To Wolfgang
with love
Mutti and Vati
1/31/1990

MODERN CARPENTRY

building construction details
in easy-to-understand form

by
WILLIS H. WAGNER
Professor, Industrial Technology
University of Northern Iowa, Cedar Falls

South Holland, Illinois
THE GOODHEART-WILLCOX CO., INC.
Publishers

Library of Congress Catalog Card Number 76-3630.
International Standard Book Number 0-87006-208-5.
23456789-76-32109876

Library of Congress Cataloging in Publication Data

Wagner, Willis H
Modern Carpentry.

Includes index.
1. Carpentry. I. Title.
TH5604.W34 1976 694 76-3630
ISBN 0-87006-208-5

INTRODUCTION

Modern Carpentry is an easy-to-understand cyclopedia of authoritative and up-to-date information on building materials and construction methods.

Modern Carpentry provides detailed coverage of all aspects of light frame construction; including site layout, foundations, framing, sheathing, roofing, windows and doors, exterior finish, and interior wall, floor and ceiling finish. Special emphasis is placed on the use of modern materials and prefabricated components in the application of interior trim, and the construction of stairs and cabinetwork. Also included is basic information covering post-and-beam construction, chimney and fireplaces, and prefabricated structures. Units are arranged in a logical sequence - similar to the order in which the various phases of building construction are performed.

Illustrations consist of more than 1400 carefully selected photos and drawings. These are accurately coordinated with written instructions and descriptions that are easy to read. A special insert in full color shows the beauty and grain characteristics of over 70 native and foreign species of wood.

Information about building materials includes size and grade descriptions and also basic technical information that covers physical properties and other important characteristics. Scientific and technological discoveries have led to the development of many new materials. The proper use and application of these materials depend on a craftsman who has considerable knowledge about the material and how it will function in the completed structure.

Modern Carpentry is designed to provide basic instruction for students in high schools, vocational-technical schools, college classes, and apprentice training programs. It can also serve as a valuable reference for students in architectural drafting classes, and for journeymen carpenters and construction supervisors. Modern Carpentry will enable do-it-yourself persons to handle many construction jobs that they would otherwise be reluctant to undertake.

Willis H. Wagner

CONTENTS

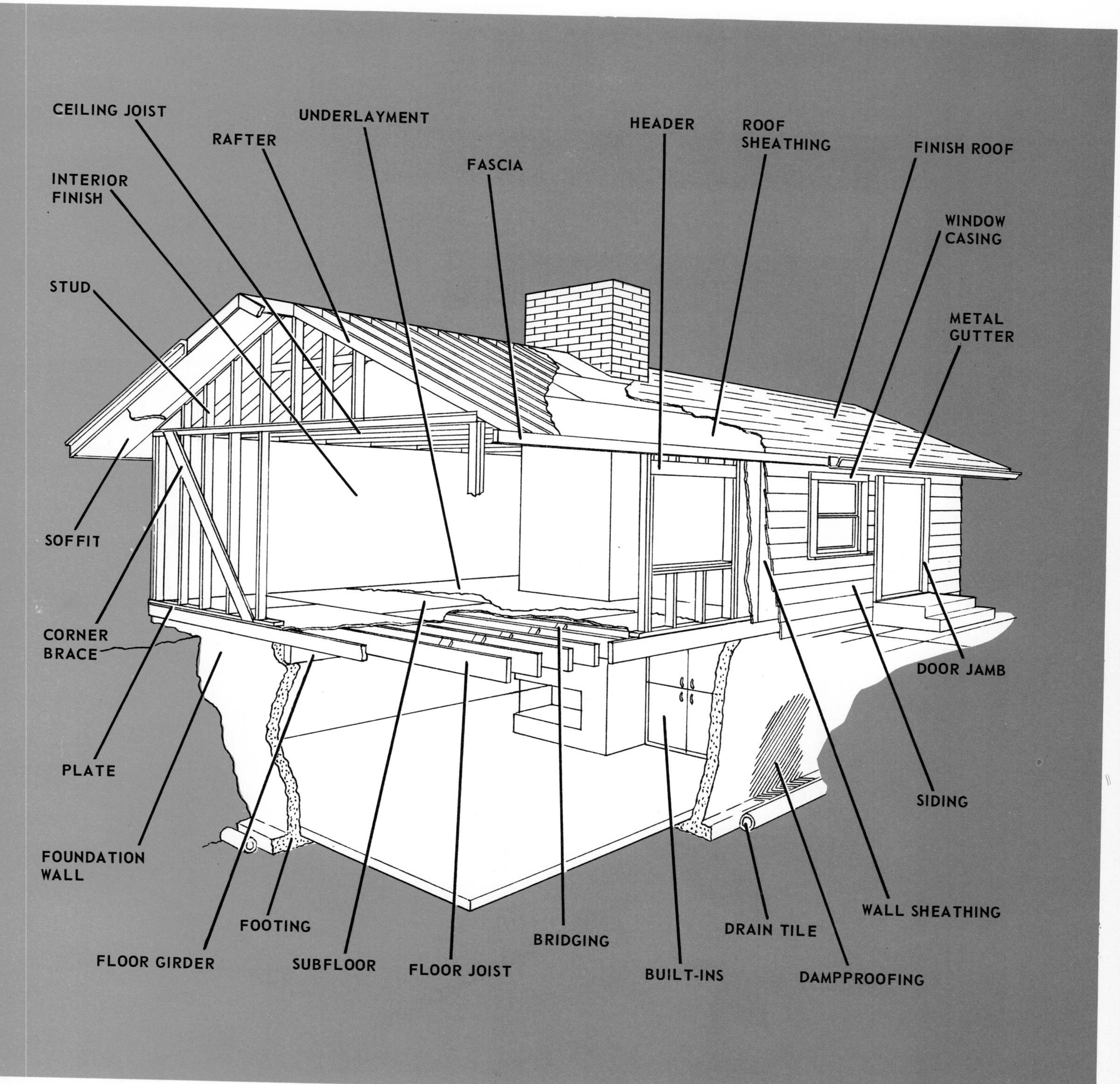

The basic parts of a house.

Fig. 1-1. Measuring and lay out tools used by the carpenter. (Stanley Tools)

Unit 1
HAND TOOLS

Hand tools are essential to every aspect of carpentry work. A great variety of tools are required because the work covers a broad range of activities.

The carpenter, a skilled craftsman, carefully selects the kind, type, and size tools that best suit his personal requirements. As he works with tools from day to day, they become an important part of his life, helping him to perform with speed and accuracy the various tasks of his trade.

Although the basic design of common woodworking tools has changed but little in many years, modern technology and industrial "know-how" have resulted in numerous improvements. Special tools have been developed to do specific jobs. The experienced carpenter appreciates the importance of having good tools and selects those that are accurately made from quality materials. He knows that such tools will enable him to do better work and will last longer.

Space is not available here for a detailed study of the selection, care, and use of all the hand tools available. Only a general description of tools most commonly used by the carpenter will be included.

Measuring and Lay Out Tools

Tools in this group are illustrated in Fig. 1-1. These tools must be handled with considerable care and kept clean if a high level of accuracy is to be maintained in their use.

The folding wood rule is indispensable. Most carpenters carry a folding rule on their person at all times. A standard size rule is 6 ft. long. Some are equipped with a metal slide that is helpful in making inside measurements. Tape rules which range in size from 6 to 12 ft. are also handy. Long tapes are available in lengths of 50 ft. and longer.

The framing square (also called a rafter square) is especially designed for the carpenter and its uses are many and varied. A description of the square and how it is used in framing is included in Unit 9. It is available in steel, aluminum, or steel with a copper-clad or blued finish.

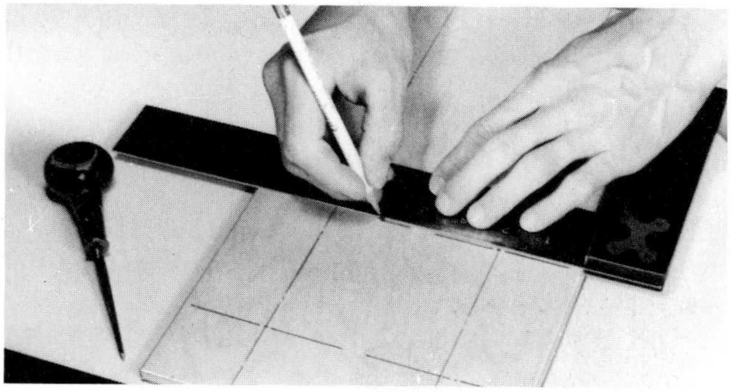

Fig. 1-2. Using a try square to lay out lines perpendicular to an edge.

Try squares are available with blades 6 to 12 in. long, with handles made of wood or metal. These are used to check the squareness of surfaces and edges and to lay out lines perpendicular to an edge, Fig. 1-2. The combination square serves a similar purpose and is also used to lay out miter joints. The adjustable blade allows it to be conveniently used as a gauging tool. The T-bevel has an adjustable blade making it possible to transfer a specified angle from one piece to another. It is useful in laying out cuts for hip and valley rafters. Fig. 1-3 illustrates how the framing square is used to set the T-bevel at a specified angle.

Wing dividers on which the legs can be locked in position, are available in several sizes, and serve a number of purposes. Given dimensions can be stepped off along a layout or transferred from one position to another. Circles and arcs can also be scribed on surfaces.

A marking gauge is used to lay out parallel lines along the edges of material. It may also be used to transfer a specified dimension from one place to another, or check sizes of material. The butt gauge can be used for about the same purposes; however, it is used mostly to lay out the gain (recess) for hinges.

In layout work a scratch awl is used to scratch

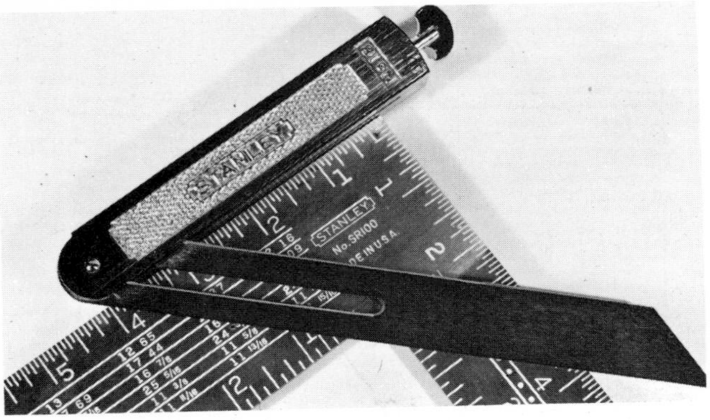

Fig. 1-3. Setting the T-bevel to a 45 deg. angle from the framing square.

lines on the surfaces of materials. It is also used to mark points and to form starter holes for small screws or nails.

Using a chalk line is an easy way to mark long, straight lines. The line (covered with chalk) is held tight and close to the surface and is snapped. This action drives the chalk onto the surface forming a distinct mark. A special reel re-chalks the line each time it is wound into the case.

The level and plumb bob are important devices for checking and laying out vertical and horizontal lines. A standard size level is 24 in. long. The body of the level may be made of wood, aluminum or special lightweight alloys. It is often used in connection with a straightedge (straight strip of wood, usually laminated to provide greater stability) when the span or height of the work is greater than the length of the level. Some carpenters prefer a level 4 to 6 ft. long since it eliminates the need for a straightedge on such work as installing door and window frames. The plumb bob is attached to a line and suspended. Its weight pulls the line in a true vertical position for lay out and checking. The point of the plumb bob is directly below the given point above.

Saws

The principal types of saws used by the carpenter are illustrated in Fig. 1-4. These are available in several different lengths as well as various tooth sizes (points per inch).

Crosscut saws, as the name implies, are designed to cut across the wood grain. They have

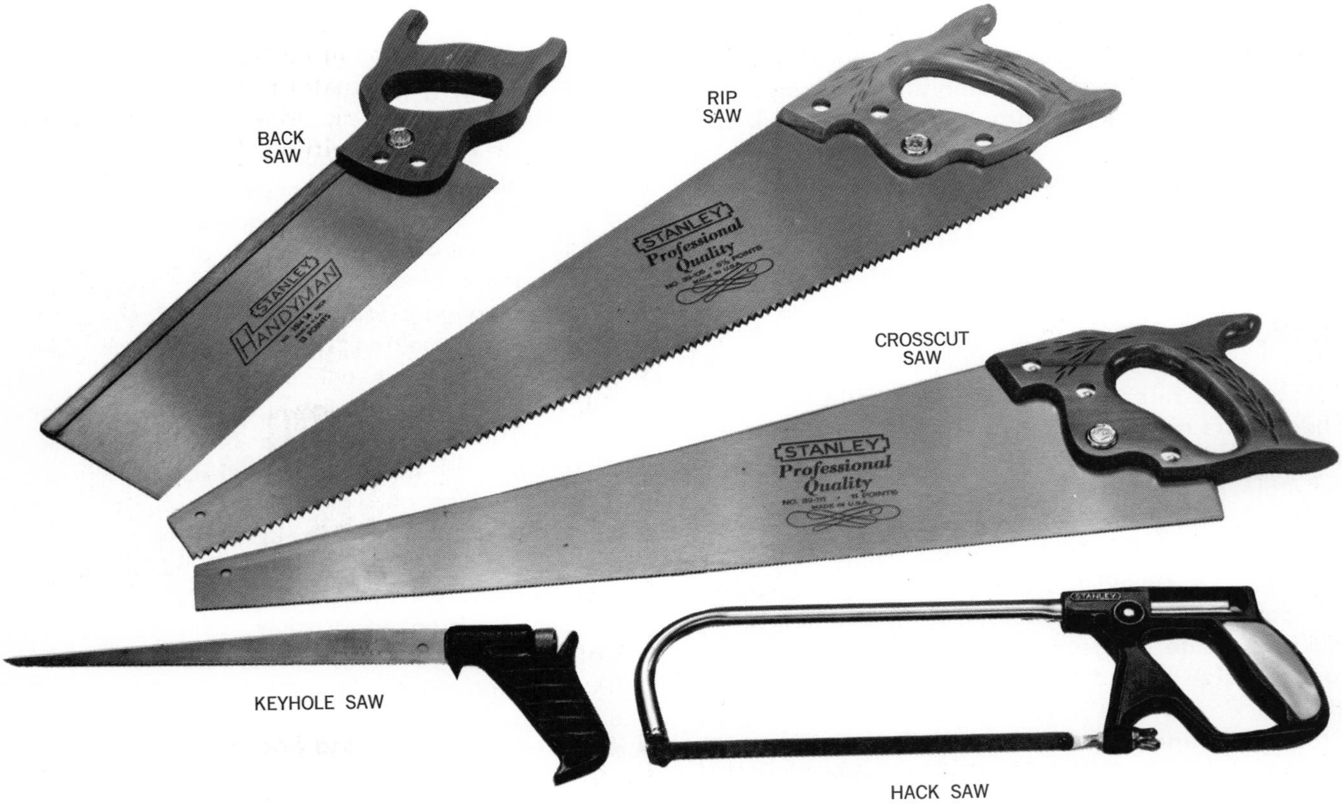

Fig. 1-4. Kinds of saws. The ripsaw has 4 1/2 teeth per in. and is referred to as a 5 1/2 point saw. The crosscut saw has 11 points.

teeth that are pointed. See Fig. 1-5. Ripsaws have chisel shaped teeth which cut best along the grain, Fig. 1-6. Note that the teeth are set (bent) alternately from side to side so the kerf (opening cut in the wood) will be large enough for the blade to run freely. Some saws have a taper, ground from the toothed edge to the back edge, which eliminates the need for a large amount of set.

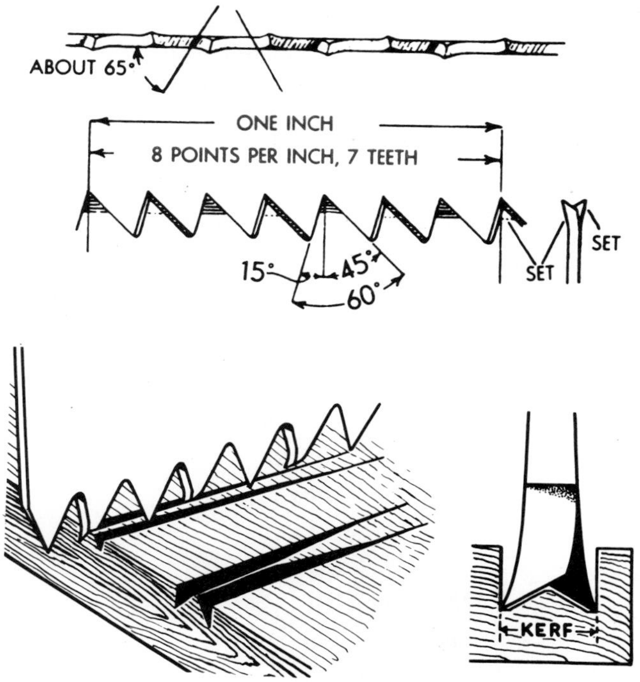

Fig. 1-5. Above. Crosscut saw teeth. Below. How crosscut teeth cut.

Every carpenter needs to include in his tool kit a good crosscut saw with a tooth size ranging from 8 to 11 points. A ripsaw is no longer considered a "must" since most ripping operations are performed with power saws.

The backsaw has a thin blade reinforced with a steel strip along the back edge. The teeth are small (14-16 points)-- thus the cuts produced are fine. It is used mostly for interior finish work. A similar type of saw is used in the miter box, Fig. 1-7, a device that accurately guides the blade when forming miters and other types of joints.

Many keyhole saws have a quick-change feature that permits the use of various sizes and kinds of blades. Although originally designed to cut keyholes, they now serve as general purpose saws for irregular cuts or work where space is limited.

Because of the wide range of work that the carpenter must be prepared to handle - a hacksaw should be included in his hand tool assortment.

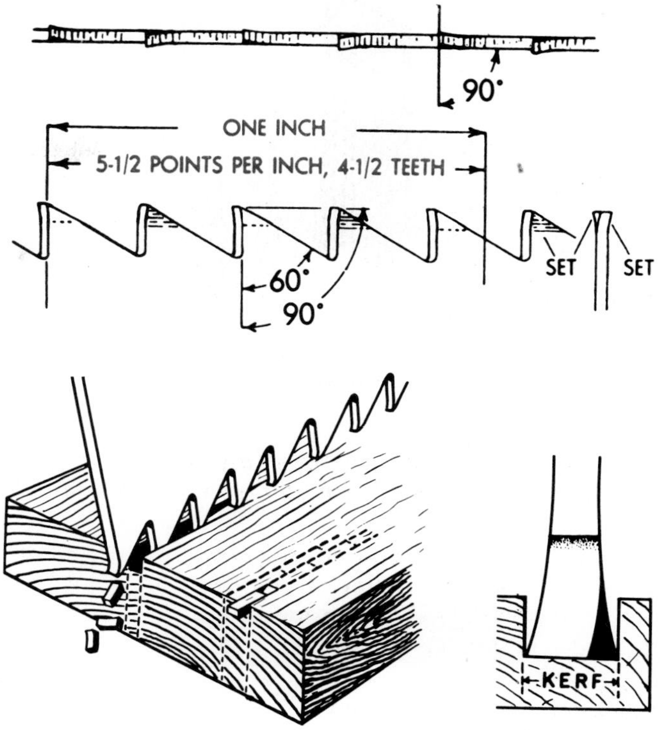

Fig. 1-6. Above. Ripsaw teeth. The 90 deg. angle is often increased slightly to give the tooth a negative rake. Below. How rip teeth cut.

This will be needed to cut nails, bolts, other metal fasteners and various metal trim used on both exterior and interior work. Most hacksaws have an adjustable frame, permitting the use of several sizes of blades.

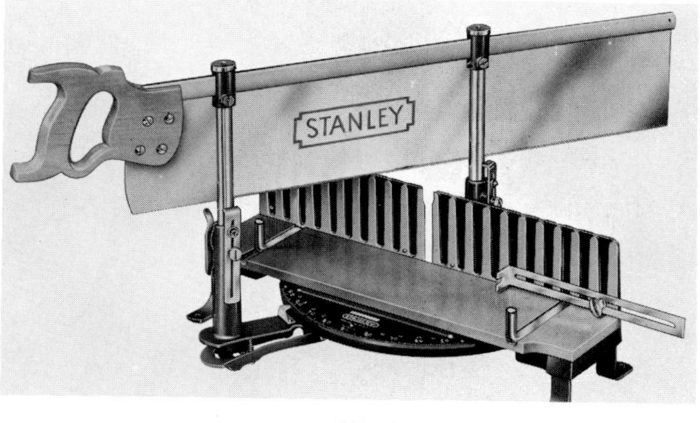

Fig. 1-7. Miter box. Valuable for making accurate cuts.

Planing, Smoothing, and Shaping

The most important tools in this group are the planes and chisels. Several other tools that depend on a cutting edge to perform the work are included in the illustration, Fig. 1-8.

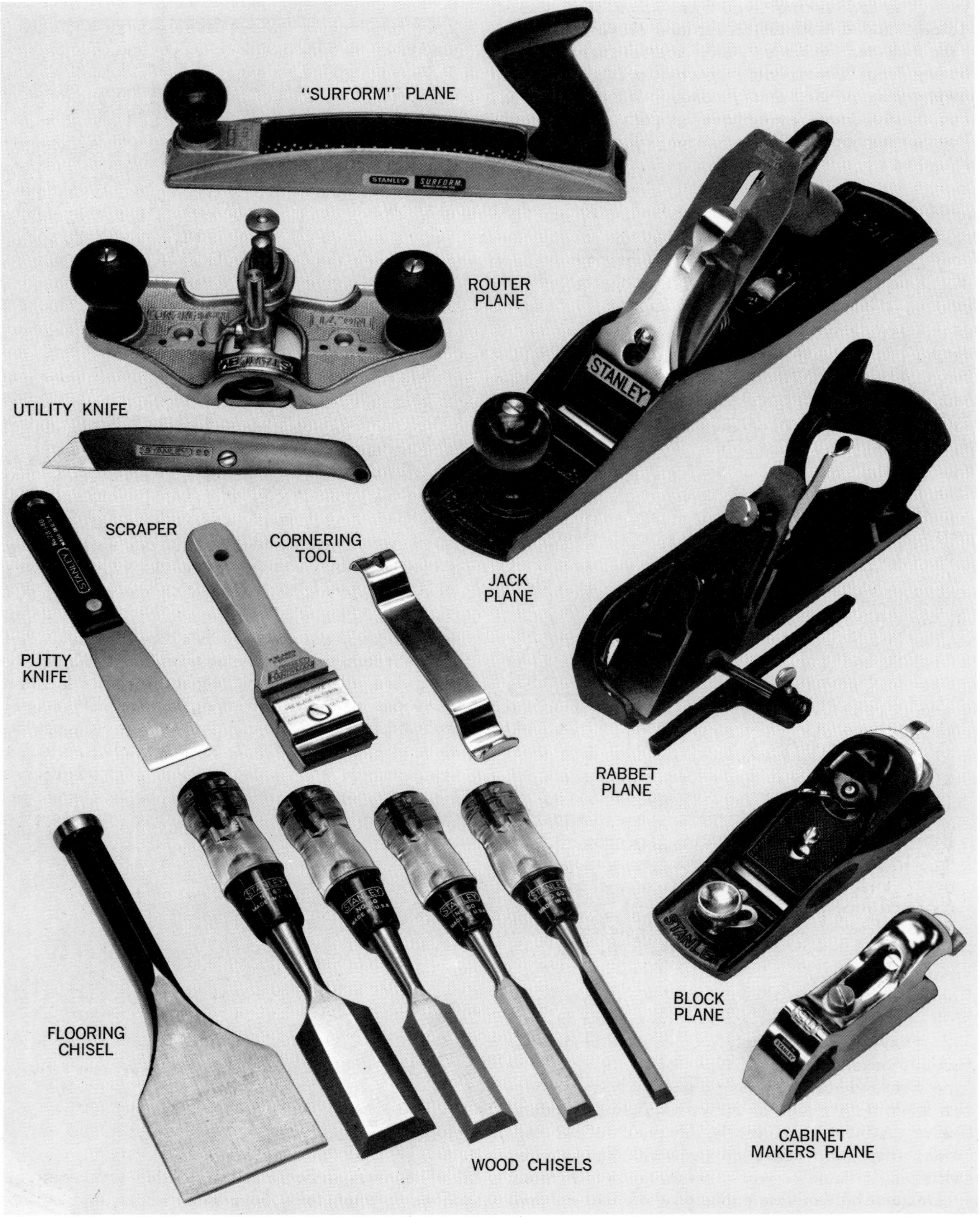

Fig. 1-8. Planes, chisels and other edge tools.
(Stanley Tools)

Standard surfacing planes include the smooth plane (8-9 in. long), jack plane (14 in. long) and the fore and jointer plane (18 to 24 in. long). These planes are constructed the same, but vary in cutter width and overall length. Fig. 1-9 illustrates the parts and how they fit together. The jack plane is commonly selected for general purpose work.

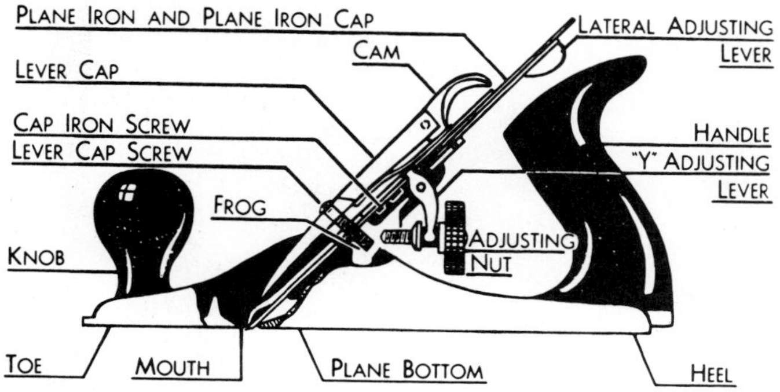

Fig. 1-9. Parts of a standard plane.

Another surfacing plane - the block plane - is especially useful for the carpenter. It is small (6-7 in. long) and can be held in one hand while the other hand is used to hold the work. The blade is mounted at a low angle and the bevel of the cutter is turned up. This plane produces a fine, smooth cut which is desirable in fitting and trimming work.

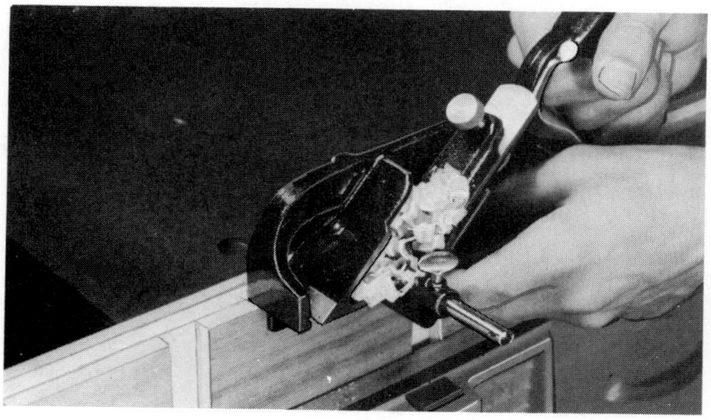

Fig. 1-10. How a rabbet plane is used.

Router and rabbet planes are designed to form dados, grooves, and rabbets. Fig. 1-10 illustrates the use of a rabbet plane. The need for these specialized hand planes has diminished in recent years due to the availability of power driven equipment. Fig. 1-11 shows the operation of a router plane.

Wood chisels are used to trim and cut away wood or composition materials to form joints or recesses. They are also helpful in paring and

Fig. 1-11. Using a router plane to smooth the bottom of a dado joint.

smoothing small, interior surfaces that are inaccessible for other edge tools. Width sizes range from 1/8 to 2 in. For general work, the carpenter usually selects 3/8, 1/2, 3/4 and 1 1/4 in. sizes. A soft-face hammer or mallet should always be used to drive the chisel when making deep cuts.

The scraper has an edge that is hooked and produces thin shaving-like cuttings. The edge is usually filed and then a tool called a burnisher is used to form the hook. There are a number of different types of scrapers. The carpenter generally prefers the type with a blade that can be quickly replaced.

Two tools used for smoothing and shaping are the tungsten carbide coated file, and the multi-blade forming tool (Surform). Each cuts rapidly and will do a considerable amount of work before it becomes dull. The surface produced is rough and requires sanding to remove the tool marks. The cornering tool, shown with the planes and chisels, is used to remove sharp corners from exposed wooden parts.

Drilling and Boring

Holes larger than 1/4 in. are made with auger bits, or bits adjustable in size, called expansive bits. The operation is called boring. Small holes are formed with hand or push drills. Tools in this group are illustrated in Fig. 1-12.

Auger bits vary in the shape and design of the twist, but all have about the same parts, as shown in Fig. 1-13. The size (diameter) of standard bits ranges from 3/16 in. to as large as 2 in. The more common range is from 1/4 to 1 in. The size is stamped on the tang or shank, often as a whole number which indicates the number of sixteenths.

Boring bits are mounted in a brace that holds and turns them into the wood. Parts of a standard ratchet brace are shown in Fig. 1-14. The ratchet

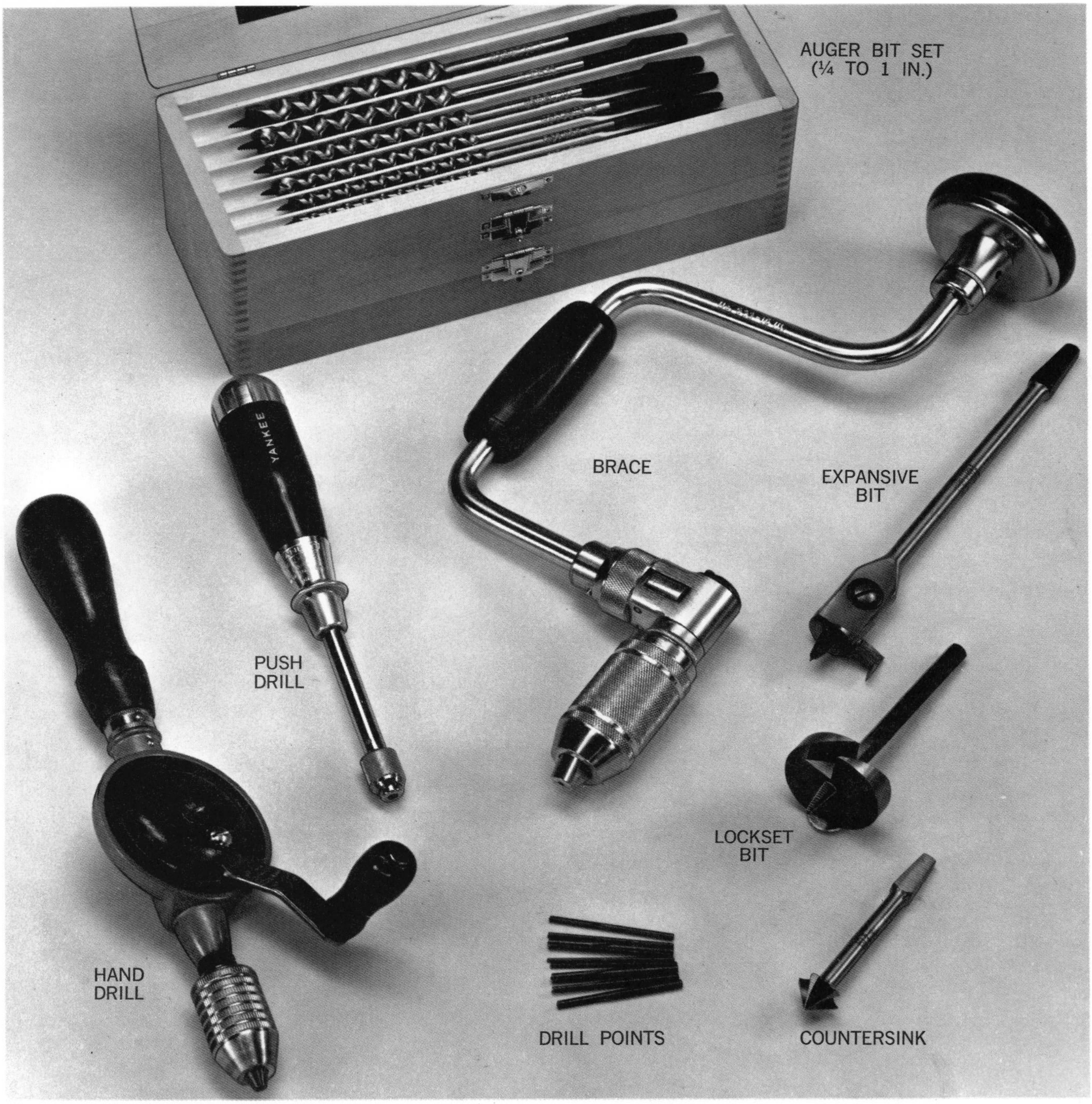

Fig. 1-12. Tools for drilling and boring holes. (Stanley Tools)

INTRODUCING "HANDY"

Hi! My name is Handy Carpenter. Just call me "Handy."

I'll be around from time to time with pointers, safety tips, and helpful suggestions.

permits boring in a location where a complete revolution of the brace is not possible. Other tools designed for the brace include the countersink which forms the recess for screw heads, and the screwdriver bit which is used to set regular wood screws.

The size of a hand drill is determined by the capacity of its chuck, usually 1/4 or 3/8 in. Regular twist drills are used in the hand drill, with a

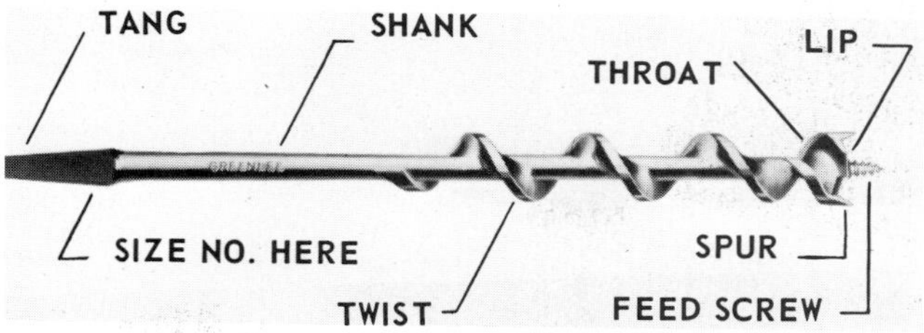

Fig. 1-13. Parts of an auger bit.

practical set for the carpenter ranging from 1/16 to 1/4 in. by thirty-seconds.

Push drills are designed to form small holes quickly. When the handle is pushed down, the drill chuck revolves. A spring inside the handle forces it back to its original position when pressure is

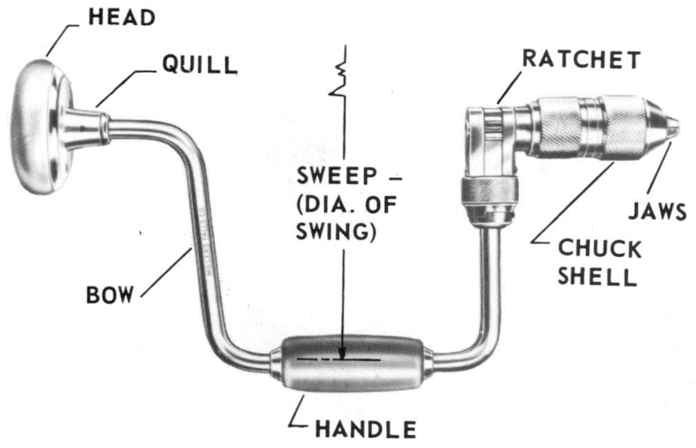

Fig. 1-14. Parts of a ratchet brace. (Millers Falls Co.)

released. Push drills use special fluted bits that range in size from 1/16 to about 3/16 in. Carpenters frequently use this type of drill to make holes for nails and screws since it can be operated with one hand. See Fig. 1-15.

Fastening Parts Together

In carpentry work, a large percent of the work consists of fastening parts together, using nails, screws, bolts and other types of connectors. Fig. 1-16 shows a group of hand tools commonly used to perform operations in this area.

Of the tools shown, the nail hammer is used most often. The carpenter usually carries a hammer in a holster attached to his belt. Two shapes of hammer heads are in common use; the curved claw and the ripping (straight) claw. The curved claw is the most common and is best suited for pulling nails. The ripping claw which may be driven between fastened pieces is used somewhat like a chisel to pry them apart.

Parts of a standard hammer are shown in Fig. 1-17. The face can be either flat or have a slightly rounded convex surface (bell face). The bell face is most often used since it will drive nails flush with the surface without leaving hammer marks on the wood. The hammer head is forged of high quality steel and is heat-treated to give the poll and face extra hardness. The size of a claw hammer is determined by the weight of its head. It is available in a range from 7 to 20 oz. The 13 oz. size is popular for general purpose work. Carpenters generally use either the 16 or 20 oz. size for rough framing.

A hammer should be given good care. It is especially important to keep the handle tight and the face clean. If a wooden handle becomes slightly loose, it can be tightened by driving the wedges deeper or installing new ones. Fig. 1-18 illustrates the procedure for pulling nails to relieve strain on the handle and protect the work.

Hatchets are used for rough work on such jobs as making stakes and building concrete forms. Some are designed for wood shingle work, especially wood shakes. See Unit 10.

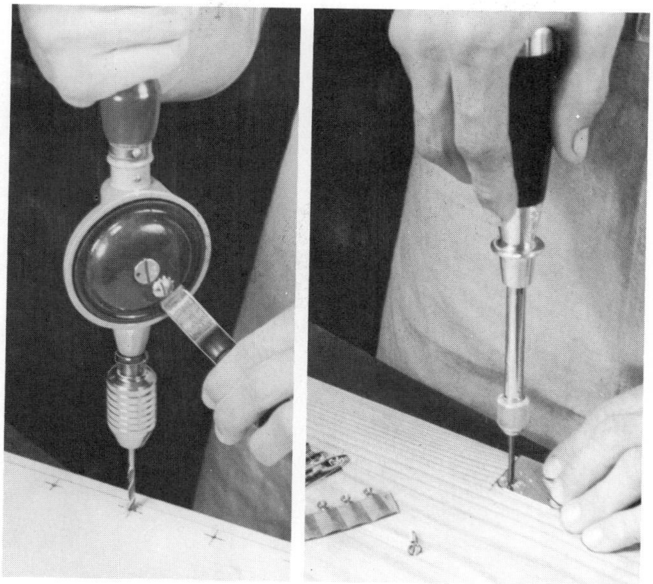

Fig. 1-15. Left. Drilling holes with a hand drill. Right. Using a push drill.

Ripping bars, also called wrecking bars, vary in length from 12 to 36 in. They are used to strip concrete forrns, disassemble scaffolding, and other rough work involving prying and nail pulling.

The nail set is designed to drive the heads of casing and finishing nails below the surface of the

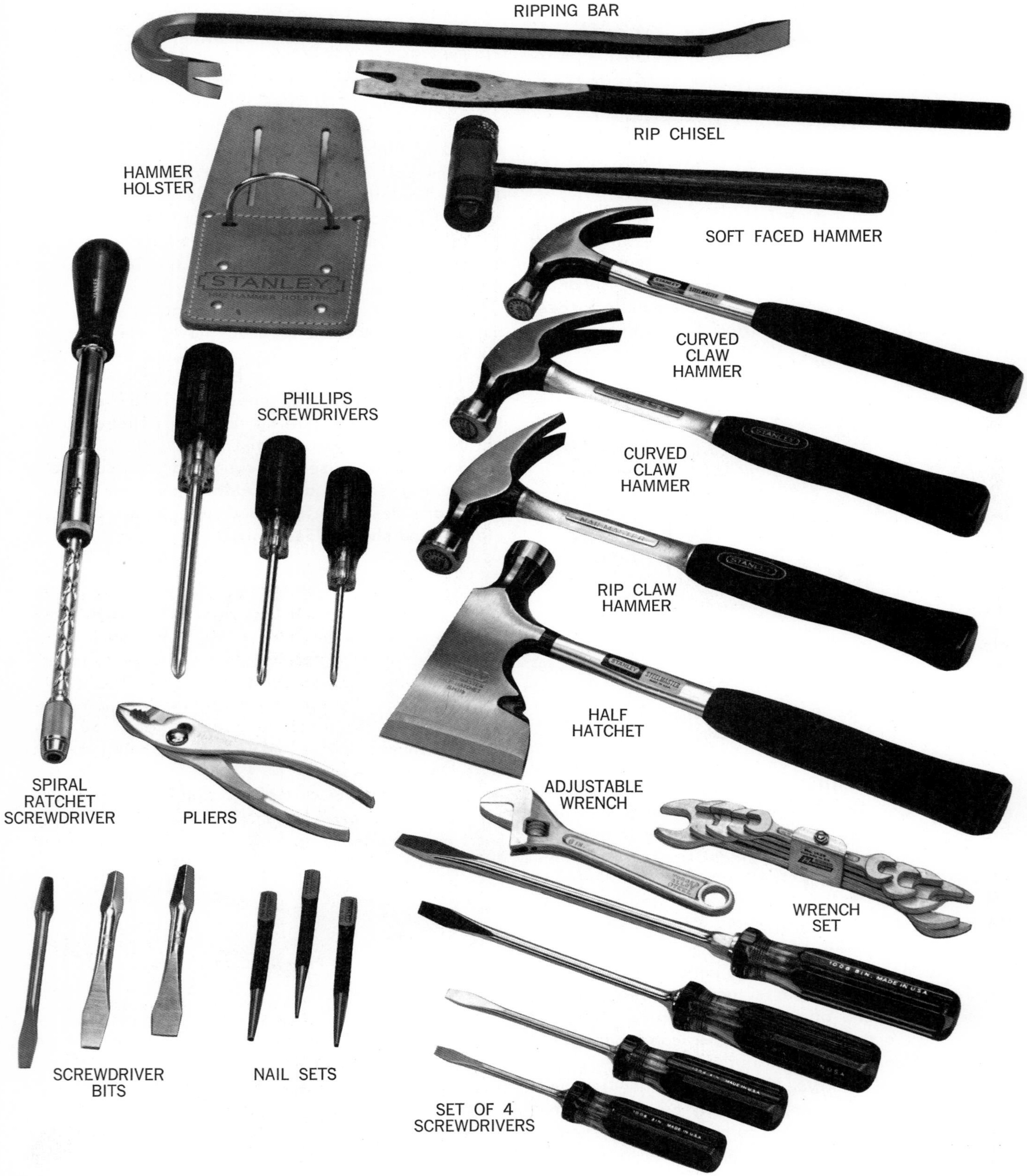

Fig. 1-16. Tools used to fasten and disassemble carpentry structures. (Stanley Tools)

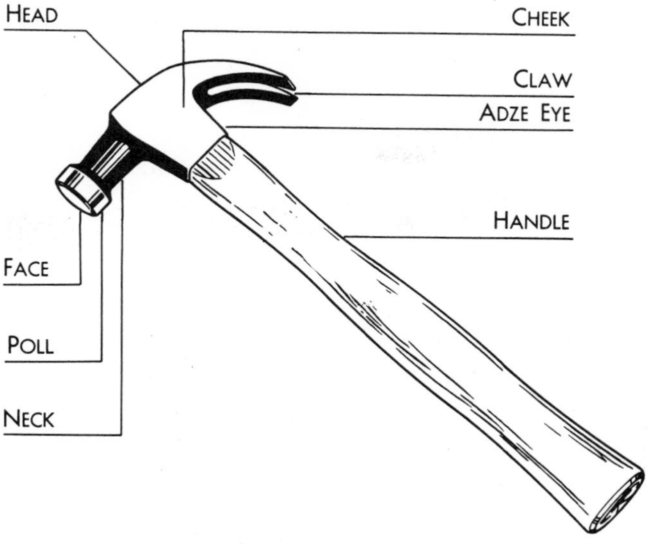

Fig. 1-17. Parts of a standard claw hammer.

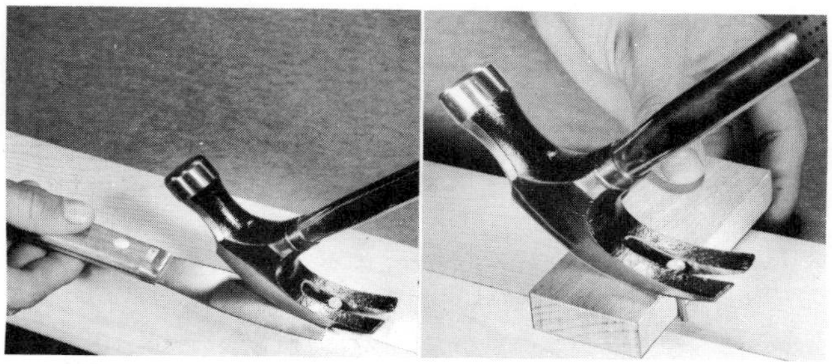

Fig. 1-18. Pulling a nail. Left. Using a putty knife to protect the surface. Right. Using a block of wood to increase leverage and protect the surface.

wood. Tips range in size (diameter) from 1/32 to 5/32 in. by thirty-seconds. Overall length is usually about 4 inches.

Parts of a standard screwdriver are shown in Fig. 1-19. A number of screwdriver sizes and styles are available. Screwdriver size is specified by giving the length of the blade, measuring from the ferrule to the tip. The most common sizes for the carpenter are shown in the group illustration; sizes 3, 4, 6 and 8 in.

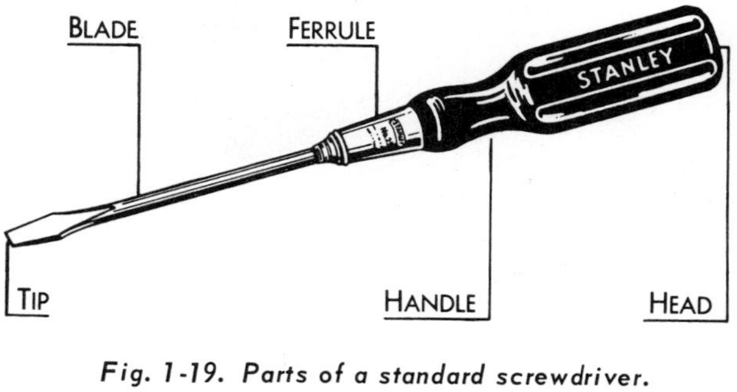

Fig. 1-19. Parts of a standard screwdriver.

The size of a Phillips screwdriver is given as a point number ranging from a No. 0, (the smallest) to a No. 4. Size numbers 1, 2, and 3 will fit most of the screws of this type used in carpentry work.

Tips of screwdrivers must be carefully shaped. For a slotted screw they must be square, of correct width and fit snugly into the screw. The width of the tip should be equal to the length of the bottom of the screw slot. The sides of the screwdriver tip should be carefully ground to an included angle of not more than 8 deg. and to a thickness that will fit the screw slot. Fig. 1-20 shows properly shaped tips for various types of screwdrivers.

Fig. 1-20. Screwdriver tips. Left. Phillips. Center. Standard. Right. Cabinet.

A handy tool for the carpenter is the spiral ratchet screwdriver. It operates somewhat like the push drill except the ratchet keeps the chuck turning in the same direction. Various bits to match screw sizes and types can be mounted in the chuck.

A kit of tools for the carpenter would not be complete without a pair or two of pliers, and several wrenches. Sizes will vary depending on the construction needs.

Today, mechanical tackers, staplers and nailers perform a variety of operations formerly done by hand nailing. See Fig. 1-21. They provide an effi-

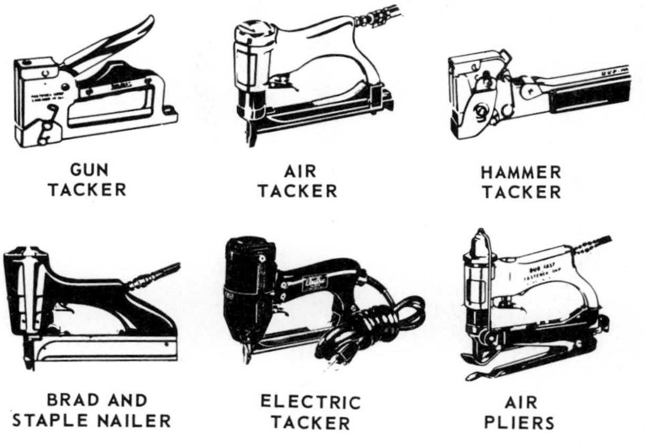

Fig. 1-21. Tackers, staplers, and nailers.

Using a 20 oz. (rip claw) hammer to nail the headers located over corner window openings. Note the temporary diagonal braces which provided rigidity until the wall sheathing is applied.

Using a spiral ratchet screwdriver to install hinges on a passage door located in a commercial building. (Stanley Tools)

cient method of attaching insulation, roofing material, underlayment, ceiling tile and many other products. An advantage of the regular stapler is that it leaves one hand free to hold the material. Fig. 1-22 illustrates a heavy-duty stapler-nailer that is operated by striking it with a mallet.

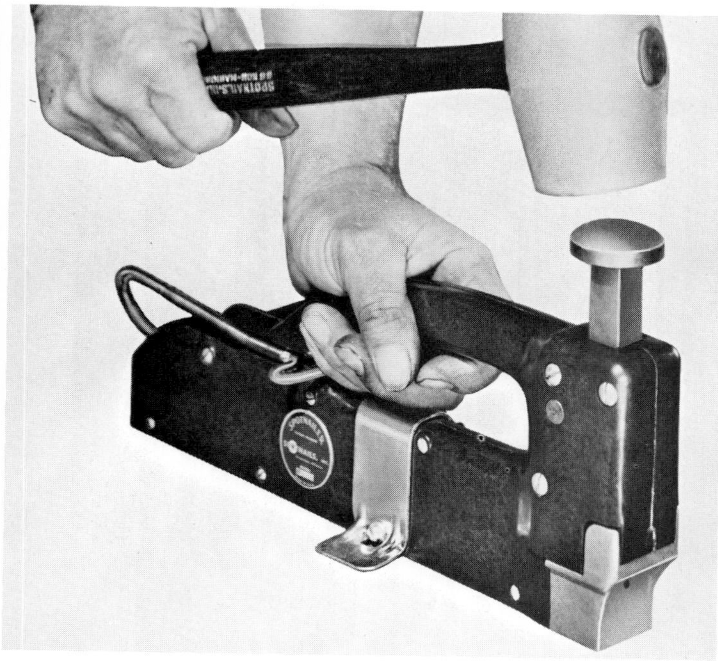

Fig. 1-22. Mallet operated nailer-tacker. (Spotnails, Inc.)

Fig. 1-23. C-clamps in use. This type of clamp is also called a carriage clamp.

Fig. 1-24. Hand screws are adaptable to woodworking.

Clamping Tools

There are several clamping tools that may prove helpful, especially in finish work. C-clamps, Fig. 1-23, are adaptable to a wide range of assemblies where parts need to be held together, while metal fasteners are being applied, or until an adhesive sets. Sometimes they can be used to clamp jigs or fixtures to machines for some special set-up. C-clamps are available in sizes (largest opening) from 1 to 12 in.

Hand screws, shown in Fig. 1-24, are ideal clamps for woodworking because the jaws are broad and distribute the pressure over a wide area. Sizes (length of the jaw) range from 4 to 24 in.

Tool Storage

Some type of tool chest or cabinet is needed to store tools and transport them to and from the job, or from one building site to another. In addition to being portable, the chest or cabinet should protect the tools from weather damage, loss, and theft.

Folding tool panels provide a practical solution to transportation and storage. Fig. 1-25 shows a design that can be carried like a large suitcase, then opened to produce a free-standing tool panel. Toolholders, especially designed to hold a given item are preferred over nails, screws, or hooks. Attaching the holder to a small subpanel and then mounting the unit on a main panel is shown in Fig. 1-26. Special locking devices may be required to hold the tools in place while the case is being transported.

One of the important advantages of an organized tool panel is the ease with which tools can be checked. At the end of the day, you can determine if any tools are missing by checking to see if there are empty holders.

Fig. 1-25. Folding tool panel which may be closed and carried like a large suitcase. (American Plywood Assoc.)

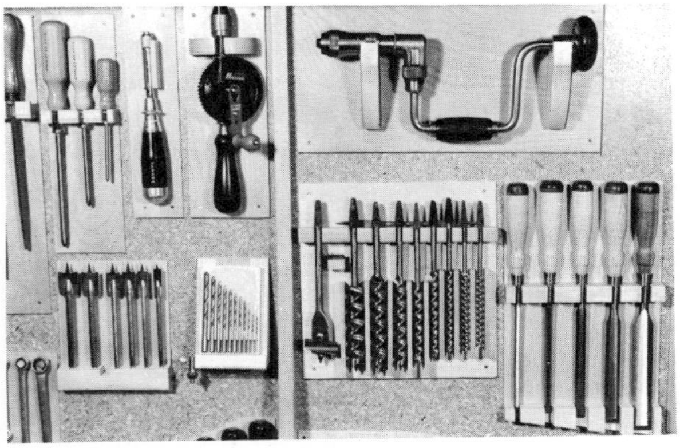

Fig. 1-26. Custom-made toolholders mounted on vertical panel.

Care and Maintenance

The experienced craftsman takes pride in his tools and keeps them sharp and in good condition. He knows that even high quality tools will not perform satisfactorily if they are dull or out of adjustment.

Tools should be wiped clean after being used. Occasionally saturate the wiping cloth lightly with oil. Some craftsmen use a lemon oil furniture polish since it has a slight cleaning action and also leaves a thin film that protects metal surfaces from rust.

Keep handles on all tools tight. When handles and fittings are broken they should be replaced.

Edge tools become dull and need to be sharpened. It is a simple matter to hone such tools on an oilstone, Fig. 1-27. For tools with single-bevel edges like planes and chisels, place the tool on the stone with the bevel flat on the surface. Raise the

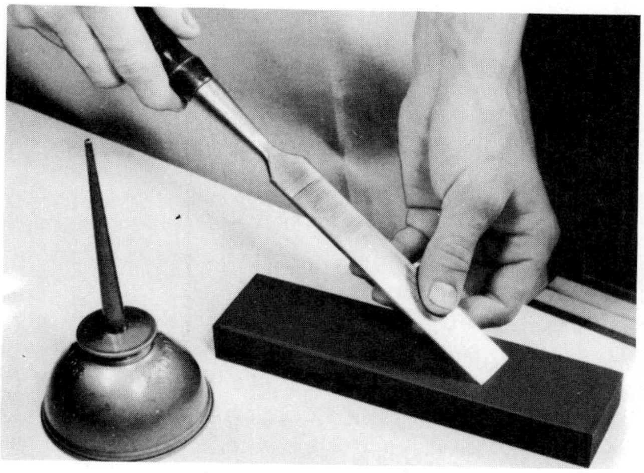

Fig. 1-27. Honing a chisel on an oilstone.

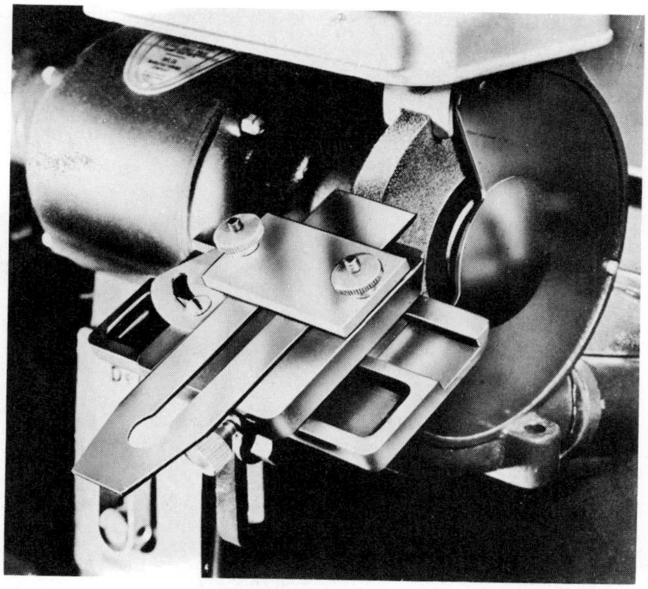

Fig. 1-28. Grinding a plane iron. Note the attachment that makes it possible to accurately form a bevel. (Rockwell International)

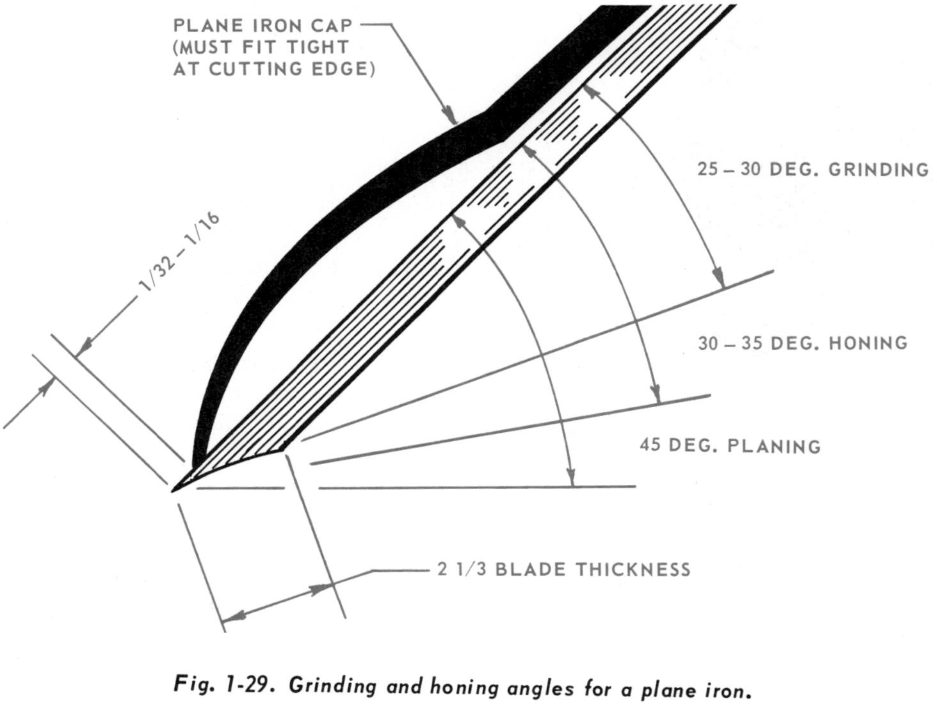

Fig. 1-29. Grinding and honing angles for a plane iron.

back edge of the tool a few degrees so just the cutting edge is in contact and then move the tool back and forth until a fine wire edge can be detected by pulling the finger over the edge. Now place the back of the tool flat on the oilstone and stroke lightly several times. Turn the tool over and again stroke the beveled side lightly. Repeat this total operation several times until the wire edge has disappeared.

If not damaged, a cutting edge can be honed a number of times before grinding is required. When the bevel becomes blunt, reshape it with a grinding operation, Fig. 1-28. The grinding angle for tools will vary somewhat depending on the work they must perform. Fig. 1-29 shows grinding and honing angles recommended for a plane iron.

Some tools may be sharpened with a file. For auger bits, it is best to use a special auger bit file. See Fig. 1-30. Sharpen the lips or cutters by stroking up through the throat. Do not file the underside. File the inside of the spurs as shown, being sure to keep them the same length.

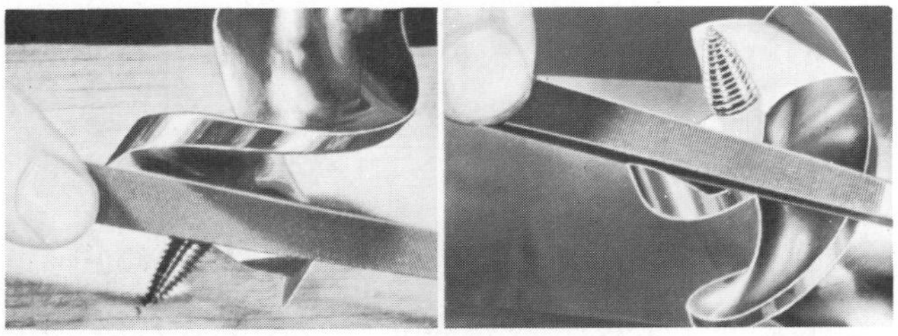

Fig. 1-30. Sharpening an auger bit. Left. Filing the cutting lip. Right. Filing the spur – inside surface only.

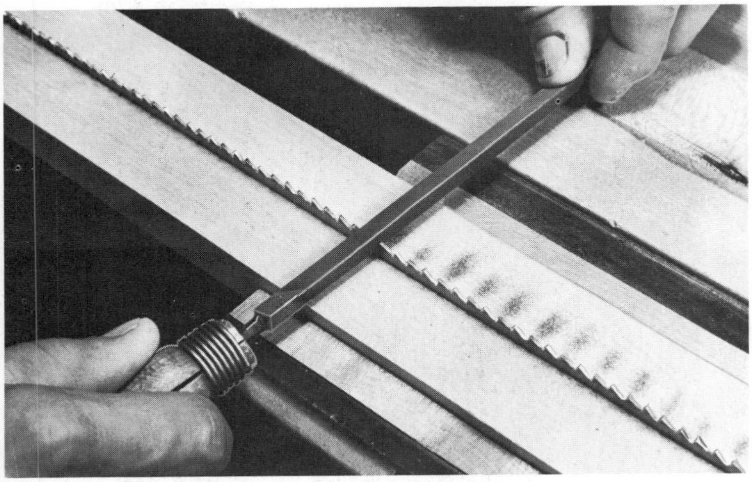

Fig. 1-31. Filing a ripsaw. The file should contact the back of a tooth set away from the operator.

Fig. 1-32. Setting the teeth. In the position shown, the saw set will bend the teeth away from the operator.

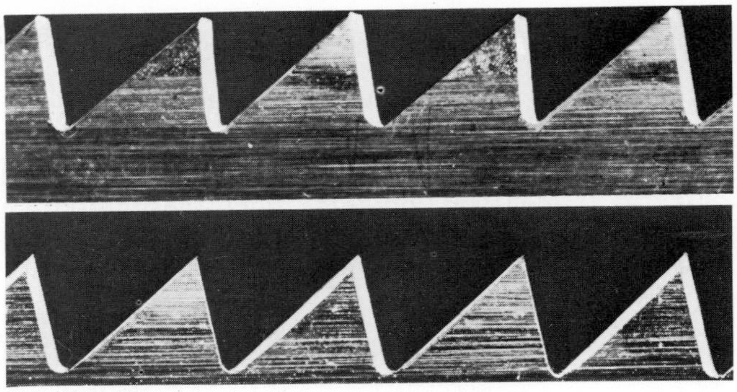

Fig. 1-33. Correctly filed saw teeth. Above. Rip teeth. Below. Crosscut teeth. (Nicholson File Co.)

Saws require filing as well as setting. Before filing they often require jointing -- where the height of the teeth are struck-off evenly. Filing a saw is a tedious operation. Most carpenters prefer to send their saws to a shop where they can be machine sharpened by an expert.

When a saw is only slightly dull, the teeth can often be made sharp with a few file strokes, as illustrated in Fig. 1-31. Use a triangular saw file. Be sure to match the original angle of the teeth. File the back of one tooth and the front of an adjacent tooth in a single stroke. Fig. 1-32 shows a saw set in operation. Fig. 1-33 shows how correctly filed teeth should appear.

Test Your Knowledge - Unit 1

Write answers on a separate sheet of paper. Do not write in this book.

1. A standard folding wood rule is _____ ft. long.
2. The blade of a framing square is 24 in. long and the tongue (other leg) is ______ in. long.
3. When checking structural members to see if they are horizontal or vertical, it is best to use a __________ .
4. A 10 point saw will have ______ teeth per in.
5. The bevel of a block plane blade is turned __________ (down, up).
6. An auger bit with the number 10 stamped on the tang or shank would bore a hole _______ in. in diameter.
7. The size of a nail hammer is determined by the __________ of its head.
8. Large screws can be driven with a screwdriver bit mounted in a __________.
9. Spurs of an auger bit are filed on the ________ (inside, outside).
10. The operation of filing off the points of saw teeth until they are all level is called _______.

Outside Assignments

1. After a study of reference books and manufacturers' catalogs, prepare a list of tools you believe the carpenter will need for rough framing and exterior finish of a typical residential structure.
2. Prepare a report on the historical development of woodworking tools used by the carpenter. Reference books on woodworking and encyclopedias will contain helpful material. If you make a report to your class, use the overhead projector or other visual aids to show the students pictures and drawings of early tools. If possible, secure actual specimens from a collector.

Skilled carpenter laying out cut on finish stair tread. Note appropriate clothing worn which includes overalls, heavy work shirt with snug fitting cuffs, work shoes, hard hat and safety glasses. (See page 25 for additional information on clothing and personal protective equipment.) The Occupational Safety and Health Administration (OSHA) specifies the wearing of approved hard hats on all construction sites. In addition to personal protective equipment, the law covers a wide range of standards to insure a safe and healthful environment in all branches of business and industry. (Stanley Tools)

BUILDER'S SUCCESS STORY . . .

from framing to "SOLD."

ASH, WHITE. Strong, stiff, and fairly heavy (42 lbs. to cu. ft.). Works fairly well with hand tools but splits easily. Heartwood is a pale tan with a texture similar to Oak. Used for millwork, cabinets, furniture, upholstered frames, boxes and crates. Used extensively for baseball bats, tennis rackets, and other sporting equipment.

WOOD IDENTIFICATION

A key element in woodworking and in carpentry is the proper identification of the wood.

This insert, which is intended as a guide and an aid to the student in learning to identify various woods, shows typical color and grain characteristics of 71 different species.

ASPEN. Soft, light, close-grained, and easy to work. White sapwood with light tan and brown streaked heartwood. Source: Europe, Western Asia and Middle Atlantic States. Used for furniture, and interior paneling.

AVODIRE. Has a distinctive odor and lustrous appearance. Medium hard, open-grain, and about the same weight and working characteristics as Mahogany. Source: Ivory Coast of West Africa. Used for fine furniture, fixtures and wall paneling.

AYOUS. Also called African Whitewood. Fairly soft, lightweight, even textured and open-grained. Source: West Africa, mainly Ivory Coast and Nigeria. Used mainly for plywood, solid stock is available.

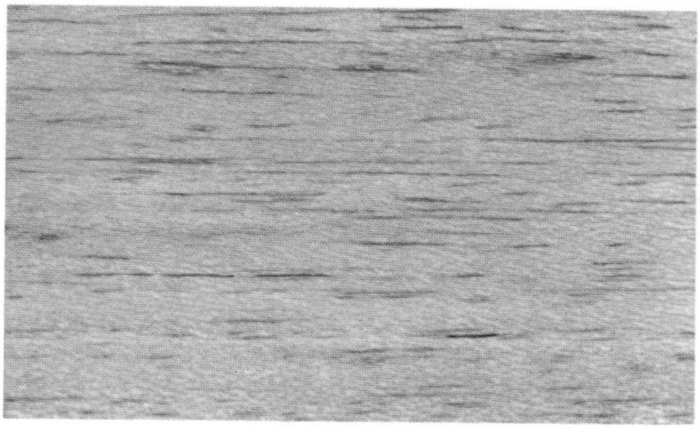

BALSA. The lightest wood available (12 lbs. to cu. ft.). Very soft and easily cut with knives. Best known for its use in model aircraft construction. Also used for life rafts, duck decoys, hat blocks, and other lightweight items. Chief source is Ecuador.

BASSWOOD. The softest and lightest (26 lbs. to cu. ft.) hardwood in commercial use. Fine, even texture with straight grain. Especially easy to work with hand tools and highly resistant to warpage. Heartwood is a light yellowish-brown. Used for drawing boards, food containers, moldings, woodenware and core stock for plywood.

BEECH. Heavy, hard and strong. (44 lbs. to cu. ft.) A good substitute for Sugar Maple but somewhat darker in color, with a slightly coarser texture. Used for flooring, furniture, brush handles, food containers, and boxes and crates.

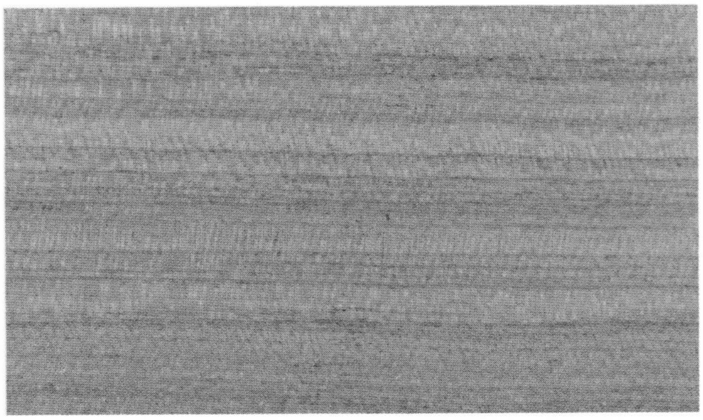

BELLA ROSA. Moderately hard, heavy and coarse textured. Striped and mottled grain patterns. Source: Philippine Islands. Available in veneer and lumber. Sample shows quartered veneer.

BENGE. A medium hard and dense wood with open grain. Heartwood is light brown with dark brown or reddish brown stripes. Source: West Africa. Used mainly for veneer in fabricating plywood. Solid stock is seldom available.

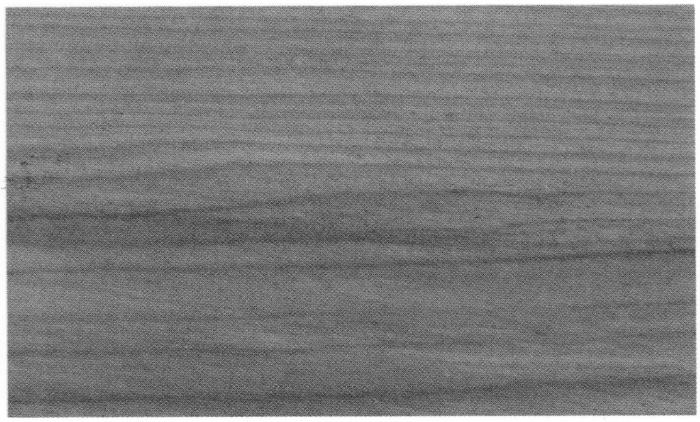

BIRCH. A hard, strong, wood (47 lbs. to cu. ft.). Works well with machines and has excellent finishing characteristics. Heartwood, reddish-brown with white sapwood. Fine grain and texture. Used extensively for quality furniture, cabinetwork, doors, interior trim, and plywood. Also used for dowels, spools, toothpicks, and clothespins.

BIRCH, WHITE. Selected sapwood rotary-cuttings of regular birch veneer. Same characteristics as birch except nearly white in color. Used to make plywood for installations requiring a very light, fine textured material.

BOXWOOD. Hard, close, and smooth grained. A very indistinct grain pattern with a light yellow color. Source: Europe, Asia, and West Indies. Used for inlays, marquetry, and instruments, especially fine rulers and scales.

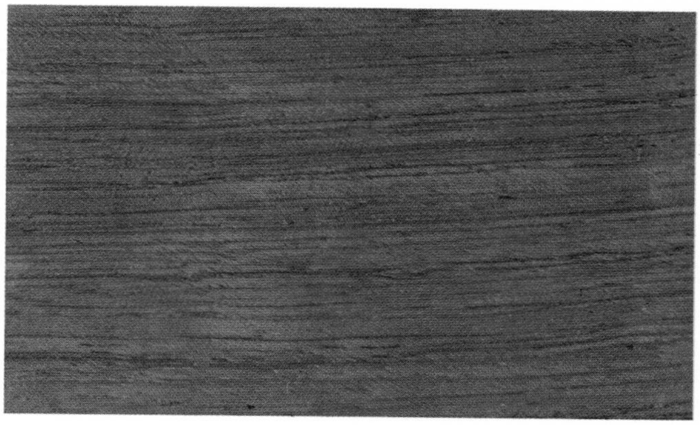

BUBINGA. Also called African Rosewood. A hard, heavy wood with open grain. The reddish brown color is often streaked with lines of dark purple. Source: West Africa. Used for fine cabinetwork in both solids and veneers.

BUCKEYE. A soft, close-grained wood that is not very strong. Nearly white in color, often blemished with dark stained lines. Source: Eastern United States. Rather scarce and usually available only in veneers.

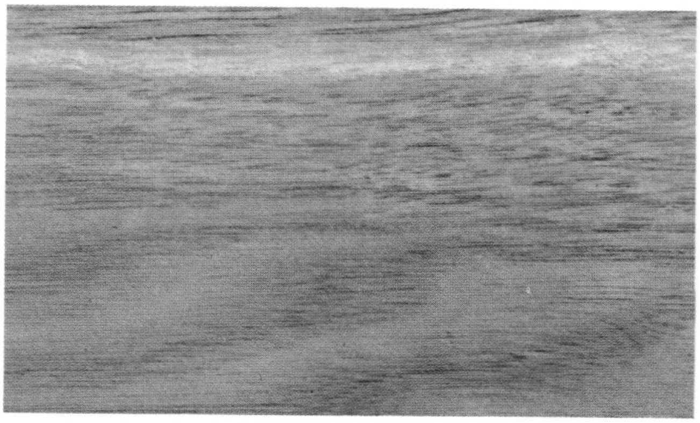

BUTTERNUT. Fairly soft, weak, light in weight (27 lbs. to cu. ft.), with a coarse texture. Grain patterns resemble walnut. Large open pores require a paste filler. Works easily with hand or machine tools. Sometimes used for interior trim, cabinetwork, and wall paneling.

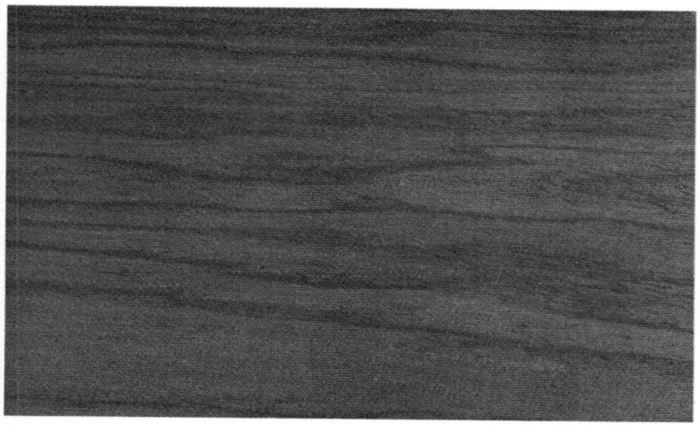

CANALLETTA. A very hard, medium textured wood. Dark brown to purplish color with very dark stripes that form interesting patterns. Source: Northern South America. Used for tool and knife handles, turnings, and walking sticks.

CEDAR, RED, EASTERN. Medium dense softwood, (34 lbs. to cu. ft.). Close grained and durable. Heartwood is red; sapwood is white. It has an aroma that inhibits the growth of moths. Knotty wood is available only in narrow widths. Used mostly for chests and novelty items.

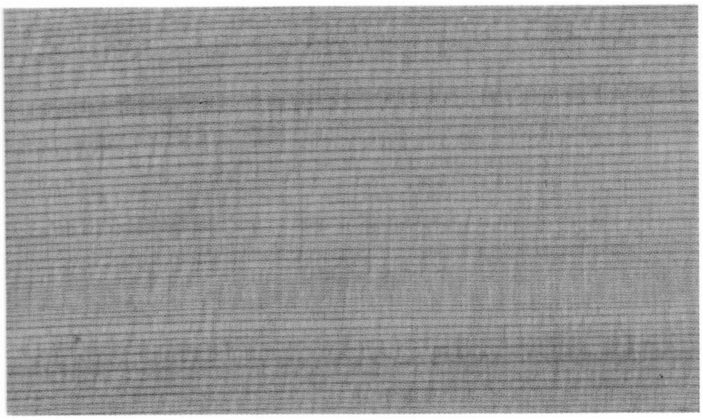

CEDAR, RED, WESTERN. A soft wood, light in weight (23 lbs. per cu. ft.). Similar to redwood except for cedar-like odor. Pronounced transition from spring to summer growth (see edge-grain sample). Source: Western coast of North America, especially Washington. Used for shingles, siding, structural timbers and utility poles.

CHERRY, BLACK. Moderately hard, strong and heavy (36 lbs. to cu. ft.). A fine, close-grained wood that machines easily and can be sanded to a very smooth finish. Heartwood is a reddish-brown with beautiful grain patterns. One of the fine furniture woods, however, there is a scarcity of good grades of lumber.

CHESTNUT. Coarse textured, open grain, and very durable. Reddish brown heartwood. Easily worked with hand or machine tools. Source: Eastern United States. Available only in a wormy grade due to the "chestnut blight."

COCOBOLA. Dense, hard and oily. A difficult wood to work because of an interwoven grain. Bright red when freshly cut, but soon darkens. Source: Central America. Used for fancy cabinetwork and knife handles.

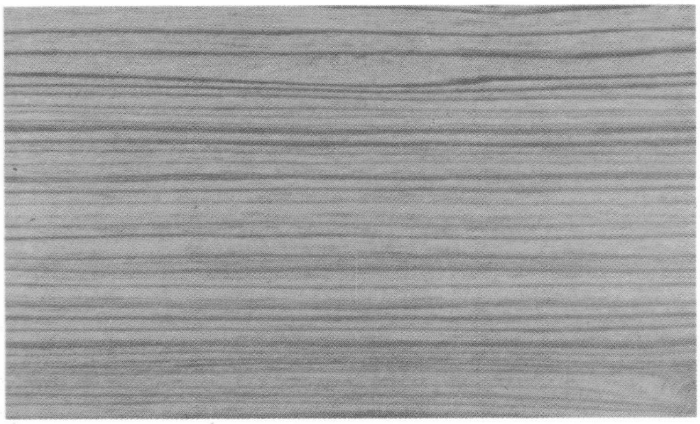

CYPRESS. Light in weight, soft and easily worked. Fairly coarse texture with annual growth rings clearly defined (sample shows edge grain). Source: Southeastern Coast of the United States. Noted for its durability against decay. Used for exterior construction and interior wall paneling.

EBONY. Extremely hard and so dense that it will not float in water. Difficult to work. Source: Ceylon, Africa and East Indies. The true black ebony comes from Ceylon and is expensive. Suitable for inlays, marquetry and small decorative articles.

ELM, AMERICAN. Strong and tough for its weight (36 lbs. to cu. ft.). Fairly coarse texture with open pores. Annular ring growth is clearly defined. Bends without breaking and machines well. Used for barrel staves, bent handles, baskets, and special types of furniture.

EMERI. Also called Ireme. A medium-hard, open-grained wood with a texture and grain pattern similar to African mahogany. Good working and finishing qualities. Source: West Africa. Used for wall panels and cabinetwork. Moderately priced. Solid stock is available.

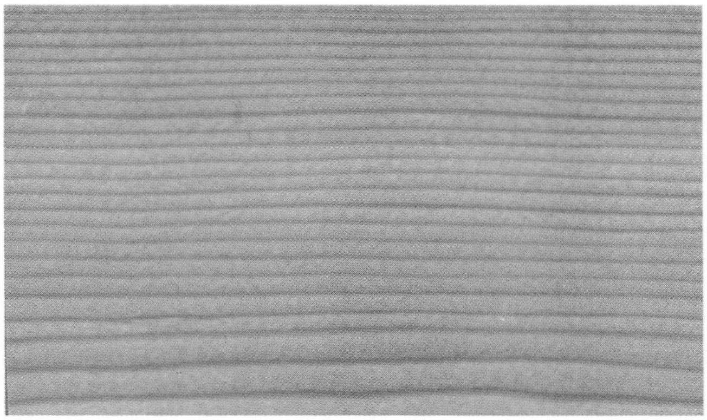

FIR, DOUGLAS. A strong, moderately heavy (34 lbs. to cu. ft.) softwood. Straight close grain with heavy contrast between spring and summer growth. Splinters easily. Used for wall and roof framing and other structural work. Vast amounts are used for plywood. Machines and sands poorly. Seldom used for finish.

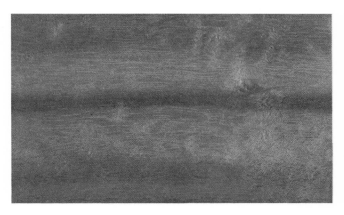

GONCALO ALVES. Strong and durable. Small open pores with a texture that varies. Fairly easy to work however, and very smooth surfaces can be produced. Source: Mexico and tropical South America. Used for furniture and cabinetwork.

GREENHEART. Extreme hardness, weight, and strength. Open pores. Light olive green to nearly black in color. Source: British Guiana and West Indies. Used for the construction of ships, docks and piling. Available only in solid stock.

GUM, SWEET. Also called Red Gum. Fairly hard and strong (36 lbs. to cu. ft.). A close grained wood that machines well but has a tendency to warp. Heartwood is reddish-brown and may be highly figured. Used extensively in furniture and cabinetmaking. Stains well, often used in combination with more expensive woods.

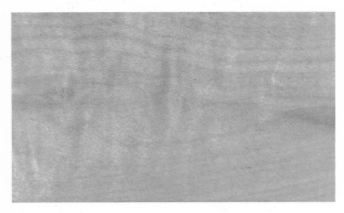

HAREWOOD. A close-grained wood nearly white in color. Grain pattern is often highly figured. Sometimes dyed to a gray color and called Silver Harewood. Source: England. Used for marquetry, inlay and paneling.

HEMLOCK. A softwood. Light in weight and moderately hard. Light reddish brown in color with a slight purplish cast. (Sample shows edge grain.) Source: Pacific Coast and Western States. Used for construction lumber and pulpwood; also for containers and plywood core stock.

HOLLY. Light, tough and very close-grained. Nearly white when cut but turns slightly brown with age and exposure. Source: South Central and Southern States. Used for inlays and marquetry and to a limited extent in cabinetwork.

IMBUYA. Also called Brazilian Walnut. Moderately hard and heavy with an open-grain and fine texture. Source: Brazil. Used for high grade furniture, cabinetwork and paneling.

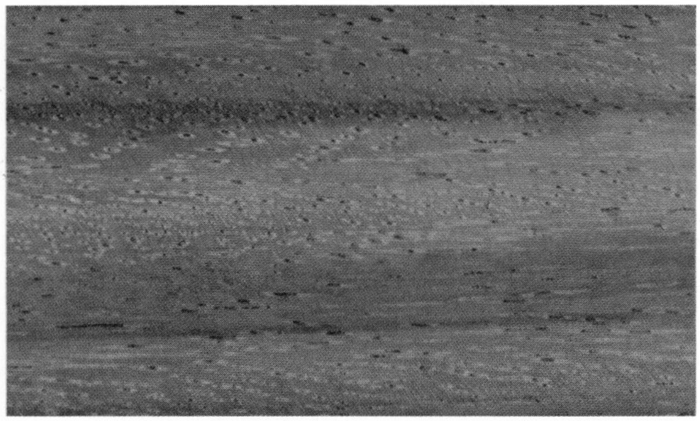

KELOBRA. Coarse texture with very large pores. Pronounced grain patterns are large and often have wavy lines. Source: Mexico, Guatemala and British Honduras. Used for furniture and cabinetry mainly in the form of veneers.

KOA. Medium hard with a texture similar to walnut. Open-grained. Color is golden to dark brown, often with dark streaks. Polishes to a lustrous sheen. The best known hardwood of the Hawaiian Islands. Used for fine furniture, art objects, and musical instruments.

LACEWOOD. Medium hard with flaky grain formed by wood rays. Usually quartered to produce small uniform flakes (as shown). Source: Australia. Occasionally used for decorative overlays on furniture and for wall paneling.

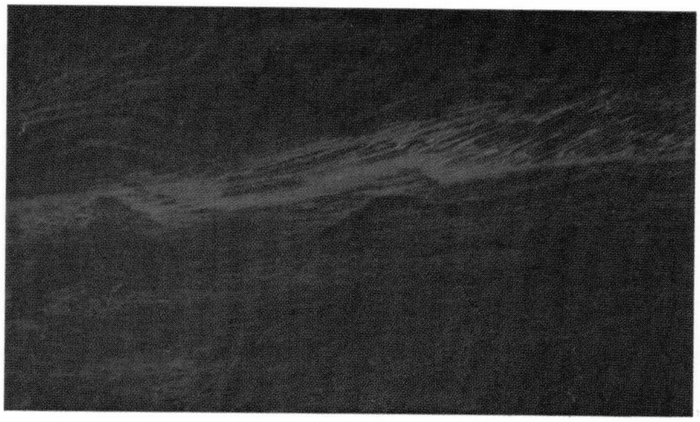

LIGNUM VITAE. Very hard and heavy (specific gravity 1.20). Oily and mildly scented, with a fine interwoven grain texture that makes it hard to work. Source: Central America, West Indies, and Northern South America. Used for bearings, pulleys, brush backs, and similar items.

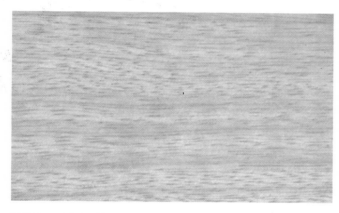

LIMBA. A light blond wood from the Congo, often sold under the trade name Korina. It has an open grain with about the same texture and hardness of Mahogany. Works easily with either hand or machine tools. Used for furniture and fixtures, especially where light tones are required.

MAHOGANY, AFRICAN. Characteristics are similar to American varieties. Slightly coarser texture and more pronounced grain patterns. Quarter sawing or slicing produces a ribbon grain effect (as shown). Source: Ivory Coast, Ghana and Nigeria. Used for fine furniture, interior finish, art objects and boats.

MAHOGANY, HONDURAS. Medium hard and dense (32 lbs. to cu. ft.). Excellent working and machining qualities. A very stable wood, with even texture, open pores, and beautiful grain patterns. Used for high grade foundry patterns and quality furniture. It turns and carves especially well.

MAHOGANY, PHILIPPINE. Medium density and hardness (37 lbs. to cu. ft.). Open grain and coarse texture. Works fairly well with hand or machine tools. Varies in color from dark red (Tanguile) to light tan (Lauan). Used for medium price furniture, fixtures, trim, wall paneling. Also, boat building and core stock in plywood.

MAKORE. Also called African Cherry and Cherry Mahogany. Somewhat like mahogany but with a finer texture and also harder and heavier. May often have a more pronounced grain pattern than sample shown. Source: West Africa, especially Nigeria. Used for furniture and cabinetry.

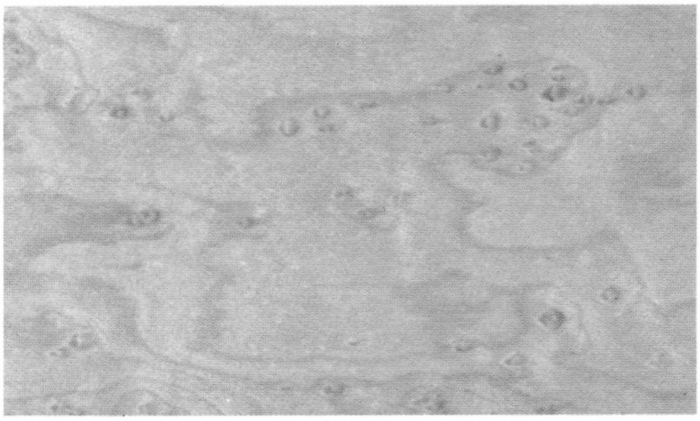

MAPLE, BIRDSEYE. Hard or Sugar Maple with tiny spots of curly grain that look like bird's eyes. The cause of this figure is not known. It may be distributed throughout the tree or located only in irregular stripes or patches. Used for highly decorative inlays and overlays.

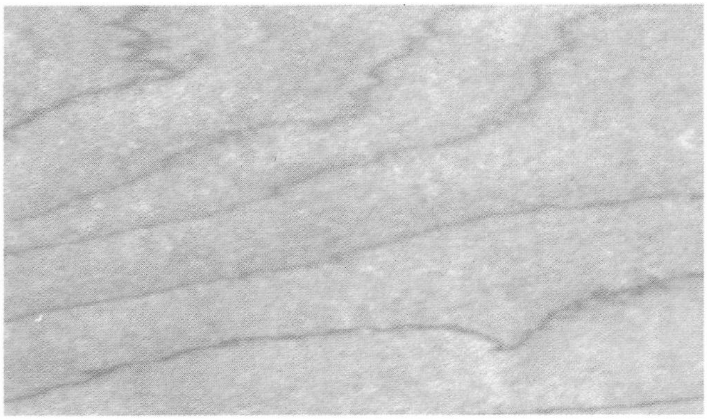

MAPLE, SUGAR. Also called Hard Maple. It is hard, strong, and heavy (44 lbs. to cu. ft.). Fine texture and grain pattern. Light tan color, with occasional dark streaks. Hard to work with hand tools but machines easily. Is an excellent turning wood. Used for floors, bowling alleys, woodenware, handles, and quality furniture.

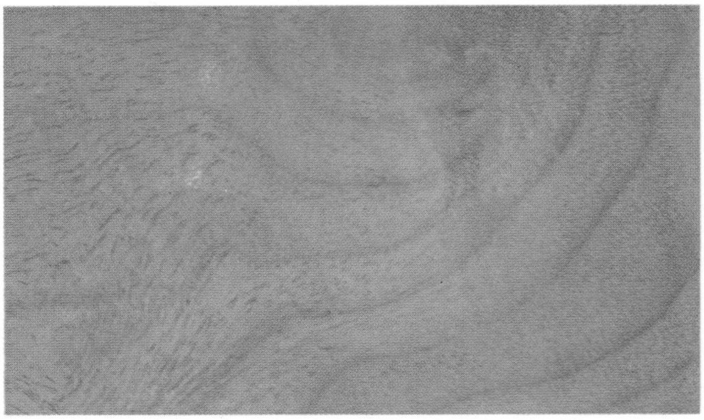

MYRTLE. Hard and strong with pore size and distribution about the same as walnut. Golden brown color with an olive green cast. Machines easily and can be polished to a high luster. Source: Southwestern Oregon. Used for decorative panels in furniture and architectural woodwork; also for art objects and novelties.

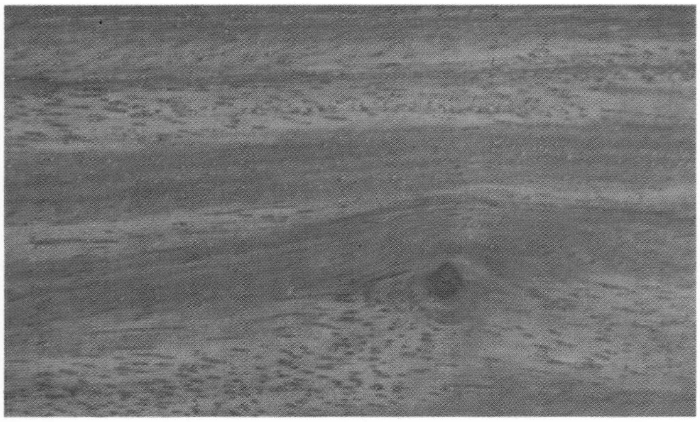

NARRA. Heavy, hard, and durable. Has a distinct grain pattern and open pores. Generally a golden yellow brown but may be found in a rose or deep red color. Source: Philippines and Malaysia. Used for high grade furniture and interior finish of ships.

OAK, ENGLISH. Open-grained with a texture similar to regular oak. Light tan to deep brown color with dark spots that create an unusual figure and grain character. Source: England. Used for fine furniture and special architectural woodwork.

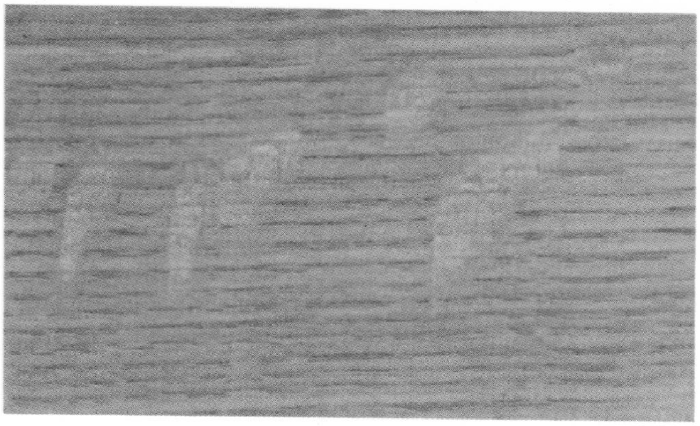

OAK, QUARTERED. Sawing or slicing oak in a radial direction results in a striking pattern as shown. The "flakes" are formed by large wood rays that reflect light. Used where dramatic wood grain effects are desired.

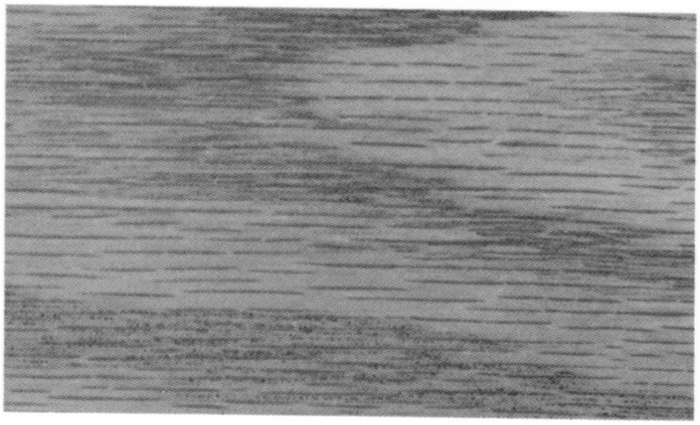

OAK, RED. Heavy (45 lbs. to cu. ft.) and hard with the same general characteristics as White Oak. Heartwood is reddish-brown in color. No tyloses in wood pores. Used for flooring, millwork and inside trim. Difficult to work with hand tools.

OAK, WHITE. Heavy (47 lbs. to cu. ft.), very hard, durable, and strong. Works best with power tools. Heartwood is greyish-brown with open pores that are distinct and plugged with a hairlike growth called tyloses. Used for high quality millwork, interior finish, furniture, carvings, boat structures, barrels and kegs.

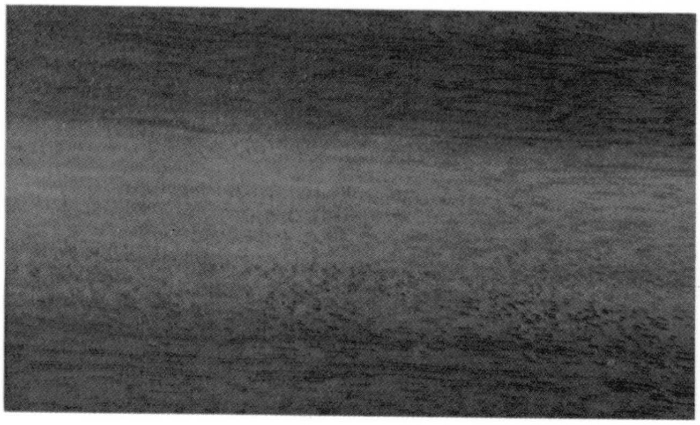

ORIENTALWOOD. Also called Australian Laurel. Medium weight with wood characteristics similar to walnut. Color is pinkish gray to brown with dark stripes as shown. Source: Australia. Used for highest quality furniture and cabinetwork.

PADOUK. Also called Vermillion. Hard firm texture with some interlocked grain that makes it difficult to work. Large open pores. Red color, may have streaks of yellow or brown. Source: Burma and West Africa. Used for art objects and novelties.

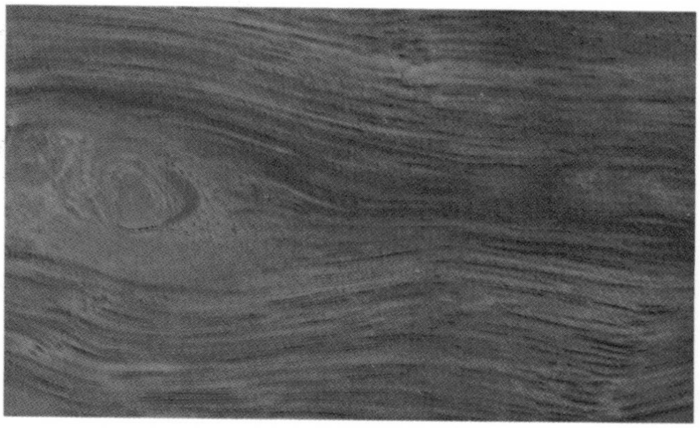

PALDAO. A fairly hard wood with large pores that are partially plugged. Grain patterns are striking and beautiful and provide an excellent example of an "exotic" wood. Source: Philippine Islands. Sometimes selected by architects for special fixtures or built-ins for public or institutional buildings.

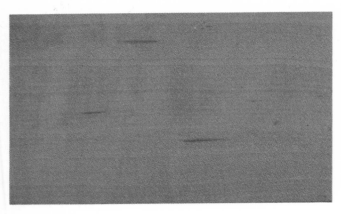

PEARWOOD. A very fine textured, close-grained wood with pores that are indistinct. Subdued grain pattern, sometimes with mottled figure. Source: United States and Europe. Used for fine furniture, marquetry, saw handles and rulers.

PINE, PONDEROSA. Lightweight (28 lbs. to cu. ft.) and soft. Straight grained and uniform texture. Not a strong wood but works easily and has little tendency to warp. Heartwood is a light reddish-brown. Change from springwood to summerwood is abrupt. Used for window and door frames, moldings, and other millwork; toys, models.

PINE, SUGAR. Lightweight, (26 lbs. to cu. ft.) soft, and uniform texture. Heartwood, light brown with many tiny resin canals that appear as brown flecks. Straight grained and warp resistant. Cuts and works very easily with hand tools. Used for foundry patterns, sash and door construction, and quality millwork.

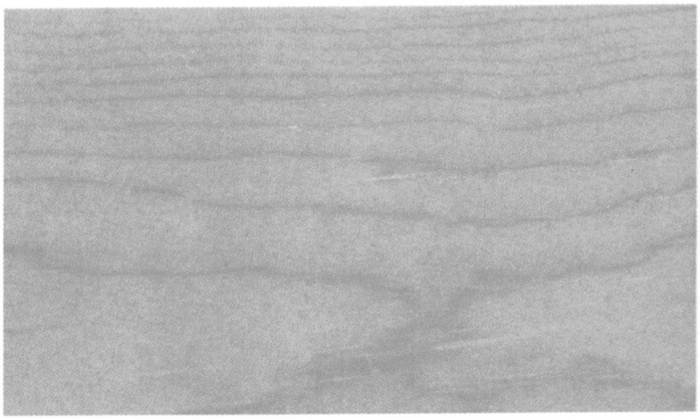

PINE, WHITE. Soft, light (28 lbs. to cu. ft.) and even texture. Cream colored with some resin canals but not as prevalent as in Sugar Pine. Used for interior and exterior trim and millwork items. Knotty grades often used for wall paneling. Works easily with hand or machine tools.

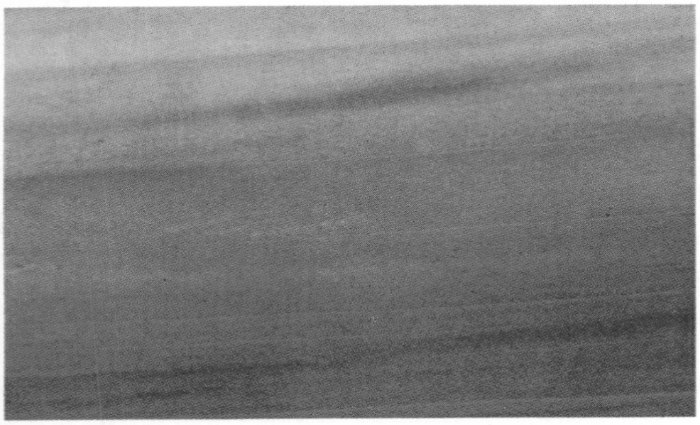

POPLAR, YELLOW. Moderately soft, light in weight (34 lbs. to cu. ft.), and even textured. Heartwood is a pale olive-brown and sapwood is greyish-white. Works well with hand or machine tools and resists warping. Used in a wide variety of products including inexpensive furniture, trunks, toys, and core stock for plywood.

PRIMAVERA. Sometimes called white or golden mahogany. Medium to coarse texture with straight and somewhat striped grain. Very similar to mahogany except for the color which is a light straw to golden yellow. Source: Southern Mexico, Guatemala, Honduras, and Salvador. One of the fine cabinet woods of the world.

REDWOOD. Soft and light in weight (28 lbs. to cu. ft.). Texture varies but is usually fine and even grained. Easy to work and durable. Heartwood is reddish-brown. Used for structures, outside finish and sometimes for interior paneling. Its durability makes it especially valuable for products exposed to water and moisture.

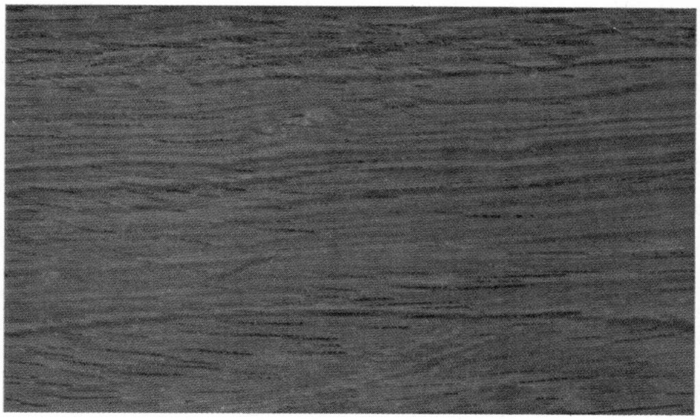

ROSEWOOD, BRAZILIAN. A very hard wood with large irregular pores. Various shades of dark brown with conspicuous black streaks. Rosewood with different colors and characteristics also comes from India, Ceylon, Madagascar and Central America. A beautiful wood used for art objects, levels, tool handles and musical instruments.

SAPELE. Grain, color, and figure are typical of African Mahogany but somewhat heavier and harder. Also not as dimensionally stable. Source: Ivory Coast and Nigeria. Used extensively in veneer and solid form for furniture, fixtures, cabinets, and boats.

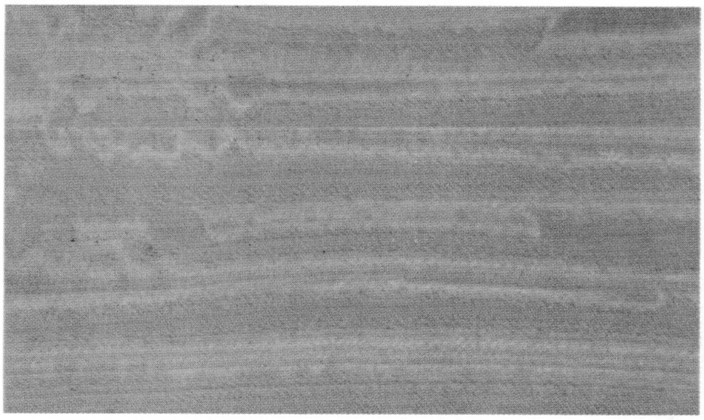

SATINWOOD. Fine grain texture, hard and heavy, with a slight oilyness. Golden yellow in color with beautiful wavy grain patterns that often give a mottled effect. Source: Puerto Rico. Honduras, Ceylon and East Indies. Used for inlays, marquetry, fine brush handles and similar items.

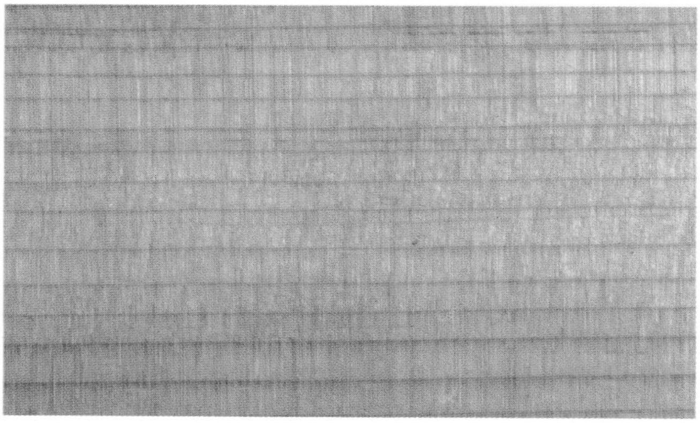

SPRUCE. A soft wood, light in weight (24 lbs. per cu. ft.). Transition from spring to summer growth is gradual (see edge grain sample). There are several species; Sitka, Englemen, and a general classification called Eastern. Source: various parts of the United States and Canada. Used for pulpwood, light construction and carpentry.

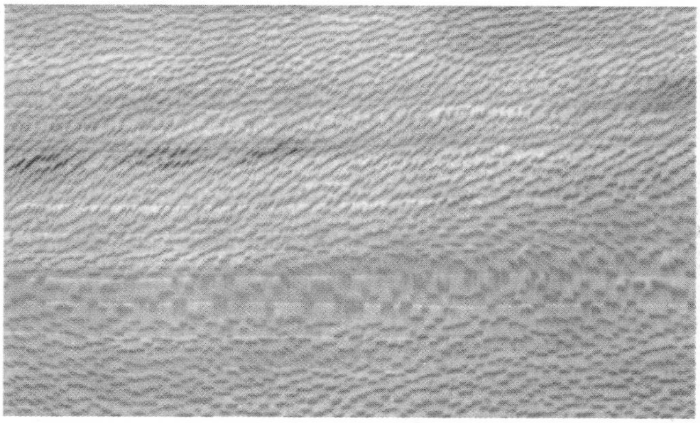

SYCAMORE. Medium density, hardness, and strength. A close-grained wood with a rather coarse texture. Easily identified by the flaky pattern of wood rays observed best in quartered stock. Source: Eastern half of United States. Used for drawer sides and lower priced furniture. Veneers are used for berry and fruit boxes.

TAMO. Also called Japanese Ash. The physical characteristics of the wood resembles American Ash. Grain patterns are extremely pronounced. Figures may consist of swirl, fiddleback or blister types. Source: Japan. Used for inlays and overlays where a highly decorative surface is required.

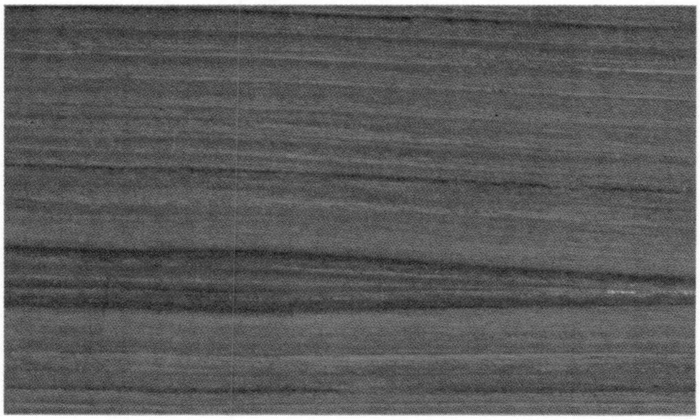

TEAK. Strong and quite hard. Resembles walnut except for a lighter tawny yellow color. Silicates and minerals in the wood give it an oily feel and dulls regular tools more quickly. Source: Burma, India, Thailand and Java. Used for fine furniture and paneling. Also used for ship building because of its great durability.

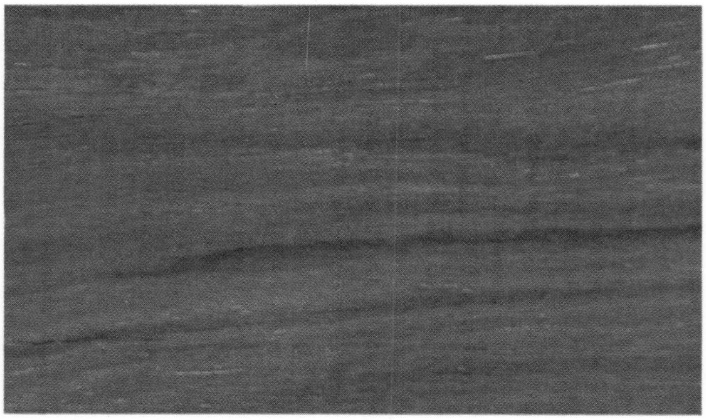

TULIPWOOD. Hard, medium textured, open-grained wood, with small pores that vary in size. Red and yellow color streaked with dark lines. Source: Northeastern Brazil. A highly decorative wood used for inlays, turnings, and novelties.

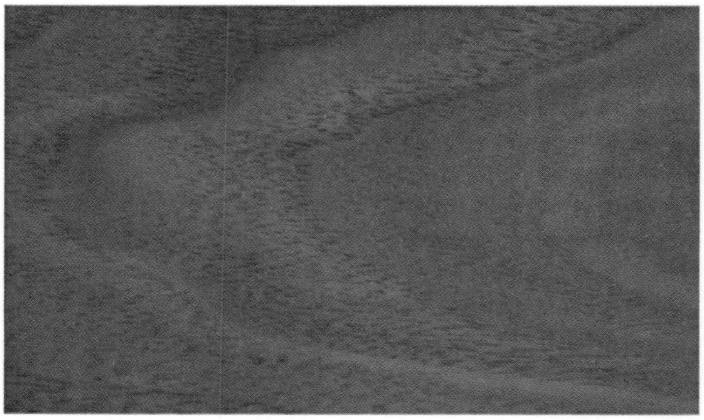

WALNUT, BLACK. Fairly dense and hard. Very strong in comparison to its weight (38 lbs. to cu. ft.). Excellent machining and finishing properties. A fine textured open grain wood with beautiful grain patterns. Heartwood is a chocolate brown with sapwood near white. Used on quality furniture, gun stocks, fine cabinetwork, etc.

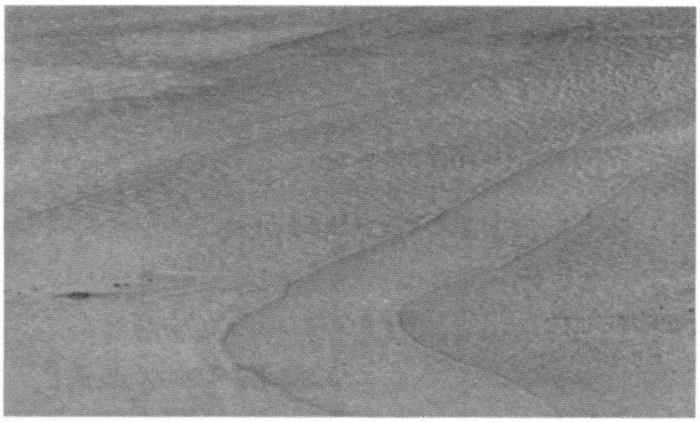

WILLOW, BLACK. Very soft and light in weight (27 lbs. to cu. ft.). Resembles basswood in workability, although there is some tendency for the machined surface to be fuzzy. Heartwood varies from light gray to dark brown. Used for some inexpensive furniture, core stock for plywood, wall paneling, toys, and novelty products.

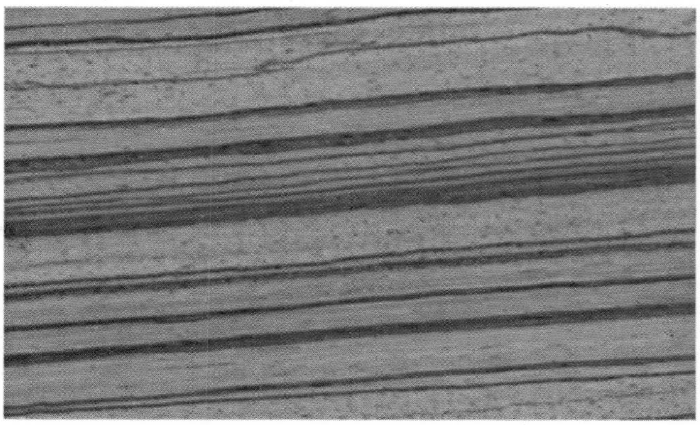

ZEBRAWOOD. Also called Zebrano. Heavy, hard, open-grained wood with a medium texture. Light gold in color with narrow streaks of dark brown in quartered stock (see sample). A highly decorative wood from Central and West Africa. Made into quartered veneer and used where a spectacular effect is desired.

GENERAL SAFETY RULES

The good carpenter recognizes that safety is an important part of his job. He knows that accidents are prevalent in building construction and that they often result in partial or total disability. Even minor cuts and bruises can be painful.

Safety is based on knowledge, skill, and an attitude of care and concern. The carpenter should know correct and proper procedures for performing the work and be familiar with the potential hazards -- how they can be minimized or eliminated.

He must develop a good attitude toward safety. This means that he believes in the importance of safety and is willing to give time and effort to a continuous study of the safest ways to perform his work. It means that he will work carefully and follow the rules.

CLOTHING: Wear clothing that is appropriate for the work and weather conditions. Trousers or overalls should fit properly and have legs without cuffs. Shirts and jackets should be kept buttoned. Sleeves should also be buttoned or rolled up. Never wear loose or ragged clothing, especially around moving machinery.

Shoes should be sturdy with thick soles that will protect your feet from protruding nails. Tennis or lightweight canvas shoes are not satisfactory. Guard against leather soles that may wear too slick to provide satisfactory traction on smooth wood surfaces.

Most carpenters prefer to wear a cloth or leather cap although a hat is also appropriate. Some construction work will require a hard hat (hat that will help prevent head injury from falling objects). Headgear should provide the necessary protection, be comfortable, permit good visibility, and shade your eyes. All clothing should be maintained in a satisfactory state of repair and not be permitted to become excessively soiled.

PERSONAL PROTECTIVE EQUIPMENT: Safety glasses should be worn whenever the work involves even the slightest hazard to your eyes. Standard specifications state that a safety lens must withstand the blow of a 1/8 in. steel ball dropped from a height of 50 in.

Safety boots and shoes are required on heavy construction jobs. They consist of special reinforced toes that will withstand a load of 2500 lbs.

Hard hats should be worn whenever you are exposed to any possibility of falling objects. Standard specifications require that such hats withstand a certain degree of denting, and resist breaking when struck with an 80 lb. ball dropped from a 5 ft. height.

Wear gloves of an appropriate type when handling rough materials, and a respirator when working in dusty areas or where finishing materials are being sprayed.

HAND TOOLS: Always select the correct type and size of tool for your work and be sure it is sharp and properly adjusted. Guard against using any tool if the handle is loose or in poor condition. Dull tools are hazardous to use because excessive force must be applied to make them cut. Oil or dirt on a tool may cause it to slip and cause an injury.

When using tools, hold them correctly. Most edge tools should be held in both hands with the cutting action away from yourself. Be careful when using your hand or fingers as a guide to start a cut.

Handle and carry tools with considerable care. Keep edged and pointed tools turned downward. Carry only a few tools at one time unless they are mounted in a special holder. Do not carry sharp tools in pockets of your clothing. When not in use, tools should be kept in special boxes, chests or cabinets.

POWER TOOLS: Before operating any power tool or machine you must be thoroughly familiar with the way it works and the correct procedures to follow. In general, when you learn to use equipment the correct way, you also learn to use it the safe way.

There are a number of general safety rules that apply to power equipment. In addition, special safety rules must be observed in the operation of each individual tool, or machine. Those that apply directly to the power tools commonly used for modern carpentry are listed in Unit 2. Study and follow them carefully.

GOOD HOUSEKEEPING: This refers to the neatness and good order of the construction site. Maintaining a satisfactory level contributes to the efficiency of the worker and is an important factor in the prevention of accidents.

Place building materials and supplies in carefully laid piles, positioned to allow adequate aisles and walkways. Rubbish and scrap should be placed in containers until disposal can be made. Do not permit blocks of wood, nails, bolts, empty cans, or pieces of wire to accumulate, since they interfere with walking and constitute a tripping hazard.

Keep tools and equipment, which are not being used, in panels or chests. This will provide protection for the tools and also the workers. In addition to improving efficiency and safety – good housekeeping helps maintain a better appearance of the construction project, which in turn will contribute to the morale of all workers.

DECKS AND FLOORS: To perform an operation safely, either with hand or power tools, the carpenter should stand on a firm, solid base. The surface should be smooth but not slippery. Do not attempt to work over rough piles of earth or on stacks of material that are unstable. Stay well away from floor openings, floor edges, and excavations as much as possible. Where this cannot be done, install adequate guardrails or barricades. In cold weather remove any accumulation of ice or cover it with sand or ice melting chemical.

EXCAVATIONS: Shoring and adequate bracing must be placed across the face of any excavation where the ground is cracked or caving is likely to occur. Inspect the excavation and shoring daily and especially after rains. Follow state and local regulations. Never climb into an open trench until proper reinforcement against cave-in has been installed – or until the sides have been sloped to the "angle of repose" of the material being excavated.

SCAFFOLDS AND LADDERS: All types of scaffolds should have a minimum safety factor of four to one. This means that the scaffold will carry a load four times greater than the load it will probably be required to support. All scaffolding should be constructed under the direction of an experienced craftsman and inspected daily before use. Ladders should be checked at frequent and regular intervals, and their use should be limited to climbing from one level to another. The performance of work while being supported on a ladder is hazardous and should be kept to a minimum.

There are many safety rules that must be observed in the use of scaffolds and ladders. These are covered in Unit 22.

FALLING OBJECTS: When working on upper levels of a structure, you should be especially cautious in handling tools and materials so there is no chance of them falling on workers below. Do not place tools on the edge of scaffolds, stepladders, window sills or on any other surface where they might be knocked off. If long pieces of lumber must be placed temporarily on end and leaned against the side of the structure, be sure they will not fall sideways. When moving through a building under construction, be aware of overhead work, and wherever possible, avoid passing directly underneath. Stay clear of materials being hoisted. Wear an approved hard hat whenever there is a possibility of falling objects.

LIFTING AND CARRYING: Injuries, such as hernias, may be caused by improper lifting or carrying heavy objects. When lifting, stand close to the load, bend your knees and grasp the object firmly -- then lift by straightening your legs and keeping your body as nearly vertical as possible. To lower the object, reverse the procedure.

When carrying a heavy load, do not turn or twist your body but make adjustments in position by shifting your feet. If the load is heavy or bulky, secure help from others. Never underestimate the weight to be moved or overestimate your own ability. Always secure assistance when carrying long pieces of lumber.

FIRE PROTECTION: Carpenters should have a good understanding of fire hazards -- their causes and methods of control. Class A fires result from burning wood and debris; Class B fires involve highly volatile materials such as gasoline, oil, paints and oil soaked rags; and Class C fires are caused by electrical wiring and equipment. Any of these fires can occur on a typical construction site.

Approved fire prevention practices should be followed throughout the construction project. Good housekeeping is an important aspect. Special precautions should be taken during the final stages of construction when heating and wiring are being installed and when highly flammable surface finishes are being applied. Always keep containers of volatile materials closed when not in use and dispose of oily rags and combustible materials promptly.

Fire extinguishers should be available on the construction site. Be sure to use the proper kind for each type of fire. Study and follow local regulations.

FIRST AID: A knowledge of first aid is important. You should understand approved procedures and be able to exercise good judgment in applying them. Remember that an accident victim may receive additional injury from extensive treatment by an unqualified person. Information of this nature can be secured from your local Red Cross.

Keep an approved first aid kit available and use it to clean and bandage even the slightest nicks and cuts.

Using a portable circular saw to make a bevel cut in 3/4 in. plywood. Note how the shoe has been tilted. When using a portable circular saw, be sure to observe the safety rules listed on page 29. (Black and Decker Co.)

Unit 2
POWER TOOLS

Modern power tools greatly reduce the time required to perform many of the operations in carpentry work. Heavy sawing, planing, and boring can be accomplished with far less human energy, and when proper tools are used in the correct way, high levels of accuracy can be maintained.

There are two general types of power tools; portable and stationary. Portable tools are light in weight, easily carried, and are held in the hands of the operator during operation. Stationary tools (also called machines) are mounted on benches or stands which rest firmly on the floor. Work is brought to the machine. In contrast, the portable tool is moved to the work. Portable tools are extensively used in nearly every stage of carpentry, especially rough framing.

Space in this Unit permits only a brief description of the kinds of power tools most commonly associated with carpentry work. This should be supplemented with woodworking textbooks and reference books devoted to power tool operation. Manufacturers' bulletins and operator's manuals are also a good source of information.

Safety

Safety must be stressed continually. Before operating any power tool, you must become thoroughly familiar with the way it works and the correct procedure to follow in its use. You must be wide awake and alert. Never operate a power tool when overly tired or ill. Think through the operation before performing it. Know what you are going to do, and what the tool will do. Make all adjustments before turning on the power and be sure blades and cutters are sharp and are of the correct type for the work.

While operating a power tool, do not allow your attention to be distracted. See Fig. 2-1. Do not distract the attention of others while they are operating power tools. Keep all safety guards in position and wear safety glasses.

Feed the work carefully and only as fast as the tool will cut it easily. Overloading is hazardous to the operator and will likely damage the tool or work. When the operation is complete, turn off the power and wait until the moving parts have stopped before leaving the machine.

Always make sure that the source of electric power is the correct voltage and that the tool switch is in the "off" position before it is plugged in. The electrical cord and plug must be in good condition and provide a ground for the tool. When

Fig. 2-1. When operating a power tool, give full attention to the work.

long extension cords are required, they too must provide for grounding and be of sufficient size to prevent excessive voltage drop. Be cautious in stringing electrical cords around the working site so they will not be damaged or interfere with other workers. Electrical shock is one of the potential hazards in operating portable tools. Always be sure that proper grounding is provided and receptacles, plugs and cords are an approved type.

Portable Circular Saws

This power tool is also called an electric hand saw or builders' saw. Its size is determined by the diameter of the largest blade it will take. Most carpenters prefer a 7 or 8 in. saw. The depth of cut is adjusted by raising or lowering the position of the base or shoe. See Fig. 2-2. On many saws it is possible to make bevel cuts by tilting the shoe.

Portable saws are often guided along the layout line "free-hand" and therefore extra clearance in the saw kerf is required. To provide this clearance, blades are usually set with about 2 1/2 gauges on each side. Fig. 2-3 shows standard types of blades. The rough cut combination blade is a popular type because it is suitable for both ripping and crosscutting. Some carpenters prefer carbide tipped teeth, Fig. 2-4, because they usually stay sharp longer than teeth of a standard blade.

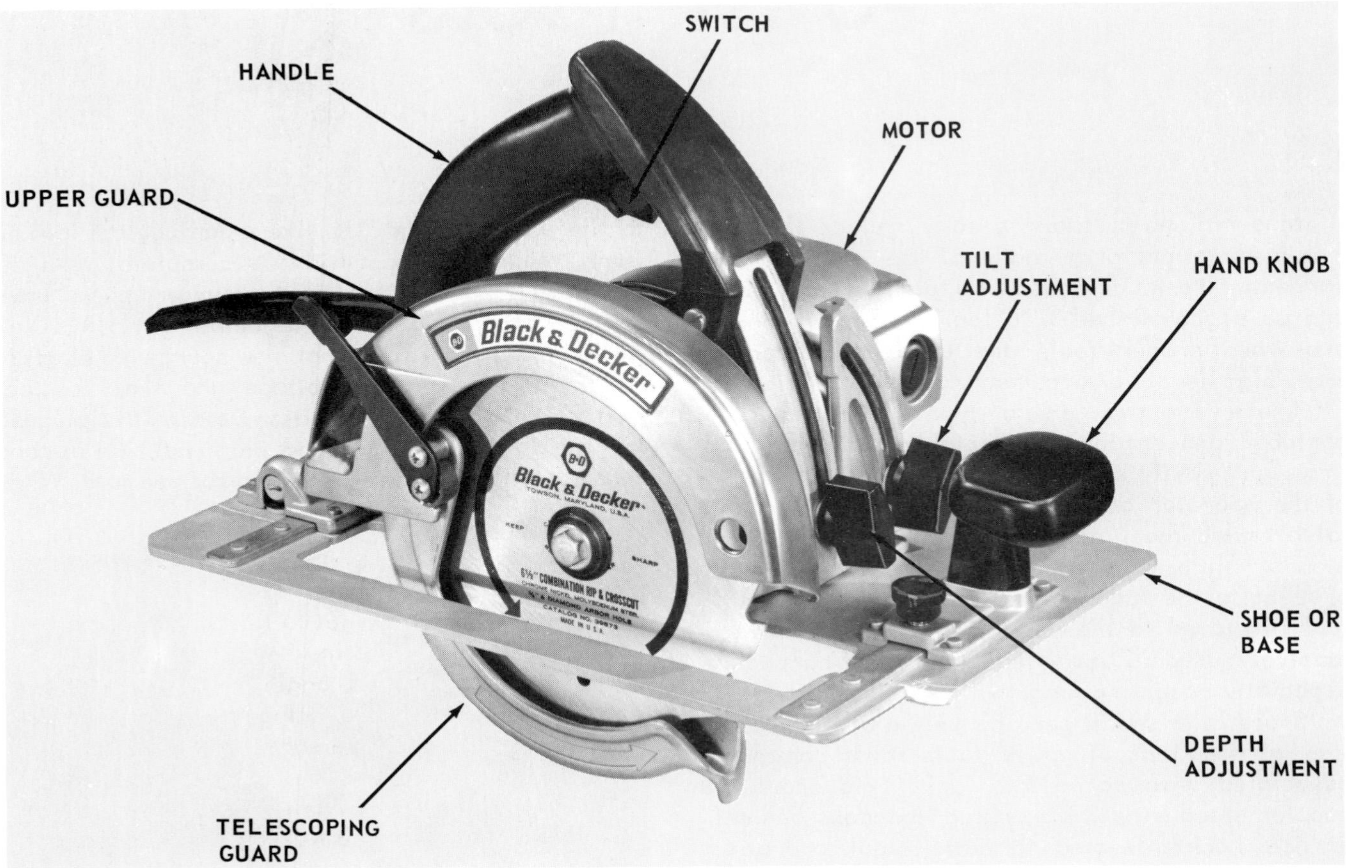

Fig. 2-2. Parts of a portable circular saw. In operation, the telescoping guard is pushed back by the stock and a spring returns it to a closed position when the cut is complete.

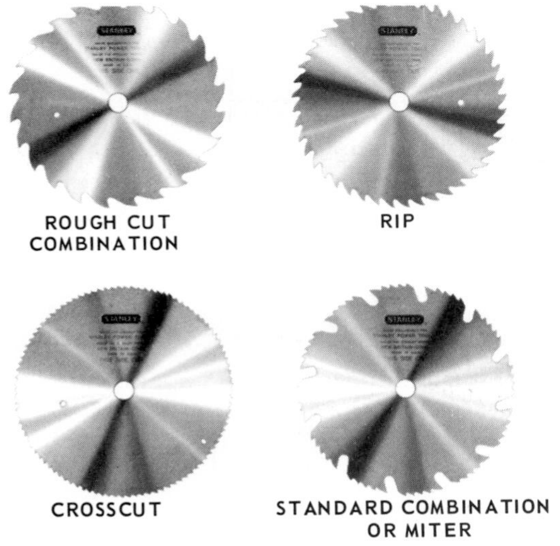

Fig. 2-3. Standard types of saw blades used on portable circular saws.

In using a portable saw, grasp the handle firmly in one hand with the forefinger ready to operate the trigger switch. The other hand should be placed on the stock, well away from the cutting line. Some saws will require both hands on the machine. Rest the base on the work and align the guide mark with the layout line. Turn on the switch, allow the motor to reach full speed and then feed it smoothly into the stock as shown in Fig. 2-5. Release the switch as soon as the cut is finished and continue to hold the saw until the blade stops.

The portable saw may be used to make cuts in assembled work. For example, flooring and roofing boards are often nailed into place and then the ends are trimmed.

Safety Rules for Portable Circular Saws

1. Stock must be well supported in such a way that the kerf will not close and bind the blade during the cut or at the end of the cut.
2. Thin materials should be supported near the cut. Small pieces should be clamped to a bench top or sawhorse.

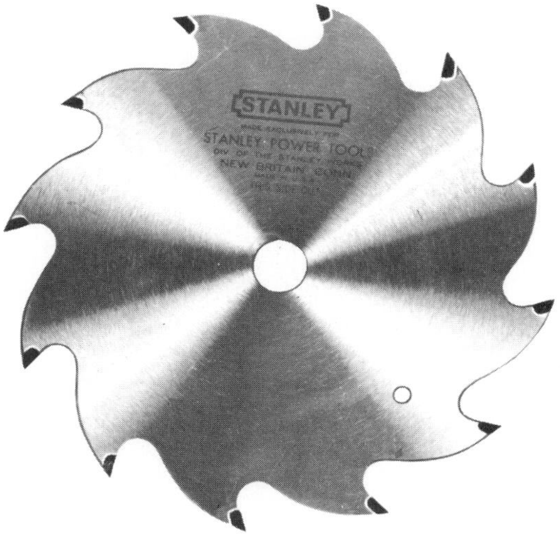

Fig. 2-4. Carbide tipped blade stays sharp longer. Handle carefully to prevent points from being broken off. (Stanley Tools)

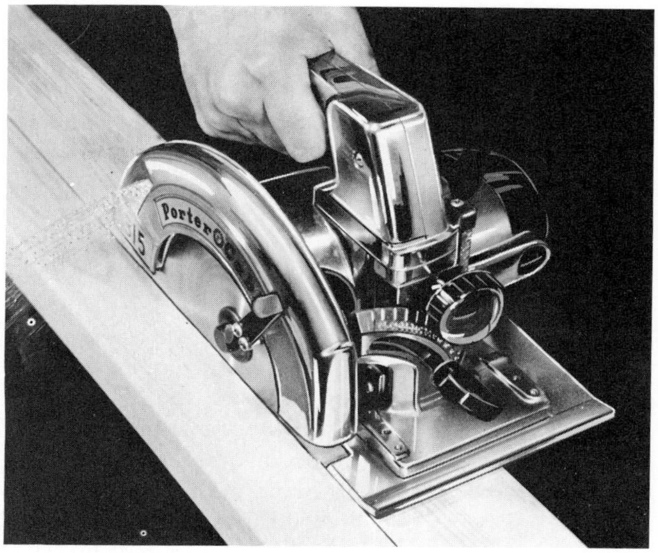

Fig. 2-5. Using a portable saw to rip along a line. Some saws have an adjustable guide that is helpful in performing this operation.

3. Be careful not to cut into the sawhorse or other supporting device.
4. Adjust the depth of cut to the thickness of the stock, plus about 1/8 in.
5. Check the base and angle adjustment to be sure they are tight. Plug the cord into a grounded outlet and be sure it will not become tangled in the work.
6. Always place the saw base on the stock with the blade clear before turning on the switch.

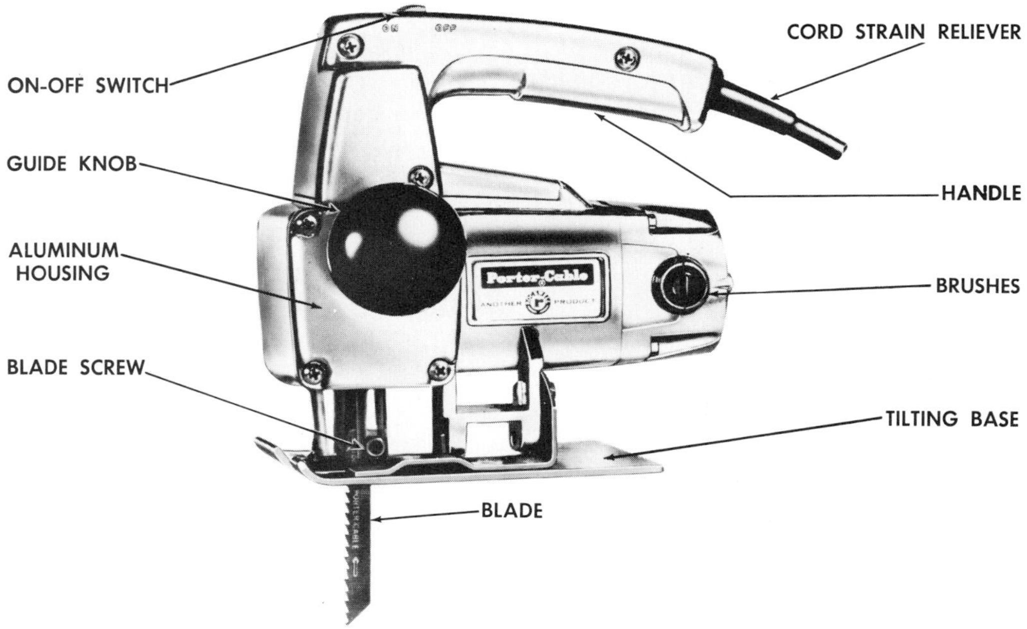

Fig. 2-6. Saber saw with principal parts identified.

7. During the cut, stand to one side of the cutting line. Never reach under the material being cut.
8. Some portable saws have two handles. In using such saws, be sure to keep both hands on the handles during the cutting operation.
9. Always unplug the machine to change blades or make adjustments.
10. Always use a sharp blade that is properly set.

Saber Saws

The saber saw, also called a bayonet saw, is a portable electric jig saw that can be used for a wide range of light work. It is used by carpenters, cabinetmakers, electricians and home craftsmen. A standard model and its basic parts are shown in Fig. 2-6. The stroke of the blade is about 1/2 in. and it operates at a speed of approximately 2500 strokes per minute.

Various blades are available for wood cutting with a range of 6 to 12 teeth per in., Fig. 2-7. For general purpose work, a blade with 10 teeth per in. is satisfactory. Always select a blade that will have at least two teeth in contact with the edge being cut. Saws will vary in the way the blade is mounted in the chuck. Follow directions listed in the manufacturer's manual. Also follow the lubrication schedule as specified in this manual.

The saber saw can be used to make straight or bevel cuts as shown in Fig. 2-8. Curves are usually cut by guiding the saw along a layout line, however, circular cuts may be made more accurately with a special guide or attachment.

The blade cuts on the up stroke so splintering will take place on the top side of the work. This must be considered when making finished cuts, especially in fine hardwood plywood. Always hold the base of the saw firmly against the surface of the material being cut.

Fig. 2-7. Saber saw blades.

Safety Rules for Saber Saws

1. Make certain the saw is properly grounded through the electrical cord. The switch must be in "off" position before connecting to power source.
2. Select the correct blade for your work and be sure it is properly mounted.
3. Disconnect the saw to change blades or make adjustments.
4. Place the base of the saw firmly on the stock before starting the cut.
5. Turn on the motor before the blade contacts the work.
6. Do not attempt to cut curves so sharp that the blade will be twisted.
7. Make certain the work is well supported and do not cut into sawhorses or other supports.

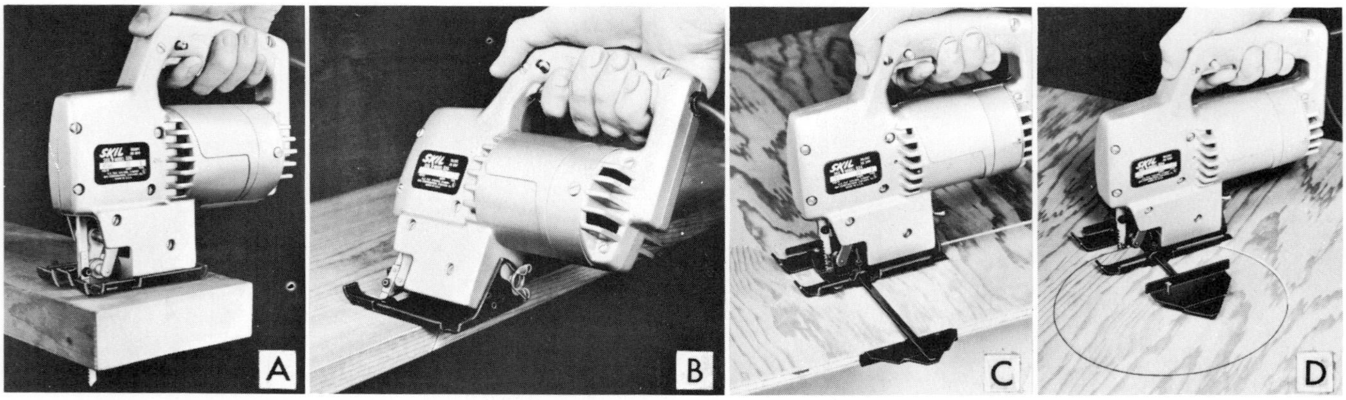

Fig. 2-8. Using a saber saw. A–Straight cutting. B–Angle cutting. C–Using a fence. D–Using a circle cutting guide. (Skil Corp.)

When cutting internal openings, a starting hole can be drilled in the waste stock,·or the saw can be held on end so the blade will cut its own opening, as shown in Fig. 2-9. This is called plunge cutting and must be undertaken with considerable care. Rest the toe of the base firmly on the work, turn on the motor, and then slowly lower the blade into the stock.

Fig. 2-9. Starting an internal cut with a saber saw.

Another portable power tool is the reciprocating saw shown in Fig. 2-10. Operation is similar to the saber saw. Fig. 2-11 shows some typical operations this type saw will perform.

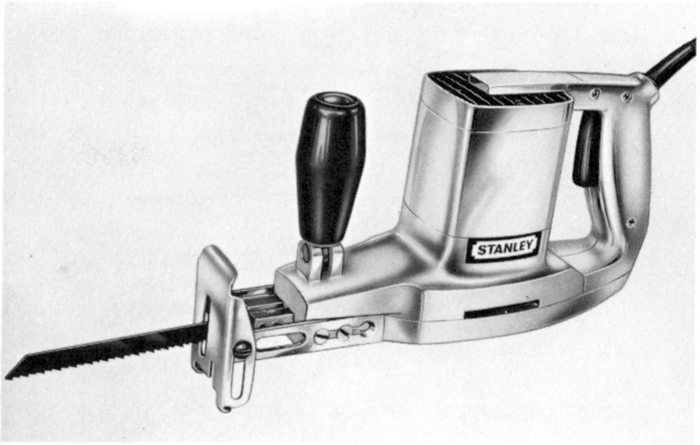

Fig. 2-10. Reciprocating saw (also called an all-purpose saw).

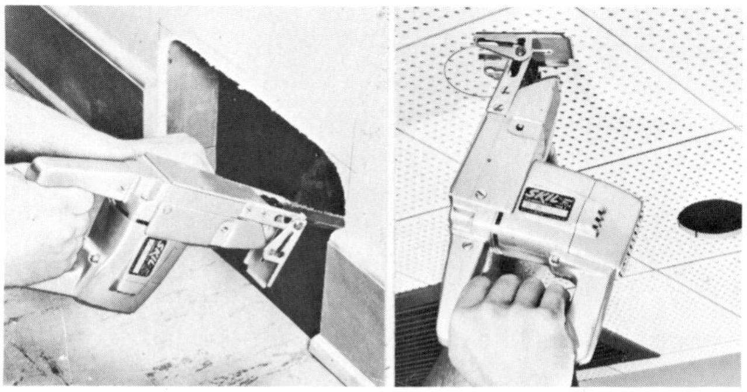

Fig. 2-11. Using a reciprocating saw. Left. Rough cutting. Right. Finish cutting.

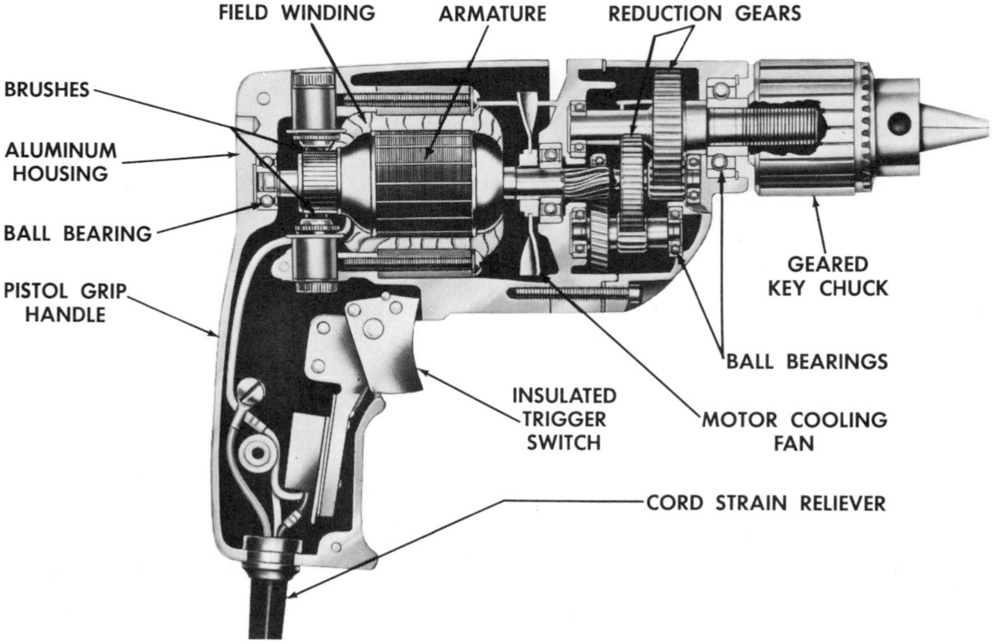

Fig. 2-12. Portable electric drill with parts identified. (Rockwell International)

Fig. 2-13. Power bits for portable electric drill.

Portable Electric Drills

Portable electric drills come in a wide range of types and sizes. The size is determined by the chuck capacity; 1/4 and 3/8 in. generally being selected by the carpenter. Fig. 2-12 illustrates the basic parts and reduction gears of a typical model. Speeds of approximately 1000 rpm are best for woodworking. Bits designed for use in a portable electric drill are shown in Fig. 2-13.

A variable speed drill is shown in Fig. 2-14. By simply depressing the trigger switch to various positions, the speed can be changed from 0 to 1000 rpm.

Fig. 2-14. Variable speed electric drill. (Stanley Power Tools)

Safety Rules for Portable Drill

1. Select the correct drill or bit for your work and mount it securely in the chuck.
2. Stock to be drilled must be held in a stationary position so it will not move during the operation.
3. Connect the drill to a properly grounded outlet with switch in the "off" position.
4. Turn on the switch for a moment to see if the bit is properly centered and running true.

Fig. 2-15. Using a hole saw in an electric drill.

Fig. 2-16. Using a cordless drill to drive a screw. (Black and Decker Mfg. Co.)

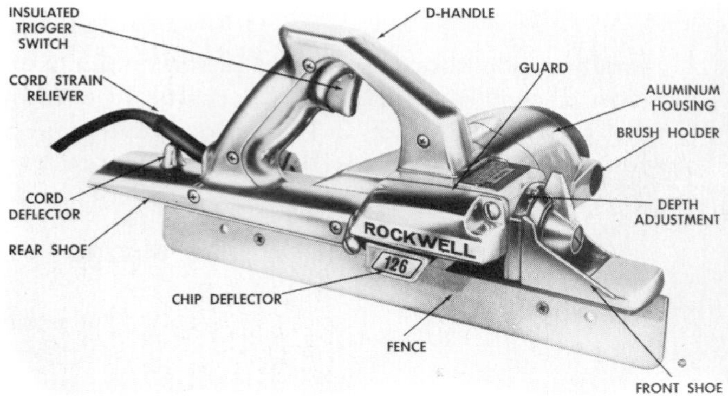

Fig. 2-17. Power plane with parts identified. (Rockwell International)

5. With the switch off, place the point of the bit in the punched layout hole.
6. Hold the drill firmly in one or both hands and at the correct drilling angle.
7. Turn on the switch and feed the drill into the work. The pressure required will vary with the size of the drill and the kind of wood. See Fig. 2-15.
8. During the operation, keep the drill aligned with the direction of the hole.
9. When drilling deep holes, especially with a twist drill, withdraw the drill several times to clear the cuttings.
10. Always remove the bit from the drill as soon as you have completed your work.

Trimming the ends of the roof joist after installation.
(Black and Decker Co.)

Using a heavy duty air-powered nailer to assemble components at a construction site. (Bostitch)

Fig. 2-18. Using the power plane to surface a board.

Cordless portable drills, Fig. 2-16, are handy for many jobs. Power is supplied by a small nickel-cadmium battery that can be recharged. Such drills are used for general maintenance work and on production jobs where there are no power lines.

Power Planes

The power plane provides the carpenter with a tool that produces finished surfaces with speed and accuracy. Fig. 2-17 shows the parts of a typical model. The motor, which operates at a speed of about 20,000 rpm, drives a spiral cutter. The depth of cut is adjusted by raising or lowering the front shoe. The rear shoe (main bed) must be kept level with the perimeter of the cutterhead.

The power plane is equipped with a fence that is adjustable for planing bevels and chamfers. For surfacing operations, it is removed as illustrated in Fig. 2-18.

Hold and operate the power plane in about the same manner as a hand plane. The work should be rigidly supported in a position that will permit the operation to be easily performed. Start the cut with the front shoe resting firmly on the work and the cutterhead slightly behind the surface, Fig. 2-19. Be sure the electric cord is kept clear. Start the motor and move the plane forward with smooth even pressure on the work. When finishing the cut, apply extra pressure on the rear shoe.

Safety Rules for Power Planes

1. Study the manufacturer's instructions for detailed information on adjustment and operation.
2. Be sure the machine is properly grounded.
3. Hold the standard power plane in both hands before you pull the trigger switch. Continue to hold it in both hands until the motor stops after releasing the switch.
4. Always be sure the work is securely clamped and held in the best position to perform the operation.
5. Do not attempt to operate a power plane designed for two hands, with one hand.
6. Disconnect the electric cord before making adjustments or changing cutters.

Fig. 2-19. Starting an edge cut with the power plane. (Black and Decker Mfg. Co.)

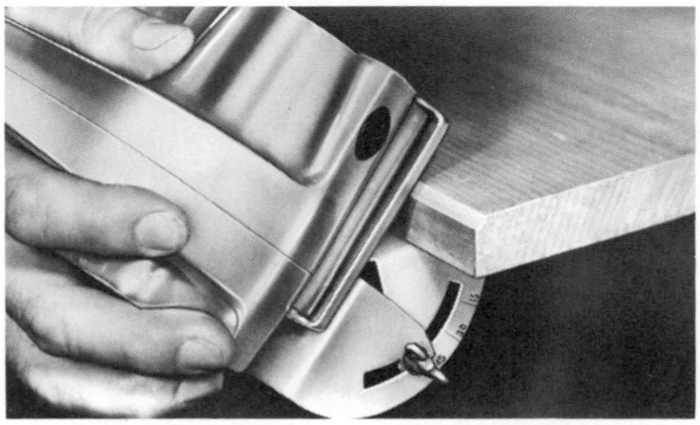

Fig. 2-20. Cutting a chamfer with a power block plane. (Rockwell International)

The power block plane, Fig. 2-20, can be used on small surfaces. It has about the same features and adjustments as the regular power plane, but is designed to be operated with one hand. When

Fig. 2-21. Portable router with motor unit mounted in an adjustable base.

using this tool, the work should be securely held or clamped in place. In planing small stock, kickbacks may occur. Be sure the hand not holding the plane is kept well out of the way.

Portable Routers

Routers are used to cut irregular shapes and to form various contours on edges, Fig. 2-21. When equipped with special guides, they can be used to cut dados, grooves, mortises and dovetail joints. Important uses in carpentry include the cutting of gains for hinges when hanging passage doors, and routing housed stringers for stair construction. See Units 16 and 17.

When mounting bits in the router, the base is usually removed as illustrated in Fig. 2-22.

Fig. 2-22. Mounting a straight bit in collet type chuck.

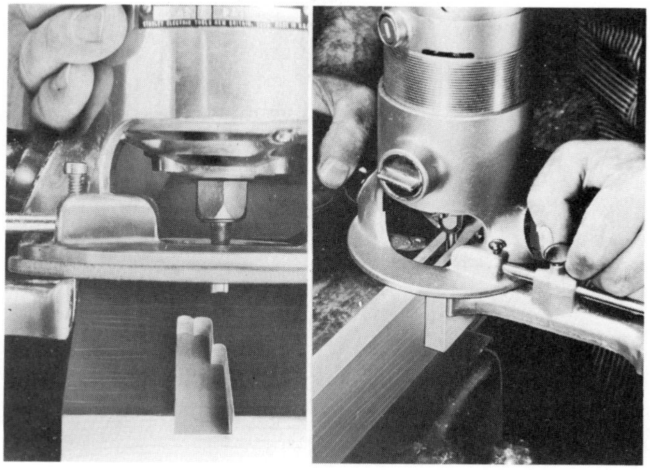

Fig. 2-23. Left. Series of cuts used to form a dado. Right. Cutting a groove. (Stanley Power Tools)

Straight bits are used when cutting dados and grooves. See Fig. 2-23.

Some bits for shaping and forming edges have a pilot tip that guides the router. The router motor revolves in a clockwise direction (when viewed from above) and should be fed from left to right when making a cut along an edge, as illustrated in Fig. 2-24. When cutting around the outside of oblong or circular pieces, always move in a counterclockwise direction.

Safety Rules for Portable Routers

1. The bit must be securely mounted in the chuck and the base must be tight.
2. Be sure the motor is properly grounded.
3. Wear eye protection.

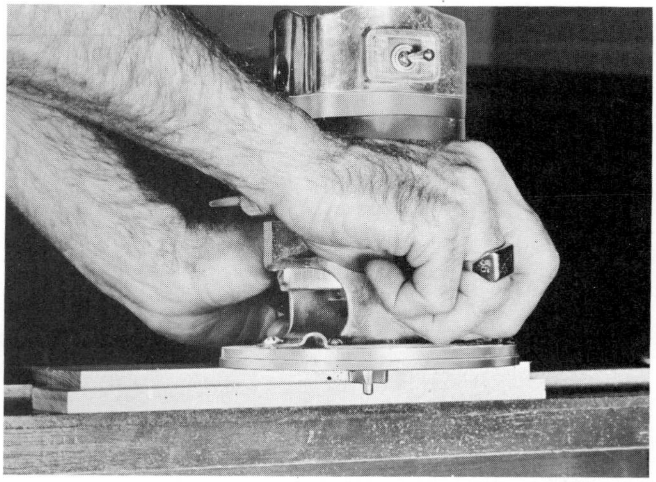

Fig. 2-24. Pilot tip on cutter controls the cut. Apply light pressure between the tip and material. Excessive pressure will cause burn marks.

4. Be certain the work is securely clamped so it will remain stationary during the routing operation.
5. Place the router base on the work, template, or guide, with the bit clear of the wood, before turning on the power. Hold it firmly when turning on the motor to overcome starting torque.
6. Hold the router with both hands and feed it smoothly through the cut in the correct direction.
7. When the cut is complete, turn off the motor. Do not lift the machine from the work until the motor has stopped.
8. Always unplug the motor when mounting bits or making adjustments.

Portable Sanders

Portable sanders include three basic types; belt, disc and finish. They vary widely in size and design. Manufacturer's instructions should be followed carefully in the mounting of abrasive belts, discs and sheets. Also follow the manufacturer's lubrication schedule.

The belt sander, Fig. 2-25, requires skillful manipulation. Stock to be sanded should be firmly supported. Be sure the switch is in the off position before plugging in the electric cord. Like all portable power tools, the sander should be properly grounded. Check the belt and make sure it is tracking properly.

Hold the sander over the work, start the motor, then lower it carefully and evenly onto the surface. Move the sander forward and backward over the surface in even strokes. At the end of each stroke, shift it sideways about one-half the width of the belt. Continue over the entire surface, holding the sander level and sanding each area the same amount. Do not press down on the sander since its weight is sufficient to provide the proper pressure for the cutting action. When work is complete, raise the machine from the surface and stop the motor.

Fig. 2-25. Using a portable belt sander. The size is the same as the belt width.

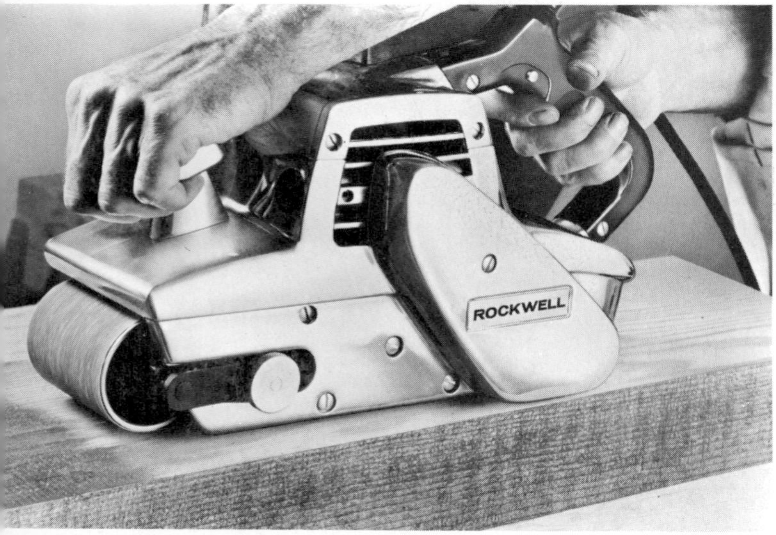

Finishing sanders, Fig. 2-26, are used for final sanding where only a small amount of material needs to be removed. They are also used for cutting down and rubbing finishing coats. There are two general types; orbital and oscillating.

Fig. 2-26. Using a finishing sander. (Stanley Power Tools)

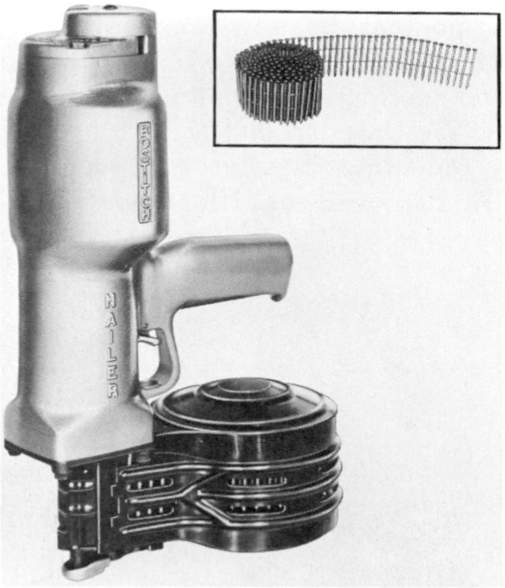

Fig. 2-27. Air-powered nailer. Drives regular headed nails as shown in detail. Operates on 60 to 90 psi.

Staplers and Nailers

A wide variety of power staplers and nailers is available. Most of them are air (pneumatic) powered. Those that are electrically operated should be properly grounded. Fig. 2-27 shows a pneu-

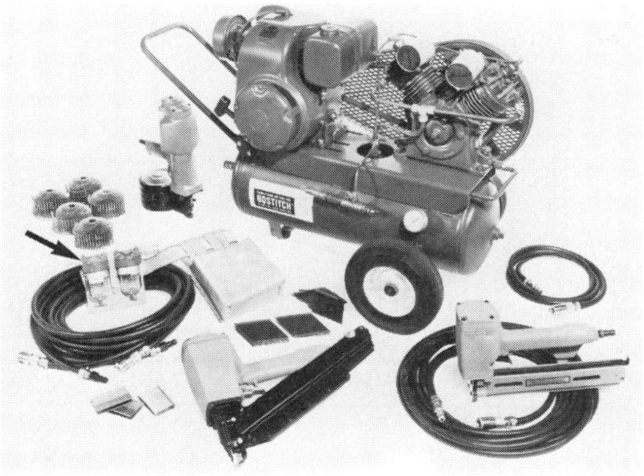

Fig. 2-28. Portable air compressor (gasoline powered) and accessories. The air filter and lubricator unit (arrow) is essential when operating pneumatic tools. (Bostitch)

matic powered nailer that drives full-headed 6d and 8d nails. The magazine holds up to 300 nails. Fig. 2-28 shows a portable air compressor.

Safety Rules for Power Staplers and Nailers

1. Study the manufacturer's operating directions and follow them carefully.
2. Use the correct type and size of fastener as recommended by the manufacturer.
3. For air-powered nailers, always use the correct pressure (seldom over 90 lbs). Be sure the compressed air is free of dust and excessive moisture.
4. Always keep the nose of the stapler or nailer pointed toward the work; never toward yourself or other workers.
5. Check all safety features and be sure they are working. Make a test, driving the staples or nails into a block of wood.
6. During use on the job, hold the nose firmly against the surface being stapled or nailed.
7. Always disconnect the tool from the air or electrical power supply when it is not being used.

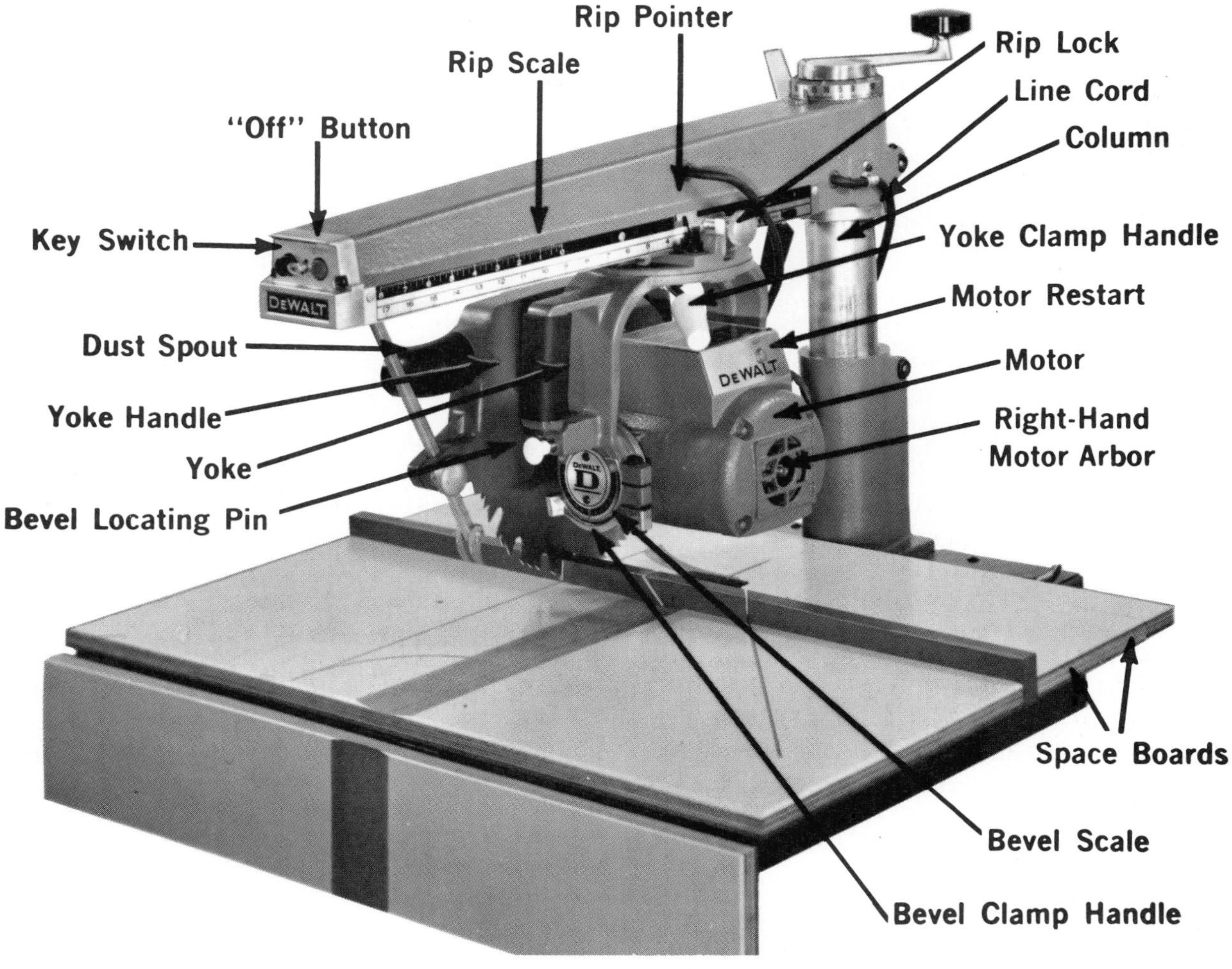

Fig. 2-29. Radial arm saw with parts identified. (Black and Decker Mfg. Co.)

Radial Arm Saws

In this power tool, the motor and blade are carried by an overhead arm and the stock is supported on a stationary table. The arm is attached to a vertical column at the back of the table. The depth of cut is controlled by raising or lowering the overhead arm. Fig. 2-29 shows the parts of a typical radial arm saw.

The motor is mounted in a yoke and may be tilted for angle cuts. The yoke is suspended from the arm on a pivot which permits the motor to be rotated in a horizontal plane. Adjustments make it possible to perform many sawing operations.

Fig. 2-30. Crosscutting on radial arm saw with lower guard in place.

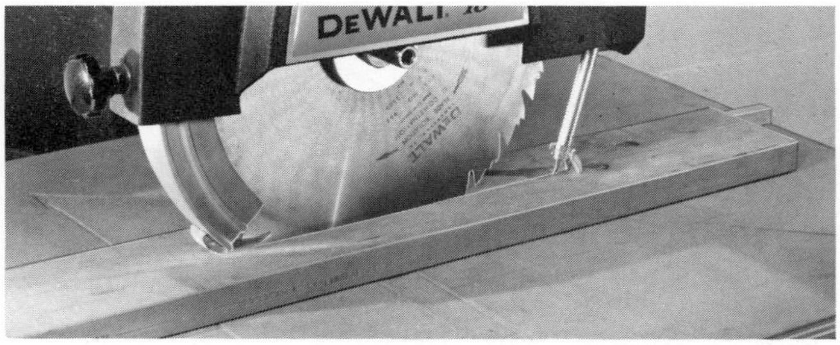

Fig. 2-31. Ripping operation. A fence (not visible) is set in the table to guide the work.

When crosscutting, mitering, beveling and dadoing, the work is held firmly on the table and the saw is pulled through the cut, Fig. 2-30. For ripping and grooving, the blade is turned parallel to the table and locked into position. Stock is then fed into the blade in somewhat the same manner as a table saw, Fig. 2-31.

For regular crosscuts and miters, first be sure the saw is against the column. Then place your work on the table and align the cut. Hold the stock firmly against the table fence with your hand at least 6 in. away from the path of the saw blade. Turn on the motor, grasp the saw handle and pull the saw firmly and slowly through the cut. See Fig. 2-32. The saw may tend to "feed itself." You must keep complete control over the rate of feed. When the cut is complete, return the saw to the rear of the table and shut off the motor.

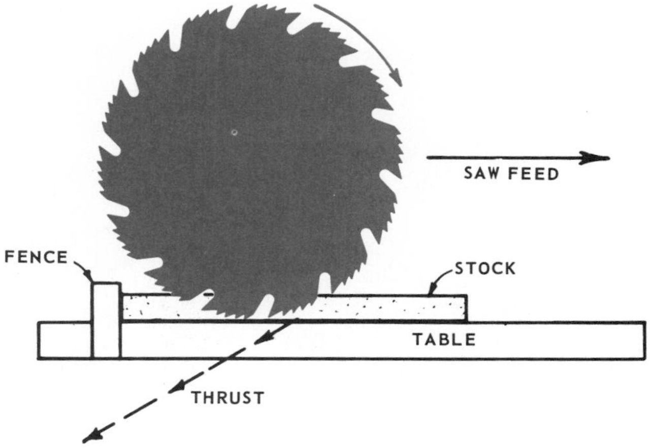

Fig. 2-32. Blade rotation and feed for crosscutting.

Safety Rules for Radial Arm Saws

1. Stock must be held firmly on the table and against the fence for all crosscutting operations. The ends of long boards must be supported level with the table.
2. Before turning on motor, be sure clamps and locking devices are tight; depth of cut is correct; and saw is slightly lower at back than front to prevent blade from "running" forward.
3. Keep the guard and anti-kickback device in position.
4. Always return the saw to the rear of the table after completing a crosscut or miter cut. Never remove stock from the table until the saw has been returned.
5. Maintain a 6 in. margin of safety. To do this you must keep your hands this distance away from the path of the saw blade.
6. Shut off the motor and wait for the blade to stop before making any adjustments.
7. Always be sure the blade has stopped before you leave the machine.
8. The table should be kept clean and free of scrap pieces and excessive amounts of sawdust. Do not attempt to clean off the table while the saw is running.
9. In crosscutting, always pull blade toward you.
10. Stock to be ripped must be flat and have one straight edge to guide it along the fence.
11. When ripping, always feed stock into the blade so that the bottom teeth are turning toward you. This will be the side opposite the anti-kickback fingers.

The radial arm saw is especially useful in cutting compound miters like cheek cuts on jack rafters. Fig. 2-33 shows a precision cut of this type. Cutting a large sheet of plywood is shown in Fig. 2-34.

Fig. 2-33. Cutting a compound miter. (Impractical to use lower guard.)

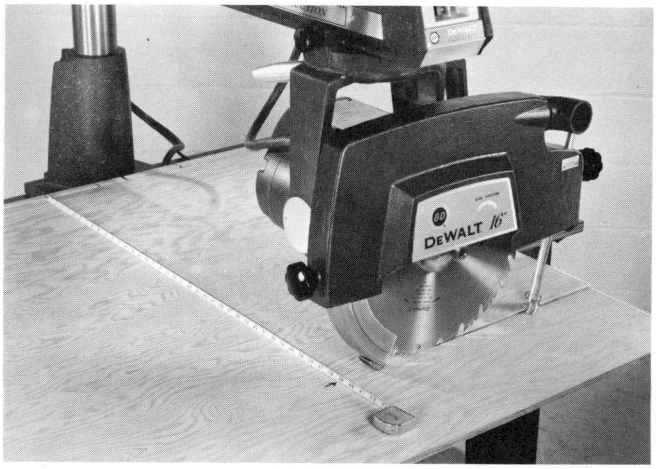

Fig. 2-34. Setup for cutting large sheets of plywood.

Table Saws and Jointers

These power tools are basic machines used in cabinetmaking and are frequently used by carpenters on projects which include on-the-job built cabinets and built-ins. Carpenters use them to some extent for cutting and fitting moldings and other inside trim work. When used for carpentry, the smaller sizes (4 - 6 in. jointers and 8 - 10 in. table saws) are usually selected because they can be easily moved from one job to another.

Fig. 2-35. Parts of a table saw. (Rockwell International)

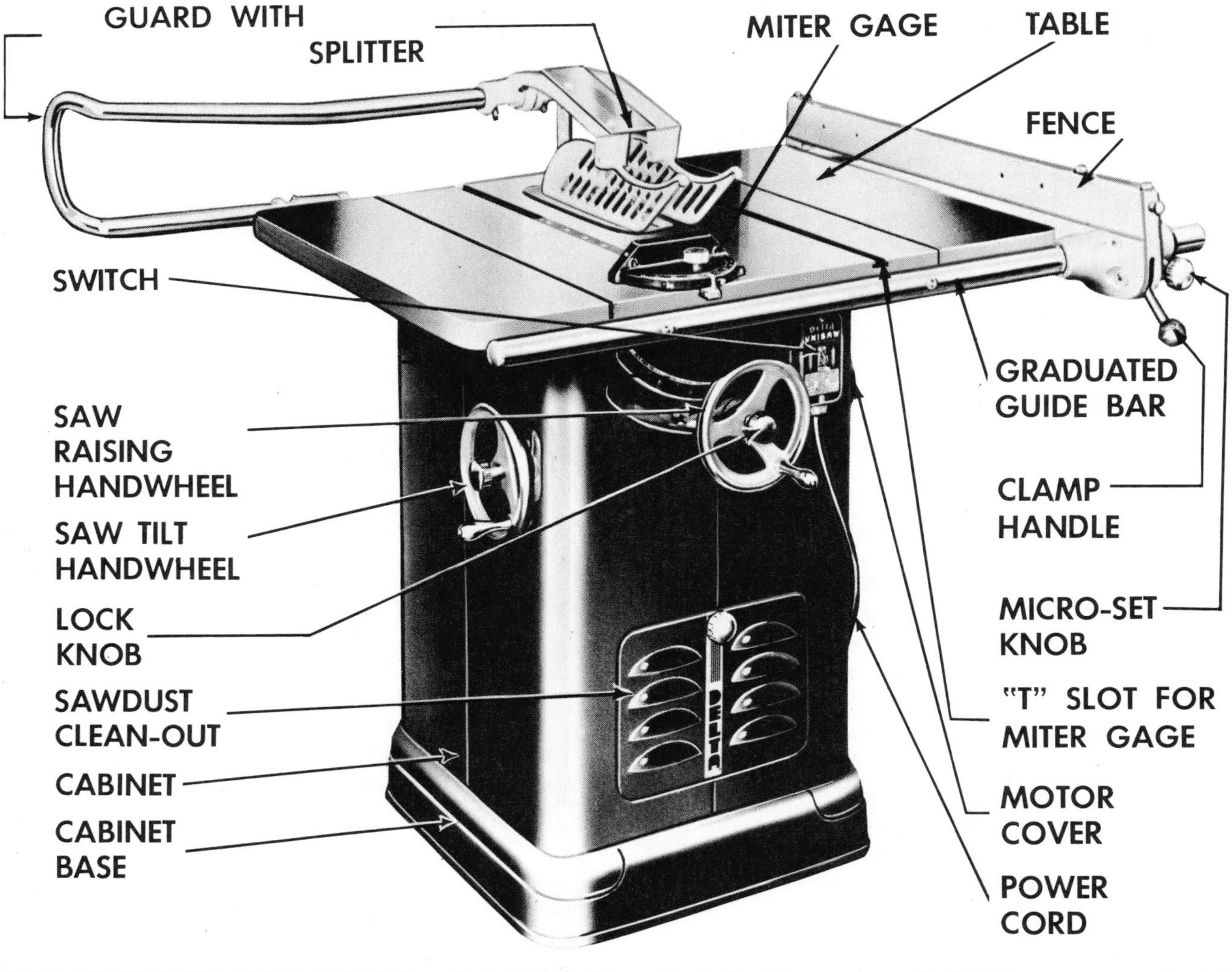

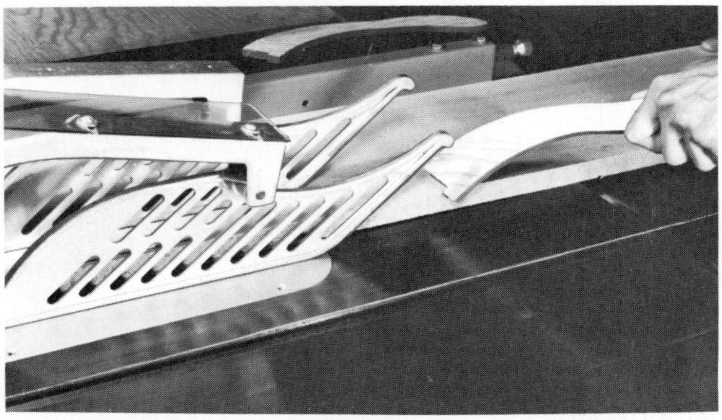

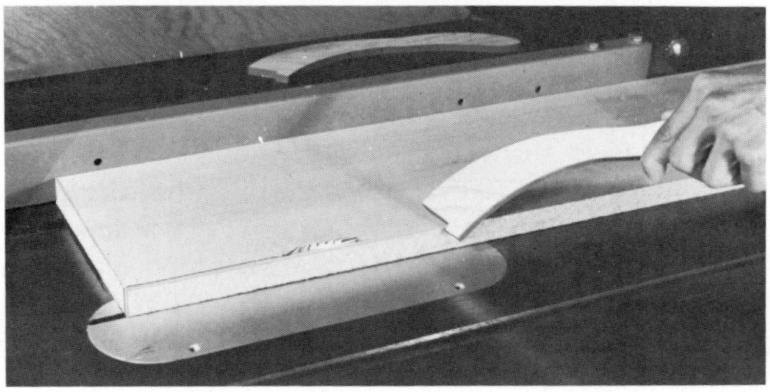

Fig. 2-36. Ripping stock to width. Above. Guard in position. Below. Guard removed to show operation.

pers. Properly set up, the table saw can be used to produce grooves, dados, rabbets and other forms basic to a wide variety of joints. The size of the saw is determined by the largest blade it will take. Fig. 2-35 shows a typical model with the parts identified.

Safety Rules for Table Saws

1. Be certain the blade is sharp and is the right one for the job at hand.
2. Make sure the saw is equipped with a guard and use it. Fig. 2-36.
3. Set the blade so it extends about 1/4 in. above the stock to be cut.
4. Stand to one side of the operating blade and do not reach across it.
5. Maintain a 4 in. margin of safety. (Do not let your hands come closer than 4 in. to the operating blade even though the guard is in position.)
6. Stock should be surfaced and at least one edge jointed before being cut on the saw.
7. The position of the stock must be controlled either by the fence or the miter gauge. Do not cut stock free hand.

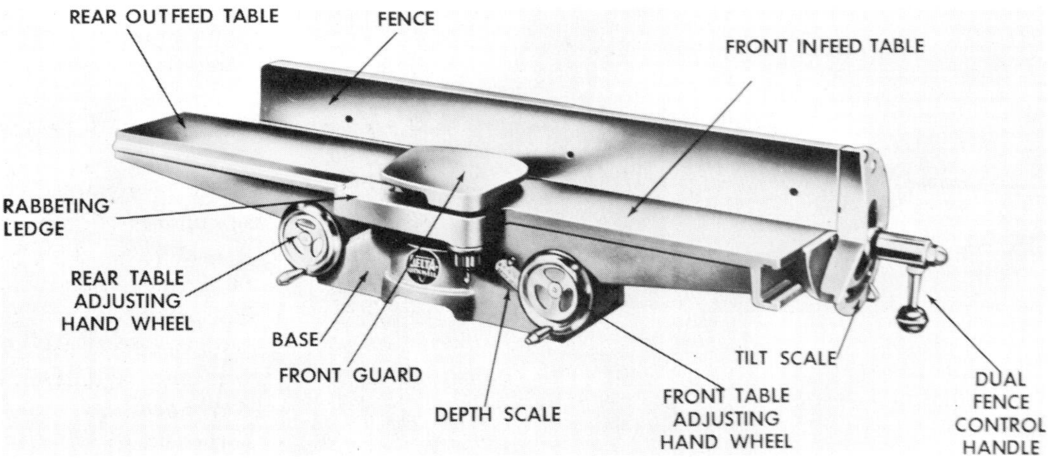

Fig. 2-37. Jointer with parts identified.

Space in this book does not permit more than a brief introduction to this equipment. Woodworking textbooks* provide a complete description of the wide variety of work they will perform along with procedures for operation.

The table saw (also called a circular saw) is used for ripping stock to width and cutting it to length. It also will cut bevels, chamfers and ta-

*Wagner, W. H., MODERN WOODWORKING, Goodheart-Willcox Co., Inc., South Holland, Ill. 60473

8. Always use push sticks when ripping short, narrow pieces.
9. Stop the saw before making adjustments.
10. Do not let small scrap cuttings accumulate around the saw blade. Use a push stick to move them away.
11. Resawing and other special setups must be carefully made and checked before the power is turned on.
12. The dado head or any special blades should be removed from the saw after use.

13. Other workers helping to "tail-off" the saw should not push or pull on the stock but only support it. The operator must control the feed and direction of the cut.
14. As work is completed, turn off the machine and remain until the blade has stopped. Clear the saw table and place waste cuttings in a scrap box.

JOINTER.

Principal parts of a jointer are shown in Fig. 2-37. The cutterhead (not shown in the photo) holds three knives and revolves at a speed of about 4500 rpm. The size of the jointer is determined by the length of these knives.

The three main parts that can be adjusted are the infeed table, the outfeed table and the fence. The outfeed table must be level with the knife edges at their highest point of rotation. This is a critical adjustment. See Fig. 2-38. If the table is too high the stock will be gradually raised out of the cut and a slight taper will be formed. If it is too low, the tail end of the stock will drop as it leaves the infeed table and cause a "bite" in the surface or edge.

The fence guides the stock over the table and knives. When jointing an edge square with a face, it should be perpendicular to the table surface, Fig. 2-39. The fence is tilted when cutting chamfers or bevels.

Safety Rules for Jointers

1. Before turning on the machine, make adjustments for depth of cut and position of fence. Be sure the guard is in place and is operating properly.
2. The maximum cut for jointing an edge on a small jointer is 1/8 in. and for a flat surface, 1/16 in.
3. Stock must be at least 12 in. long. Stock to be surfaced must be at least 3/8 in. thick unless a special feather board is used.

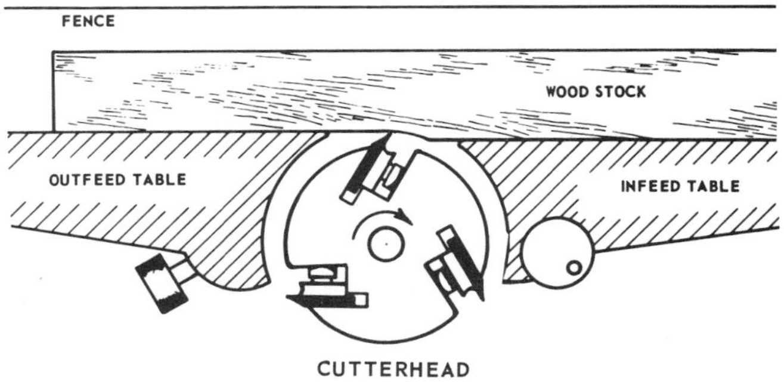

Fig. 2-38. How a jointer works. Note the direction of the wood grain.

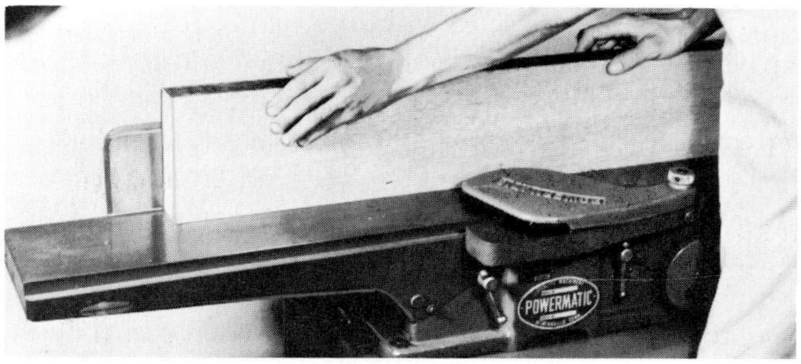

Fig. 2-39. Planing an edge. "Step" the hands along the stock so they will not bear on the stock while it passes over the cutterhead.

4. Feed the work so the knives will cut "with the grain." Use stock that is free from knots, splits and checks.
5. Keep your hands away from the cutterhead even though the guard is in position. Maintain at least a 4 in. margin of safety.
6. Use a push block when planing a flat surface. Do not apply pressure directly over the knives with your hand.
7. Do not plane end grain unless the board is at least 12 in. wide.
8. The jointer knives must be sharp. Dull knives will vibrate the stock and may cause a kickback.
9. When work is complete, turn off the machine and remain until the cutterhead has stopped.

Power Tool Care and Maintenance

Care of power tools is especially important if they are to function properly over a long period of time. Sharp blades and cutters insure accurate work and make the tool much safer to operate. The good craftsman takes pride in his tools, in their condition and also their appearance.

Most power tools are equipped with sealed bearings that seldom need attention. Follow the manufacturer's recommendations for lubrication schedules. Gear mechanisms for portable power tools usually require a special lubricant. All equipment will require a few drops of oil on controls and adjustment bearings from time to time. Clean and polish bare metal surfaces with 600 wet-or-dry abrasive paper when required. These surfaces can be kept smooth and clean by wiping them occasionally with light oil or furniture polish. Some carpenters apply a coat of paste wax to protect the surface and reduce friction.

Some power tools, especially those with a number of accessories, can be purchased with a case designed to keep the accessories organized and protected.

Cutters and blades require periodic sharpening. Most carpenters are too busy and usually do not have the equipment to accurately grind cutters or completely fit saw blades. They usually send these items to a saw shop where an expert job can be performed. The carpenter may, however, lightly hone cutters before grinding operations are required. Also he may prefer to file saw blades several times before sending them in for a complete fitting which includes jointing, gumming, setting and filing.

Test Your Knowledge - Unit 2

1. The size of a portable circular saw is determined by the _____________.
2. For general purpose work, a saber saw blade should have about _____ teeth per in.
3. When the base of the saber saw rests on a horizontal surface, the blade cuts on the _________ (up, down) stroke.
4. The depth of cut of a power plane is adjusted by raising or lowering the _________.
5. A standard router bit is held in a _______ type chuck.
6. The size of a belt sander is determined by the ____________.
7. To adjust the depth of cut of a radial arm saw, the ___________ is raised or lowered.
8. When crosscutting with the radial arm saw, the blade is ________ (pushed away, pulled toward) the operator.
9. For regular work, the ________ of the jointer should be perfectly aligned with the knife edges at their highest point.

Outside Assignments

1. Visit a builder's supply store and study the various portable circular saws on display -- also secure descriptive literature concerning the various sizes. After careful consideration, select a brand, model, and size that you believe would be best for rough framing and sheathing work on residential structures. Outline your reasons and make an oral report to your class. Include detailed specifications and the list price of the model you selected.
2. Visit with a carpenter in your locality and learn what procedures he follows in maintaining and sharpening his tools. If he uses standard saw blades - learn how he keeps them sharp. Secure his reaction to the use of hardened tooth and carbide tipped blades. If he has some of his tools sharpened at a saw shop, secure information concerning approximate prices. If a tool maintenance center is located in your area - secure information about services available and costs. Prepare your notes carefully, then make a report to your class.

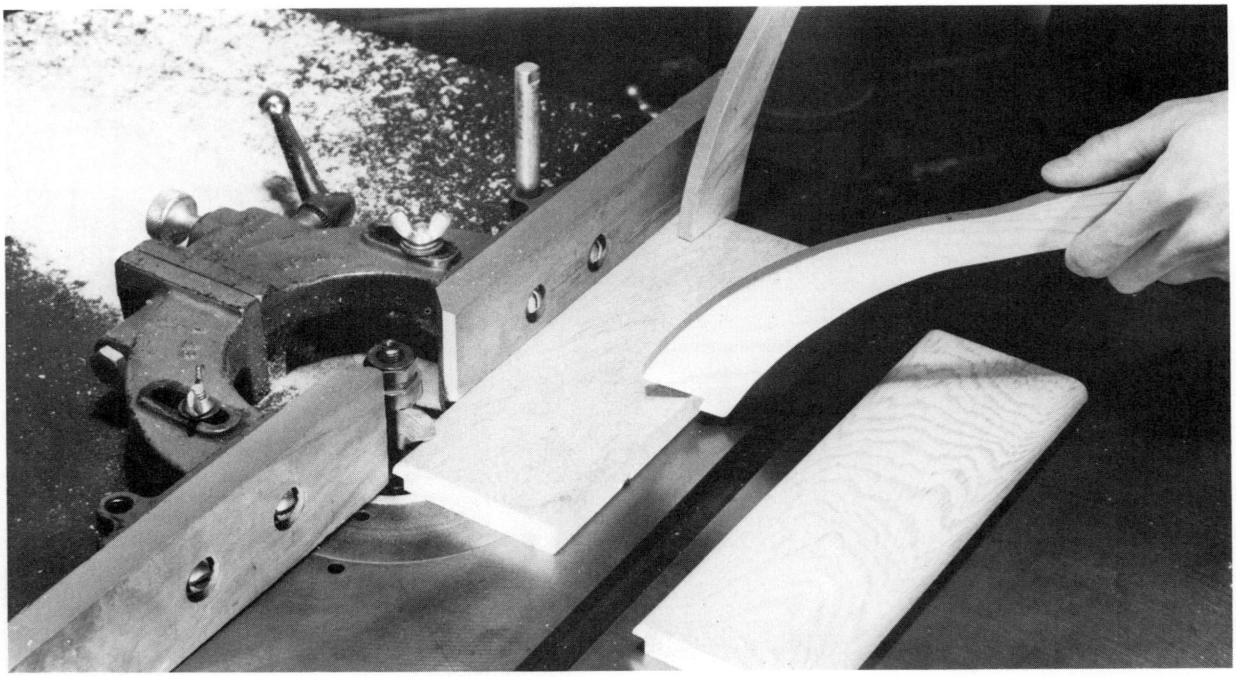

Using a shaper to form a lip on the edge of a cabinet drawer front. The use of push sticks permits the operator to keep his hands well away from the high speed cutter.

Unit 3
LEVELING INSTRUMENTS

In building construction it is important that footings and foundation walls be level, square, and the correct size.

For small structures, the carpenters' level, framing square, and rule can be used to lay out and check the foundation work. As the size increases in residential construction and in commercial and institutional buildings, special leveling instruments must be used if accuracy and efficiency are to be maintained.

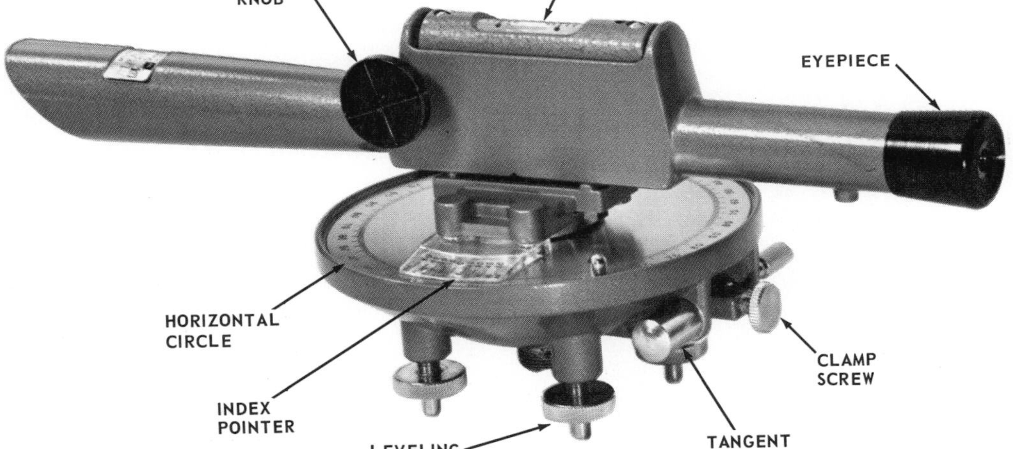

Fig. 3-1. Builders' level which is used to sight level lines and lay out or measure horizontal angles. (David White Instruments, Div. of Realist, Inc.)

Leveling Instruments

The level and level-transit are commonly used in laying out and checking construction work. These instruments can also be used for surveying and other space and land-layout jobs. When the job is too large for the chalk line, straightedge, level, and square; leveling instruments should be used. The instruments include an optical device which operates on the basic principle that a LINE OF SIGHT is a straight line that neither dips, sags or curves. Any point along a level line of sight will be the same height as any other point. In the use of these instruments, the line of sight replaces the chalk line and straightedge.

The builders' level (also called a dumpy level) is shown in Fig. 3-1. It consists of an accurate spirit level and telescope assembly that is attached to a circular base. Leveling screws are used to adjust the base after it has been mounted on the tripod. The telescope can be rotated on the base and any angle in a horizontal plane can be laid out or measured.

The level-transit, Fig. 3-2, works like the builders' level with an additional feature that permits the telescope to be pivoted up and down in a vertical plane. This makes it possible to accurately measure vertical angles or to determine if a wall is perfectly plumb (vertical). The vertical movement simplifies the operation of aligning a

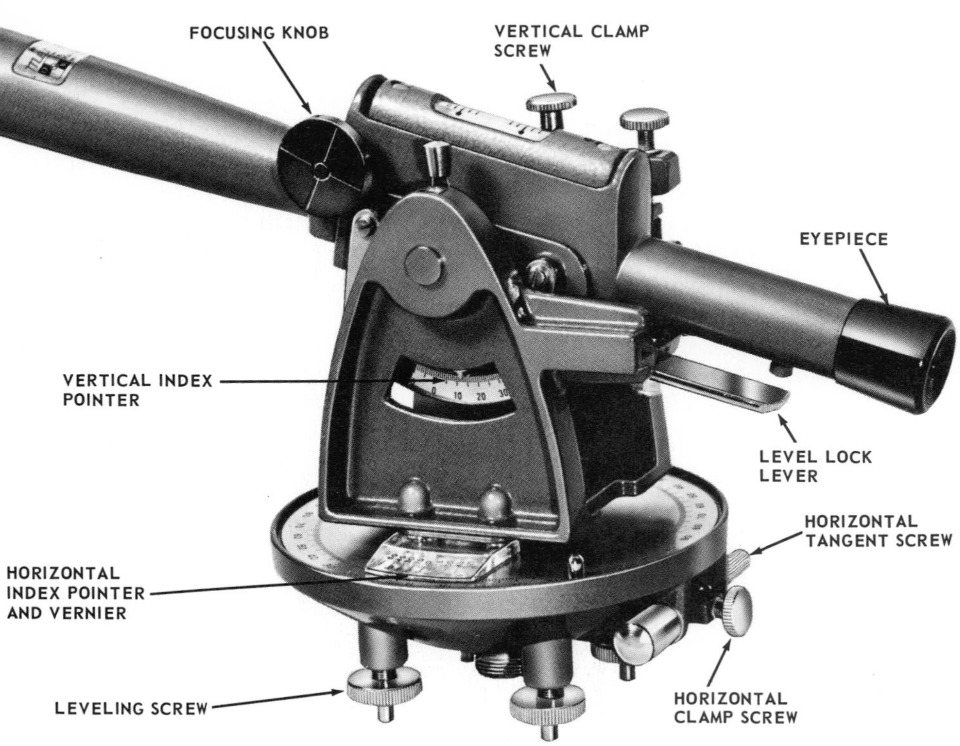

Fig. 3-2. Parts of a level-transit. This instrument can be used to lay out or check level and plumb lines. It can also be used to measure angles in either horizontal or vertical planes.

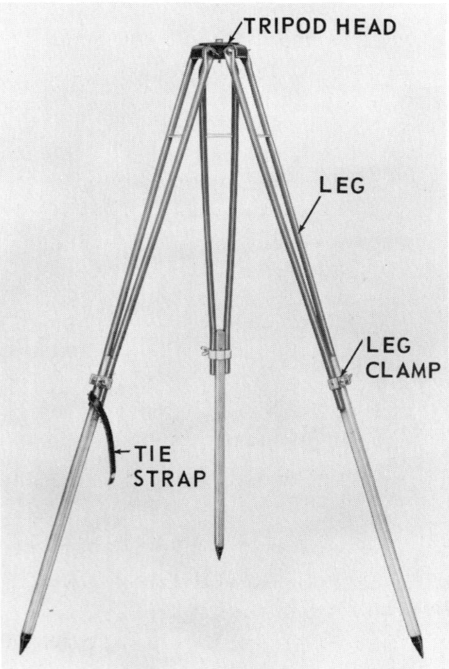

Fig. 3-3. Parts of a tripod. The legs hinge at the top for carrying and storing.

row of stakes, especially when they vary in height.

In use, both the builders' level and level-transit are mounted on tripods, Fig. 3-3. Some models, like the one shown, have adjustable legs that make it easier to set them up on sloping ground. This feature also permits the legs to be retracted for handling and storing.

When it is necessary to sight over long distances, a leveling rod is used. See Fig. 3-4. It is designed so that differences in the elevation between the position of the level and various positions where the rod is held can be easily read. The rod is especially useful for surveying work. Readings can be made by the person operating the level (the levelman), or the target can be adjusted up and down to the line of sight and then the rodman

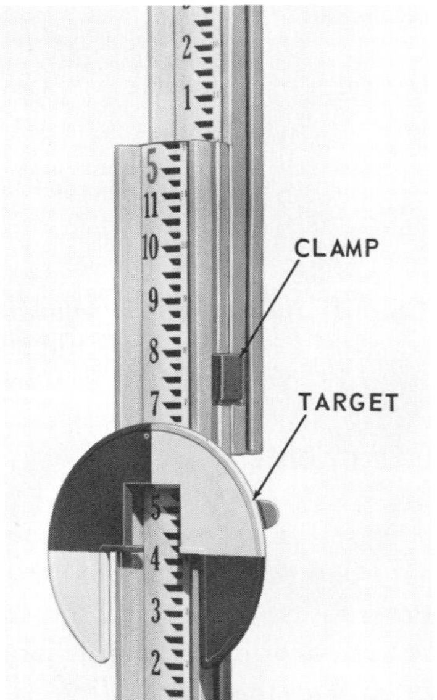

Fig. 3-4. Leveling rod. Consists of two sections that can be extended to a height of 12 feet.

(man holding the rod) can make the reading. The rod shown has graduations in feet and inches. However, for regular surveying work, the graduations are usually based on the foot and decimal parts of a foot.

When sighting short distances (100 feet or less) a regular folding rule can be held against a wood strip and read through the instrument. This procedure will be satisfactory for jobs such as setting grade stakes for a footing. Always be sure to hold the strip and rule in a vertical position.

For measurements and layouts involving long distances, steel tapes (called measuring tapes) may be used. They are contained on winding reels such as the one shown in Fig. 3-5, and are available in lengths from 50 to 300 feet. Various types or graduations may be obtained, Fig. 3-6. The carpenter will usually select one that is graduated in feet, inches and eighths like two that are shown. Surveying requires a tape that is graduated in feet and decimal parts of a foot.

Fig. 3-5. Measuring tape. Graduations in feet, inches and eighths. (Keuffel and Esser Co.)

Care of Leveling Instruments

Leveling instruments are more delicate than most other tools and equipment. Special precautions must be followed in their use so they will continue to provide accurate readings over a long period of time. Some suggestions follow:

1. Keep the instrument clean, dry, and in its carrying case when not in use.
2. When the instrument is set up, have a plastic bag or cover handy to use in case of rain.
3. It is best to grip the instrument by its base when moving it from the case to the tripod.
4. Never leave the instrument unattended when it is set up near moving equipment.
5. When going through buildings or any close quarters, carry the tripod under your arm with the instrument in front of you.
6. Never over-tighten leveling screws or any of the other adjusting screws or clamps.
7. Always set the tripod on firm ground with the legs spread well apart. When set up on floors or pavement, take extra precautions to insure that the legs will not slip.
8. For precision work, permit the instrument to reach air temperature before making readings.
9. When the lenses collect dust and dirt, clean them with a camel's hair brush, special lens paper or a clean soft rag.
10. Never use a wrench or pliers on any of the adjustments. Have the instrument checked periodically by a qualified service station or by the factory.

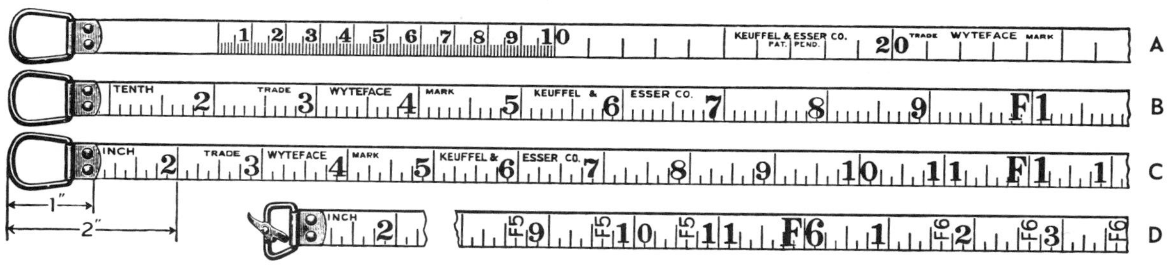

Fig. 3-6. Measuring tape graduations: A–Metric. B–Feet and decimal parts. C–Feet and inches. D–Feet and inches with the foot repeated at each inch mark.

HANDY SAYS:

"Leveling instruments and equipment will vary somewhat depending on the manufacturer. Always read and study carefully the instructions for a given brand, and learn about its features and details of operation."

Setting Up Instrument

Set up the tripod in a firm and stable position. Make sure the points are well into the ground. On hard surfaces, be sure the points will not slip. Check the tripod wing nuts and see that they are securely tightened. The legs should have a spread of about 3 1/2 feet and be adjusted so the tripod head appears to be level.

Lift the instrument carefully from its case by the base plate. Screw it firmly onto the tripod head. On some instruments the leveling screws may need to be turned up so the mounting screw can be threaded all the way into the tripod head. If the instrument is to be located over an exact point, this should be done before the final leveling.

Leveling is a very important operation in preparing the instrument for use. None of the readings taken or levels sighted will be accurate unless the instrument is level throughout the work.

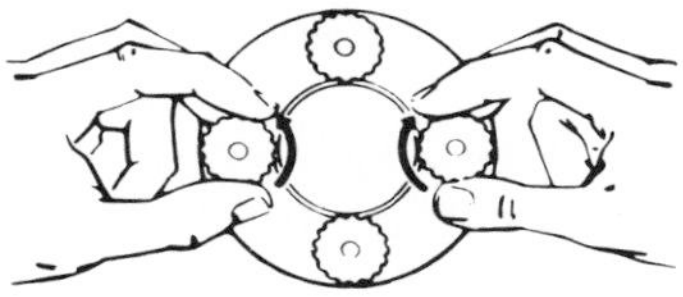

Fig. 3-7. Adjusting the leveling screws.

First, release the horizontal clamp screw and turn the telescope so it is directly over a pair of the leveling screws. Grasp the screws between the thumb and forefinger as shown in Fig. 3-7. Turn both screws uniformly with your thumbs moving toward each other or away from each other. Center the bubble of the spirit level carefully between the graduations. You will find that on most instruments the BUBBLE WILL TRAVEL IN THE DIRECTION THAT YOUR LEFT THUMB MOVES. See Fig. 3-8. Leveling screws should bear firmly on the base plate. Never tighten the screws so much they will bind.

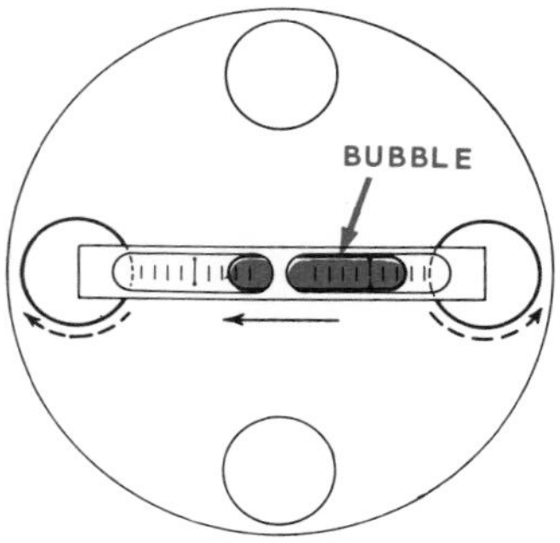

Fig. 3-8. The bubble of the spirit level moves in the same direction as the left thumb.

When the bubble is centered, turn the telescope 90 deg. so it is over the other set of leveling screws and repeat the operation. Now check and recheck the instrument over each pair of screws.

Fig. 3-9. Sighting a level line.

When the instrument is level, it will be possible to turn the telescope in a complete circle without any change in the bubble.

Sighting

The telescope will magnify or enlarge the image (object sighted). Most builders' levels have a telescope with a power of about 20X which means that the object will appear to be only one-twentieth of its actual distance. First, line up the telescope by sighting along the barrel and then look into the eyepiece, Fig. 3-9, and adjust the focusing knob until the image is clear and sharp. When the cross hairs are in approximate position on the object, Fig. 3-10, tighten the horizontal-motion clamp and make the final alignment by turning the tangent screw.

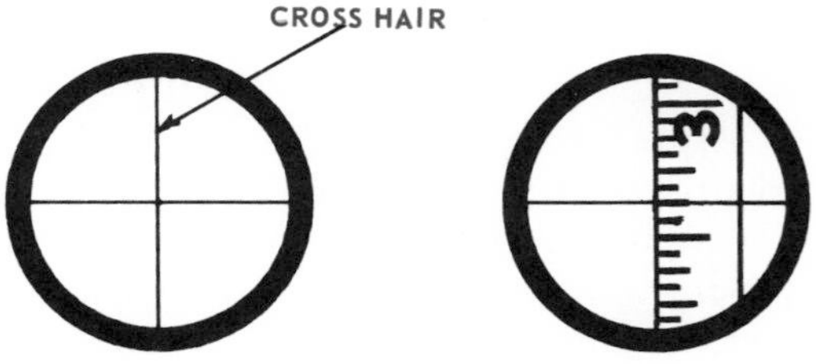

Fig. 3-10. View through the telescope. Left. Cross hairs. Right. Object in view.

Leveling

Finding the difference in the grade level between several points or transferring the same level from one point to another is called leveling. With the instrument level, the line of sight will be

Using a level-transit to plumb (make vertical) a column in a steel framework.
(David White Instruments, Div. of Realist, Inc.)

level and the readings shown in Fig. 3-11 can be used to calculate the difference in elevation. When there is a great amount of slope, the instrument can be set up between the points. The reading is taken with the rod in one position and then the instrument is carefully rotated to secure the reading at a second position.

When setting grade stakes for a footing or erecting batter boards, the instrument should be set in a central location as shown in Fig. 3-12. The distances will be about equal and it will reduce the need for changing focus on each corner. An elevation established at one corner can be quickly transferred to other corners or points in between.

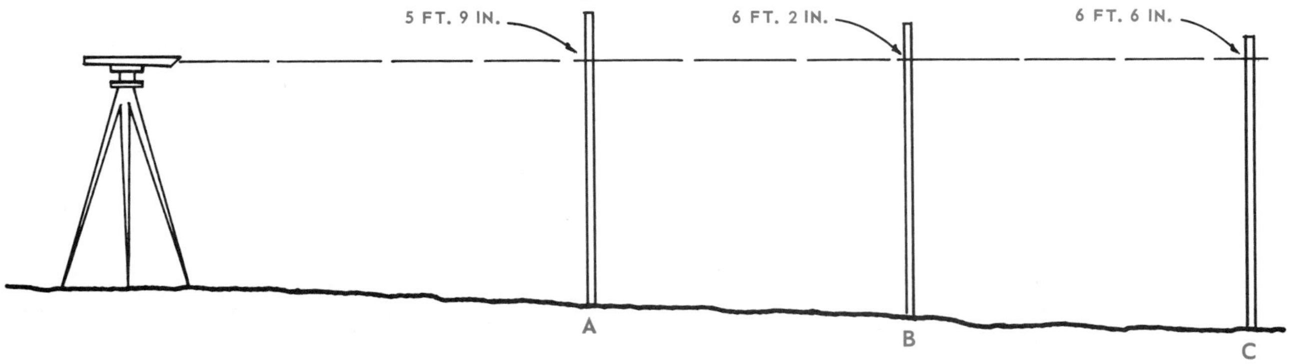

Fig. 3-11. Leveling. Point A is 9 in. higher than point C.

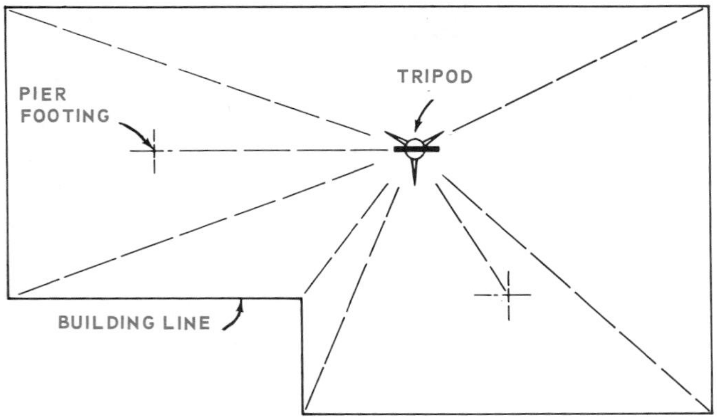

Fig. 3-12. Setting grade stakes with the instrument centrally located.

Grade stakes for footings are usually set to the approximate level by "eye" judgment. They are then carefully checked with the rod and level as they are driven deeper until the top of each stake is at the required elevation. Sometimes reference lines are drawn on construction members or stakes near the work and then transferred to the forms with a carpenters' level and rule.

There may be situations where the existing grade will not permit the setting of a stake or reference mark at the actual level of the required grade. In such cases a mark is made on the stake with the added notation of its position in relation to the required grade elevation. The letters C and F, standing for cut and fill, are generally used, Fig. 3-13.

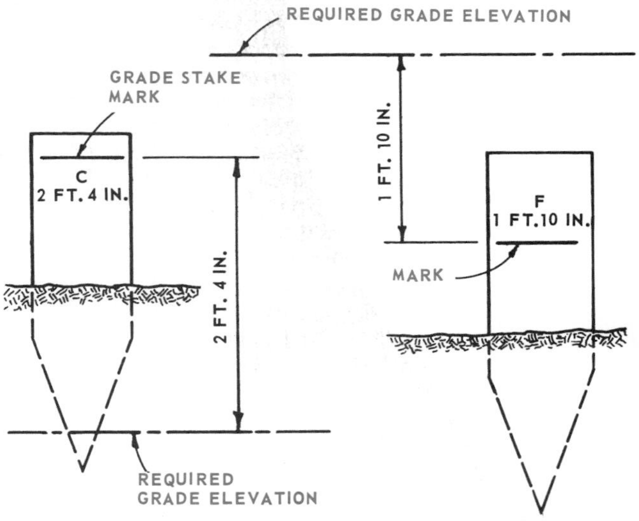

Fig. 3-13. Cut and fill stakes.

When laying out sloping building plots or "carrying" a benchmark (an officially established elevation) to the building site, it will likely be necessary to set up the instrument in several locations. Fig. 3-14 shows how reading from two positions is used to calculate or establish grade levels.

Horizontal Angles

To measure or lay out angles with the builders' level, attach a plumb bob to the center screw or hook on the underside of the instrument. Adjust the tripod so the point of the plumb bob is directly over the required point on the ground. This point is usually marked on a stake about two inches square with a nail or tack driven in at the exact spot.

Level the instrument, check the plumb bob, and then turn the telescope so the vertical cross hair is directly in line with the edge of the rod at station B. The distance is laid out with the measuring tape. See Fig. 3-15, Left. Set the horizontal circle at zero to align with the vernier zero and then swing the instrument 90 deg. (or any required angle) and position the rod so it aligns with the cross hairs. The distance to station C is also laid out with the measuring tape.

Move the instrument to station C, sight back to station A and then turn 90 deg. to locate station D after measuring the required distance. If the shape is a perfect rectangle or square, you will have completed the layout. However, you may want to move the instrument to station D and check your work.

In actual practice, you will find that it is difficult to locate a stake in a single operation. This is especially true when using a builders' level where the line of sight must be "dropped" with a plumb line to ground level. Usually it is best to set a temporary stake, see Fig. 3-16, and mark it with a line sighted from the instrument. Then with the measuring tape pulled taut and aligned with the mark, the permanent stake can be set and the exact point easily located as shown.

Major rectangles and squares of a given building line are laid out using leveling instruments in

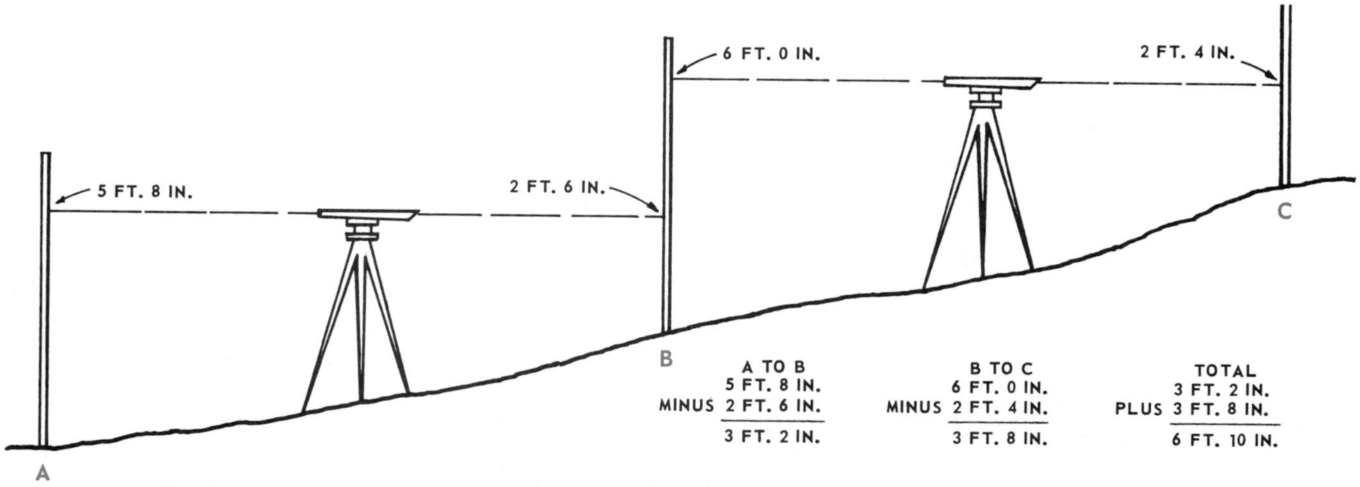

Fig. 3-14. When there is a great amount of slope or long distances are involved, the instrument will need to be set up at two or more locations.

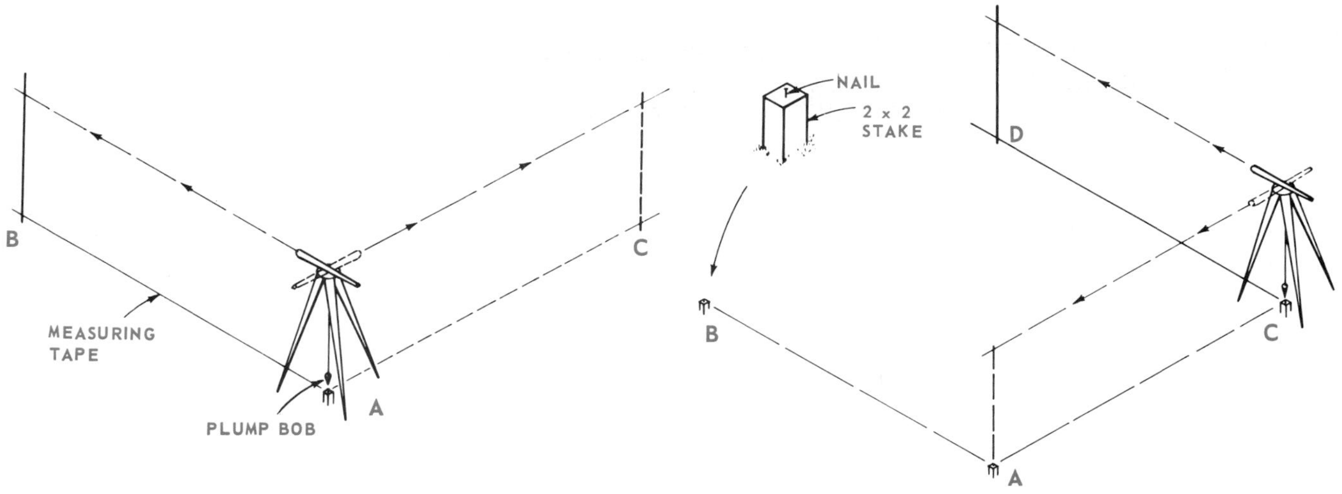

Fig. 3-15. Left. First step in laying out corners of a building. Second step. Note: the rod must be kept exactly vertical with a plumb line or carpenters' level.

the manner described. After batter boards are set and lines attached, the carpenters' level and square can be used to set stakes for small projections and irregular shapes.

Measurements of a Circle

circle	=	360 degrees
1 degree	=	60 minutes
1 minute	=	60 seconds
1 quadrant	=	1/4 of a circle or 90 degrees

When laying out or measuring angles where fractions of a degree are involved, you will need to use the vernier scale. The upper half of Fig. 3-17 shows a section of the horizontal circle and the vernier scale with the reading 0° - 0'. Now assume that you are required to lay out an angle to the left (counterclockwise). Swing the telescope to a position of 44° and then lock the horizontal-motion clamp and make the final adjustments with the tangent screw.

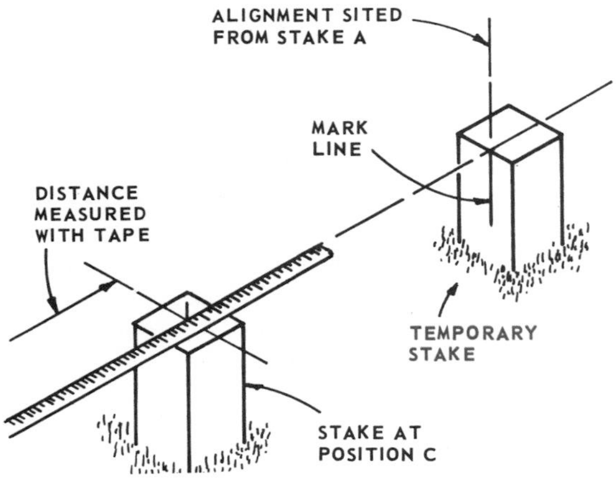

Fig. 3-16. Establishing an exact point for a lay out with the aid of a temporary stake.

For the required setting, the center point of the vernier must be moved past the 44 deg. mark. The vernier is divided into lines representing 5 min. each, thus the additional rotation is made until the fourth mark to the right of the index point lines up with a degree mark on the horizontal circle as shown in the figure. Vernier scales will not be the same on all instruments and you should study the operator's manual for instructions about the particular model you are using.

Contour Lines

Contour lines are lines that run through points of equal level. Such lines can be laid out on an area of ground by setting up the instrument at an appropriate point and directing the rodman to hold the rod at the beginning point. Sight the rod and set the target on the horizontal cross hair. The rodman then moves to the next required location and moves the rod up and down the slope until the target again aligns with the scope. Set a stake and repeat the procedure as many times as required.

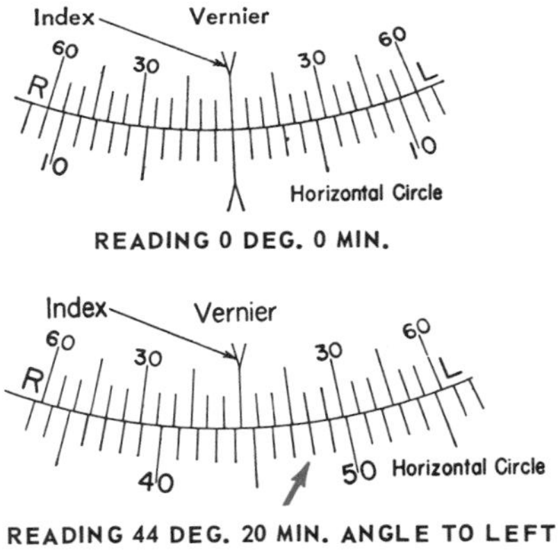

Fig. 3-17. Vernier scales. Arrow shows alignment for 20 minute reading. (Berger Scientific Supplies, Inc.)

Running Straight Lines with Transit

Although the builders' level can be used to line up stakes, fence posts, poles and roadways, more accuracy is gained with the level-transit, especially when different levels are involved.

Set the instrument directly over the reference point. Level the instrument and then release the

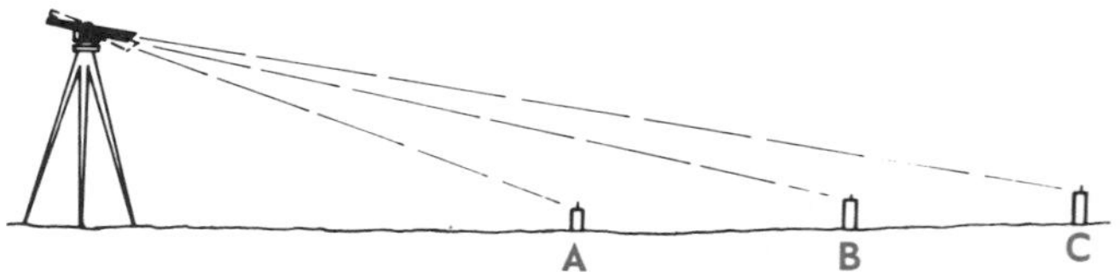

Fig. 3-18. Using the level-transit to set a row of stakes.

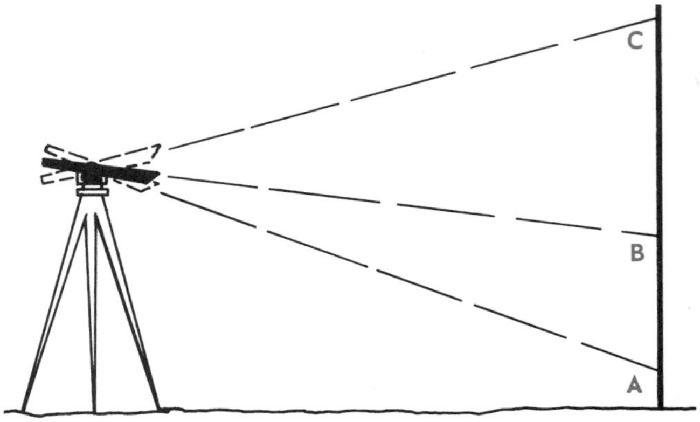

Fig. 3-19. Using a level-transit to lay out or check points in a vertical plane.

lock that holds the telescope in the level position. Swing the instrument to the required direction or until a stake is aligned with the vertical cross hair. Tighten the horizontal circle clamp so the telescope can move only in a vertical plane. Now by pointing the telescope up or down, any number of points can be located in a perfectly straight line. See Fig. 3-18.

Vertical Planes and Lines

The level-transit can be used to measure vertical angles or lay out and check building walls, flagpoles or TV antenna masts.

To establish or check vertical lines and planes, first level the instrument, then release the lever that holds the telescope in a horizontal position. Swing the instrument vertically and horizontally until a reference point in the required plane or line is sighted and then lock the horizontal clamp screw. As you rotate the telescope up and down, all of the points sighted will be located in the same vertical plane, Fig. 3-19. Plumb lines can be checked or established by first operating the instrument as shown and then moving it to a second position, usually 90° either to the right or left.

To measure or lay out angles in a vertical plane, follow the same general procedure that was used for horizontal angles.

Test Your Knowledge - Unit 3

1. In the use of leveling instruments, the ________ replaces the chalk line and straightedge.
2. The builders' level consists of a telescope assembly that is mounted on a ________ base.
3. For surveying work, a measuring tape with graduations reading in feet and ________ is usually selected.
4. The most important operation in setting up a builders' level or level-transit is the ________.
5. When sighting through the telescope, you should adjust the ________ until the image is sharp and clear.
6. When setting grade stakes for a building footing, the instrument should be set up in a _____ location.
7. To position a leveling instrument directly over a given point, a ____ ____ is used.
8. A circle is divided into degrees, ________ and ________.

Outside Assignments

1. Study the catalog of a supplier or manufacturer and develop a set of specifications for a builders' level. Be sure it includes a good carrying case. Also select a suitable tripod and measuring tape. Secure list prices for all of the items.
2. Through drawings and a written description, tell how you would proceed to lay out a baseball diamond using a level-transit.
3. Make a study of the procedures you would follow and calculations you would make to determine the height of a flag pole, tall building or mountain, using the level-transit and trigometric functions. Prepare a report for your class.

Fig. 4-1. Approximately 80 percent of our nation's homes have a structural framework of wood.

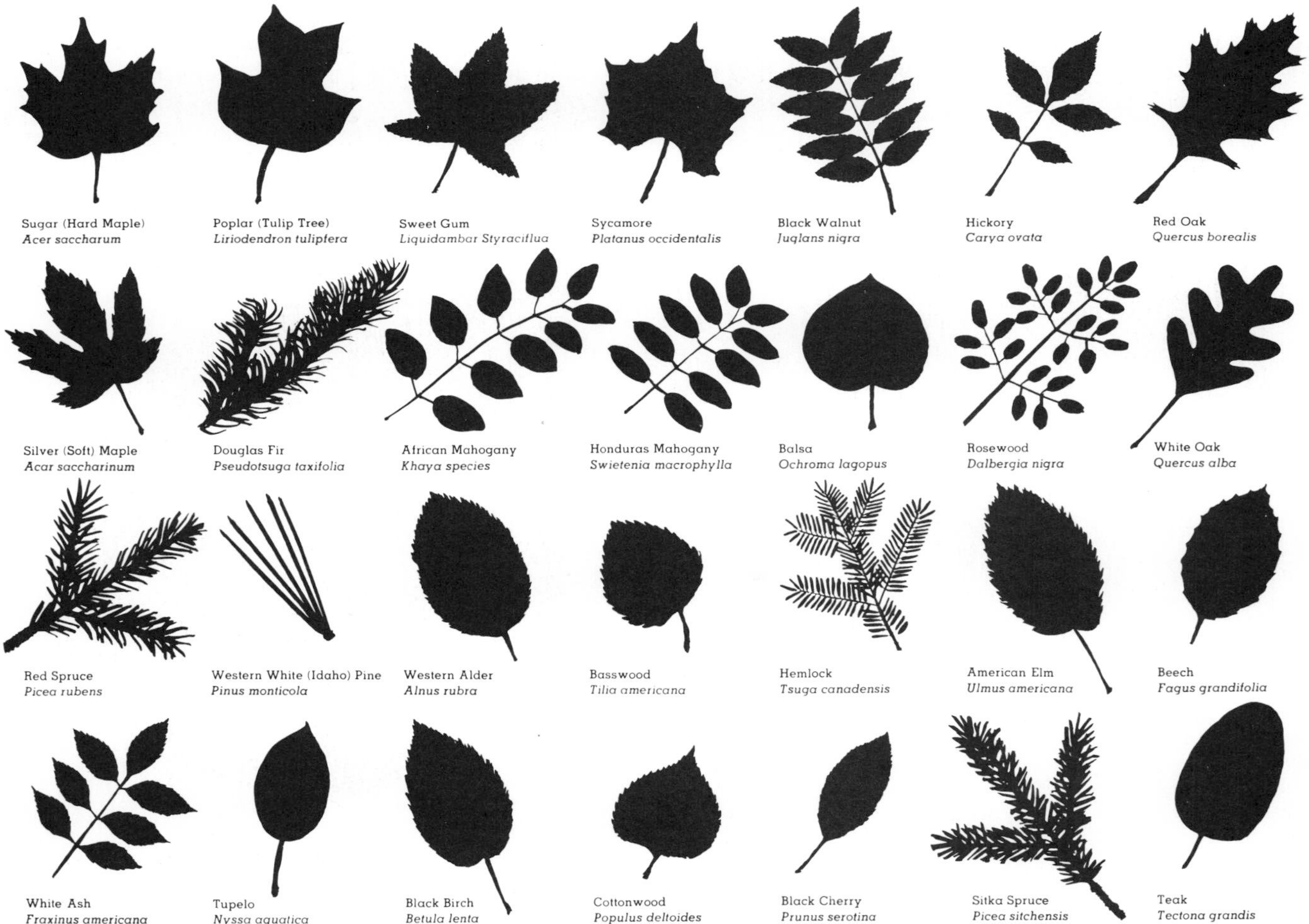

Fig. 4-2. Leaf silhouettes of 23 domestic and 5 imported species of hardwoods and softwoods. (Gamble Brothers, Inc.)

Unit 4
BUILDING MATERIALS

Wood is one of our greatest natural resources. When cut into pieces that are uniform in thickness, width and length, it becomes lumber; the material most widely used for residential construction, Fig. 4-1.

Lumber is the designation given to products of the sawmill and includes: boards used for flooring, sheathing, paneling, and trim; dimension lumber used for sills, plates, studs, rafters and other framing members; timbers used for posts, beams and heavy stringers; and numerous specialty items. The carpenter must have a good working knowledge of lumber -- kinds, grades, sizes and other aspects that apply to its selection and use. To fully appreciate and understand some of the important considerations in this area, he should know something about wood growth, its structure and characteristics.

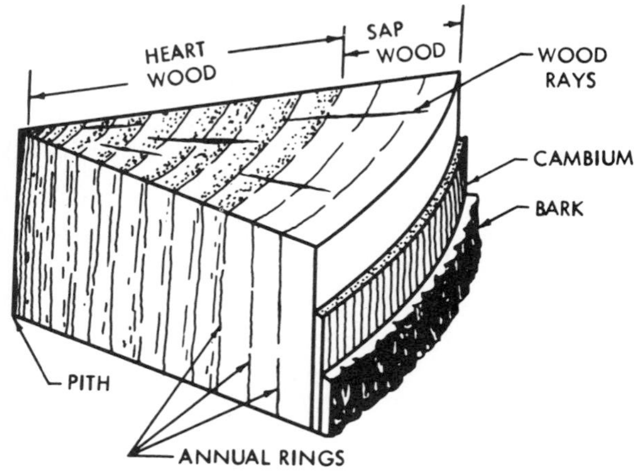

Fig. 4-3. Basic parts of a tree trunk.

Wood Structure and Growth

The basic structure of wood consists of long narrow tubes or cells (called fibers or tracheids) which are no larger around than human hair. Their length varies from about 1/25 in. in hardwoods to approximately 1/8 in. in softwoods. Tiny strands of cellulose make up the walls of the cells which are held together with a natural cement called lignin. It is this cellular structure that makes it possible to drive nails and screws into the wood and also accounts for the lightweight, low heat transmission factors, and sound absorption qualities.

The growing parts of a tree are the tips of the roots, the leaves and a layer of cells just inside the bark called the cambium. Water is absorbed by the roots and travels through the sapwood to the leaves, Fig. 4-2, where it is combined with carbon dioxide from the air. Through the miracle of photosynthesis, sunlight changes these elements to food (carbohydrates) which is then carried back to the various parts of the tree.

New cells are formed in the cambium layer, Fig. 4-3. The inside area of the layer (xylem) develops new wood cells while the outside area (phloem) develops cells that form the bark.

The growth in the cambium layer takes place in the spring and summer and forms separate layers each season. These layers are called annual rings, Fig. 4-4. Each annual ring is composed

Fig. 4-4. Annular rings formed each year indicate the age of the tree. Drought, disease or insects can interrupt the growth, causing an extra or false ring to be formed. (Forest Products Lab.)

of two layers; springwood and summerwood. In the spring, trees grow rapidly and the cells produced are large and thin walled. As the growth slows down during the summer months, the cells produced are smaller, thicker walled and appear darker in color. These annual growth rings are largely responsible for the grain patterns that are seen in the surface of boards cut from a log.

Sapwood contains living cells and may be several inches or more in thickness. The heartwood of the tree is formed as the sapwood becomes inactive and usually turns darker in color because of the presence of gums and resins. In some woods such as hemlock, spruce, and basswood, there is little or no difference in the appearance. Sapwood is as strong and heavy as heartwood but not as durable when exposed to weather.

Kinds of Wood

Lumber may be classified as either softwood or hardwood. Softwood comes from the evergreen or needle bearing trees. These are called "conifers" because many of them bear cones. See Fig. 4-5. Hardwood comes from broadleaf (deciduous) trees that shed their leaves at the end of the growing season. This classification is somewhat confusing, however, because many of the hardwood trees produce a softer wood than some of the so-called softwood trees.

Several of the more common kinds of commercial softwoods and hardwoods are:

Softwoods	Hardwoods
Douglas Fir	Basswood
Southern Pine	Willow
Western Larch	American Elm
Hemlock	*Mahogany
White Fir	Sweet Gum
Spruce	*White Ash
Ponderosa Pine	Beech
Western Red Cedar	Birch
Redwood	Cherry
Cypress	Maple
White Pine	*Oak
Sugar Pine	*Walnut

*Open grained wood

A number of hardwoods have large pores in the cellular structure (called open grain woods) and require special or additional operations in the finishing procedure.

In addition to hardness and softness, different kinds of wood will vary in weight, strength, workability, color, texture, grain pattern and odor. Check the color section, opposite page 24, as an initial step in wood identification. To further develop ability in this area, actual specimens should be studied. Several of the softwoods used in construction work are similar in appearance and considerable experience is required to make identification accurately and quickly.

Most of the samples shown in the color section were cut from plain-sawed or flat-grain stock. Edge-grain views would look different.

Availability of different species (kinds) of lumber varies somewhat throughout the country. This is especially true of framing lumber which is expensive to transport over long distances. It is usually more economical to select building materials (lumber, stone, etc.) that reflect the natural resources of the area.

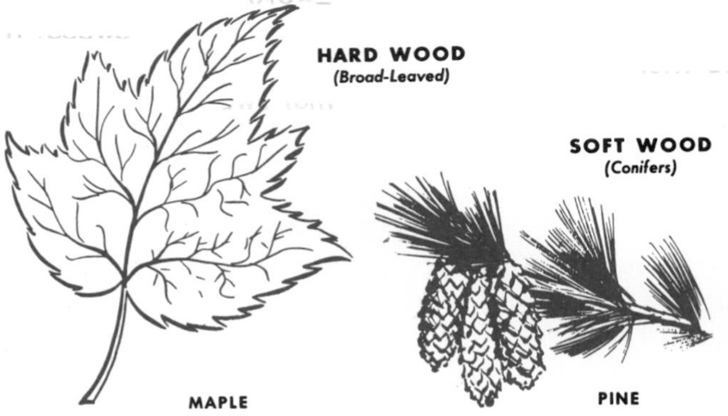

Fig. 4-5. General classification of wood. (Paxton Lumber Co.)

Cutting Methods

Most lumber is cut in such a way that the annular rings form an angle of less than 45 deg. with the surface of the board. This method produces lumber that is called FLAT-GRAINED if it is softwood, or PLAIN-SAWED if it is hardwood. Minimum waste is incurred in using this method and desirable grain patterns are obtained.

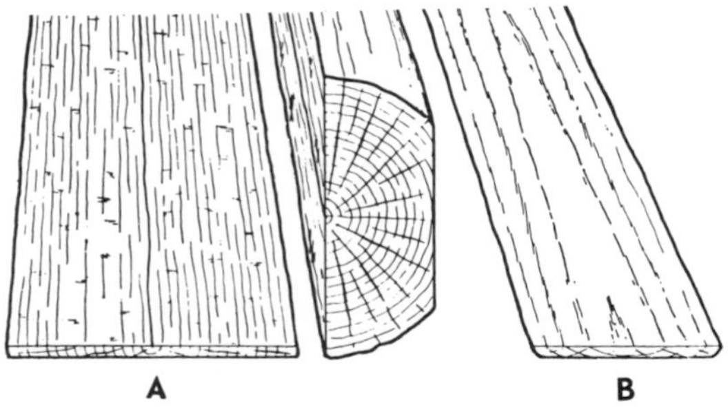

Fig. 4-6. Methods of cutting lumber. A–Edge-grain or quarter-sawed. B–Flat-grained or plain-sawed.

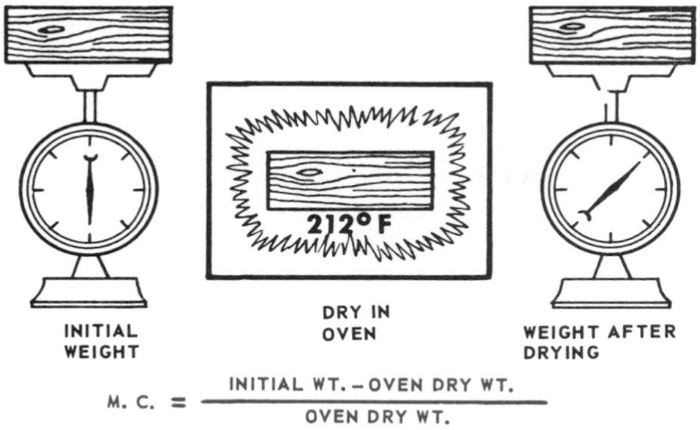

Fig. 4-7. Determining moisture content of wood.

Lumber can also be cut so the annular rings form an angle of more than 45 deg. with the surface of the board, Fig. 4-6. This method produces lumber that is called EDGE-GRAIN if it is softwood, and QUARTER-SAWED if it is hardwood. It is more difficult and expensive to use this method but it does produce lumber that swells and shrinks less in width and is not so likely to warp.

Moisture Content and Shrinkage

Before wood can be used commercially, a large part of the moisture (sap) must be removed. When a living tree is cut, more than half of its weight may be moisture. Lumber used for framing and outside finish should be dried to a moisture content of about 15 percent. Most cabinet and furniture woods are dried to a moisture content of 7 to 10 percent.

The amount of moisture or moisture content (M.C.) in wood is expressed as a percent of the oven-dry weight. To determine the moisture content, a sample is first weighed. It is then placed in an oven and dried at a temperature of about 212 deg. F. The drying is continued until it no longer loses weight. The sample is weighed again and this oven-dry weight is subtracted from the initial weight. The difference is then divided by the oven-dry weight, Fig. 4-7.

Moisture is contained in the cell cavities (free water) and in the cell walls (bound water). As the wood is dried, moisture first leaves the cell cavities. When the cells are empty but the cell walls are still full of moisture, the wood has reached a condition called the FIBER SATURATION POINT. For most woods this is about 30 percent, Fig. 4-8.

The fiber saturation point is important because wood does not start to shrink until this point is reached. As the M.C. is reduced below 30 percent, moisture is removed from the cell walls and they become smaller in size. For a one percent moisture loss below the fiber saturation point, the

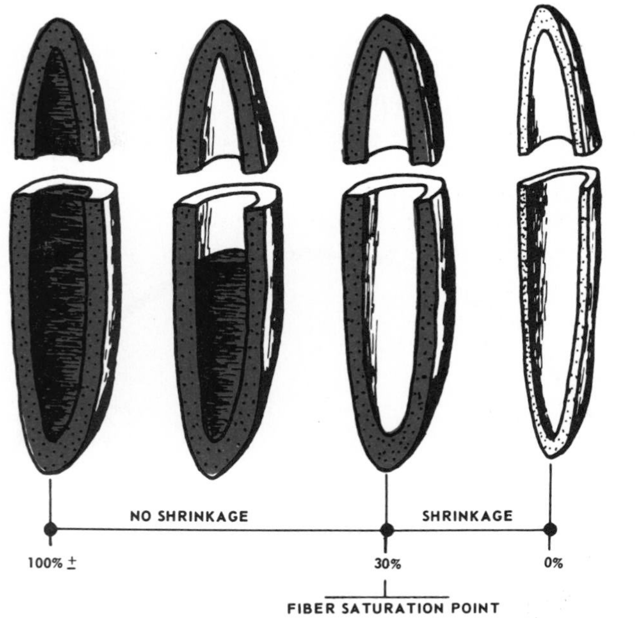

Fig. 4-8. How a wood cell dries. First the free water in the cell cavity is removed, and then the cell wall dries and shrinks.

wood will shrink about 1/30th of the total possible shrinkage. If dried to 15 percent M.C., it will have been reduced by about one-half the total shrinkage. Fig. 4-9 shows the shrinkage in a 2 x 10 joist.

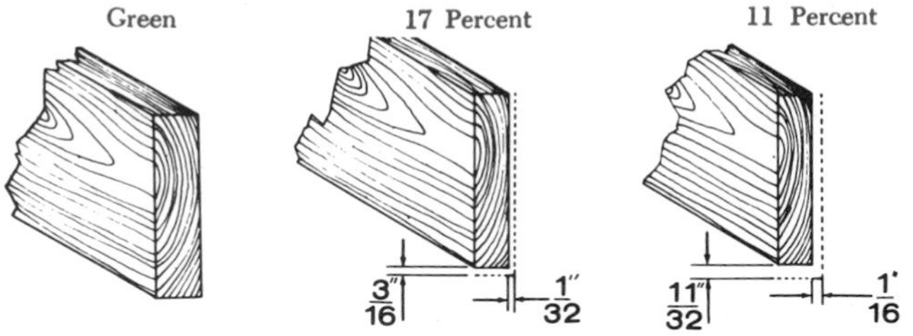

Fig. 4-9. Dimensional change in a 2 x 10 joist.

Wood shrinks most along the direction of the annual rings (tangentially) and about one-half as much across these rings. There is practically no shrinkage in the length. How this shrinkage affects lumber cut from a log is shown in Fig. 4-10. As moisture is added to wood, it swells in the same proportion that the shrinkage has taken place.

Equilibrium Moisture Content

A piece of wood will give off or take on moisture from the air around it until the moisture in the wood is balanced with that in the air. At this point the wood is said to be at equilibrium moisture content (E.M.C.). Since wood is exposed to daily and seasonal changes in the relative humidity of the air, it is continually making slight changes in its moisture content and, therefore, changes in its dimensions. This is the reason doors and drawers often stick during humid weather but work freely the rest of the year.

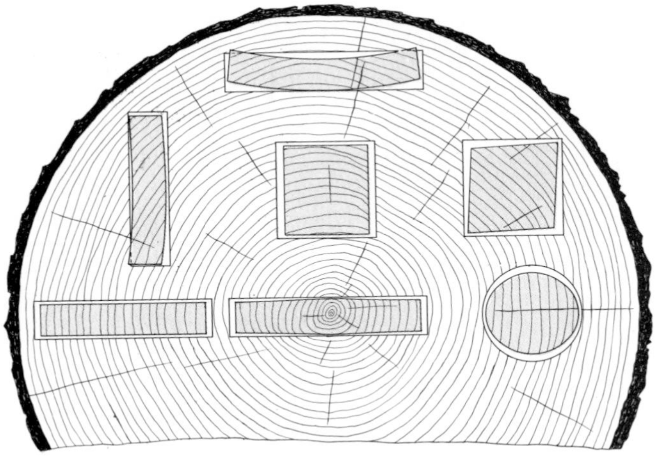

Fig. 4-10. The shrinkage and distortion of flat, square and round pieces, as affected by the direction of the annual rings.

Ideally, a wood structure should be framed with lumber at a M.C. equal to that which it will attain in service. This is not practical since lumber with such a low moisture content is seldom available and would likely gain moisture during the construction stages. Standard practice is to use lumber with a moisture content in the range of 15 to 19 percent. In heated structures, it will eventually reach a level of about 8 percent. However, this will vary in different geographical areas, Fig. 4-11.

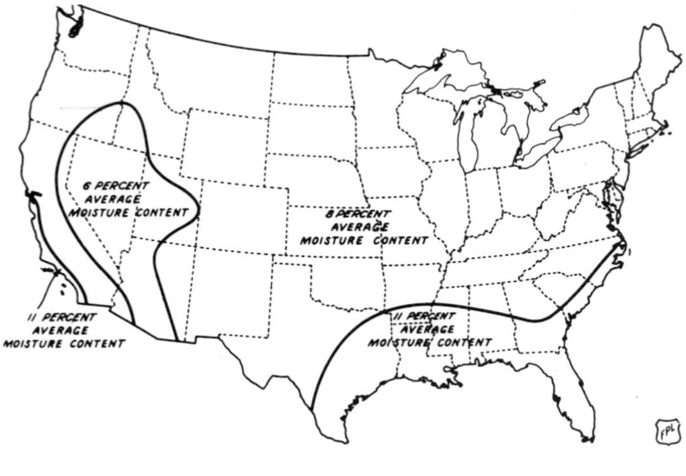

Fig. 4-11. Average moisture content of interior woodwork in various areas of the United States.

The carpenter understands that some shrinkage is inevitable and makes allowances where it will affect the structure. The first, and by far the greatest change in moisture content occurs during the first year after construction, particularly during the first heating season.

When "green" lumber (in excess of 20 percent M.C.) is used, shrinkage will be excessive and it will be almost impossible to prevent excessive warping, plaster cracks, nail pops, squeaky floors and other difficulties.

Seasoning Lumber

This is the process of reducing the moisture content to the required level specified by the grade and use. In air-drying, the lumber is simply exposed to the outside air. It is carefully stacked with stickers (wood strips) between layers so air can circulate through the pile. Boards are also spaced apart in the layers so air can move vertically. Air-drying is a relatively slow process and often creates additional defects in the wood.

Lumber is kiln-dried by placing it in huge ovens where the temperature and humidity can be carefully controlled. When the green lumber is first placed in the kiln, steam is used to keep the humidity high while the temperature is kept at a low level. Gradually the temperature is raised and the humidity reduced. Fans are used to keep the air in constant circulation over the surface of the wood. See Fig. 4-12.

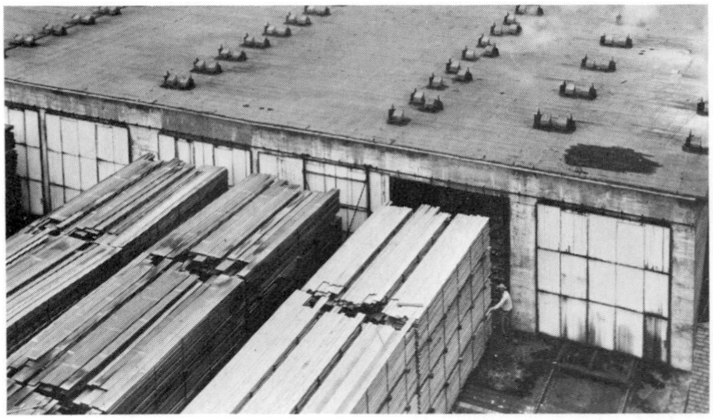

Fig. 4-12. Huge kilns used to season lumber at a modern sawmill. (Forest Products Lab.)

Moisture Meters

The moisture content of wood can be determined by oven drying a sample as previously described, or by using an electric moisture meter. Although the oven drying method is the most ac-

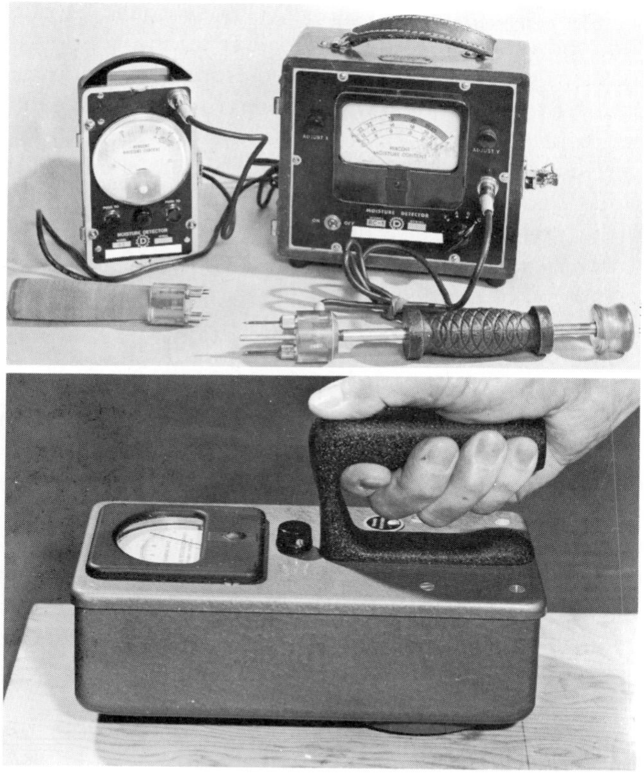

Fig. 4-13. Moisture meters. Above. Resistance type. Below. Radio-frequency type.

curate, meters are often used because readings can be secured rapidly and conveniently. The meters are usually calibrated to cover a range from 7 to 25 percent with an accuracy of plus or minus 1 percent of the moisture content.

Two types of meters are shown in Fig. 4-13. One determines the moisture content by measuring the electrical resistance between two pin-type electrodes that are driven into the wood. The other type measures the capacity of a condenser in a high-frequency circuit where the wood serves as the dielectric material of the condenser.

Lumber Defects

A defect is an irregularity occurring in or on wood that reduces its strength, durability or usefulness. It may or may not detract from appearance. For example, knots commonly considered a defect may add to the appearance of pine paneling. An imperfection that impairs only the appearance of wood is called a blemish. Some of the common defects include:

KNOTS: Caused by an imbedded branch or limb of the tree, Fig. 4-14. They are generally considered to be strength reducing - the amount depending upon the type, size and location, See Fig. 4-15.

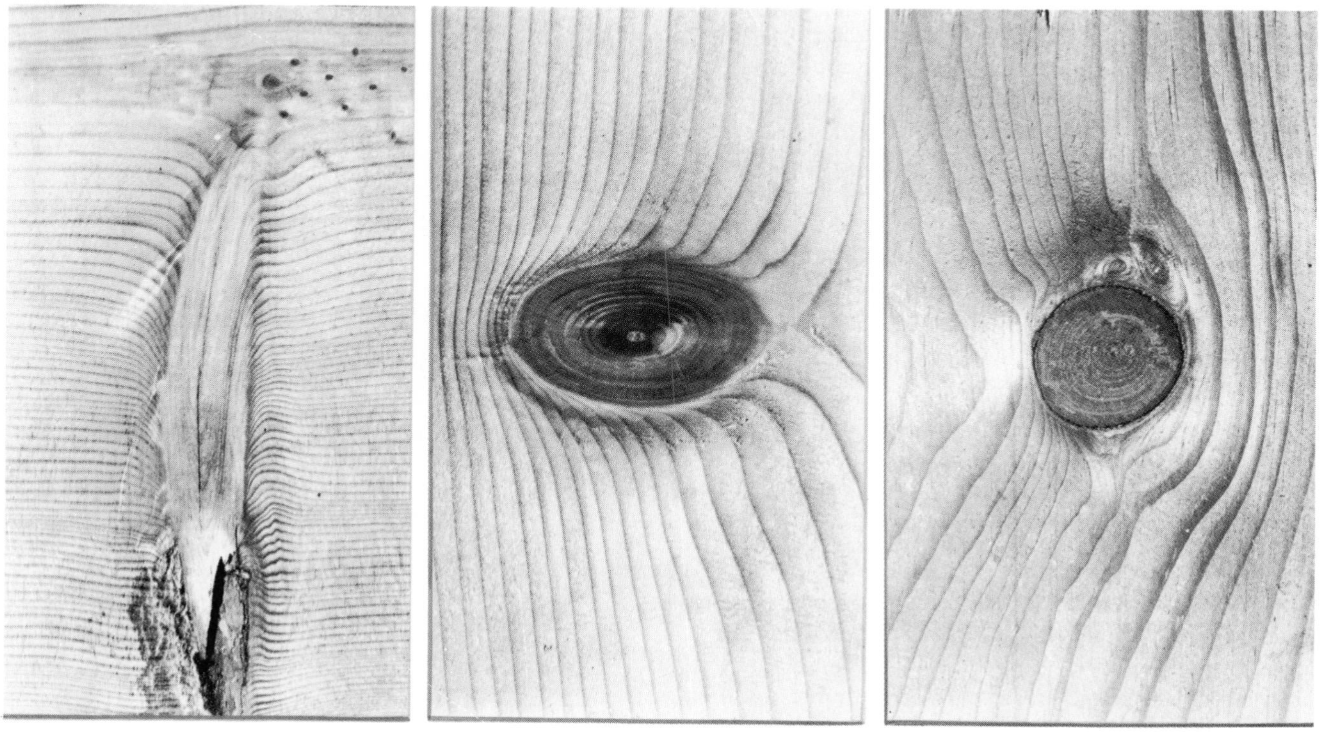

Fig. 4-14. Common kinds of knots. Left. Spike. Center. Intergrown. Right. Encased. An encased knot will usually loosen and fall out. (Forest Products Lab.)

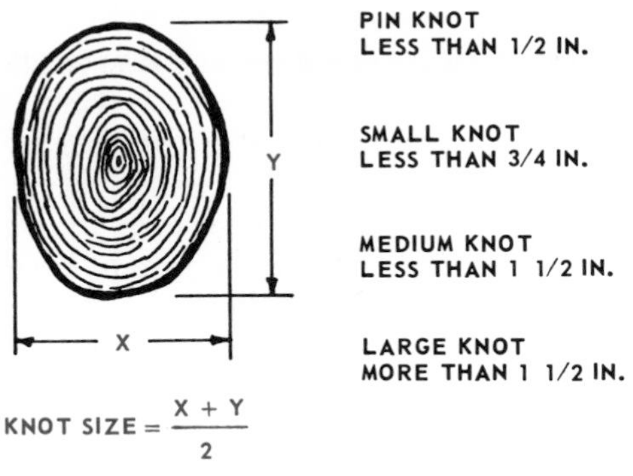

Fig. 4-15. Knot sizes.

SPLITS and CHECKS: A separation of the wood fibers along the grain and across the annular growth rings. Usually occurs at the ends of lumber - a result of uneven seasoning.

SHAKES: A separation along the grain and between the annular growth rings. Likely to occur only in species with abrupt change from spring to summer growth.

PITCH POCKETS: Internal cavities that contain or have contained pitch in either solid or liquid form.

HONEYCOMBING: Separation of the wood fibers in the interior section of the tree. May not be visible on the surface of boards.

WANE: The presence of bark or the absence of wood along the edge of the board. It forms a bevel and reduces the width.

BLUE STAIN: A discoloration caused by moldlike fungi. Objectional in appearance in some grades of lumber but has little or no effect on strength.

DECAY: A disintegration of wood fibers due to fungi. Early stages of decay may be difficult to recognize. Advanced stages result in wood that is soft, spongy, and crumbles easily.

HOLES: Holes in lumber will lower the grade. They may be caused by handling equipment or by wood boring insects or worms.

WARP: Any variation from true or plane surface. May include any one or combination of the following: cup, bow, crook, and twist (also called wind). See Fig. 4-16.

Grades (Softwoods)

Basic principles of grading lumber are formulated by the American Lumber Standards Committee and are published by the U.S. Department of Commerce. Detailed rules are developed and applied by the various associations of lumber producers -- Western Wood Products Association, Southern Pine Inspection Bureau, California Redwood Association and others. These agencies publish and distribute grading rules for the species of lumber produced in their regions and maintain qualified personnel who supervise grading standards at sawmills.

Basic classifications of softwood grading include boards, dimension, and timbers. The grades within these classifications are shown in Fig. 4-17. Another classification called FACTORY and SHOP LUMBER is graded primarily for remanufacturing purposes. It is used by millwork plants in the fabrication of windows, doors, moldings and other trim items.

The carpenter must understand that quality construction does not require that all lumber be of the best grade. Today, lumber is graded for specific uses and, in a given structure, several grades may be appropriate. The key to good economical construction is the proper use of the lowest grade which is suitable for the purpose.

Grades (Hardwoods)

Grades for hardwood lumber are established by the National Hardwood Lumber Association. FAS (firsts and seconds) is the best grade and specifies that pieces be no less than 6 in. wide by 8 ft. long -- and yield at least 83 1/3 percent clear cuttings. The next lower grade is SELECTS and permits pieces 4 in. wide by 6 ft. long. A still lower grade is designated as NO. 1 COMMON and is expected to yield 66 2/3 percent clear cuttings.

Lumber Stress Values

In softwood lumber, all dimension and timber grades except Economy and Mining are assigned stress values. Slope of grain, knot sizes and knot locations are critical considerations. There are

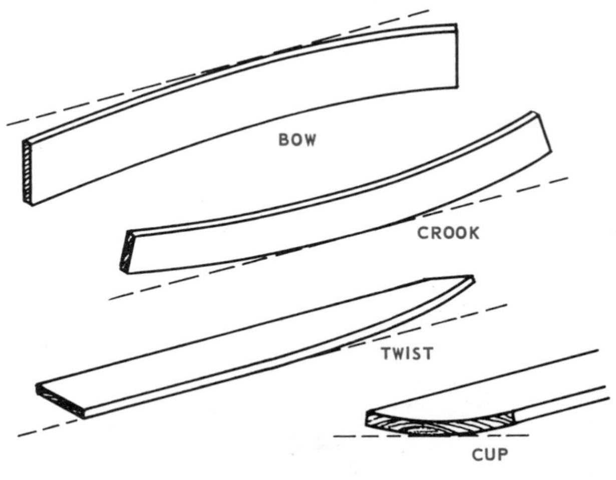

Fig. 4-16. Kinds of warp.

boards

APPEARANCE GRADES	**SELECTS**	B & BETTER (IWP—SUPREME) C SELECT (IWP—CHOICE) D SELECT (IWP—QUALITY)	
APPEARANCE GRADES	**FINISH**	SUPERIOR PRIME E	
APPEARANCE GRADES	**PANELING**	CLEAR (ANY SELECT OR FINISH GRADE) NO. 2 COMMON SELECTED FOR KNOTTY PANELING NO. 3 COMMON SELECTED FOR KNOTTY PANELING	
APPEARANCE GRADES	**SIDING** (BEVEL, BUNGALOW)	SUPERIOR PRIME	
			ALTERNATE BOARD GRADES
BOARDS SHEATHING		NO. 1 COMMON (IWP—COLONIAL) NO. 2 COMMON (IWP—STERLING) NO. 3 COMMON (IWP—STANDARD) NO. 4 COMMON (IWP—UTILITY)	SELECT MERCHANTABLE CONSTRUCTION STANDARD UTILITY

SPECIFICATION CHECK LIST

- ☐ Grades listed in order of quality.
- ☐ Include all species suited to project.
- ☐ For economy, specify lowest grade that will satisfy job requirement.
- ☐ Specify surface texture desired.
- ☐ Specify moisture content suited to project.
- ☐ Specify (W) grade stamp. For finish and exposed pieces, specify stamp on back or ends.

WESTERN RED CEDAR

FINISH PANELING AND CEILING	CLEAR HEART A B
BEVEL SIDING	CLEAR — V.G. HEART A — BEVEL SIDING B — BEVEL SIDING C — BEVEL SIDING

dimension

LIGHT FRAMING 2″ to 4″ Thick 2″ to 4″ Wide	CONSTRUCTION STANDARD UTILITY ECONOMY	This category for use where high strength values are **NOT** required; such as studs, plates, sills, cripples, blocking, etc.
STUDS 2″ to 4″ Thick 2″ to 4″ Wide	STUD ECONOMY STUD	An optional all-purpose grade limited to 10 feet and shorter. Characteristics affecting strength and stiffness values are limited so that the "Stud" grade is suitable for all stud uses, including load bearing walls.
STRUCTURAL LIGHT FRAMING 2″ to 4″ Thick 2″ to 4″ Wide	SELECT STRUCTURAL NO. 1 NO. 2 NO. 3 ECONOMY	These grades are designed to fit those engineering applications where higher bending strength ratios are needed in light framing sizes. Typical uses would be for trusses, concrete pier wall forms, etc.
STRUCTURAL JOISTS & PLANKS 2″ to 4″ Thick 6″ and Wider	SELECT STRUCTURAL NO. 1 NO. 2 NO. 3 ECONOMY	These grades are designed especially to fit in engineering applications for lumber six inches and wider, such as joists, rafters and general framing uses.

timbers

BEAMS & STRINGERS	SELECT STRUCTURAL NO. 1 NO. 2 (NO. 1 MINING) NO. 3 (NO. 2 MINING)	**POSTS & TIMBERS**	SELECT STRUCTURAL NO. 1 NO. 2 (NO. 1 MINING NO. 3 (NO. 2 MINING)

Fig. 4-17. Softwood lumber classifications and grades. Grade titles and specifications will vary among lumber manufacturers' associations and regions producing lumber.

two methods of assigning stress values, "visual" and "machine rated." In the latter method, lumber is fed into a special machine and subjected to bending forces. The stiffness of each piece (modulus of elasticity E) is measured and marked on each piece. Machine stress-rated lumber (MSR) must also meet certain visual requirements.

Lumber Sizes

When listing and calculating the size and amount of lumber, the nominal dimension is always used. Fig. 4-18 illustrates the nominal and dressed sizes for various classifications of lumber used by the carpenter. Note that nominal sizes

BOARD MEASURE

The term "board measure" indicates that a board foot is the unit for measuring lumber. A board foot is one inch thick and 12 inches square.

The number of board feet in a piece is obtained by multiplying the nominal thickness in inches by the nominal width in inches by the length in feet and dividing by 12: $\frac{(T \times W \times L)}{12}$.

Lumber less than one inch in thickness is figured as one-inch.

product classification

	thickness in.	width in.
board lumber	1″	2″ or more
light framing	2″ to 4″	2″ to 4″
studs	2″ to 4″	2″ to 4″ 10′ and shorter
structural light framing	2″ to 4″	2″ to 4″
joists & planks	2″ to 4″	6″ and wider

	thickness in.	width in.
beams & stringers	5″ and thicker	more than 2″ greater than thickness
posts & timbers	5″ x 5″ and larger	not more than 2″ greater than thickness
decking	2″ to 4″	4″ to 12″ wide
siding	thickness expressed by dimension of butt edge	
mouldings	size at thickest and widest points	

Standard lengths of lumber generally are 6 feet and longer in multiples of 1′

dimensional data / nominal, dressed, based on 1970 rules

Product	Description	Nominal Size: Thickness In.	Nominal Size: Width In.	Dressed Dimensions: Thicknesses and Widths In. Surfaced Dry	Dressed Dimensions: Thicknesses and Widths In. Surfaced Unseasoned	Lengths Ft.
FRAMING	S4S	2 3 4	2 3 4 6 8 10 12 Over 12	1-1/2 2-1/2 3-1/2 5-1/2 7-1/4 9-1/4 11-1/4 Off 3/4	1-9/16 2-9/16 3-9/16 5-5/8 7-1/2 9-1/2 11-1/2 Off 1/2	6 ft. and longer in multiples of 1′
				Thickness In.	Width In.	
TIMBERS	Rough or S4S	5 and Larger		1/2 Off Nominal		Same
		Nominal Size		**Dressed Dimensions** Surfaced Dry		
		Thickness In.	Width In.	Thickness In.	Width In.	Lengths Ft.
DECKING Decking is usually surfaced to single T&G in 2″ thickness and double T&G in 3″ and 4″ thicknesses	2″ Single T&G	2	6 8 10 12	1 1/2	5 6 3/4 8 3/4 10 3/4	6 ft. and longer in multiples of 1′
	3″ and 4″ Double T&G	3 4	6	2 1/2 3 1/2	5 1/4	
FLOORING	(D & M), (S2S & CM)	3/8 1/2 5/8 1 1 1/4 1 1/2	2 3 4 5 6	5/16 7/16 9/16 3/4 1 1 1/4	1 1/8 2 1/8 3 1/8 4 1/8 5 1/8	4 ft. and longer in multiples of 1′
CEILING AND PARTITION	(S2S & CM)	3/8 1/2 5/8 3/4	3 4 5 6	5/16 7/16 9/16 11/16	2 1/8 3 1/8 4 1/8 5 1/8	4 ft. and longer in multiples of 1′
FACTORY AND SHOP LUMBER	S2S	1 (4/4) 1 1/4 (5/4) 1 1/2 (6/4) 1 3/4 (7/4) 2 (8/4) 2 1/2 (10/4) 3 (12/4) 4 (16/4)	5 and wider (4″ and wider in 4/4 No. 1 Shop and 4/4 No. 2 Shop)	25/32 (4/4) 1 5/32 (5/4) 1 13/32 (6/4) 1 19/32 (7/4) 1 13/16 (8/4) 2 3/8 (10/4) 2 3/4 (12/4) 3 3/4 (16/4)	Usually sold random width	4 ft. and longer in multiples of 1′

ABBREVIATIONS
Abbreviated descriptions appearing in the size table are explained below.
S1S — Surfaced one side.
S2S — Surfaced two sides.
S4S — Surfaced four sides.
S1S1E — Surfaced one side, one edge.
S1S2E — Surfaced one side, two edges.
CM — Center matched.
D & M — Dressed and matched.
T & G — Tongue and grooved.
EV1S — Edge vee on one side.
S1E — Surfaced one edge.

Fig. 4-18. Standard lumber sizes. (Western Wood Products Assoc.)

coverage estimator

The following estimator provides factors for determining the exact amount of material needed for the five basic types of wood paneling. Multiply square footage to be covered by factor (length x width x factor).

	Nominal Size	WIDTH Overall	WIDTH Face	AREA FACTOR*
SHIPLAP	1 x 6	5½	5⅛	1.17
	1 x 8	7¼	6⅞	1.16
	1 x 10	9¼	8⅞	1.13
	1 x 12	11¼	10⅞	1.10
TONGUE AND GROOVE	1 x 4	3⅜	3⅛	1.28
	1 x 6	5⅜	5⅛	1.17
	1 x 8	7⅛	6⅞	1.16
	1 x 10	9⅛	8⅞	1.13
	1 x 12	11⅛	10⅞	1.10
S4S	1 x 4	3½	3½	1.14
	1 x 6	5½	5½	1.09
	1 x 8	7¼	7¼	1.10
	1 x 10	9¼	9¼	1.08
	1 x 12	11¼	11¼	1.07
PANELING PATTERNS	1 x 6	5 7/16	5 1/16	1.19
	1 x 8	7⅛	6¾	1.19
	1 x 10	9⅛	8¾	1.14
	1 x 12	11⅛	10¾	1.12
BEVEL SIDING	1 x 4	3½	3½	1.60
	1 x 6	5½	5½	1.33
	1 x 8	7¼	7¼	1.28
	1 x 10	9¼	9¼	1.21
	1 x 12	11¼	11¼	1.17

*Allowance for trim and waste should be added.

Product	Description	Nominal Size Thickness In.	Nominal Size Width In.	Dressed Dimensions Thickness In.	Dressed Dimensions Width In.	Lengths Ft.
SELECTS AND COMMONS S-DRY	S1S, S2S, S4S, S1S1E, S1S2E....	4/4 5/4 6/4 7/4 8/4 9/4 10/4 11/4 12/4 16/4	2 3 4 5 6 7 8 and wider	¾ 1 5/32 1 13/32 1 19/32 1 13/16 2 3/32 2⅜ 2 9/16 2¾ 3¾	1½ 2½ 3½ 4½ 5½ 6½ ¾ Off nominal	6 ft. and longer in multiples of 1′
FINISH AND BOARDS S-DRY	S1S, S2S, S4S, S1S1E, S1S2E ...	1 1¼ 1½	2 3 4 5 6 7 8 and wider	¾ 1 1¼	1½ 2½ 3½ 4½ 5½ 6½ ¾ off nominal	3′ and longer. In Superior grade, 3% of 3′ and 4′ and 7% of 5′ and 6′ are permitted. In Prime grade, 20% of 3′ to 6′ is permitted.
RUSTIC AND DROP SIDING	(D & M) .... If ⅜″ or ½″ T & G specified, same over-all widths apply. (Shiplapped, ⅜-in. or ½-in. lap) ..	1	6 8 10 12	23/32	5⅜ 7⅛ 9⅛ 11⅛	4 ft. and longer in multiples of 1′
PANELING AND SIDING	T&G or Shiplap.................	1	6 8 10 12	23/32	5 7/16 7⅛ 9⅛ 11⅛	Same
CEILING AND PARTITION	T&G	⅝ 1	4 6	9/16 23/32	3⅜ 5⅜	Same
BEVEL SIDING	Bevel or Bungalow Siding........ Western Red Cedar Bevel Siding available in ½″, ⅝″, ¾″ nominal thickness. Corresponding thick edge is 15/32″, 9/16″ and ¾″. Widths for 8″ and wider, ½″ off nominal.	½ ¾	4 5 6 8 10 12	15/32 butt, 3/16 tip ¾ butt, 3/16 tip	3½ 4½ 5½ 7¼ 9¼ 11¼	Same

See coverage estimator chart above for T&G widths.

MINIMUM ROUGH SIZES **Thicknesses and Widths Dry or Unseasoned All Lumber (S1E, S2E, S1S, S2S)**

80% of the pieces in a shipment shall be at least ⅛″ thicker than the standard surfaced size, the remaining 20% at least 3/32″ thicker than the surfaced size. Widths shall be at least ⅛″ wider than standard surfaced widths.

When specified to be full sawn, lumber may not be manufactured to a size less than the size specified.

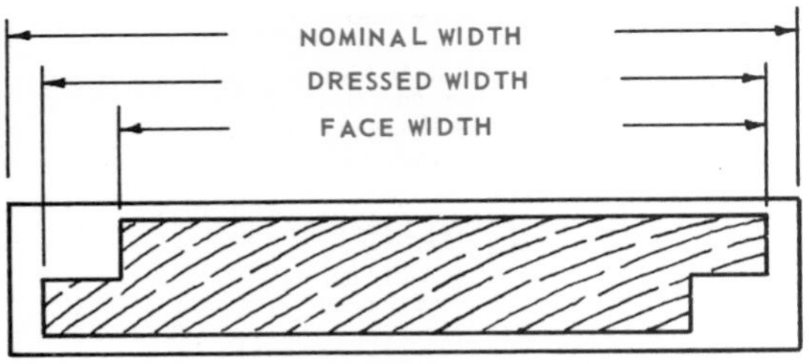

Fig. 4-19. Nominal and dressed sizes.

are sometimes listed in quarters. For example: 1 1/4 in. material is given as 5/4. This nominal dimension is its rough unfinished measurement, Fig. 4-19. The dressed size is less than the nominal size as a result of seasoning and surfacing. Dressed sizes of lumber are established by the American Lumber Standards and applied consistently throughout the industry.

Figuring Board Footage

The unit of measure for lumber is the board foot. This is a piece 1 in. thick and 12 in. square or its equivalent (144 cu. in.). Standard size pieces can be quickly calculated by visualizing the board feet included. For example: a board 1 x 12 and 10 ft. long will contain 10 bd. ft. If it were only 6 in. wide, it would be 5 bd. ft. If the original board had been 2 in. thick, it would have contained 20 bd. ft. The following formula can be applied to any size piece where the total length is given in feet:

$$\text{Bd. ft.} = \frac{\text{No. pcs.} \times T \times W \times L}{12}$$

An example of the application of the formula is shown below. Find the number of board feet in 6 pieces of lumber that measure 1" x 8" x 14':

$$\text{Bd. ft.} = \frac{\overset{1}{\cancel{6}} \times 1 \times \overset{4}{\cancel{8}} \times 14}{\underset{\underset{1}{\cancel{2}}}{\cancel{12}}} = \frac{56}{1} = 56 \text{ bd. ft.}$$

Stock that is less than 1 in. thick is figured as though it were 1 in. When the stock is thicker than 1 in., the nominal size is used. When this size contains a fraction such as 1 1/4, change it to an improper fraction (5/4) and place the numerator above the formula line and the denominator below. For example: find the board footage in 2 pieces of lumber that measure 1 1/4" x 10" x 8'.

$$\text{Bd. ft.} = \frac{\overset{1}{\cancel{2}} \times 5 \times \overset{5}{\cancel{10}} \times \overset{2}{\cancel{8}}}{\underset{\underset{1}{\cancel{2}}}{\cancel{4}} \; \underset{3}{\cancel{12}}} = \frac{50}{3} = 16\ 2/3 \text{ bd. ft.}$$

Use the nominal size of the material when figuring the footage. Items such as moldings, furring strips, and grounds are priced and sold by the lineal foot; thickness and width are disregarded.

Plywood

Plywood is constructed by gluing together a number of layers (plies) of wood with the grain direction turned at right angles in each successive layer. An odd number (3, 5, 7) of plies are used so they will be balanced on either side of a center core and so the grain of the outside layers will run in the same direction. The outer plies are called FACES or face and back. The next layers under these are called CROSS-BANDS and the other inside layer or layers are called the CORE. See Fig. 4-20. A thin plywood panel made of three layers would consist of two faces and a core.

There are two basic types of plywood; exterior and interior. EXTERIOR PLYWOOD is bonded with waterproof glues and can be used for siding, concrete forms, and other constructions that will be exposed to the weather or excessive moisture. INTERIOR PLYWOOD is bonded with glues that are not waterproof and is used for cabinets and other inside work where the moisture content of the panels will not exceed 20 percent.

Plywood can be secured in thicknesses of 1/8 in. to more than 1 in. with the common sizes being 1/4, 3/8, 1/2, 5/8 and 3/4 in. A standard panel size is 4 ft. wide by 8 ft. long. Smaller size panels are available in the hardwoods.

Plywood Grades (Softwood)

Softwood plywood for general construction is manufactured in accordance with U.S. Product Standard PSI. This standard provides a system

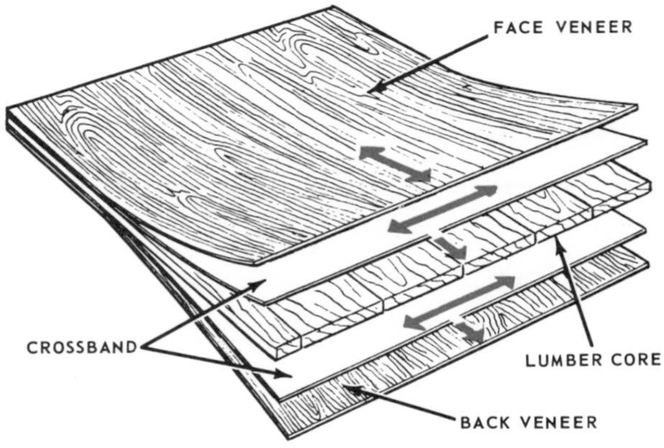

Fig. 4-20. Plywood construction. Hardwood or softwood classification depends on the kind of wood used for the face veneers.

Group 1	Group 2	Group 3	Group 4	Group 5
Apitong Beech, American Birch Sweet Yellow Douglas Fir 1[(a)] Kapur Keruing Larch, Western Maple, Sugar Pine Caribbean Ocote Pine, South. Loblolly Longleaf Shortleaf Slash Tanoak	Cedar, Port Orford Cypress Douglas Fir 2 Fir California Red Grand Noble Pacific Silver White Hemlock, Western Lauan Almon Bagtikan Mayapis Red Lauan Tangile White Lauan Maple, Black Mengkulang Meranti, Red Mersawa Pine Pond Red Virginia Western White Spruce Red Sitka Sweetgum Tamarack Yellow- poplar	Alder, Red Birch, Paper Cedar, Alaska Fir, Subalpine Hemlock, Eastern Maple, Bigleaf Pine Jack Lodgepole Ponderosa Spruce Redwood Spruce Black Engelmann White	Aspen Bigtooth Quaking Cativo Cedar Incense Western Red Cottonwood Eastern Black (Western Poplar) Pine Eastern White Sugar	Basswood Fir, Balsam Poplar, Balsam

Fig. 4-21. Classification of softwood species. (American Plywood Assoc.)

N	Smooth surface "natural finish" veneer. Select, all heartwood or all sapwood. Free of open defects. Allows not more than 6 repairs, wood only, per 4 x 8 panel, made parallel to grain and well matched for grain and color.
A	Smooth, paintable. Not more than 18 neatly made repairs, boat, sled, or router type, and parallel to grain, permitted. May be used for natural finish in less demanding applications.
B	Solid surface. Shims, circular repair plugs and tight knots to 1 inch permitted. Wood or synthetic patching material may be used. Some minor splits permitted.
C (Plugged)	Improved C veneer with splits limited to 1/8 inch width and knotholes and borer holes limited to 1/4 x 1/2 inch. Admits some broken grain. Synthetic repairs permitted.
C	Tight knots to 1-1/2 inch. Knotholes to 1 inch and some to 1-1/2 inch if total width of knots and knotholes is within specified limits. Synthetic or wood repairs. Discoloration and sanding defects that do not impair strength permitted. Limited splits allowed.
D	Knots and knotholes to 2-1/2 inch width and 1/2 inch larger within specified limits. Limited splits are permitted.

Fig. 4-22. Veneer grades of softwood plywood. (American Plywood Assoc.)

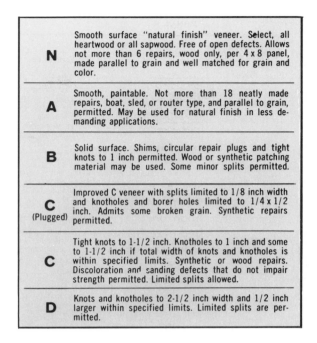

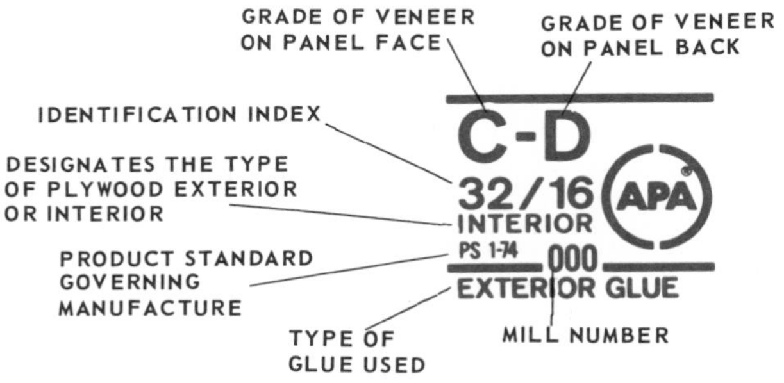

Fig. 4-23. Typical grade-trademark for plywood manufactured in compliance with U.S. Product Standard PSI-74. (American Plywood Assoc.)

for designating the species, strength, type of glue, and appearance.

Many species of softwood are used in the manufacture of plywood and are separated into five groups - based on their stiffness and strength. Group 1 lists the species with the highest level of these characteristics, Fig. 4-21. A lettering system is used to designate the quality of veneer used on the panel face, back and for the inner ply construction. The N grade is the highest grade and ranges downward through A to including D. See Fig. 4-22.

The American Plywood Association (APA) conducts a rigid testing program based on PSI-74 and member manufacturing companies are licensed to use their official grade-trademark. These grade-trademarks are stamped on each panel and cover both appearance grades and engineered (structural) classifications.

A typical grade-trademark for an engineered grade of plywood is shown in Fig. 4-23. The C-D indicates that a C grade of veneer has been used on the face of the panel and a D grade on the back. Some engineered grades include an identification index which consists of a pair of numbers separated by a slash mark (/). The number on the left indicates the maximum recommended spacing of supports in inches when the plywood is used for roof decking. The number on the right indicates the maximum recommended spacing when the plywood is used for subflooring. In general, the higher the index number, the greater the stiffness.

Fig. 4-24 lists some engineered grades of plywood and includes descriptions and most common uses. A more complete list can be secured from the American Plywood Association. A table listing appearance grades of plywood is included in the appendix.

Plywood Grades (Hardwood)

The Hardwood Plywood Institute uses a number system for grading the faces and backs of a panel. A grading specification of 1-2 would indicate a

INTERIOR TYPE

Use these terms when you specify plywood	Description and Most Common Uses	Typical Grade-trademarks	Veneer Grade			Most Common Thicknesses (inch) (2) (3)				
			Face	Back	Inner Plies					
C-D INT-APA (1) (4)	For wall and roof sheathing, subflooring, industrial uses such as pallets. Also available with intermediate glue or exterior glue. Specify intermediate glue if moderate construction delays are expected; exterior glue for better durability in somewhat longer construction delays, and for wood foundations.	C-D 48/24 INTERIOR PS 1-74 000 (APA)	C	D	D	5/16	3/8	1/2	5/8	3/4
STRUCTURAL I C-D INT APA (4) and STRUCTURAL II C-D INT-APA (4)	Unsanded structural grades where plywood strength properties are of maximum importance: structural diaphragms, box beams, gusset plates, stressed-skin panels, containers, pallet bins. Made only with exterior glue.	STRUCTURAL I C-D 32/16 INTERIOR PS 1-74 000 EXTERIOR GLUE (APA)	C	D	D	5/16	3/8	1/2	5/8	3/4
C-D PLUGGED INT-APA (1) (4) (5)	For built-ins, wall and ceiling tile backing, cable reels, walkways, separator boards. Not a substitute for Underlayment, as it lacks Underlayment's punch-through resistance. Touch-sanded.	C-D PLUGGED GROUP 2 INTERIOR PS 1-74 000 (APA)	C Plgd.	D	D	5/16	3/8	1/2	5/8	3/4
2•4•1 INT-APA (1) (6)	Combination subfloor-underlayment. Quality base for resilient floor coverings, carpeting, wood strip flooring. Use 2•4•1 with exterior glue in areas subject to moisture. Unsanded or touch-sanded as specified.	2•4•1 GROUP 1 INTERIOR PS 1-74 000 (APA)	C Plgd.	D	C & D	(available 1-1/8" or 1-1/4")				

EXTERIOR TYPE

C-C EXT-APA (4)	Unsanded grade with waterproof bond for subflooring and roof decking, siding on service and farm buildings, wood foundations, crating, pallets, pallet bins, cable reels.	C-C 42/20 EXTERIOR PS 1-74 000 (APA)	C	C	C	5/16	3/8	1/2	5/8	3/4
STRUCTURAL I C-C EXT-APA (7) (4) and STRUCTURAL II C-C EXT-APA (7) (4)	For engineered applications in construction and industry where full Exterior type panels are required. Unsanded. See (5) for Group requirements.	STRUCTURAL I C-C 32/16 EXTERIOR PS 1-74 000 (APA)	C	C	C	5/16	3/8	1/2	5/8	3/4

(1) Also available with exterior or intermediate glue (check dealer for availability of intermediate glue in your area).
(2) All grades available tongue-and-grooved in panels 1/2" and thicker.
(3) Panels are standard 4x8-foot size. Other sizes available.
(4) May be grade-trademarked with Identification Index.
(5) Also available in Structural I (all plies limited to Group 1 species) and Structural II (all plies limited to Group 1, 2, or 3 species).
(6) Available in Group 1, 2, or 3 only.
(7) Also available in Exterior sanded grades.

Fig. 4-24. Selected list of engineered grades of softwood plywood. Appearance grades are listed in the appendix. (American Plywood Assoc.)

good face with grain carefully matched and a good back but without careful grain matching. A number 3 back would permit noticeable defects and patching but would be generally sound. A special or PREMIUM grade of hardwood is known as "architectural" or "sequence-matched." This usually requires an order to a plywood mill for a series of matched plywood panels.

For either softwood or hardwood plywood, it is common practice to designate in a general way the grade by the symbol G2S (good two sides) or G1S (good one side).

In addition to the various kinds, types, and grades, hardwood plywood is made with different core constructions. The two most common are the veneer core and the lumber core, as shown in Fig. 4-25. VENEER CORES are the least expensive and are fairly stable and warp resistant. LUMBER CORES are easier to cut, the edges are better for shaping and finishing, and they hold nails and

screws better. Plywood is also manufactured with a particle board core. It is made by gluing veneers directly to the particle board surface.

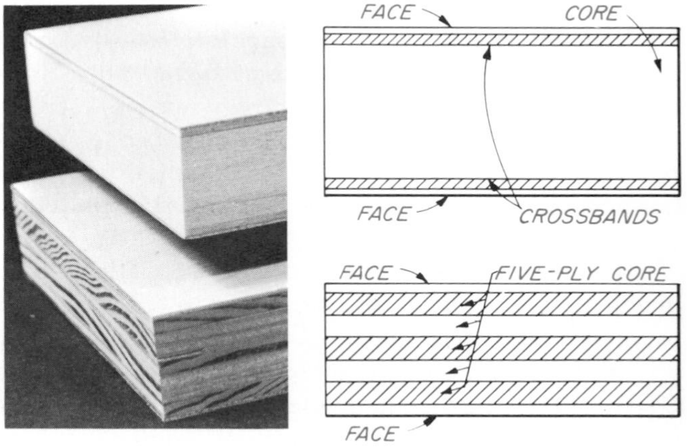

Fig. 4-25. Hardwood plywood. Above. Lumber core. Below. Veneer core.

Hardboard and Particle Board

Hardboards and particle boards are used extensively in modern construction for siding and interior wall surfaces. In cabinetwork they serve as appropriate materials for drawer bottoms and concealed panels in cases, cabinets and chests. Fig. 4-26 shows some of the various types of hardboards and particle boards that are available. These are manufactured by many different companies and are sold under various trade names.

Hardboard is made of refined wood fibers, pressed together to form a hard, dense material (50 - 80 lbs. per cu. ft.). There are two types; standard and tempered. Tempered hardboard is impregnated with oils and resins that make it harder, slightly heavier, more water resistant and darker in appearance. Hardboard is manufactured with one side smooth (S1S) or both sides smooth (S2S). It is available in thicknesses from 1/12 in. to 5/16 in. with the most common thicknesses being 1/8, 3/16 and 1/4 in. Panels are 4 ft. wide and come in standard lengths of 8, 10, 12 and 16 ft.

Particle board is made of wood flakes, chips and shavings bonded together with resins or adhesives. It is not as heavy as hardboard (about 40 lbs. per cu. ft.) and is available in thicker pieces. Particle board may be constructed of layers made of different size wood particles; large ones in the center to provide strength and fine ones at the surface to provide smoothness. Extensive use is made of particle board as a base for plastic laminates and as a core for plywood. It is available in a wide range of thicknesses from 1/4 in. to 1 7/16 in. The most common panel size is 4 x 8 ft.

Fig. 4-26. Manufactured materials. Left. Hardboard. Right. Particle board. (Weyerhaeuser Co.)

The unit of measure for plywood, hardboard and particle board is the square foot (sq. ft.). A standard 4 ft. x 8 ft. panel contains 32 sq. ft. Prices are quoted per square foot on the basis of full panel purchase and vary widely depending on the kind, thickness and grade.

Wood Treatments

Wood and wood products can be protected from attack by fungi, insects and borers by the application of special chemicals or wood preservatives. The degree of protection depends on the effectiveness of the chemical and how thoroughly it penetrates the material. Millwork plants employ extensive treatment processes in the manufacture of such items as door frames and window units.

There are two general classes of wood preservatives: oils, such as creosote, and petroleum solutions of pentachlorophenol; and certain salts that can be dissolved in water. The selection of a preservative should be based on its effectiveness in protecting the wood and also on any side effects that may result - discoloration of painted surfaces or objectionable odors.

A number of commercially prepared preservatives are available for on-the-job application.

Study the manufacturer's directions and recommendations carefully. Use special precautions in handling solutions since some contain toxic chemicals and may also create a fire or explosion hazard during application.

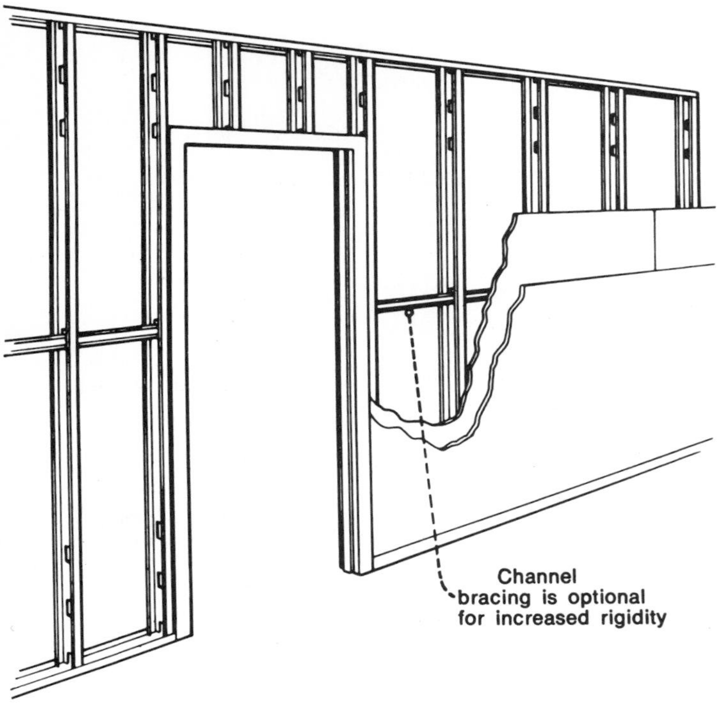
Channel
bracing is optional
for increased rigidity

Fig. 4-27. Metal stud partition system. (National Gypsum Co.)

Nonwood Materials

The carpenter works with a number of materials other than the lumber and wood-base products previously discussed. Some of the more common items include gypsum lath, wallboard, and sheathing; insulation boards and blankets; shingles made of mineral fiber and asphalt; metal lath and flashing units; and a wide range of resilient flooring materials.

METAL STUDS are another example of a nonwood material. Originally designed for commercial and institutional construction, their use is now being extended to residential structures. A typical stud consists of a metal channel with openings through which electrical and plumbing lines can be installed. See Fig. 4-27. These are attached to base and ceiling channels with metal screws or clips. Wall surface material is attached to the stud using self-drilling drywall screws. Some web-type studs have a special metal edge into which nails can be driven. Metal stud systems are usually designed for nonload bearing walls and partitions.

Additional specialized building materials are described in detail in other sections of this book.

Handling and Storing

Building materials are expensive and every precaution should be taken to maintain them in good condition. After they are delivered to the construction site, this becomes the responsibility of the carpenter.

Piles of framing lumber and sheathing should be laid on level skids raised at least 6 in. above the ground. Be sure all pieces are well supported and are lying straight. Cover the material with canvas or waterproof paper, laid to shed water. Polyethylene film provides a watertight covering, Fig. 4-28.

Exterior finish materials, door frames and window units should not be delivered until the structure is partially enclosed and the roof surface complete. In cold weather, the entire structure should be enclosed and heated before interior finish and cabinetwork are delivered and stored.

Fig. 4-28. Using polyethylene film to cover framing materials.

Fig. 4-29. Storing plywood. (E. L. Bruce Co.)

When finish lumber is received at a higher or lower moisture content than it will attain in the structure, it should be open stacked with wood strips so air can circulate freely around each piece.

Plywood, especially the fine hardwoods, must be handled with care. Sanded faces become soiled and scarred if not protected. In storing, the panels should be laid flat as illustrated in Fig. 4-29.

Metal Fasteners

Nails, the metal fasteners commonly used by carpenters, are available in a wide range of types and sizes. Basic kinds are illustrated in Fig. 4-30.

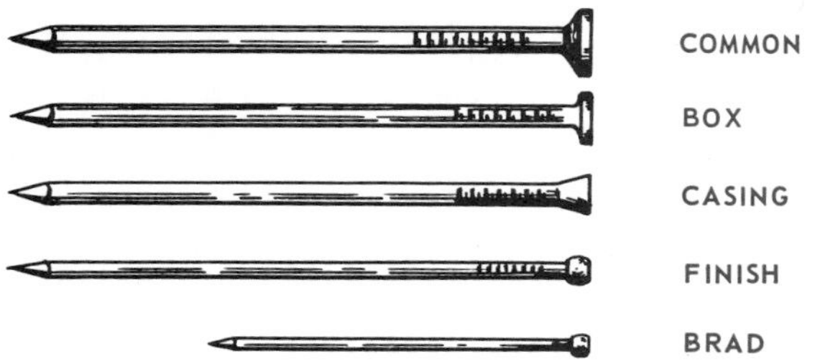

Fig. 4-30. Basic types of nails.

The common nail has a heavy cross section and is designed for rough framing. The thinner box nail is used for toe nailing in frame construction and light work. The casing nail is the same weight as the box nail, but has a small conical head. It is used in finish carpentry work to attach door and window casings and other wood trim. Finishing nails and brads are quite similar and have the thinnest cross section and the smallest head.

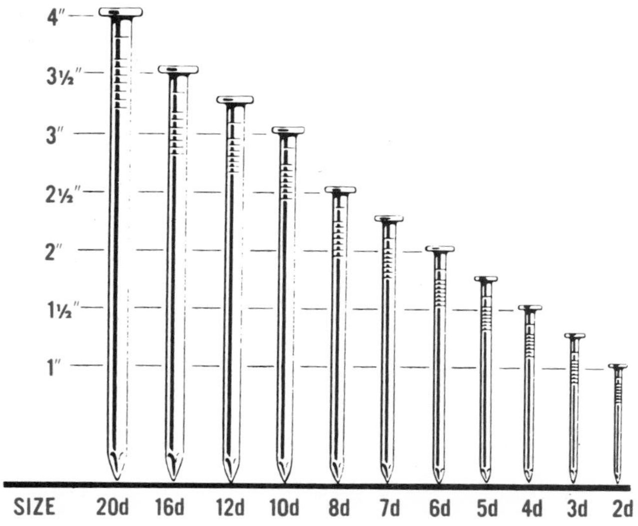

Fig. 4-31. Nail sizes. (United States Steel)

The nail size unit is called a "penny" and is abbreviated with the lower case letter d. It indicates the length of the nail. A 2d (2 penny) nail is 1 in. long. A 6d (6 penny) nail is 2 in. long. See Fig. 4-31. This measurement applies to common, box, casing and finish nails. Brads and small box nails are specified by their actual length and gauge number.

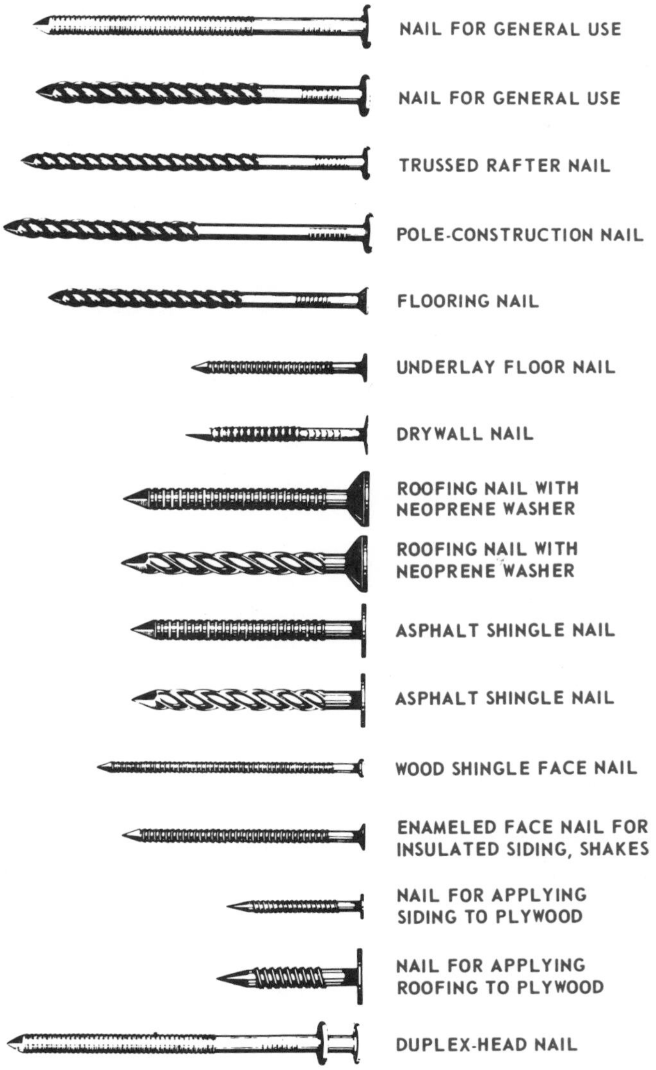

Fig. 4-32. Annular and spiral threaded nails designed for special purposes. (Independent Nail and Packing Co.)

Fig. 4-32 shows a few of the many available specialized nails. Each is designed for a special purpose, with either annular or spiral threads that greatly increase the holding power. Some nails have special coatings of zinc, cement or resin. Coating or threading increases its holding power. Nails are made from such materials as iron, steel, copper, bronze, aluminum and stainless steel.

Wood screws have greater holding power than nails and are often used for interior construction and cabinetwork. Their size is determined by the length and diameter (gauge number). Screws are classified according to the shape of head, surface finish and the material from which they are made. See Fig. 4-33. Wood screws are available in

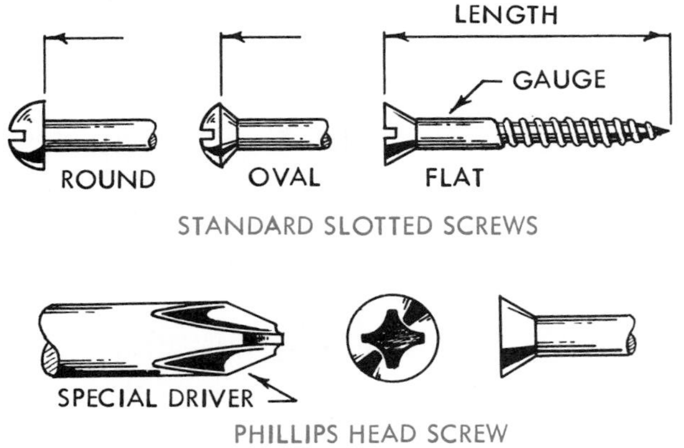

Fig. 4-33. Kinds of common wood screws.

lengths from 1/4 to 6 in., and in gauge numbers from 0 to 24. The gauge number can vary for a given length of screw. For example, a 3/4 in. screw is available in gauge numbers of 4 through 12. The No. 4 would be a thin screw, while the No. 12 would have a large diameter. From one gauge number to the next, the size of the wood screw changes by 13 thousandths (.013) of an inch.

Most wood screws used today are made of mild steel with a zinc chromate finish. They are labeled as F. H., which stands for flat head. Nickel and chromium plated screws, also screws made of brass, are available for special work. Wood screws are usually priced and sold by the box. Each box contains 100 screws.

Additional fasteners the carpenter will find useful include lag screws, hanger bolts, carriage bolts (especially designed for woodwork), corrugated fasteners and metal splines. Specialized metal fasteners are described in other sections of this book where their application is covered.

Adhesives

The adhesives the carpenter uses may be classified as glue and mastics. Research and development have created many new products in this area; some are highly specialized and designed

Fig. 4-34. Using polyvinyl resin emulsion glue to assemble stairs.

for a specific material and/or application procedure. Brief descriptions of several of the commonly used glues follow:

POLYVINYL RESIN EMULSION GLUE (generally called polyvinyl or white glue) is excellent for interior construction. It comes ready to use in plastic squeeze bottles, Fig. 4-34, and is easily applied. This glue sets up rapidly, does not stain the wood or dull tools, and holds wood parts securely.

Polyvinyl glue hardens when its moisture content is removed through absorption into the wood or through evaporation. It is not waterproof and therefore is not suitable for assemblies that will be subjected to high humidities or moisture. The vinyl-acetate materials used in the glue are thermoplastic, which means that under heat they will soften; and should not be used in constructions where the temperature may rise above 165 deg.

UREA-FORMALDEHYDE RESIN GLUE (usually called urea resin) is available in a dry powder form which contains the hardening agent or catalyst. It is mixed with water to a creamy consistency for use. Urea resin is moisture resistant, dries to a light brown color and holds wood securely. It hardens through chemical action when water is added and sets at room temperatures in from 4 to 8 hours.

CONTACT CEMENT is applied to each surface and allowed to dry until a piece of paper will not stick to the film. The surfaces are then pressed firmly together and bonding takes place immediately. The pieces must be carefully aligned for the initial contact because they cannot be changed after they touch. The bonding time is not critical and can usually be performed any time within one hour.

Contact cement is made with a neoprene rubber base and is an excellent adhesive for applying plastic laminates or joining parts that cannot be

Fig. 4-35. Mounting gypsum wallboard on a concrete block wall. Note the beads of adhesive that have been applied with a caulking gun. (National Gypsum Co.)

clamped together easily. It works well for applying thin veneer strips to plywood edges and can also be used to join combinations of wood, cloth, leather, rubber and plastics. Contact cement usually contains volatile, flammable solvents and the work area where it is applied must be well ventilated.

CASEIN GLUE is made from milk curd, hydrated lime and sodium hydroxide. It is supplied in powder form and is mixed with cold water for use. After mixing, it should set for about 15 minutes before it is applied. It is classified as a water resistant glue.

Casein glue is used for structural laminating and works well where the moisture content of the wood is high. It has good joint filling qualities and is therefore often used on materials that have not been carefully surfaced. Casein is used for gluing oily woods such as teak, padouk and lemon wood. Its main disadvantages are that it stains the wood, especially such species as oak, maple and redwood, and has an abrasive effect on tool edges.

Mastics

Mastics are a heavy, pasty type of adhesive that have revolutionized the methods used in the application of wallboards, wood paneling and some types of floors. They vary in their characteristics and application methods, and are usually designed for a specific type of material. Some are waterproof and others must be used where there is no excessive moisture.

One application method consists of placing several gobs on the surface of the material and then pressing the unit firmly in place causing the mastic to spread over a wider area. Some mastics are spread over the surface with a notched trowel while still others are designed for caulking gun application, Fig. 4-35.

Mastics are usually packaged in metal containers or gun cartridges ready for application. Always follow the directions of the manufacturer of the product being applied.

Test Your Knowledge - Unit 4

1. The natural cement that holds wood cells together is called ____________.
2. New wood cells are formed in the ____________ layer.
3. Which of the following kinds of wood is classified as a hardwood? Hemlock, redwood, willow, spruce.
4. When a softwood log is cut so the annular rings form an angle greater than 45 deg. with the surface of the boards, the lumber is called ______________.
5. What is the moisture content of a board if a test sample that originally weighed 11.5 oz. was found to weigh 10 oz. after oven drying?
6. The fiber saturation point is about ______ M.C. for nearly all kinds of wood.
7. The letters E.M.C. are an abbreviation for the term ____________ moisture content.
8. A large knot is defined as one that is over ____ in. in size.
9. Basic classifications of softwood lumber include yard lumber, structural lumber, and ____________ lumber.
10. The best grade of finish softwood lumber is ____________.
11. The best available grade of hardwood lumber is ____________.
12. How many board feet of lumber is contained in a pile of 24 pieces of 2" x 4" x 8'?

Outside Assignments

1. Prepare a visual aid that shows various metal fasteners used in carpentry. Include nails, screws, carriage bolts, and other items. Try to include a good representation of spiral, ring groove and coated nails. Label each item or group of items, giving the correct name, size and other information.
2. Secure a group of softwood samples that will be representative of the species of lumber used in your locality. Instead of writing the proper name on each piece, use a number that corresponds with your master list. This will permit you to give a wood identification quiz to members of your class. As the samples

are passed around, the students can record the numbers and their answers on a sheet of paper.

3. Visit a local building supply center and secure information and literature concerning the various grades and species of lumber normally carried in stock. Prepare written descriptions of the defects permitted in several of the grades commonly selected by builders in the area. Secure list prices of these grades to gain some understanding of the savings that can be gained by using a lower classification. Make a summary report to your class.
4. After a study of reference materials, prepare a paper on stress-rated lumber. Include information on grade stamps and their interpretation. Also define the f (fiber stress in bending) and the corresponding E (stiffness) rating. Try to include a description of modern equipment used in the grading process.

Rotary cutting a Douglas fir log mounted in a giant lathe. A long knife peels a continuous ribbon of veneer at speeds up to 600 lineal feet per minute. Softwood veneers are cut to thicknesses ranging from 1/10 in. to 1/4 in. (Georgia-Pacific)

Fig. 5-1. Included in this Unit are drawings from plans which provide constructional details on this attractive home.

Unit 5
PLANS, SPECIFICATIONS, CODES

In building construction, a good plan and a well defined contract are important. "Early understandings make long friendships."

A basic qualification for every carpenter is the ability to "read" and understand architectural drawings (plans, blueprints) and correctly interpret the information included in written specifications.

To provide the vast amount of information needed to build a modern home, a number of sheets are required. These sheets, when bound together, make up a set of plans.

The original plan drawings and specifications are reproduced so the owner, contractor, and various tradesmen may have copies. Lumber dealers and equipment suppliers must also have access to the plans.

What Set of Plans Includes

A set of house plans usually includes a plot plan, foundation or basement plan, floor plans, elevations (drawings of front, rear and sides of building), and various detail drawings. Included also are drawings which show the electrical, plumbing, heating and air conditioning layouts.

Stock Plans

Today a wide range of ready-drawn house plans (called stock plans) are available.

In this Unit, we will use drawings from a set of stock plans* for illustrative purposes, and we will discuss house plan details. See Fig. 5-1.

Scale

Architectural plans must be drawn to scale. Residential plan views are generally drawn to a one-fourth inch scale (1/4" = 1' - 0"). This means that for each 1/4 in. on the plan, the building dimension will be 1 foot.

When certain parts of the structure need to be shown in greater detail, they are drawn to a larger scale. Details such as framing plans are frequently drawn to a smaller scale (1/8" = 1' - 0").

Floor Plans

Floor plans show the size and outline of the building -- and provide considerable additional information. Extensive use is made of dimension

*Plan 102 – L. F. Garlinghouse Company, Topeka, Kansas.

lines (explained under heading of Dimensions, page 80) to show the location and size of interior partitions, doors, windows and stairs. Plumbing fixtures and various appliance and utility installations can also be shown in this view. Fig. 5-2 shows the floor plan of the residence illustrated in Fig. 5-1.

Foundation plans are similar to floor plans and are often combined with basement plans. The foot-

of the structure on the building site. It includes lot lines and the outside lines of the building. See Fig. 5-4.

Elevations

Elevations show the outside of the structure. These are scaled so all elements will appear in true relationship. Generally the various elevations

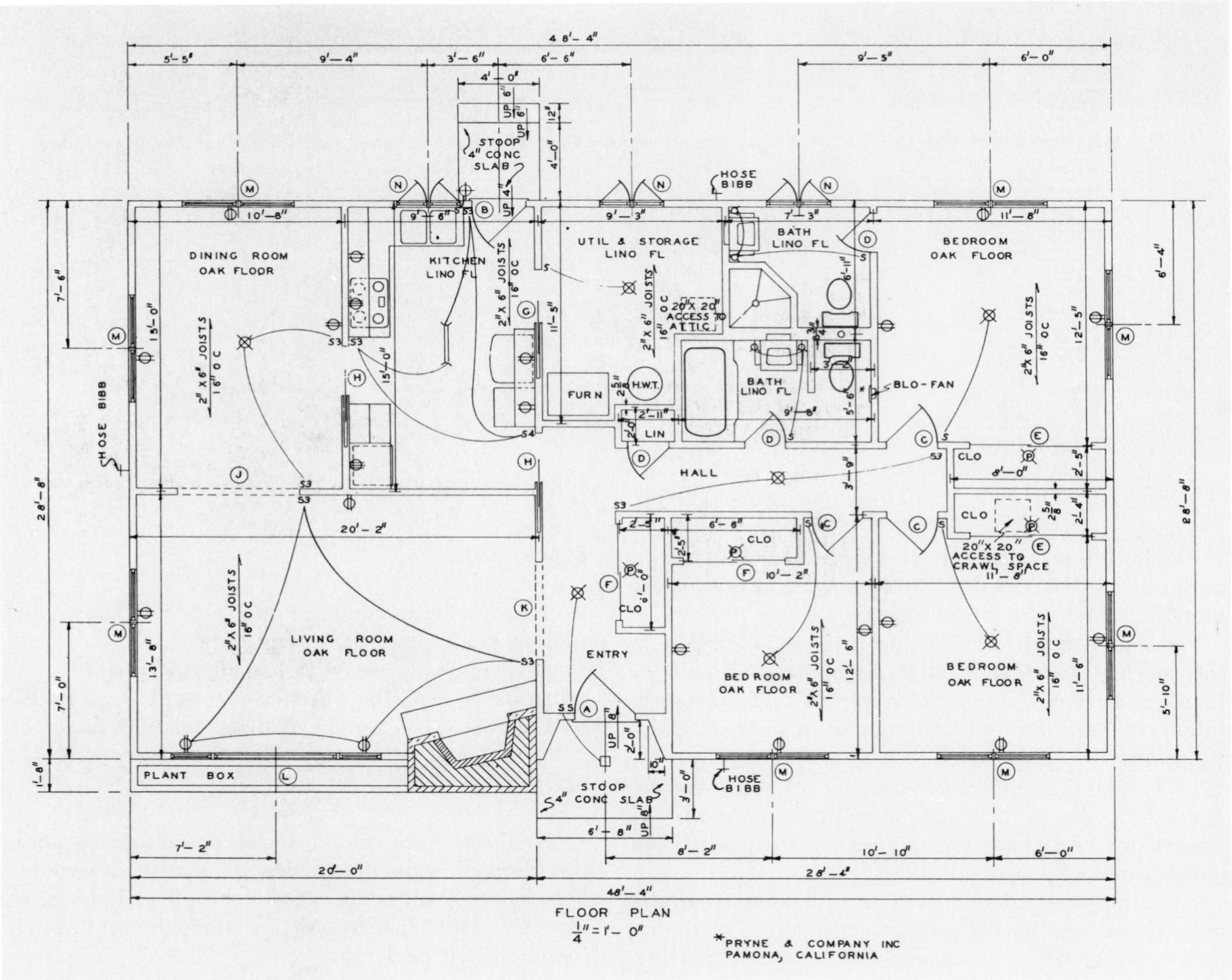

Fig. 5-2. Floor plan for residence shown in Fig. 5-1. This is a reduced-size reproduction of the original floor plan drawing, which was drawn to a scale of 1/4" = 1'-0".

ings, if shown, are represented by a dotted line since it is assumed that the basement floor is in place and that the grade covers the footings on the outside, Fig. 5-3.

A complete set of architectural drawings usually includes a plot plan which shows the location

are connected to the site by listing them according to the directions they face. When plans are not designed for a given location, the names front, rear, left-side and right-side are used. See Figs. 5-5 and 5-6.

Some of the information that the carpenter can

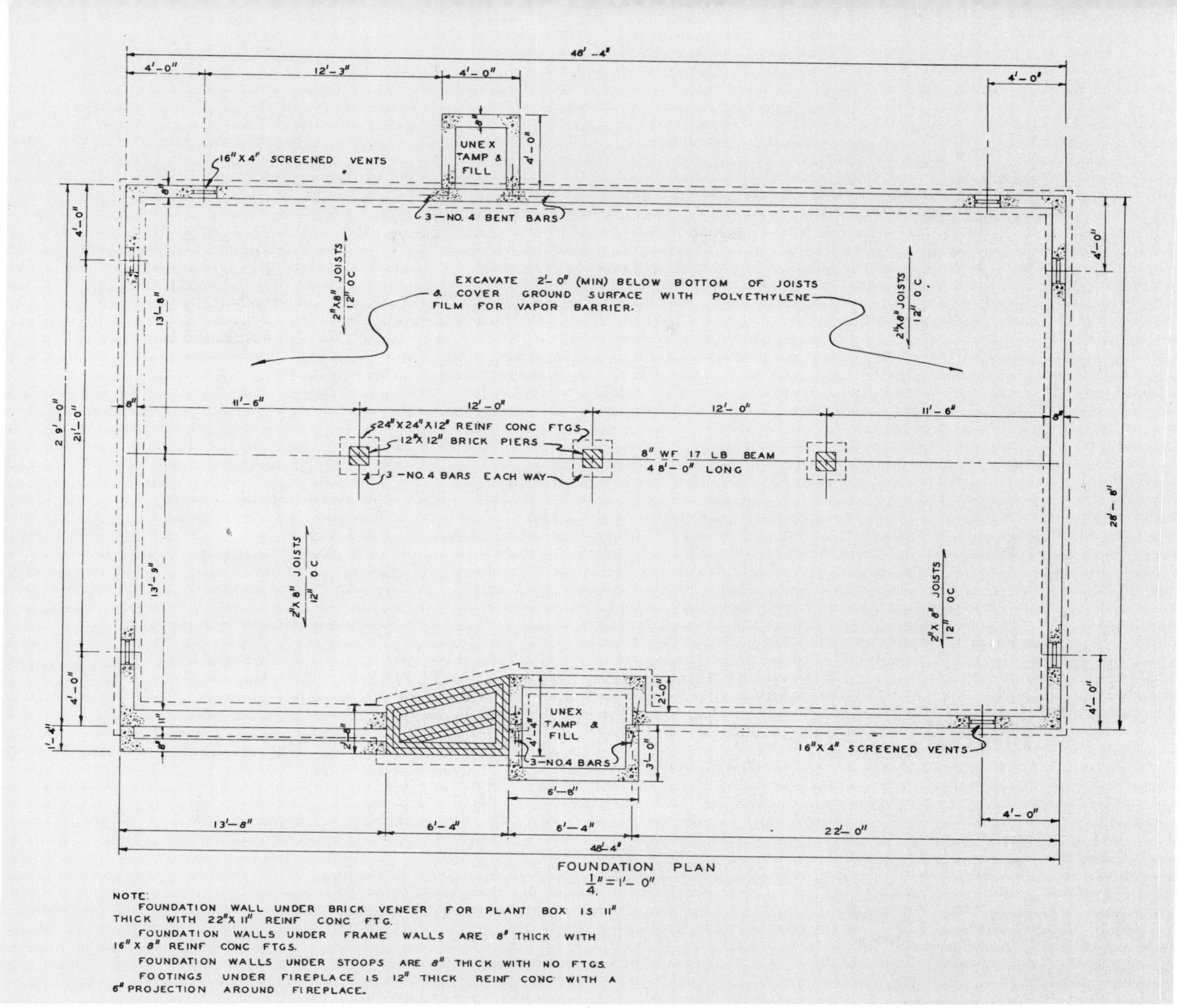

Fig. 5-3. Foundation plan for residence, shown in Fig. 5-1.

Fig. 5-4. Plot plan. Note spaces for fill-in of lot dimensions.

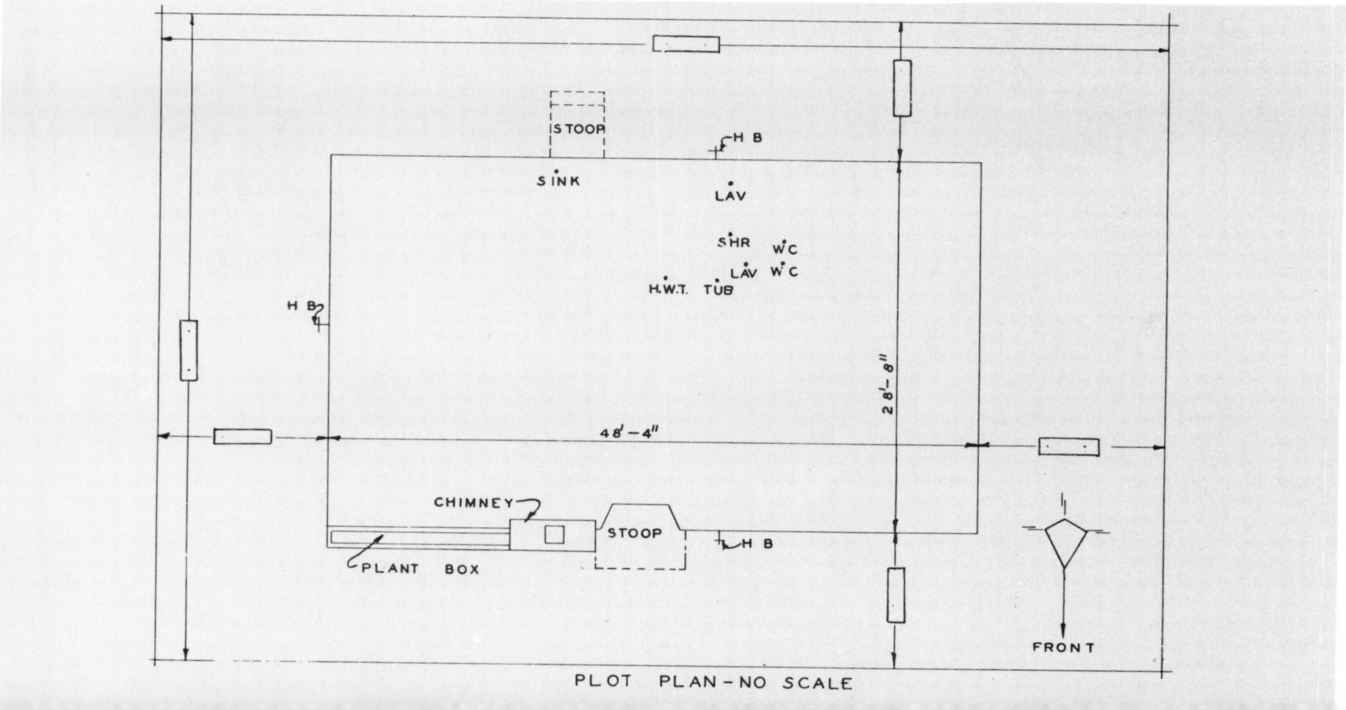

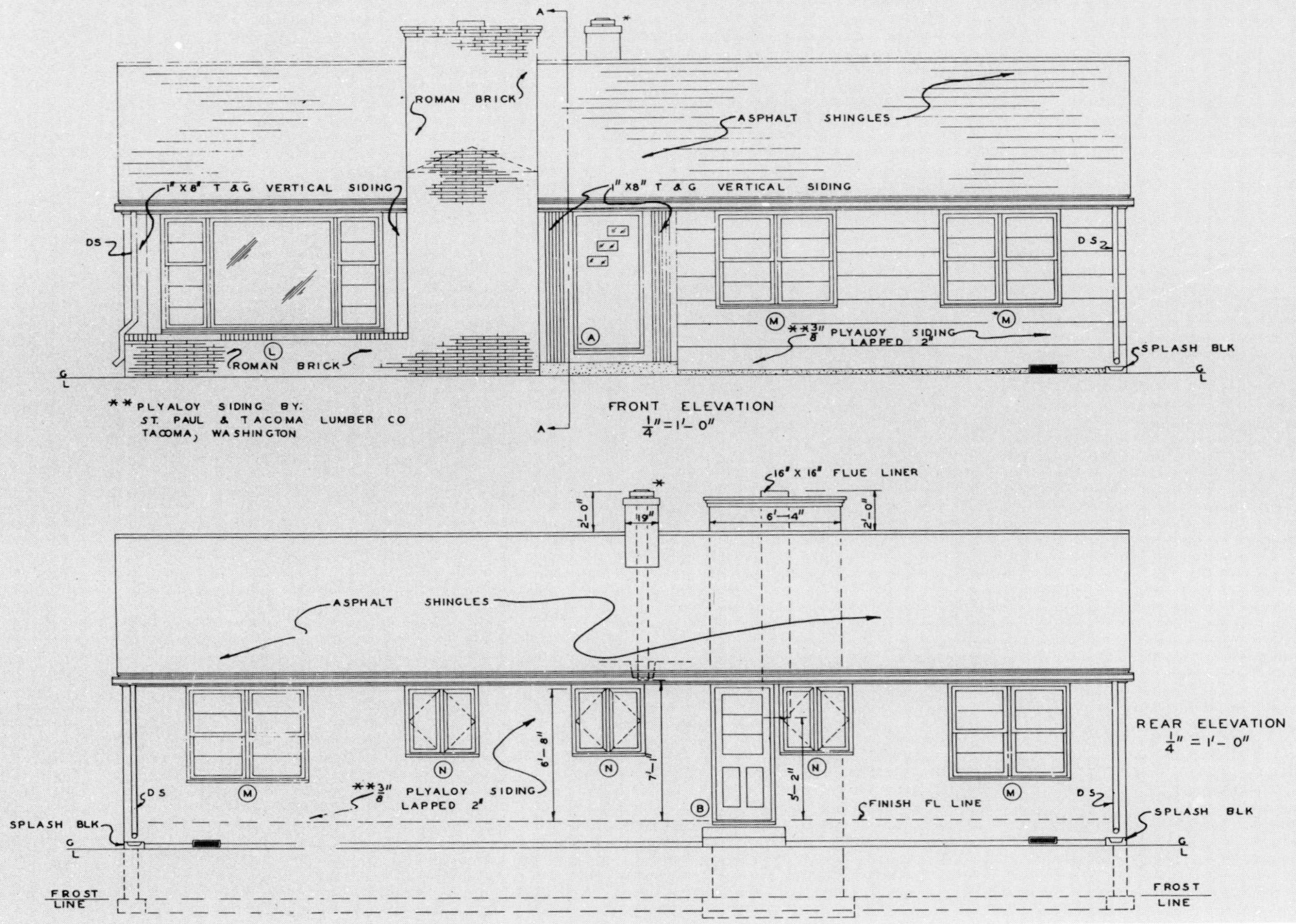

Fig. 5-5. Above. Front elevation. Below. Rear elevation.

Fig. 5-6. Above. Right elevation. Below. Left elevation.

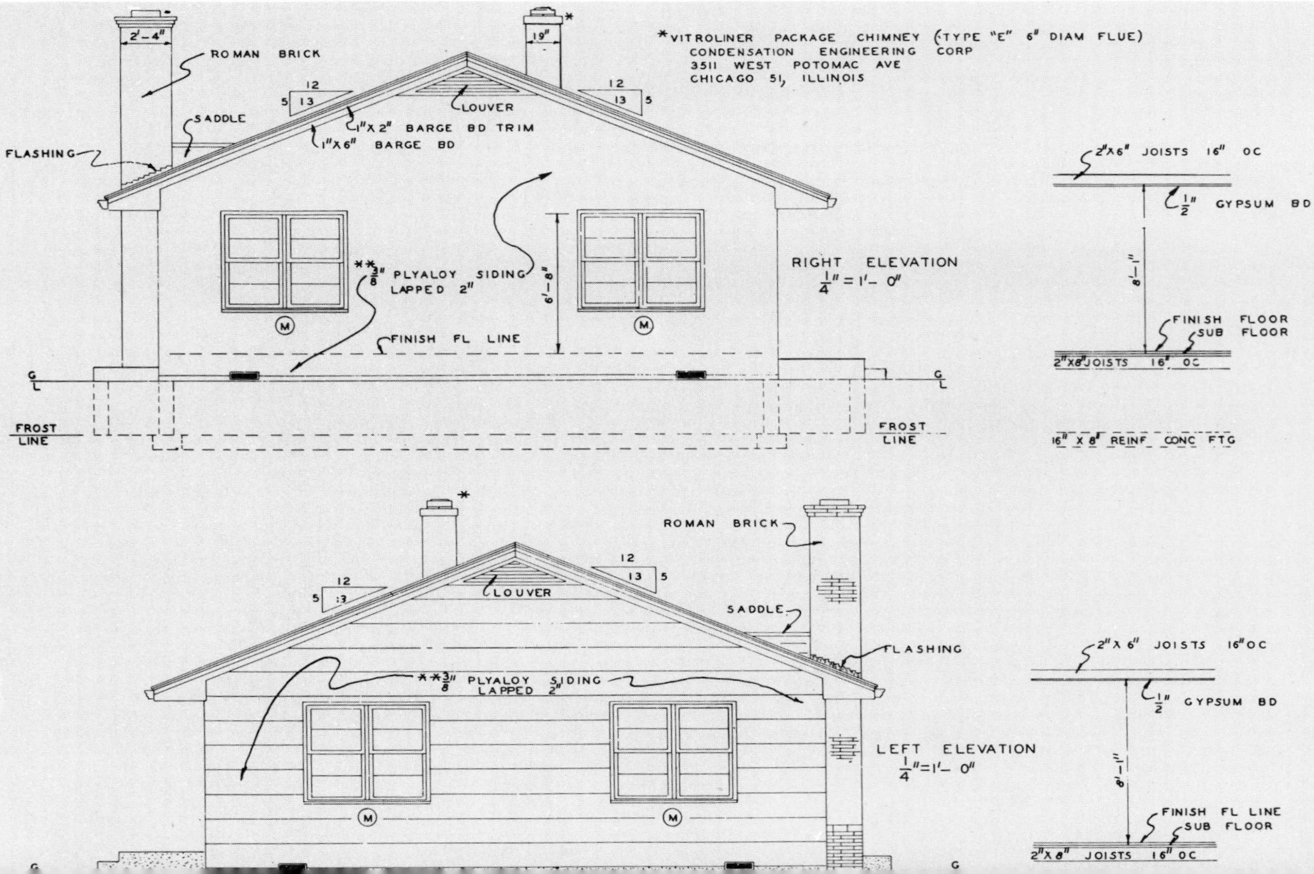

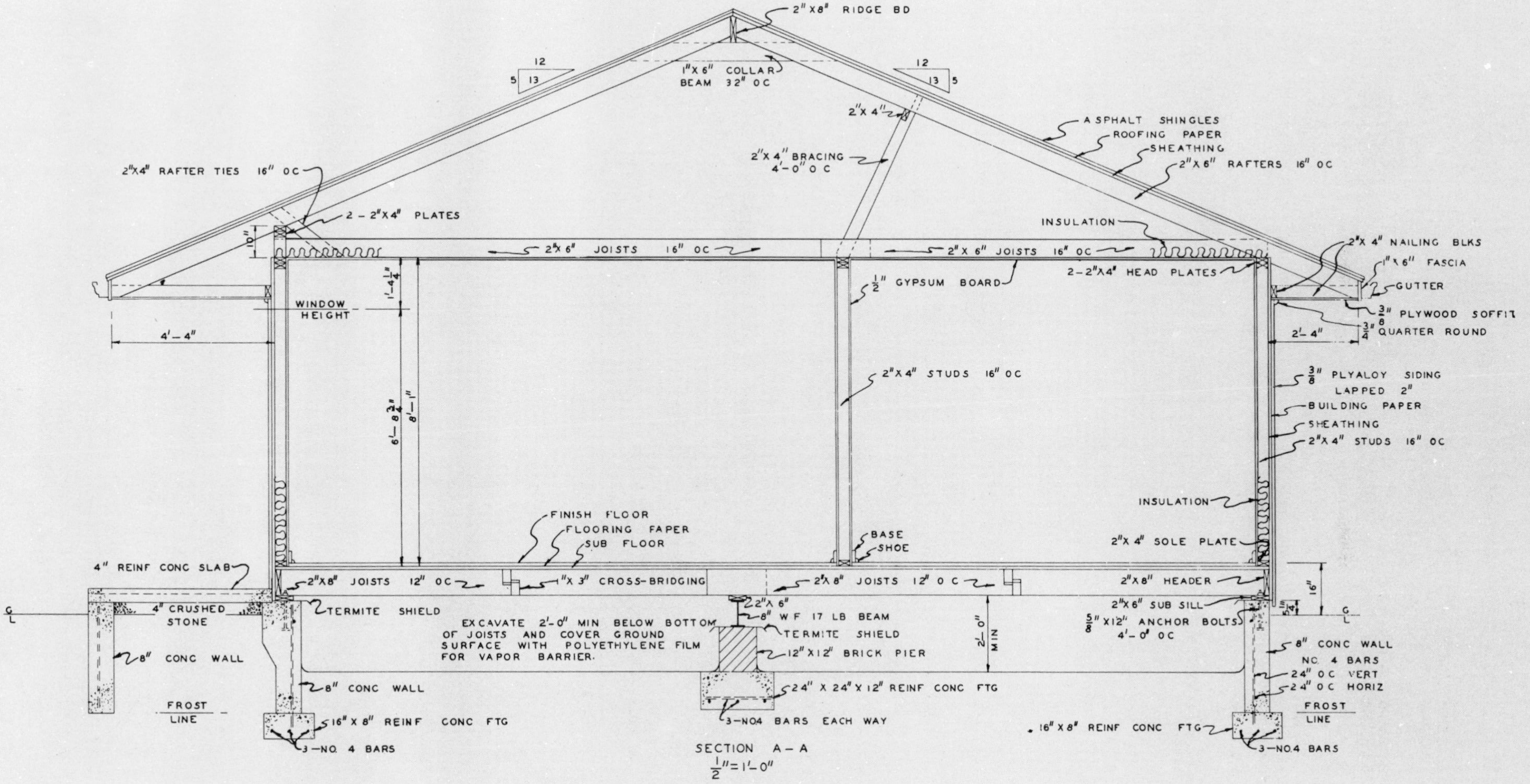

Fig. 5-7. *Section view A-A. See front elevation, Fig. 5-5, for A-A location. Note that this part of the original plan drawing was done to a scale of 1/2" = 1'-0".*

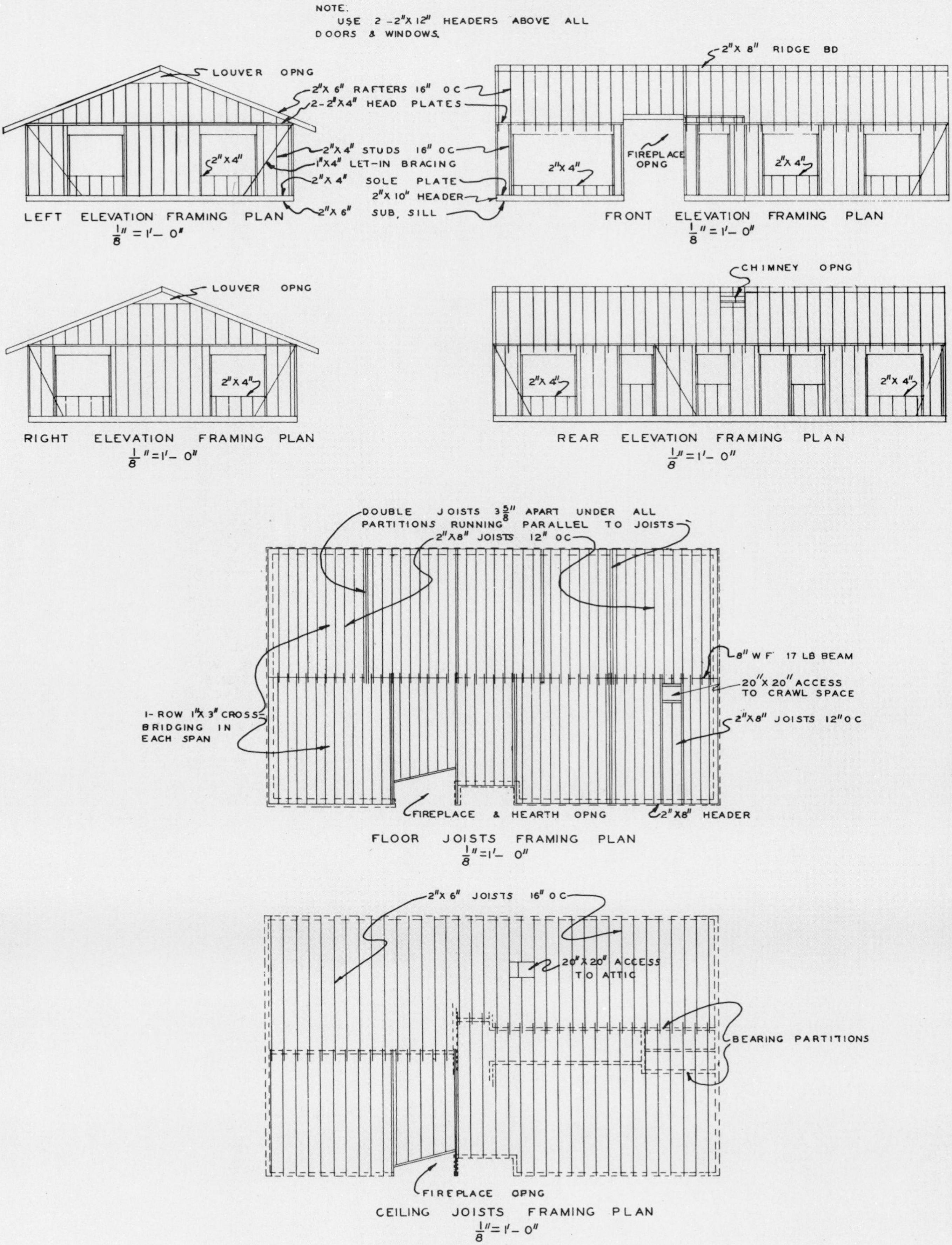

Fig. 5-8. Framing plans. Front, rear, left, right elevations; floor and ceiling joists. (Original plan drawing was 1/8" = 1'-0".)

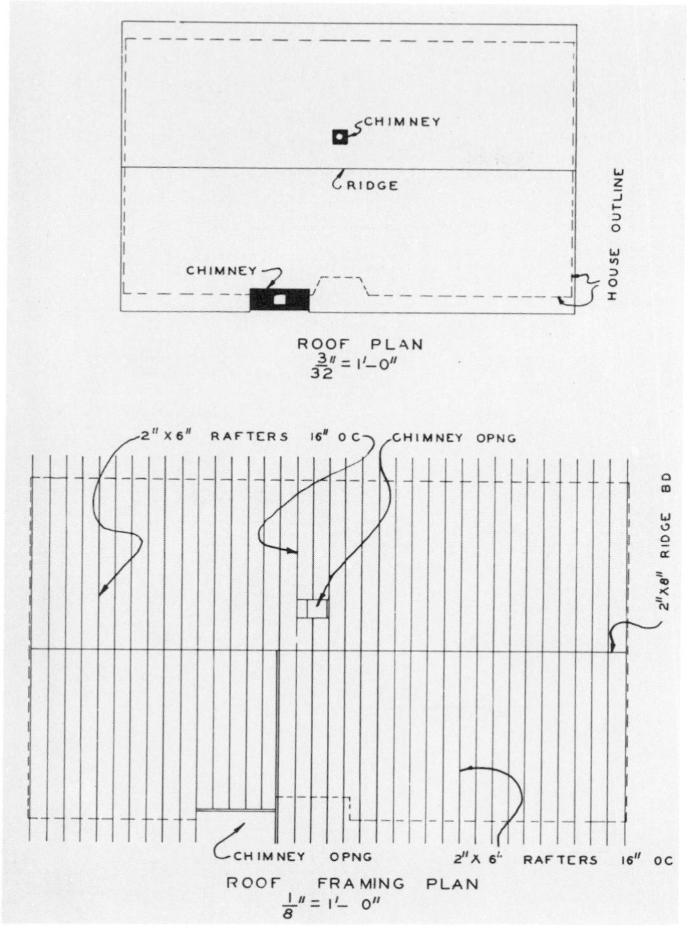

Fig. 5-9. Above. Roof plan. Below. Roof framing plan.

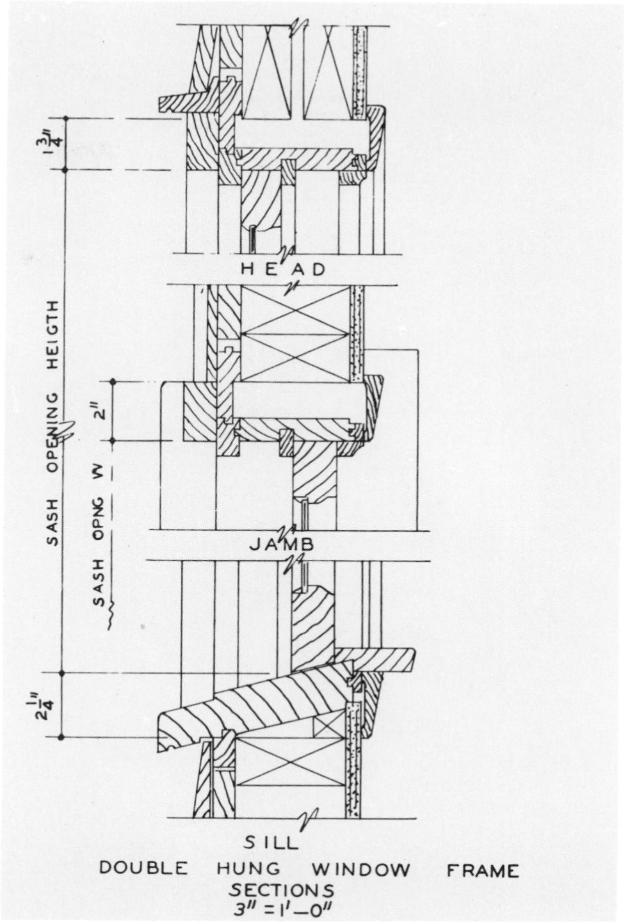

Fig. 5-11. Window frame sections.

WINDOW SCHEDULE

CODE	QUAN	NO. LTS	GLASS SIZE	SASH SIZE	ROUGH OPENING	REMARKS
L	1	1 5	68" X 61" 20" X 24"/36"	(1) 6'– 0" X 5'–6" (2) 2'– 0" X 5'–6"	10'– 8" X 5'–10"	1–PICTURE WINDOW FLANKED EACH SIDE BY 1–DOUBLE HUNG WINDOW 2 – 2" MULLIONS
M	8	4	28" X 24"	2'– 8" X 4'– 6"	5'– 10" X 4'–10"	DOUBLE HUNG WINDOWS, DOUBLE UNITS, 2" MULLION
N	3	2	14" X 33"	3'– 2" X 3'–2"	3'–3" X 3'–3 3/8"	#C–21–27 CURTIS CONVERTIBLE WINDOWS, CASEMENT TYPE

DOOR SCHEDULE

CODE	QUAN	SIZE	THK	ROUGH OPENING	JAMB SIZE	TYPE	DESIGN	REMARKS
A	1	3'–0" X 6'–8"	1 3/4"	3'– 3" X 6'– 10 1/4"	1 5/16" X 4 7/8"	HINGED	3 LTS, SOLID CORE	FRONT ENTRANCE DOOR
B	1	2'–8" X 6'–8"	1 3/4"	2'–11" X 6'– 10 1/4"	1 5/16" X 4 7/8"	HINGED	3 LTS, 2 PANELS	REAR SERVICE DOOR
C	3	2'–6" X 6'–8"	1 3/8"	2'–8 1/2" X 6'–10 1/2"	3/4" X 4 5/8"	HINGED	FLUSH HOLLOW CORE	INTERIOR DOORS
D	3	2'–0" X 6'–8"	1 3/8"	2'–2 1/2" X 6'–10 1/2"	3/4" X 4 5/8"	HINGED	FLUSH HOLLOW CORE	INTERIOR DOORS
E	2	(2) 3'–0" X 6'–8"	1 3/8"	6'–1 1/2" X 6'–11 1/4"	3/4" X 4 5/8"	SLIDING	FLUSH HOLLOW CORE	BY–PASSING CLOSET DOORS
F	2	(2) 2–6" X 6'–8"	1 3/8"	5'–1 1/2" X 6'–11 1/4"	3/4" X 4 5/8"	SLIDING	FLUSH HOLLOW CORE	BY–PASSING CLOSET DOORS
G	1	2'–8" X 6'–8"	1 3/8"	5'– 6" X 7'– 0"		SLIDING	FLUSH HOLLOW CORE	RECESSED DOOR
H	2	2'–6" X 6'–8"	1 3/8"	5– 2" X 7'– 0"		SLIDING	FLUSH HOLLOW CORE	RECESSED DOORS
J	1	6'–0" X 6'–8"		6'–2 1/2" X 6'–10 1/2"	3/4" X 4 5/8"			CASED OPENING
K	1	5'–0" X 6'–8"		5'–2 1/2" X 6'–10 1/2"	3/4" X 4 5/8"			CASED OPENING

Fig. 5-10. Window and door schedules.

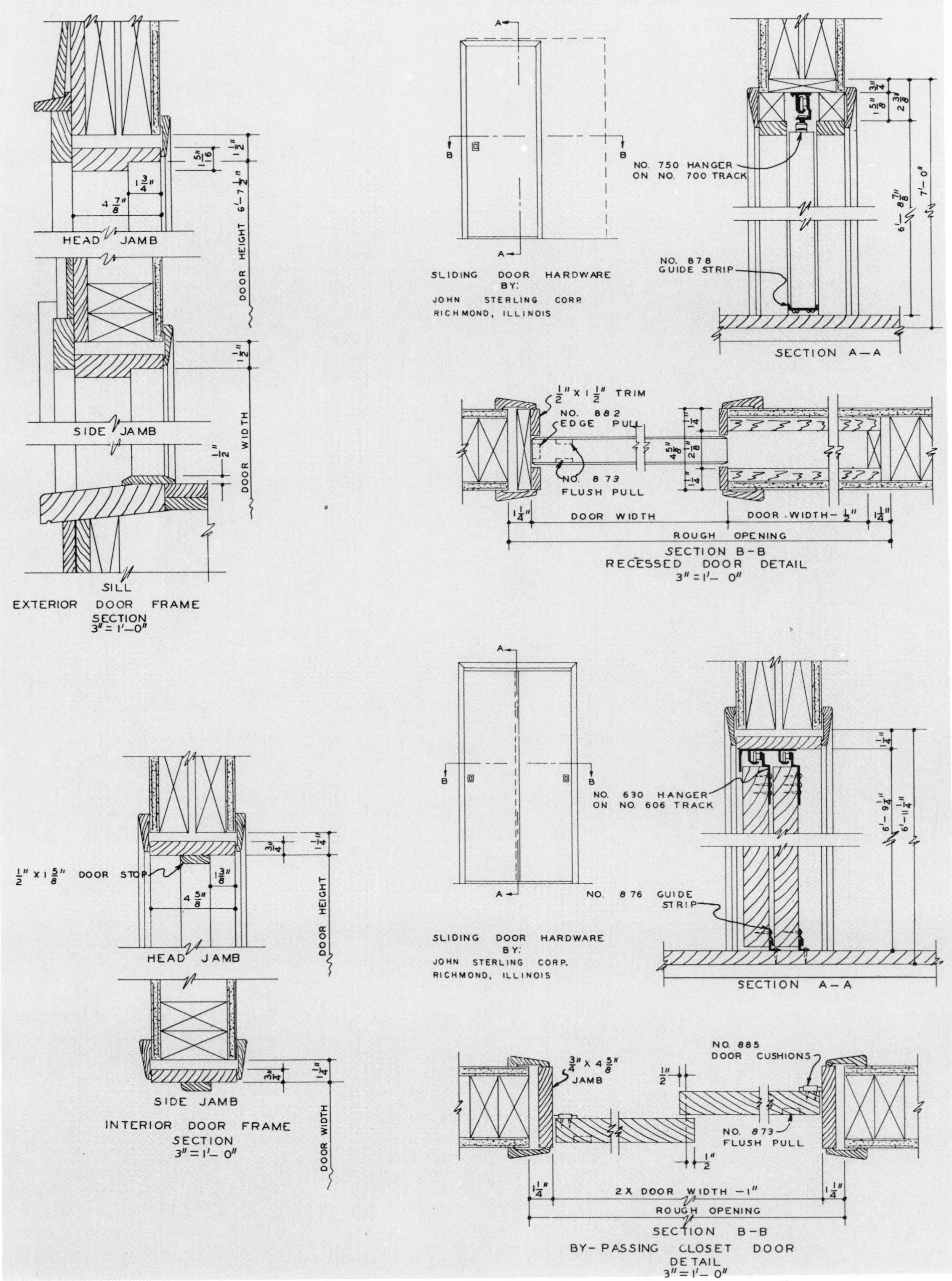

Fig. 5-12. Door frame sections.

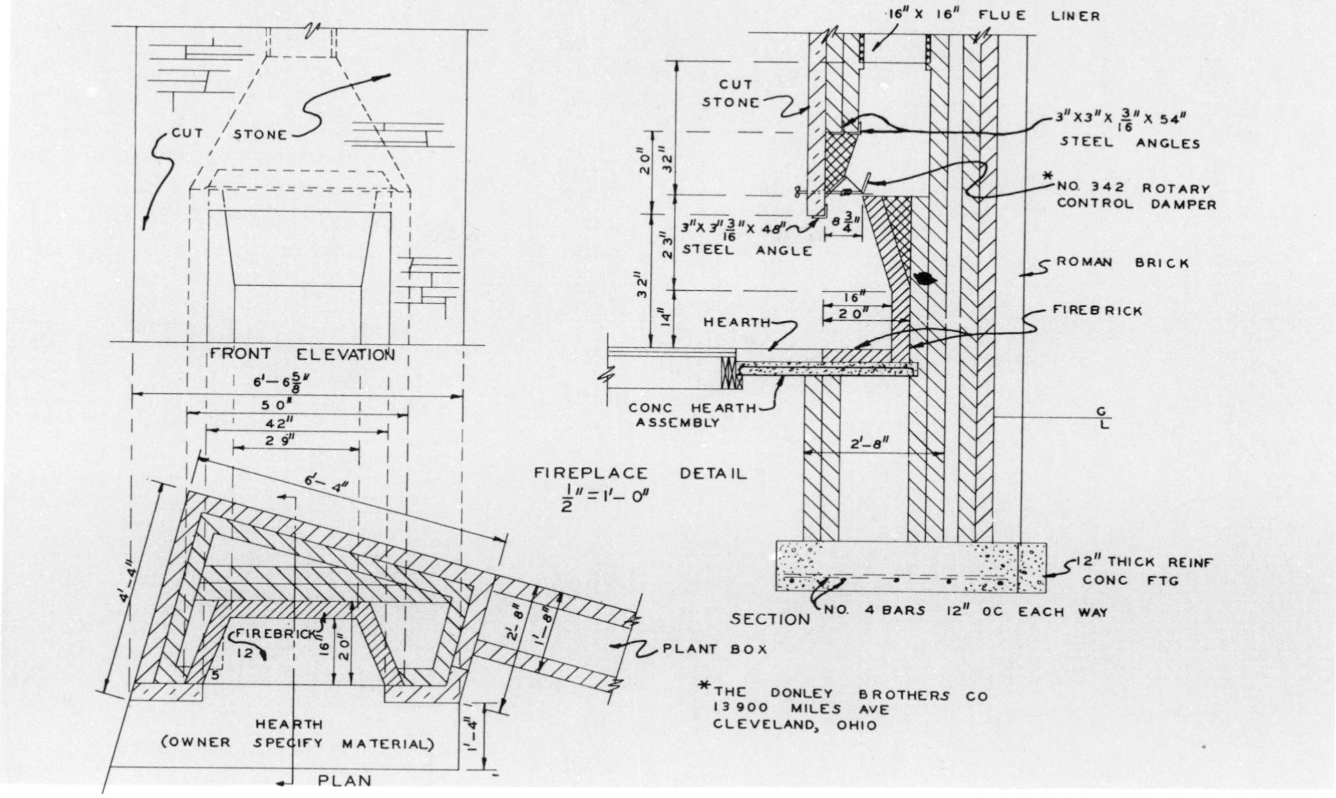

Fig. 5-13. Fireplace details.

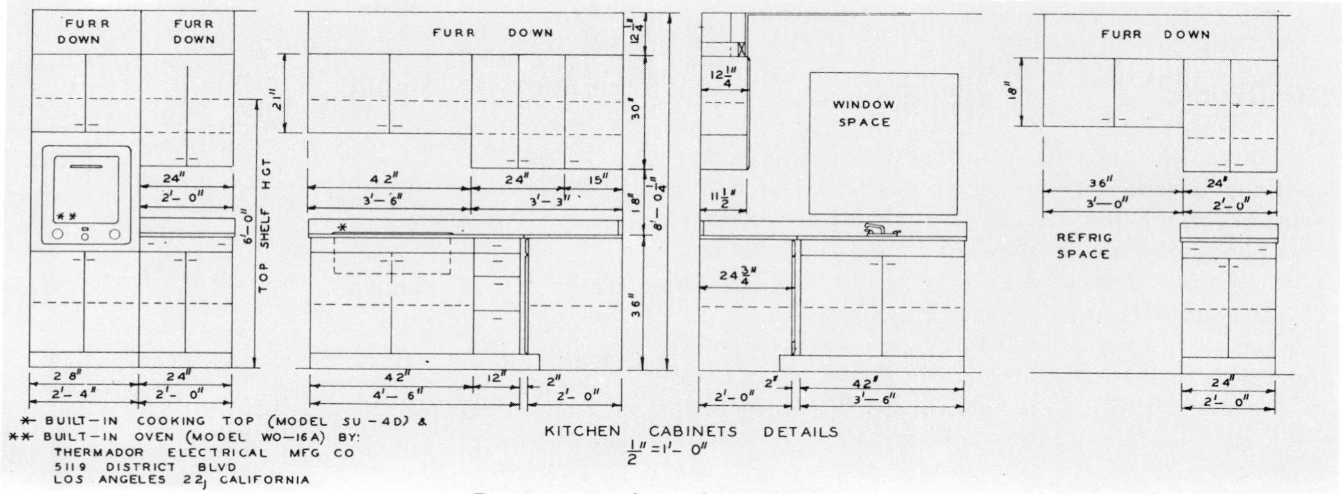

Fig. 5-14. Kitchen cabinet details.

secure from the exterior views: floor levels, grade lines, window and door heights, roof slopes, and the kinds of material used on the surface of walls and roof. Foundation and footing lines, located below grade, are indicated with broken lines.

Sections

To show the various parts of a structure and how they fit together, a section view is included. This provides information on footing and foundation design and also construction details of the sill and cornice. Sizes of framing lumber are listed along with the types and kinds of material to be used for sheathing, outside and inside wall surfaces, location of insulation, etc.

A complicated structure may require a number of wall sections to completely show all the details of construction.

Details

To assist the carpenter, house plans usually show: Elevation, floor, ceiling and roof plans, framing plans, window and door schedules and frame sections; also details of important items such as fireplaces and built-in kitchen cabinets. See Figs. 5-8 through 5-15.

$2\frac{1}{4}''$

CASING

$\frac{3}{4}''$

$\frac{1}{2}''$

SHOE

$3\frac{5}{8}''$

$\frac{3}{4}''$

BASE

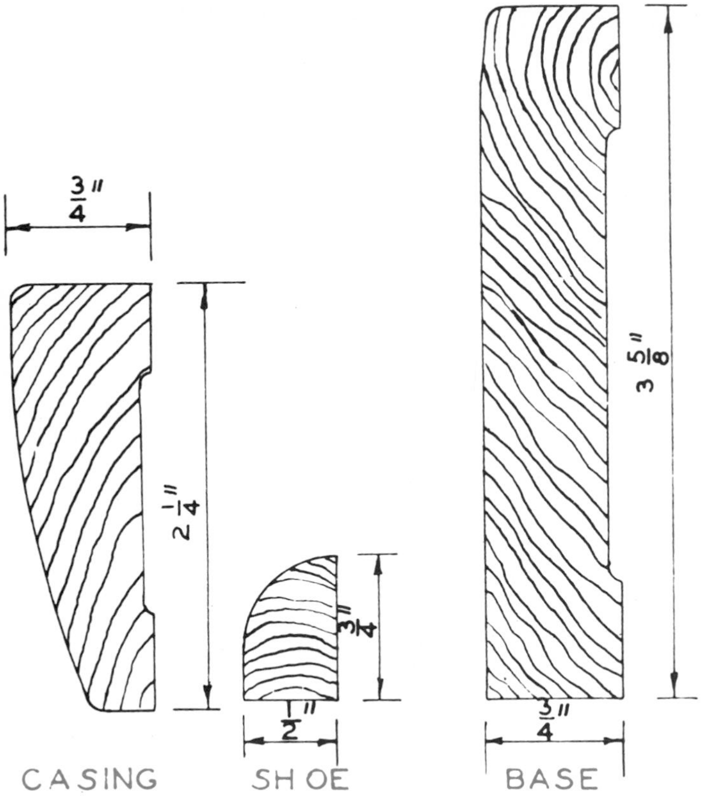

Fig. 5-15. Casing, shoe and base. Full size details.

Dimensions

Dimension lines on architectural drawings are continuous lines with the size placed above the line near or at the middle. In general, all dimensions over one foot are expressed in feet and inches. For example: standard ceiling height is given as 8' – 0" rather than 96".

The carpenter prefers to work with feet and inches since the measurement in this form is usually easier to visualize and apply. When working with architectural plans and laying out various distances, he often needs to add and subtract dimensions. Procedures for making the calculations are:

Addition
$$
\begin{array}{r}
6' - 8'' \\
4' - 6'' \\
2' - 4'' \\
1' - 2'' \\
\hline
13' - 20'' = 14' - 8''
\end{array}
$$

Subtraction
$$
\begin{array}{r}
7' - 16'' \\
*8' - 4'' \\
6' - 10'' \\
\hline
1' - 6''
\end{array}
$$

*(change 8' – 4" to 7' – 16")

Symbols

Since architectural plans are drawn to a small scale, materials and constructions can seldom be shown as they actually appear. Also it would require too much time to produce drawings of this

	PLAN	ELEVATION	SECTION
WOOD	FLOOR AREAS LEFT BLANK	SIDING PANEL	FRAMING FINISH
BRICK	FACE / COMMON	FACE OR COMMON	SAME AS PLAN VIEW
STONE	CUT / RUBBLE	CUT RUBBLE	CUT RUBBLE
CONCRETE			SAME AS PLAN VIEW
CONCRETE BLOCK			SAME AS PLAN VIEW
EARTH	NONE	NONE	
GLASS			LARGE SCALE / SMALL SCALE
INSULATION	SAME AS SECTION	INSULATION	LOOSE FILL OR BATT / BOARD
PLASTER	SAME AS SECTION	PLASTER	STUD / LATH AND PLASTER
STRUCTURAL STEEL		INDICATE BY NOTE	
SHEET METAL FLASHING	INDICATE BY NOTE		SHOW CONTOUR
TILE	FLOOR	WALL	

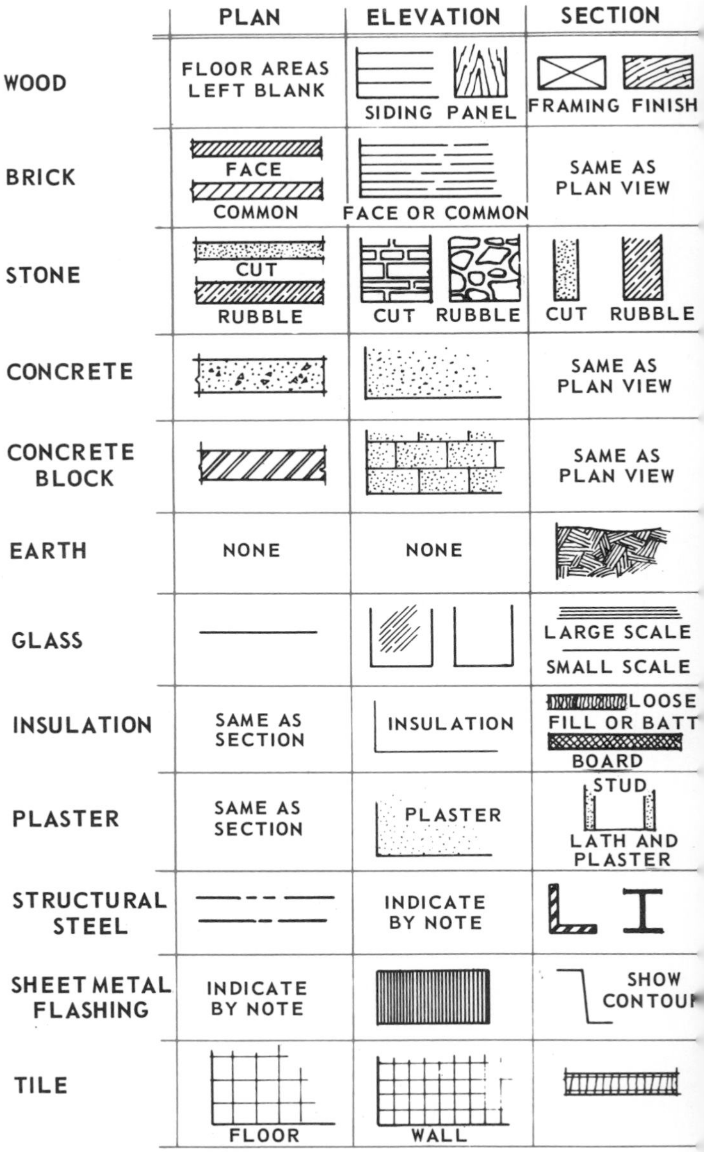

Fig. 5-16. Symbols for materials.

nature. The architect, therefore, uses symbols to represent materials and other items and certain approved short-cuts (called conventional representations) to simplify the illustration of assemblies and other elements of the structure. Generally accepted symbols are illustrated in Figs. 5-16 through 5-19.

Abbreviations are commonly used on plans to save space. Refer to the appendix for a listing.

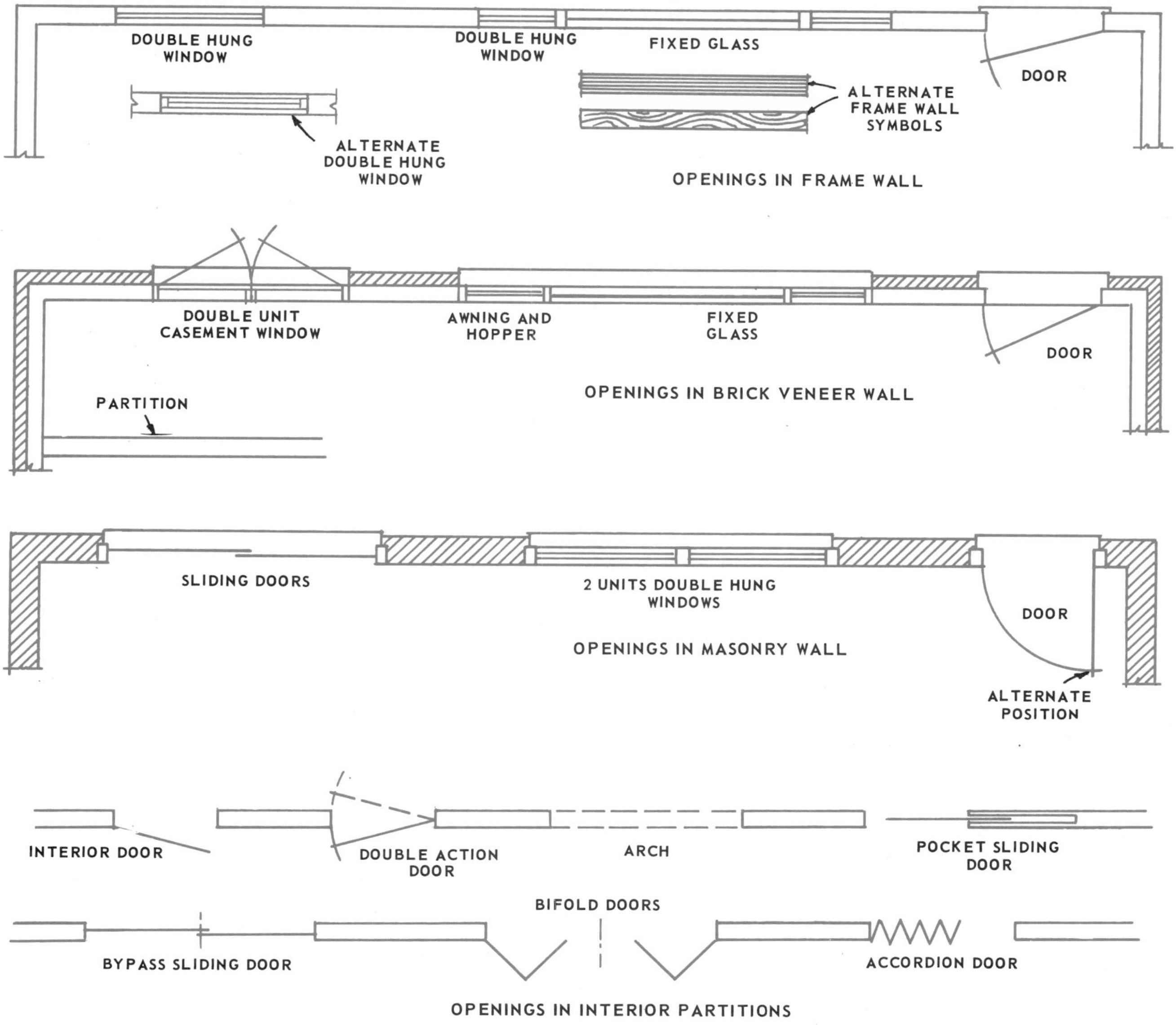

Fig. 5-17. How openings are represented in plan views.

How to Scale a Drawing

Architectural plans include dimension lines that show many distances and sizes, but the carpenter may require a dimension that is not shown. To get this dimension, he will need to scale the drawing. An architect's scale, Fig. 5-20, may be used for this purpose. Each division of an architect's scale represents 1 ft. The foot division is divided into 12 parts, each part being equal to 1 in.

Another way to scale a plan is to use a regular folding rule and calculate the distance, as shown in Fig. 5-21.

Changing Plans

Minor changes in plans, desired by the owner as the job progresses, such as changing the size or location of a window or making a revision in the design of a built-in cabinet, can usually be handled by the carpenter. Sketches or notations should be recorded on each set of plans so there will be no misunderstandings.

Major changes such as the relocation of a load-bearing wall, or stairs, may generate a "chain-reaction" of problems and should be undertaken only after the necessary plan changes have been made by an architect and approved by the owner.

Specifications

Although the working drawings show many of the requirements for a structure -- certain supplementary information is best presented in written form or specifications (commonly called Specs). The carpenter should check and carefully follow these specs.

Headings generally included in specifications for a residential structure are listed below:

1. General Requirements, Conditions and Information
2. Excavating and Grading
3. Masonry and Concrete Work
4. Sheet Metal Work
5. Rough Carpentry and Roofing
6. Finish Carpentry and Millwork
7. Insulation, Caulking and Glazing
8. Lath and Plaster or Drywall
9. Schedule for Room Finishes
10. Painting and Finishing
11. Tile Work
12. Electrical Work
13. Plumbing
14. Heating and Air Conditioning
15. Landscaping

Carefully prepared specifications are valuable to the contractor, estimator, tradesmen and the building supply dealer. They protect the owner and help to insure good workmanship. In addition to the items previously described, the specifications may include information and requirements regarding building permits, contract payment provisions, insurance and bonding, and provisions for making changes in the original plans.

A set of house plans generally includes also, a complete lumber and mill list (doors, windows, built-ins, etc.).

Modular Construction

The modular coordination (construction) concept is based on the use of a standard grid divided into 4 in. squares. See Fig. 5-22. Actually each

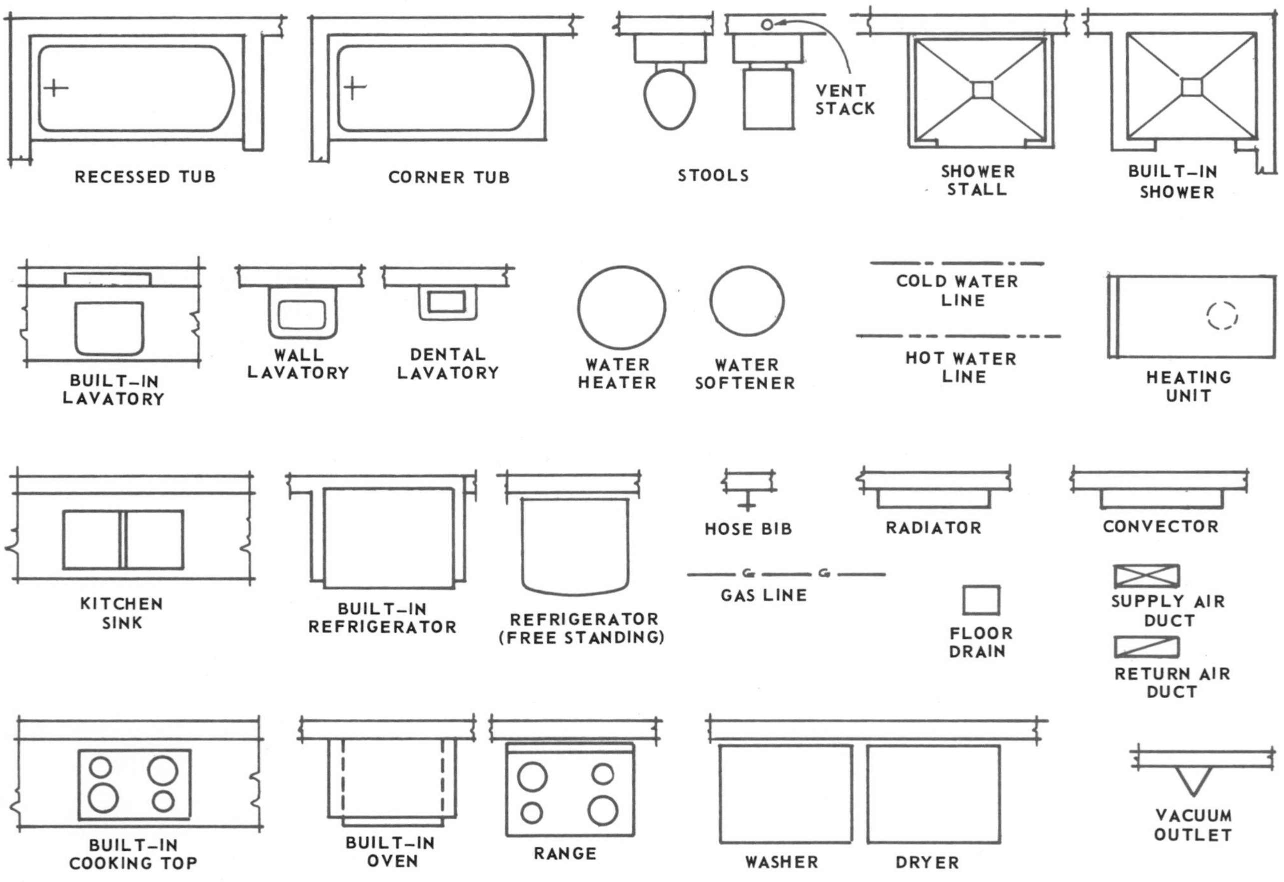

Fig. 5-18. Symbols for plumbing fixtures, appliances and mechanical equipment.

Under each of the headings the content is usually divided into sections concerning: Scope of work, specifications of materials to be used, application methods and procedures, and guarantee of quality and performance.

individual square (module) should be considered to be the base of a cube so it can be applied to elevations as well as horizontal planes.

All dimensions are based on multiples of 4 in. including 16 in., 24 in. and 48 inches -- the last

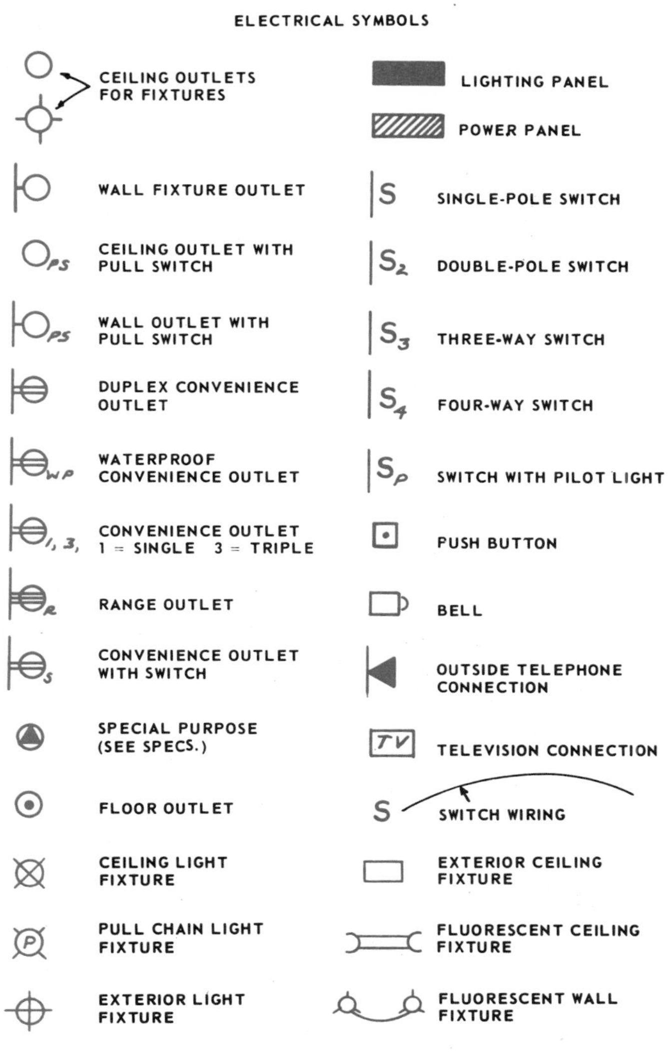
ELECTRICAL SYMBOLS
CEILING OUTLETS FOR FIXTURES
WALL FIXTURE OUTLET
CEILING OUTLET WITH PULL SWITCH
WALL OUTLET WITH PULL SWITCH
DUPLEX CONVENIENCE OUTLET
WATERPROOF CONVENIENCE OUTLET
CONVENIENCE OUTLET 1 = SINGLE 3 = TRIPLE
RANGE OUTLET
CONVENIENCE OUTLET WITH SWITCH
SPECIAL PURPOSE (SEE SPECS.)
FLOOR OUTLET
CEILING LIGHT FIXTURE
PULL CHAIN LIGHT FIXTURE
EXTERIOR LIGHT FIXTURE
LIGHTING PANEL
POWER PANEL
SINGLE-POLE SWITCH
DOUBLE-POLE SWITCH
THREE-WAY SWITCH
FOUR-WAY SWITCH
SWITCH WITH PILOT LIGHT
PUSH BUTTON
BELL
OUTSIDE TELEPHONE CONNECTION
TELEVISION CONNECTION
SWITCH WIRING
EXTERIOR CEILING FIXTURE
FLUORESCENT CEILING FIXTURE
FLUORESCENT WALL FIXTURE

Fig. 5-19. Electrical symbols.

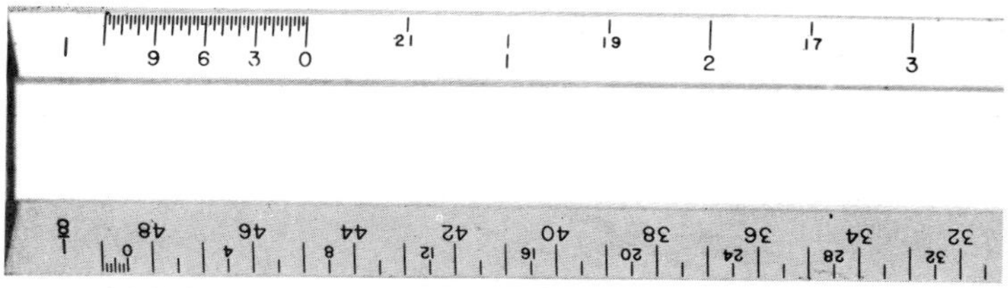

Fig. 5-20. Architect's scale. Each division represents 1 ft. The foot division is divided into 12 parts, each part being equal to 1 in.

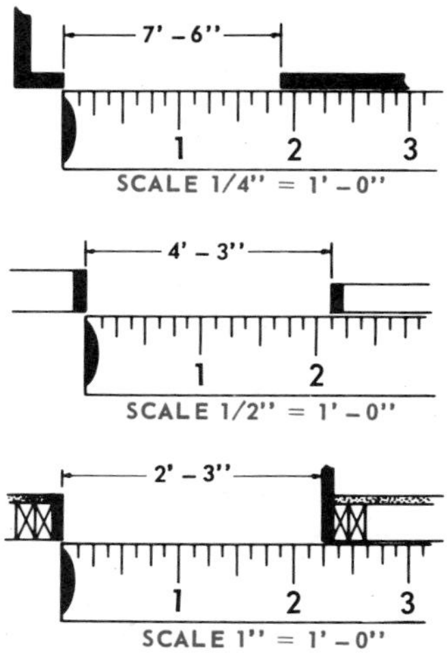
7' – 6"
SCALE 1/4" = 1' – 0"
4' – 3"
SCALE 1/2" = 1' – 0"
2' – 3"
SCALE 1" = 1' – 0"

Fig. 5-21. Using a folding rule to scale a plan.

two of which are sometimes called the minor and major module. Many building materials and fabricated units are manufactured to coordinate with modular dimensions. This helps to eliminate costly cutting and fitting during construction. A good

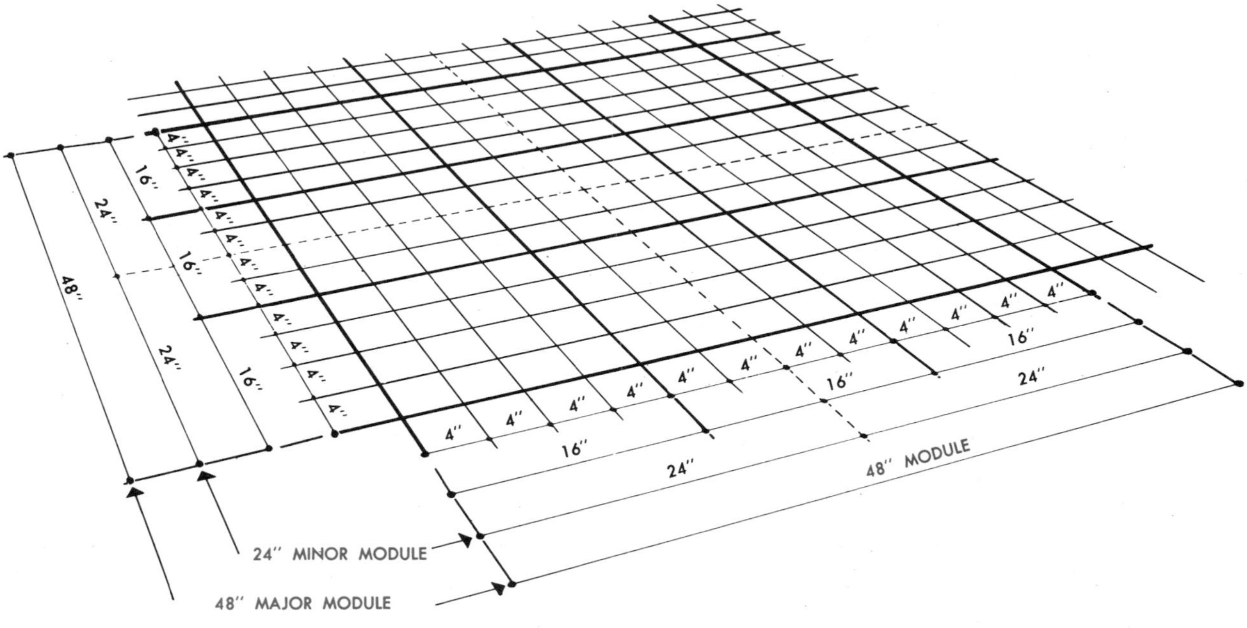

Fig. 5-22. Plans for modular structures, to be built with components or by conventional framing methods, are designed to exact grid sizes. (National Forest Products Assoc.)

example of this system is illustrated in standard concrete blocks which are manufactured in nominal sizes of 8 x 8 x 16. The actual size is 3/8 in. less in each dimension to allow for bonding (mortar joints). See Unit 6.

Modular dimension standards for manufactured components have been developed by the National Lumber Manufacturers Association. The system is called Unicom which stands for "uniform manufacture of components." The Unicom system helps to make it possible to apply modern mass production methods to the field of building construction.

Building Codes

A building code is a collection of laws listed in booklet form that apply to a given community. The code covers all important aspects of the erection of a new building and also the alteration, repair, and demolition of existing buildings. The basic purpose is to provide for the health, safety and general welfare of the occupants of the home being built and other people in the community.

HANDY SAYS:

"In doing any kind of carpentry work, the importance of closely following all building codes applicable to the job cannot be overemphasized."

A code deals with standards of performance and material specifications, also minimum requirements concerning such design factors as room size, ceiling heights, lighting and ventilation. In the area of materials and methods, many codes include detailed directions covering installations.

Building codes issued by some cities have been criticized because they are not up-to-date and do not permit the use of new materials and methods. Another complaint is the lack of uniformity.

Some items in a code necessarily must be adjusted to local conditions. In northern climates, footings need to be deeper than in Southern States; structures in "hurricane belts" require extra bracing.

Modern research and development have resulted in so many improvements in building construction that it now becomes a tremendous task to prepare and continually up-date building codes. Because of this, many communities have adopted model codes. Today, four major organizations provide a service of this nature.

The UNIFORM BUILDING CODE, published by the International Conference of Building Officials, has been widely accepted. The organization provides annual revisions and the entire code is republished every three years. A short form is available which covers buildings not over two stories in height and containing less than 6000 sq. ft. of ground floor area.

Another organization, the Building Officials and Code Administrators International, Inc. has developed the BOCA - BASIC BUILDING CODE. An abridged form, designed for residential construction, includes plumbing and wiring standards.

One of the first model codes was introduced by The American Insurance Association (successor to the National Board of Fire Underwriters). This publication is now known as the NATIONAL BUILDING CODE. An abbreviated edition is also available.

A model building code called the STANDARD BUILDING CODE, used in Southern States, covers problems in this region. It is prepared under the direction of the Southern Building Code Congress International, Inc.

In addition to building codes adopted by the local community (cities, towns, counties), the carpenter must be informed of certain laws at the state level that govern buildings. Several states have developed building codes for adoption by their local communities; however, for the most part, state codes deal mainly with fire protection and special needs for public buildings.

A carefully prepared and up-to-date code is not sufficient in itself to insure safe and adequate buildings. All codes must be properly administered by officials that are experts in the field. Under these conditions, the owner can be assured of a well constructed building and the carpenter will be protected against the unfair competition of those who are willing to sacrifice quality for an excessive margin of profit.

Standards

Building codes are based on standards developed by manufacturers, trade associations, government agencies, professionals and tradesmen, all of whom are seeking a desirable level of quality through efficient means. A particular material, method, or procedure is technically described through specifications. Specifications become standards when their use is formally adopted by broad groups of manufacturers and builders and/or recognized agencies and associations.

Organizations devoted to the establishment of standards, many of which are directly related

City of Cedar Falls, Iowa

BUILDING DEPARTMENT

INSPECTION RECORD

OFFICE 266-1761, Ext. 20

DON'T COVER OR REMOVE THIS CARD

JOB ADDRESS ....

NATURE OF WORK ....

USE OF BUILDING ....

BUILDING PERMIT NO. .... DATE ISSUED ....

OWNER ....

INSPECTIONS

FOOTINGS and BASEMENT WALLS
Date ....
Approved by.................... Building Official
Do not back-fill until above has been signed

ROUGH PLUMBING, Approved by.................... Plumbing Inspector

ROUGH WIRING, Approved by .... Electric Inspector

FRAMING
Date ....
Approved by.................... Building Official
Cover no work until above has been signed

LATHING
Date ....
Approved by.................... Building Official
Do not plaster until above has been signed

FINAL
Date ....
Plumbing approved by.................... Plumbing Inspector

FINAL
Date....................
Electrical approved by .... Electrical Inspector

FINAL
Date....................
Furnace, hot water tank approved by.................... Utility Official

FINAL
Date....................
Building approved by.................... Building Official
Do not occupy until above has been signed

REMARKS ....

Fig. 5-23. Typical inspection record card, which is posted at the building site.

to the field of building construction, include: The American Society for Testing and Materials (ASTM); American National Standards Institute (ANSI); and Underwriters' Laboratories, Inc. (UL).

Commercial standards are developed by the Commodity Standards Division of the U.S. Department of Commerce. The chief purpose of the agency is to establish quality requirements and approved methods of testing, rating and labeling. These standards are designated by the initials CS, followed by a code number and the year of the latest revision.

Building Permits and Inspections

To secure a building permit, the contractor or owner must file a formal application and submit drawings and specifications for the proposed structure. Building officials review the plans and specs to determine if they meet the requirements of the local code. If all conditions are satisfactory, a building permit is issued. Sometimes it may be necessary to submit supporting data concerning the correctness of the design to indicate how the code requirements will be met.

When construction is started, an inspection card, Fig. 5-23, is posted at the building site. This is filled out and signed by the inspector as work progresses. Work on the structure should not proceed beyond the point indicated in each successive inspection. The carpenters on the job must give close attention to this record and make certain that mechr .ical aspects (heating, plumbing and electrical) are not enclosed before the installations have been approved by the building inspector. In some communities, a final inspection must be made and an occupancy permit obtained prior to occupying the house.

Test Your Knowledge - Unit 5

1. A set of house plans usually includes:______ ______________________________. and __________drawings.
2. Residential plans are usually drawn to a scale of____in. = _____ft.____in.
3. Floor plans show the_______and outline of the building.
4. The plot plan shows the ______________________________________.
5. Elevation drawings show the__________of the structure.
6. A section view shows the________of a structure and how _____ _____ ____________.
7. Dimension lines are ______________lines with the size being placed_________the line near the ____________.
8. Draw symbols which represent these materials and items:
   a. Concrete:
   b. Double hung window:
   c. Interior door:
   d. Refrigerator:
   e. Wall lavatory:
   f. Three-way switch:
   g. Range outlet:
   h. Wall fixture outlet:
9. To obtain a plan dimension not shown, an __________ scale may be used.

10. Working drawings (plans) provide much information required by the builder. Supplementary information is supplied by written _____, commonly called __________ .
11. The modular construction concept is based on the use of a standard grid divided into ____in. squares.
12. A building code covers all important aspects of the erection of a building. (True or False).

Outside Assignments

1. Secure a complete set of plans for an average size residence and make a careful study of the views shown (try to borrow a set from a local builder). Make a list of symbols, notes and abbreviations that you do not understand and then refer to reference books and architectural standards books to secure the information. Secure assistance from your instructor if you have difficulty with some of the views.
2. Make a thorough study of the building code in your community. Become familiar with the various sections that are covered and especially note requirements that apply to residential work. Submit a general outline of the material you feel is most important.
3. Make a trip to your city offices and visit with the commissioner or director of building. Be sure to call for an appointment in advance of your trip. During your visit secure information concerning building permits and inspection procedures. Learn the cost and what plans and specifications need to be submitted. Also secure information about zoning restrictions and other public ordinances that apply to residential construction. Prepare carefully organized notes and make an oral report to your class.

Architectural draftsman prepares a floor plan. Lines must be carefully drawn and distances accurately scaled. (Stenson, Warm and Grimes Architects, Inc.)

Unit 6
FOOTINGS AND FOUNDATIONS

The carpenter is an important man in the field of building construction. Even in modern commercial and institutional structures built of steel and reinforced concrete as in Fig. 6-1, there is a large amount of work he must do. The carpenter must continually coordinate his efforts with those of the other trades.

Fig. 6-1. Carpenters and other tradesmen at work on modern office building. (Construction Craftsman Magazine)

In the construction of single family dwellings and other smaller structures, the carpenter must not only work with other tradesmen, but he is often the key man in the operation. He works closely with the architect and owner in carrying out and coordinating the total plan.

On some jobs the carpenter may be required to lay out the building lines and supervise the excavation. He builds the forms for footings and foundation walls. The carpenter needs a working knowledge of standards and practices in concrete work. In this Unit, some of the material presented deals with masonry. It is included because of its close relationship to carpentry.

Laying Out Building Lines

After the site is cleared, lot lines should be located and checked. To protect the owner and builder, this should be done by or with the help of a registered engineer or licensed surveyor. The assistance may include the establishment of building lines and grade levels. The carpenter should be familiar with local building codes.

It is best to establish building lines with the use of leveling instruments following the procedure described in Unit 3. Lines can however be pulled from lot markers. When following such a procedure, it is important that distances be laid out perpendicular to existing lines and that building lines be square. To establish a right angle, the 6-8-10 method (based on the pythagorean theorem) can be used, as shown in Fig. 6-2.

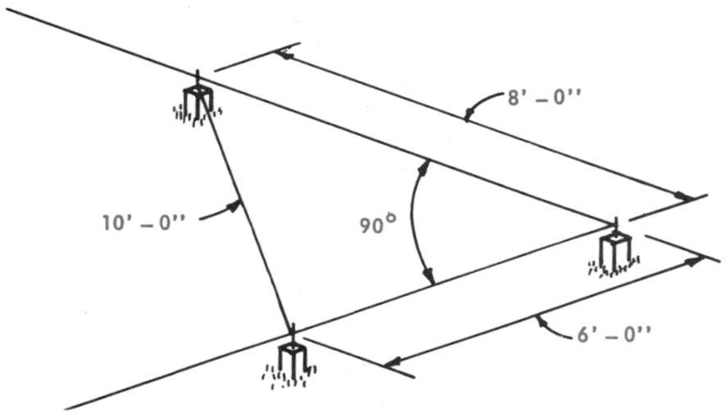

Fig. 6-2. Using the 6–8–10 method to lay out or check a right angle.

Locate corners formed by the intersection of the outside surfaces of foundation walls. Mark the positions by driving stakes; then set tacks in the stake tops at the exact spot.

After all building lines are established, check them carefully. Measure the length and, even though they were laid out with a transit, measure

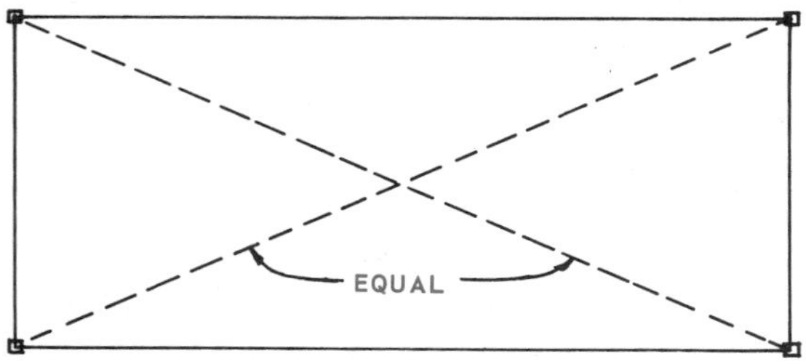

Fig. 6-3. Diagonals of a square or rectangle will be equal in length. Always apply this final check to building lines.

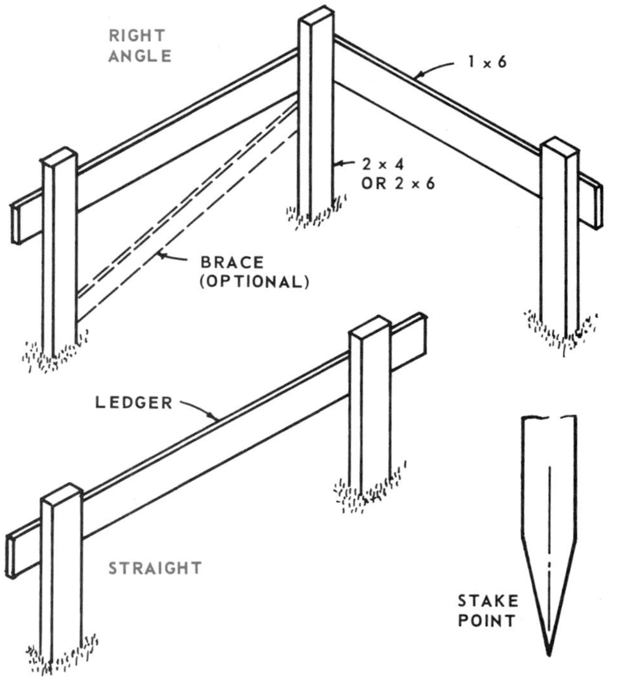

Fig. 6-4. Batter boards. When the soil is loose or they must be higher than 3 ft., braces should be used.

the diagonals of squares and rectangles, Fig. 6-3. An out-of-square foundation can result in continuous problems throughout construction.

Batter Boards

Batter boards, Fig. 6-4, are set up around the building layout stakes. Use 2 x 4 pieces for the stakes and 1 x 6 or wider pieces for the ledgers. Locate the batter boards 4 ft. or more away from the building lines.

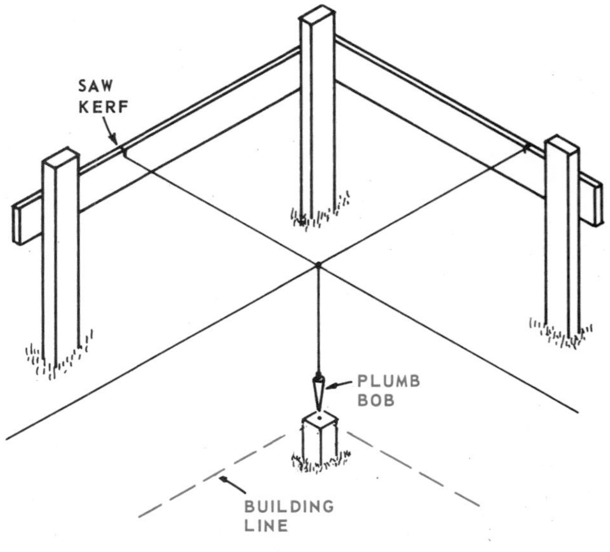

Fig. 6-5. Setting up building lines on batter boards.

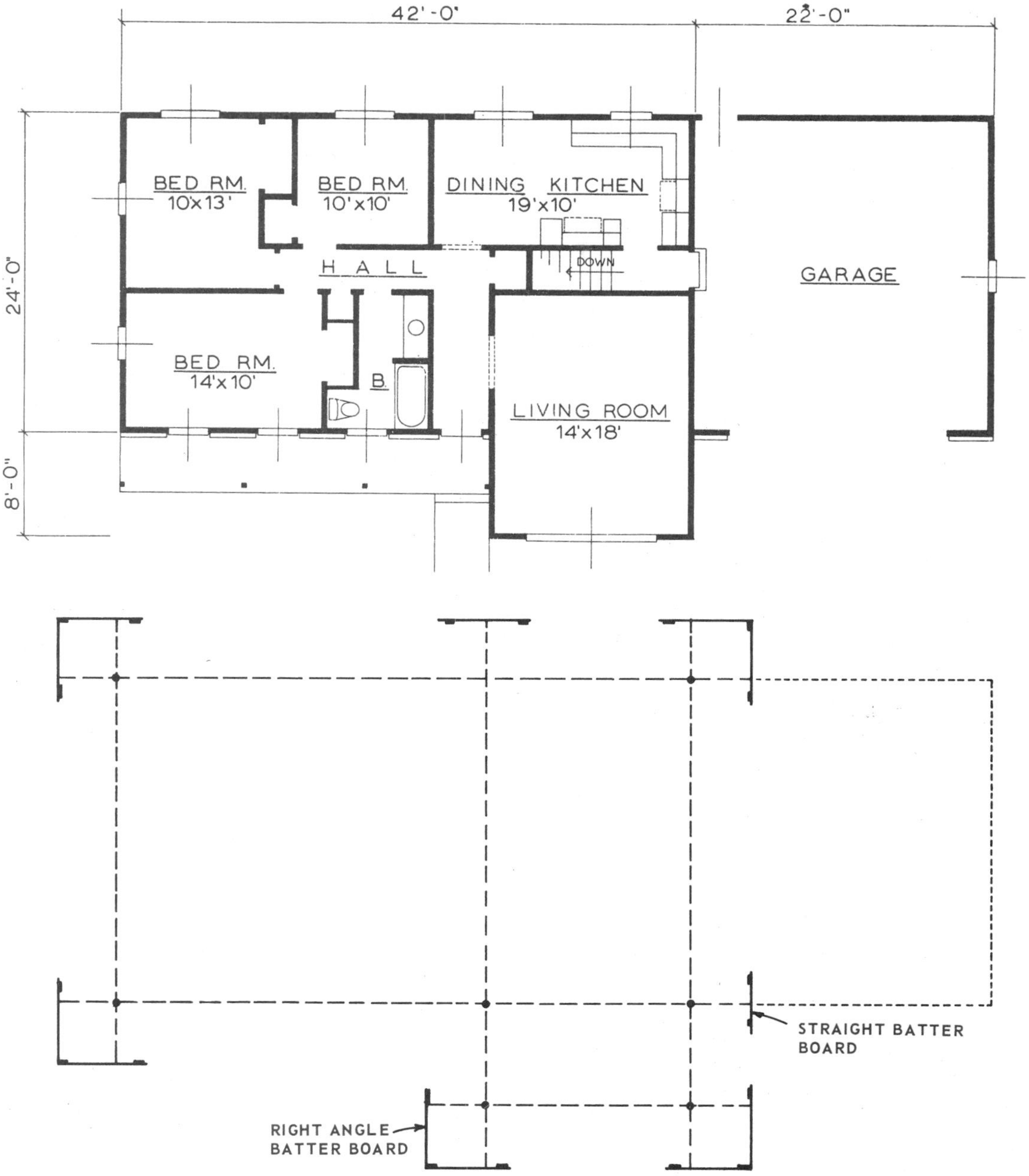

Fig. 6-6. Above. Plans showing overall dimensions. Below. Building lines set up around excavated area. The actual excavation should be at least 2 ft. outside of these lines.

Nail the ledger boards to the stakes in a level position and at a convenient working height, preferably slightly above the top of the foundation. The batter boards should be approximately level with each other. Also be sure that the ledger boards are long enough to extend well past each corner.

Using lines and a plumb bob, pull the lines so they pass directly over the layout stakes. Mark the top of the ledger boards where the lines cross. Make a shallow saw kerf as shown in Fig. 6-5. Pull the lines tight and fasten them to nails driven in the back of the ledger, or wrap them around the ledger and run them through the saw kerf several times.

Excavation

Building sites on steep slopes or rugged terrain should be rough graded before the building is laid out. Top soil should be removed and piled where it will not interfere with construction. This can be used for the finished grade after the building is complete.

Where no grading is needed, the site can be laid out and batter boards erected, as shown in the lower drawing, Fig. 6-6. Stakes marking the perimeter of the rough excavation are set and the lines are removed from the batter boards during the work. For regular basement foundations, the excavation should extend beyond the building lines by at least 2 ft. to allow clearance for formwork. Foundations for structures with a slab floor or crawl space will need little excavating beyond the trench for footings and walls.

The depth of the excavation can be calculated from a study of the vertical section views of the architectural plans.

In cold climates, it is important that foundations be located below the frost line. If the moisture in the soil under the footing freezes, this may force the foundation wall upward and cause cracks and damage that is difficult to repair. Local building codes usually cover these requirements.

It is common practice to establish the depth of the excavation and, consequently, the height of the foundation, by using the highest elevation on the perimeter of the excavation as the control point, Fig. 6-7. Foundations should extend above the finished grade (about 8 in.) so the wood finish and framing members will be adequately protected from soil moisture. The finished grade should be sloped from all sides of the structure so surface water will run away from the foundation.

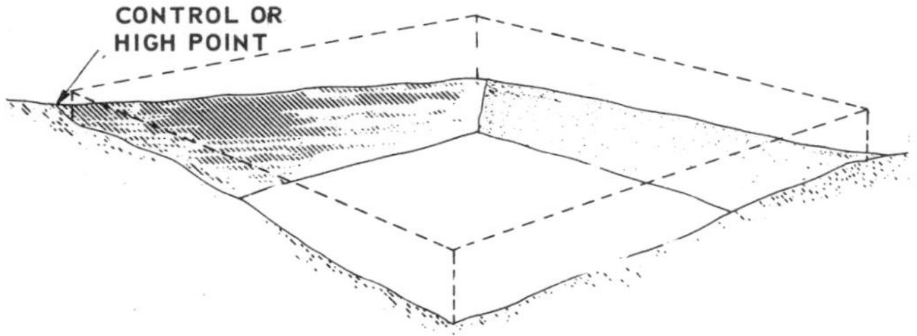

Fig. 6-7. Method of establishing depth of excavation.

The depth of the excavation may be further controlled or limited by the overall levelness of the site, slope to the street or adjacent property, and the elevation of sewer lines. Normally, solving these problems is the responsibility of the architect. Information on grade, foundation and floor levels is usually included in the working drawings.

Footing Design

The size and shape of footings needed depends largely on soil conditions. Footings must be wide enough to spread the load over sufficient area

Type of soil	Capacity, tons per sq.ft.
Soft clay	1
Wet sand or firm clay	2
Fine, dry sand	3
Hard, dry clay or coarse sand	4
Gravel	6

Fig. 6-8. Load carrying capacities of various types of soil.

so there is no possibility of settlement. Load-bearing capacities of soils vary considerably. See Fig. 6-8.

In residential and smaller building construction, no great load is placed on the wall footings and a safe design is usually secured by making the width twice as wide as the foundation wall. See Fig. 6-9. The average thickness of a footing is about 8 in.

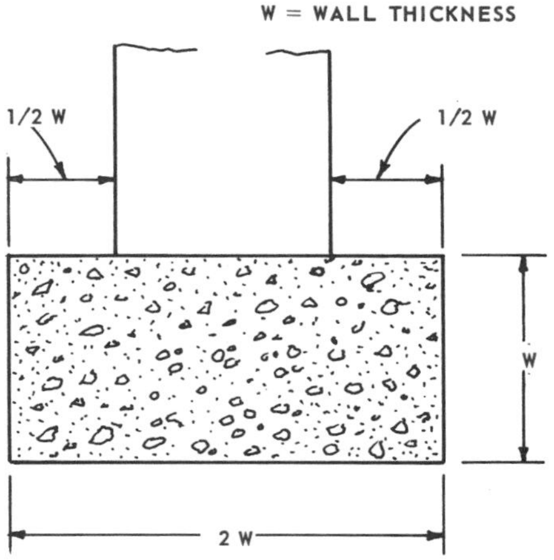

Fig. 6-9. Standard design for a footing in residential construction.

Footings under columns and posts carry heavy, concentrated loads and are usually from 2 to 3 ft. square. The thickness should be about 1 1/2 times the distance from the face of the column to the edge of the footing.

Reinforced footings are used in regions subject to earthquakes or where the footings must extend over areas of the foundation bed containing poor load-bearing material. Some structural designs may also require the use of reinforcing. The common practice is to use two No. 5 (5/8 in.) bars for 12 x 24 in. footings. At least 3 in. of concrete should cover the reinforcing at all points.

In a single story dwelling where chimney footings are independent of other footings, the footings should have a minimum projection of 4 in. on each side. For a two-story house, chimney footings should have a minimum thickness of 12 in. and a minimum projection of 6 in. on each side. Exact dimensions will vary somewhat according to the weight of the chimney and the nature of the soil. Where chimneys occur in outside walls or inside bearing walls, chimney footings should be constructed as part of the wall footing. Concrete for the chimney and for the wall footings should be placed at the same time.

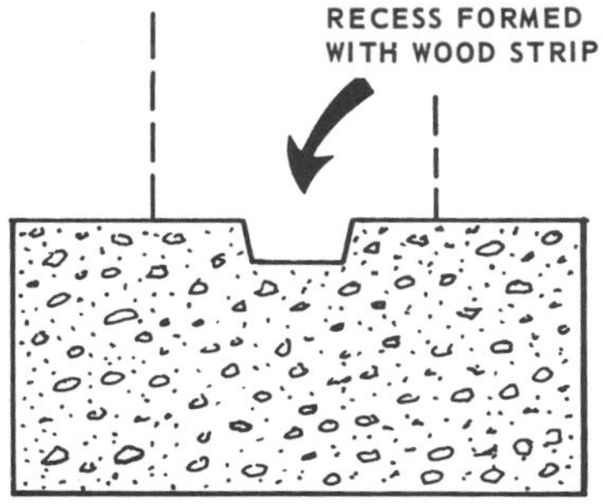

Fig. 6-10. Footing for concrete foundation with key formed by using wooden strip placed in the concrete.

Footings that will support cast-in-place concrete walls may be formed with a recess which provides a keyed joint as illustrated in Fig. 6-10. A number of typical footing designs are shown in Fig. 6-11. These are not working drawings and will need to be adapted to local conditions and legal requirements.

Forms for Footings

After the excavation is complete, the footings are laid out and constructed. Lines are replaced on the batter boards and corner points are dropped with a plumb bob to the bottom of the excavation.

HANDY SAYS:

After the excavation is complete, check the batter boards carefully. They may have been disturbed by the excavating equipment. Make necessary adjustments before proceeding with the footing layout.

Drive stakes and establish points at the corners of the foundation walls. Set up a builders' level at a central point in the excavation and drive a number of grade stakes (level with the top of the footing) along the footing line and at approximate points where column footings are required. Corner stakes can also be driven to the exact height of the top of the footing. Connect the corner stakes with lines tied to nails in the top of the stakes.

Working from these building lines, construct the outside form for the footing. The form boards will be located outside the building lines by a distance equal to the footing extension (usually 4 in. for an 8 in. foundation wall). See Fig. 6-12. The top edge of the form boards must be level with the grade stakes. Transfer the measurement from the grade stakes to the form. Use a carpenter's level. After the outside form boards are in place, it will be relatively easy to set the inside sections. See Fig. 6-13.

Forms constructed of 1 in. boards should be held in place with stakes placed 2 to 3 ft. apart. Stakes may be placed farther apart when 2 in. material is used. Using a number of spacers or spreaders to locate the inside form will save measuring time, Fig. 6-14.

Forms for column footings are usually set after the wall footing forms are complete. These are located by direct measurements from the building lines and are leveled to stakes previously set.

Some hand digging and leveling of the excavation will probably be necessary as forms are set. LOOSE DIRT AND DEBRIS MUST ALWAYS BE REMOVED FROM THE GROUND THAT WILL BE LOCATED UNDER A FOOTING EVEN THOUGH THE RESULTING DEPTH WILL BE GREATER THAN REQUIRED. The top of the footing must be level. The bottom may vary somewhat as long as the minimum thickness is maintained.

Form boards are temporarily nailed to stakes and to each other. Double-headed nails may be used. If regular nails are used, they should be driven only part way into the wood. Consider problems in form removal. Nail through stakes into the form boards. Do not nail from the inside.

When the forms are complete and checked for sturdiness and accuracy, the line, line stakes and grade stakes are removed and the concrete is placed.

Concrete

Concrete is made by mixing cement, fine aggregate (sand), coarse aggregate (gravel or crushed stone), and water in proper proportions. When the coarse aggregate is eliminated and the mix consists of cement, sand and water, it is called mortar or grout. A common error is to refer to either concrete or mortar as cement. It is incorrect to call a concrete floor a cement floor.

To prepare concrete, the aggregate and cement are first mixed together and then the water is added. The water causes a chemical action (called hydration) to take place and the mass hardens. The hardening process is not a result of drying out. Instead of drying out, the concrete should be kept moist during the initial hydration process.

The compressive strength of concrete is high, but its tensile strength (stretching, bending or

Fig. 6-11. Footing designs for various foundation walls. The grid over the drawings is formed with 4 in. squares. (Portland Cement Assoc.)

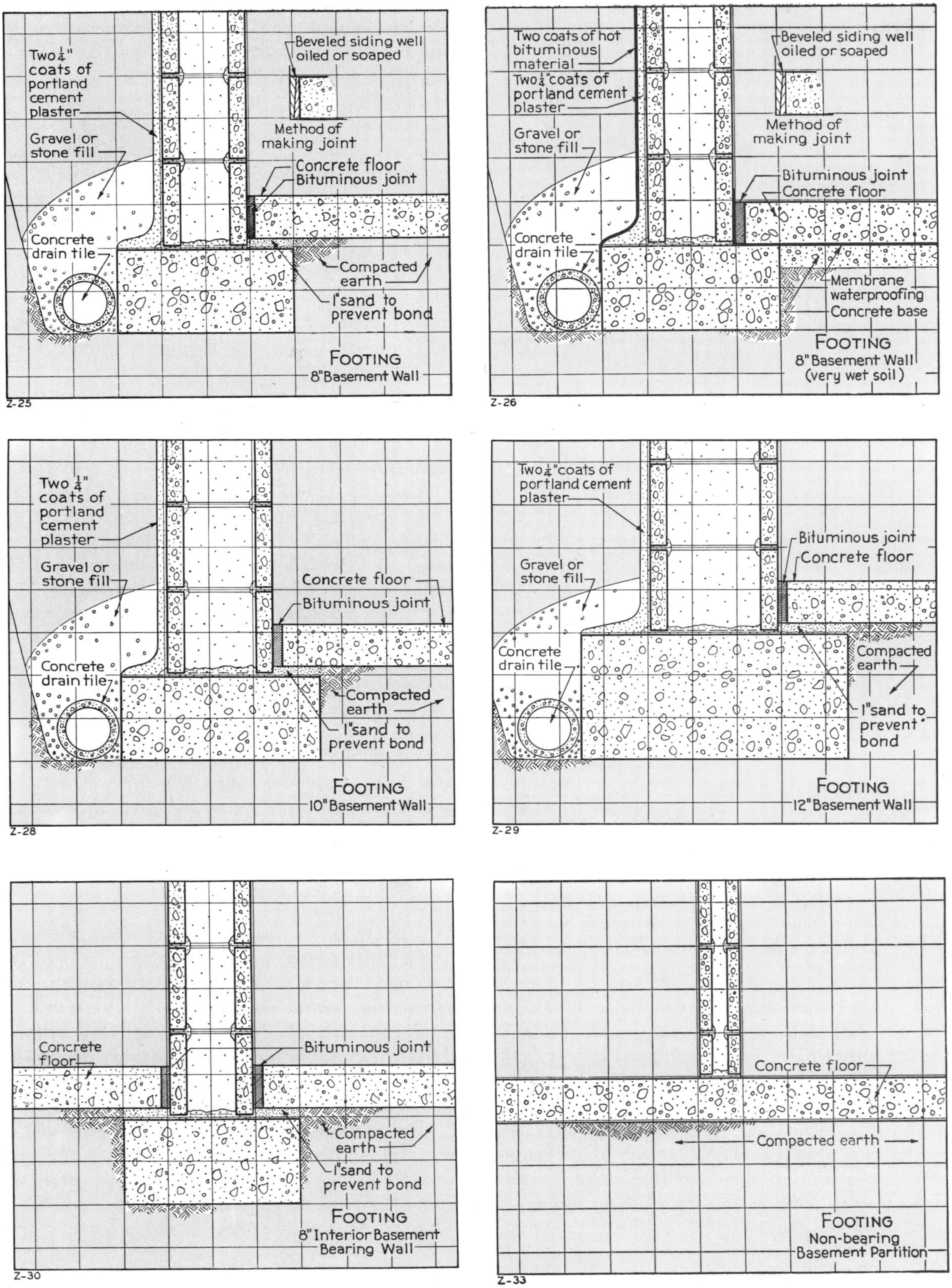

twisting) is relatively low. Consequently, when concrete is used for beams, columns and girders, it must be reinforced with steel. When it must resist compression forces only, reinforcement is usually not included.

Cement

Most cement used today is Portland cement which is usually manufactured from limestone mixed with shale, clay or marl. The properly proportioned raw materials are pulverized and fed into kilns where they are heated to a temperature of about 2700 deg. F. and maintained at that temperature for a certain time. As a result of chemical changes produced by the heat, the material is transformed into a clinker. The clinker is then finely ground to form a powder.

Each sack of Portland cement holds 94 lbs. - equal to one cubic foot in volume. Cement should be a free-flowing powder. If it contains lumps that cannot be pulverized easily between thumb and fingers, it should not be used.

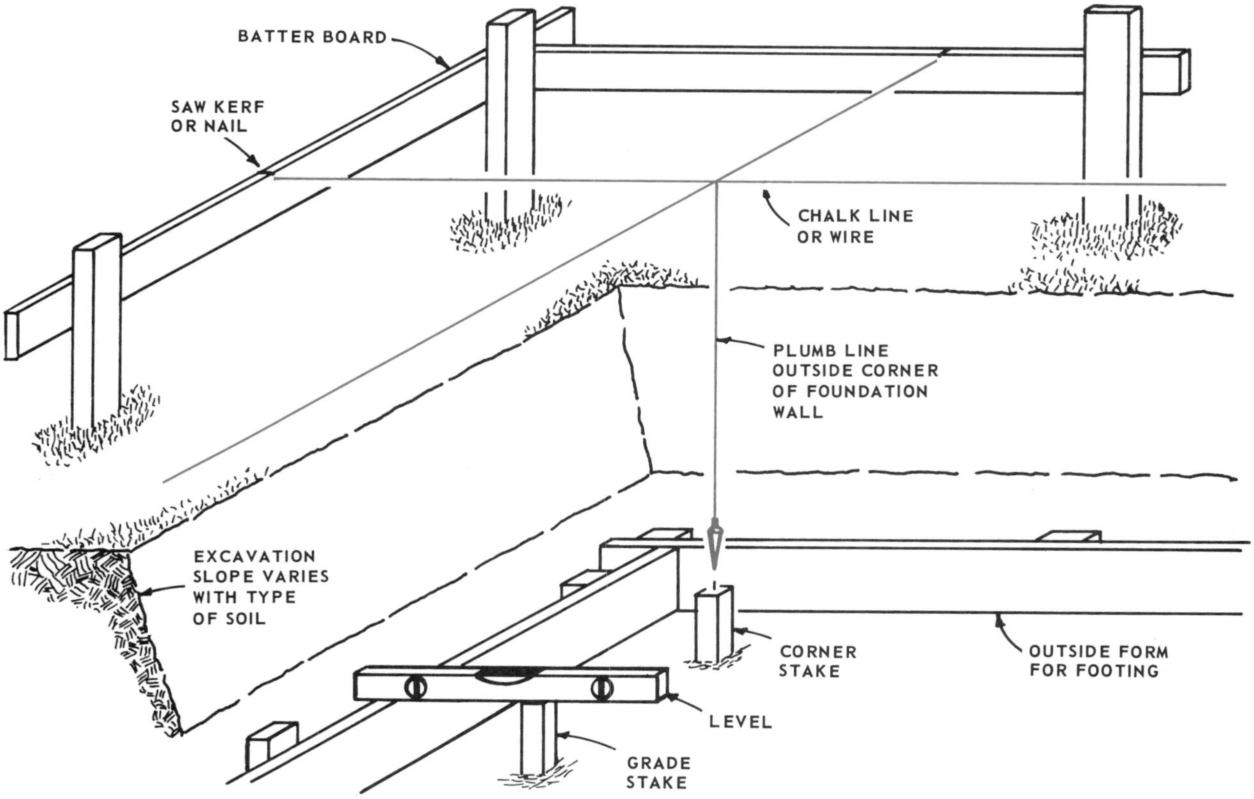

Fig. 6-12. Laying out forms for footings. The outside forms are built first.

Fig. 6-13. Measuring and setting footing forms. (Portland Cement Assoc.)

Aggregates

The inert material which is combined with the cement and water is called the aggregate. It may consist of sand, crushed stone, gravel or lightweight materials such as expanded slag, clay or shale.

The large coarse aggregate (seldom over 2 1/2 in. in size) particles form the basic structural members of the concrete. The voids between these particles are filled with smaller particles, and the voids between these smaller particles are filled with still smaller particles until a dense mass is formed. The cement and water paste, which thoroughly coats each particle, then binds them solidly together when it hardens.

Good quality concrete is formed when aggregate size and proportion are carefully determined and controlled. In specifying a concrete mix, the first number refers to the proportion of Portland cement, the second number the sand, and the third the gravel or crushed stone. In a 1-2-3 mix, for example, we would have 1 part Portland cement, 2 parts sand and 3 parts gravel.

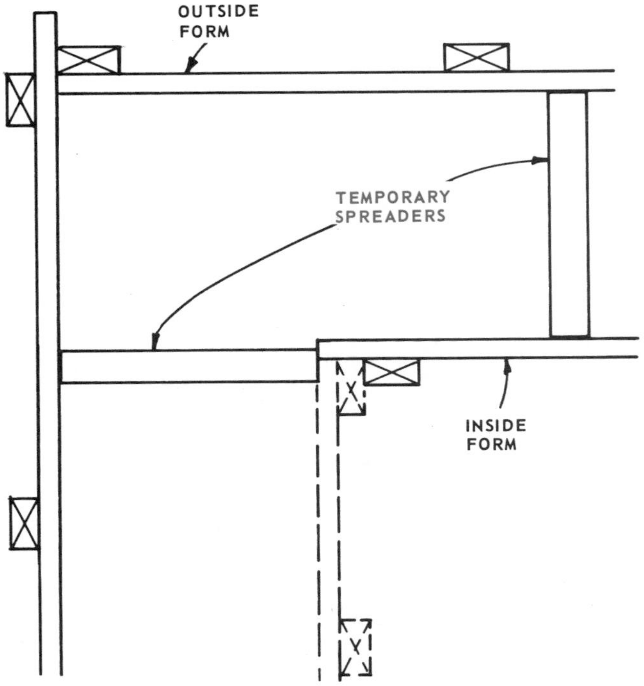

Fig. 6-14. Using precut spreaders to locate the inside form boards saves time.

Today, practically all concrete is delivered to the building site in ready-mix trucks. Ready-mix concrete is purchased by the cubic yard (27 cu. ft.) and is available in a number of psi ratings. A minimum order is usually 1 cu. yd. Fractional parts (1/4, 1/2, 1/3) can be furnished beyond this amount. An average size ready-mix truck has a capacity of about 5 cu. yds.

Wall Forms

Many different types of wall forming systems are available. Regardless of the details of form construction and methods of erection, there are certain basic considerations that should be understood and applied to all systems.

For quality work the forms used must be tight, smooth, defect-free and properly aligned. Joints

Fig. 6-15. Using a wheelbarrow and chute to place concrete in a narrow form. (American Plywood Assoc.)

between form boards or panels should be tight to prevent the loss of the cement paste which will tend to weaken the concrete and result in honeycombing.

Wall forms must be strong and well braced to resist the side pressure created by the plastic concrete, Fig. 6-15. This pressure increases tremendously as the height of the wall is increased. Regular concrete weighs about 150 lbs. per cu. ft. If it were immediately poured into a form 8 ft. high, it would create a pressure of about 1200 lbs. per sq. ft. along the bottom side of the form.

In actual practice, this amount of pressure is reduced through compaction and hardening of the concrete and it tends to support itself. Thus, the lateral pressure will be related to the amount of

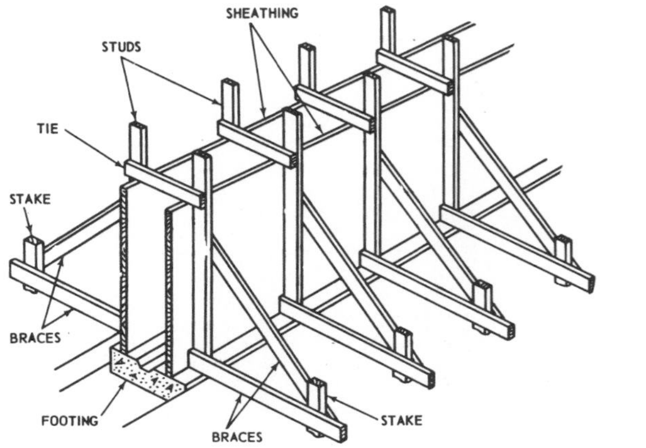

Fig. 6-16. Suggested design for low wall forms -- 3 ft. or less in height.

concrete placed per hour, the outside temperature and the amount of mechanical vibration.

Low wall forms, up to about 3 ft. in height, can be assembled from 1 in. sheathing boards or 3/4 in. plywood, supported by two-by-four studs spaced 2 ft. apart, Fig. 6-16. The height can be increased somewhat if the studs are spaced closer together.

For walls over 4 ft. in height, the studs should be backed with wales to provide greater strength, Fig. 6-17.

Form Ties

Wire ties and wooden spreaders formerly used have been largely replaced with various manufactured devices in which the function of the spreader and tie are combined. Fig. 6-18 shows a type called a snap tie. This is available for various wall thicknesses. The rod goes through small holes bored through the sheathing and studs. Holes through the wales can be larger for easy assembly or the wales can be doubled as shown. The

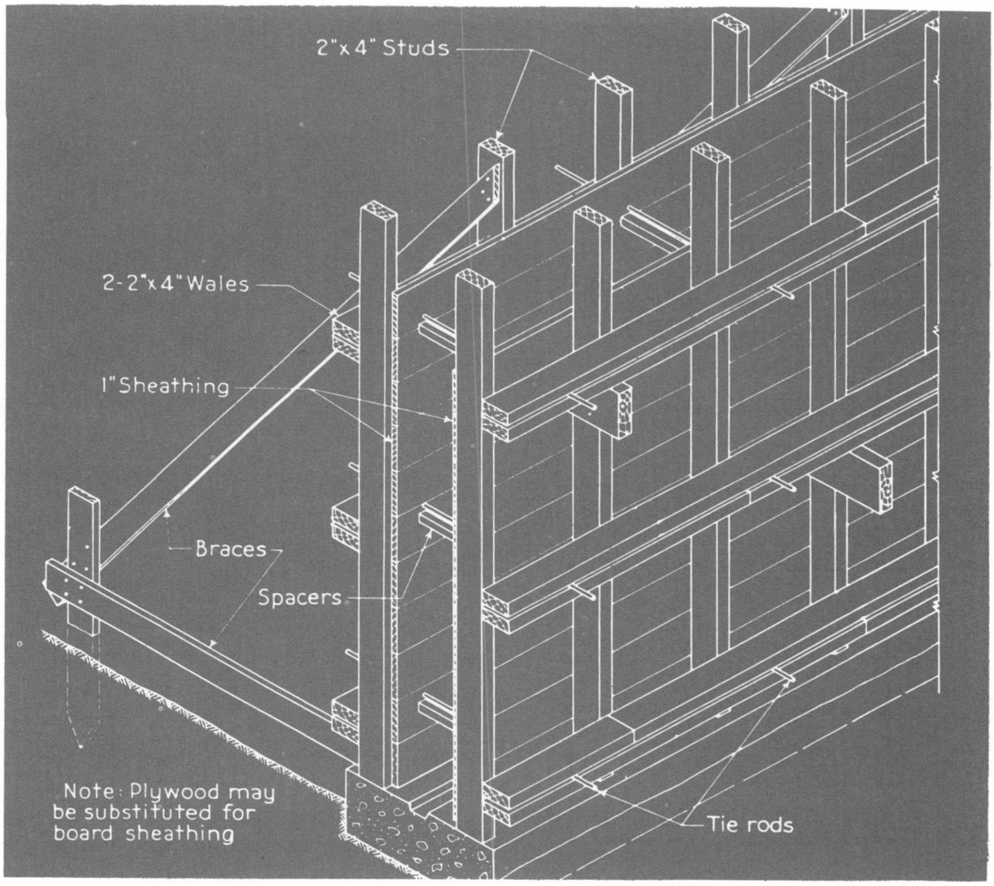

Fig. 6-17. A built-in-place form using tie rods to keep the sides from spreading.

Fig. 6-18. Section through a wall form held in position with a patented wall tie. (Universal Form Clamp Co.)

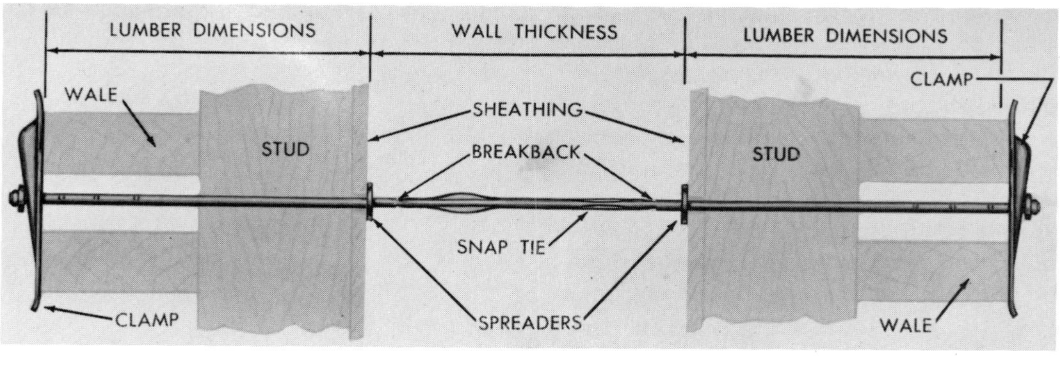

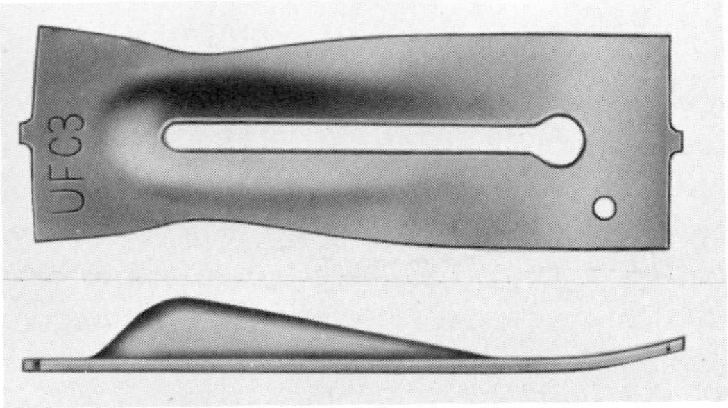

Fig. 6-19. Wall tie clamp slips over the tie rod and is tapped downward to tighten the assembly.

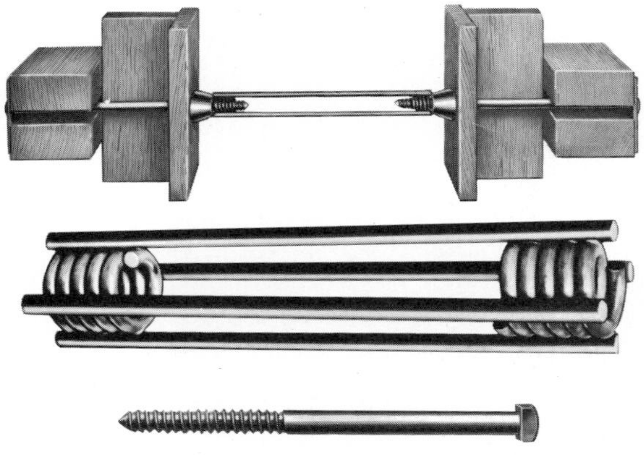

Fig. 6-20. Top. Heavy-duty wall tie assembly. Center. Internal member. Consists of helical coils welded to two or four high-strength steel rods. Below. Lag screw which threads into helical coils. (Richmond Screw Anchor Co.)

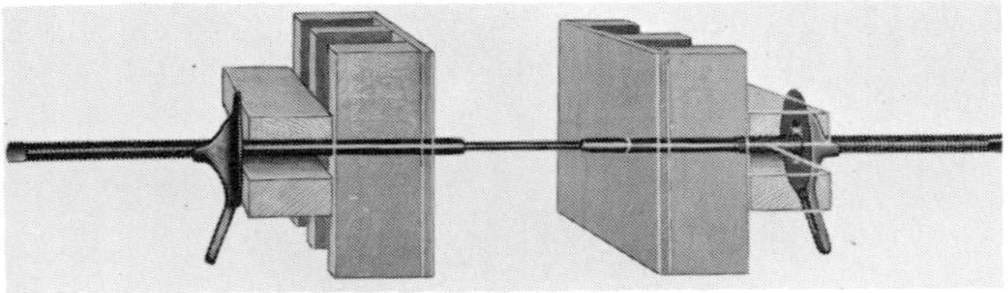

Fig. 6-21. Threaded rod wall tie assembly.

spreader washer is rigidly set on the tie rod, thus holding the forms apart. Clamps, Fig. 6-19, are placed over the rod and tapped down to tighten and hold the assembly together.

After the concrete has set, the clamps can be quickly removed and the forms stripped. A special wrench is used to break off the outer sections of the rod. The rod breaks at a small indentation located about 1 in. below the concrete surface. The hole in the concrete is patched with grout or mortar, completely embedding the remaining portion of the tie and providing a smooth surface.

Other types of patented wall ties are shown in Fig. 6-20 and Fig. 6-21. The coil type spreader is assembled with a cone of wood, plastic or metal and a lag screw. The cone provides smooth contact with the form and leaves a recess that is easy to fill. The other type consists of threaded rods that attach to the center tie section and are then screwed out after the forms are stripped. Pressure is applied with either a nut-washer or a tilt-lock clamp. The latter allows for rapid assembly and disassembly since the threads do not engage except when the clamp is perpendicular with the bolt.

Fig. 6-22. Assembling prefabricated form panels. (Georgia-Pacific)

Panel Forms

Today, prefabricated panels are used for most wall forming. The panels, made from a special grade of plywood, are attached to wood or metal frames. See Fig. 6-22. Some carpenters build their own panels, using 3/4 in. plywood and 2 x 4 studs to form 4 ft. by 8 ft. units, Fig. 6-23. For standard columns and other units, prefabricated forms, such as shown in Fig. 6-24, will save time.

When erecting panel forms, the units may be fastened together as shown in Fig. 6-25. Wall ties are installed as the panels are assembled and then wales and bracing are added to align and strengthen the completed form. Be sure to select straight lumber for the wales so they will hold the panels in proper alignment.

HANDY SAYS:

Check the form work carefully before placing the concrete. A form that fails during the pouring will waste material and cause a lot of extra work.

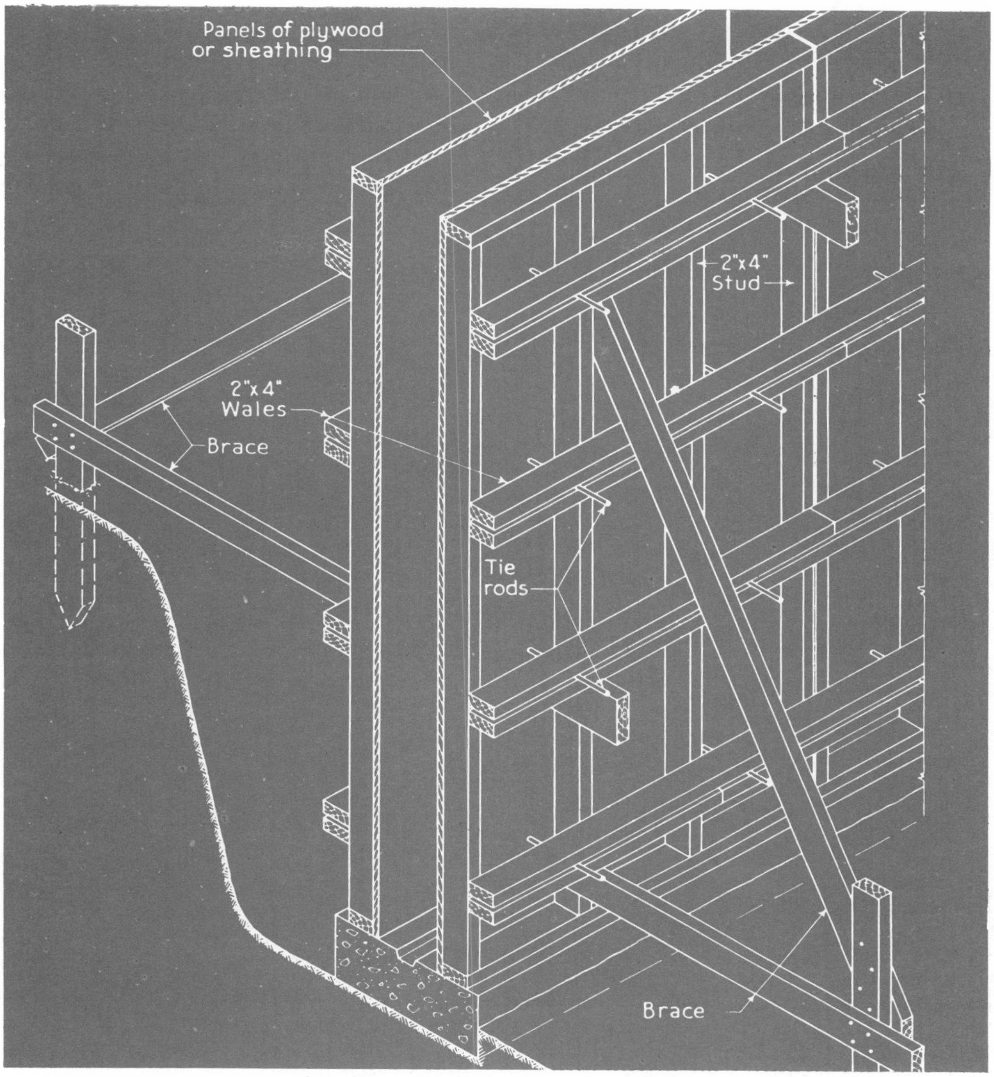

Fig. 6-23. Form assembled from prefabricated panels. The panels are designed so they can be readily removed from one job and reassembled for use on subsequent work.

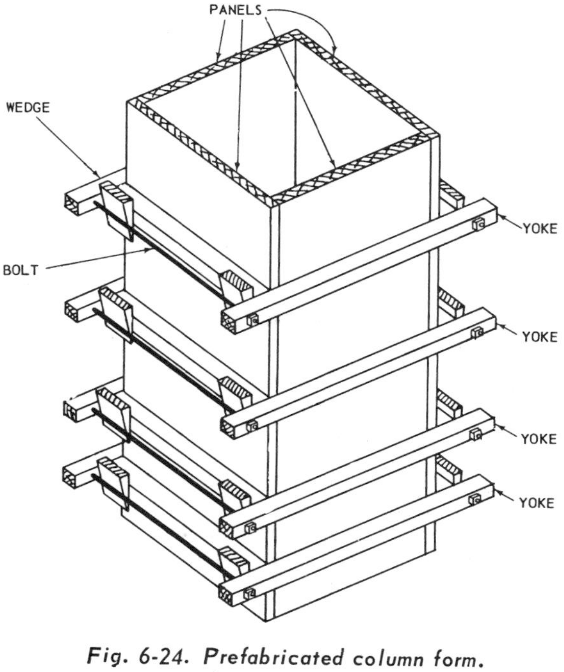

Fig. 6-24. Prefabricated column form.

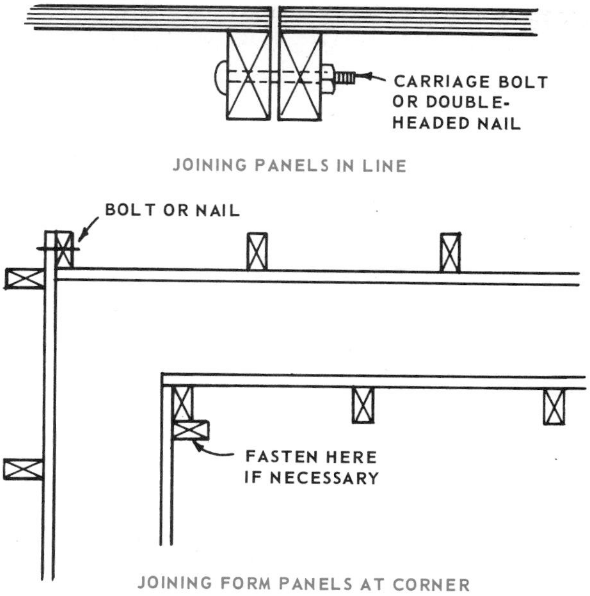

Fig. 6-25. Methods of attaching form panels constructed of plywood and 2 x 4 studs.

Since panel forms are designed to be used many times, they should be treated to prevent the concrete from sticking to the surfaces using special form coatings which are available.

An on-the-site view, Fig. 6-26, shows a typical panel form assembly after the concrete has been placed and stripping has just been started. Care should be exercised in removing form components.

Fig. 6-26. Form panels ready to be stripped. Note that one of the wales has already been removed.

Fig. 6-27. Manufactured forms made from a special grade of plywood mounted on steel frames. Patented corners and form ties insure rapid and accurate assembly. (Universal Form Clamp Co.)

They should be thoroughly cleaned and then carefully sorted and stacked for movement to the next job or storage.

Manufacturers have developed many forming systems to replace or supplement panel forms built by the carpenter. For residential work they usually consist of steel frames and exterior grade plywood panels. Sometimes the plywood is coated with a special plastic material to create a smooth finish on the concrete and prevent it from sticking to the surface. The panel units are relatively light in weight and can be easily handled and transported from one building site to another. Specially designed devices are used to assemble and space the components quickly and accurately. See Fig. 6-27.

Windows and Doors

Several procedures are followed in forming openings in foundation walls for doors and windows. In poured walls, forms or stops are built into the regular forms. Nailing strips may be at-

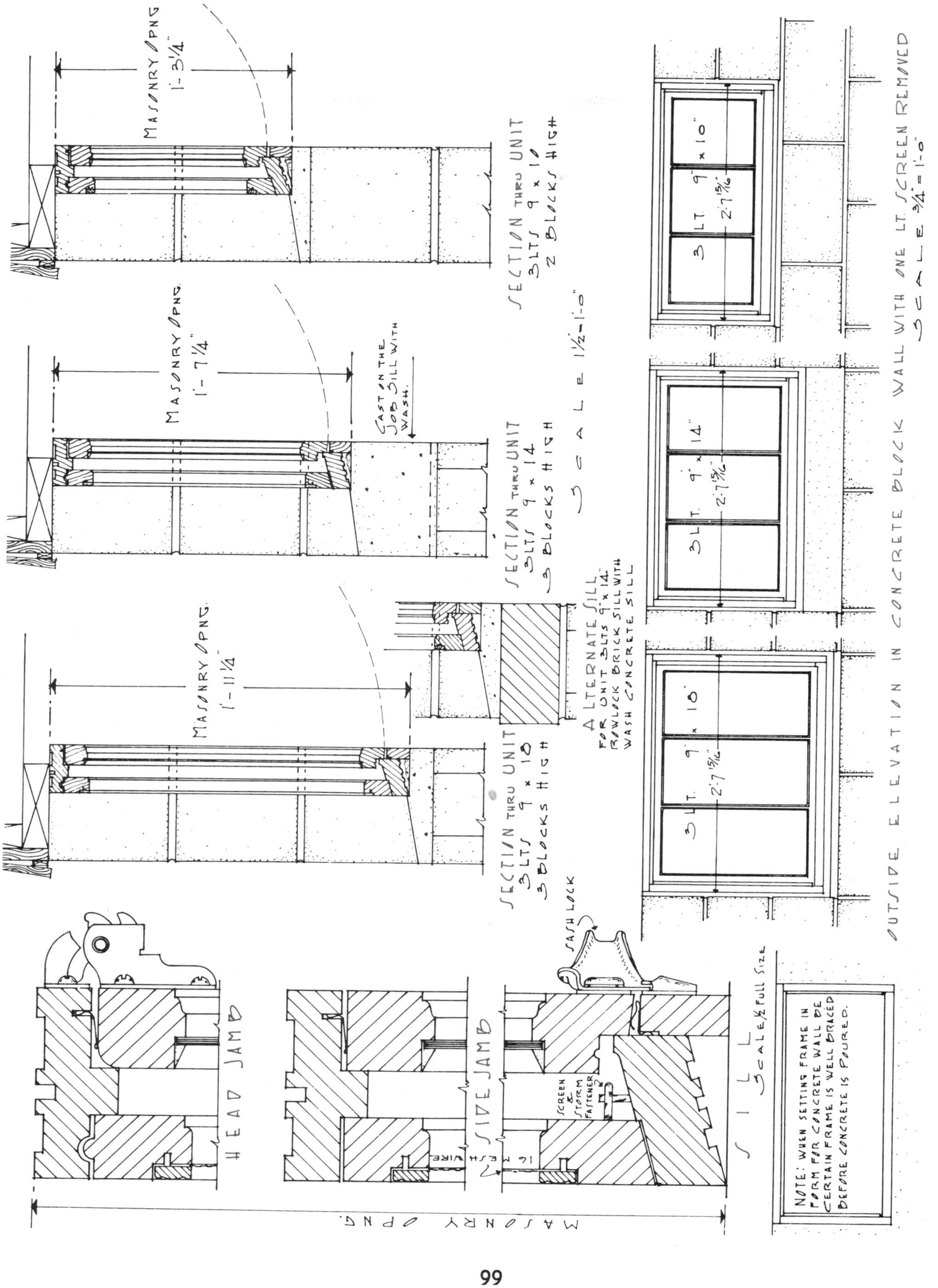

Fig. 6-28. Detail of basement window units that can be installed in a concrete or masonry foundation wall.

tached to the form and cast into the concrete. Frames are then secured to these strips after the forms are removed.

In concrete block construction, door and window frames are set in place and the masonry units are constructed around them. The outside surface of the frames have grooves into which the mortar flows, forming a key. Basement windows are usually located level with the top of the foundation wall and the sill carries the weight of the structure across the opening, Fig. 6-28.

Lintels

Masonry that is carried across the top of openings is supported by a structural unit called a lintel. This can be a precast concrete unit that includes metal reinforcing bars, or steel angle irons can be used as shown in Fig. 6-29. Another method consists of laying a course of lintel blocks across the opening (supported by a frame) and then filling the blocks with reinforcing bars and concrete, Fig. 6-30.

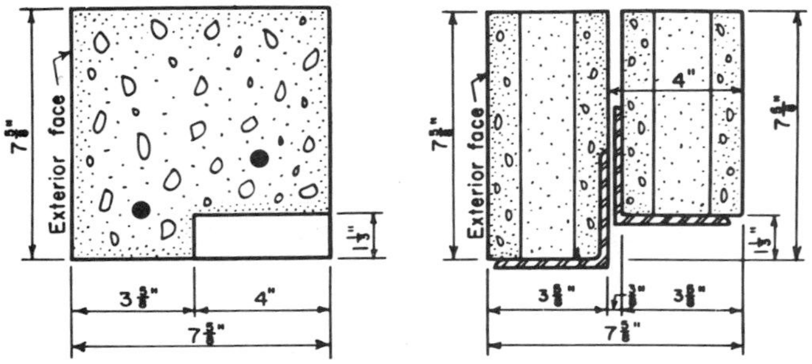

Fig. 6-29. Lintels for masonry walls. Left. Precast unit. Right. Steel angle irons.

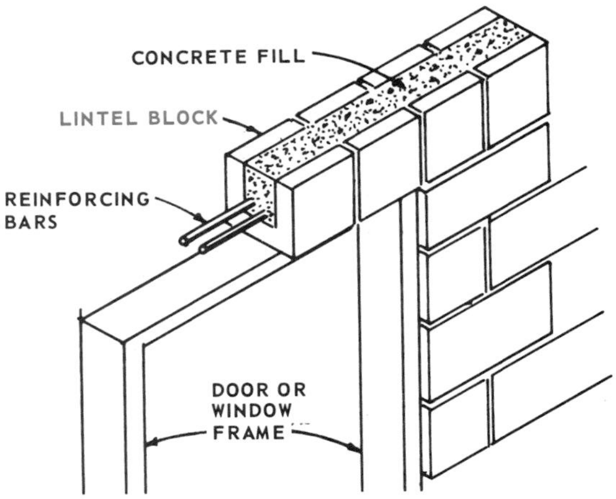

Fig. 6-30. How lintel blocks are used to provide support across an opening.

Placing Concrete

Usually most of the concrete can be poured directly from the ready-mix truck into the forms, Fig. 6-31. To move the concrete to other areas not accessible to the truck, a wheelbarrow is generally used.

Concrete should be placed as near to its final position as possible. To avoid segregation, it

Fig. 6-31. Concrete being placed directly into the forms from a ready-mix truck.

should not be placed in large quantities at a given point and allowed to run or be worked over a long distance. Concrete should be placed in forms promptly after mixing.

In general, concrete for walls should be placed in the forms in horizontal layers of uniform thickness not exceeding 6 to 12 in. deep. As the concrete is placed, it is spaded or vibrated enough to compact it thoroughly and produce a dense mass. Working the concrete next to the form tends to produce a smooth surface, free from honeycombing, along the form faces. A spade or thin board, Fig. 6-32, may be used for this purpose. Thus the large aggregates are forced away from the forms and any air that may have been trapped along the form face is released. Mechanical vibrators are effective in consolidating concrete. However, vibrators create an increased pressure on the forms and this factor must be considered in the form design. The vibrator should not be held in one location long enough to draw a pool of grout from the surrounding concrete.

Anchor Bolts

Wood plates are fastened to the top of foundation walls with 1/2 in. anchor bolts spaced not more than 4 ft. apart. In concrete walls, they are set in place as soon as the pour is completed and leveled off. Anchor bolts are set in the cores of a concrete block wall and should be about 18 in. long. A piece of metal lath is placed in the second horizontal joint below the top of the wall to hold the grout or mortar. After the wall is complete, the core is filled and the bolt installed as shown in Fig. 6-33.

Wall forms can usually be stripped after one or two days.

Concrete Block Foundations

In some localities, concrete blocks are used extensively for foundation walls and other masonry construction. The standard weight block, made from Portland cement and aggregates such as sand, fine gravel or crushed stone, weighs about 40-50 lbs. (8 x 8 x 16 size). Lightweight units are made from Portland cement and natural or manufactured aggregates including volcanic cinders, pumice and foundry slag. A lightweight unit weighs between 25 and 35 lbs. and usually has a much lower U factor (measurement of the heat flow or heat transmission through materials).

Blocks used should comply with specifications provided by the American Society for Testing Materials. ASTM specifications for a Grade A

Fig. 6-32. As concrete is placed, it should be spaded along both sides of the form to eliminate honeycombing. (Portland Cement Assoc.)

Fig. 6-33. Installing anchor bolts in concrete block wall.

load-bearing unit requires that the compression strength equal 1000 lbs. per sq. inch; thus an 8 x 8 x 16 unit must withstand about 128,000 lbs. or 64 tons.

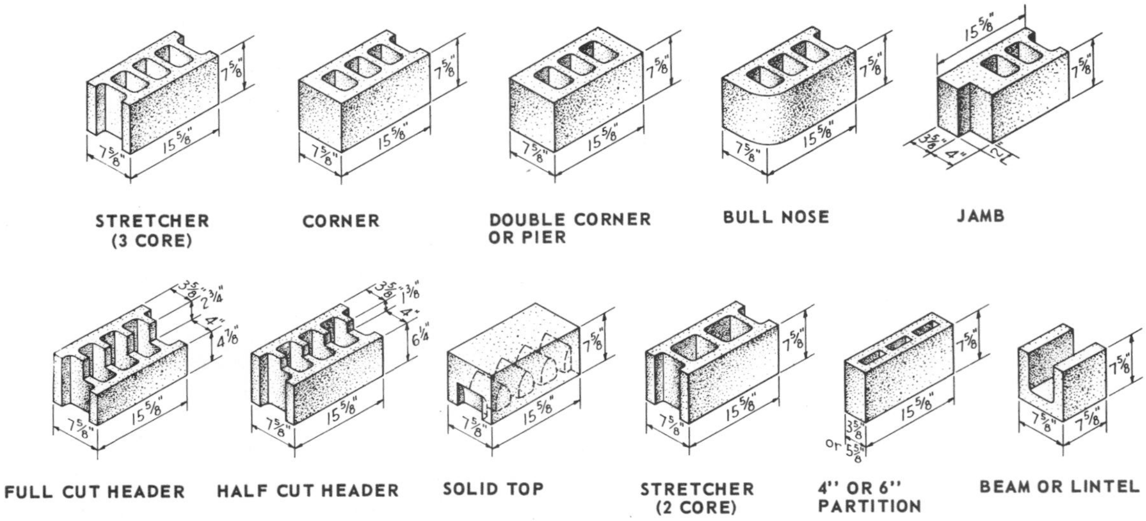

Fig. 6-34. Shapes and sizes of typical concrete masonry units. (Portland Cement Assoc.)

Sizes and Shapes

Blocks are classified as solid or hollow. A solid unit is one in which the core area is 25 percent or less of the total cross-sectional area. Blocks are usually available in 4, 6, 8, 10 and 12 in. widths and 4 and 8 in. heights. Fig. 6-34 shows some of the sizes and shapes with names that indicate use.

Sizes are actually 3/8 in. shorter than their nominal (name) dimensions to allow for the mortar joint. For example: the so-called 8 x 8 x 16 block is actually 7 5/8 x 7 5/8 x 15 5/8. With a standard 3/8 mortar joint, the laid-in-the-wall height will be 8 in. and the width 16 in. as shown in Fig. 6-35.

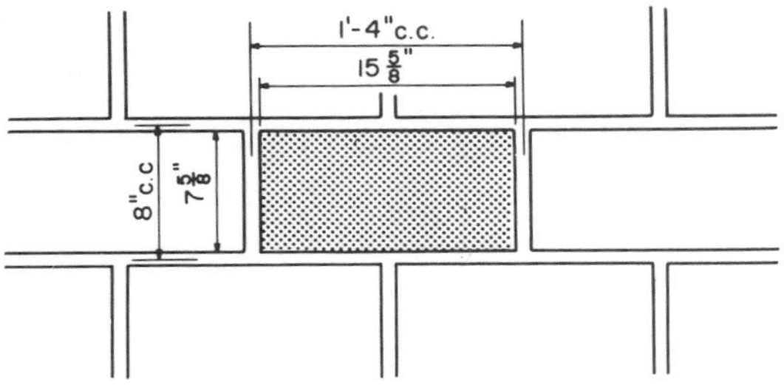

Fig. 6-35. How 8 x 8 x 16 block is laid in wall with standard 3/8 in. mortar joint.

Insulating Foundation Walls

In northern climates, when residential plans include a finished basement, insulating the foundation walls may be desirable. There are several ways to reduce the heat flow. Lightweight masonry units that have a lower U factor may be used. The cores of regular or lightweight blocks can be filled with an insulating material as shown in Fig. 6-36. Other methods include cavity wall construction or the use of various forms of rigid or flexible insulation applied to the interior surface. Additional information about insulation is included in Unit 13.

Waterproofing

In most localities, the outside of poured concrete or concrete masonry basement walls should be waterproofed below the finished grade and drain tile installed. See Fig. 6-37.

Masonry (block) walls may be waterproofed by an application of cement plaster. The wall surface should be clean and dampened with a water spray just before the first coat of plaster is ap-

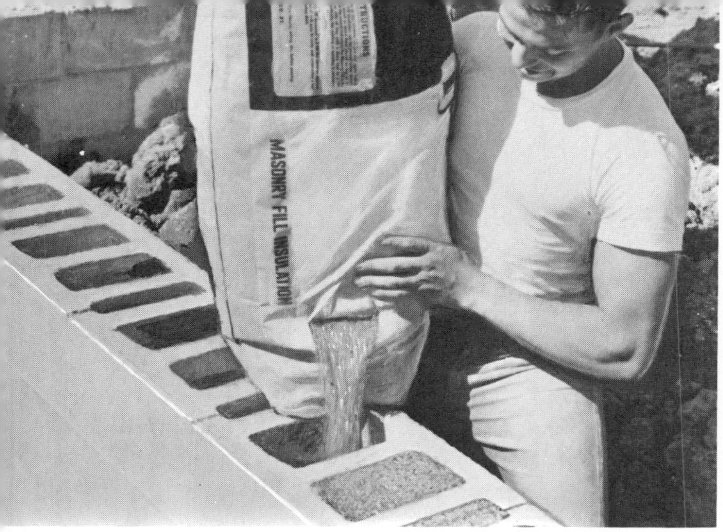

Fig. 6-36. Insulating concrete blocks by filling the cores with water-repellent vermiculite. (Vermiculite Institute)

Fig. 6-38. Applying plaster to the outside of a basement wall. The concrete blocks have been dampened. (Portland Cement Assoc.)

plied, Fig. 6-38. The plaster can be made of cement and sand (1 to 2 1/2, mix by volume) or mortar used to lay concrete blocks may be used. When the first coat has partially hardened, it should be roughened with a scratcher to provide better bond for the second coat.

After the first coat has hardened at least 24 hours, the second coat is applied, Fig. 6-39. Again dampen the surface just before applying the plaster. Both the plaster coats should extend from about 6 in. above the finished grade to the footing.

Fig. 6-39. Applying second coat of plaster. Observe how the first coat was roughened.

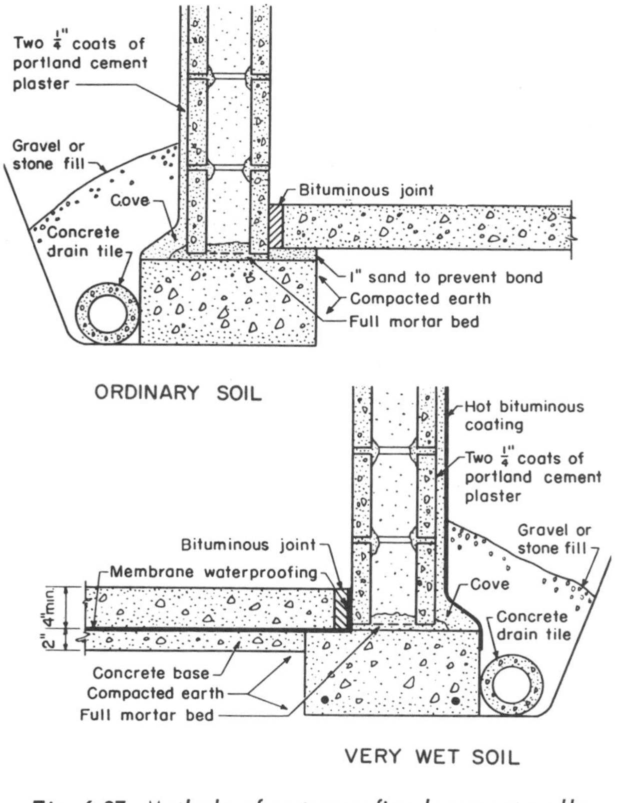

Fig. 6-37. Methods of waterproofing basement walls.

Fig. 6-40. Laying drain tile along the footing. Note the wall has been coated with asphalt. Perforated plastic pipe is also used to make this installation.

A cove of plaster should be formed between the footing and wall to prevent water from collecting and seeping through the joint. The second coat should be kept damp for at least 48 hours.

In poorly drained soils, or when it is important to secure added protection against moisture penetration, the plaster may require coating with asphalt waterproofing or hot bituminous material.

To waterproof poured walls, bituminous waterproofing is generally used without cement plaster.

Fig. 6-40 shows a drain tile being laid along the side of the footing. The drain should lead away from the foundation to an outlet that always remains open. In some localities, a drain may be connected to the sanitary sewer lines. The drain tile is usually placed at a slope of about 1 in. in 20 ft. and spaced 1/4 in. apart with strips of tar paper being used to cover the joint. After the tile is in place, it should be covered with a 6 to 8 in. layer of coarse gravel or crushed stone.

Backfilling

After the foundation has cured sufficiently and waterproofing has been completed, Fig. 6-41, the excavation outside the walls is filled with earth. If tile lines have been installed, care must be taken to protect the lines from movement. When heavy power equipment is used to backfill, the operator must be careful not to damage the walls.

Fig. 6-41. Foundation wall complete and ready for backfilling. Pressed steel areaways have been set in place over the basement windows.

Slab-On-Ground Construction

Today many commercial and residential structures are built without basements and the main floor is formed by placing concrete directly on the ground. Footings and foundations are somewhat similar to those for basements except they need to extend down only to solid soil and below the frost line.

In slab-on-ground construction, insulation and moisture control are essential. Fig. 6-42 shows details. The earth under the floor is called the subgrade and must be firm and completely free of sod, roots and debris.

A fill, usually coarse granular (4 in. minimum), is placed over the finished subgrade. The fill should be brought to the desired grade and thoroughly compacted. This granulary fill may be slag, gravel or crushed stone, preferably ranging from 1/2 in. to 1 in. in size. The material should be of uniform size particles, without fines, to insure a maximum volume of air space in the fill. Air spaces will add to insulating qualities and reduce capillary attraction of subsoil moisture (action by which soil moisture passes through fill.)

In areas where the subsoil is not well drained, a line of drain tile may be required around the outside edge of the exterior wall footings.

While preparing the subgrade and fill, various mechanical installations should be made. Under floor ducts, where used, are usually embedded in the granular fill. Water service supply lines, if placed under the floor slab, should be installed in trenches deep enough to prevent possible damage by freezing. Connections to utilities should be brought to a point above the finished concrete floor level prior to concreting.

After the fill has been compacted and brought to grade, a vapor barrier should be placed over the sub-base to stop the movement of water

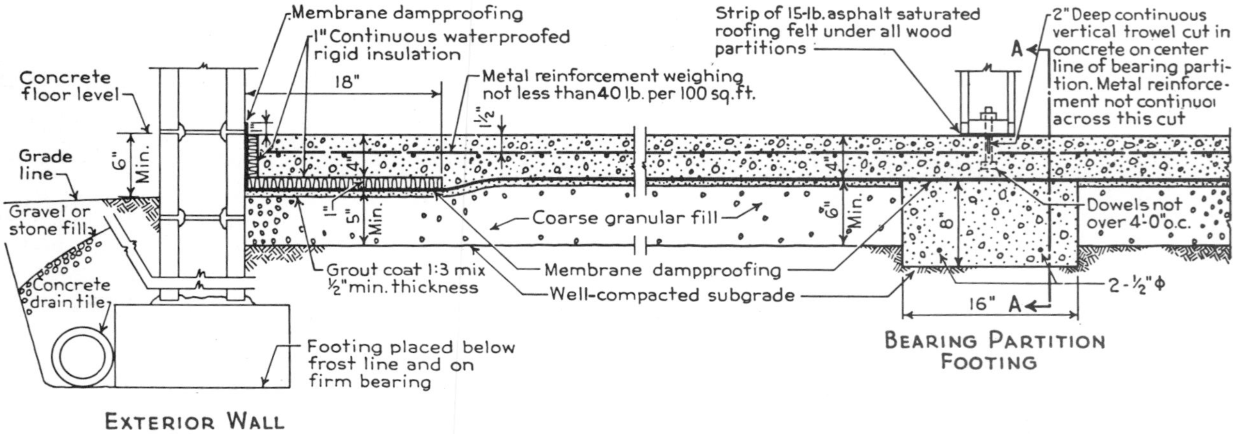

Fig. 6-42. Details for slab-on-ground construction.

into the slab. Among materials widely used as vapor barriers are 55 lb. roll roofing, 4-mil polyethylene film and asphaltic-impregnated kraft papers. Strips should be lapped 6 in. to form a complete seal. A vapor barrier is essential under every section of the floor. Instructions supplied by the manufacturer should be carefully followed.

Perimeter insulation is important since it reduces heat loss from the floor slab to the outside. The insulation material must be rigid and stable in the presence of wet concrete. Fig. 6-43 shows a foundation design and the position of perimeter insulation typical of residential frame construction. Thickness of the insulation varies from 1 to 2 in. depending on outside temperature and type of heating. The insulation can be placed either horizontally as shown, or vertically along the foundation wall. Refer to Unit 13 for additional information.

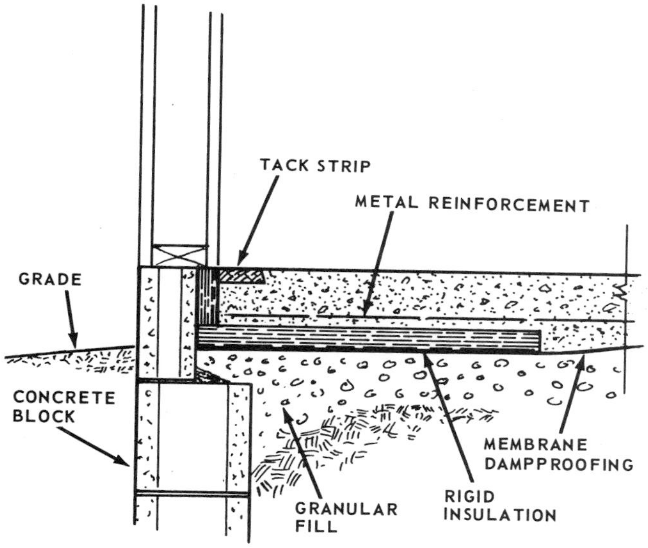

Fig. 6-43. Suggested construction for slab-on-ground when used with wood frame structure.

When the insulation, vapor barrier, and all mechanical aspects are complete, reinforcing mesh is laid in position. Check local building codes for requirements. Usually a 6 x 6 x 10 ga. mesh is sufficient for residential work. This should be located from 1 to 1 1/2 in. below the top surface of the concrete. Fig. 6-44 shows the concrete being placed for a typical slab-on-ground residential floor. For industrial and commercial buildings, the floor is thicker and heavier reinforcement is used, Fig. 6-45.

Terrazzo, ceramic tile, asphalt tile, wood flooring, linoleum and wall-to-wall carpeting are coverings appropriate for use on concrete floors.

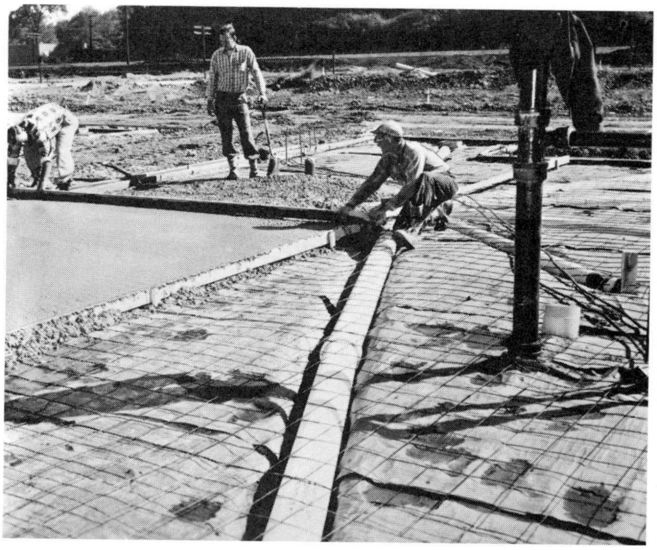

Fig. 6-44. Placing concrete for a residential floor. Note the heating ducts, plumbing and water service lines and vapor barrier. (Portland Cement Assoc.)

When linoleum, asphalt tile or similar resilient-type flooring materials are to be applied, the concrete surface is usually given a smooth steel-troweled finish. Information on finished wood floors is provided in Unit 15.

Basement Floors

Many of the considerations previously listed for slab-on-ground floors also apply to basement floors. Basement floors are poured later in the building sequence; sometime after the framing and roof are complete and after the plumber has in-

Fig. 6-45. Placing concrete for a slab-on-grade floor for an industrial building. The vapor barrier is a heavy polyethylene film that is especially designed for this purpose. (St. Regis Paper Co.)

stalled waste plumbing and water service lines. The concrete may be brought in through basement windows or openings for stairs. Sometimes the rough opening for a fireplace located on an outside wall will provide easy access.

Main walks leading to front entrances should be at least 4 ft. wide while those to secondary entrances may be 3 ft. or slightly less. In most areas, sidewalks are 4 in. thick and the formwork is constructed with 2 x 4 lumber. Walks and drives

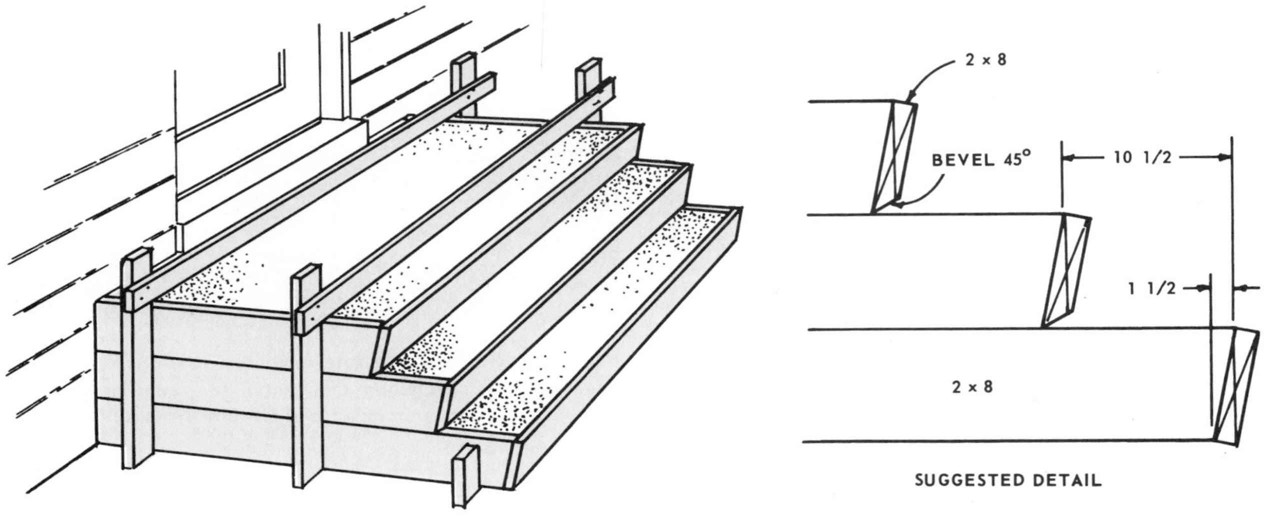

Fig. 6-46. Forms for entrance platform and steps.

Entrance Platforms and Steps

Entrance platform foundations should be constructed as a part of the main foundation, or firmly attached to the main foundation. Reinforcing bars, placed in the wall when it is constructed, can provide the connection.

Steps may be included and poured as an integral part of the platform, Fig. 6-46. When the steps are more than 3 ft. wide, 2 in. stock should be used to prevent the risers from bulging. Detailed information on stair construction is included in Unit 16. In the illustration, 2 x 8 in. riser boards are set at an angle of about 15 deg. to provide a slight overhang (nosing). Also, the bottom edges of the boards are beveled to permit the mason to trowel the entire surface of the tread.

If concrete stairs are to be poured against a wall, or between two existing walls, forms can be constructed in a manner similar to that shown in Fig. 6-47.

Sidewalks and Drives

Sidewalks and drives are usually not laid until the finished grading is complete. If considerable earth has been moved and there is extensive fill, it is best to delay construction until the grade has thoroughly settled.

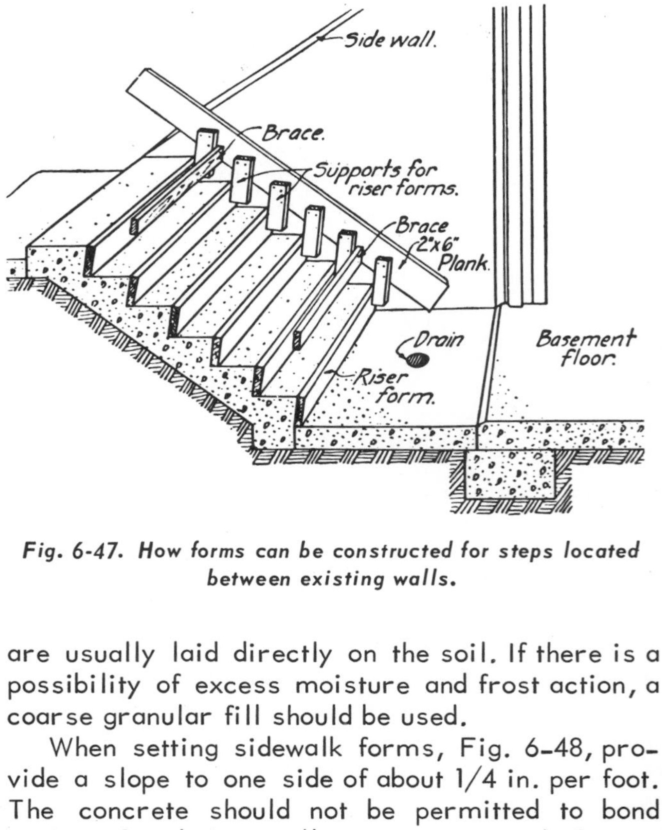

Fig. 6-47. How forms can be constructed for steps located between existing walls.

are usually laid directly on the soil. If there is a possibility of excess moisture and frost action, a coarse granular fill should be used.

When setting sidewalk forms, Fig. 6-48, provide a slope to one side of about 1/4 in. per foot. The concrete should not be permitted to bond against foundation walls or entrance platforms, but should be permitted to "float" on the ground. A 1/2 in. thick strip of asphalt impregnated com-

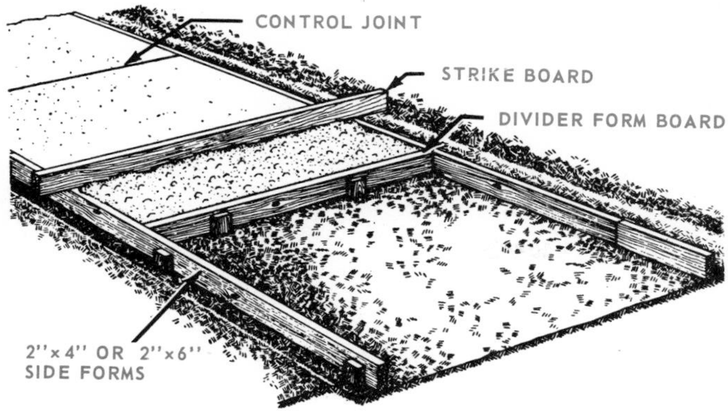

Fig. 6-48. Forms for a sidewalk. They are usually 4 in. thick. When subjected to heavy vehicles, the thickness should be increased to 6 in. or reinforcing should be added. Note the wood strike board that is used to level the concrete between the forms. It is worked back and forth and gradually moved ahead.

Fig. 6-49. Placing concrete near its final position in a concrete driveway. Forms consist of patented steel channels held in place with steel pins.

Fig. 6-50. Close-up view shows polyethylene film underlayment that prevents water in the concrete from soaking into the subgrade. Note the hook (arrow) being used to raise the reinforcing into the concrete.

position board is commonly used to form this separation.

Driveways should be 5 or 6 inches thick and include reinforcing mesh. A single drive should be at least 10 ft. wide and a double drive a minimum of 16 ft. wide. They need to have a minimum cross slope of 1/4 in. per foot.

Concrete should be placed between the forms so it will be close to its final position. Guard against overworking the concrete while it is still plastic since this will tend to bring excess water and fine material to the surface, and cause scaling and dusting after the concrete has cured.

After the concrete is placed and roughly spread between the forms, it should immediately be screeded. This is accomplished by moving a straightedge back and forth in a saw like motion across the top of the forms. A small amount of concrete should always be kept ahead of the straightedge.

When the screeding operation is complete, a float made of wood, aluminum or magnesium is moved over the surface. When skillfully performed, this operation removes high spots, fills depressions, and smooths other irregularities. As the concrete stiffens and the water sheen disappears from the surface, edges can be finished and control joints cut. These joints should extend to a depth of at least one-fifth of the thickness of the concrete. For sidewalks and drives the distance between the control joints is usually about equal to the width of the slab. They can be formed with a groover and straightedge or cut with a power saw equipped with a masonry blade. Sawing is done 18 to 24 hours after the concrete is poured.

Edges of walks and driveways should be rounded by working an edger tool along the side forms. The edging tool can also be used to finish the control joints. Surface finishing operations should be performed after the concrete has hardened enough to become somewhat stiff. For a rough finish that will not become slick during rainy weather, the surface can be stroked with a stiff bristle broom. When a finer texture is desired, the surface should be steel troweled and then lightly brushed with a soft bristle broom. Several key steps in the pouring and finishing of a concrete drive are shown in Figs. 6-49 to 6-53.

Proper curing of the concrete requires that it be protected against moisture lost during the early stages of hardening (hydration). Covers of waterproof paper or polyethylene film are commonly used. A convenient procedure consists of spraying the concrete with a plastic-based curing compound that forms a continuous membrane over the surface.

Fig. 6-51. Using a power screed to strike-off and compact the concrete. The engine is belted to a small off-center flywheel that vibrates the wooden frame.

Fig. 6-52. Using a lightweight aluminum float to smooth the concrete surface.

Wood Foundations

The All-Weather Wood Foundation System is so named because it can be installed during nearly any condition of temperature and precipitation. It provides comfortable living space in basement areas because the stud wall can be fully insulated. All wood parts are pressure treated with a solution of chemicals that make the fibers useless as a food for insects and the fungus growth that causes decay. Foundation sections of 2 in. lumber and exterior plywood can be panelized in fabricating plants or constructed on the building site. Pressure treated wood foundations, Fig. 6-54, have been approved by major code groups and accepted by FHA, HUD, and FmHA (Farmers Home Adm.).

For a regular basement, the site is excavated to the required depth. Plumbing lines are installed and provisions made for foundation drainage according to local requirements. Some soils will require a sump (pit for water collection) which is connected to a storm sewer, pump, or other positive drainage. The subgrade is then covered with a 4 to 6 in. layer of porous gravel or crushed

Fig. 6-53. Using a power troweler to produce a smooth, hard surface. Note the edger tool (arrow) being used along the form.

Fig. 6-54. All-Weather Wood Foundation. Lumber and plywood are pressure treated according to standards developed by the American Wood Preservers Institute. (American Plywood Assoc.)

stone and carefully leveled. Footing plates of 2 x 6 or 2 x 8 material are installed directly on this base and the wall sections erected. Nails and other fasteners should be made of either silicon bronze, copper, or hot-dipped zinc coated steel. Special caulking compounds are used to seal all joints in the plywood sheathing.

Before pouring the basement floor, the porous gravel or crushed stone base is covered with a polyethylene film (6 mil. thick) and a screed board is attached to the foundation wall. See Fig. 6-55.

The first floor frame is installed on the double top plate of the foundation wall with special attention given to methods of attachment so that inward forces will be transferred to the floor structure. Where joists run parallel to the wall, blocking should be installed between the outside joist and

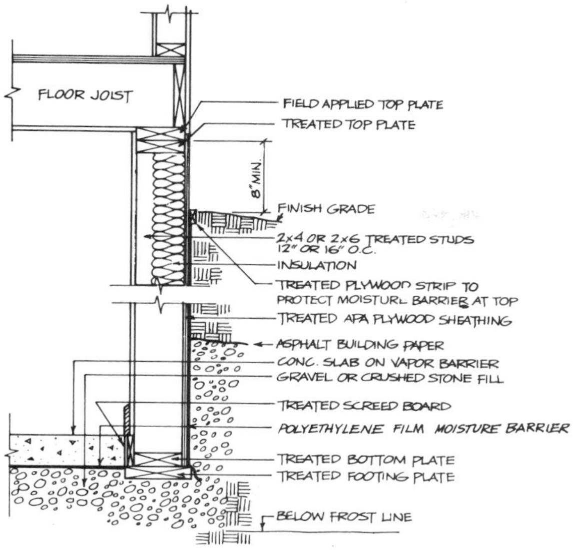

Fig. 6-55. General details of a wood foundation. (American Plywood Assoc.)

the first interior joist.

Immediately before backfilling, the outside surface of the foundation wall below grade is covered with a 6 mil. thick polyethylene film. The top edge is bonded to the plywood with a special adhesive. The adhesive is also used to seal all vertical lapped joints (6 in. min.) in the film. A strip of treated wood or other approved material is installed along the top edge to cover and protect the film.

As with any foundation system, satisfactory performance demands full compliance with recommended standards covering design, fabrication, and installation. Standards for wood foundations are contained in a Manual published by the National Forest Products Association, 1619 Massachusetts Avenue, N.W., Washington, D.C. 20036.

Estimating Materials

The standard unit of measure of concrete is the cubic yard which contains 27 cu. ft. The following formula can be used to determine the cubic feet for any square or rectangular area when all the dimensions are given in feet:

$$\text{cu. yds.} = \frac{\text{width} \times \text{length} \times \text{thickness}}{27}$$

For example: the concrete needed to pour a basement floor 4 in. thick that was 30 ft. x 42 ft. would be figured as follows:

$$\text{cu. yds.} = \frac{30 \times 42 \times 1/3}{27}$$

$$= \frac{30 \times 42 \times 1}{27 \quad 3}$$

$$= \frac{\overset{10}{\cancel{30}} \times \overset{14}{\cancel{42}} \times 1}{\underset{9}{\cancel{27}} \quad \underset{1}{\cancel{3}}} = \frac{140}{9}$$

$$= 15.56 \text{ cu. yds.}$$

For estimating, it is easier to keep all the dimensions in feet or fractions of a foot. For example: a footing with a total perimeter of 162 ft. and a cross section of 8 in. x 16 in. would be figured as follows:

$$\text{cu. yds.} = \frac{162 \times 2/3 \times 1\ 1/3}{27}$$

$$= \frac{162}{27} \times \frac{2}{3} \times \frac{4}{3}$$

$$= \frac{\overset{2}{\cancel{\overset{\cancel{6}}{\cancel{162}}}}}{\underset{1}{\cancel{27}}} \times \frac{2}{\underset{1}{\cancel{3}}} \times \frac{4}{3} = \frac{16}{3}$$

$$= 5\ 1/3 \text{ cu. yds.}$$

The amount of concrete determined by the formula does not allow for waste or slight variations in the cross sections of the form. An additional 5 to 10 percent is usually figured to cover these factors.

The number of concrete masonry units can be estimated by first determining the number of units required in each course and then multiplying by the number of courses between the footing and plate. For example: find the number of 8 x 8 x 16 in. blocks required to construct a foundation wall with a total perimeter of 144 ft. and laid 11 courses high.

$$\text{Total number} = \frac{\text{perimeter}}{\text{unit length}} \times \text{number of courses}$$

$$= \frac{144}{1\ 1/3\ (16'')} \times 11$$

$$= \frac{144}{\frac{4}{3}} \times 11$$

$$= \frac{144 \quad 3}{4} \times 11$$

$$= \frac{\overset{36}{\cancel{144}} \quad 3}{\underset{1}{\cancel{4}}} \times 11$$

Wall thickness	For 100 sq.ft. of wall		For 100 concrete block
in.	Number of block*	Mortar** cu.ft.	Mortar** cu.ft.
8	112.5	2.6	2.3
12	112.5	2.6	2.3

* Based on block having an exposed face of 7⅝ x 15⅝ in. and laid up with ⅜-in. mortar joints.

** With face shell mortar bedding—10 per cent wastage included.

Fig. 6-56. Quantities of concrete blocks and mortar. (Portland Cement Assoc.)

= 108 x 11

= 1188 blocks

Another method is to figure the face area of the wall in square feet and divide by 100. This figure is then multiplied by 112.5 which is the number of 8 x 8 x 16 in. blocks required to construct 100 sq. ft. of wall. This same figure can be multiplied by 2.6 to secure the cu. ft. of mortar required. See the table in Fig. 6-56.

Test Your Knowledge - Unit 6

1. In laying out building lines, a right angle may be established by using the _____ _____ _____ method.
2. Batter boards should be located _____ feet or more away from the building lines.
3. Building sites on steep slopes or rugged terrain should be rough graded before the building is laid out. True or False?
4. In cold climates, foundations should be located below the ____________ line.
5. In residential construction, a safe design is usually obtained by making the width of the footing ________ as wide as the foundation wall.
6. Foundation forms constructed of 1 in. boards should be held in place with stakes placed ____ to ____ ft. apart.
7. Loose dirt and debris (should or should not) be removed from the ground under a footing.
8. Concrete is made by mixing _______, _______, _________ and water in proper proportions.
9. Is it correct to call a concrete floor a cement floor?
10. Concrete hardens by a chemical action called ________________.
11. Each sack of Portland cement holds _____ lbs.
12. In mixing concrete, a 1, 2, 3 mix would specify 1 part _______, 2 parts _______, and 3 parts __________.
13. Ready-mix concrete is purchased by the _____ ___________.
14. When placing concrete in forms, working the concrete next to the forms tends to produce a __________ surface along the form faces.
15. Wood plates are fastened to the top of a foundation wall with ________ ________.
16. A concrete block specified as an 8 x 8 x 16 block is actually _____ x _____ x _____.
17. A concrete basement wall may be waterproofed by using an application of __________ ________ or _________ waterproofing.
18. In slab-on-ground construction, a _____ _____ should be laid over the sub-base to stop the movement of ________ ________ into the concrete slab.
19. In most areas, sidewalks are ___ in. thick.
20. How many 8 x 8 x 16 concrete blocks are required to lay 100 square feet of wall surface?

Outside Assignments

1. Visit a ready-mix concrete plant in your locality and study the operations. Secure information concerning the following: source of aggregates, handling and storing cement and aggregates, equipment used to measure and proportion mixtures, size of truck-mounted mixers, distance trucks can travel without extra charge, cost of a cubic yard of concrete in various psi ratings and fractional parts of a cubic yard that can be ordered. Prepare a written report.
2. Secure a set of house plans that includes a fireplace located on an outside wall. Prepare a scaled (1 1/2" = 1' - 0") drawing of the formwork you would use for the footings under the fireplace wall. Show individual form boards and stakes. Include one or more section views to describe the shape of the footing as well as the form materials.
3. Study reference materials and booklets prepared by such organizations as the Portland Cement Association or the Perlite Institute. Secure information about air entrained concrete, slump tests, lightweight aggregates, ultra-lightweight concrete, thermal conductivity, compression tests, reinforcing, and prestressed concrete units. Prepare an outline and make an oral report to the class.
4. Study the building code in your area and learn about such requirements as: building placement from property lines, design of footings for residential structures, minimum depths for footings and foundations and basic construction of concrete and masonry foundation walls. Summarize your findings in a written report for your class.

Unit 7
FLOOR FRAMING

When the foundation is complete and the concrete has properly set up, work can be started on the floor framing. Before starting with the framing, most carpenters like to have the area outside the walls backfilled and the ground brought to rough-grade level. This makes it easier to deliver and stack lumber in convenient locations on the building site and it provides the carpenter with easier access to the building.

Types of Framing

The type and method of framing used in a structure will be determined to a large extent by the basic design. See Fig. 7-1. In addition to this, methods may vary because of conditions peculiar to a certain locality, materials available, and the personal experience and preference of the builder.

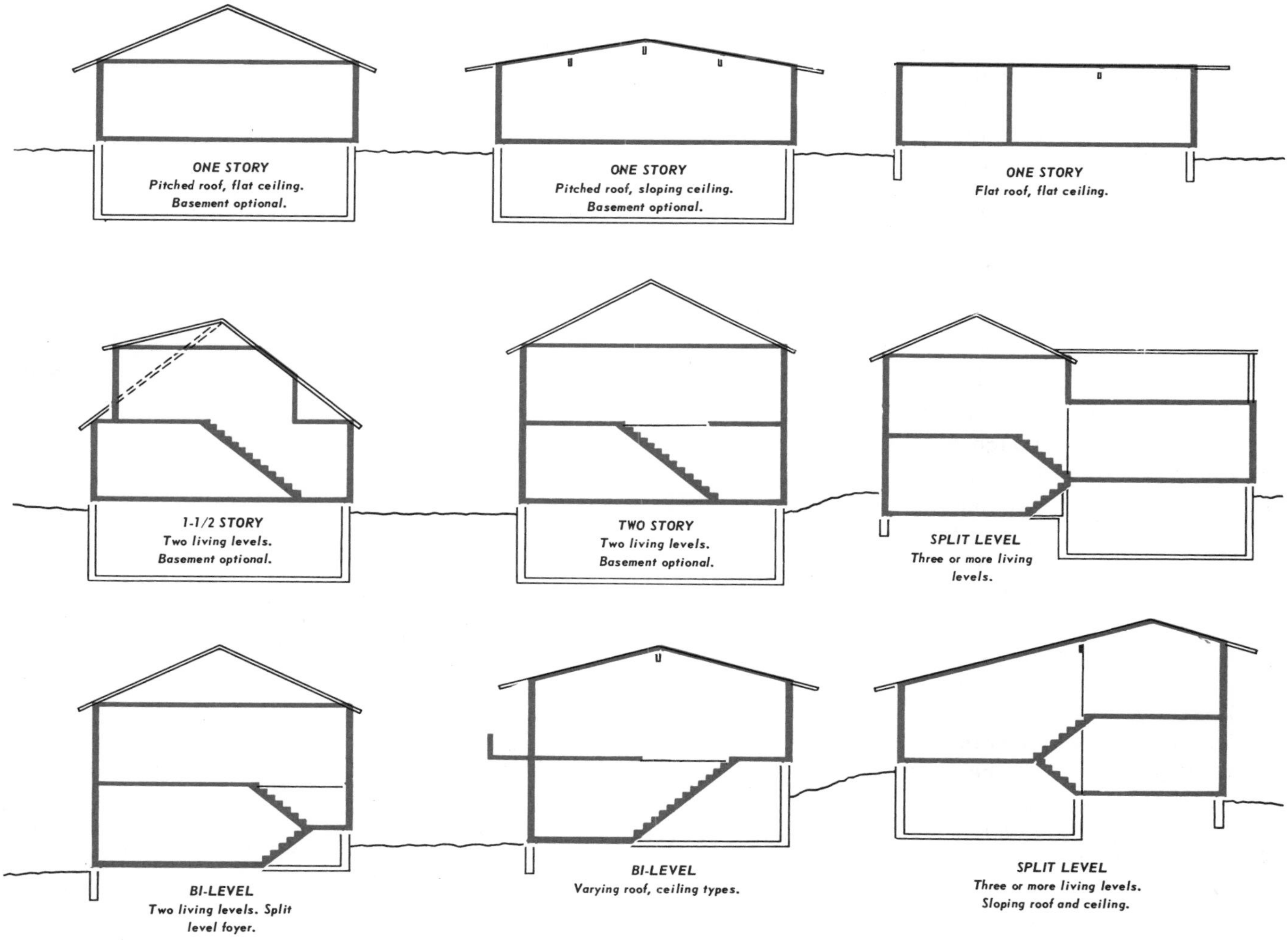

Fig. 7-1. Basic types in modern residential construction. Methods and framing systems depend on the type of space and its orientation, and the kind of materials used.

In some parts of the country, buildings must

be constructed with special resistance to winds and rains; in other parts, earthquakes may be the greatest hazard; while in still other areas, heavy loads of damp snow may require special design. Regardless of the area, all structures should be built to reduce shrinkage and warping to a minimum and be constructed so they will be resistant to the hazard of fire.

Fig. 7-2. Platform type of framing applied to a multi-story apartment building. Note that floor joists on upper levels extend to the outside edge of the double plates with blocking between, see arrow. This method of assembly provides better support for the ends of the joists. (Western Wood Products Assoc.)

Fig. 7-3. Platform framing applied to a modern two-story structure. (National Forest Products Assoc.)

Fig. 7-4. Platform framing details: A–First floor level. B–Second floor level.

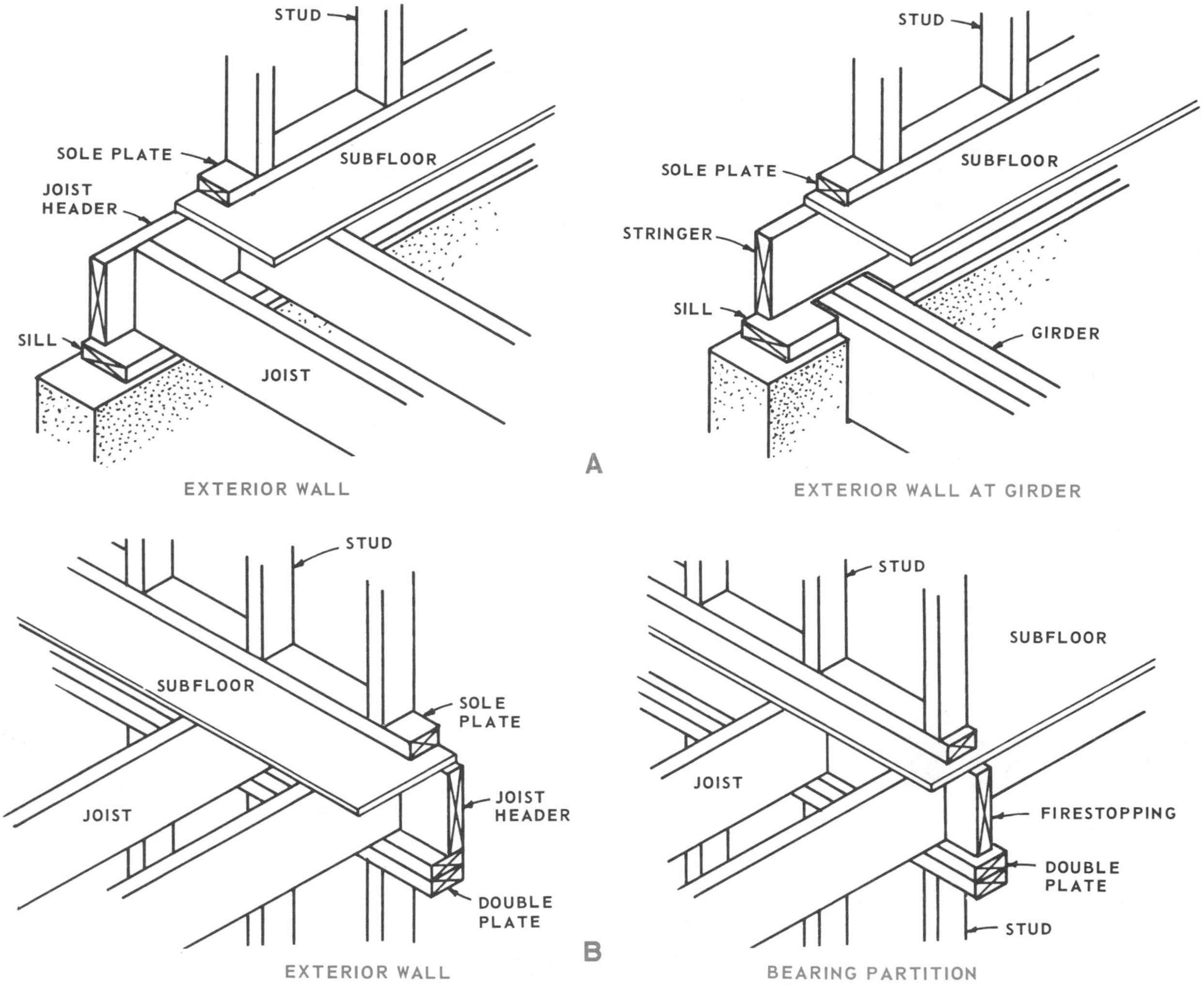

Two basic types of framing are platform framing (also called western framing) and balloon framing. Joists, studs, plates and rafters are the common structural members in both types of framing. Material which has a nominal 2 in. thickness is used. Plank and beam framing (also called post and beam) is considerably different, utilizing heavy structural members 4 in. and more in thickness. This is covered in Unit 20.

Platform Framing

In platform framing, which is used for most modern residential construction, the first floor is built on top of the foundation walls as though it were a "platform." See Fig. 7-2. The floor provides a base upon which the carpenter can assemble wall sections safely and accurately and then raise them into place. The wall sections (one story high) and partitions support a platform for the second floor where the wall sections and partitions are again built and erected. Each floor is framed separately.

Platform framing is satisfactory for both one story and multi-story structures, Fig. 7-3. Fire stops are automatically provided and since the carpenter performs nearly all the work on a solid platform, a high safety factor is an additional feature. Settlement, due to shrinkage, occurs in an even and uniform manner throughout the structure.

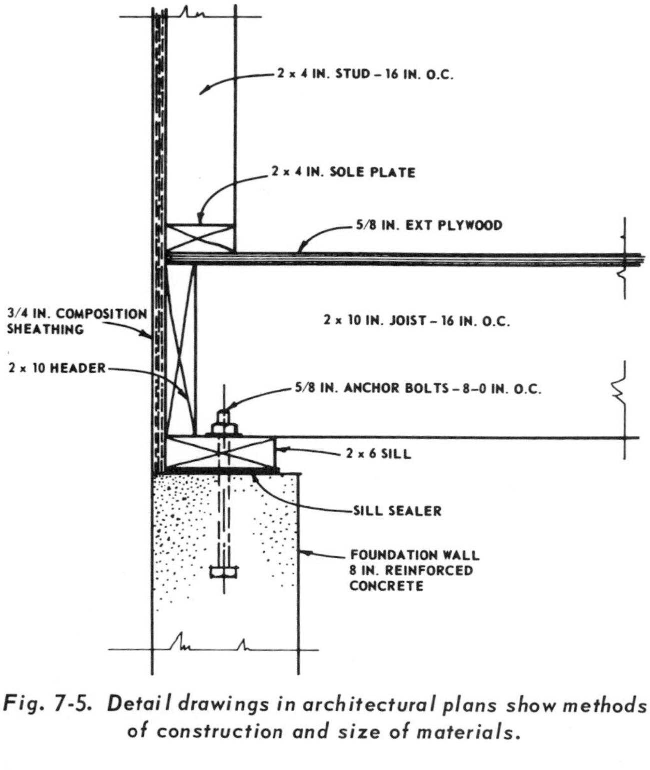

Fig. 7-5. Detail drawings in architectural plans show methods of construction and size of materials.

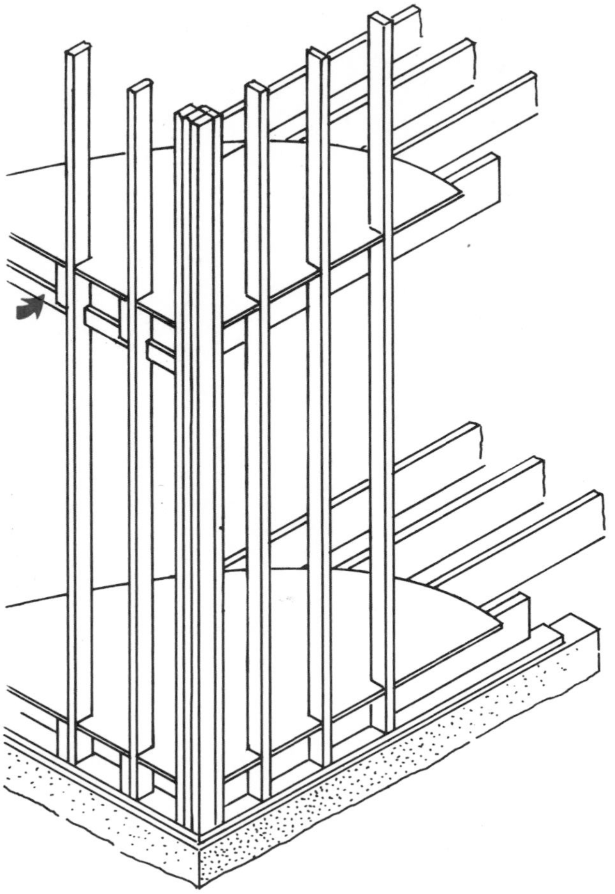

Fig. 7-6. Balloon framing. Second floor joists rest on a ribbon (arrow) set into the studs.

Typical construction methods used at first and second floor levels are illustrated in Fig. 7-4. The only firestopping needed is built into the floor frame at the second floor level where it prevents the spread of fire in a horizontal direction. Here it also serves as solid bridging, holding the joists in a plumb (vertical) position.

The type of framing to be used is usually specified in the architectural plans through various sectional views of floors, walls and ceilings. A typical detail drawing of first floor framing would include not only the type of construction but also the size and spacing of the various members. See Fig. 7-5.

Balloon Framing

The distinguishing feature of the balloon type of framing, Fig. 7-6, (now seldom used) is that the studs are continuous from the foundation to the rafter plate. Ends of the second floor joists are supported on a ribbon and are spiked to the stud. Since the space between the studs is unobstructed between the two levels, firestopping must be included. This space, which also occurs in load-bearing partitions, permits easy installation of service pipes and wiring.

In balloon framing, shrinkage is reduced because the amount of cross-sectional lumber is minimized. Wood shrinks across its width but practically no shrinkage occurs lengthwise. Thus

ilar to those in a chimney flue. In case of fire, flames and heat would move unobstructed through the open spaces spreading the fire rapidly through the structure.

Firestopping, when properly placed, also provides extra bracing and can form edges for attaching flooring and wall surface material.

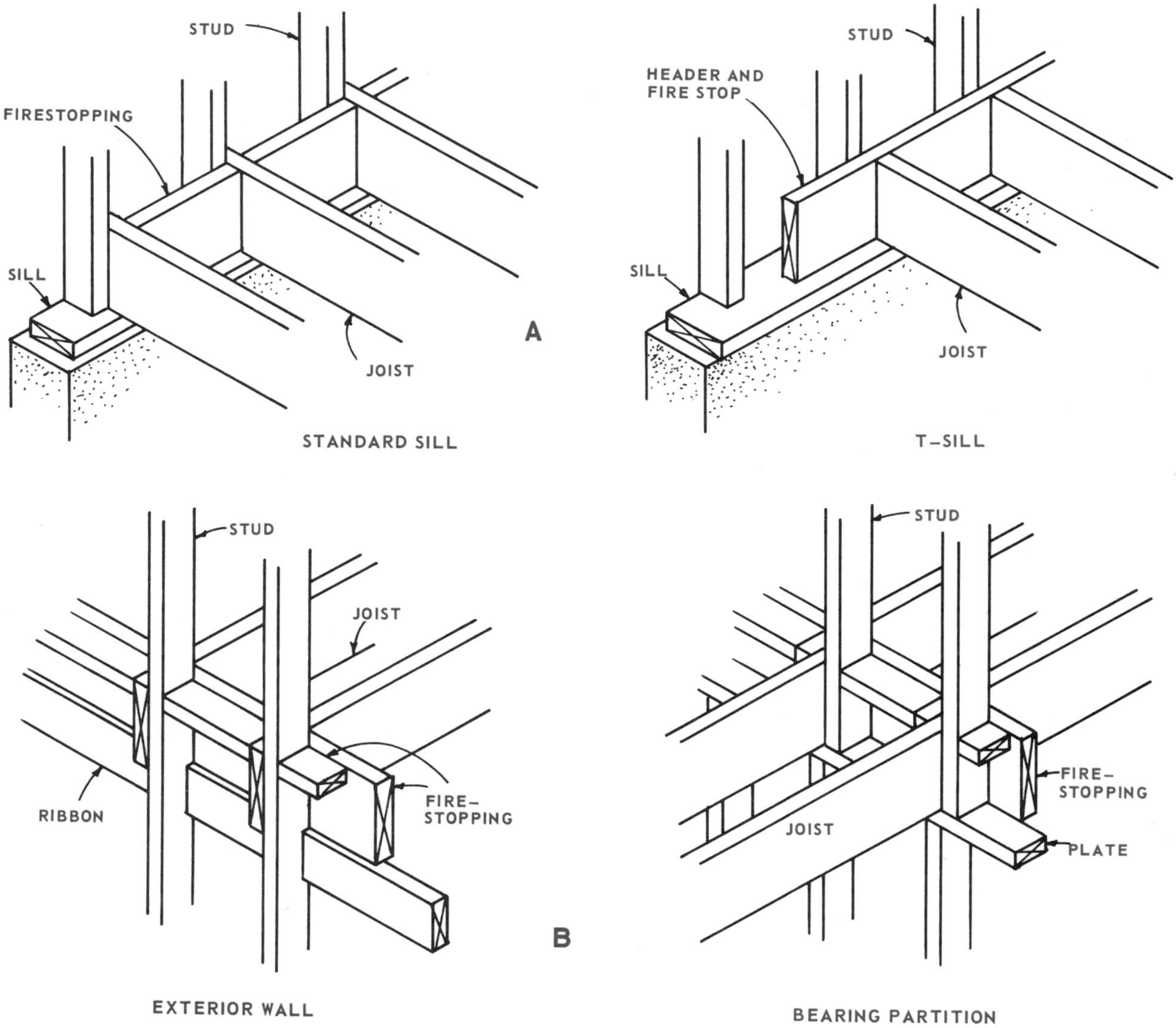

Fig. 7-7. Balloon framing details: A–First floor level. B–Second floor level.

the additional stability (vertical) of the balloon frame makes it adaptable to two-story structures, especially those where masonry veneer or stucco will be used for the exterior wall finish.

Fig. 7-7 shows typical methods of constructing the floor frame at first and second floor levels. Note how the spaces between the studs (vertical framing members) are closed with firestopping. If left open, these spaces will permit drafts sim-

Girders and Beams

Joists are the supporting members of the floor frame and rest on top of the foundation walls. Usually the distance between these walls (called the span) is so great that additional support must be provided. Girders (also called beams), resting on the foundation walls and on posts or columns, provide this additional support. The girders may consist of solid timbers, built-up members, or steel beams. Sometimes a load-bearing partition can replace a girder or beam.

To determine the size of a girder, it is necessary to: 1. Find the distance between girder supports (span). 2. Find the girder load width. A girder must carry the weight of the floors on each side to the mid-point of the joists which rest upon it. 3. Find the "total floor load" per square foot carried by joists and bearing partitions to girder. This will be the sum of loads per square foot listed in the diagram, Fig. 7-8, with the exception of the roof loads which are carried on the outside walls unless braces or partitions are placed under the rafters, in which case a portion of the roof load is carried to the girder by joists and partitions. 4. Find the total load on the girder. This is the product of girder span x girder load width x total floor load. 5. Select proper size of girder from the table, Fig. 7-9, which indicates safe loads on standard size girders for spans from 6 ft. to 10 ft. Shortening the span is usually the most economical way to increase the load that a girder will carry.

Built-up girders can be made of three or four pieces of 2 in. lumber nailed together with 20d nails as shown in Fig. 7-10. Joints should occur only over columns or posts.

In many localities, steel beams are used instead of wood girders. Sizes required depend upon the load to be carried and may be figured the same as specified for wood girders. After the

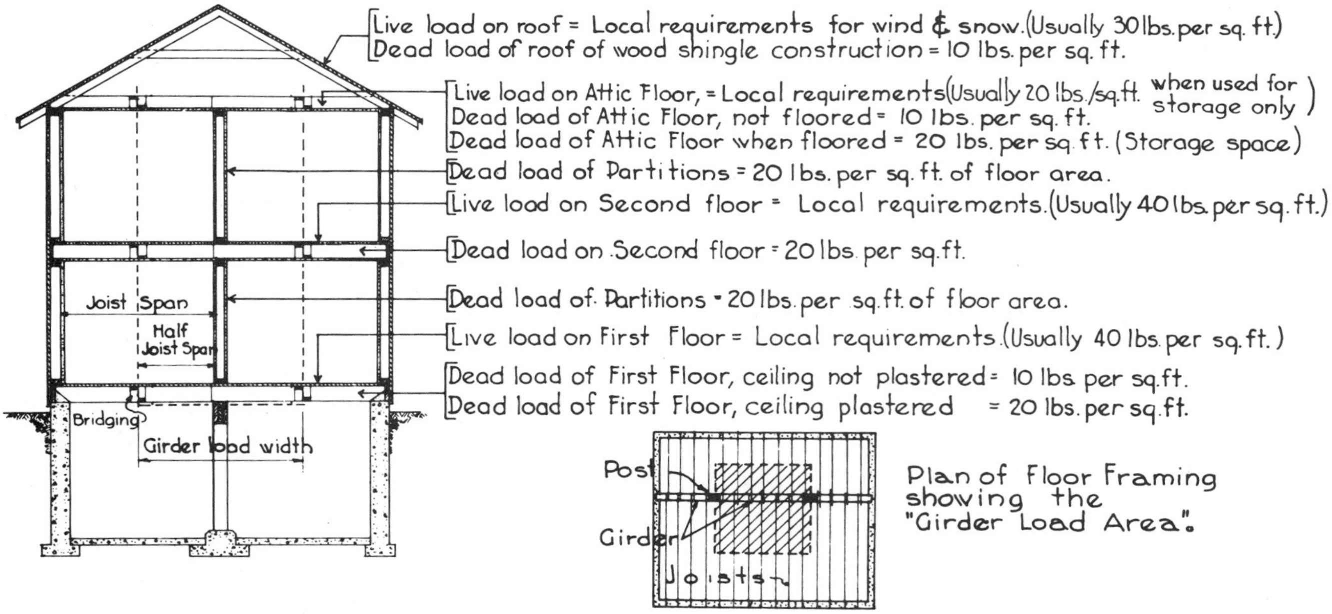

Fig. 7-8. Diagram showing method of figuring loads for residential framing.

GIRDERS	SAFE LOAD IN LBS. FOR SPANS FROM 6 TO 10 FEET				
SIZE	6 FT.	7 FT.	8 FT.	9 FT.	10 FT.
6 x 8 SOLID	8,306	7,118	6,220	5,539	4,583
6 x 8 BUILT-UP	7,359	6,306	5,511	4,908	4,062
6 x 10 SOLID	11,357	10,804	9,980	8,887	7,997
6 x 10 BUILT-UP	10,068	9,576	8,844	7,878	7,086
8 x 8 SOLID	11,326	9,706	8,482	7,553	6,250
8 x 8 BUILT-UP	9,812	8,408	7,348	6,544	5,416
8 x 10 SOLID	15,487	14,732	13,608	12,116	10,902
8 x 10 BUILT-UP	13,424	12,768	11,792	10,504	9,448

Fig. 7-9. Table indicating safe loads on standard size wood girders for spans of 6 – 10 ft.

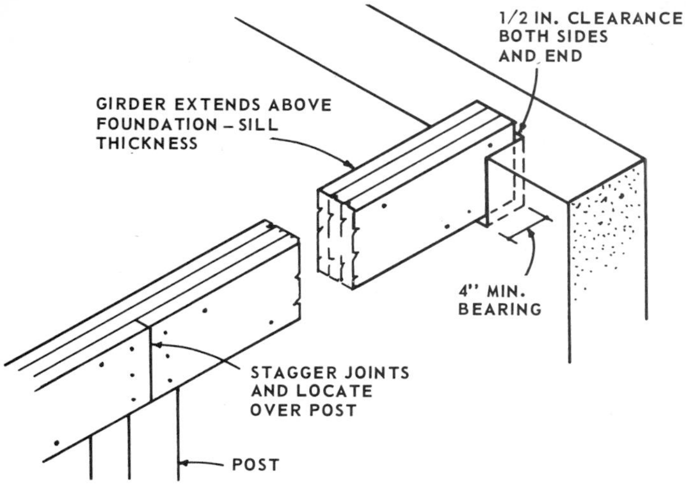

Fig. 7-10. Built–up wood girder. Nail spacing must not be greater than 32 in. along the top and bottom edge. A metal bearing plate should be placed under the girder at the foundation wall.

approximate weight has been determined, the correct beam size can be selected from the table, Fig. 7-11.

Wood beams vary in width, depth, species and grade. Steel beams vary in weight, depth and thickness of web and flanges. While all weights shown in the table are manufactured, frequently only the lighter weights are carried in stock at local sources of supply. The lightest weight approved for a given depth is usually selected for residential construction.

Posts and Columns

For ordinary wood posts (not longer than 9 ft. or smaller than 6 in. x 6 in.), it will be safe to assume that a post whose greatest dimension is

[Standard I-beams. Allowable fiber stress 18,000 pounds]

Span in feet	4 inches deep by—				5 inches deep by—			6 inches deep by—			7 inches deep by—		
	7.7	8.5	9.5	10.5	10.0	12.25	14.75	12.5	14.75	17.25	15.3	17.5	20.0
4	9.0	9.5	10.1	10.7	14.5	16.2	18.0	21.8	23.8	26.0	31.0	33.4	36.0
5	7.2	7.6	8.0	8.5	11.6	13.0	14.4	17.4	19.0	20.8	24.8	26.7	28.7
6	6.0	6.3	6.7	7.1	9.7	10.8	12.0	14.5	15.9	17.3	20.7	22.2	24.0
7	5.1	5.4	5.7	6.1	8.3	9.3	10.3	12.5	13.6	14.9	17.7	19.1	20.5
8	4.5	4.7	5.0	5.3	7.3	8.1	9.0	10.9	11.9	13.0	15.5	16.7	18.0
9	4.0	4.2	4.5	4.7	6.5	7.2	8.0	9.7	10.6	11.6	13.8	14.8	16.0
10	3.6	3.8	4.0	4.3	5.8	6.5	7.2	8.7	9.5	10.4	12.4	13.3	14.4
11	—	—	—	—	5.3	5.9	6.5	7.9	8.7	9.5	11.3	12.1	13.1
12	—	—	—	—	—	—	—	7.3	7.9	8.7	10.3	11.1	12.0
13	—	—	—	—	—	—	—	6.7	7.3	8.0	9.5	10.3	11.1
14	—	—	—	—	—	—	—	6.2	6.8	7.4	8.9	9.5	10.3
15	—	—	—	—	—	—	—	—	—	—	8.3	8.9	9.6
16	—	—	—	—	—	—	—	—	—	—	7.7	8.3	9.0
17	—	—	—	—	—	—	—	—	—	—	—	—	—
18	—	—	—	—	—	—	—	—	—	—	—	—	—
19	—	—	—	—	—	—	—	—	—	—	—	—	—
20	—	—	—	—	—	—	—	—	—	—	—	—	—

{Beam Weight lb/ft
{Safe Load = 36,000 lb

Span in feet	8 inches deep by—				9 inches deep by—				10 inches deep by—			
	18.4	20.5	23.0	25.5	21.8	25.0	30.0	35.0	25.4	30.0	35.0	40.0
4	42.7	45.2	48.2	51.1	56.6	60.9	67.6	74.2	73.3	80.1	87.5	94.8
5	34.1	36.1	38.5	40.9	45.3	48.7	54.1	59.4	58.6	64.1	70.0	75.8
6	28.5	30.1	32.1	34.1	37.7	40.6	45.1	49.5	48.8	53.4	58.3	63.2
7	24.4	25.8	27.5	29.2	32.3	34.8	38.6	42.4	41.9	45.8	50.0	59.2
8	21.3	22.6	24.1	25.5	28.3	30.5	33.8	37.1	36.6	40.1	43.7	47.4
9	19.0	20.1	21.4	22.7	25.2	27.1	30.0	33.0	32.6	35.6	38.9	42.1
10	17.1	18.1	19.3	20.4	22.6	24.4	27.0	29.7	29.3	32.0	35.0	37.9
11	15.5	16.4	17.5	18.6	20.6	22.2	24.6	27.0	26.6	29.1	31.8	34.5
12	14.2	15.1	16.1	17.0	18.9	20.3	22.5	24.7	24.4	26.7	29.2	31.6
13	13.1	13.9	14.8	15.7	17.4	18.7	20.8	22.8	22.5	24.6	26.9	29.2
14	12.2	12.9	13.8	14.6	16.2	17.4	19.3	21.2	20.9	22.9	25.0	27.1
15	11.4	12.0	12.8	13.6	15.1	16.2	18.0	19.8	19.5	21.4	23.3	25.3
16	10.7	11.3	12.0	12.8	14.2	15.2	16.9	18.6	18.3	20.0	21.9	23.7
17	10.0	10.6	11.3	12.0	13.3	14.3	15.9	17.3	17.2	18.8	20.6	22.3
18	9.5	10.0	10.7	11.4	12.6	13.3	15.0	16.5	16.3	17.8	19.4	21.1
19	9.0	9.5	10.1	10.8	11.9	12.8	14.2	15.6	15.4	16.9	18.4	20.0
20	8.5	9.0	9.6	10.2	11.3	12.2	13.5	14.8	14.7	16.0	17.5	19.0

NOTE.—If the reading in the table above is 12.4, the safe load is 12,400 pounds.

Example: If the total load on the girder is 13,500 pounds uniformly distributed and the span or distance between basement piers is 9 feet, then a 7 inch by 15.3 pound I-beam is the proper size to use, for by the table it will carry 13,800 pounds (13.8 in the table). This figure will be found on the line of 9-foot span and in the 15.3-pound column of the 7-inch beam.

These figures are taken from A. I. S. C. Manual.

Fig. 7-11. Steel I–beams. Safe loads in thousands of pounds, uniformly distributed along the beam. Beam weight is based on one linear foot.

equal to the width of the girder it supports will carry the girder load. For example, a 6 x 6 in. post would be suitable for a girder 6 in. wide; for a girder 8 in. wide, a 6 in. x 8 in. or 8 in. x 8 in. post should be used.

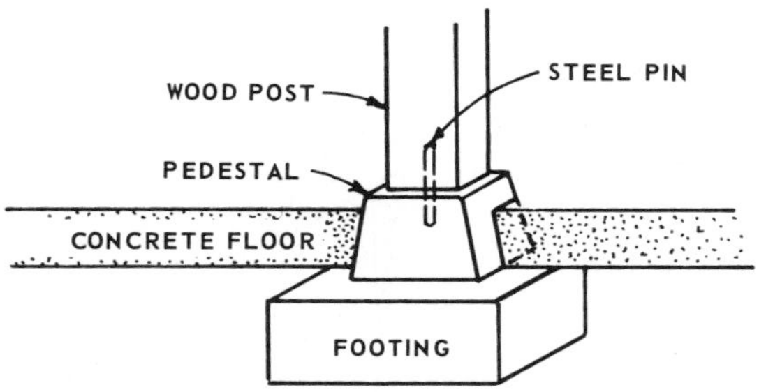

Fig. 7-12. Footing design for wood post.

Adequate footings must be provided for girder posts and columns. Wood posts should be supported on footings which extend above the floor level, as shown in Fig. 7-12. To make sure the posts will not slide off their footing, pieces of 1/2 in. diameter reinforcing rod or iron bolts of that size should be inserted in the footing before the concrete sets and should be allowed to project about 3 in. into holes bored in the bottom of the posts.

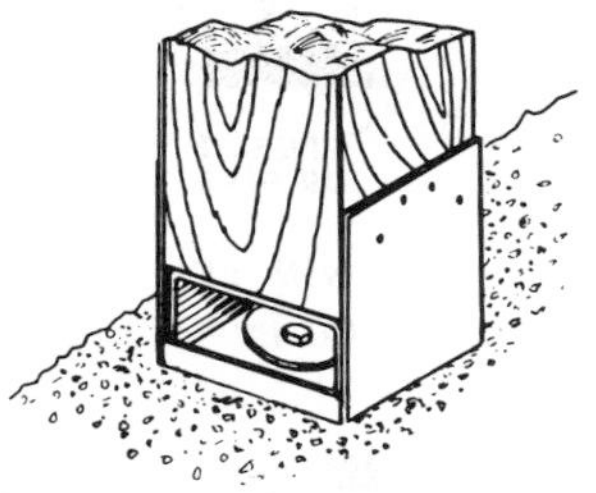

Fig. 7-13. Manufactured post anchor that permits lateral adjustment. (Timber Engineering Co.)

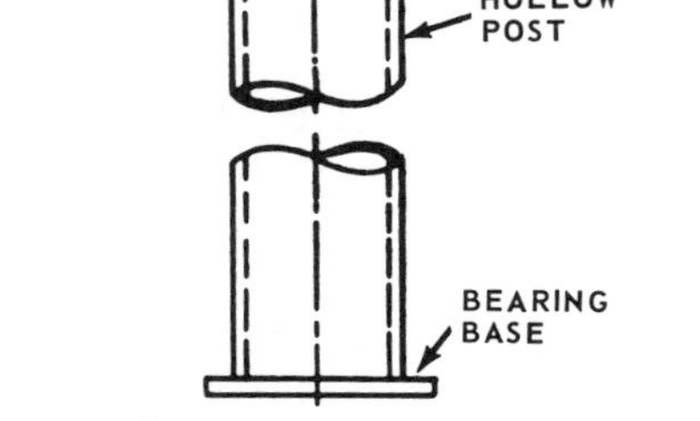

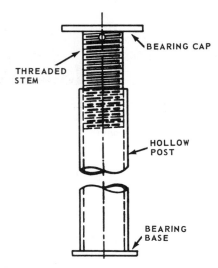

Fig. 7-14. Steel post with threaded section permits easy installation and adjustments.

A post anchor, Fig. 7-13, is available that holds the wood post securely in place. It spaces the bottom of the post above the floor thus protecting the wood from dampness. The bracket can be adjusted for position if the anchor bolt was improperly placed.

When a wood column supports a steel girder, fitting the end of the column with a metal cap is desirable. If wood supports wood, a metal cap should be provided to give an even bearing surface and prevent end grain of the post from crushing the horizontal grain of the wood girder.

HANDY SAYS:

"Be sure the tops of posts and columns and also the seats in foundation walls are flat so the girder or beam is well supported with its sides plumb."

A built-up wood post may be made by spiking together three 2 x 6's. The pieces used should be free from defects and securely nailed together as excessive loading may cause the members to buckle away from each other and fail individually instead of acting together.

Today, steel posts are used extensively for girder and beam support. The post should be capped with a steel plate to provide a suitable bearing area. A steel post designed especially for this purpose is shown in Fig. 7-14. This has a threaded area inside the top end. A heavy stem threads into the top so the post becomes adjustable in length. This makes it possible to adjust the post to the exact length required at the time of installation. Later, as wooden structural members shrink, the post can be lengthened to provide needed adjustment.

Framing Over Girders and Beams

A common method of framing joists over girders and beams is shown in Fig. 7-15. The steel beam is placed level with the foundation wall and the 2 in. wood pad then carries the joists level with the sill. When a wooden girder is used, it is usually set so the top is level with the sill.

If additional headroom (space between floor and girder) is required, the joists can be notched and carried on a ledger, Fig. 7-16. When it is necessary for the lower side of the girder to be

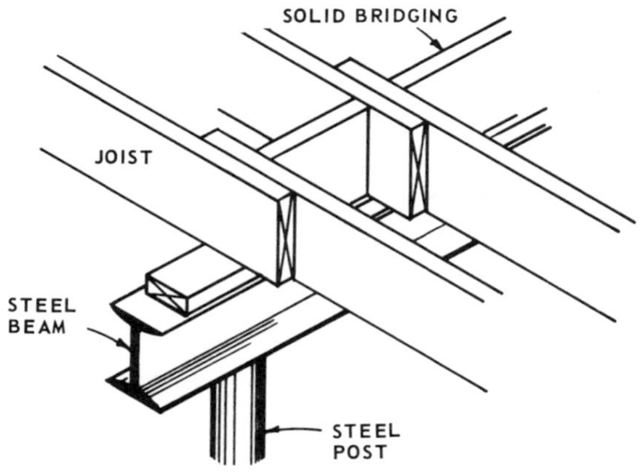

Fig. 7-15. Joists supported on a steel beam. The top of the steel beam is set flush with the top of the foundation wall.

flush with the joists to provide an unbroken ceiling surface, the joists should be supported with hangers or stirrups, Fig. 7-17.

Framing joists to steel beams at various levels can be accomplished with special hangers in somewhat the same manner as suggested for wood girders. Special consideration must be given to the fact that the joists will likely shrink while the steel beam will remain the same size. For average work with a 2 x 10 in. joist, an allowance of 3/8 in. above the top flange of the steel girder or beam is usually sufficient. A suggested method of attaching joists is shown in Fig. 7-18. Notching the joists so they rest on the lower flange of the beam is not recommended because the sloping surface does not provide sufficient bearing. Another solution to the problem would be to use an H-beam in place of an I-beam.

Sill Construction

After girders and beams are set in place, the next step is to attach the sill to the foundation wall. This is the part of the side walls or floor frame that rests horizontally on the foundation. It is also called the sill plate or the mudsill. The latter term originates from the procedure of correcting irregularities in the masonry work by embedding the sill in a layer of fresh mortar or grout.

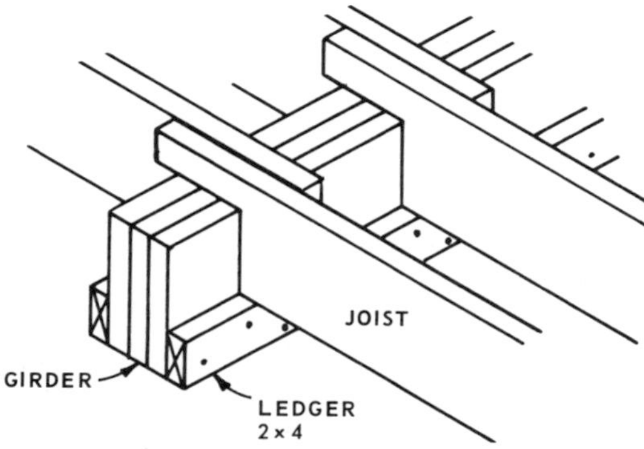

Fig. 7-16. Wood girder raised to provide additional headroom. Joists should bear on the ledger strip and not on the top of the girder. Use 16d nails to attach the ledger to the girder.

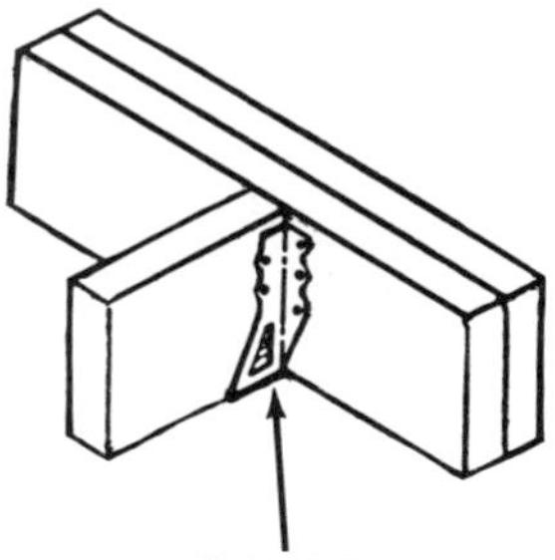

Fig. 7-17. Joist and beam hanger. Available in a wide range of sizes. Special nails are furnished. (Timber Engineering Co.)

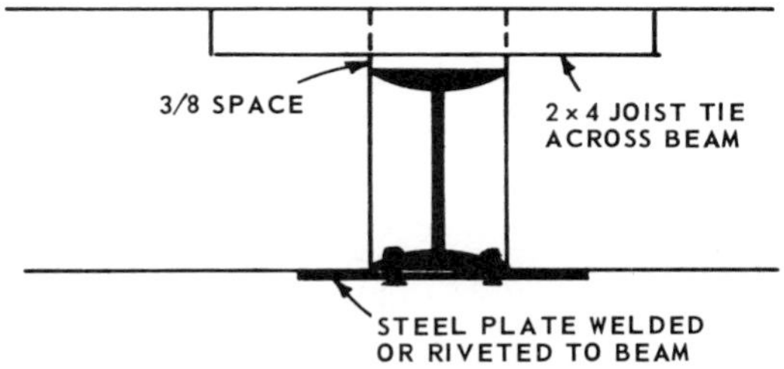

Fig. 7-18. Supporting joists with steel I-beam. The 3/8 in. space provides for joist shrinkage.

Fig. 7-19. Sill Sealer. Available in 50 ft. rolls. Nominal thickness is 1 in. but compresses to as little as 1/32 in. when installed. Seals against air, dirt and insects. (Owens-Corning Fiberglas Corp.)

Sills usually consist of 2 x 6 in. lumber; however the width may vary depending on the type of construction. The sills are attached to the foundation wall with anchor bolts, the size and spacing of which is specified in local building codes. Fig. 7-19 shows a resilient waterproof material called sill sealer which may be used under the sill before it is bolted in place.

Termite Shields

If termites are prevalent in your locality, special shields should be provided to prevent destruction from this insect. Termites live underground and come to the surface to feed on wood. They may burrow through inferior masonry or build earthen tubes on the sides of masonry walls to reach the wood structure.

The wood sill should be located at least 8 in. above the ground and be protected by a metal shield (not less than 26 gauge) that extends out over the foundation wall as shown in Fig. 7-20. In areas where termite damage is great, additional precautions should be observed. Sometimes it is necessary to use lumber for lower framing members that has been treated with chemicals, or to poison the soil around the foundation and under the structure.

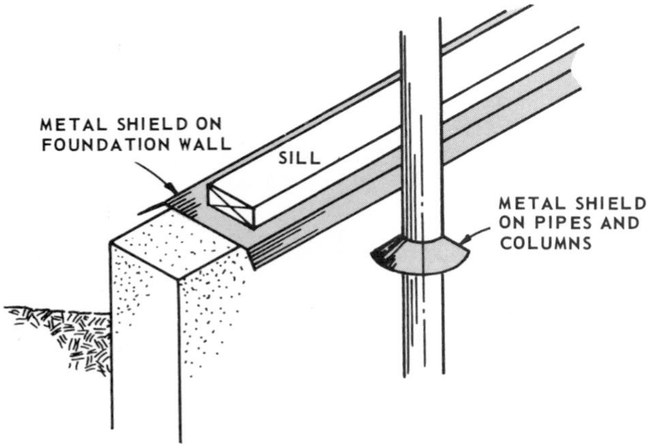

Fig. 7-20. Termite Shields. Use galvanized iron or other suitable metal, 26 gauge or heavier.

Installing Sills

Remove the washers and nuts from the anchor bolts and lay the sill along the foundation wall, Fig. 7-21. The end of the sill is set back from the outside of the foundation a distance equal to the sheathing thickness. Now draw lines across the sill on each side of the bolts as shown. Measure the distance from the center of the bolt to the outside of the foundation and subtract the thickness of the sheathing. Use this distance to locate the position of the bolt holes from the edge of the sill. You will probably need to make separate measurements for each anchor bolt.

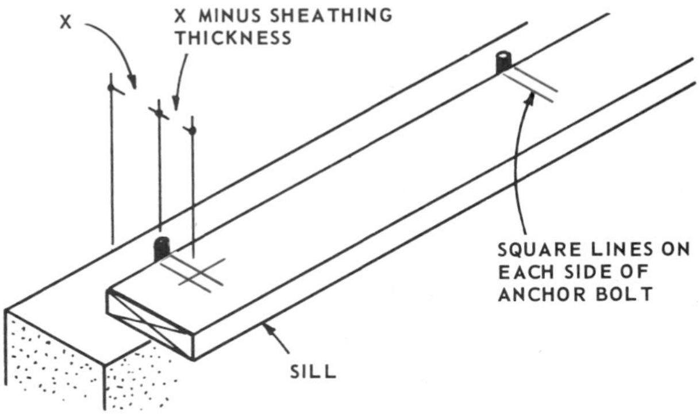

Fig. 7-21. Laying out anchor bolt holes.

After all the holes are located, place the sill on sawhorses and bore the holes. Most carpenters prefer to bore the hole about 1/4 in. larger than the diameter of the bolts to allow for slight inaccuracies in the layout and to permit adjustments when the sill is anchored in place. As each section is laid out and holes bored, place it in position over the bolts.

SPAN OF JOISTS — Span Calculations provide for carrying the live loads shown and the additional weight of the joists and double flooring.

Size	Spacing	20# L.L.	30# Live Load		40# Live Load		50# Live Load		60# Live Load	
		Plaster Clg.	Plaster Clg.	No Plaster	Plaster Clg.	No Plaster	Plaster Clg.	No Plaster	Plaster Clg.	No Plaster
2×4	12"	8'-8"								
	16"	7'-11"								
	24"	6'-11"								
2×6	12"	13'-3"	11'-6"	14'-10"	10'-8"	13'-2"	10'-0"	12'-0"	9'-6"	11'-1"
	16"	12'-1"	10'-6"	12'-11"	9'-8"	11'-6"	9'-1"	10'-5"	8'-7"	9'-8"
	24"	10'-8"	9'-3"	10'-8"	8'-6"	9'-6"	8'-0"	8'-7"	7'-7"	7'-10"
2×8	12"	17'-6"	15'-3"	19'-7"	14'-1"	17'-5"	13'-3"	15'-10"	12'-7"	14'-8"
	16"	16'-0"	13'-11"	17'-1"	12'-11"	15'-3"	12'-1"	13'-10"	11'-5"	12'-9"
	24"	14'-2"	12'-3"	14'-2"	11'-4"	12'-6"	10'-7"	11'-4"	10'-1"	10'-6"
2×10	12"	21'-11"	19'-2"	24'-6"	17'-9"	21'-10"	16'-8"	19'-11"	15'-10"	18'-5"
	16"	20'-2"	17'-6"	21'-6"	16'-3"	19'-2"	15'-3"	17'-5"	14'-6"	16'-1"
	24"	17'-10"	15'-6"	17'-10"	14'-3"	15'-10"	13'-5"	14'-4"	12'-8"	13'-3"
2×12	12"	26'-3"	23'-0"	29'-4"	21'-4"	26'-3"	20'-1"	24'-0"	19'-1"	22'-2"
	16"	24'-3"	21'-1"	25'-10"	19'-7"	23'-0"	18'-5"	21'-0"	17'-5"	19'-5"
	24"	21'-6"	18'-8"	21'-5"	17'-3"	19'-1"	16'-2"	17'-4"	15'-4"	16'-0"
3×8	12"	20'-0"	17'-7"	24'-3"	16'-4"	21'-8"	15'-4"	19'-10"	14'-7"	18'-4"
	16"	18'-6"	16'-1"	21'-4"	14'-11"	19'-1"	14'-1"	17'-4"	13'-4"	16'-0"
	24"	16'-5"	14'-3"	17'-9"	13'-2"	15'-9"	12'-4"	14'-4"	11'-9"	13'-3"
3×10	12"	25'-0"	22'-0"	30'-2"	20'-6"	27'-1"	19'-3"	24'-10"	18'-4"	23'-0"
	16"	23'-2"	20'-3"	26'-8"	18'-10"	23'-10"	17'-8"	21'-9"	16'-10"	20'-2"
	24"	20'-7"	17'-11"	22'-3"	16'-7"	19'-10"	15'-7"	18'-1"	14'-10"	16'-8"

Fig. 7-22. Table showing maximum safe spans for highest quality wood joists. Values will vary for different species of lumber. Always check local building codes for exact data.

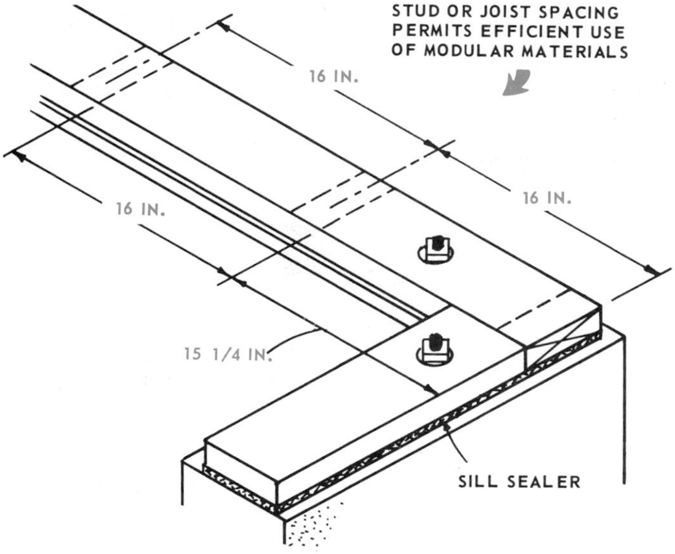

Fig. 7-23. Sill attached to the foundation and spaces laid out for studs or joists.

When all sill sections are fitted, remove them from the anchor bolts, install the sill sealer, and then replace them. As nuts (be sure to include washers) are tightened, see to it that the sills are properly aligned and are in the correct position from the edge of the foundation wall. The sill must be level and straight. Low spots in the foundation can be shimmed with wooden wedges; however it is better to use grout or mortar.

Joists

Floor joists are the framing members that carry the weight from the various areas of the floor to the sills and girders. In residential construction, they generally consist of 2 in. lumber placed on edge. In heavier construction, steel bar joists and reinforced concrete joists are used.

The most common spacing of wooden joists is 16 in. O.C. (on center) but 12, 20 and 24 in. centers are also used under certain conditions. The table in Fig. 7-22 lists the safe spans for joists under various live loads with plastered and non-plastered ceilings below. It is necessary to know the span, the live load (usually 40 pounds per square foot) and whether or not the ceiling below is plastered to select the proper size joists for ordinary load conditions. Joists must not only be strong enough to carry the load that rests on them, but they must also be stiff enough to prevent undue bending or vibration. Building codes usually specify that the deflection (bending downward at the center) must not exceed 1/360th of the span with a normal live load. This would equal 1/2" for a 15' - 0" span.

Laying Out Joists

Study the architectural plans carefully and note the direction the joists are to run. Also become familiar with the various openings that show the location of posts, columns and supporting partitions. These may also show the center lines of girders.

The position of the floor joists can be laid out directly on the sill as illustrated in Fig. 7-23. The position of an intersecting framing member may be laid out by marking a single line and then placing an X to indicate the position of the part, Fig. 7-24.

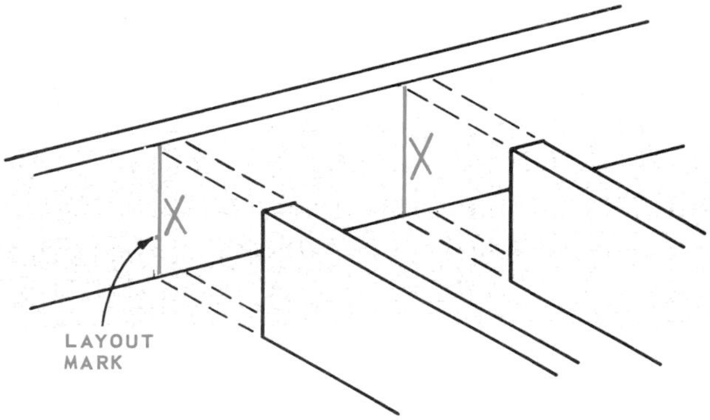

Fig. 7-24. How to lay out the position of framing members. The line is marked with a square.

On platform construction, the joist spacing is usually laid out on the joist header rather than the sill. Instead of measuring each individual space around the perimeter of the building, efficiency and accuracy can be gained by making a master layout on a strip of wood (called a rod) and then transferring the layout to the headers or sill. The strip is then used to make the layout along girders and the wall on the opposite side. When the joists

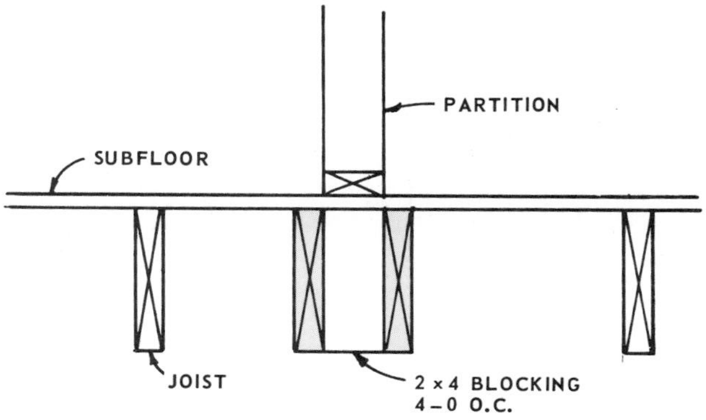

Fig. 7-25. Double joists spaced apart under a partition to permit installation of heating or plumbing.

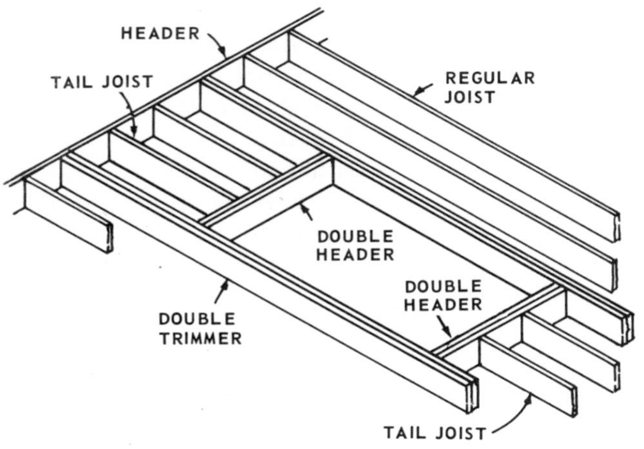

Fig. 7-26. Framing members around a floor opening.

are lapped at the girder, the X (position of the joists) is marked on the other side of the layout line for the opposite wall.

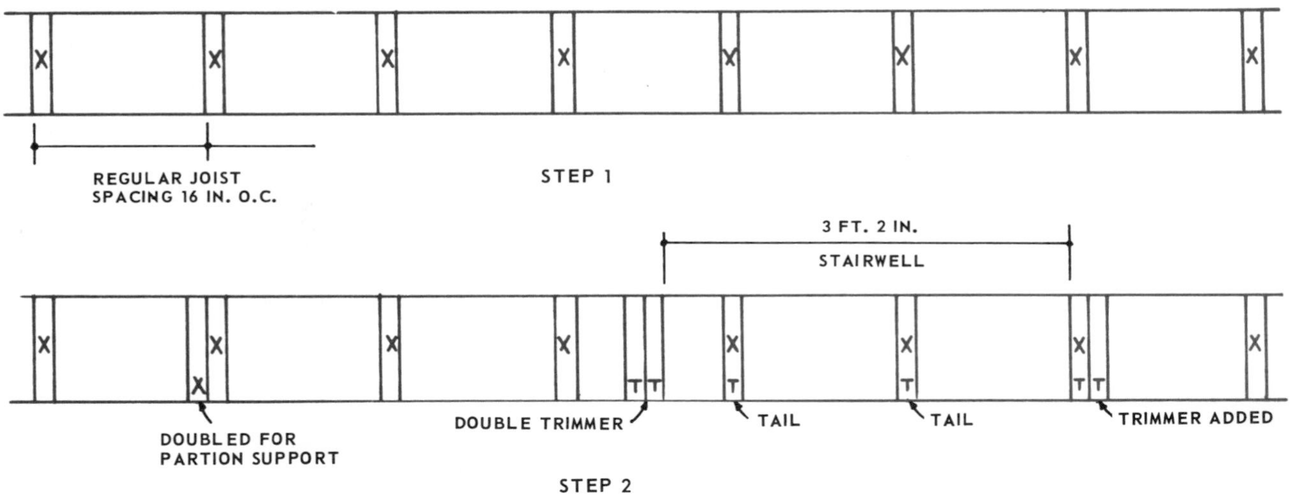

Fig. 7-27. Header joist layout. Use a rod to lay out the regular joist spacing, and then add the position of double and trimmer joists.

Joists are doubled where extra loads must be supported. When a partition runs parallel to the joists, a double joist is placed underneath. Partitions which are to carry plumbing or heating pipes are usually spaced apart to permit easy access, Fig. 7-25. Joists must also be doubled around openings in the floor frame for stairways, chimneys, and fireplaces. These joists are called trimmers and support the headers which carry the tail joists. See Fig. 7-26. The carpenter must become thoroughly familiar with the plans at each floor level so adequate support can be provided.

Select straight lumber for the header joist and lay out the standard spacing along its entire length, Fig. 7-27. Add the position for any doubled joists under partitions and also trimmer joists that will be required along openings. Where regular joists will become tail joists, change the X mark to a T as shown.

Installing Joists

After the header joists are laid out, place them in position and toenail them to the sill. Place all full length joists in position, Fig. 7-28, with the crown (slight warpage called crook) turned up. Hold the end tightly against the header and along the layout line so the sides will be plumb. Attach the joists to the header using a nailing pattern consisting of three 16d nails.

Fig. 7-28. Full length joists in position.

HANDY SAYS:

"The Uniform Building Code states that in standard framing, nails should not be spaced closer together than one-half their length; nor closer to the edge of a framing member than one-fourth their length."

Repeat the operation for the joist along the opposite wall. If the joists butt at the girder (join end to end without overlapping), they should be joined with an overlay board or metal fastener. If they lap, they can be nailed together using 10d nails. Also use 10d nails to toenail the joist in the proper position along the girder.

Nail doubled joists together using 12 or 16d nails spaced about 1 ft. along the top and bottom edge. First, drive several nails straight through to pull the two surfaces tightly together and clinch the protruding ends. Finish the nailing pattern by driving the nails at a slight angle, Fig. 7-29, which will prevent them from going all the way through and also increase their holding power. Fig. 7-30 shows a pneumatic powered nailer being used to fasten joists together.

Framing Openings

Place boards or sheets of plywood across the joists to provide a temporary working deck to install header and tail joists. First set the trimmer

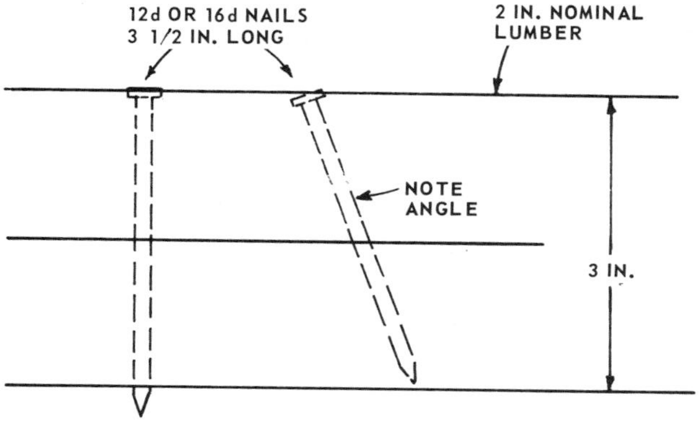

Fig. 7-29. Drive the 12d or 16d nail at a slight angle when nailing double joists and headers together.

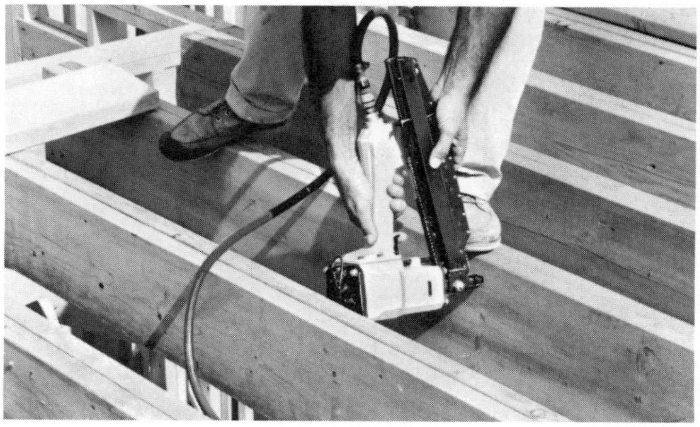

Fig. 7-30. Using a pneumatic powered nailer to fasten double joists together. (Duo-Fast Fastener Corp.)

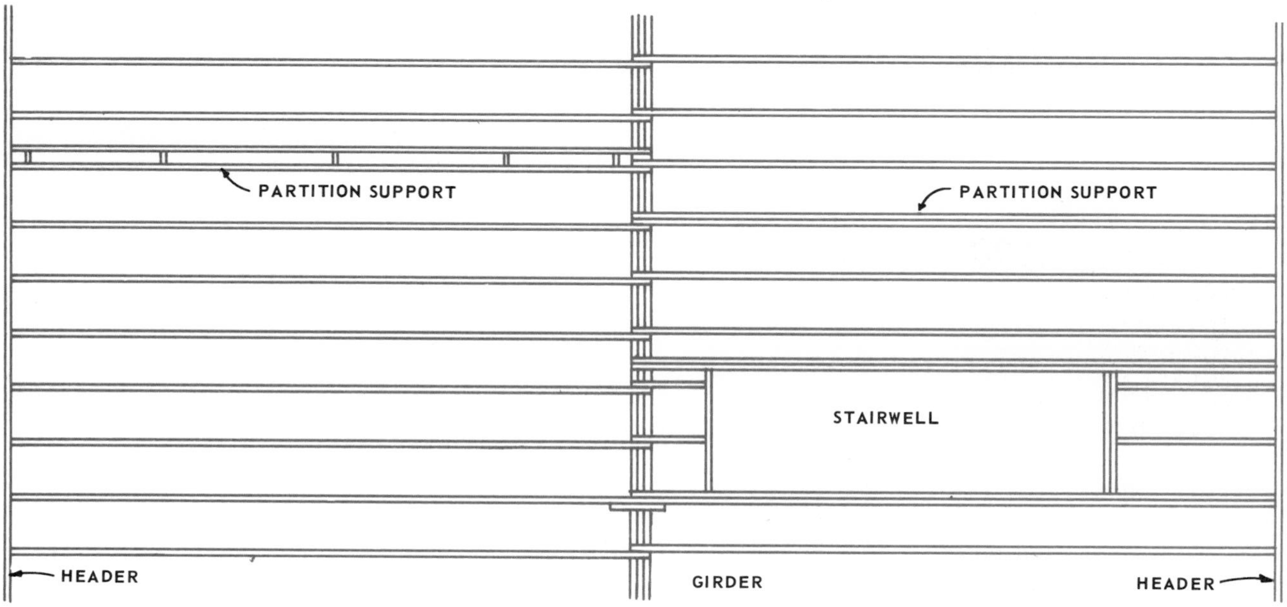

Fig. 7-31. Completed floor frame showing stairwell opening. Note that single header is used at the end of the opening, next to the girder, since the span of the tail joists and header is short.

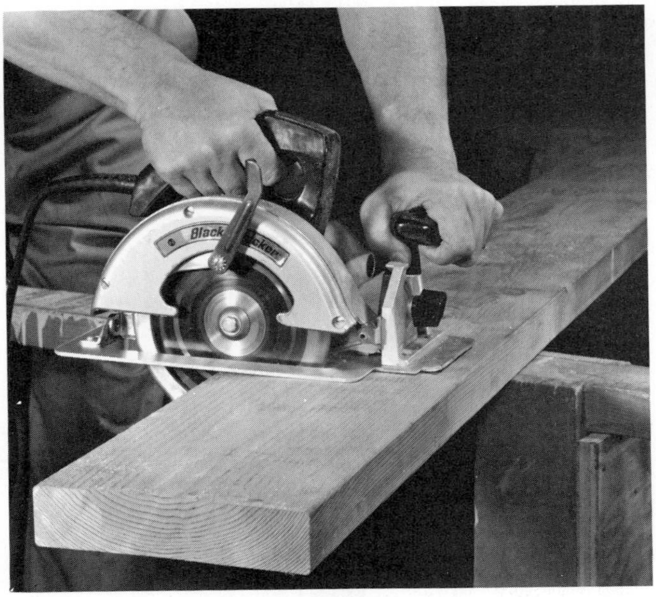

Fig. 7-32. Using a portable electric saw to cut header.

joists in place. Sometimes a regular joist will be located in such a position that it can serve as the first trimmer. Fig. 7-31 shows a plan view of how the finished assembly will appear.

The length of the headers can be secured from the layout on the main header joist. Cut the required number of headers and tail joists to length, Fig. 7-32. Make the cuts square and true. Considerable strength will be lost in the finished assembly if the members do not fit tightly together. Lay out the position of the tail joists on the headers by transferring the marks made on the main header in the initial layout.

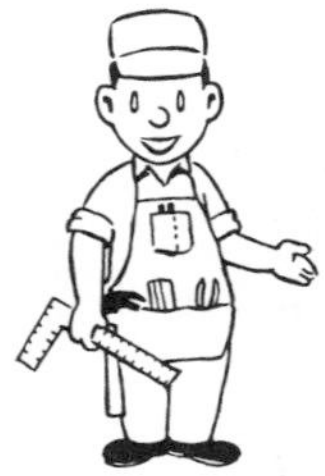

HANDY SAYS:

"Maintain a high level of accuracy in laying out and in cutting floor framing members. The strength of the assembly depends on all the parts fitting tightly together."

When the assembly of tail joists and first headers is small, they are sometimes nailed together and then set in place. Usually, however, the headers are installed and then the tail joists are nailed into position. One of the tail joists can be temporarily nailed to each trimmer to accurately locate the header and hold it while it is being nailed.

Fig. 7-33 illustrates the procedure to follow in assembling the members of an opening with nails. After the first header and tail joists are in position between the first trimmers, the second or double header is nailed in place. Be sure to nail through the first trimmer into the second header using three 16d nails at each end. Finally the

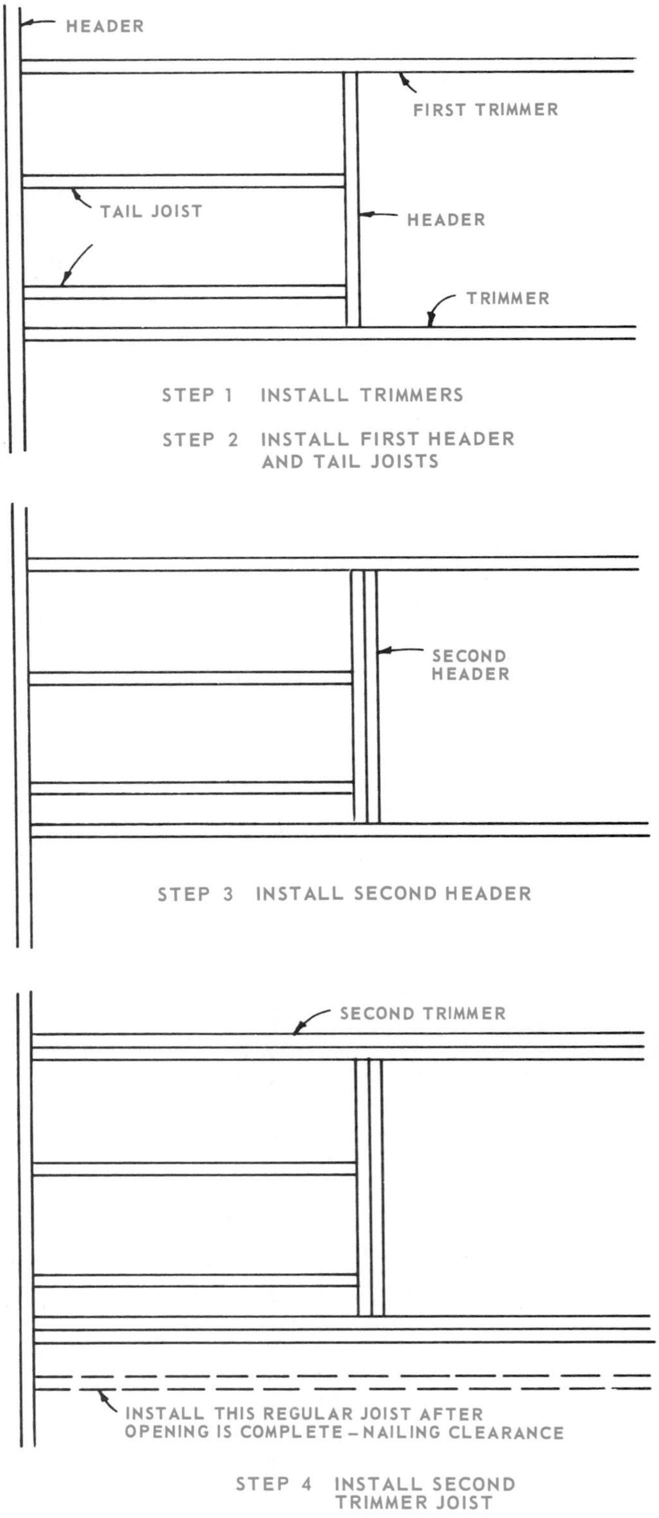

Fig. 7-33. Procedure to follow using nails in assembling framing members of an opening.

Framing a four-story apartment building. Foreground shows second-story walls and partitions on which the third-story floor frame will be constructed. Code requirements include firewalls and sprinkler system.
(Western Wood Products Assoc.)

second trimmer is placed and nailed to the first trimmer. A suggested nailing pattern for the entire assembly is shown in Fig. 7-34.

This nailing pattern will support a concentrated load of 300 lbs. at any point on the floor, or uniformly distributed load of 50 lbs. per square foot

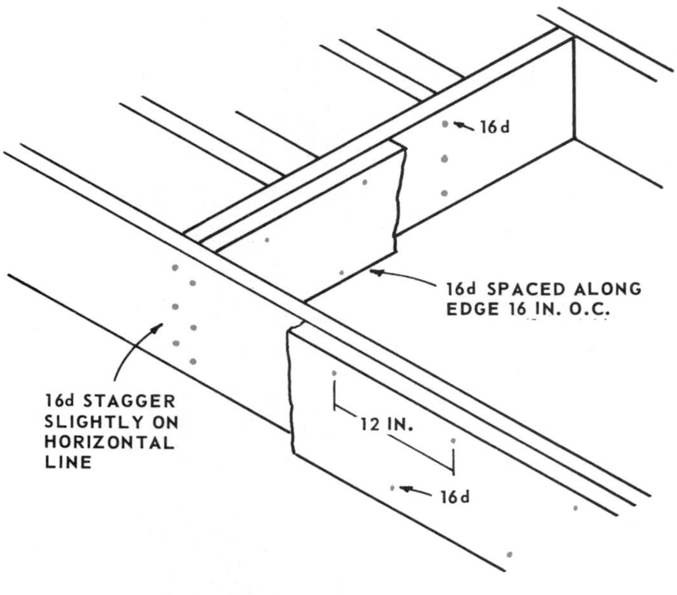

Fig. 7-34. Nailing pattern for floor openings.

with any spacing and span of tail beams ordinarily used in residential construction, provided the length of floor opening is parallel to the length of the joist. If length of opening is at right angles to joists, excessive loading may be carried to the junction of headers with trimmers. Anticipated loads should be checked and increased nailing or additional supports provided at these junctions when needed.

Today, metal framing anchors are often used to assemble headers, trimmers and tail joists, Fig. 7-35. They are manufactured from 18 gauge zinc coated sheet steel in a variety of sizes and shapes. Special nails for attaching the anchors are also available. The National Forest Products Association recommends the use of framing anchors or ledger strips to support tail joists that are over 12 feet long.

Bridging

Some recent studies have shown that where joists are properly secured at the ends, and subflooring is adequate and carefully nailed, bridging may be eliminated. However, many local building codes list requirements in this area and general standards suggest that bridging be installed at intervals not to exceed 8 feet.

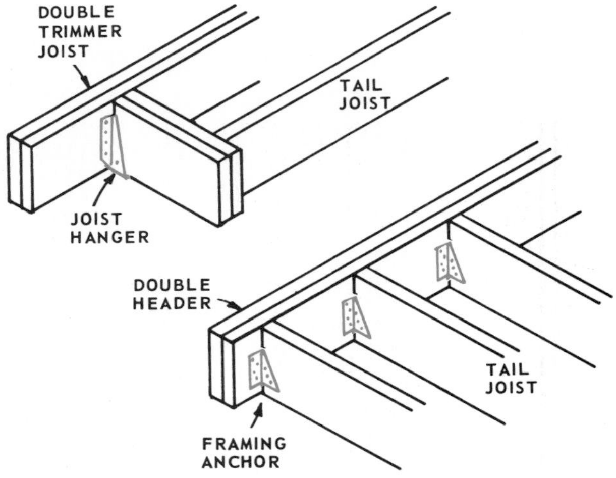

Fig. 7-35. Using joist hangers and framing anchors to assemble members around opening.

Regular bridging, sometimes called herringbone or cross bridging, is composed of pieces of lumber set in a diagonal position between the joists to form an X. Its purpose is to hold the joists in a vertical position and transfer the load from one joist to the next. Fig. 7-36 shows how the carpenters framing square can be used to lay out a pattern for bridging. Pieces can be cut rapidly on the radial arm saw or a jig can be set up to use the portable electric saw or a hand saw.

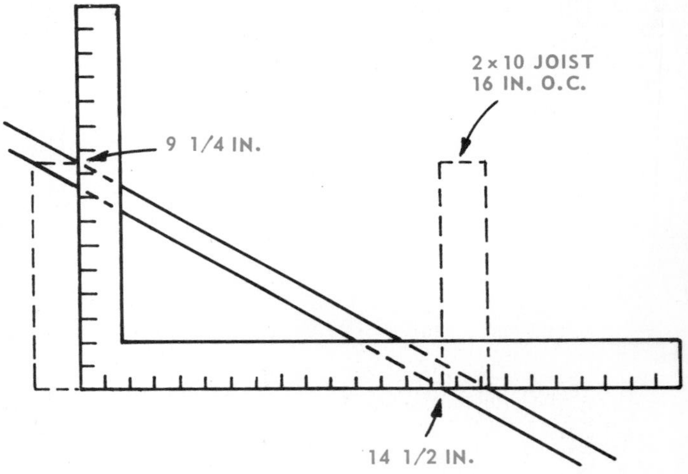

Fig. 7-36. Laying out bridging with the framing square. The line for the lower cut can be secured by shifting the square to a second position along the stock.

To install bridging, first determine the position of the run and then snap a chalk line across the tops of the joist. Start two 8d nails into the ends of all the bridging and then attach a piece to each side of every joist. Alternating the position of the two pieces, first on one side of the chalk line and then on the other side at the next joist, will make the installation easy and correct, Fig. 7-37. The lower ends of the bridging are not

Fig. 7-37. Installing cross bridging at the second floor level. Note the use of blocking in small spaces and over supporting partitions. Also note the framing around the stair opening. (Weyerhaeuser Co.)

Fig. 7-38. Installing solid bridging or blocking over a girder.

nailed until the subflooring is complete or until the under surfaces of the floor are to be enclosed.

Solid bridging, as the name implies, consists of solid blocks set between the joists. This is often used to fit in odd size spaces in a run of cross bridging since the required piece can be cut quickly. Solid bridging, also called blocking,

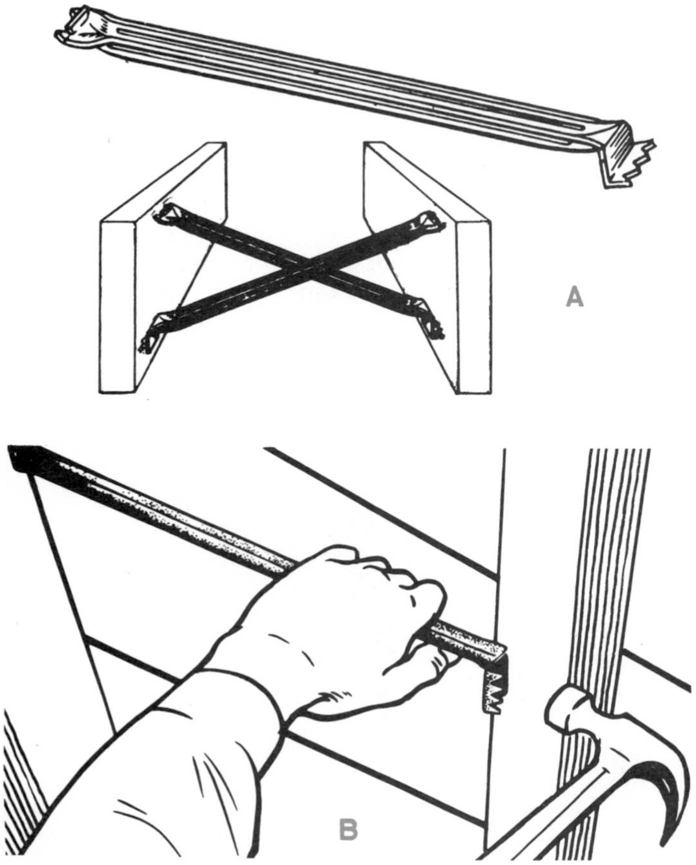

Fig. 7-39. A–Steel bridging and completed installation. B–Method of installation. (Timber Engineering Co.)

is used for an installation like the one shown in Fig. 7-38. Here the chief purpose is to hold the joist in a vertical position. However, it also results in added rigidity to the total assembly.

Several types of prefabricated steel bridging are available that can be installed quickly. The type shown in Fig. 7-39 is manufactured from sheet steel with a V cross section that makes it rigid. No nails are required and it is quickly driven into place with a regular hammer. This meets FHA Minimum Property Standards, and is approved by the Uniform Building Code.

After the bridging is installed, the floor frame should be checked carefully to see that nailing patterns have been completed in all members. After this is done, Fig. 7-40, the frame is ready to receive the subflooring.

Fig. 7-40. Floor frame complete and ready for subflooring. (National Forest Products Assoc.)

Special Framing Problems

In modern residential construction, the design may include a section of floor that overhangs (projects beyond) a lower floor or basement level. When the floor joists run perpendicular to the projected section, the framing is comparatively easy since it is only necessary to use longer joists. If, however, the floor joists run parallel to the wall, the construction must be formed with cantilever joists as illustrated in Fig. 7-41. The exact spacing and length of the members will depend on the weight of the outside wall that will be supported. Usually cantilevered joists should extend inward at least twice as far as they project over the supporting wall. Note that since the force at the inside double header is upward -- the ledger strip must be positioned at the top.

Entrance halls, bathrooms and other areas are often finished with tile or stone that is installed on a concrete base. To provide room for this base, the floor frame must be lowered. When the area is not large, this can be accomplished by doubling joists of a smaller dimension, Fig. 7-42. Additional support can be secured by reducing the spacing. When area is large, steel or wood girders and posts should be incorporated into structure.

Bathrooms must support unusually heavy loads; heavy fixtures and often the additional weight of a tile floor. The fixed dead load imposed by a tile floor will average around 30 lbs. per square foot and the load from bathroom fixtures from 10 to 20 lbs. per square foot, making a total of between 40 and 50 lbs. dead load. In addition to this, it is frequently necessary to cut joists to provide for water service and waste pipes. Special precautions must therefore be taken in framing bathroom floors to provide adequate support.

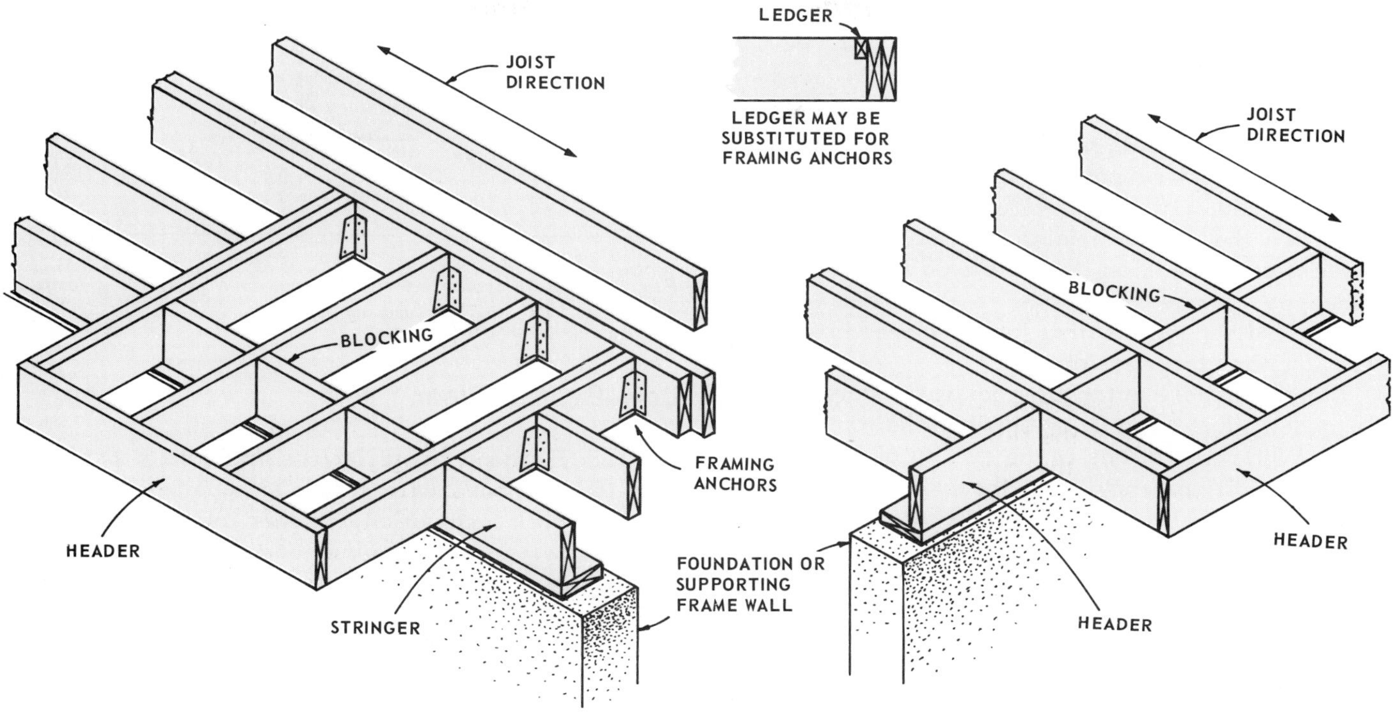

Fig. 7-41. Floor framing for overhangs and projections. Design depends on the direction the joists run. Blocking holds the joists vertical, adds rigidity, and closes the space.

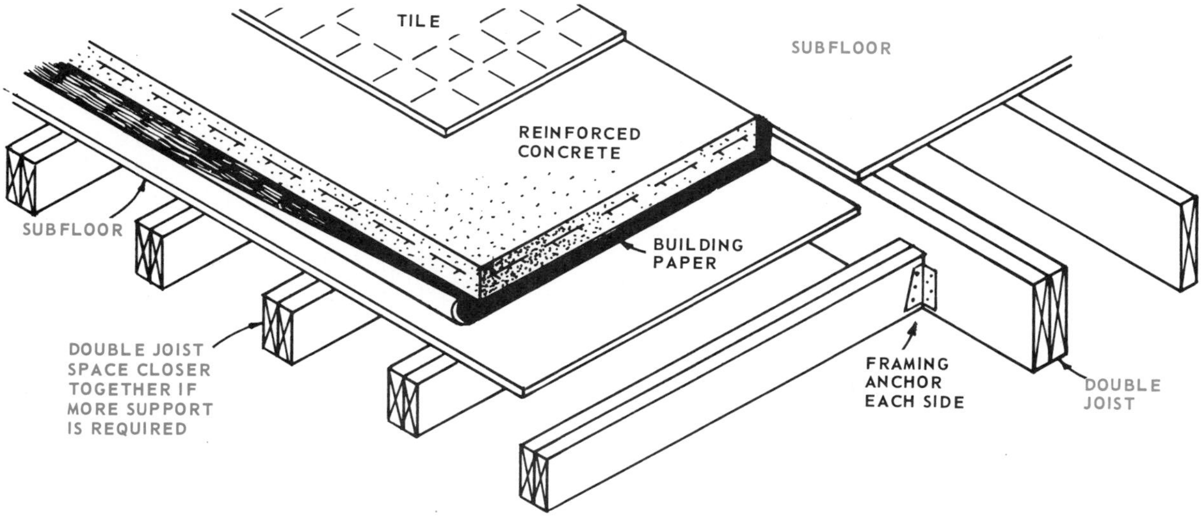

Fig. 7-42. Lowering the floor frame to accommodate a concrete base for tile or stone surfaces.

Cutting Floor Joists

If a joist is supported on edge at the ends between two sawhorses and it is desired to cut the joist in two at the center, the saw will bind. The sides of the cut will close in and bind the saw, compressing the top of the joist. Suppose the cut is started from underneath. With this type of cut the saw will not bind; in fact as the sawing progresses, the opening made by the saw will become larger. The lower part of the joist will be pulled or stretched.

When the top of a joist is in compression and the bottom in tension, there is a point at which the stresses change from one to the other. At this point, called the neutral axis, there is neither tension nor compression.

In the usual rectangular joist, the neutral axis is assumed to be halfway between the top and bottom. While variations in the quality of lumber and other conditions may shift the axis slightly, this assumption is accurate enough for all practical considerations.

If there is neither compression nor tension at the neutral axis or center of the piece, it is obvious that a hole, provided it is not too large (not over one-fourth the total depth of the joist), would have but little effect on the strength.

Weight is most effective in producing bending if it is at the center of the span. Therefore a weakness is more likely to reduce the strength of a joist or beam if it is near the center of the span. Considering this and the previous principle, the following precautions in cutting joists should be observed:

1. Whenever possible, holes cut in the joist should be made at or close to the neutral axis. If limited to one-fourth of the total depth, no material reduction in strength will result.
2. In cases where it is necessary to cut joists, the cuts should be made from the top. For example: if a 2 x 8 joist is cut to a depth of 4 in., its strength will be reduced to that of a 2 x 4. If the cut from the top is 2 in., it will be equivalent to a 2 x 6. When a joist is cut, the loss in strength must be compensated for by providing headers and trimmers or by adding extra joists. One way of solving the problem for large plumbing pipes is shown in Fig. 7-43.
3. If the cut is made elsewhere than at the center of span, the weakening effect will not be as great, but it is advisable to provide fully as much compensating strength as lost by the cut.

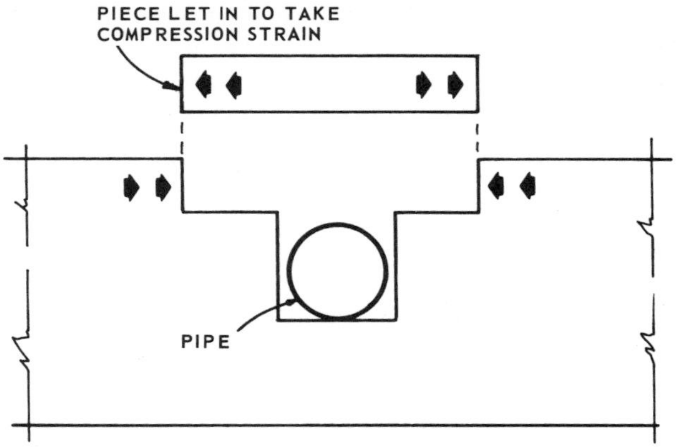

Fig. 7-43. Reinforcing an opening cut in a joist for a large plumbing pipe.

Low-Profile Floor Frames

Today, many home buyers indicate a preference for a house with a low silhouette. Standard wood floor construction, whether over a basement or a crawl space, requires adequate distance between the framing members and the ground, and thus places the first floor level at a considerable distance above the finished grade.

Various framing systems have been devised to lower the floor in relationship to the outside grade. Basically these consist of designing the foundation in such a way that the floor frame is surrounded and protected by the wall, Fig. 7-44. In construction of this type, special precautions must be taken to assure that the joists have adequate bearing surface and that allowance is made for shrinkage of the wood members.

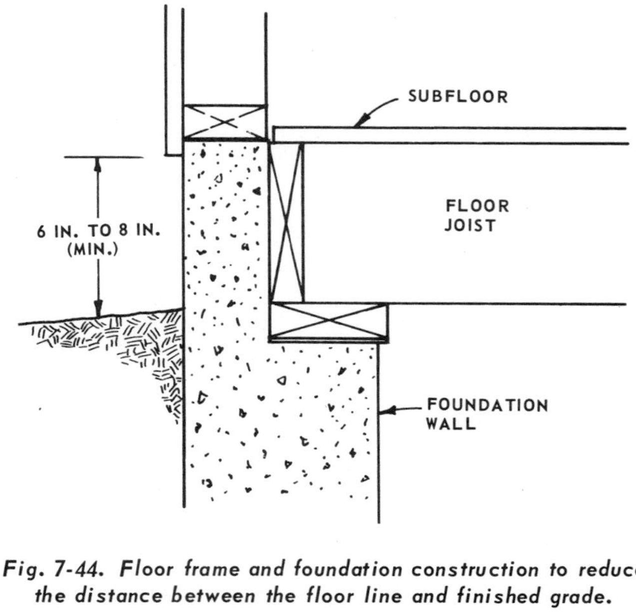

Fig. 7-44. Floor frame and foundation construction to reduce the distance between the floor line and finished grade.

Fig. 7-45 shows a low-profile floor framing system located over a crawl space. The underfloor space was reduced to a minimum and used as a plenum (enclosed space for air under slightly greater pressure than that surrounding the enclosure) for heating and air conditioning. Economy was secured through the use of smaller joists and beams, supported on appropriately spaced piers. Standard requirements for ventilation and insulation of conventional crawl space is included in Unit 13.

Subfloors

The laying of the subfloor is the final step in completing the work on the floor frame. Either plywood, shiplap, tongue-and-groove flooring, or common boards can be used. The main purpose of the subfloor is to add rigidity to the structure and provide a base for the application of finished flooring materials. It also provides a surface upon which the carpenter can lay out and construct additional framing.

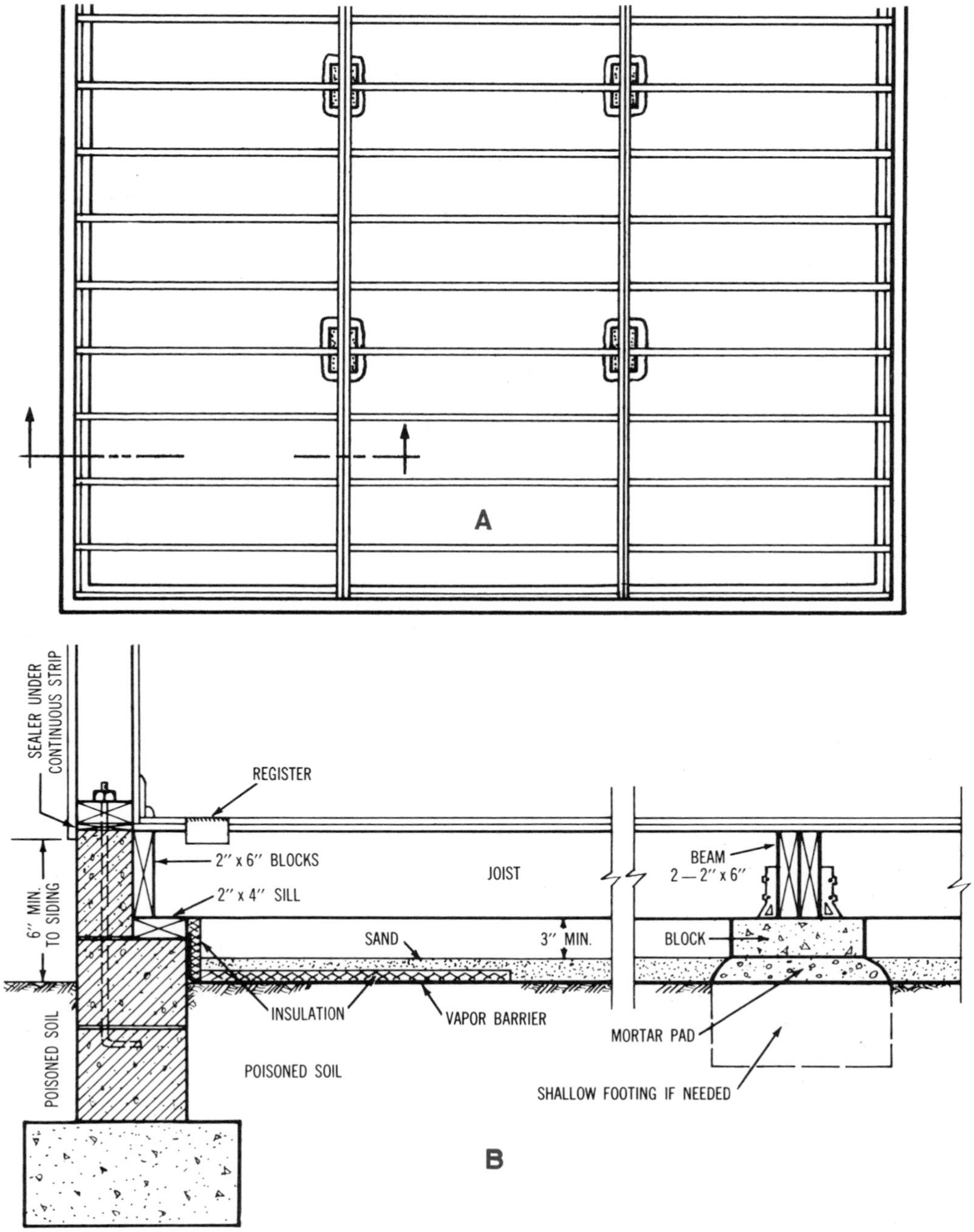

Fig. 7-45. A–Plan view of floor frame over crawl space. B–Section view of floor system. Under-floor space serves as a heating and air conditioning plenum. (National Forest Products Assoc.)

Fig. 7-46. Using shiplap for subflooring. Note that the boards are laid diagonally across the joists. (Laying diagonally is considered obsolete in many areas.)

Shiplap or common boards may be laid at a 45 deg. angle with the joists. This provides a bracing action and permits a conventional tongue-and-groove finish flooring to be laid either parallel or perpendicular to the joists. When laying shiplap at an angle, it is best to start some distance from a corner and then lay back to the corner after short cuttings have accumulated, Fig. 7-46. It is considered good practice to let ends extend out over walls and openings and then saw them off after the boards have been nailed in place.

Subflooring of the board or shiplap type, Fig. 7-47, is nailed at each joist with 8d nails. Use two nails in each board when the width is under 6 in. and three nails for widths of 6 in. and over.

If subfloor is tongued and grooved on ends and

Fig. 7-47. Subflooring boards being laid perpendicular to the joist. The carpenter in the foreground is filling in the full nailing pattern. (National Forest Products Assoc.)

edges, Fig. 7-48, end joints need not be made over joists. Subfloor is preferably laid without cracks between boards. If accumulation of water on the subfloor during construction is likely, it may be desirable to leave sufficient cracks to permit drainage and avoid swelling of the subfloor with resultant cupping and warping.

Plywood

In modern construction, a large amount of plywood is used for subflooring. It provides a smooth even base and acts as a horizontal diaphragm that gives added strength to the building. Plywood can be installed rapidly and usually insures a squeak-free floor.

Fig. 7-48. Laying side and end matched flooring. End joints do not need to occur over joists. (Weyerhaeuser Co.)

Although 1/2 in. plywood over joists spaced 16 in. O.C. meets the minimum FHA requirements, many builders use 5/8 in. plywood. The long dimension of the sheet should run perpendicular to the joists and joints should be broken in successive courses, Fig. 7-49. For 5/8 in. plywood, use 8d nails spaced 6 in. along edges and 10 in. along intermediate members.

Using pneumatic nailers and staplers to install a plywood subfloor reduces the installation time. See Fig. 7-50.

Fiberboard

In recent years fiberboard products have been perfected that meet the requirements for subflooring materials. These are composed of wood fibers bonded together with special adhesives. When installing these products, be sure to follow the instructions supplied by the manufacturer.

Estimating Materials

If you are required to estimate the number and size of floor joists on a job, first scale the plan and determine the lengths that will be needed. Be sure to allow sufficient length for full bearing on girders and partitions. Average residential structures will require several different lengths. Mul-

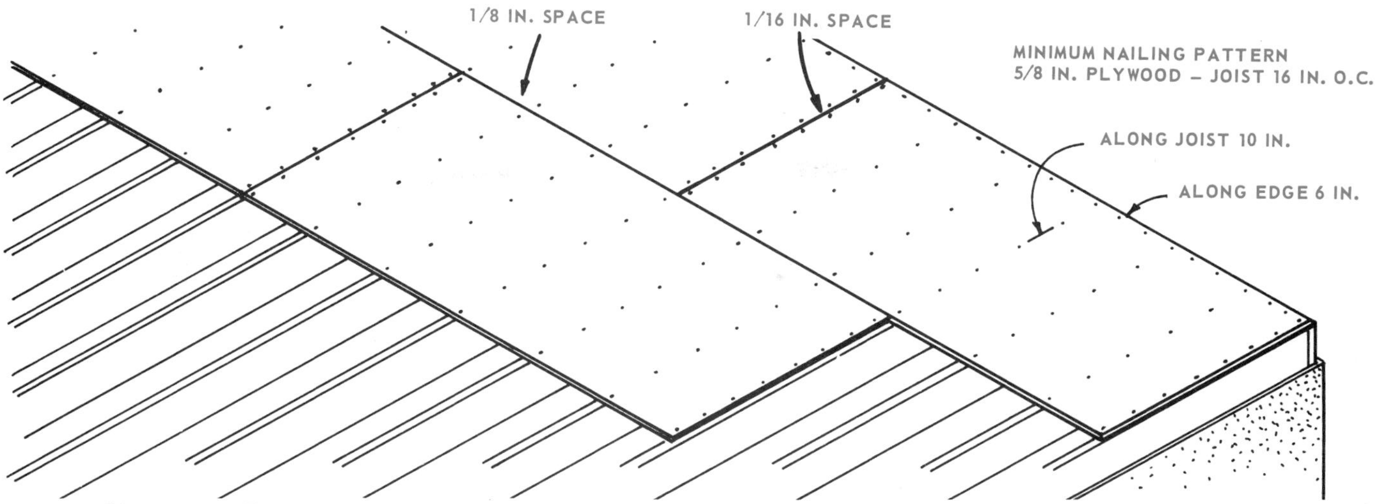

Fig. 7-49. Using plywood for the subfloor. Note the nailing pattern and recommended spacing between sheets.

tiply the length of the wall that carries the joists by 3/4 for spacing 16" O.C. (3/5 for 20, 1/2 for 24) and add one more. Also add extra pieces for doubled joists under partitions and trimmer joists and headers at openings.

No. of joists = length of wall x 3/4 + 1 + extras

Some carpenters figure one joist for every foot of wall upon which the joists rest (16" O.C. spacing); the over-run allowing extra pieces needed for doubles, trimmers and headers.

Header joists are usually figured separately and added to the above figures. Your listing should include the cross section size and the number of pieces of each length.

Example:

Joists (Floor)

40 pcs - 2 x 10 x 16' - 0"
36 pcs - 2 x 10 x 14' - 0"

Joists (Headers)

4 pcs - 2 x 10 x 16' - 0"
2 pcs - 2 x 10 x 12' - 0"

Procedures for estimating the subflooring will vary somewhat depending on the type of material used. Usually the area is figured by multiplying the overall length and width, and then subtracting major areas that will not be covered. These include breaks in the wall line and openings for stairs, fireplaces and other items.

This will give the net area and the basic amount of material needed. To this must be added waste and certain other extras. For example: when 8 in. shiplap is used, multiply the basic figure by 1.15 and then add another 15 percent for waste. If the shiplap is laid diagonally to the joist, another 5 percent should be added. Individual boards are not specified and the amount is simply listed in board feet (equal to the square footage need) along with a description of the material.

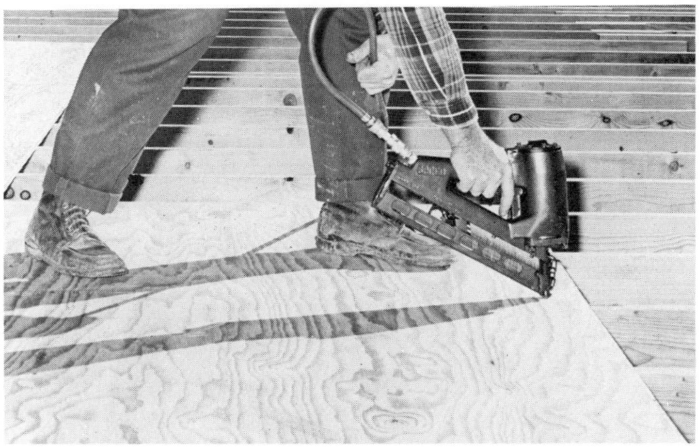

Fig. 7-50. Using a pneumatic nailer to apply a 5/8 in. plywood subfloor. Note that the long dimension of the plywood sheet runs perpendicular to the floor joists. (Senco Products, Inc.)

When using plywood, there is practically no waste and the basic figure is divided by 32 (sq. ft. in a 4 x 8 sheet) and rounded out to the next whole number. This will be the number of pieces of plywood required and is listed as follows:

38 pcs - 5/8 x 4 x 8 Plywood - Standard 32/16

A newer and more complete method of specifying plywood as recommended by the American Plywood Association is included in Unit 4.

Test Your Knowledge
Unit 7

1. The type of framing used in most one-story construction is ______________.
2. When requirements call for the joists to be framed flush with the underside of a wood girder, it is best to use __________.
3. Standard construction usually requires that the sill be spaced back from the foundation wall a distance equal to the _______ _______.
4. The studs of a balloon type frame run continuously from the __________ to the rafter plate.
5. In a balloon frame, second floor joists rest on a horizontal member called a _____________.
6. In residential construction, the deflection of first floor joists under normal live loads should not exceed ________ of the span.
7. A member of the floor frame that runs from the main header to a header for an opening is called a ____________ joist.
8. When framing a floor opening, the double header should be nailed in place before the second ___________ is installed.
9. Cantilevered joists should extend inward at least ______ times the distance that they overhang the supporting wall.
10. Large holes bored through joists for pipes or wiring should be made at the ________ (top, bottom, center).
11. Shiplap of a nominal width of 8 in. should be applied with ____ (2,3,4) 8d nails at each joist.
12. When plywood is used for the subflooring, the short dimension of the panel should run __________ (parallel, perpendicular) to the joist.

Outside Assignments

1. Secure a set of architectural plans for a house with a conventional basement. Study the methods of construction specified in the sections and detailed drawings and then prepare a first-floor framing plan. Start by tracing the foundation walls and supports shown in the basement or foundation plans and then add all joists, headers and other framing members. Your drawing should be similar to the one in Fig. 7-31.
2. Working from a set of architectural plans for a single-story house, develop a list of materials required to frame the floor. Select the type of subflooring, if not specified, and estimate the amount of material needed.
3. Obtain a copy of the local building code in your area and study the requirements that apply directly to floor framing. Prepare a list of these items along with sketches that might clarify complicated written descriptions. Make an oral report to your class.

On-site construction of a glued floor system using a gun operated with compressed air. A 1/4 in. bead of glue is applied to the joist and panels are then nailed in place. Structural tests have shown that stiffness is increased (about 25 percent with 2 x 8 joist and 5/8 in. plywood). In addition the system insures a squeak-free construction and reduces labor costs. (American Plywood Assoc.)

Unit 8

WALL AND CEILING FRAMING

Wall framing includes the assembling of vertical and horizontal members that form the exterior and interior walls of the structure. The framework supports the upper floors, ceilings and roof and serves as a nailing base for inside and outside wall-covering materials. The inside walls are called partitions.

In one-story structures, studs are sometimes placed 24 in. O.C. (on center); however, 16 in. spacing is more commonly used. Fig. 8-1 shows a typical wall frame with openings for a window and door. Note the extra studs used at the corner, at the sides of the openings and where an interior partition joins the outside wall.

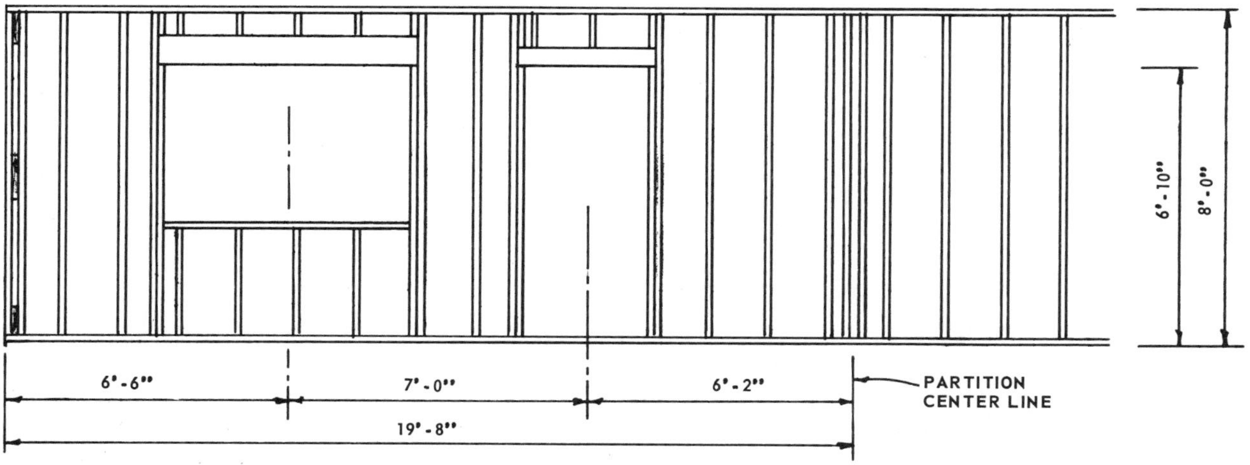

Fig. 8-1. Elevation view of a typical wall frame section. The 8' – 0" height is extended sufficiently by the second top plate to allow for floor and ceiling surface materials.

The term "system" is commonly used to designate methods and materials of a given construction. It is used in connection with floors, ceilings, and roofs as well as walls and includes not only the design of the framework but also the surface-covering materials and the methods by which they are applied. For example: a floor system would include the details of the sill construction, size and spacing of joists, and the kind of subflooring and application requirements.

Parts of the Wall Frame

The wall-framing members used in conventional construction includes sole plates, top plates, studs and headers (also called lintels). Studs and plates are made from 2 x 4 lumber while headers usually require heavier material. Ribbons of 1 x 4 stock are used for bracing that must be built into a wall where the sheathing does not provide sufficient rigidity.

The conventional stud spacing of 16 in. or 24 in. spacing has evolved from years of established practice and is based more on accommodating the wall-covering materials than on the actual calculation of imposed loads.

Fig. 8-2 illustrates in more detail the various parts and how they fit together. Full length studs become cripple studs when they are terminated because of the opening. Trimmer studs not only stiffen the sides of the opening but also bear the direct weight of the header. The regular stud spacing is continuous along the wall, irregardless of openings, so that modular wall-surface materials can be applied with a minimum of cutting.

Wall-framing lumber must be strong, have good nail-holding characteristics and be free of warp. The latter is especially important in construction systems that include a dry-wall interior finish. Stud and No. 3 grades are approved and used throughout the country, including such species as douglas fir, larch, hemlock, yellow pine and spruce. See Unit 4 for additional information.

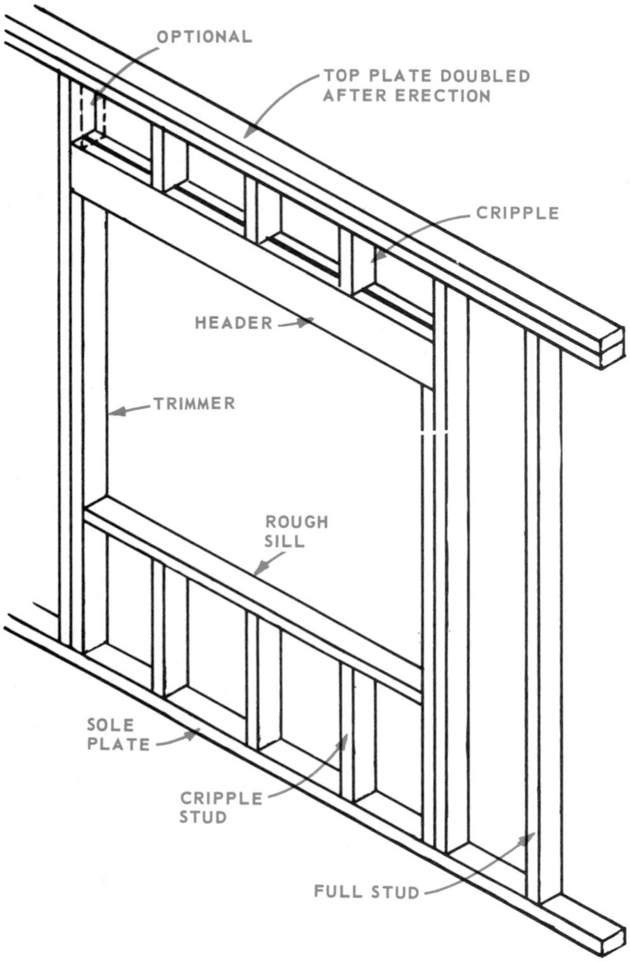

Fig. 8-2. Parts of a typical wall frame at a window opening. Door openings are framed the same way, with the rough sill and lower cripple studs removed.

Corners

There are several methods that can be used to form the outside corners of the wall frame.

In platform construction the wall frame is usually assembled in sections on the rough floor and then raised into position. Corners are formed when the section of a side wall and end wall are brought together. The method most commonly used is to include a second stud in the side wall frame -- spaced from the end stud with three or four blocks. When the end wall is raised into position, the completed corner is formed, Fig. 8-3. An alternate method is to turn the position of the extra stud as shown. Only straight studs should be selected for corners. Assemble with 10d nails, spaced 12 in. apart and stagger from one edge to the other as shown. Include extra nails where the filler blocks occur.

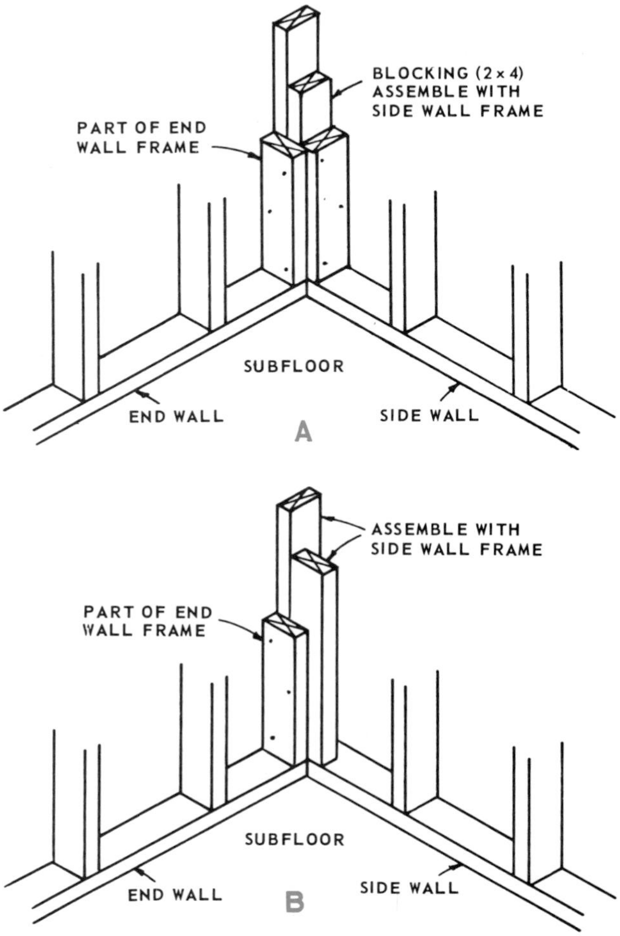

Fig. 8-3. Corner design in platform construction. A–Formed with three full studs and blocking. B–Formed with three full studs. No blocking required.

Some carpenters prefer to build the corners for platform construction as separate units and set them in place (carefully plumbed and braced) before the wall sections are raised. This procedure makes it easier to plumb and straighten the wall sections but does not permit the application of sheathing while the frame is still on the floor deck. In balloon construction, the corners are made up and installed as a separate unit, Fig. 8-4.

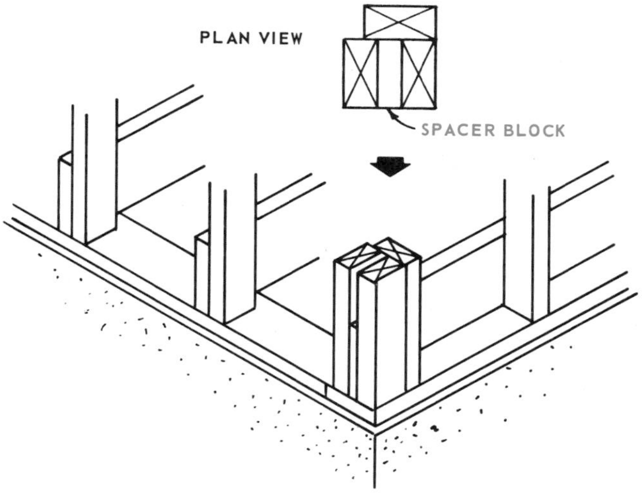

Fig. 8-4. Corner construction in balloon framing.

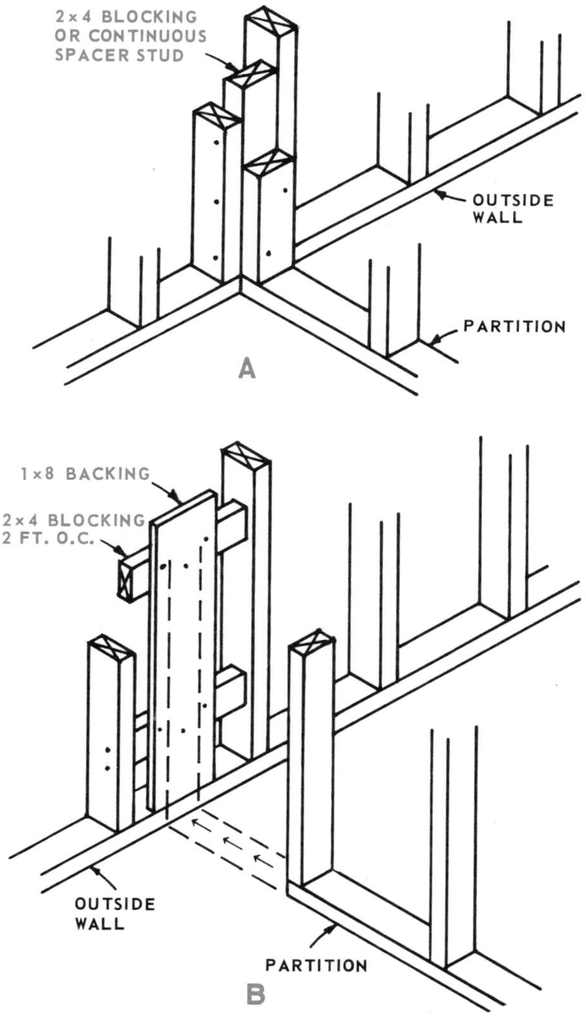

Fig. 8-5. Partition intersections. A–Using extra studs. Nail the wall studs to the spacers with 16d nails spaced 12 in. O.C. Use 10d nails to attach the partition stud. B–Using blocking between the regular stud spaces. Install the 2x4 blocking with 16d nails. Use 8d nails to attach the backing board.

Partition Intersections

Solid anchorage is essential where partitions intersect outside walls. Here, like outside corners, the design of the framing should result in the forming of inside corners to which the wall-covering material can be nailed.

Several methods are commonly employed. One consists of including extra studs in the outside wall, Fig. 8-5, to which the inside partition is attached. Another method utilizes the regular stud spacing with blocking and nailers inserted to provide the necessary attachment to the partition.

Rough Openings

Study the architectural drawings to learn the size and position of the rough openings (R.O.). The position will be given in plan views with dimension lines -- usually running from corners and/or intersecting partitions to the center lines of the openings. The height of rough openings can be secured from elevations and sectional views. Their size will be shown on the plan view or listed in a table called a door and window schedule. The size of the rough openings is always listed with the width given first and the height second. Study Unit 11 for additional requirements concerning window and door openings.

Headers carry the weight of the ceiling and roof across door and window openings. They are formed by nailing two members together and placing them on edge. A 1/2 in. plywood spacer is inserted between the pieces to make the thickness equal to that of the wall, Fig. 8-6.

The length of the header will be equal to the rough opening plus two trimmers (3 in.). The size of the lumber used in the header depends on the width of the opening. Local building codes may include requirements for headers. The table below gives the size of headers normally required for various R.O. widths under several load conditions.

HEADER SPANS

Material on Edge	Supporting one floor, ceiling, roof	Supporting only ceiling and roof
2 x 4	3' - 0"	3' - 6"
2 x 6	5' - 0"	6' - 0"
2 x 8	7' - 0"	8' - 0"
2 x 10	8' - 0"	10' - 0"
2 x 12	9' - 0"	12' - 0"

Headers will also be required across openings in load-bearing partitions. When the load is especially heavy or the span is unusually wide, a truss may be required, Fig. 8-7. The design of special

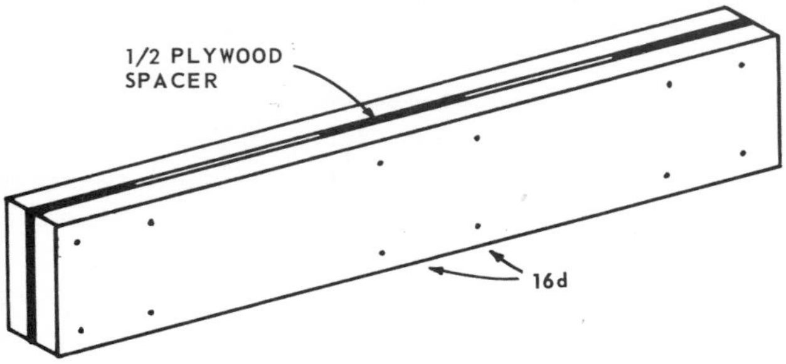

Fig. 8-6. Header construction. The plywood spacers are placed 16 in. to 24 in. O.C.

trusses is usually included in the architectural plans and should be based on careful calculations by a competent architect or engineer.

In modern platform construction, extra studs are included wherever the rough opening occurs, as shown in the standard assembly, Fig. 8-8. The stud and trimmer at the side of the opening furnish support for the header and also provide a nailing base for inside and outside window and door casing members. Some carpenters also prefer to double the rough sill to add nailing base for window stools and aprons.

In large window openings the size of the header will reduce the length of the upper cripple studs to a point where they cannot be easily assembled and should be replaced with flat blocking. Another solution is to increase the header size to completely fill the space to the plate. Some builders follow this practice and extend it to include all openings, regardless of the span. They have found that the cost of labor required to cut and fit the cripple studs is usually greater than the cost of the larger headers. A disadvantage of such construction is the extra shrinkage which, without special precaution in the application of interior wall finish, may result in cracks above doors and windows.

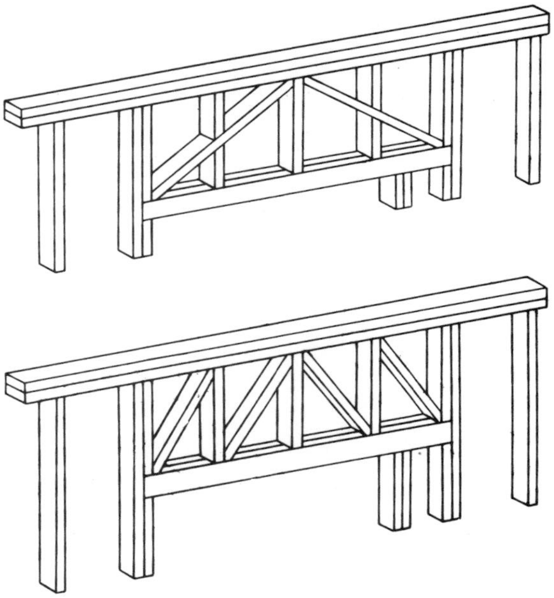

Fig. 8-7. Suggested designs for trussed headers.

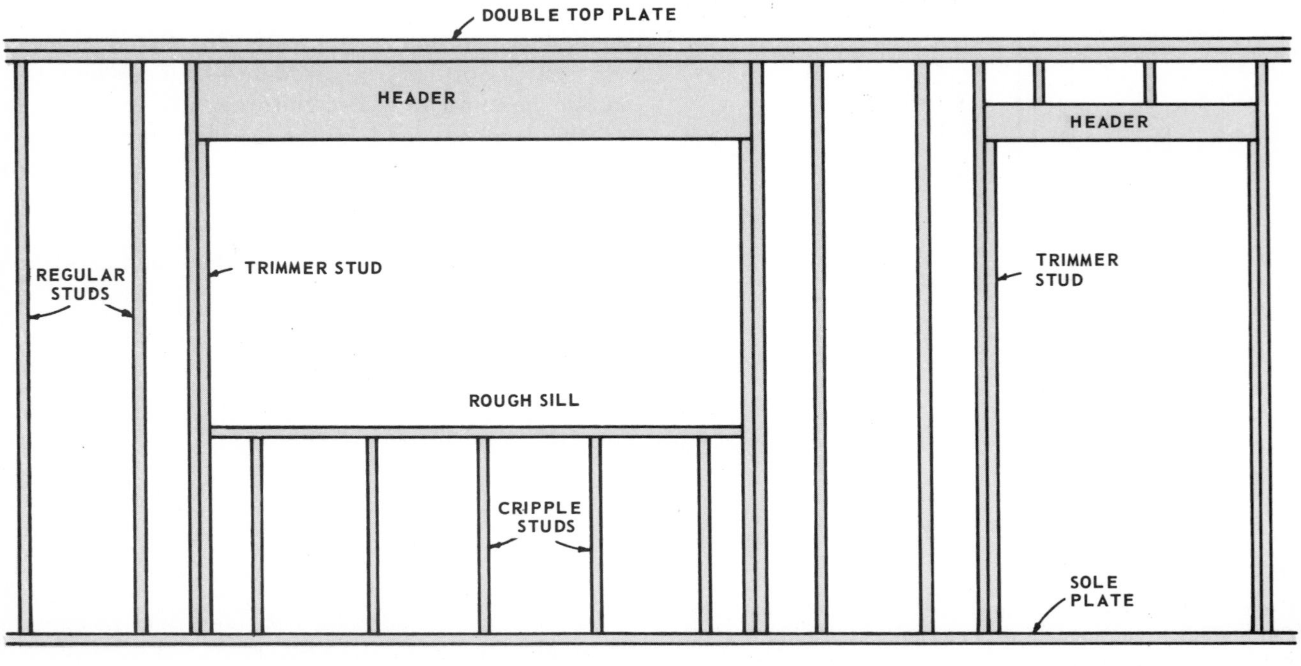

Fig. 8-8. Framed openings in platform construction. Note the header size is larger than required, but fills the space completely thus saving labor in cutting and assembling cripples.

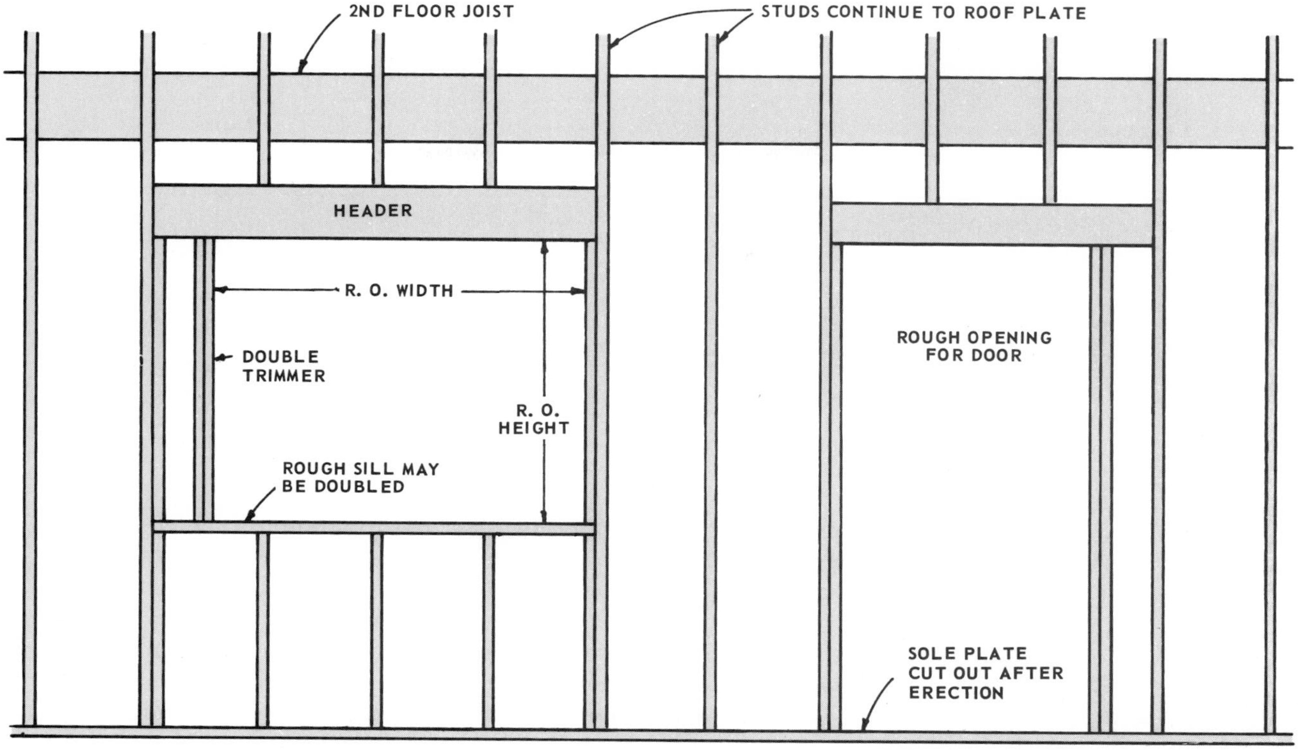

Fig. 8-9. Framed openings in balloon construction.

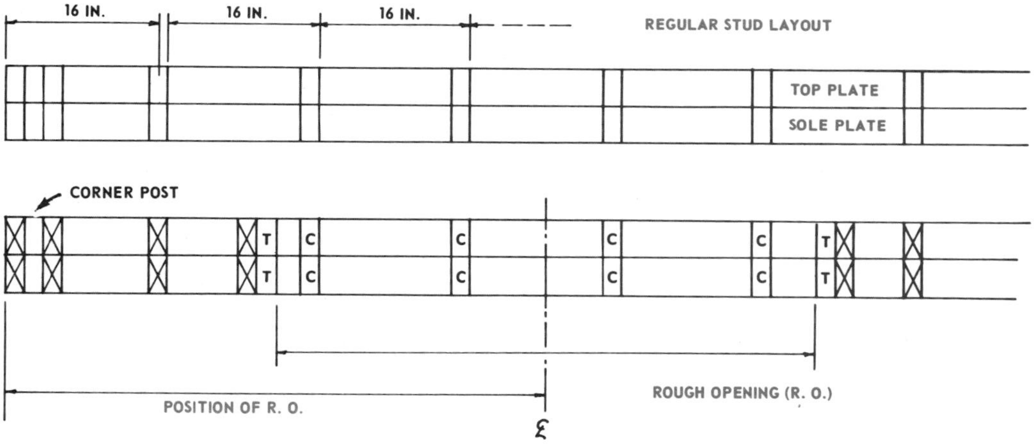

Fig. 8-10. Layout of sole and top plates. Above. Regular stud spacing marked. Below. Converted for window opening.

Several different procedures may be followed in framing openings in balloon construction. Sometimes it is best to erect all of the studs first and then lay out and cut openings. Another procedure is to erect only the full length studs and then fill in the vertical area with the assembly of cripple studs, rough sills and headers. In balloon construction, where the studs extend from the sole plate to the roof plate, the common practice is to extend headers beyond the opening to the next regular stud as shown in Fig. 8-9.

Plate Layout

Select straight 2 x 4 stock and lay two pieces of equal length along the perimeter of the floor. The length of these pieces, which will form the separate wall sections, should be determined by the rough openings included and the procedure that will be used to raise the sections into place.

Lay out the plates along the main side walls first. Align the ends with the floor frame and then mark the regular stud spacing all the way along

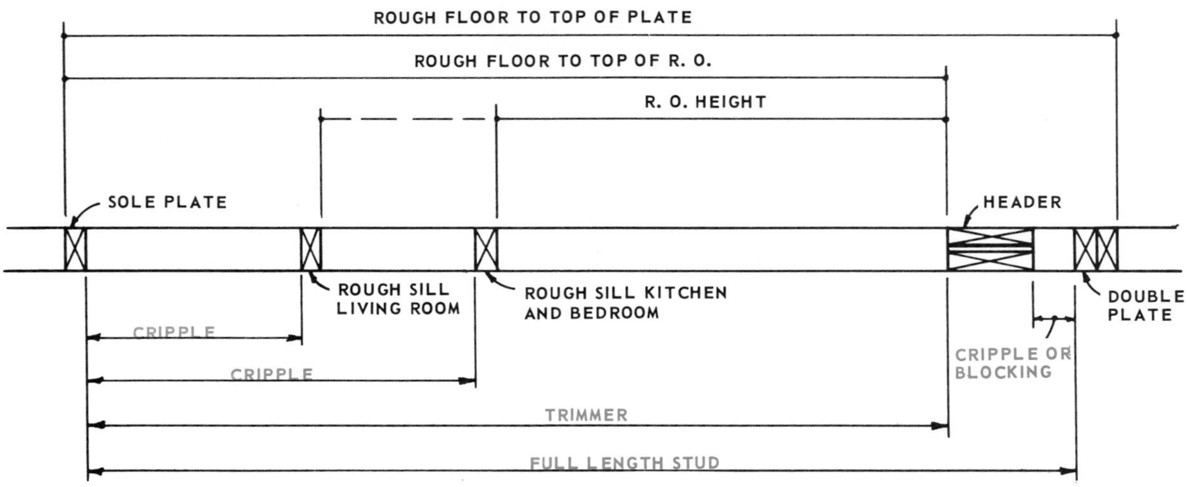

Fig. 8-11. Master stud pattern. The size of the various studs can be secured directly from this layout.

both plates, Fig. 8-10. Some carpenters prefer to secure the pieces to the floor with several nails so they will not slip while the layout is being made.

Refer to the architectural plans and lay out the center lines for each door and window opening. Measure on each side of the center line one-half the width of the opening and mark for trimmer studs outside of these points. Just outside the trimmer stud include a full length stud. Identify the positions with the letter T for trimmer studs and X for full length studs. Now mark all of the original stud spaces located between the trimmers with the letter C, thus designating them as cripple studs.

Lay out the center lines of intersecting partitions and add full length studs if required by the method of construction. When blocking between regular studs will be used, the center line will be needed to position the backing strip. Plan the layout of wall corners carefully so they will fit together correctly when the wall sections are erected.

HANDY SAYS:

"Check over your rough opening layouts carefully, since errors may be difficult to recognize at this point."

Master Stud Layout

In any sizeable construction project, the carpenter usually develops a master story pole. This is made on a smooth straight board of appropriate width and is actually a full size drawing of a typical wall section in simplified form. It includes floor levels, ceiling heights, window and door heights, as well as the thickness of the various materials used in the structure.

The carpenter lays out the story pole (sometimes called a rod) not because he wishes to avoid mathematical calculations but because he knows that it is an efficient means of measuring to full scale and storing this information during the construction. It becomes a handy reference and when using it, mistakes in measurements are less likely to occur. In modern split-level designs, a story pole becomes a necessity.

A master stud pattern is similar to a story pole but does not include as much information. The layout can be made on either a 1 x 4 or 2 x 4 that is straight. First lay out the distance from the rough floor to the ceiling. This dimension can be taken directly from the story pole or secured from the architectural plans. Draw (full size) the position of the sole plate and double top plate, Fig. 8-11. Now lay out the position and size of the header. When several header sizes are used, they can be superimposed on each other.

Lay out the height of the rough openings, measuring from the bottom side of the header and then draw the rough sill in the correct position. The length of the various studs (regular, trimmer, cripple) can now be taken directly from this full-size layout. When the header height of the doors is different from that of the windows, use the other side of the pattern to keep them separate. In multi-story or split-level structures, a master stud layout will probably be required for each level.

Constructing Wall Sections

Working from the master stud layout, cut the various stud lengths. In modern construction, it is seldom necessary to cut standard full length studs. These are usually precision end trimmed (P.E.T.) at the mill and delivered to the construction site ready to assemble. Cut and assemble the headers. Their length and also the length of the rough sill can be taken directly from the plate layout.

The sequence to be followed in assembling wall sections, especially the rough opening, will vary among carpenters. The following is presented as one of several procedures that could be used:

Move the top plate away from the sole plate about a stud length. Turn both plates on edge with the layout marks inward. Place a full-length stud at each position where specified, Fig. 8-12. Nail the top plate and sole plate to the studs using two 16d nails at the end of each stud.

Set the trimmer studs in place on the sole plate and nail them to the full-length studs. Now place the header in position so it is tight against the end of the trimmer and nail through the full-length stud into the header using 16d nails, Fig. 8-13. The upper cripples can be installed after the header is placed.

For window openings, transfer the position of the cripple studs from the sole plate to the rough sill and then make the assembly using 16d nails. Some carpenters prefer to erect the wall section and then install the lower cripples and rough sill.

Fig. 8-12. Assembling full-length studs between the sole plate and top plate. (National Forest Products Assoc.)

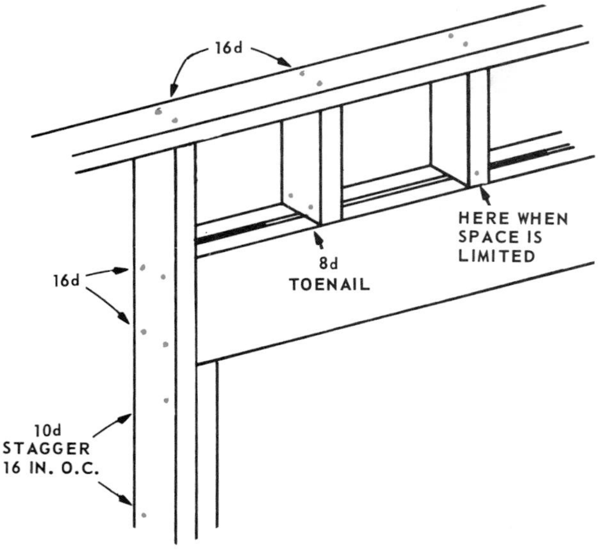

Fig. 8-13. Nailing pattern for wall assembly.

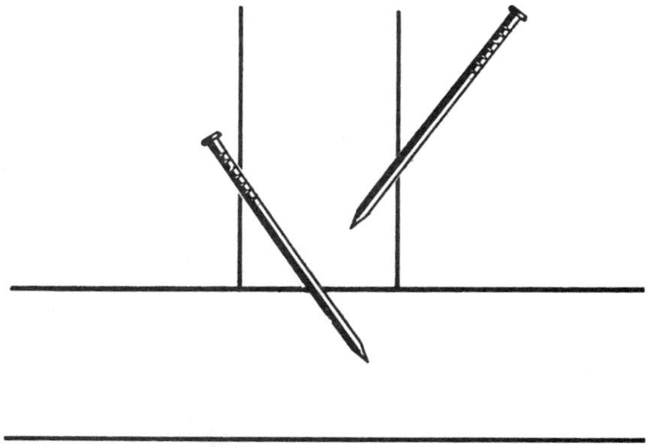

Fig. 8-14. Toenailing. Use four 8d nails.

When this procedure is followed, the lower ends of the cripple studs are toenailed to the sole plate, Fig. 8-14.

Add studs or blocking at positions where partitions will intersect outside walls. Also install any wall bracing that may be required for special installations, remembering that the inside surface of the frame is turned down.

In modern construction, wall sheathing is often applied to the frame before it is raised into position, Fig. 8-15. Make certain that the framework is square before starting the application. Check the diagonals across the corners -- they must be equal. To hold the frame square while the sheathing is being applied, fasten a diagonal brace across one corner or nail two edges of the frame temporarily to the floor.

Fig. 8-15. Applying sheathing to the wall frame before it is raised into position. (National Forest Products Assoc.)

Erecting Wall Sections

Most one-story wall sections can be raised by hand, Fig. 8-16. Larger structures will require the use of a crane or other equipment.

Before raising a section, be sure it is in the correct location and that bracing is at hand and ready to be applied. If the section is large, check to see that sufficient help is available and that each worker knows what to do. When raising frames to which sheathing has not been applied, Fig. 8-17, it is good practice to install a temporary diagonal brace across the assembly if regular bracing is not included.

Fig. 8-16. Raising a wall section that has been sheathed.

Immediately after the wall section is in a vertical position, secure it with braces attached near the top and running to the subfloor at about a 45 deg. angle. Make final adjustments in the position of the sole plate. Be sure it is straight; then nail it to the floor frame using 20d nails driven through the subfloor and into the joists.

Loosen the braces one at a time and carefully plumb the corners and mid-point along the wall. This can be accomplished with a plumb line, but on one-story construction, a carpenters' level is generally used. If the wall is slightly warped, use a straightedge, Fig. 8-18.

HANDY SAYS:

"When plumbing a wall with a carpenters' level, hold the level so you can look straight in at the bubble. If the wall framing member or surface is warped, you should hold the level against a long straightedge that has a spacer lug at the top and bottom."

After one section of the wall is in place, proceed to other sections. No particular sequence needs to be followed although most carpenters prefer to erect main side walls first and then tie in end walls and smaller projections. Procedures must be determined on each individual project, depending on the design and construction methods. Fig. 8-19 shows a building site with wall construction in progress.

Partitions

When the outside wall frame is complete, partitions are constructed and erected. Since the goal at this stage of construction is to enclose the structure and make the roof watertight, only those partitions that support the ceiling and/or roof (bearing partitions) are usually installed. Nonbearing partitions that simply enclose space and provide a framework for wall-covering materials can be postponed until later. Roof trusses, often used in modern construction, are supported entirely by the outside walls. When they are used, inside partition work is seldom started until the roof is complete.

The center lines of the partitions are established from a study of the plans and then marked

Fig. 8-17. Raising a wall framework. Studs are spaced on a 24 in. module. This spacing can be used for studs up to 10 ft. long in single-story construction. In two-story structures, stud length on the lower level should be limited to 8 feet. (Western Wood Products Assoc.)

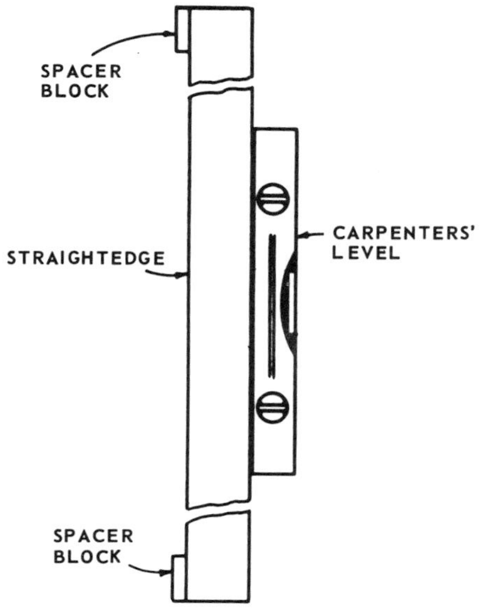

Fig. 8-18. Level and straightedge used to plumb a wall.

on the floor with a chalk line. Plates are laid out, studs and headers are cut and the partitions are assembled and erected in the same way as outside walls, Fig. 8-20. Erect long partitions first, then cross partitions, and finally short partitions that form closets, wardrobes and alcoves.

The corners and intersections are constructed similar to those described for outside walls. The size and amount of blocking, however, can be reduced especially in nonbearing partitions. The chief concern is to provide nailing surfaces at inside and outside corners to which wall-covering material can be attached.

Nonbearing partitions do not require headers. Many openings can be framed with single pieces of 2 x 4 lumber as shown in Fig. 8-21. Most carpenters, however, include trimmers around openings because they provide more rigidity and additional framework for attaching casing and trim. Door openings in partitions (and outside walls)

*Fig. 8-19. Wall construction in progress.
(Weyerhaeuser Co.)*

are framed with the sole plate included at the bottom of the opening. After the framework is erected, the sole plate is cut out with a hand saw. Rough door openings are generally made 2 1/2 in. wider than the finished door size.

Partitions around bathrooms carry plumbing pipes and may need to be thicker than those provided by regular 2 x 4 stock. Instead of using 2 x 6 studs, added thickness is sometimes gained by nailing strips on the edge of the regular frame.

The soundproofing of partitions between noisy areas and quiet areas is receiving increased emphasis in modern homes and may require a special method of framing. See Unit 13 for information on this subject.

Fig. 8-20. Raising a main bearing partition. Note that a section of the double plate has been used to splice the top plate.

Fig. 8-21. Nonbearing partitions.

View at center of rear elevation. Note that floor joists on the upper level extend to the outside edge of the double plate with blocking between. This is good practice since it provides better support for the joists.

Interior view after sheathing has been applied. Living room studs (arrow) extend continuously from floor to ceiling since they are not required to support the upper level floor frame. Note the fiberglass shower/tub unit partially visible in upper right.

Front view of structure with wall and partition framing complete and sheathing applied. Stairwell framing can be partially seen through rough opening for living room window.

Split-level framing. Also study Fig. 8-31 on page 150.

Small alcoves, wardrobes and partitions in closets are often framed with 2 x 2 material or by turning 2 x 4 stock sideways, thus conserving space. This is usually satisfactory where the thinner constructions are short and intersect regular walls, Fig. 8-22.

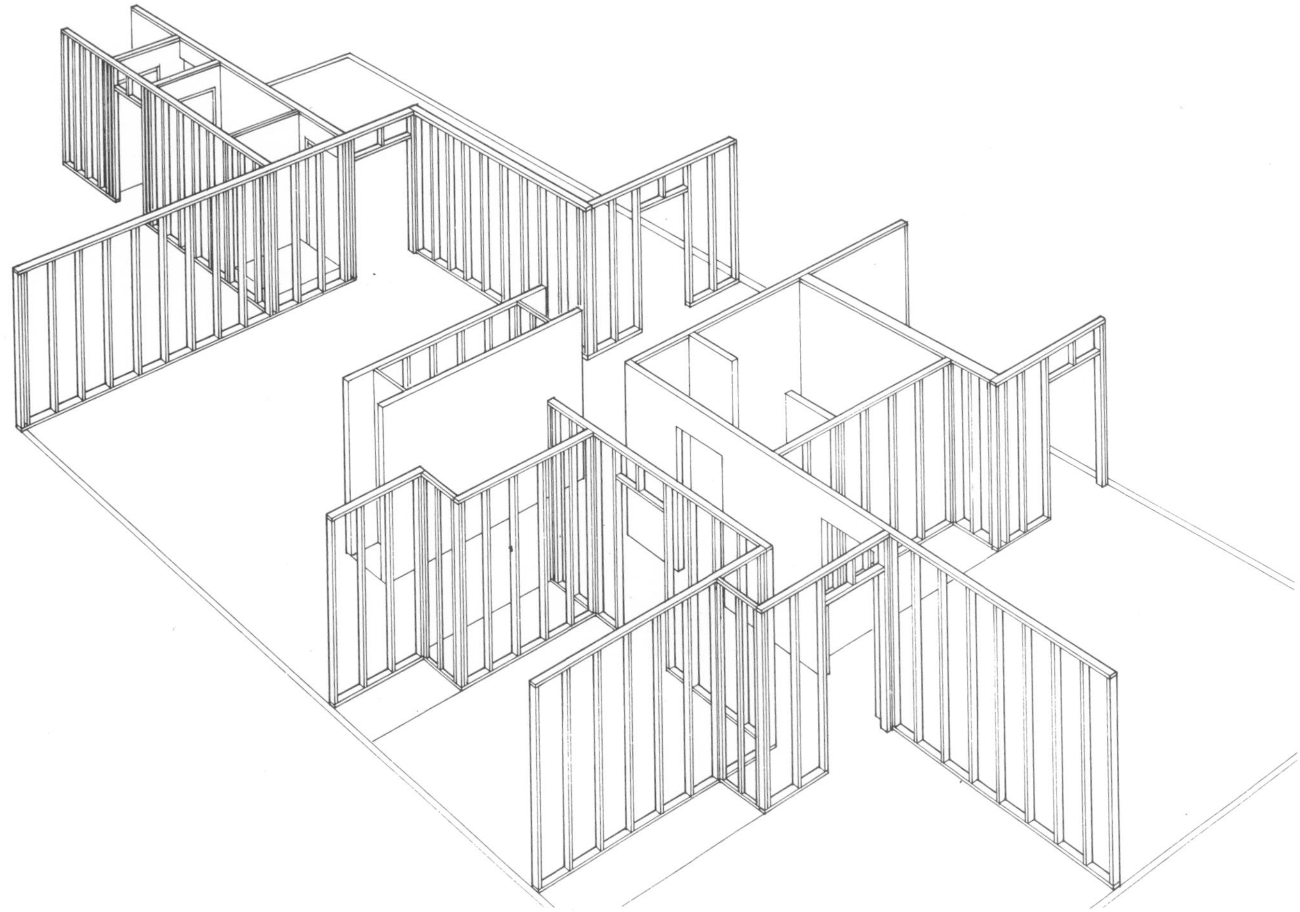

Fig. 8-22. Partitions in a modern residential structure. The short partitions in closets and wardrobes are made of 2 x 2 material to conserve space. (National Forest Products Assoc.)

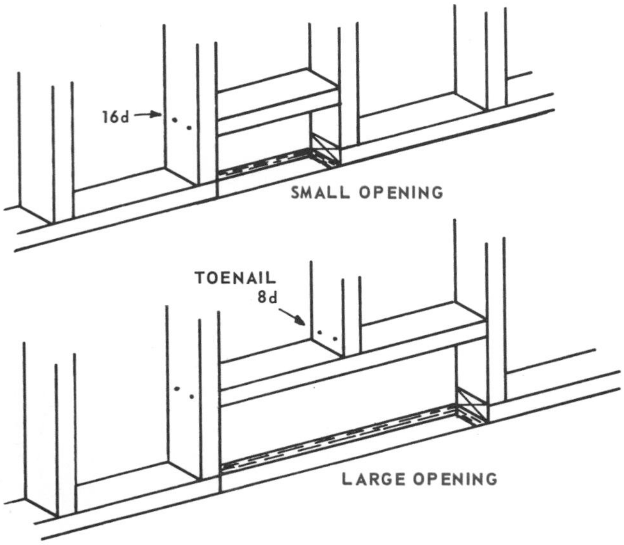

Fig. 8-23. Framing openings for heating ducts.

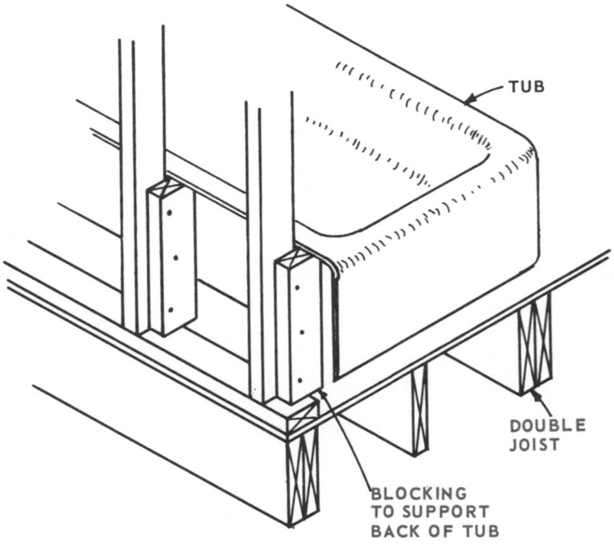

Fig. 8-24. Extra support for tub installation.

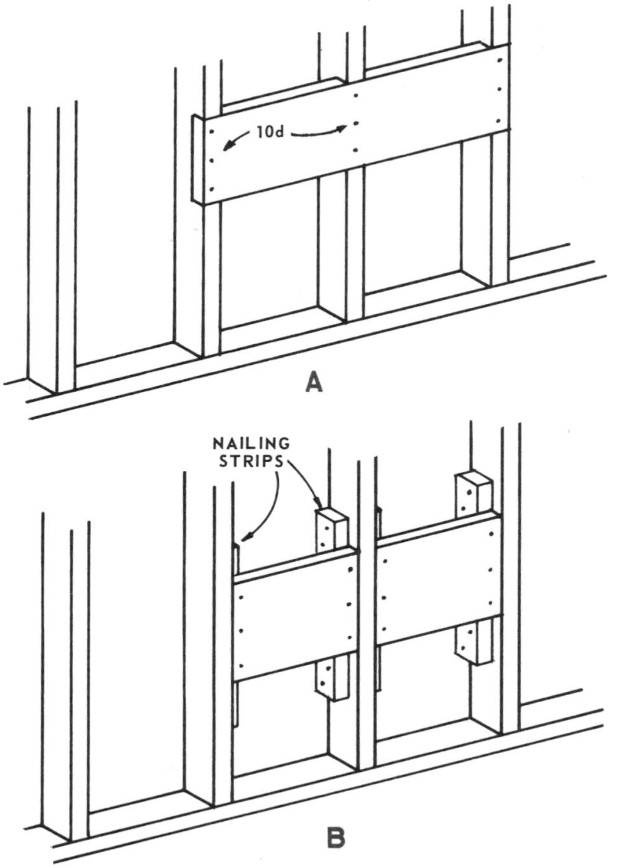

Fig. 8-25. Wall backing for mounting various fixtures and appliances. A–Backing cut into studs. B–Backing attached to nailing strips.

During wall and partition framing, various important details can be added. Openings for the installation of heating ducts, Fig. 8-23, are easily cut and framed at this time. Bath tubs and plumbing fixtures usually require extra support, Fig. 8-24. Framing for recessed cabinets and backing for towel bars, Fig. 8-25, can be included at this time or later but must be completed before wall-covering materials are applied. Small items, not critical to the structure, can often increase efficiency and quality of work during the finishing stages. For example: corner blocks, Fig. 8-26, will make it possible to nail back from the end of the baseboard, thus eliminating the possibility of splitting the wood.

HANDY SAYS:

"The carpenter must continually study and plan the sequence of the job, so that neither the weather nor work of other tradesmen will cause slow-downs or bottlenecks."

Bracing

Exterior walls usually need some type of bracing to resist lateral stresses. Some applications of material such as plywood provide sufficient rigidity and bracing can be eliminated. Always check the exact requirements of the local building code.

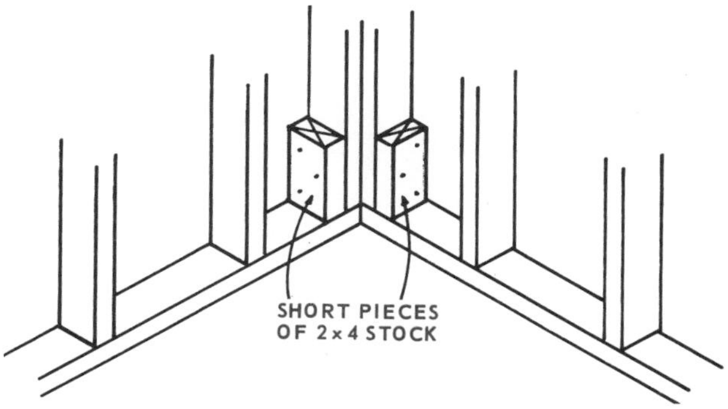

Fig. 8-26. Additional blocking in corners provides better nailing support for baseboards.

A standard installation of let-in corner bracing is shown in Fig. 8-27. Note how the bracing is applied when an opening interferes with the diagonal run from top plate to sole plate.

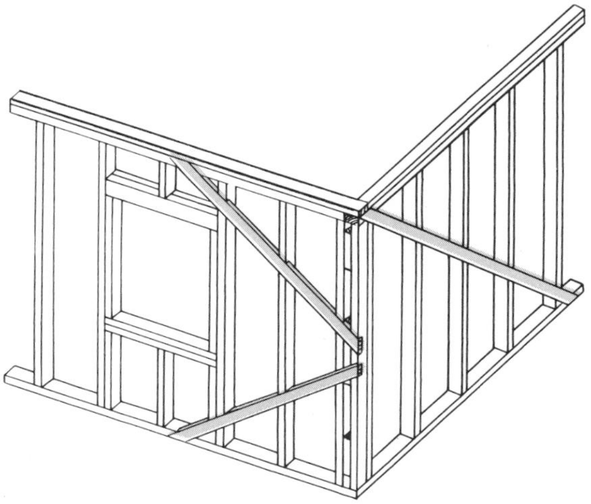

Fig. 8-27. Bracing set into walls to add rigidity to the framework.

Make the braces from solid 1 x 4 lumber cut to the correct length with an angle at the bottom end. Tack (nail temporarily) each brace in position and mark the layout on each stud. Remove the brace and gauge the depth of cut required. Make the side cuts with a saw and remove the wood between the cuts with a chisel. When all cuts are cleared, nail the brace in place with two 8d nails at each stud and trim off the top end.

Double Plate

To add support under joists and rafters, the top plate is doubled. This also serves to further tie the wall frame together. Select long straight lumber. Install the double plate with 10d nails; two nails near the ends of each piece and the others staggered 16 in. apart, Fig. 8-28.

Fig. 8-28. Doubling the top plate after walls and partitions are erected. (Western Wood Products Assoc.)

Joints should be located at least four feet from those in the lower top plate. At corners and intersections, the joints are lapped as shown in Fig. 8-29.

Special Problems

In addition to the standard framing thus far described, the carpenter will often be required to build structures that include special design features. Preferably, these special designs should be carefully engineered by the architect and complete details included in the plans. Occasionally, only the shape and size description is included and the construction details must be developed by the carpenter. For example: architectural plans may include only the size and shape of a bay window, Fig. 8-30, with rough opening sizes included. Here the carpenter must visualize the construction details, lay out and construct the floor frame to carry the projection, and then build the wall and roof frame. If he thoroughly understands regular framing requirements and procedures, he generally has little difficulty in applying them to special problems.

The tri-level or split-level house has gained wide acceptance and presents some extra problems in wall framing. Generally, a platform type of construction, Fig. 8-31, is used although the floor joists for upper levels could be carried on ribbons cut into the studs. The architectural plans will prescribe the type of construction and should also include careful calculations of distances between floor levels. When working with split-level designs, the good carpenter should prepare accurate story poles which show full-size layouts of vertical distances and actual sizes of the construction materials.

Wall Sheathing

If the wall sections were not covered with sheathing before erection, then this application should be made after they are erected and before the roof framing is started. The sheathing adds a great amount of rigidity and strength to the framework.

Today, plywood and fiberboard are used extensively for this purpose. The large sheets can be applied rapidly and usually add sufficient lateral strength to wall sections that diagonal bracing can be eliminated in one-story structures.

Fiberboard sheathing is made largely from wood fibers with added weather-proofing ingredients. It is commonly available in 4 x 8 ft. sheets. However, sheets as large as 8 x 14 ft. are manufactured. Regular fiberboard sheathing is also available in a 2 x 8 ft. size with tongue and groove

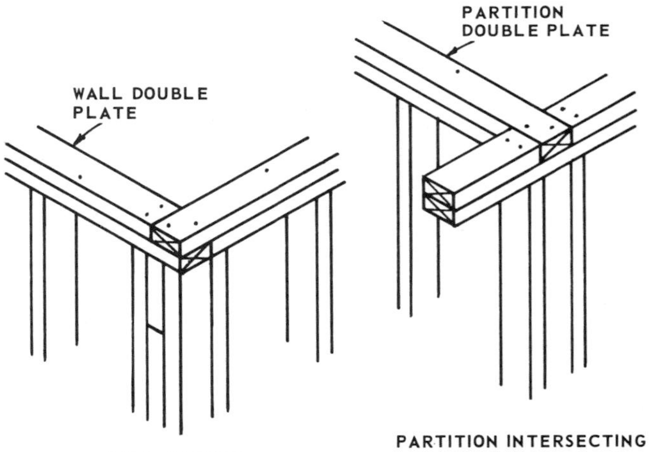

Fig. 8-29- Double plate installation at corners and intersections. Use 10d nails.

or shiplap joints. It is applied horizontally and supplementary corner bracing is required. When 4 x 8 sheets are applied vertically, Fig. 8-32, bracing is usually not required. Regular fiberboard sheathing cannot be used as a nailing base

Plywood sheathing may be either of the interior or exterior type in a structural grade. It should be at least 5/16 in. thick for studs spaced 16 in. on centers and 3/8 in. thick for studs spaced 24 in. on centers. A 1/2 in. thickness is often used

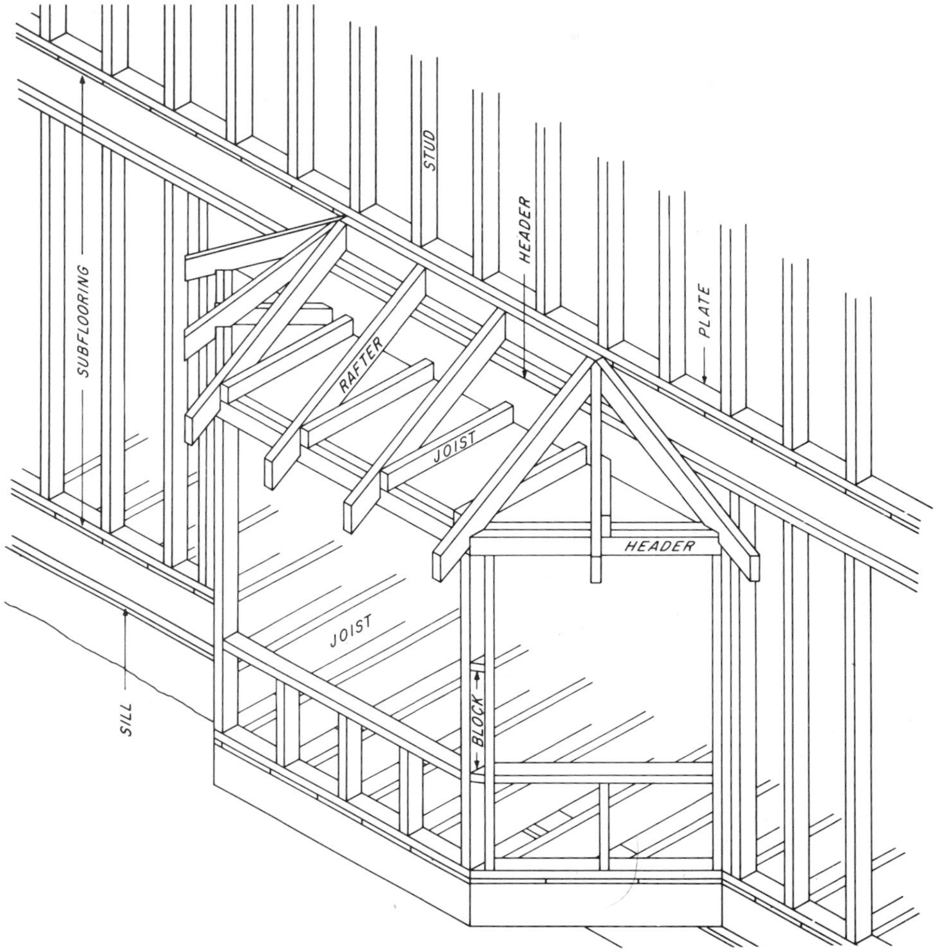

Fig. 8-30. Framing for a bay window.
(National Forest Products Assoc.)

for exterior wall finish materials. However, a special nail-base fiberboard is available. See the table in Unit 4 for further information.

Standard thicknesses of fiberboard sheathing are 1/2 and 25/32 in. The 1/2 in. sheathing is fastened with 1 1/2 in. roofing nails and the 25/32 in. requires 1 3/4 in. roofing nails. Space the nails 3 in. on centers around the edge and 6 in. on center at the intermediate supports. They should never be placed closer than 3/8 in. to the edge.

When gypsum sheathing (usually 1/2 in. thick) is used, it is necessary to include corner bracing. See Fig. 8-32A. This type of sheathing is usually installed with 1 3/4 in. nails spaced 4 in. around the edge and 8 in. on intermediate supports.

so that the exterior finish material can be fastened directly to the plywood. The sheets can be applied vertically or horizontally, Fig. 8-33. Fasten the edges with 6d nails spaced 6 in. on centers and 12 in. on centers along the intermediate members. Corner bracing is usually not required when plywood panels are used for wall sheathing.

Wood boards are still commonly used for sheathing, especially in certain localities. They should be at least 3/4 in. thick and not over 12 in. wide. When applied diagonally (procedure is still followed in some areas) in opposite directions from each corner, no additional corner bracing is required. The boards must be cut with the end joints occurring over the centers of the studs. Use

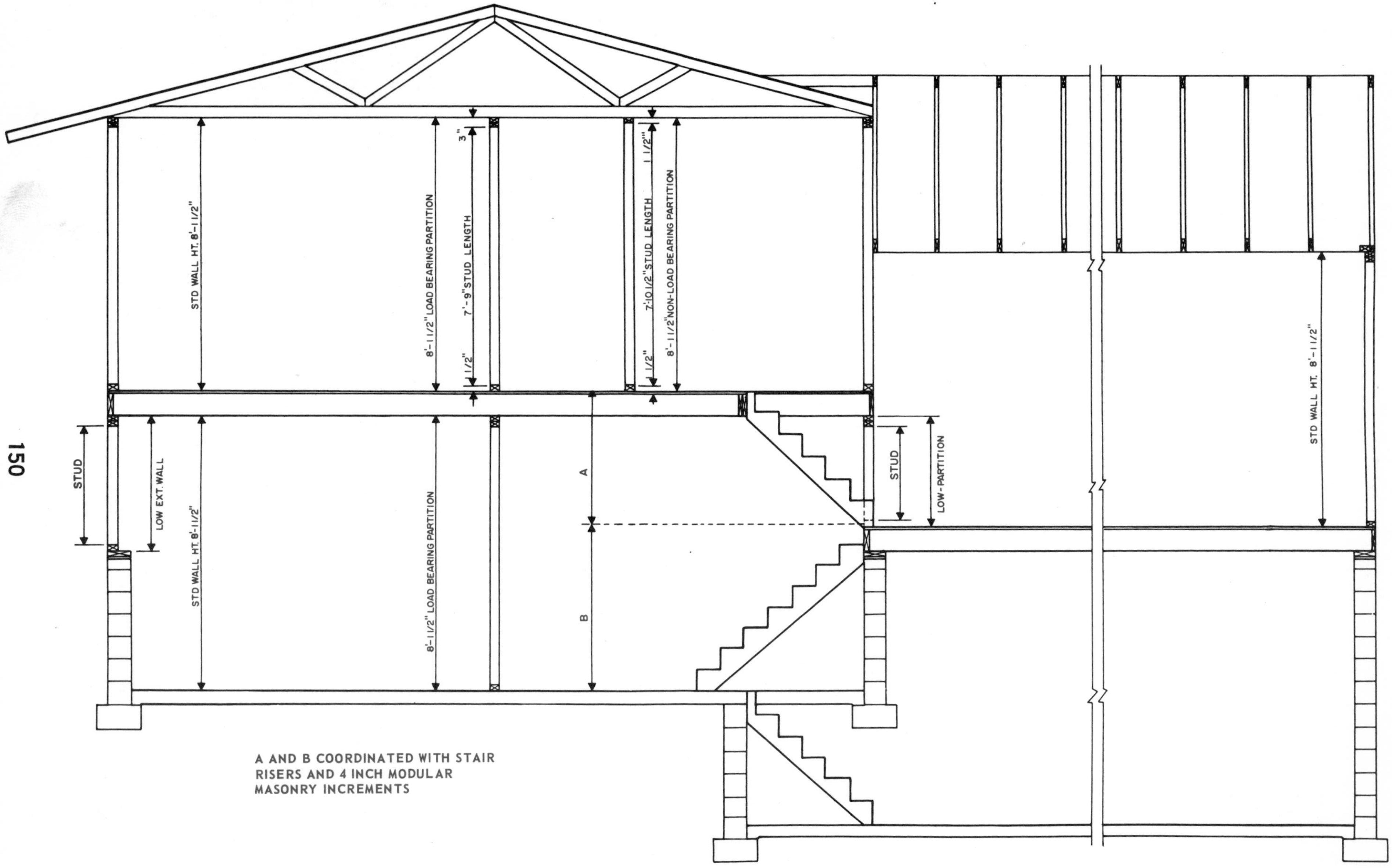

Fig. 8-31. Modern split-level design utilizes platform construction. (National Forest Products Assoc.)

Fig. 8-32. In applying 4 ft. wide sheathing, leave 1/8 in. space between adjoining panels and at the ends of panels.

Fig. 8-32A. Using gypsum wall sheathing. Note use of plywood at corners.

two 8d regular nails or 7d threaded nails for 8 in. nominal widths and under. Use three nails in widths over 8 in. Fig. 8-34 shows an inside view of a wall covered with shiplap boards. Note particularily the diagonal bracing-- required when the installation is made horizontally.

When using boards with side and end matching (tongue and groove) for sheathing, the joints do not need to occur over a stud, Fig. 8-35. This type of sheathing can be applied rapidly since little end cutting and fitting is required.

Fig. 8-33. Installing plywood sheathing. Corner bracing is usually not required. (American Plywood Assoc.)

Fig. 8-34. Interior view of a wall sheathed with 8 in. shiplap. Note the cut-in corner brace. (Western Wood Products Assoc.)

Fig. 8-35. Applying end matched sheathing at a diagonal with the studs. Bracing not required. (Weyerhaeuser Co.)

Ceiling Frame

A ceiling frame can be defined as the assembly just below the roof that carries the ceiling surface. On other levels, the ceiling is carried on the floor joists. Basic construction is similar to floor framing; the main difference being that lighter members are used and header joists are not included around the outside.

SIZE IN.	SPACING IN.	GROUP A FT. IN.	GROUP B FT. IN.	GROUP C FT. IN.	GROUP D FT. IN.
2x4	12	9–5	9–0	8–7	4–1
	16	8–7	8–2	7–9	3–6
2x6	12	14–4	13–8	13–0	9–1
	16	13–0	12–5	11–10	7–9
2x8	12	19–6	18–8	17–9	14–3
	16	17–9	16–11	16–1	12–4
2x10	12	24–9	23–8	22–6	19–6
	16	22–6	21–6	20–5	16–10

Fig. 8-36. Spans for ceiling joists, figured for a normal dead load and a live load of 20 lbs. per sq. ft. –– permitting the attic to be used for storage. (National Building Code)

The main framing members are called joists and, like floor joists, their size is determined by the length of span and the spacing used. To coordinate with walls and permit the use of a wide range of surface materials, a spacing of 12 in. and 16 in. O.C. is commonly used. Size and quality requirements must also be based on the type of ceiling finish (plaster or dry-wall) and what use will be made of the attic space, Fig. 8-36. The architectural plans will usually include specifications and these requirements should be checked with local building codes.

Ceiling joists usually run across the narrow dimensions of a structure. However, some may be placed to run in one direction and others at right angles as shown in Fig. 8-37. By running joists in different directions, the length of the span can often be reduced.

In large living rooms, the mid-point of the joists may need to be supported by a beam. This beam can be located below the joists or installed flush with the joists. In the latter installation, the joists are carried on a ledger as shown in Fig. 8-38. Joist hangers, like those described in the unit on floor framing, can also be used. Sometimes a beam is installed above the joists, in the attic area, and tied to the joists with metal straps.

At the outer end, the upper corner of the joists must be cut to match the slope of the roof. To lay out the pattern for this cut, you may use the framing square as illustrated in Fig. 8-39. When the amount of stock to be removed is small, it may be accomplished after the joists and rafters are in place, using a hatchet or saw.

When the ceiling joists run parallel to the edge of the roof, the outside member will likely interfere with the roof slope. This will often oc-

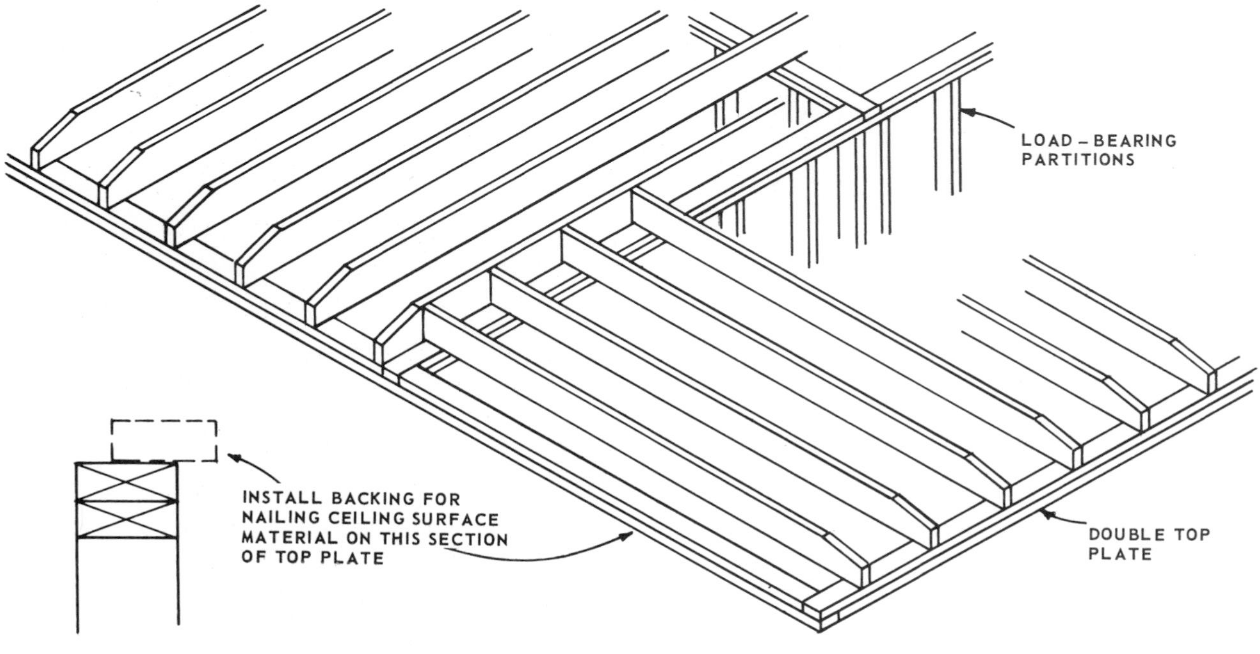

Fig. 8-37. Ceiling frame. Joists in foreground are turned at a right angle to reduce the span.

Fig. 8-38. Ceiling joists supported by a ledger strip nailed to a flush beam. (Western Wood Products Assoc.)

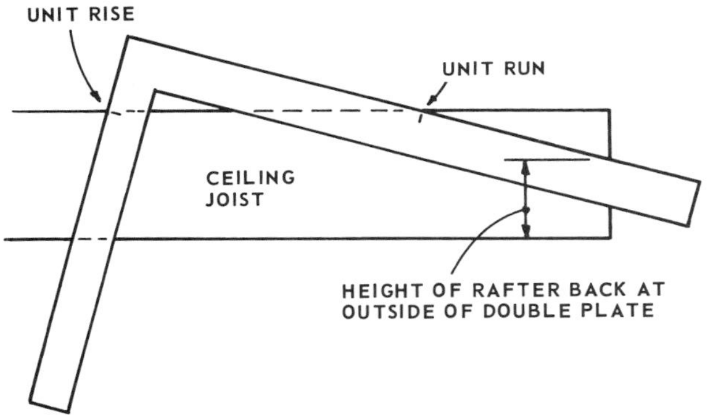

Fig. 8-39. Laying out the trim cut on the end of a ceiling joist to match the slope of the roof.

cur in low-pitched hip roof designs. The ceiling frame in this area should be constructed with stub joists running perpendicular to the regular joists, Fig. 8-40.

Lay out the position of the ceiling joists along the top plate using a rod in about the same manner as for floor joists. When a double plate is used, the joists do not need to align with the studs in the wall. This layout should, however, be coordinated with the position of the roof rafters so that some of the joists can be nailed directly to them. Ceiling joists are installed before the rafters and are toenailed to the plate using two 10d nails on each side.

Partitions or walls that run parallel to the joists must be fastened to the ceiling frame and a nailing strip to carry the ceiling material included. Various size materials can be installed in

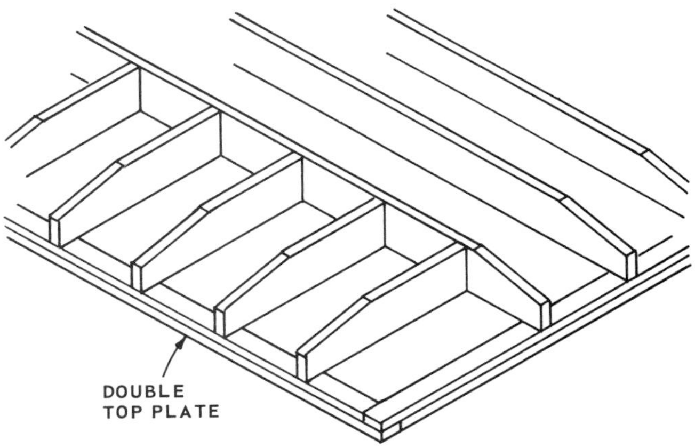

Fig. 8-40. Stub ceiling joists along the end wall. Required for a low-pitched hip roof.

a number of ways, the chief requirement being that they provide adequate support. Fig. 8-41 shows a typical method of making such an installation.

An access hole (also called a scuttle hole) must be included in the ceiling frame to provide an entrance to the attic area. Fire regulations and building codes usually list minimum size requirements. The architectural plans generally include size description and show where it should be located. The opening is framed following the procedure used for openings in the floor. If the size of the opening is relatively small (2 to 3 ft. square) doubling of joists and headers is not required.

Estimating Materials

To estimate framing materials in this area of work, first determine the total lineal feet by adding together the length of each wall and partition. The architectural plans will include the dimensions of outside walls. These can be added together. Partitions, especially those that are short, may not be dimensioned and you will need to scale the drawing. It is a good idea to place a colored pencil check mark on each wall and partition as its length is added to the list.

For plates, multiply the total figure by three (1 sole plate + 2 top plates) and add about 10 percent for waste. Order this number of linear feet of lumber in random lengths or convert to the number of pieces of a specific length. Three pieces of 2 x 4 one foot long equals 2 board feet; thus the total lineal feet of walls and partitions can be quickly converted to board measure if required.

Example:

Wall & Partition Length = 240
Total Plate Material = Length x 3 + 10%
= 240 x 3 + 10%
= 792 Lineal ft.
or 57 pcs - 2 x 4 x 14' - 0"

The total length of all walls and partitions is also used to estimate the number of studs required. When studs are spaced 16 in. O.C. multiply the total length by 3/4 and then add two studs for each corner, intersection and opening. Using the preceding figure for the wall length and assuming there are 12 corners, 10 intersections and 20 openings, the following example is included:

Total Studs = Total Length x 3/4 + 2 (corners + intersections + openings)

$$= \frac{240 \times 3}{4} + 2\ (12 + 10 + 20)$$

$$= \frac{\overset{60}{\cancel{240}} \times 3}{\cancel{4}} + 2\ (42)$$

$$= 60 \times 3 + 84$$

$$= 264$$

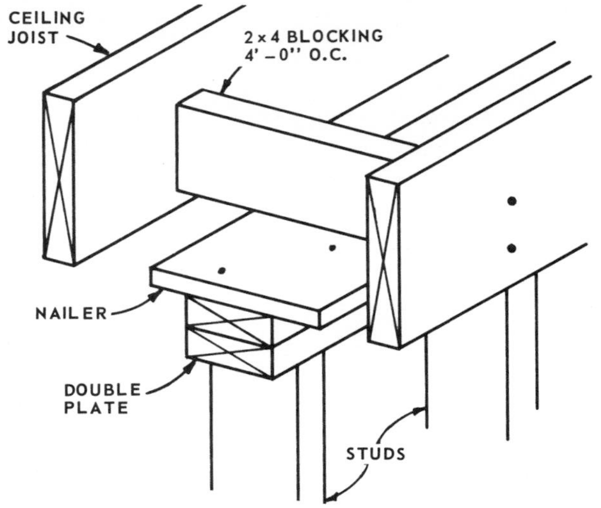

Fig. 8-41. Anchoring partitions to the ceiling frame when they run parallel to the joists.

Many carpenters estimate the number of studs by simply figuring one stud (spaced 16 in. O.C.) for each lineal foot of wall space and adding about 10 percent for waste. They figure the over-run on spacing will provide the extras needed for corners and openings. This method is rapid and fairly accurate. It will not allow for sufficient studs in a small house cut up into many rooms, while too many studs will likely be figured for a large house with wide windows and open interiors.

Lumber for headers must be calculated by analyzing the requirements for each opening. Use the R.O. width plus the thickness of the trimmers.

Ceiling joists are estimated by about the same method as is used for floor joists. Since ceilings will be relatively free of openings, no extras or waste needs to be included. Because of this, the short method should not be applied to ceiling joists. Use the following formula and include the size of the joists required:

Number of Ceiling Joists = Wall Length x 3/4 + 1

To estimate the amount of wall sheathing, first find the total perimeter of the structure. Multiply this figure by the wall height measured from the top of the foundation when the sheathing extends over the sill construction. The product will be the gross square footage of the wall surface. Now calculate the area of each major opening (windows and doors) and subtract this total from the original figure. Round the opening sizes downward to the nearest foot.

Net Area = Perimeter x Height - Wall Openings

Add to the net area to be sheathed allowances for waste and other extras when common boards or shiplap are used. See Unit 7 for more information. When using fiberboard or plywood for the sheathing material, there is only slight waste and the net area can be divided by the square footage per sheet to secure the number of pieces required. For example: if the net area to be sheathed is 1060 sq. ft. and the fiberboard sheets selected are 4' x 9', the calculations would be as follows:

$$\text{Net Area} = 1060$$

$$\text{Fiberboard Sheet Size} = 4 \times 9 = 36$$

$$\text{Number of Sheets} = \frac{1060}{36}$$

$$= 29 +$$

$$= 30$$

HANDY SAYS:

"Most problems in estimating, although based on simple formulas, usually contain so many variables that a lot of good judgment must be applied to their solution. This judgment is acquired through experience."

Test Your Knowledge - Unit 8

1. In modern platform construction, the horizontal member located at the bottom of a wall frame is called a __________ __________.
2. How many studs would be required for a plain wall panel 8' - 0" long if they are spaced 16 in. O.C.?
3. Trimmer studs stiffen the sides of an opening and carry the weight of the __________.
4. The first layout to be marked on the plates is the ________ __________spacing.
5. A master stud pattern is laid out somewhat like a __________ __________.
6. The layout of the cripple studs on the rough sill can be marked directly from the ________ __________.
7. Most carpenters prefer to erect the ____(side, end) walls first.
8. Standard let-in bracing is commonly made from ________(1 x 4, 2 x 4) lumber.
9. Joints formed along the doubled top plate should be at least ________apart.
10. Regular fiberboard sheathing __________(can, cannot) be used as a nailing base for exterior wall finish materials.
11. The position of the ceiling joists along the double plate should be coordinated with the __________.
12. The first step in estimating the number of studs required is to figure the total length of all__________ and __________.

Outside Assignments

1. Secure a set of architectural plans for a single-story house. Study the details of construction -- especially typical wall sections. Prepare a scale drawing of the framing required for the front walls. Be sure the rough openings are the correct size and in their proper location. Your drawing should look somewhat like the one in Fig. 8-1.
2. Working from the same set of plans, develop an estimated list of materials for the wall frame and sheathing. Include the number and size of studs; the number and size of lumber for headers; the material for plates, and type and amount of sheathing. Secure prices from your local supplier and figure the total cost of the materials.
3. Obtain descriptive literature about the various types of fiberboard and gypsum sheathing. Secure this material from local lumber dealers or write directly to manufacturers of these products. Also study books and other reference materials. Prepare a report based on the information you obtain. Include: grades, manufacturing processes, characteristics, and application requirements. Also include current prices and purchase information.

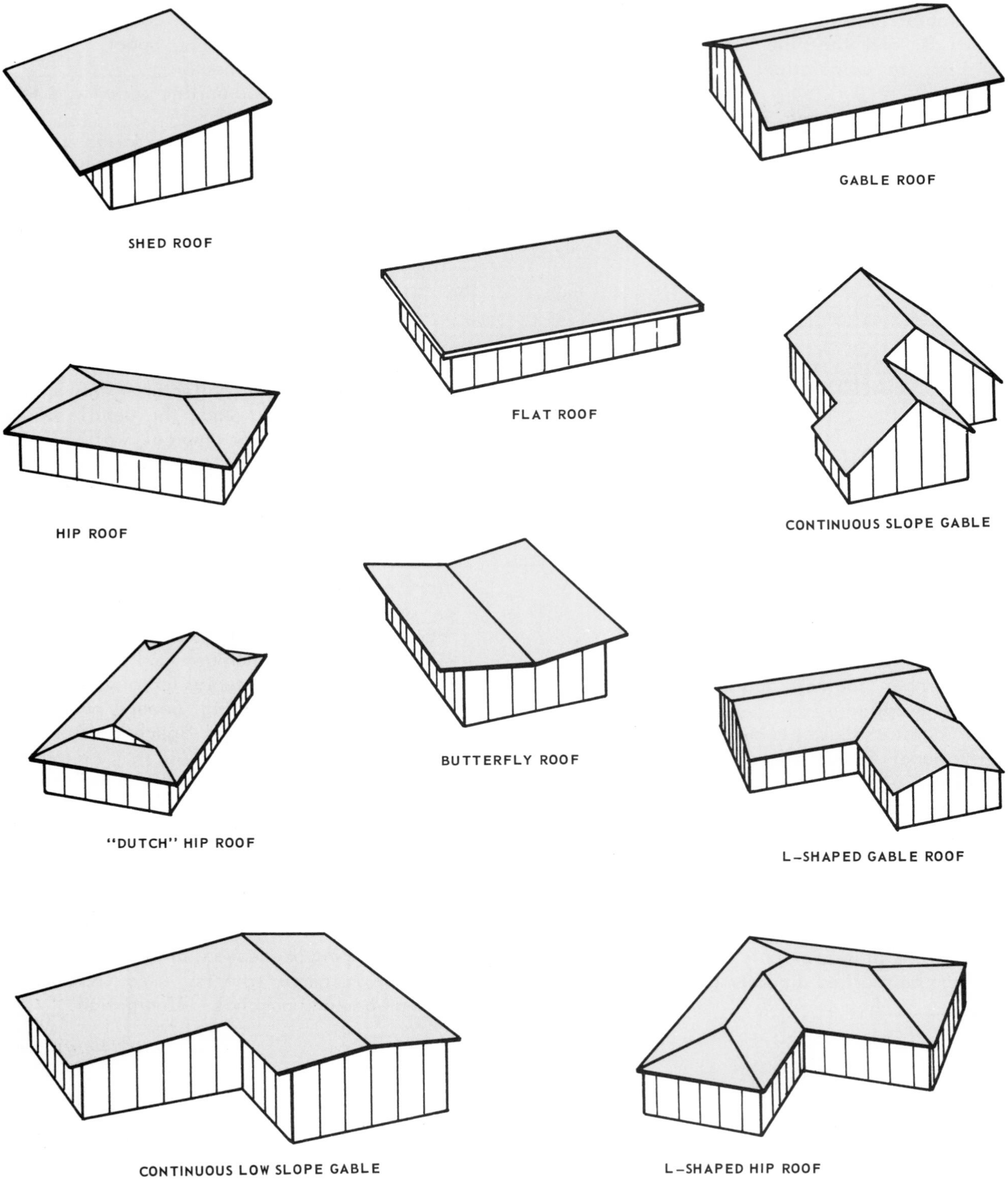

Fig. 9-1. Types of roofs used in residential construction.

Unit 9
ROOF FRAMING

Roof framing provides a base to which the roofing materials will be attached. It must be strong and rigid. Besides this, the roof -- if carefully designed and proportioned -- can contribute a distinctive and decorative feature to the structure.

In framing various types of roofs, you will encounter very little difficulty if you give careful attention to the principles involved and thoroughly understand the basic layouts and procedures presented in this Unit.

Roof Types

Although there is a wide variation in roof shapes, most of them can be classified among the following types, Fig. 9-1.

Flat roof: In this type, the roof is supported on joists that also carry the ceiling material. It may have a slight slope to provide drainage.

Shed roof: The simplest type of pitched (sloped) roof. Sometimes called a lean-to-roof since it is often a part of a larger structure. Used in contemporary designs where the ceiling is attached directly to the roof framing.

Butterfly roof: Sheds water toward the middle. Its use has become practical with the advent of improved methods in flashing, drainage, and waterproofing. Since this type tends to accumulate ice and snow, its use is usually limited to warmer climates.

Gable roof: Two surfaces slope from the center line of the structure forming gables on each end. Because of the simplicity of its design and relative low cost, the gable roof is used extensively for residential construction.

Hip roof: Consists of four sloping sides. The line where the adjacent sides intersect is called the hip. One advantage of this type results from the protective overhang formed on end walls as well as side walls.

Gambrel roof: A variation of the gable roof where each slope is broken, usually near the center. This style is used on two-story construction and permits more efficient use of the second floor level. Dormers, Fig. 9-2, above, are usually included. Typical of colonial America and the period immediately following.

Mansard roof: Similar to the hip roof except that each of the four sides has a double slope. The lower slope approaches the vertical while the upper slope is more or less flat. Like the gambrel roof, the main advantage is the additional space formed in the rooms on the upper level. The name is derived from its originator, architect Francois Mansart (1598-1666). Fig. 9-2, below.

Fig. 9-2. Traditional roof designs. Above. Gambrel. Below. Mansard.

Parts of Roof Frame

The plan view of a roof, Fig. 9-3, is a composite of several roof types with the kinds of rafters identified.

The COMMON rafters are those that run at a right angle (plan view) from the wall plate to the ridge. A plain gable roof consists entirely of rafters of this kind. HIP rafters also run from the plate to the ridge, but at a 45 deg. angle. They form the intersections of the adjacent slopes of a hip roof. VALLEY rafters extend diagonally from the plate to the ridge in the hollow formed by the intersection of two roof sections.

Parts of a Rafter

Rafters are formed by laying out and making various cuts. Fig. 9-4 shows the cuts for a common rafter and the sections formed. The ridge cut allows the upper end to fit tightly against the ridge. The bird's-mouth is formed by a seat cut

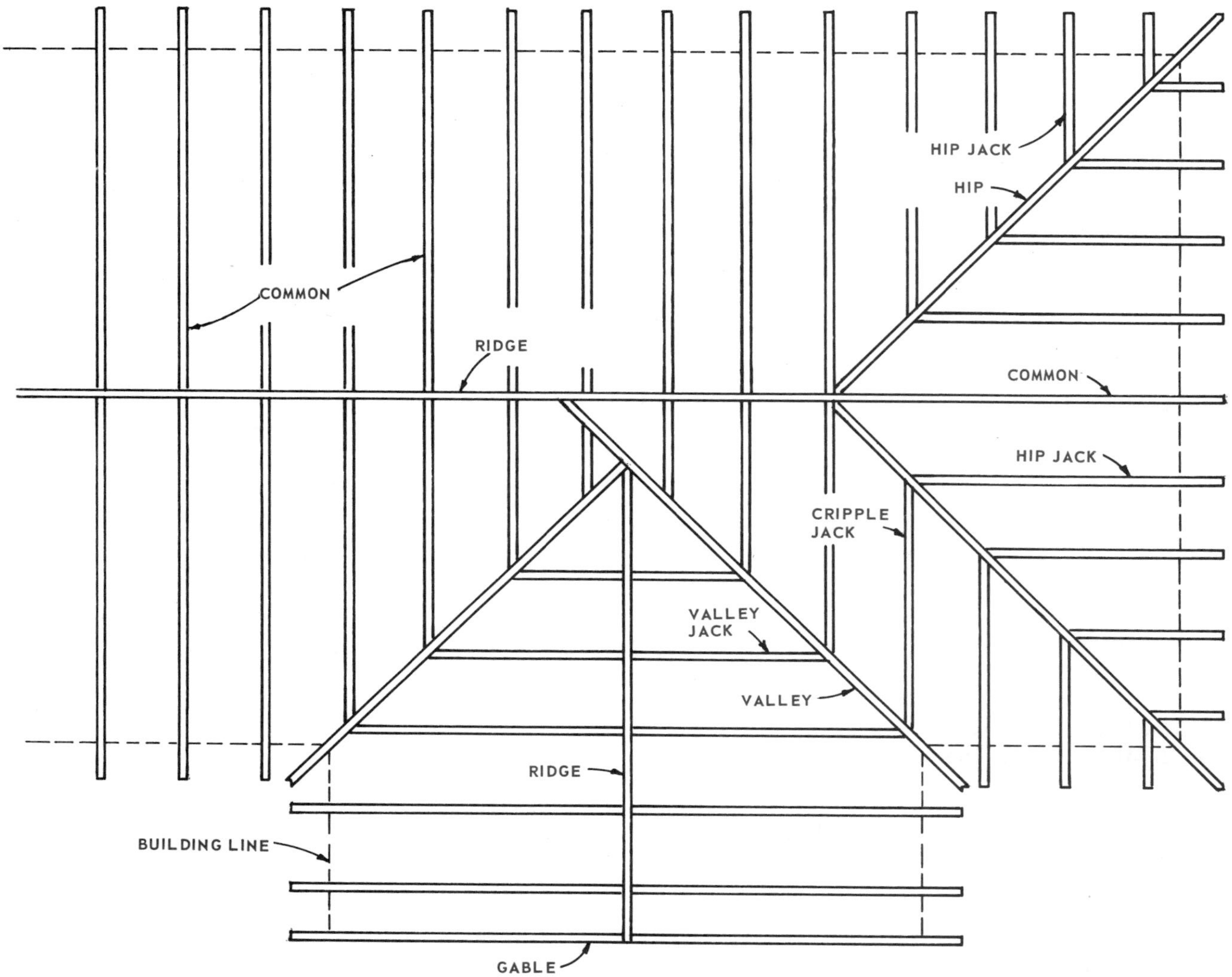

Fig. 9-3. Plan view of roof frame, showing ridges and various kinds of rafters.

Three kinds of jack rafters are: the HIP JACK which is the same as the lower part of a common rafter but intersects a hip rafter instead of the ridge. The VALLEY JACK is the same as the upper end of a common rafter but intersects a valley rafter instead of the plate. The CRIPPLE JACK rafter, also called a cripple rafter, intersects neither the plate or the ridge and is terminated at each end by hip and valley rafters. The cripple jack rafter may be further defined as a hip-valley cripple jack or a valley cripple jack.

and plumb (vertical) cut when the rafter extends beyond the plate. This extension is called the overhang or tail. When there is no overhang, the bottom of the rafter is terminated by a seat cut and a plumb cut that extends upward.

Layout Terms and Principles

Roof framing is a practical application of geometry; the area of mathematics that deals with the relationships of points, lines, and surfaces. It

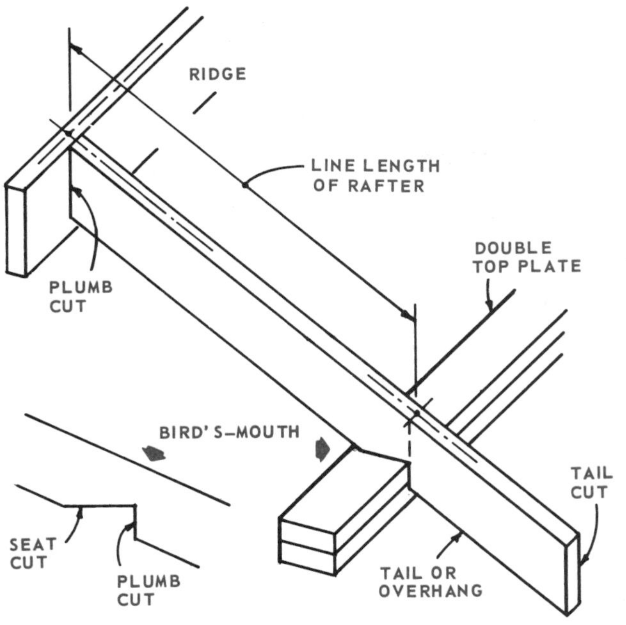

Fig. 9-4. Parts of a rafter.

is based largely on the properties of the right triangle where the horizontal distance is the base; the vertical distance is the altitude; and the length of the rafter is the hypotenuse.

If any two sides of a right triangle, Fig. 9-5, are known, the third side can be found mathematically. The formula used is: $H^2=B^2+A^2$, where H is the hypotenuse, B is the base and A is the altitude. The solution involves extracting the square root which is a rather time consuming process. The answer could also be found through the application of trigonometric functions.

In on-the-job use, the carpenter uses the tables on the framing square or a direct layout method that is rapid and practical for his work.

In rafter layout, the base of the right triangle is called the run. It is measured from the outside of the plate to the center of the ridge. The altitude or rise is the total distance the rafter extends above the plate. Other layout terms and the relationship of the parts of the roof frame are illustrated in Fig. 9-6.

Slope and Pitch

Slope indicates the incline of a roof as a ratio of the vertical rise to the horizontal run. It is properly expressed as X distance in 12. For example: a roof that rises at the rate of 4 in. for each foot of run, is designated as having a 4 in 12 slope. A triangular symbol above the roof line in the architectural plans is used to convey this information. The slope of a roof is sometimes called the "cut of the roof."

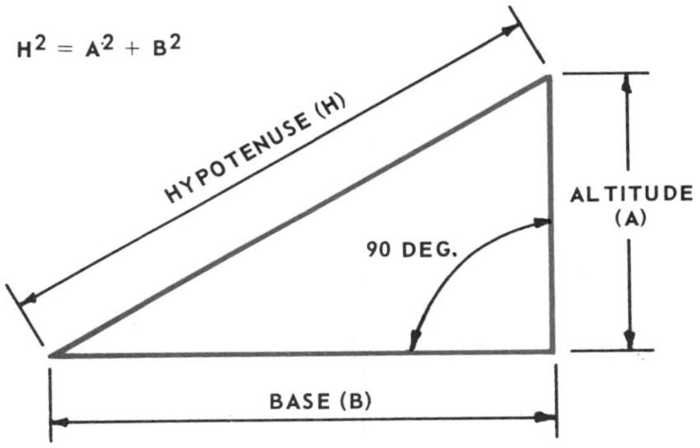

Fig. 9-5. Mathematical solution of a right triangle.

Pitch indicates the incline of the roof as a ratio of the vertical rise to the span (twice the run). It is expressed as a fraction. For example: if the total roof rise was 4 ft. and the total span was 24 ft., the pitch would be 1/6.

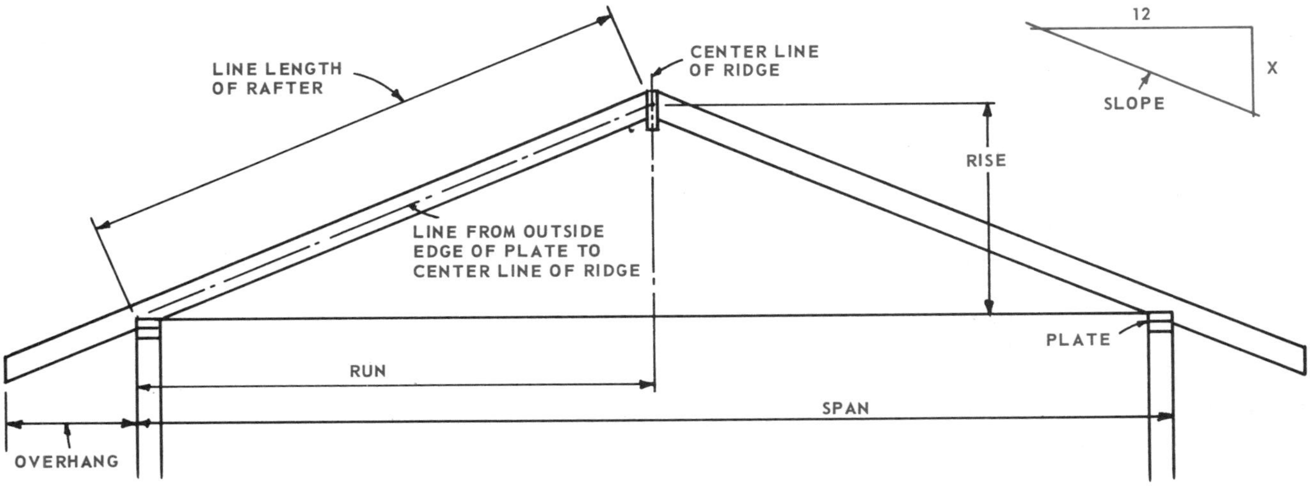

Fig. 9-6. Basic terms and dimensions used in rafter layout.

Unit Measurements

The framing square, also called a steel square or carpenters' square, is the basic layout tool in roof framing. The side with the manufacturer's name is called the face and the opposite side the back. The longer arm (24 in.) is called the body or blade while the shorter one (16 in.) is called the tongue.

Because this measuring instrument (framing square) is not large enough to make the rafter layout at one setting, it is necessary to use smaller divisions that are called units. The foot (12 in.) is the standard unit for horizontal run and the unit rise is always based on this distance. The unit run and rise is used to lay out plumb and level cuts, Fig. 9-7, required at the ridge, bird's-mouth and tail of the rafter.

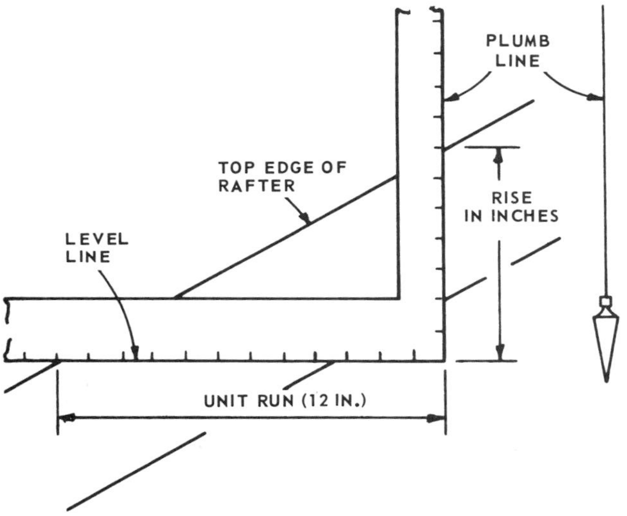

Fig. 9-7. Unit measurements used to lay out horizontal and plumb lines.

Framing Plans

When working with simple designs, the carpenter can easily visualize the roof framing. Since the wall framing is already in place, the only additional information needed will be the slope of the roof and the amount of overhang required; items which are included in the architectural plans. When the structure consists of a complicated roof design, the architect often includes a roof framing plan. If, however, a framing plan is not included, then the carpenter should prepare his own plan. This can be accomplished by making a scaled drawing on tracing paper laid directly over the floor plans. Include ridges, overhang and all the various rafters. The drawing may be made similar to the one in Fig. 9-3, although satisfactory results may be obtained by simply using a single line to represent the framing members.

SIZE OF RAFTER (Inches)	SPACING OF RAFTER (Inches)	MAXIMUM ALLOWABLE SPAN (Feet and Inches Measured Along the Horizontal Projection)			
		Group I	Group II	Group III	Group IV[2]
2 x 4	12	10-0	9-0	7-0	4-0
	16	9-0	7-6	6-0	3-6
	24	7-6	6-6	5-0	3-0
	32	6-6	5-6	4-6	2-6
2 x 6	12	17-6	15-0	12-6	9-0
	16	15-6	13-0	11-0	8-0
	24	12-6	11-0	9-0	6-6
	32	11-0	9-6	8-0	5-6
2 x 8	12	23-0	20-0	17-0	13-0
	16	20-0	18-0	15-0	11-6
	24	17-0	15-0	12-6	9-6
	32	14-6	13-0	11-0	8-6
2 x 10	12	28-6	26-6	22-0	17-6
	16	25-6	23-6	19-6	15-6
	24	21-0	19-6	16-0	12-6
	32	18-6	17-0	14-0	11-0

Fig. 9-8. Sample table showing maximum runs allowed for rafters sloped 4 to 12 or greater. Groups refer to species of wood. Be sure to secure this kind of information from your local building code.

Accurate spacing should be maintained between rafters with hips and valleys drawn at a 45 deg. angle and jack rafters drawn parallel to the common rafters.

Size of Material

As in floor and ceiling framing, the cross section size of the rafter is determined by the spacing, and span or length. Requirements will vary for different localities and local building codes must be consulted. A table listing rafter sizes for various loads and spacings is shown in Fig. 9-8. Stock for hips, valleys, and ridges is usually larger than that of the other framing members.

For purposes of estimating and ordering material, the rafter lengths can be determined with fair accuracy by making a scaled layout with the framing square. Use the back of the square where the outside edge of the blade and tongue are divided into inches and twelfths. Assume the inches to be feet and each division an inch.

First draw a triangle, using the unit run and unit rise specified by the slope of the roof. If the total run is more than 12 ft.-- extend the length of the base and hypotenuse. Now position the square along the base line until the total run aligns with the acute angle and mark the point where the tongue crosses the sloping line, Fig. 9-9. Use either the blade or tongue of the square and measure the hypotenuse to secure the rafter length. Be sure to add the overhang and allow extra material, if needed, for cuts at the ridge and tail.

Laying Out Common Rafters

Rafters can be laid out by the step-off method or their length can be calculated by using figures located in the table on the framing square. Most carpenters prefer the step-off method and then use the rafter tables to check their work. By either method, a pattern rafter is first carefully laid out, checked, and cut. This is then used to mark other rafters of the same size and kind.

For the pattern layout, select a piece of lumber that is straight and true. Place the stock on a pair of sawhorses. Usually the carpenter will stand on the side that will be the top edge of the rafter because it is easier to hold and manipulate the framing square. In order to provide illustrations describing rafter layouts that can be quickly understood, this position has been reversed -- with the rafter shown in the position it will have when installed in the roof frame.

HANDY SAYS:

"As you lay out rafters, try to visualize how each will appear when in its final position in the completed roof frame. Forming the habit of visualizing the rafter in its proper place will help to eliminate errors."

To lay out a rafter by the step-off method, place the framing square on the stock and align the figures with the top edge of the rafter -- unit run (12 in.) on the blade and the unit rise on the tongue. Patented clips are available that will maintain the square in this position or a pair of handscrews can be clamped to the blade and tongue. To insure accuracy, the figures on the square must be positioned exactly over the edge of the stock each time a line is marked. Be sure to use a sharp pencil to make the layout lines.

Start at the top of the rafter; hold the square in position and draw the ridge line. Continue to hold the square in the same position and mark the length of the odd unit (8 in. used in the example). Now shift the square along the edge of the stock until the tongue is even with the 8 in. mark. Draw a line along the tongue and mark the 12 in. point on the blade for a full unit, Fig. 9-10.

Move the square to the 12 in. point just laid out and repeat the marking procedure. Continue until the required number of full units are laid out. This number will be equal to the number of feet in the total run (6 used in the example).

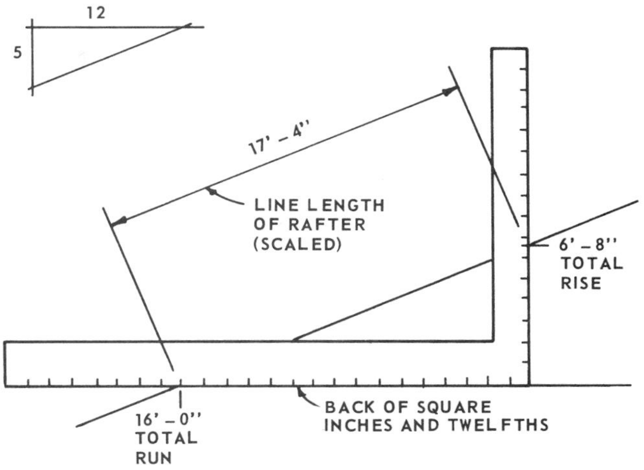

Fig. 9-9. Estimating rafter lengths with framing square.

Form the bird's-mouth by drawing a horizontal line (seat cut) to meet the building line so the surface will be about equal to the width of the plate. The size of the bird's-mouth may vary depending upon the design of the overhang. In the illustration, Fig. 9-10, note that the position of the square has been turned over to mark these cuts and also to lay out the overhang. This may or may not be necessary depending on the length of the rafter blank.

To lay out the overhang, start with the plumb cut of the bird's-mouth and mark full units first and then any odd unit that remains. The tail cut may be plumb, square, or a combination of plumb and level. Check the cornice details shown in the architectural plans for exact requirements.

The final step in the layout consists of shortening the rafter at the ridge. With the square in position, draw a plumb line spaced from the ridge line a horizontal distance equal to one-half the thickness of the ridge, Fig. 9-10. Now make the cuts you have laid out and label the rafter as a pattern, indicating the roof section to which it belongs.

Using the Rafter Table

You can calculate the length of a common rafter using the table on the framing square. See Fig. 9-11. Under the full scale number that corresponds with the unit rise (example: 5) secure the number in the first line. This is the line length of

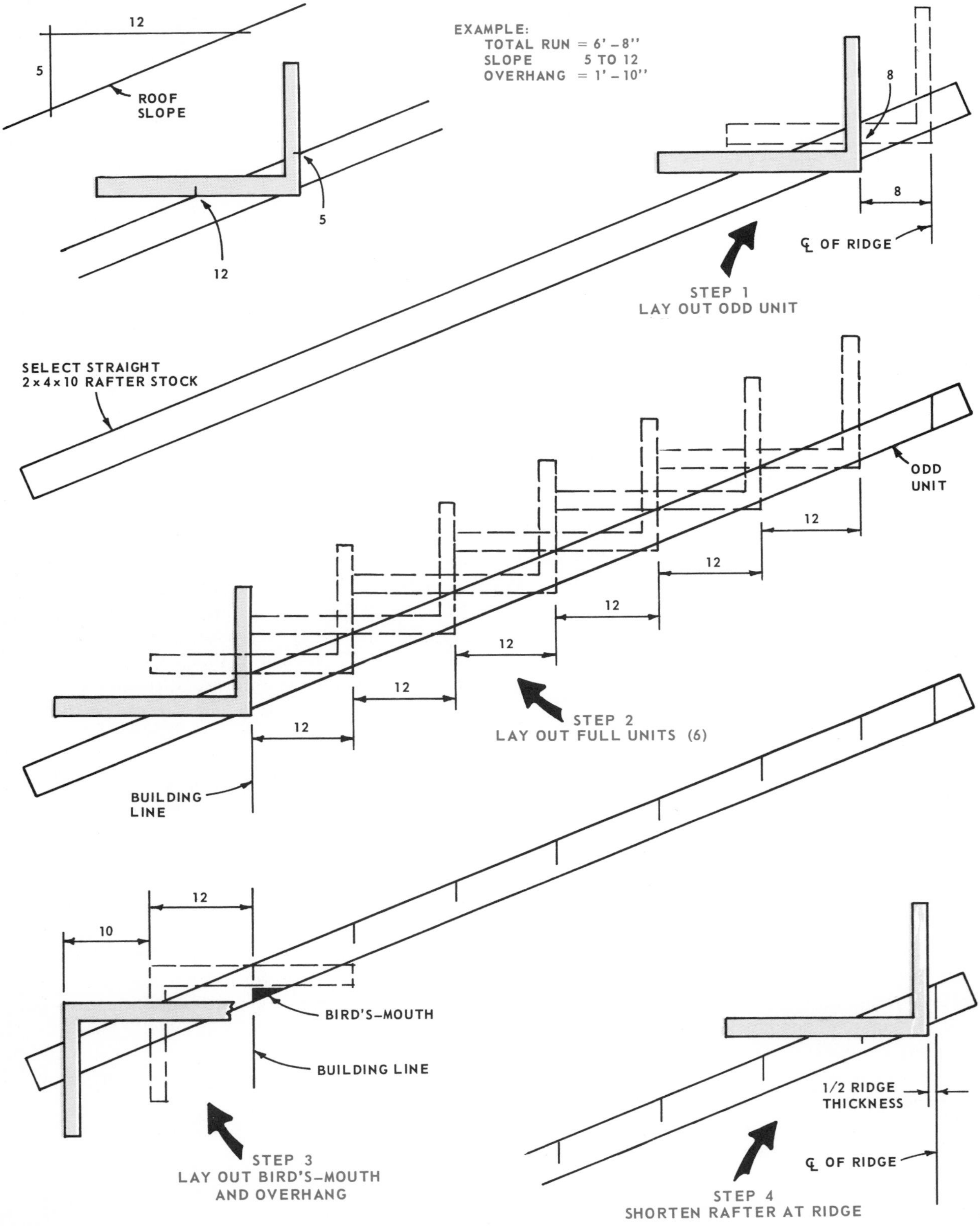

Fig. 9-10. Procedure for laying out a common rafter by the step-off method.

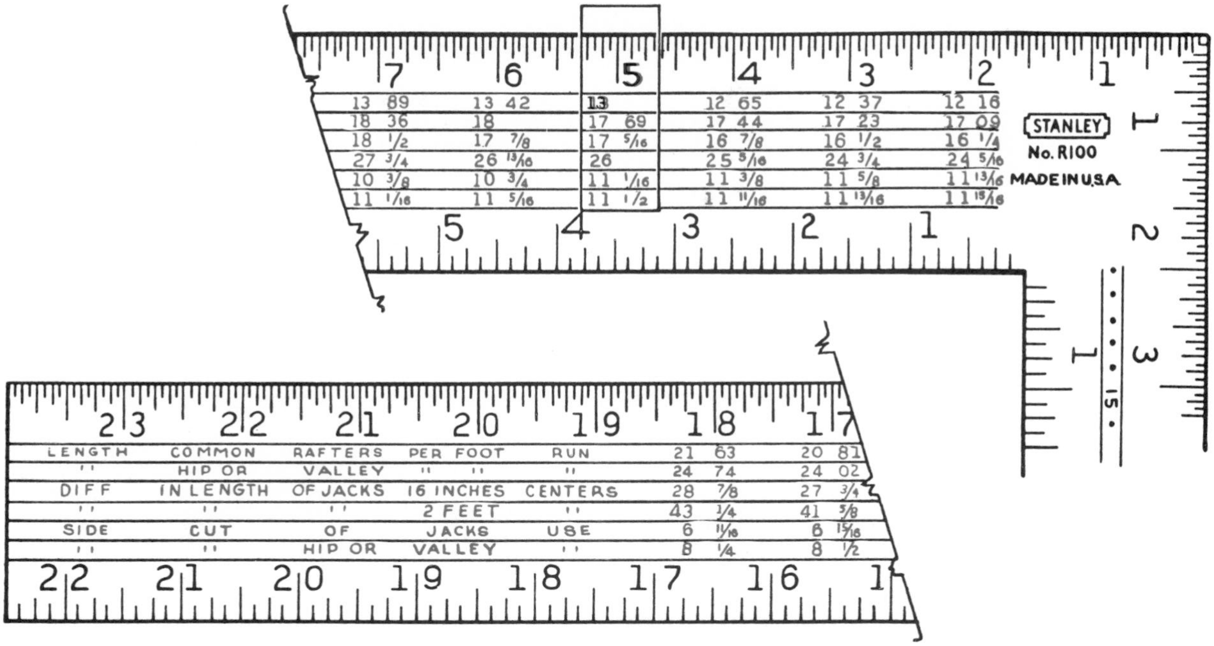

Fig. 9-11. Rafter table on framing square.

the rafter in inches for one foot of run. To find the length of the rafter from the building line to the center of the ridge, multiply the units of run by the figure from the table as shown below:

Example No. 1 Run = 6' - 8" Slope - 5 to 12

Run = 6' - 8"	= 6 2/3 units
Table No.	= 13
Rafter Length	= 6 2/3 units x 13"
	= 86 2/3"
	= 7' - 2 2/3"

Example No. 2 Run = 10' - 4" Slope - 4 to 12

Run = 10' - 4"	= 10 1/3 units
Table No.	= 12.65
Rafter Length	= 10 1/3 x 12.65"
	= 130.72"
	= 10' - 10.72"
	= 10' - 10 3/4"

The above calculations give the line length of the rafter -- running from the center of the ridge to the outside of the plate. If used to make the pattern layout, the overhang will need to be added and the rafter will also need to be shortened at the ridge.

Using the rafter pattern, cut the number required, Fig. 9-12. Some carpenters prefer to stand the completed rafters along the outside wall so they will be easily accessible during the assembly of the roof frame.

Fig. 9-12. Cutting bird's-mouth in common rafter using special cutterhead on radial arm saw.

Erecting a Gable Roof

In conventional framing, it is considered good practice to lay out the rafter spacing along the wall plate at the same time the layout is made for the ceiling joists. When rafters are spaced 2 ft. O.C. and ceiling joists are spaced 16 in. O.C., the layout is coordinated as shown in Fig. 9-13. The plate layout is important and the roof framing plan should be carefully followed.

Select straight pieces of ridge stock and lay out the rafter spacing by transferring the marking directly from the plate or a layout rod. Joints in the ridge should occur at the center line of a rafter. Cut the pieces that will make up the ridge and lay them across the ceiling joists, close to where they will be assembled with the rafters.

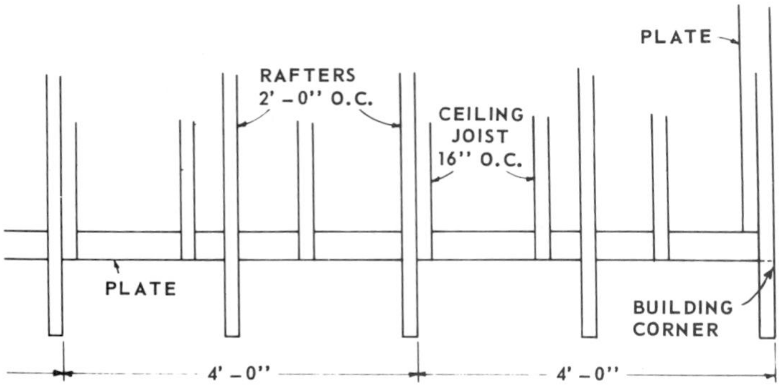

Fig. 9-13. Plan view of ceiling joist and rafter layout. A joist is nailed to every other rafter, and serves as a tie to prevent the wall from spreading.

In preparing to assemble a roof frame, be sure that rafters, ridge boards, and temporary bracing are readily accessible. Select straight rafters for the gable end and nail one in place at the plate as shown in Fig. 9-14. Install a rafter on the oppo-

Fig. 9-14. Nailing the first rafter to the plate. (Western Wood Products Assoc.)

site side with a workman at the ridge continuing to support both rafters. Now place the ridge between the two rafters and nail it temporarily in place. Move about 5 rafter spaces from the end and install another pair of rafters. Plumb and brace the assembly and make any adjustments necessary in the nailing of the first rafters. Fig. 9-15 shows the assembly and nailing pattern at the plate. Special framing anchors, Fig. 9-16, are often used.

To make the initial assembly, some carpenters prefer to first attach the ridge onto several rafters on one side. This assembly is then raised, the rafters nailed to the plate, and then several rafters are installed on the opposite side.

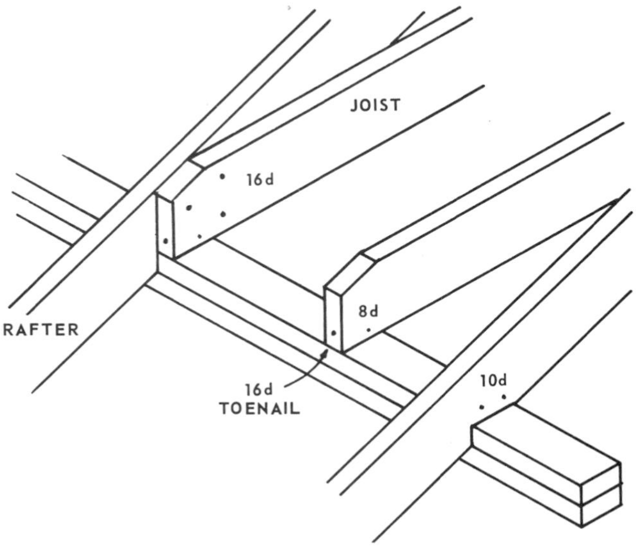

Fig. 9-15. Nailing pattern for joists and rafters at wall plate. Toenailing is duplicated on the opposite side.

Fig. 9-16. Framing anchors, designed especially for joining rafters to the plate. (Timber Engineering Co.)

Install the intervening rafters. First nail the rafter at the plate and then at the ridge. As shown in Fig. 9-17, drive 16d nails through the ridge into the rafter. The rafters on the opposite side

Fig. 9-17. Assembling a rafter to the ridge with 16d nails. (Western Wood Products Assoc.)

of the ridge are toenailed. Installing only a few rafters on one side before placing matching rafters on the opposite side will make it easier to keep the ridge straight.

Continue to add sections of ridge and assemble the rafters, Figs. 9-18 and 9-19. Sight along the ridge to see that it is straight and level. Add bracing when required. Always install rafters with the crown (curve or warp) turned upward.

HANDY SAYS:

"Use extra care when framing a roof to prevent a fall. Erect solid scaffolding wherever it will be helpful. Avoid working directly above another person."

Gable End Frame

Square a line across the end wall plate directly below the center of the gable. If a ventilator is to be installed, measure one-half of the opening size on each side of the center line, and mark for the first stud. Lay out the balance of the stud spacing.

Stand a stud upright at the first space and plumb it with a level. Mark across the edge of the stud at the underside of the rafter. Repeat the operation at the second space. The distance between the two lengths, Fig. 9-20, will be the common difference and can be used to lay out the length of all the other studs. When the spacing is laid out from a center line as described -- the studs can be cut in pairs and fitted into the frame.

The common difference of the stud lengths can also be secured from a manipulation of the framing square, Fig. 9-21. Set the square for the unit run and rise, and mark a line along the blade.

Fig. 9-18. Roof assembly. Adding ridge section.

Fig. 9-19. Common rafters on a residential structure that will require a complicated roof frame. (National Forest Products Assoc.)

Now slide the square along this line until the stud spacing (Example, 16" O.C.) aligns with the edge. The measured distance along the tongue of the square will be the common difference.

Hip and Valley Rafters

Hip roofs or intersecting gable roofs consist of some or all of the following rafters: common, hip, valley, jack, and cripple.

First cut and frame the common rafters and ridge boards. The ridge of a hip roof is cut to a length equal to the length of the building minus twice the run and plus the thickness of the rafter stock. It should intersect with the common rafters as shown. See Fig. 9-23. From the corners of the building, lay out along the side walls a distance equal to one-half the span. These points will be the center of the first common rafters. All other rafters, both common and jack, are laid out from this position.

Where the two roof surfaces slant upward from adjoining walls, they will meet on a sloping line called a hip. The rafter supporting this intersection is known as a hip rafter. In a plan view, the hip rafter will be observed as a diagonal of a square, Fig. 9-24. Two common rafters form two sides of the square while the outside walls form the other two sides. The diagonal of this square is the total run of the hip rafter.

Since the unit run of the common rafter is 12 in., the unit run of the hip rafter will be the diagonal of a 12 in. square. When calculated accurately, this is 16.97 but for actual application is rounded-out to 17 in. To lay out a hip rafter, therefore, follow the same procedure used for a common rafter, with the exception that the 17 in. mark is used on the blade of the square instead of the 12 in. mark, Fig. 9-25. The odd unit must also be adjusted. Its length is found by measuring the diagonal of a square, the sides of which are equal to the length of the odd unit.

Valley rafters will also be the diagonal of a square; the sides of which are formed by ridges and common rafters. The layout is made in the same manner as described for hip rafters with 17 in. used for the unit run.

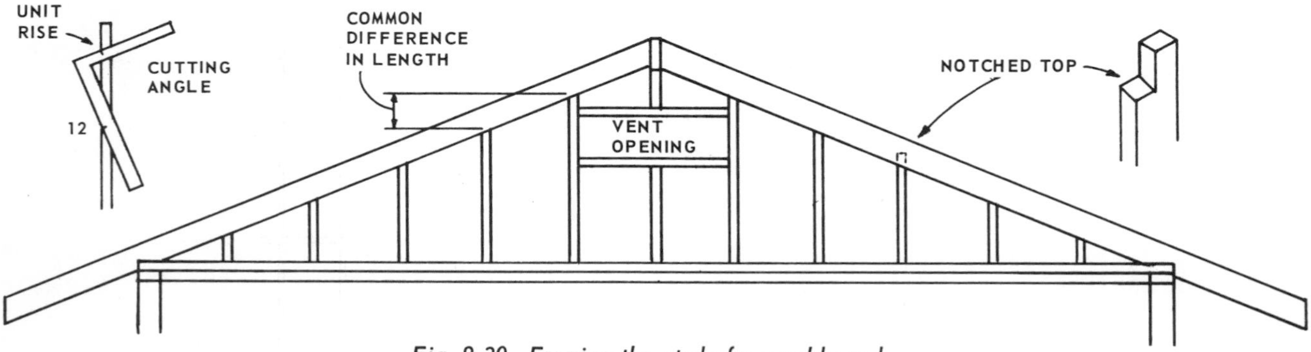

Fig. 9-20. Framing the studs for a gable end.

In modern residential construction, roof designs often include an extended rake (gable overhang). Typical framing as illustrated in Fig. 9-22, requires the construction of the gable end frame before the roof frame is completed. The architectural plans will usually include requirements for the size and spacing of frame members.

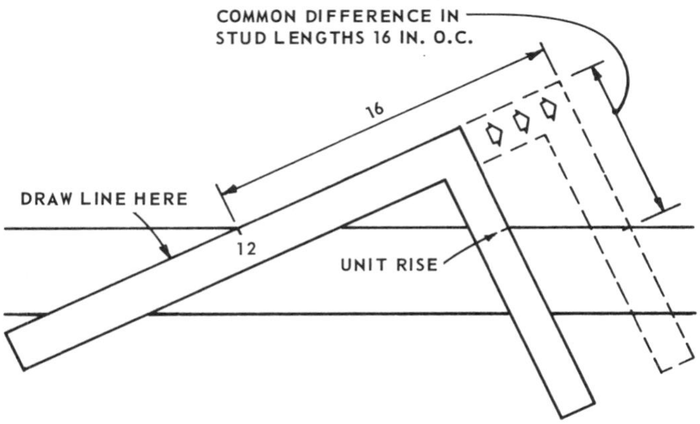

Fig. 9-21. Using the framing square to find the common difference in stud lengths for a gable.

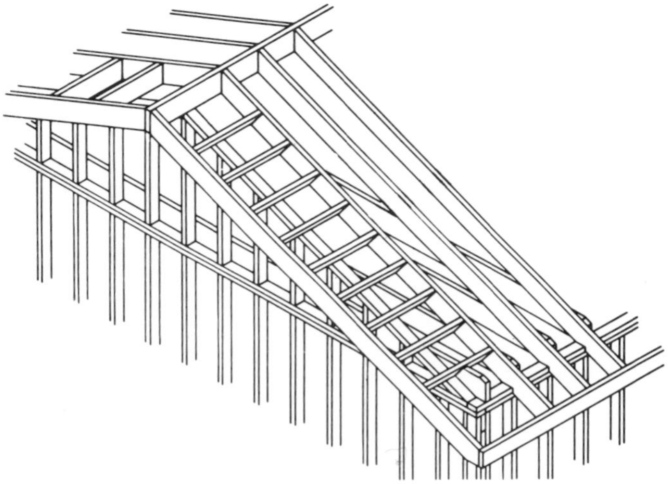

Fig. 9-22. Typical framing for an extended roof rake at a gable end.

The line length of hip and valley rafters can be determined from rafter tables just like common rafters. Using the same figures as used in a previous example, the calculations would be as follows:

Run = 6' - 8"	Slope 5 to 12
Run = 6' - 8"	= 6 2/3 units
Table No. 2	= 17.69
Hip or Valley Length	= 17.69 x 6 2/3
	= 117.93
	= 9' - 9 15/16"
	= 9' - 10"

Hip and valley rafters must be shortened at the ridge by a horizontal distance equal to one-half of the 45 deg. thickness of the ridge, Fig. 9-26. The side cuts are then laid out as shown in the illustration, or by using figures from the sixth line of the rafter table (Example, 11 1/2). All the figures in the table are based on or related to 12, so place 12 on the blade and 11 1/2 on the tongue along the edge of the rafter as shown and draw the

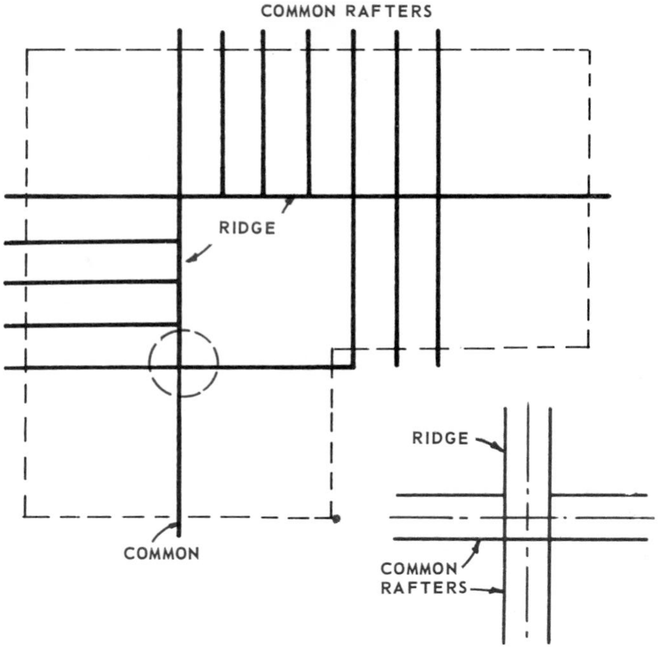

Fig. 9-23. The first step in framing a hip or intersecting roof is to install the common rafters and ridges.

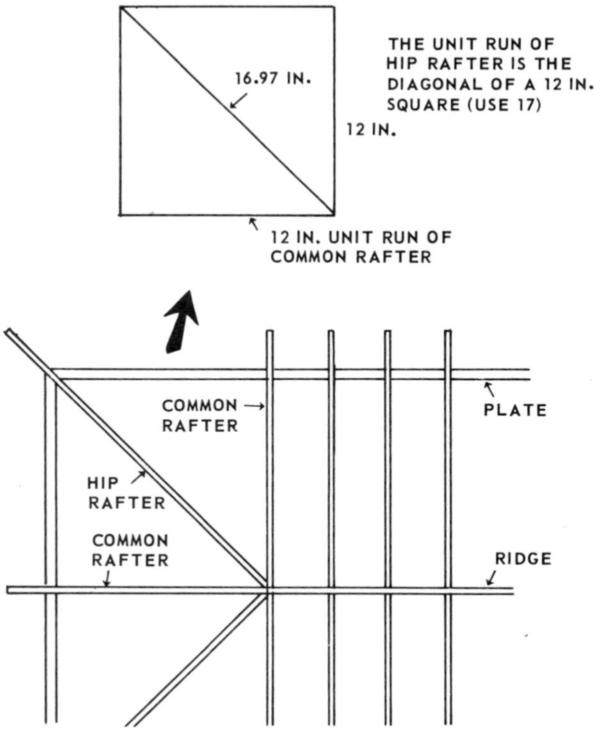

Fig. 9-24. The hip rafter is a diagonal of a square formed by the walls and common rafters.

angle for each cut, Fig. 9-27. Now draw plumb lines, using 17 on the blade and the unit rise on the tongue. The tail cuts at the end of the rafters are laid out using the same angle.

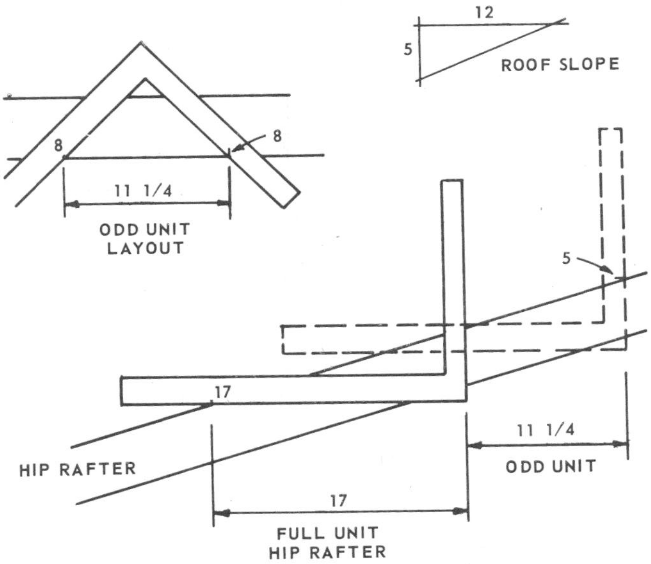

Fig. 9-25. Starting the layout of a hip rafter. Slope and odd unit size are the same as used in the example for the common rafter layout.

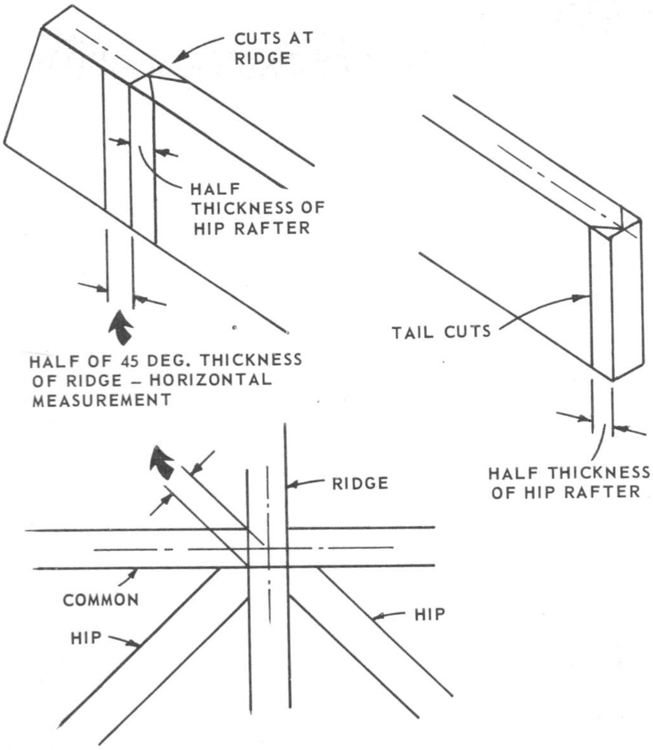

Fig. 9-26. Shortening a hip rafter and making side cuts. With the exception of the tail cut, the layout can also be applied to a valley rafter.

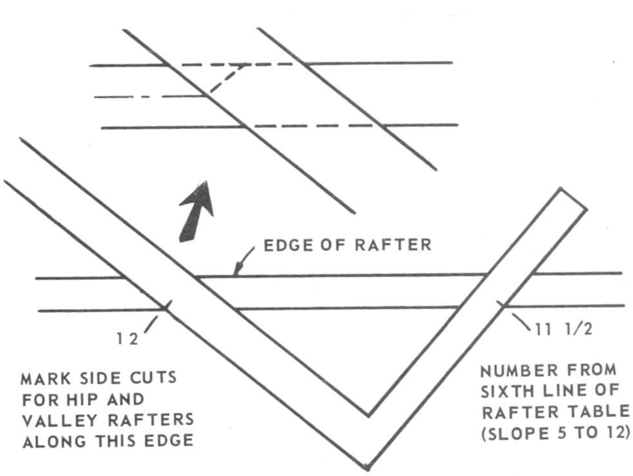

Fig. 9-27. Using the framing square to lay out side cuts on hip and valley rafters.

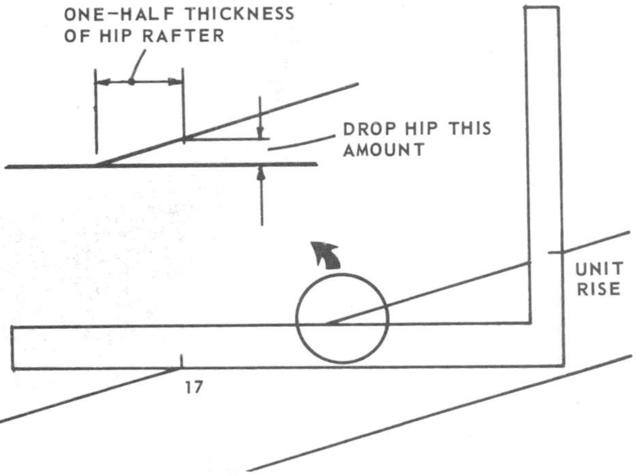

Fig. 9-28. Layout to determine distance to drop a hip rafter.

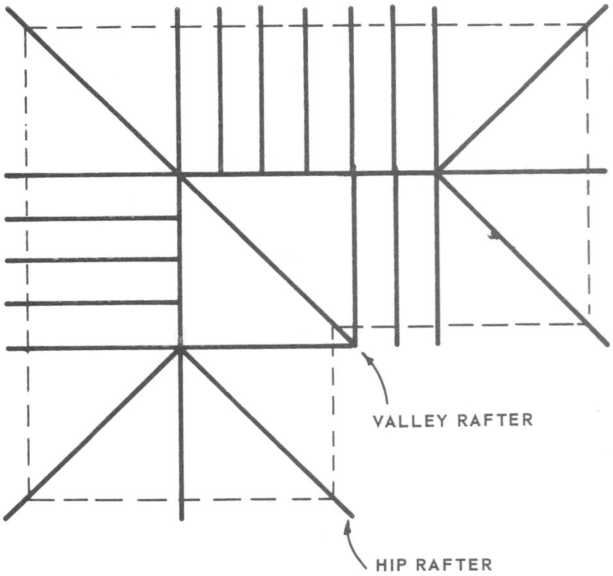

Fig. 9-29. Hip and valley rafters added to the roof frame.

A center line along the top edge of a hip rafter is where the roof surfaces actually meet. The corners of the rafter extend slightly above this line and some adjustment must be made. They could be planed off. However, it is easier to make the seat cut slightly deeper thus lowering the entire rafter, Fig. 9-28.

The plumb cut of the bird's-mouth for valley rafters must be trimmed so it will fit into the corner formed by the walls. Although side cuts of approximately 45 deg. could be made, it is more practical to move the plumb cut toward the tail of the rafter by a horizontal distance equal to the 45 deg. thickness of the rafter. The tail cut of the valley rafter must be shortened by this same amount to provide space for the intersection of the fascia board of the cornice.

After hip and valley rafters are laid out and cut, they are added to the roof frame as shown in Fig. 9-29.

Jack Rafters

Hip jack rafters have the same tail and overhang as common rafters. Since they intersect the hip rafter, their length from the bird's-mouth to this point of intersection varies. When they are equally spaced along the plate, this variance is consistent and is referred to as the common difference, Fig. 9-30.

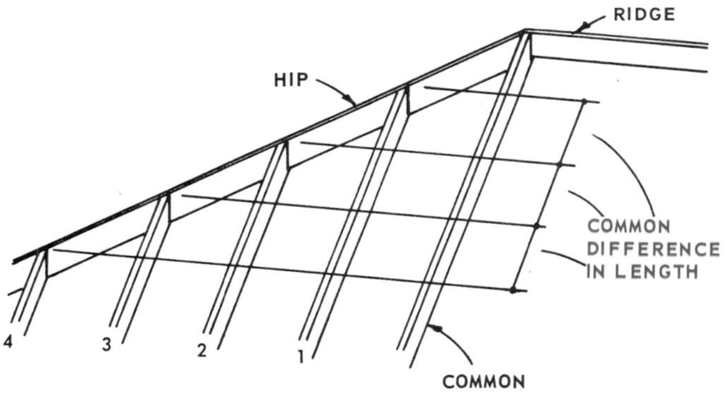

Fig. 9-30. When evenly spaced, hip jack rafters have a common difference.

The difference in length (common difference) can be secured from the third or fourth line of the rafter table, Fig. 9-11. For a roof slope of 5 to 12, with rafters spaced 24 in. O.C., the figure from the table is 26 in. The common difference can also be found through a manipulation of the framing square as illustrated in Fig. 9-31. Hold the square along the edge of a smooth piece of lumber according to the unit run and rise of the roof. Draw a line along the blade and then slide the square along this line to a point equal to the rafter spacing. The distance thus laid out along the edge of the lumber will be the required difference in length.

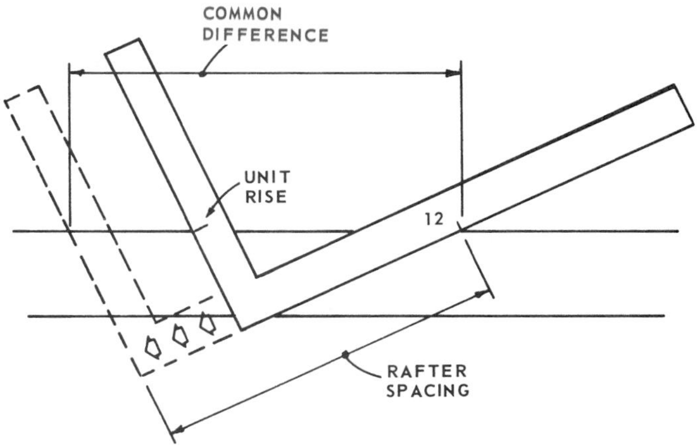

Fig. 9-31. Using the framing square to determine the common difference of jack rafters.

To make the layout for jack rafters, select a piece of straight lumber and lay out the bird's-mouth and overhang from the common rafter pattern. Next, lay out the line length of a common rafter -- the distance from the plumb cut of the bird's-mouth to the center line of the ridge. For the first jack rafter down from the ridge, lay out the common difference in length. Now shorten the jack rafter by one-half the 45 deg. thickness of the hip, Fig. 9-32. Square this line across the top of the rafter and mark the center point. Through this point, lay out the side cut as shown in the illustration, or by using the number from the fifth line of the rafter table, Fig. 9-33. Mark the plumb lines that will be followed when the cut is made.

Move down the rafter the common difference and mark the cutting line for the next hip jack. A sliding T-bevel will be a good tool to use. Continue until they are all laid out and then use this pattern to mark the jack rafters required. Each hip in the roof assembly requires one set of jack rafters made up of matching pairs. A pair consists of two rafters of the same length with the side cuts made in opposite directions.

Similar procedures are followed in laying out valley jack rafters. For these, however, it is usually best to start the layout at the building line, Fig. 9-34, and move toward the ridge. The longest valley jack will be the same as a common rafter except for the side cut at the bottom. Use the common rafter pattern and extend the plumb cut of the bird's-mouth to the top edge. Lay out the

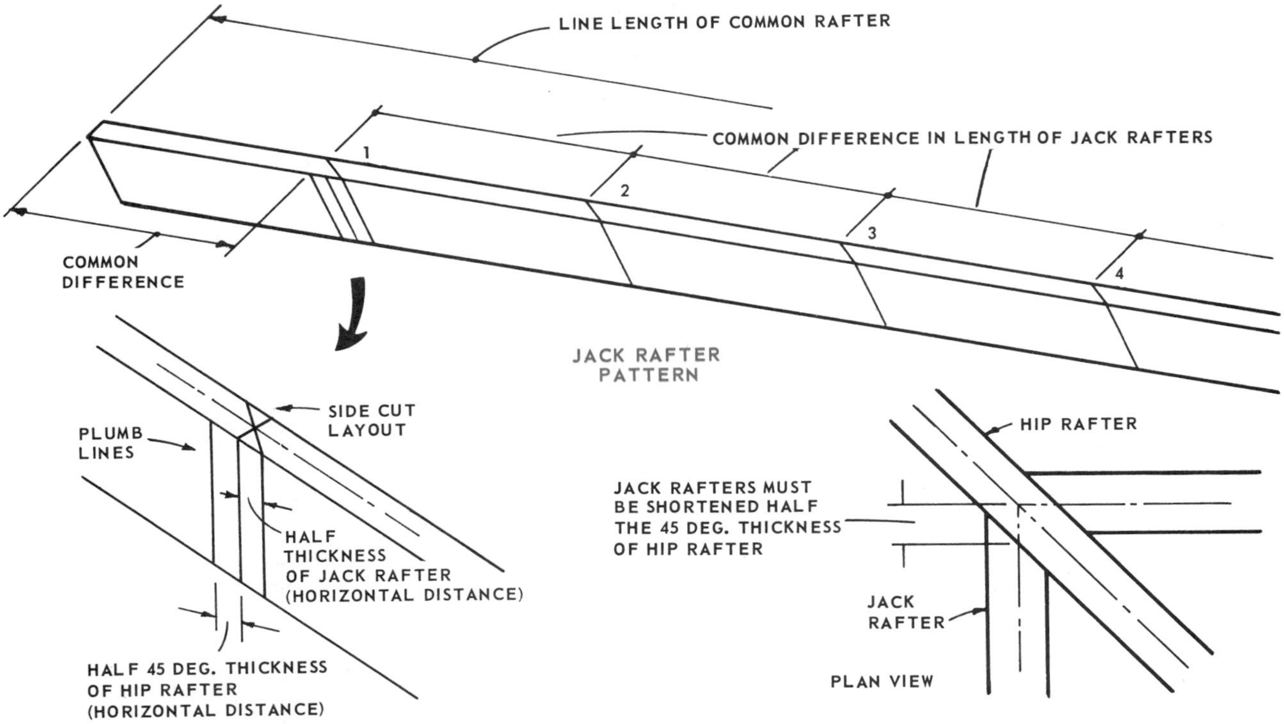

Fig. 9-32. Laying out a pattern for jack rafters.

side cut by marking (horizontally) one-half the thickness of the rafter or by using the framing square and applying the numbers located in the fifth line of the rafter table.

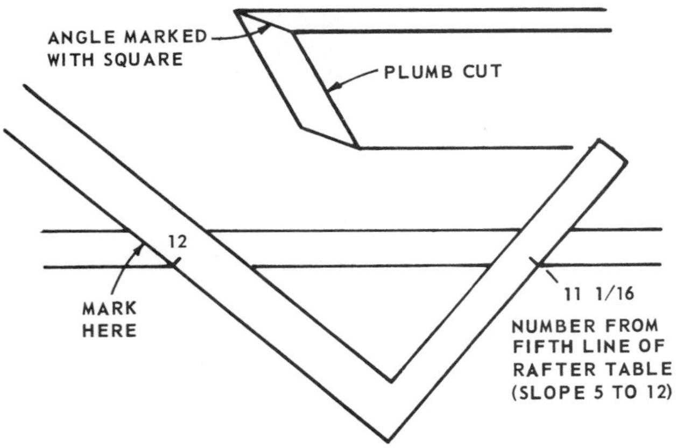

Fig. 9-33. Using framing square to lay out side cuts for jack rafters.

The common difference for valley jack rafters is obtained by the same procedure used for hip jacks. Lay out this distance from the longest valley jack to the next -- continuing along the pattern until all lengths are marked. Now use this pattern to cut all the valley rafters. They are cut in pairs in the same way as hip jacks.

When all jack rafters are laid out, they should be carefully cut. It requires a great deal of skill to make side cuts with a hand saw and it is usually better to use a portable electric saw with an adjustable base and guide. Radial arm saws are designed to do accurate cutting and are well suited for this kind of work, Fig. 9-35.

HANDY SAYS:

"The strength of a roof frame depends a great deal on the quality of the joints. Use special care on the side cuts of jack rafters so the joining surfaces will fit tightly together."

Erecting Jack Rafters

When all of the various jack rafters are cut, assemble them into the roof frame, Fig. 9-36. Nailing patterns will depend on the size of the various members. Use 10d nails, spaced so they will be near the heel of the side cut as they go from the jack into the hip or valley rafter.

Jack rafters should be erected in pairs to prevent the hip and valley rafters from being pushed out of line. It is good practice to first place a pair about halfway between the plate and ridge, while carefully sighting the hip or valley straight and true. Temporary bracing could also be used for this purpose. Be sure the outside walls running parallel to the ceiling joists are securely tied into the ceiling frame before hip jack rafters are installed, since these rafters tend to push outward when they are nailed in place.

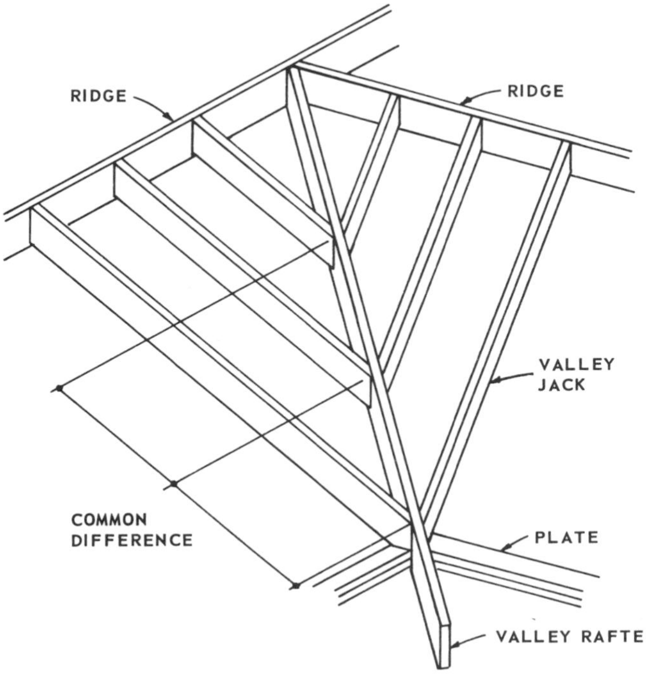

Fig. 9-34. Valley jack rafters. For a given slope and spacing, the common difference will be the same as that for hip jacks.

After all rafters have been erected and securely nailed in place, check over the frame carefully. If some rafters are bowed sideways, they can be held straight with a strip of lumber located across the center of the span. Each rafter is sighted, moved as needed, and a nail driven through the strip to hold it in place. After the roof has been sheathed to this point, the strip is removed.

Fig. 9-35. Side cuts for jack rafters can be accurately made with radial arm saw.

Special Problems

When framing intersecting roofs where the spans of the two sections are not equal, the ridges will not meet. To support the ridge of the narrow section, one of the valley rafters is continued to the main ridge. See Fig. 9-3, at the beginning of this Unit. The length of this extended or supporting valley is found by the same method used in the layout of a hip rafter. It is shortened at the ridge just like the hip but only a single side cut is required. The other valley rafter is framed against the supporting valley with a square, plumb cut.

A rafter framed between the two valley rafters is called a valley cripple jack. The angle of the side cut at the top is the reverse of the side cut at

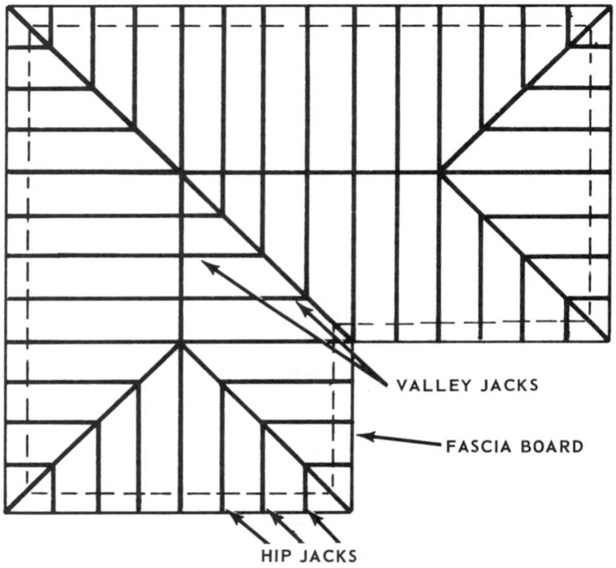

Fig. 9-36. Hip and valley jack rafters assembled in the roof frame.

the lower end. The run of the valley cripple is one side of a square, Fig. 9-37. This run is equal to twice the distance from the center line of the valley cripple jack to the intersection of the center lines of the two valley rafters. Lay out the length of the cripple by the same method used for a common rafter and shorten each end, one-half the 45 deg. thickness of the valley rafter stock. Make the side cuts in the same way used for regular jack rafters.

Rafters running between hips and valleys are called hip-valley cripple jacks. They require side cuts on each end like those on regular jack rafters. Since hip and valley rafters are parallel to each other, all cripple rafters running between them in a given roof section will be the same length.

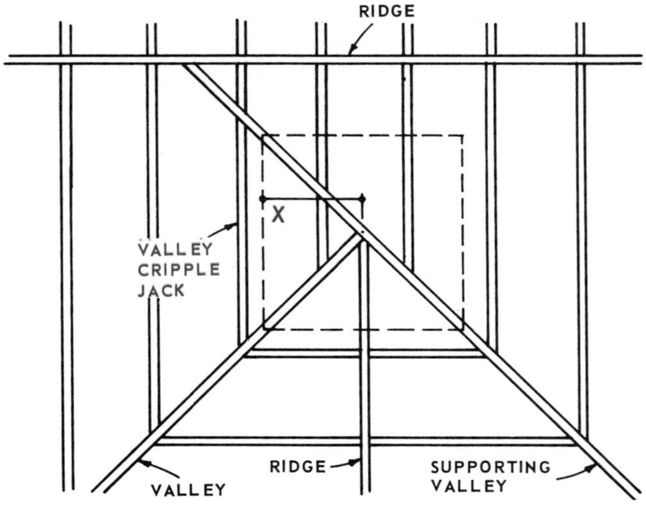

Fig. 9-37. The run of a valley cripple jack is twice the X distance.

The run of a hip-valley cripple rafter will be equal to the side of a square, Fig. 9-38, the size of which is determined by the length of the plate between the hip and valley rafter. Use this distance and lay out the cripple, in the same manner used for a common rafter. Shorten each end by an amount equal to half the 45 deg. thickness of the hip and valley rafter stock. Now lay out and mark the side cuts, following the same procedure used for hip and valley jacks. Side cuts, required on each end, form parallel planes.

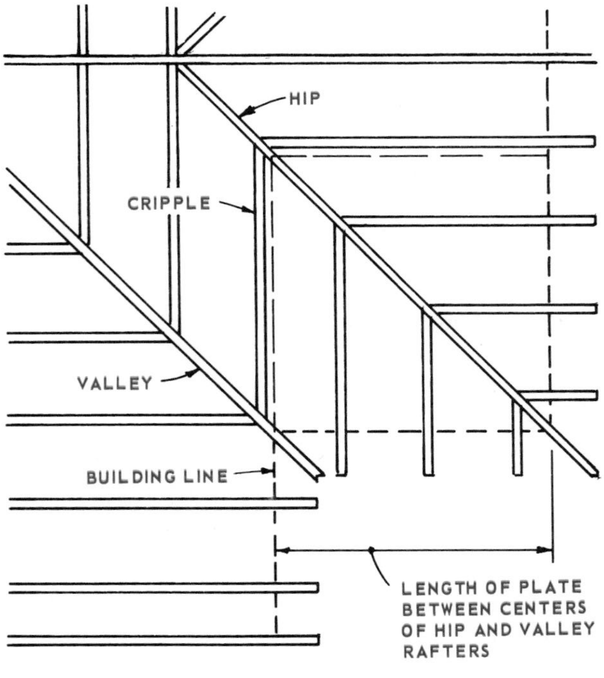

Fig. 9-38. The run of a hip-valley cripple jack is equal to the length of the wall plate.

Roof Openings

Some openings may be required in the roof for chimneys, skylights, or other items. For large openings, follow about the same procedure used in floor framing. To construct small size openings, the entire framework is first completed and then the opening is laid out and framed.

For a chimney opening, Fig. 9-39, use a plumb line to make the layout on the rafters from openings already formed in the ceiling or floor frame. Nail a temporary strip across the top of the rafters to be cut and at least two on each side of the opening. This will support the ends of the rafters while the opening is being formed. Now cut the rafters and nail the headers in place. If the size of the opening is large -- double the headers and also add a trimmer rafter to each side.

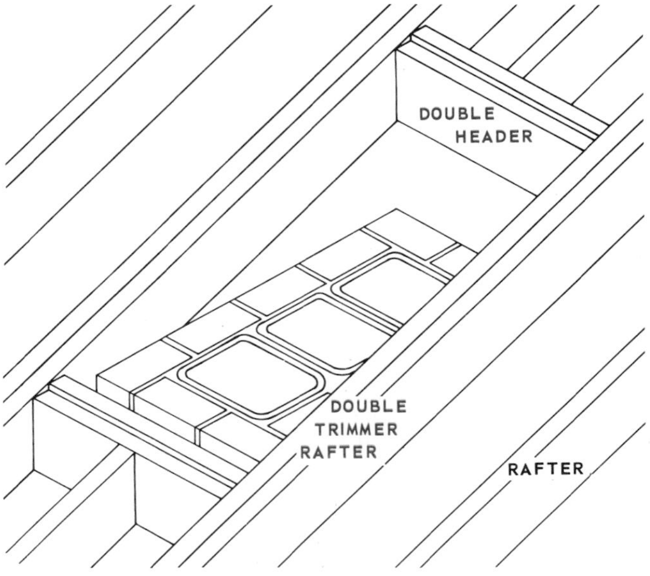

Fig. 9-39. Framing an opening for a chimney. Provide 2 in. of clearance on each side and end. Note that the headers are set plumb.

Roof Anchorage

Rafters usually rest only on the outside walls of a structure. They lean against each other at the ridge, thus providing mutual support. This causes an outward thrust along the plate that must be considered in the framing design.

Side walls are normally well secured by the ceiling joists which are also tied to some of the rafters. End walls, however, will be parallel to the joists and need extra support, especially when located under a hip roof. Fig. 9-40 shows an approved method of reinforcing roof framing

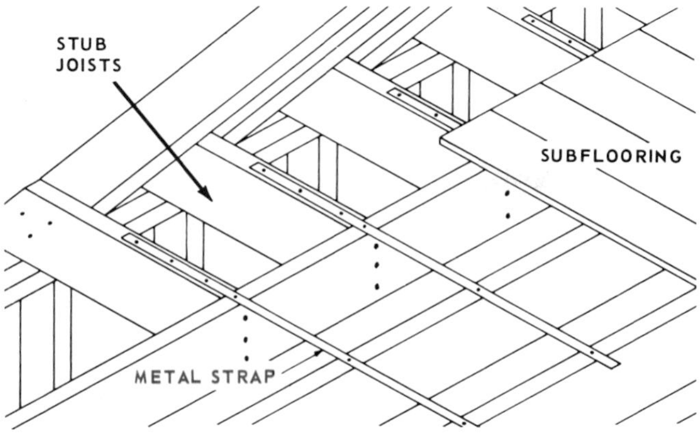

Fig. 9-40. Anchorage for a wall using stub joists and metal straps. (National Forest Products Assoc.)

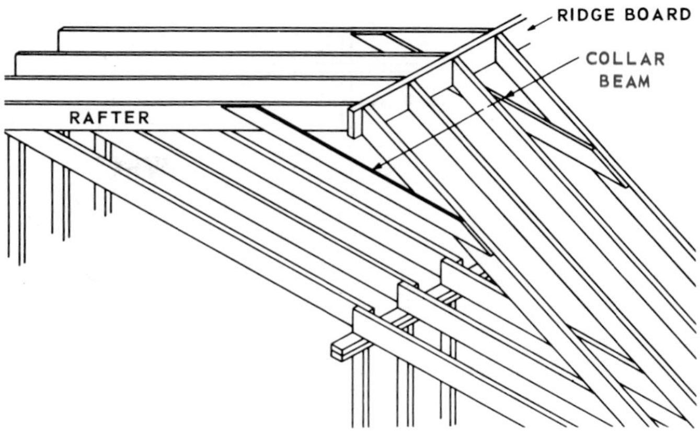

Fig. 9-41. Reinforcing a roof frame with collar beams.

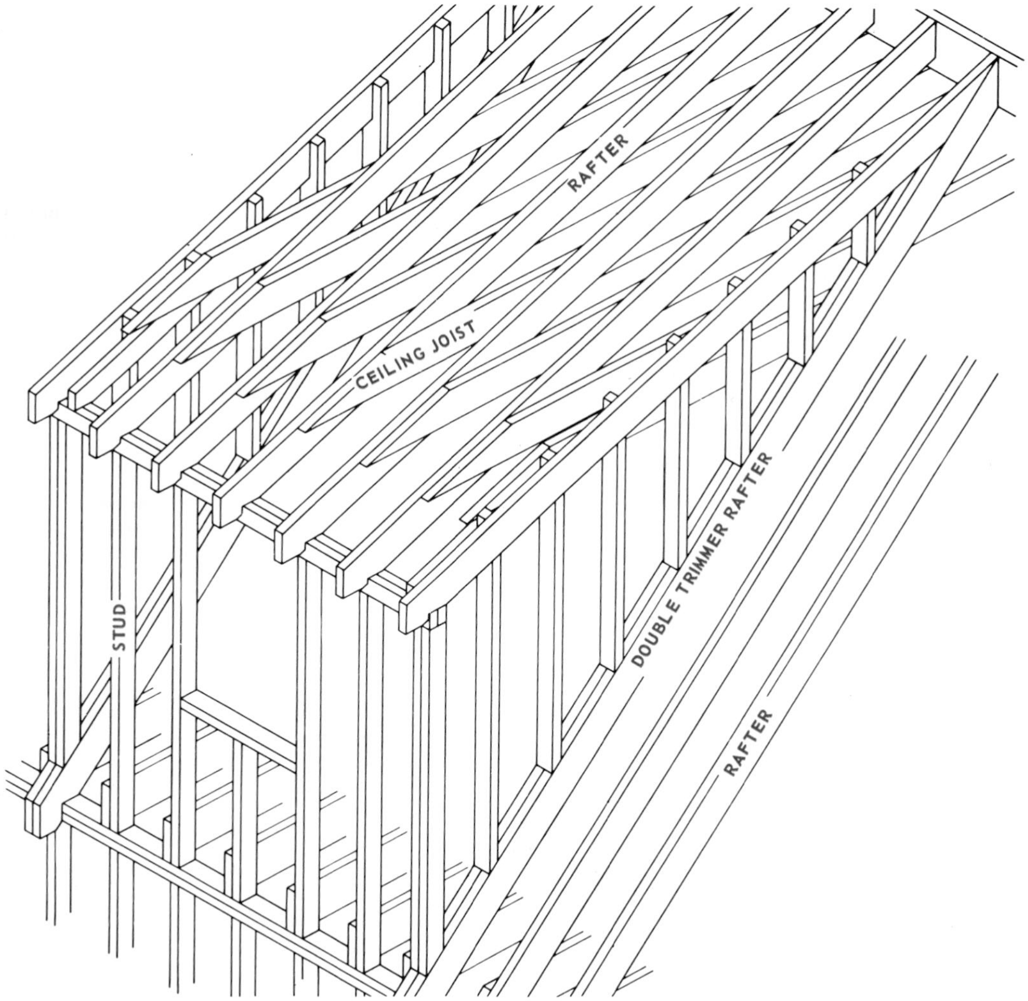

Fig. 9-42. Typical framing for a shed dormer. A nailer strip is added along the double trimmer to carry the roof sheathing.

through the use of stub ceiling joists and metal straps. Framing anchors can be substituted for the metal straps when sub-flooring is included in the assembly.

Collar Beams

Collar beams are ties between rafters on opposite sides of a roof, Fig. 9-41. When the attic

Conventional wall framing is used for the second floor. Rough openings (arrow denotes header) are for dormer alcoves instead of doors or windows. Note that floor joists extend outward beyond the first floor wall plate.

Upper roof level is framed in the same manner as a hip roof with the lower end of the rafters extending over the plate. This extension provides a connection with the mansard rafters (arrow).

Close-up view shows framing complete with sheathing applied to the upper roof section. The face frames for the dormers were cut and assembled inside and then installed. Note how the lower mansard rafters run from the upper rafters to the 2 x 4 plate installed along the floor joists.

Sheathing being applied to lower roof. Rough fascia board (arrow) is installed along ends of floor joists. The rounded roofs of the dormers are constructed from closely spaced blocking and then sheathed with flexible composition board. Shingle bundles help hold underlayment in place on upper roof.

Shingles have been applied and dormers are being finished. The faces of the dormers were covered with a primed composition board. Weatherproof metal strips are being applied to outside corners (arrow). Intersections between the dormer and roof require extensive flashing applied by an expert.

Dormers and roof complete. A finished fascia will be installed along with the soffit of the overhang. The wall finish will consist of brick veneer. Before soffit is applied, insulation batts will be placed under the floor of dormer alcoves.

Framing, sheathing and finishing a mansard roof designed to provide an overhang for the first floor level.

space is finished, they can serve as ceiling supports, if properly spaced. Collar beams do not support the roof but provide bracing and stiffening, and help hold the ridge and rafters together. A standard application consists of 1 x 6 boards, installed at every third pair of rafters.

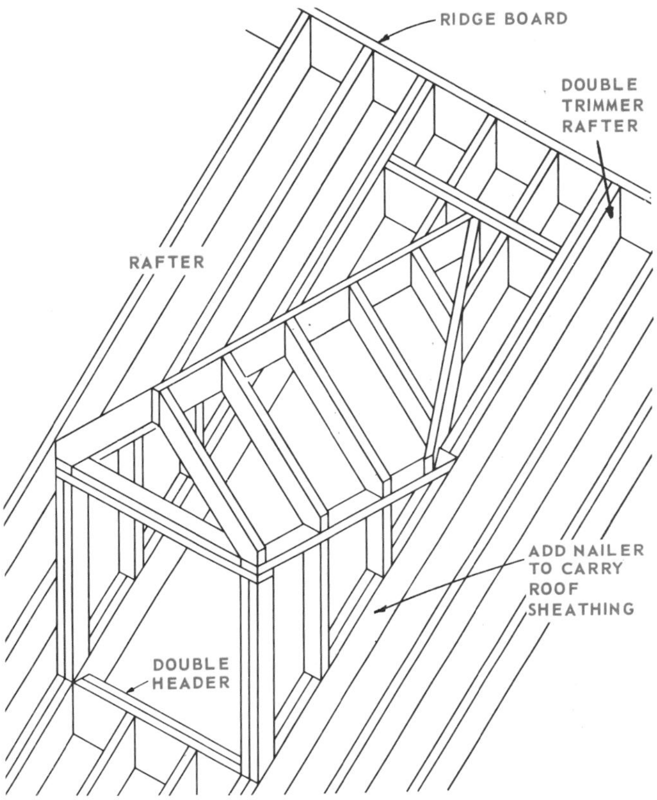

Fig. 9-43. Typical dormer framing. Another method consists of extending the wall framing from the floor through the roof opening.

Purlins

If the span of the rafters exceeds the maximum allowed, additional support must be provided. A purlin, usually a 2 x 4, is attached to the underside of the rafters and this member is then supported by bracing, also 2 x 4 stock, resting on a plate over a supporting partition. The bracing under the purlin may be placed at various angles to carry from the mid-point of the rafter to the support below. See Fig. 9-46.

Dormers

A dormer is a framed structure projecting above a sloping roof surface, and normally contains a vertical window unit. Although its chief purpose is to provide light, ventilation and additional interior space, it should also enhance the exterior appearance of the structure.

The shed dormer's width is not restricted by the roof design and is used where a large amount of additional interior space is required, Fig. 9-42. In the simplest construction, the front wall is extended straight up from the main wall plate while double trimmer rafters are used to carry the side wall. The rise of the roof is figured from the top of the dormer plate to the main roof ridge, while the run will be equal to that of the main roof. Be sure to provide sufficient slope for the dormer roof.

Gable dormers, Fig. 9-43, are designed to provide openings for windows and are located at various positions between the plate and ridge of the main roof. Double headers and trimmers are framed into the required roof opening as shown.

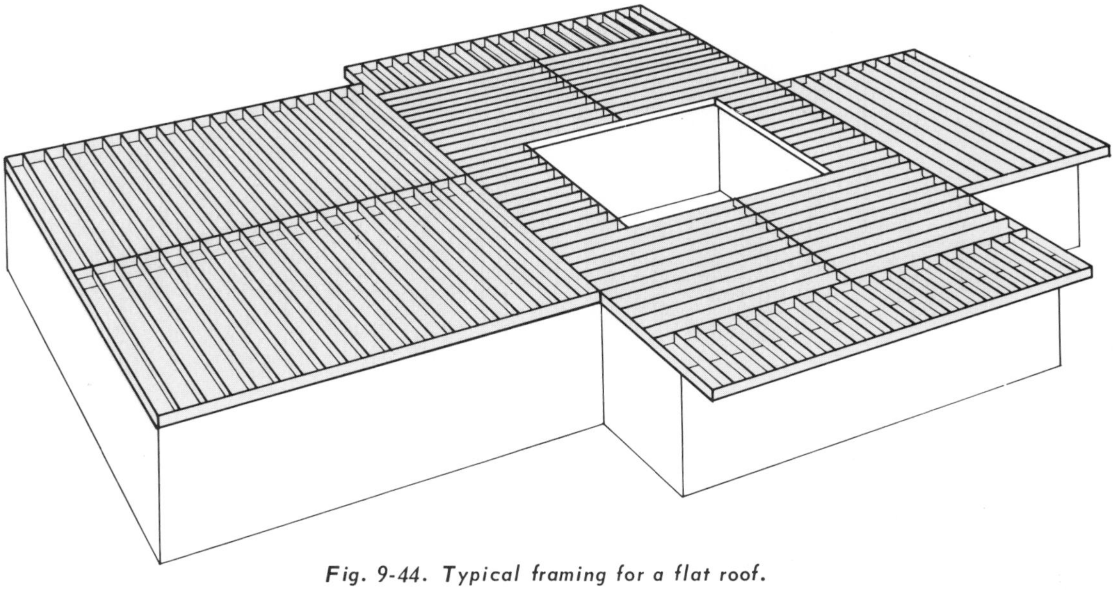

Fig. 9-44. Typical framing for a flat roof.

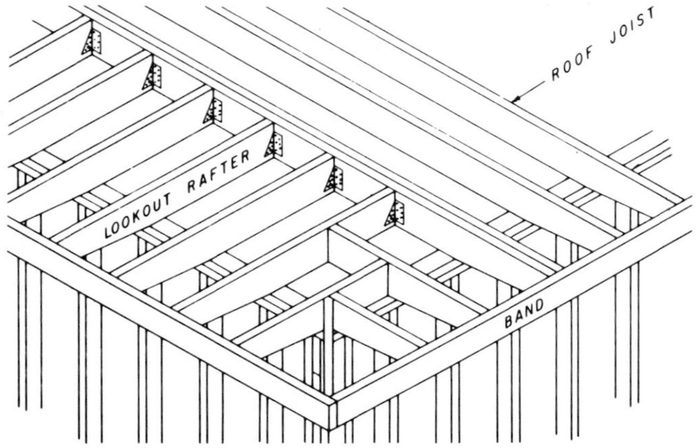

Fig. 9-45. Flat roof framing to form an overhang.

The gable roof is constructed in a conventional manner using the same slope as that of the main structure.

Flat Roofs

Today, flat roofs are often used in residential construction. They provide the long, low appearance desired in contemporary designs. Improvements in roofing surface materials and methods of application make this type of roof practical.

Fig. 9-44 shows the similarity between a flat roof frame and floor frame. The main members are called roof joists and support both the roof and ceiling. Because of this combined load, 2 x 10's or 2 x 12's -- spaced 16 in. O.C. are generally used. Requirements will vary in different localities. Local building codes must be checked.

Methods and procedures used to frame flat roofs are about the same as those followed in constructing a floor. Most designs will require an overhang with the ends of the joists tied together with a header or band. Cantilevered rafters are

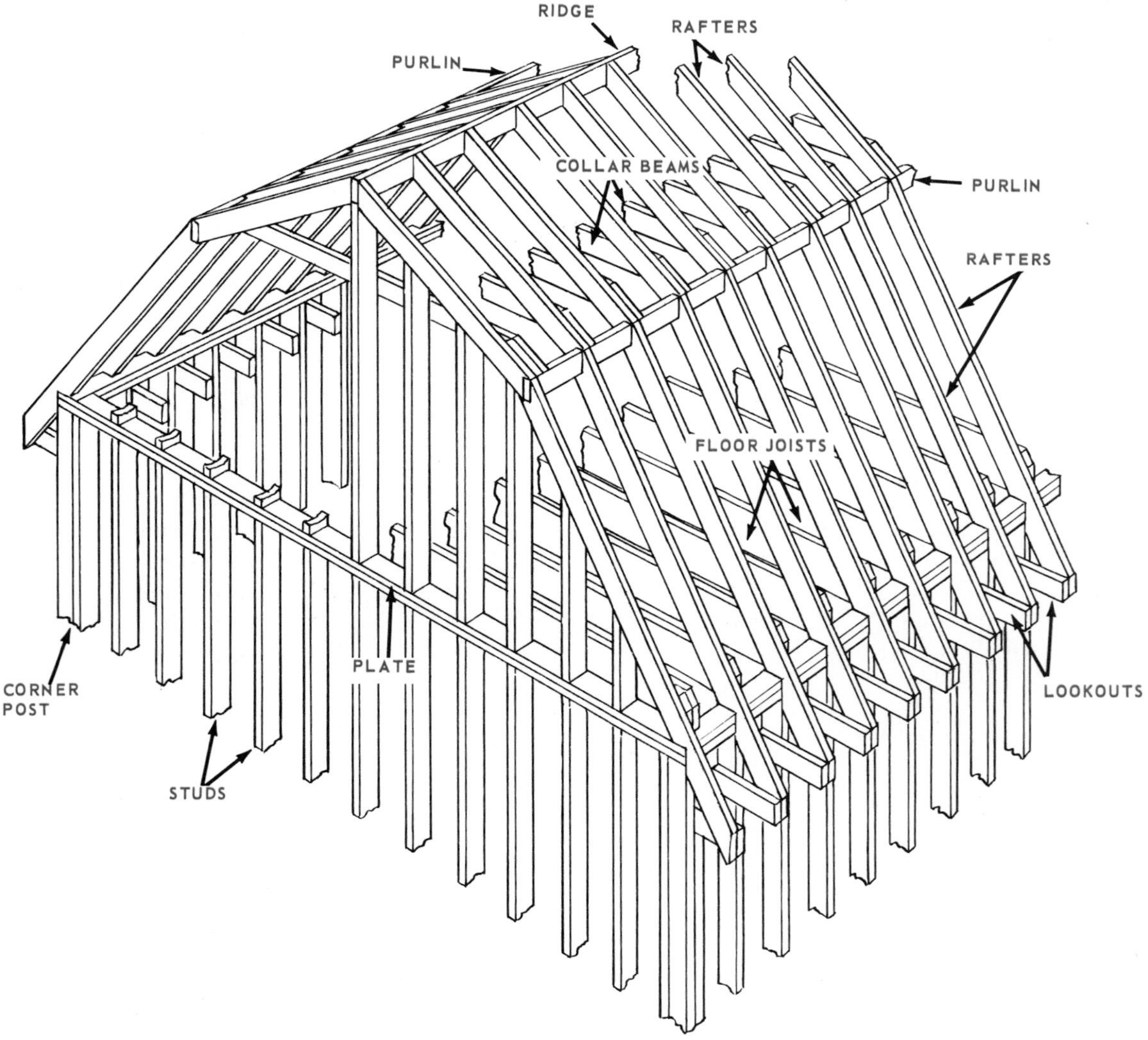

Fig. 9-46. Framing a gambrel roof.

tied to doubled roof joists as illustrated in Fig. 9-45. Corners can be formed as shown or carried on a longer diagonal joist that intersects the doubled joist.

Since no load bearing partitions are required, more freedom in the planning and division of interior space is possible. They permit larger rooms without extra beams and supports. Another advantage which may reduce labor costs, is the opportunity to apply surface materials to outside walls, ceilings, and floors before partitions are constructed.

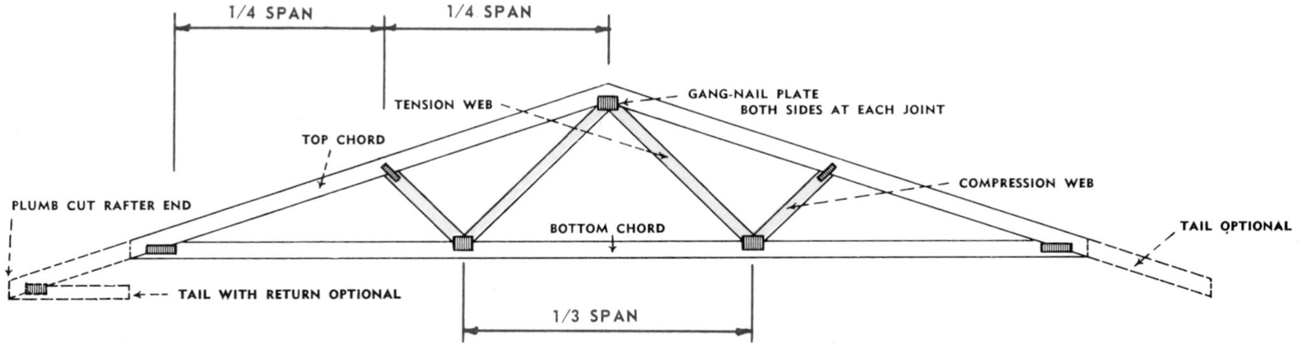

Fig. 9-47. Standard W or Fink truss commonly used in residential construction.

Gambrel Roof

The gambrel roof is somewhat similar to the gable roof. It consists of two separate roof surfaces on each side of the ridge, Fig. 9-46. The upper section usually forms about a 30 deg. angle with a horizontal plane while the lower section forms about a 60 deg. angle.

In residential construction, this type of roof is usually framed with a purlin located where the two surfaces meet. The rafters are notched to receive the purlin which is supported on partitions and/or tied to another purlin on the opposite side of the building with collar beams.

Procedures used to frame a gable roof can be applied to the gambrel roof. The rise and run of each surface is secured from the architectural plans. The two sets of rafters are laid out in the same way as previously described for a common rafter. It may be desirable to make a full-size sectional drawing (if not included in the plans) at the intersection of the two slopes to better visualize and proportion the end cuts of the rafters.

Roof Truss Construction

A truss is a framework that is designed to carry a load between two or more supports. The principle used in its design is based on the rigidity of the triangle. Triangular shapes are built into the frame in such a way that the stresses of the various parts are parallel to the members making up the structure.

Roof trusses are frames that carry the roof and ceiling surfaces. They rest on the exterior walls and span the entire width of the structure.

There are many types and configurations of roof trusses. One commonly used in residential construction is the W or Fink truss illustrated in Fig. 9-47. Trusses should always be constructed according to designs developed from engineering data. There are several sources for such material which usually includes not only detailed construction drawings but also specifications concerning materials and fasteners. Tables listing the exact dimensions of the various members and suggested material lists, Fig. 9-48, will be helpful to the builder.

Roof trusses must be made of structurally sound lumber and assembled with carefully fitted joints. Although the carpenter is seldom required to determine the sizes of truss members or the type of joints, he should have sufficient understanding of their design to appreciate the necessity of first-class workmanship in their construction.

Trusses are pre-cut and assembled at ground level and then raised into position as a unit. Spacing of 24 in. O.C. is commonly used. However, 16 in. O.C. and other spacing may be required in some designs. When the truss is in position and loaded, there will be a slight sag. To compensate for this, the lower member (called the bottom chord) is raised slightly during fabrication of the unit. This adjustment is called camber and is measured at the mid-point of the span. A standard truss, 24 feet long, will usually require about 1/2 in. of camber.

SPAN L	DIMENSIONS A	B	C	DESIGN STRESSES U_1	U_2	L_1	L_2	V_1	D_1
20'-0"	5'-5"	5'-3⅝"	2'-7⅝"	1756	1450	1614	1472	430	430
22'-0"	5'-11½"	5'-10¹⁄₁₆"	2'-10¹³⁄₁₆"	1932	1595	1775	1619	473	473
24'-0"	6'-6"	6'-4⁷⁄₁₆"	3'-2"	2108	1740	1936	1766	516	516
26'-0"	7'-0½"	6'-10⅞"	3'-5¼"	2282	1885	2097	1913	559	559
28'-0"	7'-7"	7'-5¼"	3'-8⁷⁄₁₆"	2459	2030	2258	2060	602	602
30'-0"	8'-1½"	7'-11¹¹⁄₁₆"	3'-11⅝"	2634	2175	2420	2207	645	645
32'-0"	8'-8"	8'-6¹⁄₁₆"	4'-2¹³⁄₁₆"	2810	2320	2581	2354	688	688

Span	LUMBER 2" x 6" No.	2" x 6" Length	2" x 4" No.	2" x 4" Length	Total F.B.M.
20'-0"	2	12'-0"	2 2	12'-0" 10'-0"	53
22'-0"	2	14'-0"	2 2	14'-0" 10'-0"	60
24'-0"	2	14'-0"	2 2	14'-0" 12'-0"	63
26'-0"	2	16'-0"	2 2	16'-0" 12'-0"	70
28'-0"	2	16'-0"	2 2	16'-0" 14'-0"	73
30'-0"	2	18'-0"	2 2	18'-0" 14'-0"	79
32'-0"	2	20'-0"	2 2	18'-0" 14'-0"	83

HARDWARE

No.	Item	Size
11	Split Rings	2½" Diam.
2	Trip-L-Grip	Type A
1	Bolt	½" x 7½"
2	Bolts	½" x 6"
4	Bolts	½" x 4"
14	Washers	½"

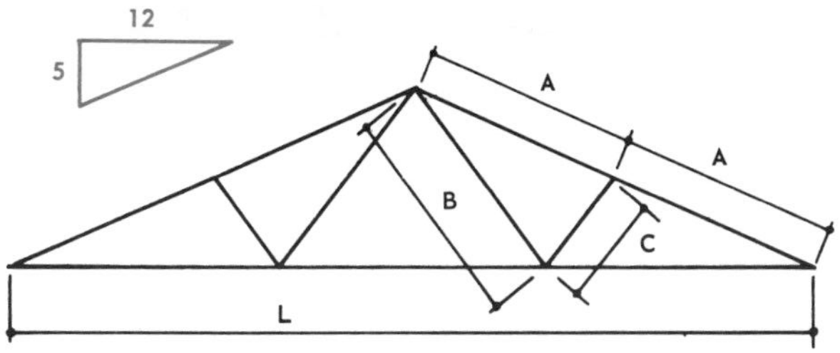

LUMBER--Lumber shall be of a good grade of sufficient quality to permit the following allowable unit stresses:
c = 900#/□" Compression parallel to grain.
f = 900#/□" Extreme fiber in bending.
E = 1,600,000#/□" Modulus of elasticity.

CONNECTORS--Timber connectors shall be 2-1/2" diameter split rings and Trip-L-Grip framing anchors.

BOLTS--Bolts shall be 1/2" diameter machine bolts with 2" x 2" x 1/8" plate washers, 2-1/8" diameter cast or malleable iron washers, or ordinary cut washers.

DIMENSIONS--Dimensions shown will provide approximately 1/2" camber at bottom chord panel points. Utilize full uncut length of bottom chord pieces by increasing the spacing of connectors in the splice.

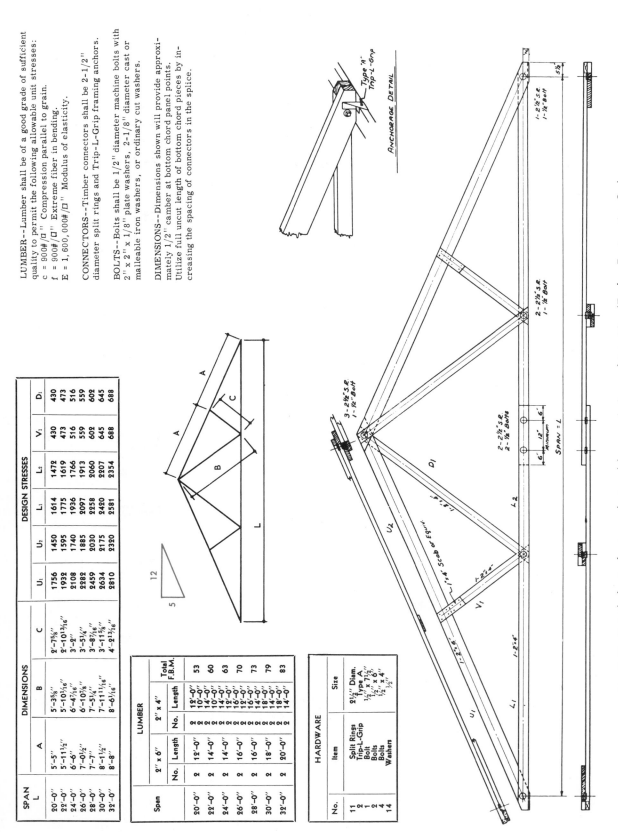

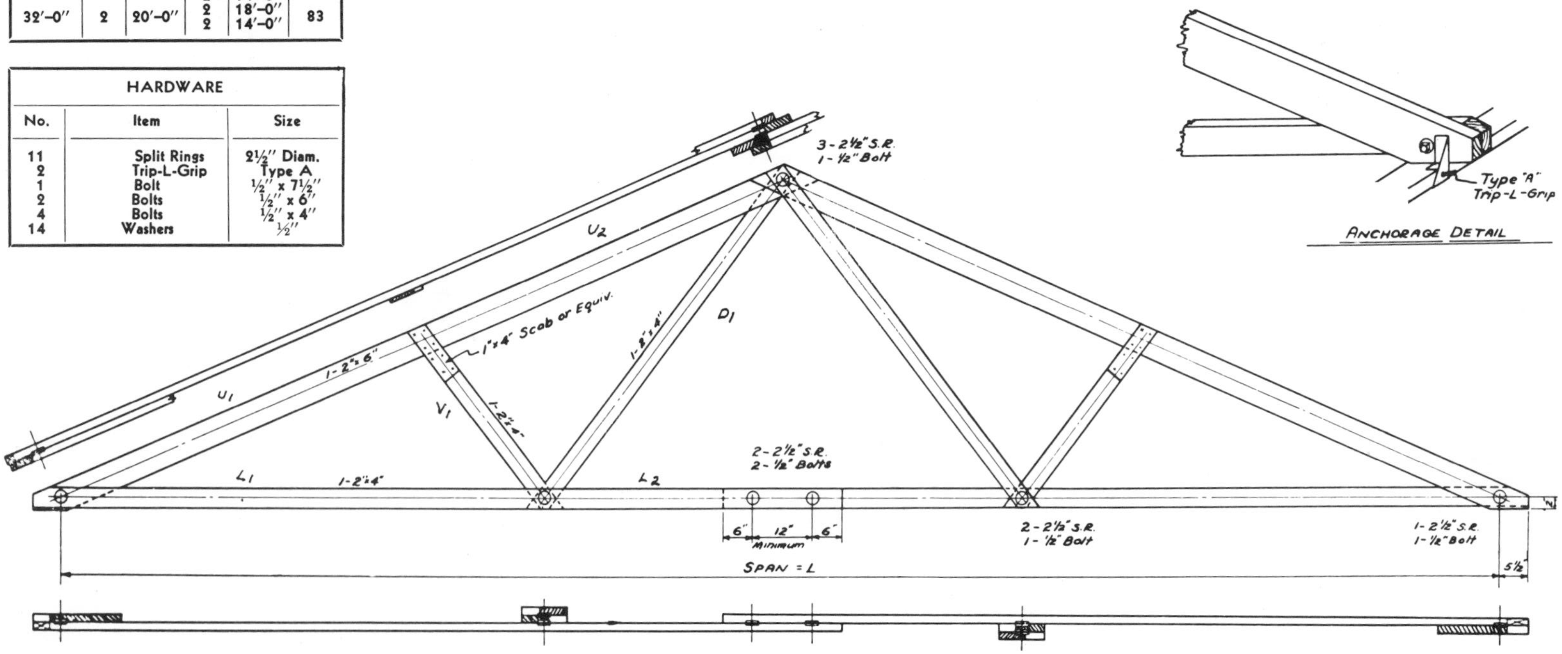

Fig. 9-48. Detail sheet of a truss designed for a roof slope of 5 in 12. (Timber Engineering Co.)

In the construction of trusses, it is essential that joint slippage be held to a minimum. Regular nailing patterns are usually not satisfactory and special connectors must be used. Various kinds are available that hold the joint securely and are easy to apply, Fig. 9-49. Plywood gussets applied with glue and nails to both sides of the joint are shown in Fig. 9-50.

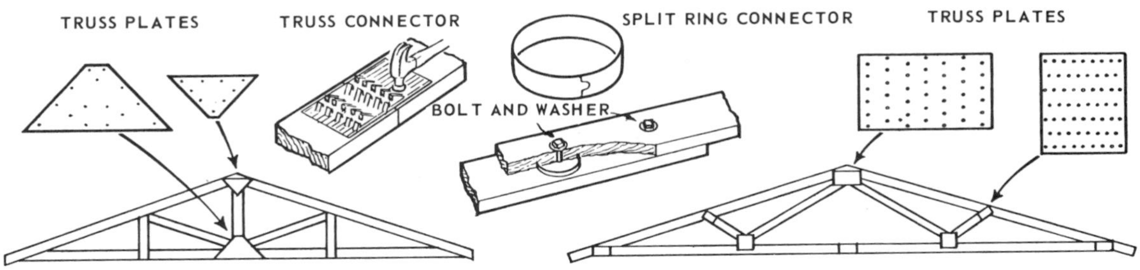

Fig. 9-49. Plates and connectors for roof trusses. Truss plates are made in many sizes, shapes and types; some are drilled for nailing, others are punctured so the protruding metal bites into the wood when installed. Some have nails formed in the plate. Ring-type truss connectors fit into grooves bored in the wood to supplement the shear strength of the bolt and distribute stress.

When the number of trusses required is small, they can be laid out and constructed on any clear floor area. First make a full-size layout on the floor, snapping chalk lines for long line lengths and using straightedges to draw shorter lines. Carefully follow the data provided.

Fig. 9-50. Assembling a roof truss with a plywood gusset. (American Plywood Assoc.)

Align the lumber with the layout to mark the size and then cut the pieces accurately. Work first with the top and bottom chords and then the web members. Cut enough material from straight lumber for a single truss and one extra of each piece to use as a pattern.

Align all members and check the fit. If ring connectors are used, bore the holes required. Tack all joints together and then apply the truss plates or other types of connectors. Turn the truss over and complete the opposite side. This truss will serve as a pattern. Check it over carefully and then nail it to the floor.

Using the pattern pieces, cut the amount of material needed to build the additional trusses required. Assemble the trusses, one at a time, by clamping the parts in position over the pattern truss as shown in Fig. 9-51. Be sure the joints are tight together. Use extra clamps on pieces that are warped. When all members are in place, apply the connectors, Fig. 9-52. Unclamp the truss, flip it over, and apply fasteners to the opposite side.

Fig. 9-51. Clamping lower chord member to pattern truss.

When a large number of trusses are required for a major building project, it is worthwhile to use a portable truss assembly unit, Fig. 9-53. This kind of equipment will insure accurate assemblies and also raise the construction to a more comfortable working height.

Fig. 9-52. Using an air-driven nailer to fasten metal plates to a roof truss. (Bostitch)

Fig. 9-53. Portable truss assembly unit can be adjusted to various sizes and designs. (Bostitch)

Fig. 9-54. Special sized trusses installed on a roof deck to connect an intersecting roof. (National Forest Products Assoc.)

Trusses for residential structures can normally be erected without special equipment. Each truss is simply placed upside down on the walls at the point of installation and then the peak is turned upward, revolving the unit into position.

HANDY SAYS:

"Use extra precaution when raising roof trusses. The first truss should be held with guy wires and all succeeding trusses carefully braced to prevent overturning."

Roofs framed with trusses need not be limited to gable types. Hip roofs can be framed through the use of special trusses or with a combination of trussed and conventional rafter-and-joist framing. In one method of truss construction, the main roof is assembled and sheathed and then special trusses are mounted on the roof deck to form the intersecting roof frame, Fig. 9-54.

In modern construction, roof trusses are very often cut and assembled in nearby manufacturing plants and then delivered to the job ready to install, Fig. 9-55. These prefabricated units are usually very satisfactory because accuracy and quality can be carefully controlled. Refer to Unit 20 for more information.

Roof Sheathing

When the roof frame is complete, check it carefully to see that all members are secure and that nailing patterns are adequate.

The frame is then ready for the sheathing which provides a nailing base for the roof covering and also adds to its strength and rigidity. Sheathing materials include plywood, shiplap, and common boards. Other special materials are also

Fig. 9-55. Prefabricated roof trusses being delivered to building site ready to install.

Fig. 9-56. Sheathing a hip roof frame using panels formed with solid boards, bonded together with heavy kraft paper. Panels shown are 2 ft. wide and 16 ft. long. The panels are first nailed in place and then cut along the hip. Those that are long enough are turned over and used on the adjacent slope. (Western Wood Products Assoc.)

Shiplap and common boards must be applied solid if asphalt shingles or other composition materials are used for the finished roof surface. For wood shingles, metal sheets, or tile, board sheathing may be spaced according to the course arrangement. They should be nailed in place with two 8d nails at each rafter, with joints located over the center of the rafter. To obtain maximum rigidity, use long boards, particularly at roof ends.

When end-matched boards are used, Fig. 9-58, the joints may be made between rafters. Joints in adjacent boards must not occur in the same rafter space and no boards should be used that are not long enough to be carried on at least two rafters.

Fit sheathing boards carefully at valleys and hips and nail them securely. This will insure a solid, smooth base for the installation of flashing materials. Around chimney openings, the boards should have a 1/2 in. clearance from the masonry.

Fig. 9-57. Applying plywood sheathing to roof frame.

available. For example: one product consists of panels formed with solid wood boards, bonded together with heavy kraft paper, Fig. 9-56.

Before starting the sheathing, erect the necessary scaffold that will make it easy and safe to install the boards or panels along the lower edge of the roof, Fig. 9-57. This scaffold can also be used later to build the cornice work after the roof surface is complete.

Framing members must have a 2 in. clearance. Always nail boards securely around openings.

Plywood is an ideal material for roof sheathing. It can be installed rapidly, holds nails well, resists swelling and shrinkage, and, because of the large panels, adds considerable rigidity to the roof frame. It is laid with the face grain perpendicular to the rafters, with end joints formed directly over the center of the rafter. Small pieces

Fig. 9-58. Installing end-matched sheathing. The species of wood is hemlock. (Weyerhaeuser Co.)

can be used but they should always cover at least two rafter spaces.

For wood or asphalt shingles with a rafter spacing of 16 in., 5/16 in. plywood is usually recommended. For a 24 in. span, a 3/8 in. thickness should be used. Slate, tile, and asbestos-cement shingles require 1/2 in. plywood for 16 in. rafter spacing and 5/8 in. plywood for 24 in. spacing. It should be nailed to rafters with 6d nails, spaced 6 in. apart on edges and 12 in. elsewhere. If wood shingles are used and the plywood sheathing is less than 1/2 in. thick, 1 by 2 in. nailing strips, spaced according to shingle exposure, should be nailed to the plywood. For a flat deck under built-up roofing, use 1/2 in. thickness.

When handling large sheets of plywood, use extra precaution. They may slide off of a roof if they are not properly secured. A special plywood rack, Fig. 9-59, will be helpful in moving sheets from the ground to the roof and will also serve for storage until they are nailed in place.

Fig. 9-59. Using a ladder rack in handling sheathing. (American Plywood Assoc.)

HANDY SAYS:

"Use special care in handling sheets of plywood on a roof, especially if there is a wind. You may be thrown off balance or the plywood may be blown off the roof and strike someone below."

Estimating Materials

The number of rafters required for a plain gable roof is easy to figure. Simply multiply the length of the building by 3/4 for spacing 16" O.C. (3/5 for 20, 1/2 for 24) and add one more. Double this figure for the other side of the roof. To determine the length of the rafter, use the 12th scale on the framing square as previously described in this Unit.

Example:

Building Size 28 x 40. Roof slope 4 to 12.
Overhang 2' - 0". Rafter spacing 24" O.C.

Total Rafter Run	= 16' - 0"
Total Rafter Length	= 16' - 11"
Nearest Std. Length	= 18' - 0"
Number of pieces	= 2 (Length of wall x 1/2 + 1)
	= 2 (40 x 1/2 + 1)
	= 2 x 21
	= 42

Rafter Estimate: 42 pcs. 2 x 8 x 18' - 0"

When estimating a hip roof, it is not necessary to figure each jack rafter. The number of jack rafters required for one side of a hip is counted to obtain the number of pieces of common rafter stock. This will normally supply sufficient rafter material for the other side of the hip. For a short method on a plain hip roof, proceed as if it were a gable roof. Add one extra common rafter for each hip and also figure and add the hip rafters required.

With complicated roof frames, it is best to work from a complete framing plan. Apply the methods described for plain roofs to the various sections. Make colored check marks on the rafters as they are figured so you will not double up on some areas and skip others. In estimating material for the total roof frame, remember to include material for ridges, collar ties, and bracing.

To estimate the roof sheathing, first figure the total surface and then apply the same procedures as used for subflooring and wall sheathing. Since the total area of the roof surface will also be needed to estimate shingles, building paper, or other roof surface materials, it is worth the extra time required to figure the area accurately.

For a plain gable roof, multiply the length of the ridge by the length of a common rafter and double the amount. Figure a plain hip roof as though it were a gable roof but instead of multiplying the length of a common rafter by the ridge, multiply it by the length of the building plus twice the overhang.

When working with complicated plans and intersecting roof lines, first determine the main roof areas; common rafter length times the length of the ridge times two. Now add the triangles that make up the other sections (located over jack rafters). Remember that the area of a triangle equals one-half the base times the altitude. The altitude of most triangular roof areas will be the length of a common rafter located in or near the perimeter of the triangle. A plan view of the roof lines will be helpful since all horizontal lines (roof edges and ridges) will be seen true length and can be scaled.

Always add an extra percentage for waste when estimating sheathing requirements for roofs that are broken up by an unusually large number of valleys and hips.

Model Construction

Students of carpentry can often gain worthwhile experiences through construction of scale model framing. Work of this kind requires much time so it is often best to construct only a part or section of a given structure. See Fig. 9-60.

Using a scale of 1 1/2" = 1' - 0" will usually make it possible to apply regular framing procedures in the construction of a model that is not too large to handle and store. Cut framing members to their nominal (name) size. For example: a 2 x 4 cut to this scale would actually measure

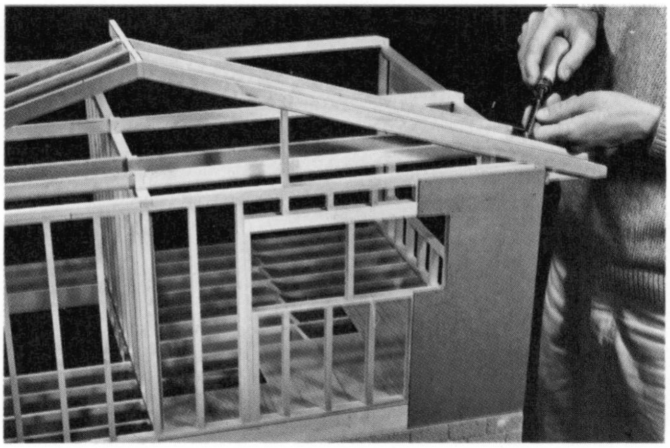

Fig. 9-60. Model carpentry construction. Above. Using a brad pusher to install a rafter. Below. Model construction based on modular layout. Roof trusses were constructed in a jig.

1/4" x 1/2" while a 2 x 10 would measure 1/4" x 1 1/4". Make all framing materials from clear white pine or sugar pine which has sufficient strength and is easy to work. Use small brads and fast setting glue to make the assembly.

Materials other than wood can often be simulated from a wide range of items. For example: foundation work can be built of rigid foamed plastic (Styrofoam) and then brushed with a creamy mixture of Portland cement and water.

Test Your Knowledge - Unit 9

1. A type of sloping roof that simplifies the construction of an overhang for all outside walls is called a ____________ roof.
2. The pitch of a roof is indicated by a fraction formed by placing the rise over the _________.
3. The tongue of a framing square is____in. long.
4. When laying out a rafter for a run that includes an odd unit, the ____________(full unit, odd unit) is laid out first.

5. The bird's-mouth is formed by a ________cut and a plumb cut.
6. The final step in laying out a common rafter is to shorten it at the ____________.
7. When assembling a roof frame, joints in the ridge should occur at the ______________of a rafter.
8. The part of a gable roof that extends beyond the end walls is called the _________.
9. When laying out a hip or valley rafter, use ___ in. for the unit run.
10. Figures used to make side cuts for hip and valley rafters are found in the _______line of the rafter table on the framing square.
11. Hip jack rafters have the same tail and overhang as ___________rafters.
12. Jack rafters should be erected in _________to keep the hip or valley rafters straight.
13. Horizontal ties between rafters on opposite sides of the ridge and usually located in the upper half of the frame are called ____ _____.
14. The two general types of dormers are _____and gable.
15. In residential construction, the gambrel roof is usually framed with a ___________located where the two surfaces of different slopes are joined together.
16. The adjustment in the lower chord of a roof truss to compensate for sag is called _______.
17. The sheathing on a roof frame provides a nailing base for shingles and also adds _________.
18. End joints in the sheathing boards can be made between rafters when ________ lumber is used.
19. The thickness of plywood required for a sheathing application will vary for different roofing materials and different _____ _______.
20. To calculate the area of a plain gable roof, multiply the length of the ridge by the length of a ________ __________and then double the product.

Outside Assignments

1. Secure a set of house plans where the design includes a hip roof and/or intersecting sections and which does not include a roof framing plan. Study the elevations and detail sections and then prepare a roof framing plan.

   Overlay the floor plan with a sheet of tracing paper and then trace the wall and draw all roof framing members to accurate scale. Be sure to include openings for chimneys and other items that would be helpful to the carpenter.
2. Prepare an estimate of the framing materials required for the above roof. Include the dimensions for all lumber needed. Refer to the detail drawings or specifications to find lumber size requirements. If this information is not included in the plans, secure it from the local building code.
3. Working from a set of architectural plans for a residential structure, make a layout for a common rafter in one of the roof sections. Use a good straight piece of stock. If dimension lumber is not available, a piece of 1 in. material may be used. Make the layout by the step-off method and cover all operations including the shortening at the ridge. When completed, present a brief demonstration to the class, showing them the procedure you followed.
4. Make a study of the various types of roof trusses. Learn their names and the basic design patterns. List the advantages and disadvantages of each and find out where they are most commonly used. Prepare a display board with line drawings of about eight of the most practical types and label each with an appropriate caption.

Fig. 10-1. Application of roofing materials. Chimneys and other projections must be completed before installation of finished roofing materials. (Weyerhaeuser Co.)

Unit 10
ROOFING MATERIALS

Roofing materials protect the structure and its contents from the sun, rain, snow, wind, and dust. In addition to weather protection, a good roof should offer some measure of fire resistance and have a high durability factor. Due to the large amount of surface that is usually visible, especially in sloping roofs, the materials can contribute to the attractiveness of the building. Roofing materials can add color, texture, and pattern.

Roof construction and finish consists of a number of operations, most of which must follow a definite sequence. After the framing is complete, the fascia (vertical trim board attached to the rafter ends) is installed and the roof sheathing is applied. All items that will project through the roof should be built or installed; including chimneys, vent pipes, and special facilities for electrical and communications service. Performing any of this work after the finished roof is applied may result in damage to the roof. In larger structures, Fig. 10-1, many of these operations may be underway at the same time.

Types of Material

Materials used for sloping roofs include asphalt, wood, and mineral fiber shingles; slate and tile. Sheet materials such as roll roofing, galvanized iron, aluminum, and copper are sometimes used.

For flat roofs and low-sloped roofs, a membrane system is used. It consists of a continuous watertight surface, usually obtained through built-up roofs or seamed metal sheets. Built-up roofs are fabricated on the job by laminating roofing felts with asphalt or coal tar pitch and then coating the surface with crushed stone or gravel. Metal roofs of this type are assembled from flat sheets with a special seam that is soldered or sealed with special compounds to insure watertightness.

The selection of roofing materials is influenced by such factors as; the initial cost, maintenance costs, durability, and appearance. The slope of the roof limits the selection. Low-sloped roofs require a more watertight system than steep roofs, Fig. 10-2. Materials such as tile and slate require heavier roof frames. Local building codes may prohibit the use of certain materials because of the fire hazard or because they will not resist the high winds or other elements prevalent in a certain locality.

shingles. Depending on the type of material and method of application, the shingles may furnish one (single coverage), two (double coverage) or even three (triple coverage) thicknesses of material on the roof.

Exposure: The shortest distance in inches between the edges of adjacent courses measured at right angles to the ridge.

Head Lap: The shortest distance in inches from the lower edge of an overlapping shingle or sheet, to the roof deck, Fig. 10-3.

Side Lap: The shortest distance in inches which horizontally adjacent elements of roofing overlap each other.

Shingle Butt: The lower exposed edge of shingle.

Preparing the Roof Deck

The roof sheathing should be smooth, securely attached to the frame, and provide an adequate base to receive and hold the roofing nails and fas-

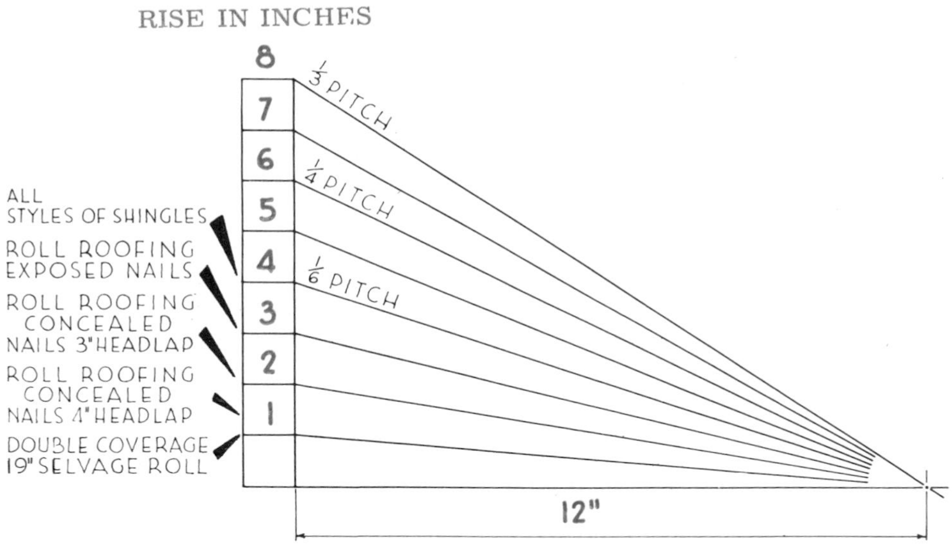

Fig. 10-2. Slope of roof limits the range of materials that can be used. The portion of the individual units (shingles) exposed to the weather must be reduced on low-sloped roofs.

Roofing Terms

Slope and pitch have already been defined in Unit 9. Several other terms commonly used include:

Square: Roofing materials are estimated and sold by the square. This is the amount of a given type of material needed to provide 100 sq. ft. of finished roof surface.

Coverage: This indicates the amount of weather protection provided by the overlapping of the

teners. All types of shingles can be applied over solid sheathing. Spaced sheathing is sometimes used for wood shingles. When solid boards are used for sheathing, they should not be over 6 in. wide.

It is also important that the attic space be properly ventilated to minimize condensation of moisture after the building is completed and ready for use. Sometimes moisture vapor from the lower stories, rising to the attic, will be chilled below its dew point and will condense on the underside of the roof deck, causing sheathing boards

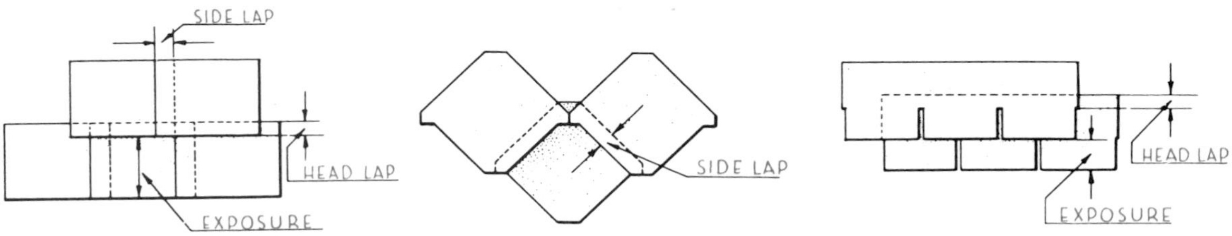

Fig. 10-3. Terms used in the application of roofing materials.

Fig. 10-4. General specifications and installation data for common asphalt roofing products. (Asphalt Roofing Manufacturers Assoc.)

PRODUCT	Approximate Shipping Weight		Sqs. Per Package	Length	Width	Side or End Lap	Top Lap	Exposure	Underwriters' Listing
	Per Roll	Per Sq.							
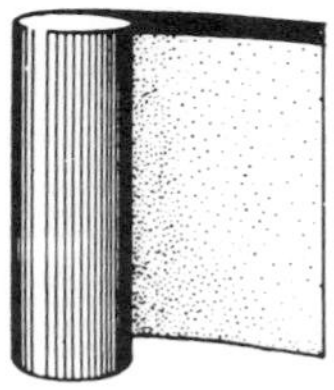 Mineral Surface Roll	75# to 90#	75# to 90#	One	36' 38'	36" 36"	6"	2" 4"	34" 32"	C
		Available in some areas in 9/10 or 3/4 Square rolls.							
 Mineral Surface Roll Double Coverage	55# to 70#	55# to 70#	One Half	36'	36"	6"	19"	17"	C
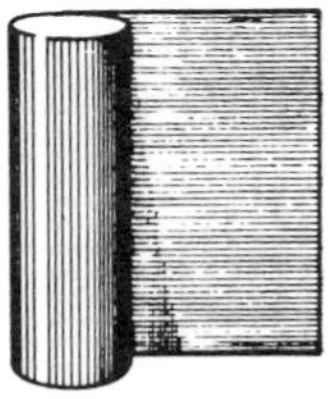 Coated Roll	50# to 65#	50# to 65#	One	36'	36"	6"	2"	34"	None
 Saturated Felt	60# 60# 60#	15# 20# 30#	4 3 2	144' 108' 72'	36" 36" 36"	4" to 6"	2"	34"	None

PRODUCT	Configuration	Per Square			Size		Exposure	Underwriters' Listing
		Approximate Shipping Weight	Shingles	Bundles	Width	Length		
Wood Appearance Strip Shingle More Than One Thickness Per Strip 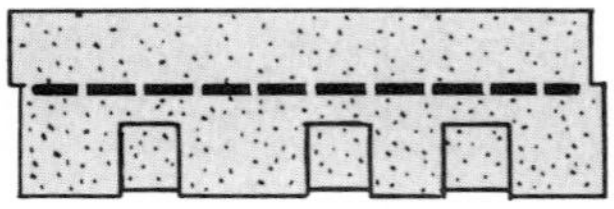 Laminated or Job Applied	Various Edge, Surface Texture & Application Treatments	285# to 390#	67 to 90	4 or 5	11-1/2" to 15"	36" or 40"	4" to 6"	A or C - Many Wind Resistant
Wood Appearance Strip Shingle Single Thickness Per Strip 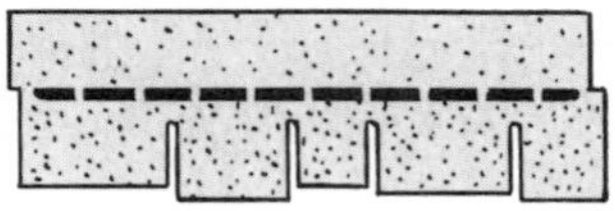	Various Edge, Surface Texture & Application Treatments	Various 250# to 350#	78 to 90	3 or 4	12" or 12-1/4"	36" or 40"	4" to 5-1/8"	A or C - Many Wind Resistant
Self-Sealing Strip Shingle 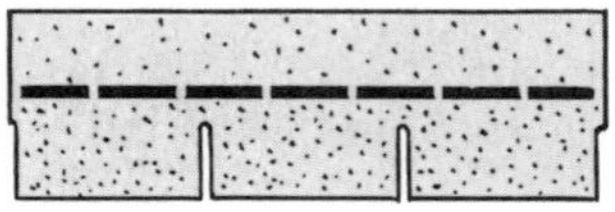	Conventional 3 Tab	205#-240#	78 or 80	3	12" or 12-1/4"	36"	5" or 5-1/8"	A or C - All Wind Resistant
	2 or 4 Tab	Various 215# to 325#	78 or 80	3 or 4	12" or 12-1/4"	36"	5" or 5-1/8"	
Self-Sealing Strip Shingle 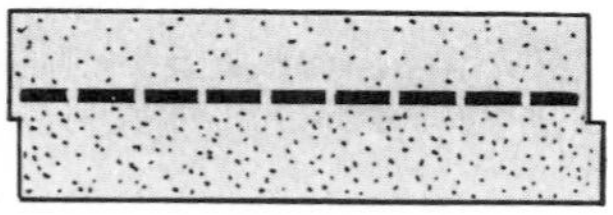 No Cut Out	Various Edge and Texture Treatments	Various 215# to 290#	78 to 81	3 or 4	12" or 12-1/4"	36" or 36-1/4"	5"	A or C - All Wind Resistant
Individual Lock Down 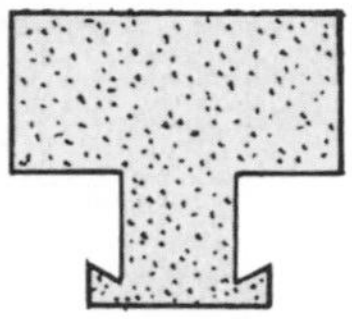 Basic Design	Several Design Variations	180# to 250#	72 to 120	3 or 4	18" to 22-1/4"	20" to 22-1/2"	-	C - Many Wind Resistant

to warp and buckle. To avoid this, louvered openings should be constructed high up under the eaves in the gable ends or at such locations as will insure adequate ventilation. Louvers should have a total effective area equivalent to 1/2 sq. in. per square foot of attic space. Refer to Unit 13 for additional information.

Inspect the roof deck to see that nailing patterns are complete and that there are no protruding nails. Joints should be smooth and free of sharp edges that might cut through the roofing materials. Repair large knot holes (over 1" diameter) by covering with a piece of sheet metal. Clean the roof surface of any chips or other scrap material.

Asphalt Roofing Products

Asphalt roofing products are widely used in modern construction. They can be broadly classified in three groups: saturated felts, roll roofing, and shingles.

Saturated felts, used under shingles for sheathing paper and for laminations in constructing a built-up roof, consist of dry felt impregnated with asphalt or coal tar. Saturated felt is made in different weights, the most common being 15 lb. The weight indicates the amount necessary to cover 100 sq. ft. of roof surface with a single layer.

Roll roofing is made by adding a coating of weather-resistant asphalt to a felt which has been impregnated with a saturant asphalt. Some types of roll roofing are surfaced with mineral granules to produce desired colors. The granules also help make the roofing fireproof and protect it from the rays of the sun.

Asphalt shingles are surfaced with mineral granules. They are available in many patterns; some are individual shingles and some come in strips. Data concerning shingles and other general groupings of asphalt products is given in Fig. 10-4. There are additional products within each group that differ as to weight and dimension. For example: the 3-tab square-butt strip shingle is available in many qualities and colors and in weights from 205 to 240 lbs. per square.

HANDY SAYS:

"Safety considerations are very important in roofing work. Be sure to erect a secure scaffold that will support the worker at a waist-high level with the eaves. Study Unit 22 for information and directions in this area."

Underlayment

When the roof deck is completed and dry, it should be covered with asphalt-saturated felt or other material that has a low vapor resistance. This underlayment protects the sheathing from moisture until the shingles are laid; provides additional weather protection by preventing the entrance of wind driven rain and snow; and prevents direct contact between the shingles and resinous areas in the sheathing.

Materials such as coated sheets or heavy felts which might act as a vapor barrier should not be used because they would permit moisture and frost to accumulate between the covering and the roof deck. Although 15 lb. roofer's felt is commonly used for this purpose, requirements will vary depending on the kind of shingles and the roof slope.

General application standards for underlayment suggest a 2 in. toplap at all horizontal joints and a 4 in. sidelap at all end joints, Fig. 10-5. It should be lapped at least 6 in. on each side of the center line of hips and valleys.

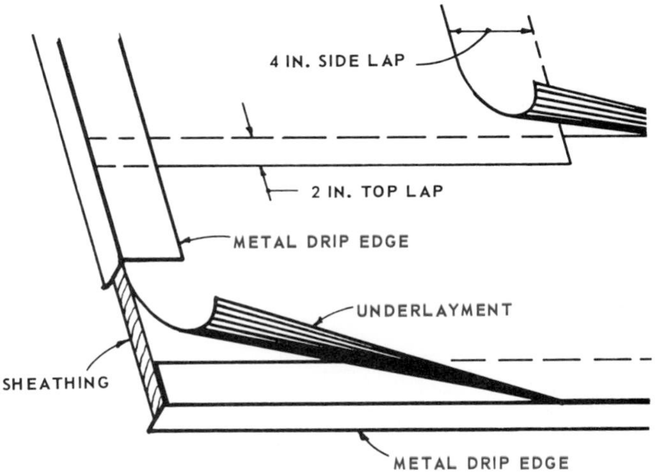

Fig. 10-5. Application of underlayment and metal drip edge. Note that the underlayment is laid over the drip edge along the eaves.

Drip Edge

For efficient water-shedding at the roof edges and along the eaves and rake, it is recommended that a metal drip edge be installed. Various shapes, formed from 26 ga. galvanized steel, are available. They extend back about 3 in. from the roof edge and are bent downward over the edge, causing the water to drip free of underlying cornice construction.

Flashing at Eaves

In some work it is highly recommended that an eaves flashing strip be installed over the underlayment and metal drip edge. This strip may be smooth or mineral surfaced roll roofing. It should be wide enough to extend from the edge of the roof to a point along the roof that is about 12 in. inside the wall line.

The purpose of this flashing is to prevent the penetration of water resulting from the thawing and freezing of melting snow or the backing-up of frozen slush in the eaves troughs. The lower edge of this strip should be placed even with the drip edge.

Valley Flashing (Open Type)

The installation of roofing materials is complicated by the intersection of other roofs, adjoining walls, and such projections as chimneys and soil stacks. Making these areas watertight requires a special construction that is called flashing. Materials used for flashing include: tin-coated metal, galvanized metal, copper, lead, aluminum, asphalt shingles, and roll roofing.

Valley flashing, where two sloping roofs meet, is one of the most critical constructions. Water drainage is concentrated at this point and leakage would create a serious problem. For asphalt shingles, the recommended flashing material is 90 lb. mineral surfaced asphalt roll roofing, installed as illustrated in Fig. 10-6A.

The first strip, at least 18 in. wide, is centered in the valley and laid with the mineral surface

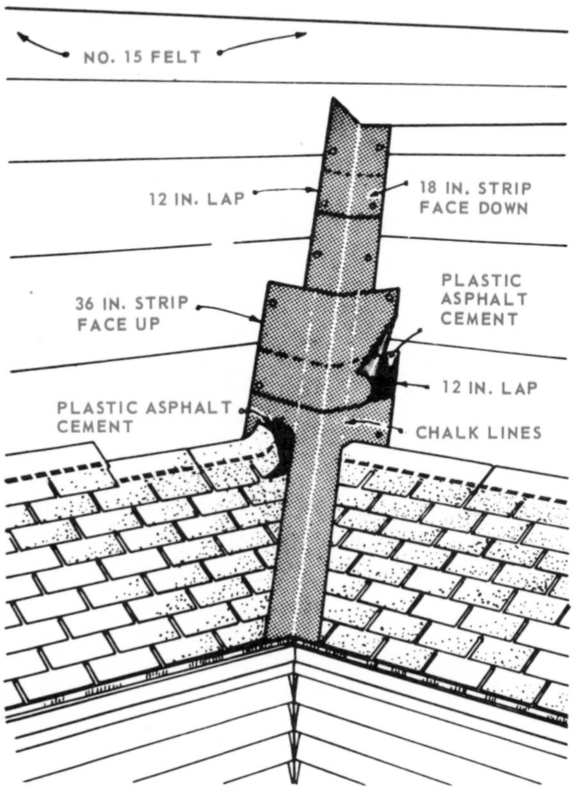

Fig. 10-6A. Open valley flashing using 90 lb. roll roofing.

down. After nailing this strip, a second strip 36 in. wide is laid in place with the mineral side up. When it is necessary to join either strip, they are lapped at least 12 in. and secured with plastic asphalt cement. As each strip is laid, first nail one edge and then press the material firmly into the valley as the second edge is attached.

Before applying the shingles, snap a center chalk line in the valley and then one on each side. The outside lines will mark the width of the waterway which should be 6 in. wide at the ridge and the lines should diverge at a rate of 1/8 in. per foot as they approach the eave. Thus a valley 8 ft. long would be 7 in. wide at the eave.

When a course of shingles meets the valley, the chalk line serves as a guide in trimming the last unit. After the shingle is trimmed, cut off the upper corner at about a 45 deg. angle with the valley line and then cement the end of the shingle in place.

Woven and Closed-Cut Valleys

Some applicators of asphalt shingles prefer to use a woven or closed-cut valley design, especially on reroofing work. Strip shingles are the only type that can be used. It is essential that a single unit be of sufficient width to cross the lowest point of the valley and continue upward on each roof surface a minimum of 12 in. To provide this position for shingles that form the center of the valley, it is necessary to cut some of the preceding shingle strips. Nails or other fasteners must be spaced at least 6 in. away from the valley centerline.

In a reroofing application, it is sometimes necessary to build up the trough of the existing valley to the average level of the existing roof surface. This can usually be accomplished with a beveled, wooden strip.

The first step in constructing either type of valley is to apply a 36 in. width of 50 lb. or heavier roll roofing. For a woven valley, the first course is laid along the eave of one of the roof surfaces. It is then extended across the valley a distance of at least 12 in. The first course is then laid along the intersecting roof and extended across the valley over the previously applied shingle. Succeeding courses are alternated; first along one roof area and then the other as shown in Fig. 10-6B. One or both of the roof areas could be partially laid before constructing the valley.

When laying the shingles across the valley, be sure to press them firmly into the valley - and position the nails at least 6 in. on either side of

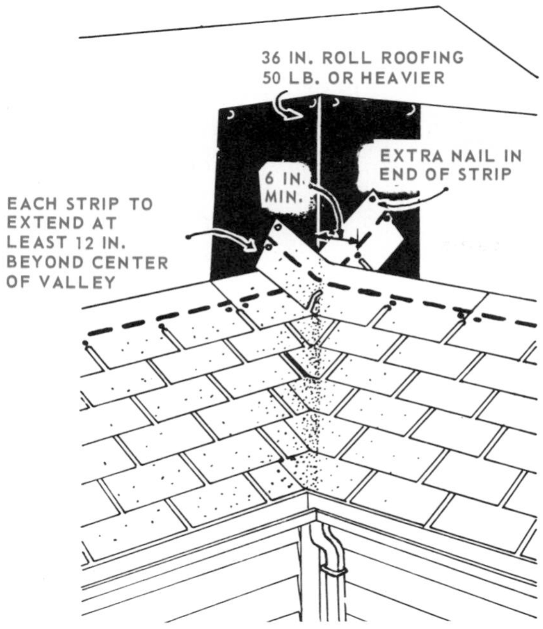

Fig. 10-6B. Woven valley design. Commonly used in reroofing applications.

the centerline. Use two nails at the end of each terminal strip as shown in the diagram.

To construct a closed-cut valley, install the roll roofing and then apply all the shingles on one roof surface. Carry each strip across the valley and onto the adjoining roof, following the procedure previously described for the woven valley. When the roof surface is complete, apply the first course of shingles along the eaves of the intersecting roof. Where this course meets the valley, trim the shingle along a line 2 in. back from the centerline of the valley. Also trim off the upper corner of the shingle to prevent water from running back along the top edge. Embed the end of the shingle in a 3 in. wide strip of plastic asphalt cement. Succeeding courses are applied and completed as shown in Fig. 10-6C.

Either the open, woven, or closed-cut valley construction can be used between a main roof and a gable dormer. The main roof area is first laid up to a point just above the lower end of the valley - and carefully flashed where it meets the dormer walls. The valley lining is then installed and extended about 1/4 in. below the edge of the dormer roof section. Shingle units are then applied. Before laying the shingles on the main roof, it is good procedure to measure and, if necessary, make slight adjustments so the courses will align with those of the dormer section.

Flashing at a Wall

Where the roof joins a vertical wall, it is best to install metal flashing shingles over the end of each course, Fig. 10-7. They should be about 6 in. long and at least 2 in. wider than the exposed surface of the regular shingle. The metal shingle is bent so about 2 in. extends horizontally and the balance extends upward along the wall. As each course of shingles is laid, a metal flashing shingle is set in place and secured with one nail in the top corner. When the roof is complete, the finished siding is installed over the flashing with sufficient clearance to paint the lower edges.

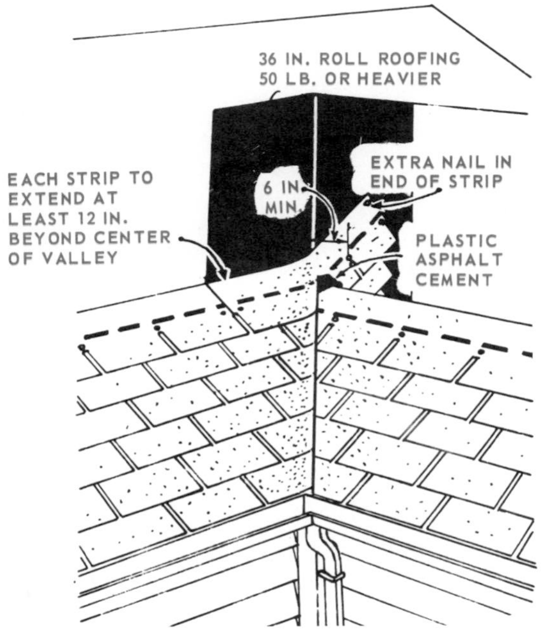

Fig. 10-6C. Closed-cut valley. Shingles on the right are cut along a line spaced 2 in. from the valley centerline.

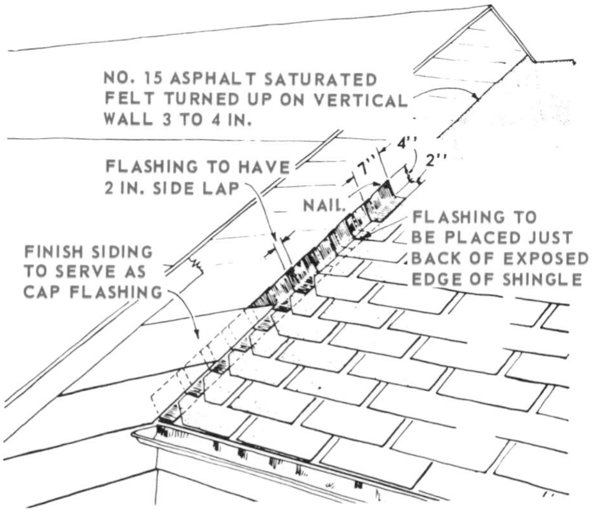

Fig. 10-7. Metal flashing shingles used to waterproof joint between sloping roof and vertical wall. Generally referred to as step-flashing.

Note: Drawings and diagrams showing the application of asphalt shingles furnished by Asphalt Roofing Manufacturers Assoc.

Strip Shingles

On small roofs, strip shingles may be laid starting at either end. When the roof surface is over 30 ft. in length, it is usually best to start at the center and work both ways from a line perpendicular to the eaves and ridge. Asphalt shingles will vary slightly in length (about plus or minus 1/4 in. in a 36 in. strip) and there may be some variations in width. Thus to control the proper placement so shingles will be accurately aligned horizontally and vertically, chalk lines should be used.

When making the application from the center of the roof toward the ends, snap a number of chalk lines between the eaves and ridge to serve as reference marks for starting each course. Space them according to the type of shingle and laying pattern. These lines are used in the same way that the rake edge of the roof is used when the application is started at the roof end as shown in the illustrations. The shingles do not need to be cut. Instead, full shingles are aligned with the chalk lines to form the desired pattern.

Chalk lines, parallel to the eaves and ridge, will be helpful in maintaining straight horizontal lines along the butt edge of the shingle. Usually, only about every fifth or sixth course should be checked in this way when the shingles are skillfully applied. Inexperienced workers may need to set up chalk lines for every second or third course.

When roofing materials are delivered to the building site, they should be handled with care and protected from damage. Try to avoid handling asphalt shingles in extreme heat or cold.

HANDY SAYS:

"To secure the best performance from any roofing material, always read and study the manufacturer's direction sheets and make the installation as specified."

Nailing and Fastening

Nails used to apply asphalt roofing must have a large head (3/8 in. to 7/16 in. dia.) and a sharp point, Fig. 10-8. Most manufacturers recommend 11 or 12 ga. galvanized steel nail with barbed shanks. Aluminum nails are also used. The length should be sufficient to penetrate nearly the full thickness of the sheathing.

The number of nails and correct placement are both vital factors in proper application of a roofing material. For three-tab square-butt shingles, use a minimum of 4 nails per strip as shown in the application diagrams. Align each shingle care-

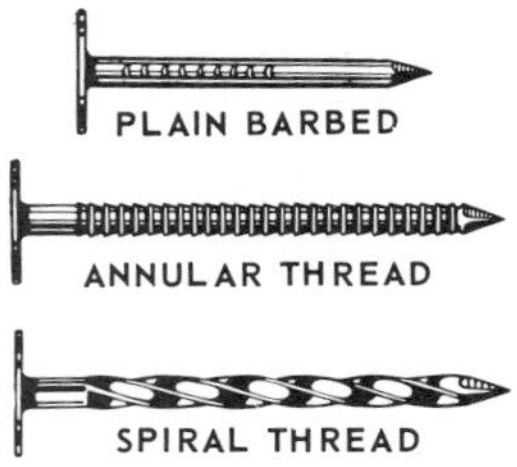

Fig. 10-8. Nails used for asphalt shingles. Use 1 1/4 in. nails for new roofs, and 1 3/4 in. nails for reroofing. Average shingles will require 480 nails per square or 2 1/4 lbs. of 1 1/4 in. size – 2 3/4 lbs. of 1 3/4 in. size.

fully and start the nailing from the end next to the one previously laid and proceed across the shingle. This will prevent buckling. Drive nails straight so the edge of the head will not cut into the shingle. The nail head should be driven flush, not sunk into the surface. If for some reason the nail fails to hit solid sheathing--drive another nail in a slightly different location.

In modern construction, pneumatic powered staplers are often used to install asphalt shingles, Fig. 10-9. Special staples with an extra wide crown should be used. Always check and follow manufacturer's recommendations when using specialized equipment.

Fig. 10-9. Installing three-tab square butt asphalt shingles with pneumatic powered stapler. (Senco Products Inc.)

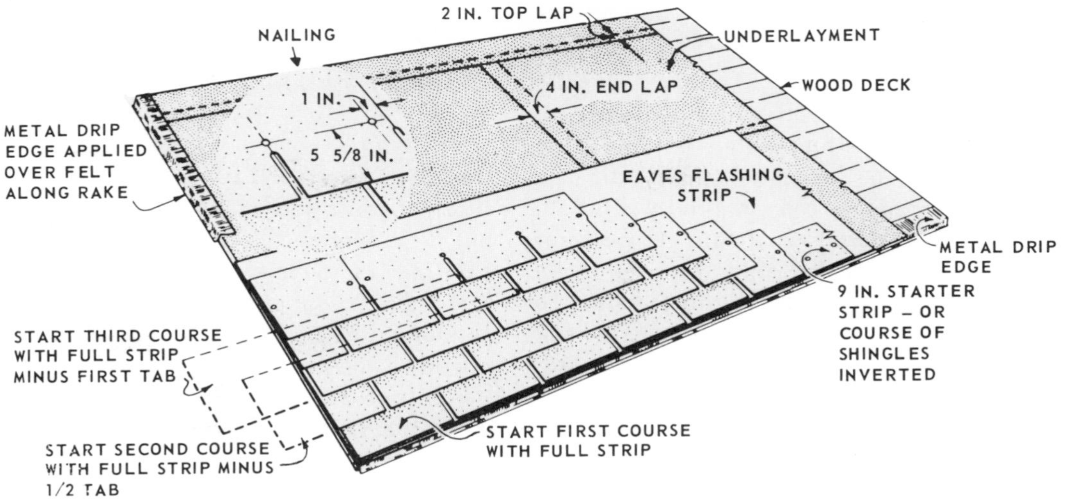

Fig. 10-10. Three-tab square butt shingles laid so the cutouts are centered over the tabs in the course directly below.

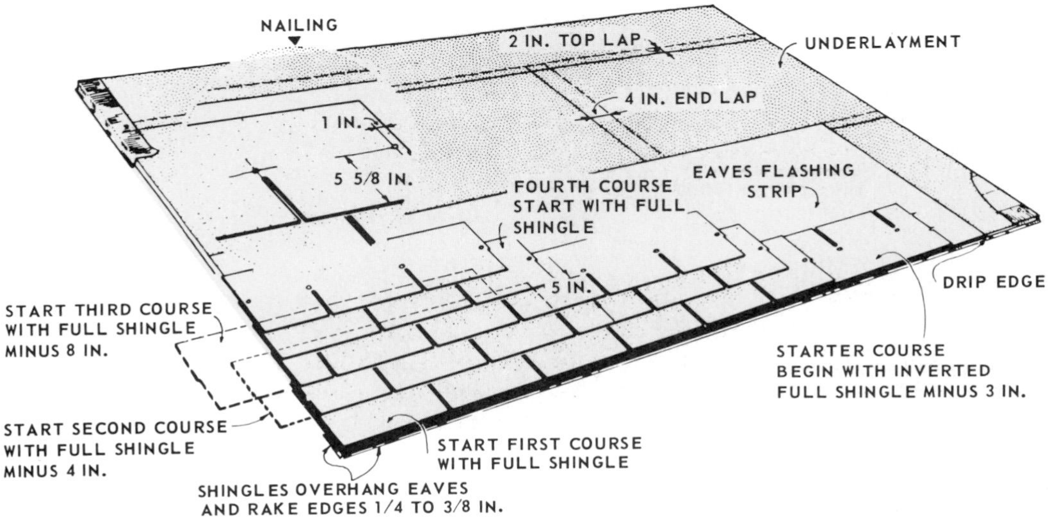

Fig. 10-11. Three-tab square butt shingles laid with cutouts breaking joints on thirds.

Starter Strip

The purpose of a starter strip is to back up the first course of shingles and fill in the space between the tabs. Use a strip of mineral surfaced roofing (9 in. or wider) of a weight and color to match the shingles. Apply the strip so it overhangs the drip edge slightly and secure it with nails spaced 3 to 4 in. above the edge. Space the nails so they will not be exposed at the cutouts between the tabs of the first course of shingles. Sometimes an inverted row of shingles is used instead of the starter strip.

First and Succeeding Courses

The first course is started with a full shingle. Succeeding courses are then started with either full or cut strips, depending upon the type of shingle and the laying pattern.

Three-tab square-butt shingle strips are commonly laid so the cutouts are centered over the tab in the course directly below, thus the cutouts in every other course will be exactly aligned, Fig.

Fig. 10-12. Approved nailing pattern for three-tab square butt shingles. (Asphalt Roofing Manufacturers Assoc.)

10-10. For this type of pattern, start the second course with a strip from which 6 in. has been cut. The third course is started with a strip with a full tab removed and the fourth with one-half of a strip. Continue to reduce the length of shingle accordingly for subsequent courses. A pair of tin snips can be used to cut the shingles.

The diagram in Fig. 10-11 shows the procedure to follow for a pattern where the cutouts break joints on thirds. The second course is started with a strip shortened by four inches; the third by 8 in. The fourth course is started with a full strip.

Using an approved nailing pattern for three-tab shingles is very important in securing the best appearance and full weather protection. Manufacturers recommend that four nails be used as shown in Fig. 10-12. When shingles are applied with an exposure of five inches, nails should be placed 5/8 in. above tops of cutouts. One nail should be placed above each cutout and one nail spaced in 1 in. from each end. Nails should not be placed in or above the factory applied adhesive strip.

Hexagonal Shingles

This type of strip shingle is available in two-tab and three-tab units. These shingles do not permit spacing variations as previously described for square-butt shingles. Begin the application with a starter strip or a course of inverted shingles as shown in Fig. 10-13. The first course starts with a full unit and the second course with a full unit less one-half tab as shown. Each course is applied so the lower edge of the tabs align exactly with the cutouts of the preceding course.

Chimney Flashing

Flashing where the chimney projects through the roof must allow for some movement that will likely result from settlement or shrinkage of the building frame. To provide for this movement, the flashing is divided into two parts; the base flashing that is secured to the roof deck and the cap flashing (also called counter flashing) that is secured to the chimney.

Before the base flashing is applied, the shingles are laid up to the front face of the chimney. When this is complete, lay out and cut, from 90 lb. mineral surfaced roofing, the front section of flashing as shown in Fig. 10-14. A similar section can be cut for the back if there is no saddle (small auxiliary roof deck built above the chimney to divert water to either side). Such a structure is not required when the chimney is small and located high on the roof near the ridge. See Unit 19 for more information.

Cement the front base flashing into place and then cut and apply the side pieces as shown. The base flashing at the back of the chimney is applied last. All the sections are cemented together as they are applied. An alternate method of applying base flashing is shown in Fig. 10-15. Sheet metal is often used for base flashing and applied by the step method previously described in the section on wall flashing.

Cap flashing consists of metal sheets set into the mortar joints of the chimney when it is constructed. The final chimney flashing operation is to simply bend the metal strips down over the base flashing. The metal is set into the mortar

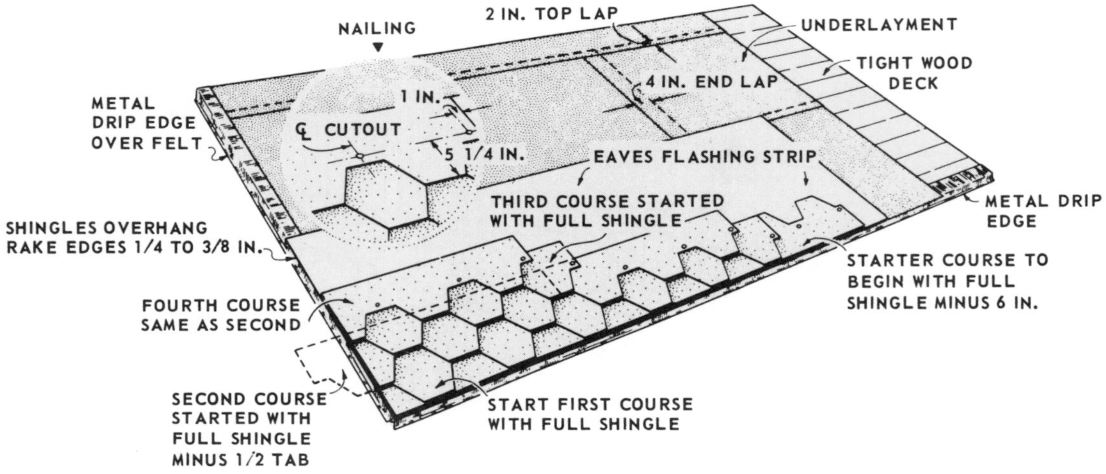

Fig. 10-13. Standard application of three-tab hexagonal shingles.

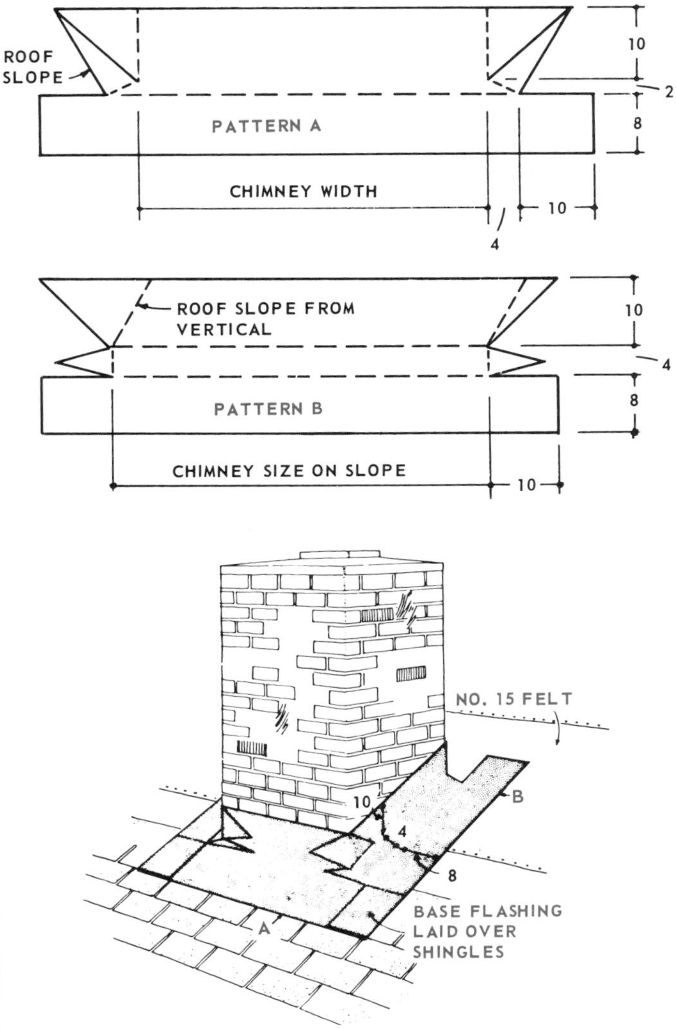

Fig. 10-14. Base flashing for a chimney. Above. Pattern layouts. Below. Flashing cut and applied.

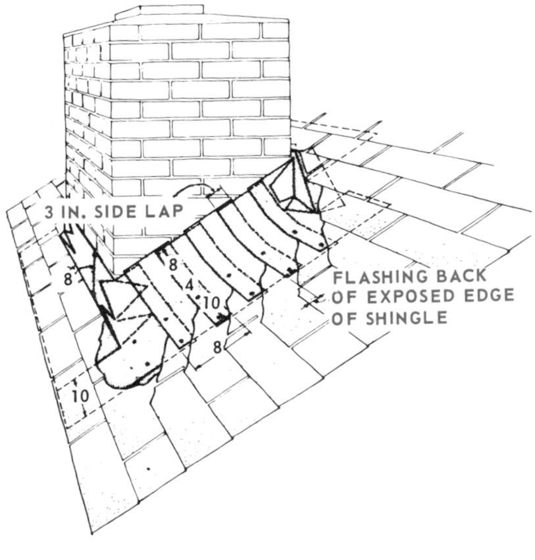

Fig. 10-15. Alternate method of applying base flashing to the side of a chimney.

joints to a depth of 1 1/2 in. Cap flashing on the front of the chimney can be one continuous piece while on the sides it must be stepped-up in sections to align with the roof slope, Fig. 10-16.

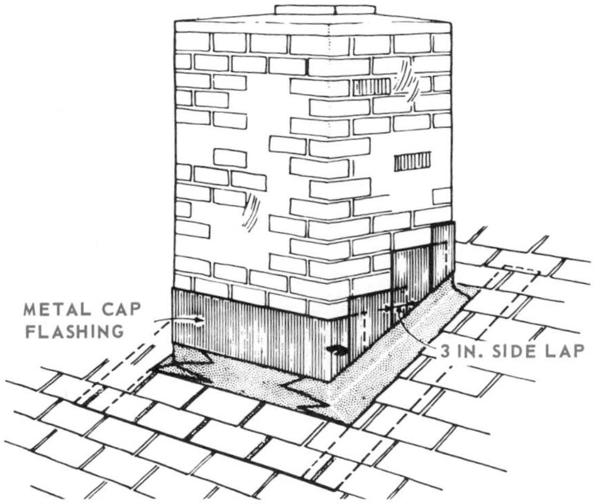

Fig. 10-16. Metal cap flashing set in mortar joints of the chimney covers base flashing.

Vent Stack Flashing

Pipes projecting through the roof must be carefully flashed. Various prefabricated flanges are available for this purpose. Asphalt products can also be used successfully for the flashing.

The roofing is first applied up to where the stack projects. Shingles are cut and fitted around the stack, Fig. 10-17. A flange is then carefully cemented in place and the roof shingles laid over the top as shown. The flange must be large enough to extend at least 4 in. below, 8 in. above, and 6 in. on each side.

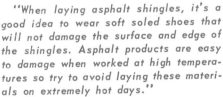

HANDY SAYS:

"When laying asphalt shingles, it's a good idea to wear soft soled shoes that will not damage the surface and edge of the shingles. Asphalt products are easy to damage when worked at high temperatures so try to avoid laying these materials on extremely hot days."

Hips and Ridges

Special hip and ridge shingles are usually available from the manufacturer. The special shingles can be easily made, however, by cutting pieces 9 in. by 12 in. from either square-butt shingle strips or mineral surfaced roll roofing that matches the color of the shingles.

After the shingles are cut, bend them lengthwise in the center line. In cold weather, the shingle should be warmed before bending to prevent cracks and breaks. Begin at the bottom of the hips or one end of the ridge. Lap the units to provide a 5 in. exposure as illustrated in Fig. 10-18. Secure with one nail on each side, 5 1/2 in. back from the exposed end and 1 in. from the edge.

Metal ridge roll is not recommended for asphalt shingles since corrosion may discolor the roof.

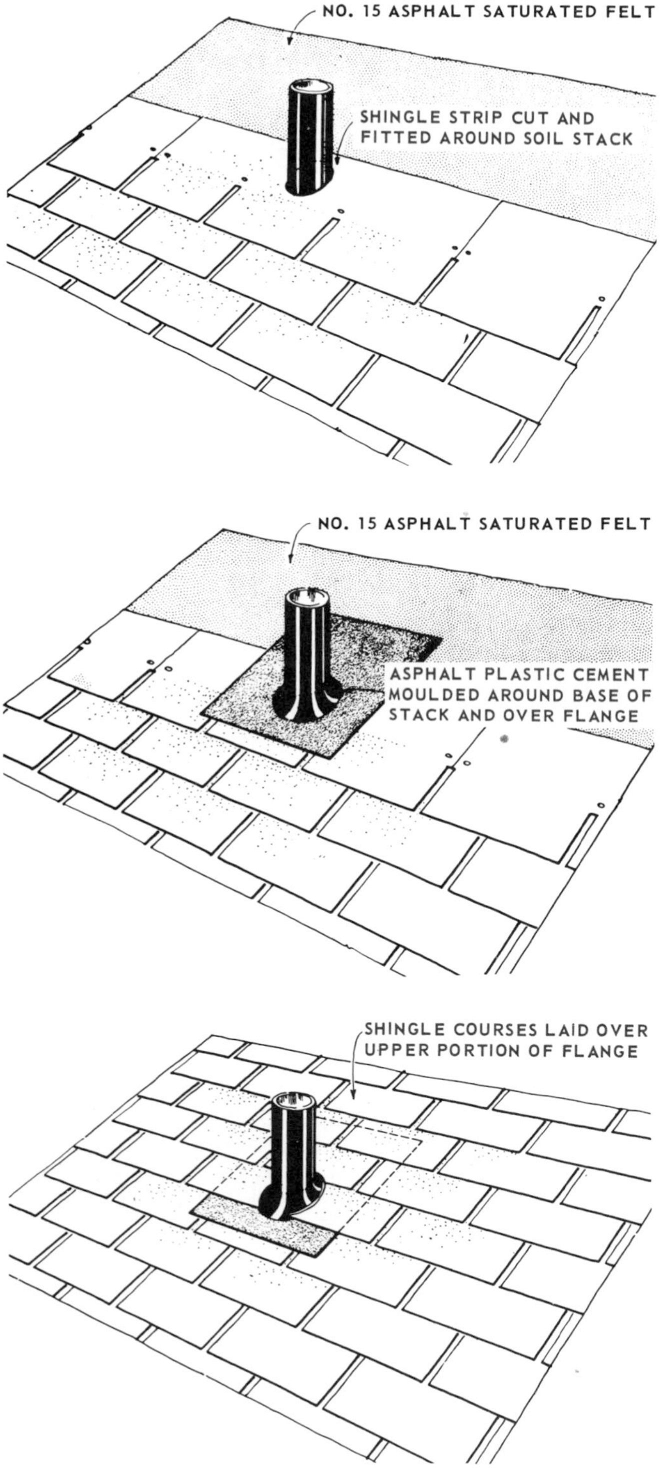

Fig. 10-17. Flashing vent stack. Top. Shingles laid up to stack and last course fitted. Center. Flange in place. Bottom. Shingles applied over flange.

Wind Protection

Shingles that are provided with factory applied adhesive under each tab are available for use in localities where high winds are frequent. After installation, only a few warm days are needed to thoroughly seal the tabs to the course below and thus prevent them from being blown up by strong winds. This precaution is especially important on low sloping roofs where it is easier for the wind to "get under" the shingles.

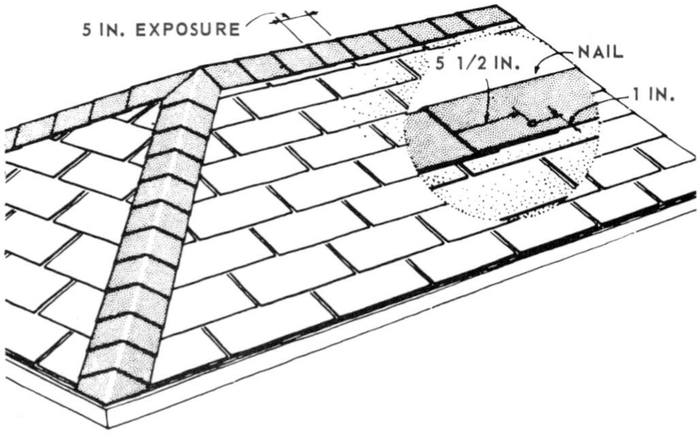

Fig. 10-18. Application of hip and ridge shingles.

If regular shingles are used, the tabs can be cemented as shown in Fig. 10-19. Apply a spot of special tab cement about 1 in. square with a putty knife or caulking gun and then press the tab down. Avoid lifting the tab any more than necessary while applying the cement.

A variety of interlocking shingles are designed to provide resistance against strong winds. They are used for both new construction and reroofing. Details of the interlocking devices and methods of application vary considerably. Always study and follow the manufacturer's directions when installing all types of shingles.

Individual Asphalt Shingles

Roof surfaces may be laid with an individual asphalt shingle. There are several sizes and designs available. One commonly used is 12 in. wide and 16 in. long. Several patterns can be used in its

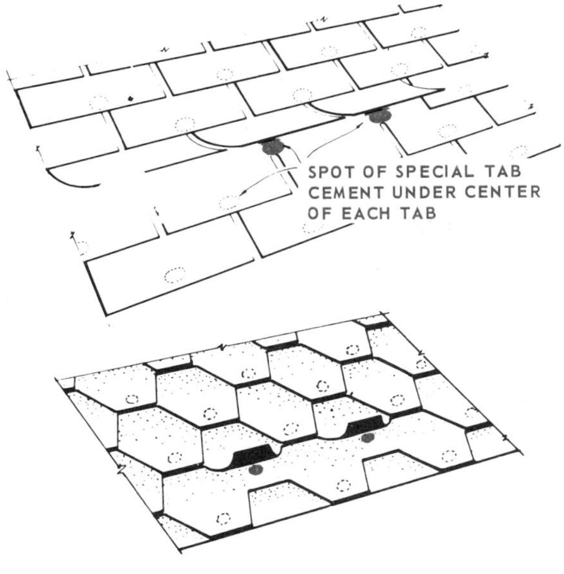

Fig. 10-19. Asphalt shingle tabs may be cemented down for wind protection. Above. Square butt shingles. Below. Hexagonal shingles.

application, one of which is illustrated in Fig. 10-20. Follow the same procedure that was described for strip shingles. Horizontal and vertical chalk lines should be used to insure accurate alignment.

the roof to a point 24 in. inside the interior wall line of the building. See Fig. 10-21.

3. Shingles provided with factory applied adhesive and manufactured to conform to the Underwriters Laboratories Standard for Class "C" Wind Resistant shingles, or, if "free" tab square-butt strips are used, cement all the tabs as previously described.
4. Follow application methods shown in Fig. 10-22.

Reroofing

When a reroofing job is being considered, a choice must be made between removing the old roofing and leaving it in place. It is usually not necessary to remove old wood shingles, old asphalt shingles, or old roll roofing before putting on a new asphalt roof provided that:

1. The strength of the existing deck and framing is adequate to support the weight of workers and additional new roofing, as well as snow and wind loads.
2. The existing deck is sound and will provide good anchorage for the nails used in applying new roofing.

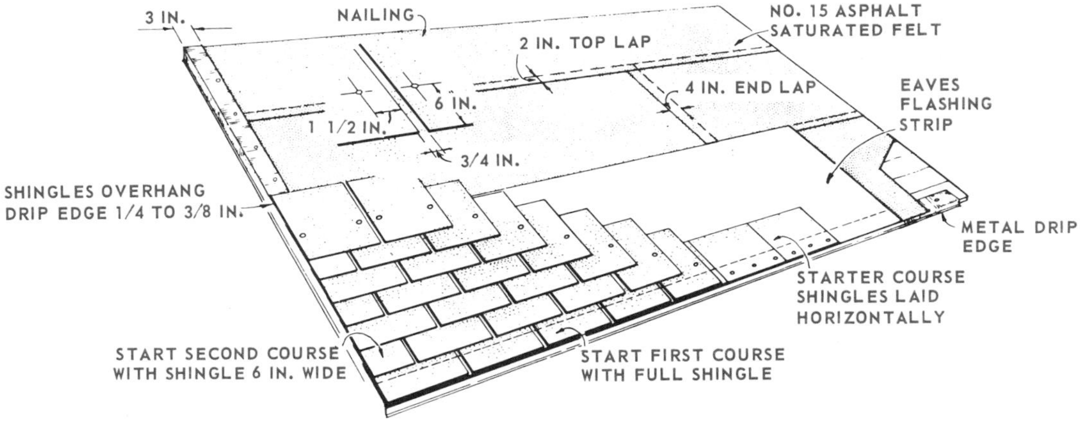

Fig. 10-20. Application of giant individual asphalt shingles by the American method.

Low-Slope Roofs

When applying asphalt shingles to slopes less than 4 in 12, certain additional application procedures should be followed. Slopes as low as 2 in 12 can be made watertight and windtight if the installation includes:

1. A double thickness felt underlayment. Lap each course over the preceeding one 19 in., starting with a 19 in. strip.
2. In areas where the January daily average temperature is 25 deg. F. or less, cement the two felt layers together from the eave up

If it is decided to put on new roofing over old wood shingles, all loose or protruding nails should be removed and the shingles renailed in new locations; loose shingles nailed down; warped shingles split and segments renailed; missing shingles replaced; and shingles at eaves and rakes cut back far enough to allow the application of 4 in. to 6 in. nominal 1 in. thick strips. These strips should be nailed in place allowing their outside edges to project beyond the roof deck the same distance as the old wood shingles.

To provide a smooth surface for asphalt roofing, it is often advisable that a "backer board" be

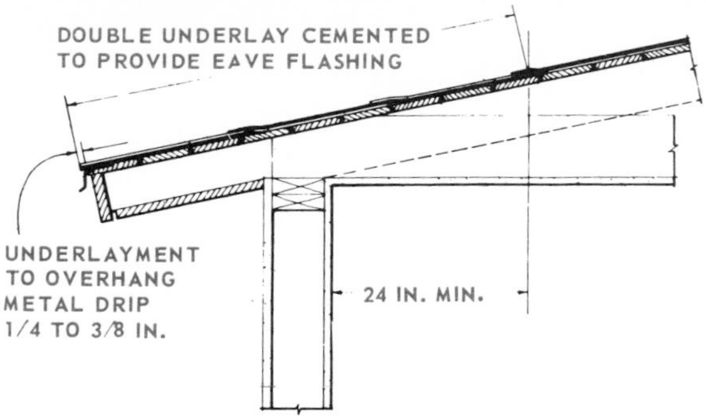

Fig. 10-21. Underlay plies cemented together to form eaves flashing for low-sloped roofs.

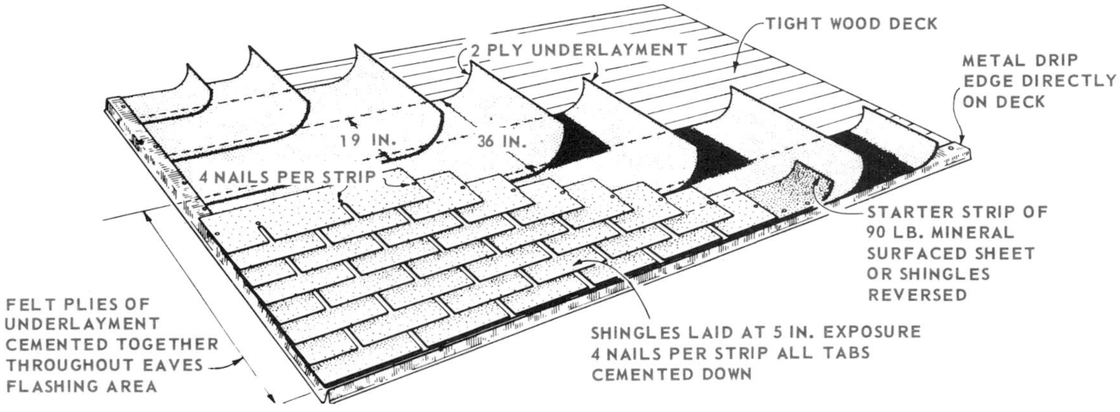

Fig. 10-22. Application of shingles on low-slope roofs.

applied over the wood shingles or that beveled wood "feathering strips" be used along the butts of each course of old shingles.

When the old roof consists of square butt asphalt shingles with a 5 in. exposure, a new application of self-sealing strip shingles can be applied as shown in Fig. 10-23. This application pattern will insure a smooth, even appearance.

The joint between a vertical wall and roof surface should be sealed in a reroofing application. First apply a strip of smooth roll roofing about 8 in. wide. Nail each edge firmly with a spacing of about 4 in. O.C. As the shingles are applied, asphaltic plastic cement is spread on the strip and the shingles are thoroughly bedded. To insure a tight joint, use a caulking gun to apply a final bead of cement between the ends of the shingles and the siding.

When old shingles are to be removed before applying a new roof, it is common practice to use a flat-bladed shovel as shown in Fig. 10-24. Both asphalt and wood shingles can be removed in this manner. A shovel can also be used to remove the old underlayment.

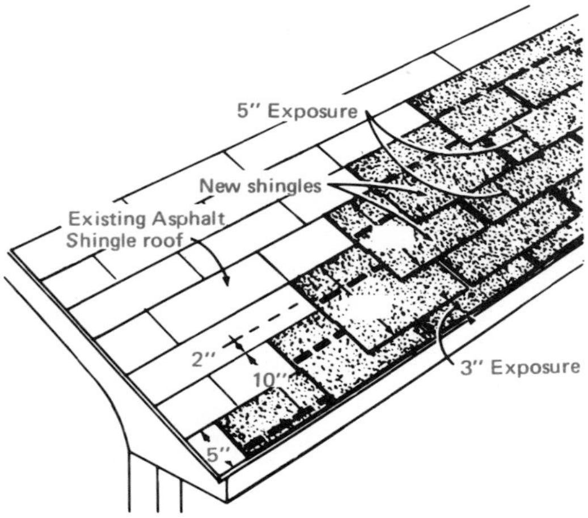

Fig. 10-23. Recommended application pattern over existing asphalt shingle roof. (Asphalt Roofing Manufacturers Assoc.)

Fig. 10-24. Using a shovel to remove old asphalt shingles. (Asphalt Roofing Manufacturers Assoc.)

Built-Up Roofing

A flat roof or a roof with very little slope must be covered with a watertight membrane. Most flat roofs today are covered with built-up roofing which, if properly installed, is quite durable. The components are delivered to the job. The roofing itself is constructed directly on the deck in a series of layers or plies.

On a wood deck, a layer of roofing felt is first nailed in place with galvanized nails, Fig. 10-25. Each succeeding layer is then mopped in place with hot asphalt or hot coal tar pitch. The top layer is coated and covered with a layer of gravel, crushed stone or marble chips which provides a weathering surface and may also improve the appearance. This mineral covering is usually applied at the rate of 300 to 400 lb. per 100 sq. ft.

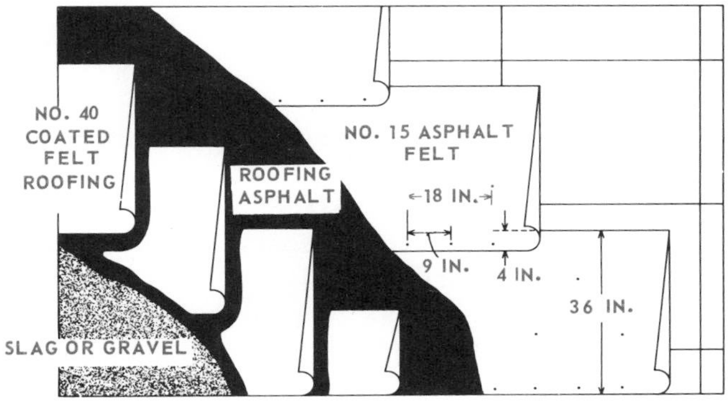

Fig. 10-25. Construction of built-up roofing. (Ruberoid Co.)

Built-up roofs for residential structures are normally 3 or 4 plies (layers) and are limited to a slope of not over 2 in 12. Asphalt moppings are used on sloping roofs while coal-tar pitch, which has a "cold flow" must be limited to flat surfaces.

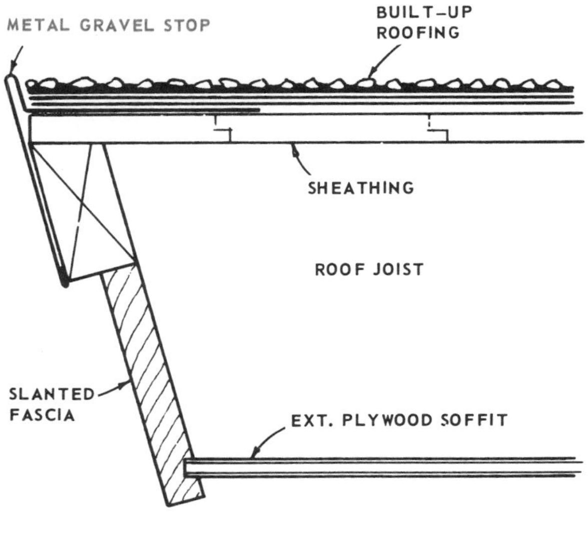

Fig. 10-26. Section through edge of flat roof overhang showing metal gravel stop.

A metal gravel stop, preferably made of copper, is fastened to the edge of the roof deck to serve as a trim member and keep the gravel and pitch in place, Fig. 10-26.

When a leak occurs in a built-up roof, flashings at parapet walls, chimneys, and vents should be inspected carefully, because the first roof failures usually occur at these locations. Bituminous flashings are made of saturated felt and flashing cement, Fig. 10-27. Flashing cement should be forced behind the felt if it has separated from the wall at the upper edge, and the edge sealed with a strip of bituminous-saturated cotton fabric 4 in. wide, embedded and coated with flashing cement.

Bare spots on a built-up roof where the mineral surfacing is not properly embedded should be repaired. First clean the area, apply a heavy coating of hot asphalt and then spread more gravel or slag. Felts which have disintegrated should be cut away and replaced with new felt. The new felt should be mopped in place, allowing at least one additional layer of felt to extend not less than 6 in. beyond the other layers.

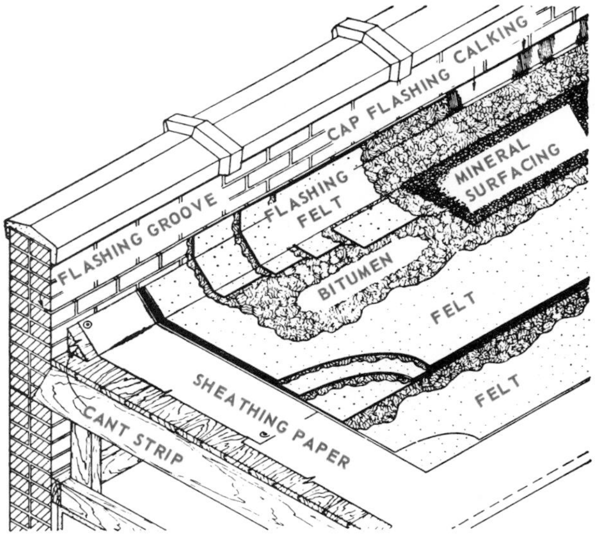

Fig. 10-27. Flashing construction for built-up roofs.

Wood Shingles

Wood shingles have been used for many years in residential construction. A disadvantage is that wood shingles, unless treated, have little resistance to fire. Building codes often prohibit their use. Since wood weathers to a soft, mellow color after exposure, wood shingles provide an appearance that is desired by many home owners. When properly installed, they also provide a very durable roof.

Grade	Length	Thickness (at Butt)	No. of Courses Per Bundle	Bdls/Cartons Per Square	Shipping Weight	Description
No. 1 BLUE LABEL	16" (Fivex) 18" (Perfections) 24" (Royals)	.40" .45" .50"	20/20 18/18 13/14	4 bdls. 4 bdls. 4 bdls.	144 lbs. 158 lbs. 192 lbs.	The premium grade of shingles for roofs and sidewalls. These top-grade shingles are 100% heartwood, 100% clear and 100% edge-grain.
No. 2 RED LABEL	16" (Fivex) 18" (Perfections) 24" (Royals)	.40" .45" .50"	20/20 18/18 13/14	4 bdls. 4 bdls. 4 bdls.	144 lbs. 158 lbs. 192 lbs.	A good grade for all applications. Not less than 10" clear on 16" shingles, 11" clear on 18" shingles and 15" clear on 24" shingles. Flat grain and limited sapwood are permitted in this grade.
No. 3 BLACK LABEL	16" (Fivex) 18" (Perfections) 24" (Royals)	.40" .45" .50"	20/20 18/18 13/14	4 bdls. 4 bdls. 4 bdls.	144 lbs. 158 lbs. 192 lbs.	A utility grade for economy applications and secondary buildings. Not less than 6" clear on 16" and 18" shingles, 10" clear on 24" shingles.
No. 1 or No. 2 REBUTTED-REJOINTED	16" (Fivex) 18" (Perfections) 24" (Royals)	.40" .45" .50"	33/33 28/28 13/14	1 carton 1 carton 4 bdls.	60 lbs. 60 lbs. 192 lbs.	Same specifications as above but machine trimmed for exactly parallel edges with butts sawn at precise right angles. Used for sidewall application where tightly fitting joints between shingles are desired. Also available with smooth sanded face.
No. 4 UNDER-COURSING	16" (Fivex) 18" (Perfections)	.40" .45"	14/14 or 20/20 14/14 or 18/18	2 bdls. 2 bdls. 2 bdls. 2 bdls.	60 lbs. 72 lbs. 60 lbs. 79 lbs.	A utility grade for undercoursing on double-coursed sidewall applications or for interior accent walls.

Fig. 10-28. Wood shingles. Grades and specifications. (Red Cedar Shingle and Handsplit Shake Bureau)

Wood shingles are made from western red cedar, redwood, and cypress, all of which are highly decay resistant. They are taper sawed and graded No. 1, No. 2, and No. 3, plus a utility grade. The best grade is cut in such a way that the annular rings are perpendicular to the surface. Butt ends vary in thickness from 1/2 to about 3/4 in. as shown in Fig. 10-28. Wood shingles are manufactured in random widths and in lengths of 16, 18, and 24 inches. They are packaged in bundles; four bundles containing enough shingles to cover one square (100 sq. ft.) when a standard application is made.

The exposure of wood shingles is dependent on the slope of the roof. When the slope is 5 in 12 or greater, standard exposures of 5, 5 1/2, and7 1/2 inches are used for 16, 18, and 24 inch sizes respectively. On roofs with lower slopes, the exposure should be reduced to 3 3/4, 4 1/4, and 5 3/4 in. which will provide a minimum of four layers of shingles over the entire roof area. In any type of construction there should be a minimum of three layers at any given point to insure complete protection against heavy wind-driven rain.

Fig. 10-29. Open or spaced sheathing may be used for wood shingles or shakes. (Western Wood Products)

Sheathing

Solid sheathing for wood shingles may consist of matched or unmatched 1 in. boards, shiplap, or plywood. Open or spaced sheathing, Fig. 10-29, is sometimes used because it costs less and permits the shingles to dry out quickly. One reason for using solid sheathing is to gain the added insulation and resistance to infiltration that such a deck offers.

One method of applying spaced roof sheathing is to space 1 by 3 in., 1 by 4 in. or 1 by 6 in. boards the same distance apart as the anticipated shingle exposure and to nail each course of shingles to a separate board. Another is to use 1 by 6 in. lumber as sheathing boards with two courses of shingles nailed to each one. Fig. 10-30 shows the recommended sheathing placement and spacing for various shingle exposures when the latter method is used.

Underlayment

Normally an underlayment is not used for wood shingles, when applied on either spaced or solid sheathing. If it is desirable to use roofing paper to prevent air infiltration, the roof may be covered with rosin-sized building paper or "dry" unsaturated felts. Saturated building paper is usually not recommended because of the condensation trouble it may cause.

Flashing

In areas where outside temperatures drop to 0 deg. F. or colder and there is a possibility of ice forming along the eaves and causing water to back up, an eaves flashing strip is recommended. Follow the same procedure for making the installation as described for asphalt shingles.

The importance of using good materials for valleys and flashings cannot be overemphasized. Materials used for this purpose include tin plate, lead-clad iron, galvanized iron, lead, copper and aluminum sheets.

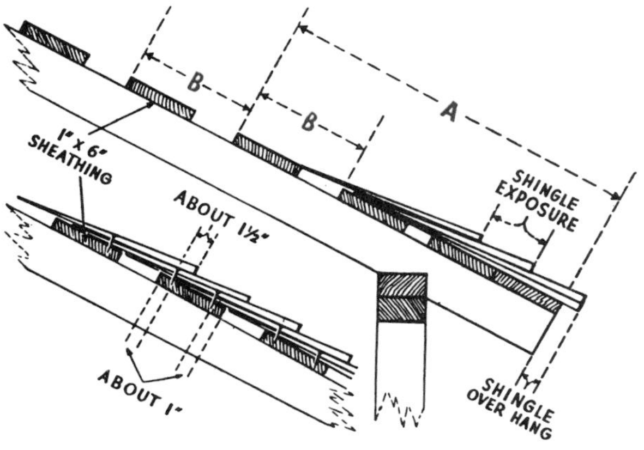

SHEATH-ING SIZE	SHINGLE EXPO-SURE	COLUMN 1 (A) Distance between upper edge of arbitrarily placed key sheathing board and lower edge of starting sheathing board at the eave	COLUMN 2 (B) Proper spacing of sheathing boards (upper edge to upper edge) below and above the key sheathing board indicated in Column 1.
1 x 6" . .	3¾"	27½"	7½" Edge to Edge
	4¼"	30½"	8½" Edge to Edge
	4½"	32 "	9 " Edge to Edge
	5 "	30½"	10 " Edge to Edge

EXAMPLE—4½" shingle exposure, 1 x 6 sheathing, and 1½" shingle overhang. Apply first sheathing board where desired at lower edge of roof; then attach upper edge of 1 x 6 sheathing board a distance of 32" (Column 1) from lower edge of first sheathing board at eave-line. Next, nail sheathing boards 9" apart (upper edge to upper edge as shown in Column 2) below this board until tight sheathing is encountered and above this board until peak or ridge of roof is reached. Starting course of shingles should be given an overhang of 1½"; shingles should be nailed 1½" above butt line of next course to be applied and nails should strike sheathing about 1" from each edge.

Fig. 10-30. Spacing of 1 x 6 in. sheathing boards for various shingle exposures.

If galvanized iron (mild steel coated with a layer of zinc) is selected, 24 or 26 ga. metal should be used. Tin, or galvanized sheets with less than 2 oz. of zinc per square foot, should be painted on both sides with white lead and oil paint and allowed to dry before being used. When making bends, care should be taken not to crack the zinc coating. On roofs of 1/2 pitch or steeper, the valley sheets should extend up on both sides of the center of the valley for a distance of at least 7 in. On roofs of less pitch, wider valley sheets should be used, with a minimum extension of at least 10 in. on both sides, Fig. 10-31. The open portion of the valley is usually about 4 in. wide and should gradually increase in width toward the lower end.

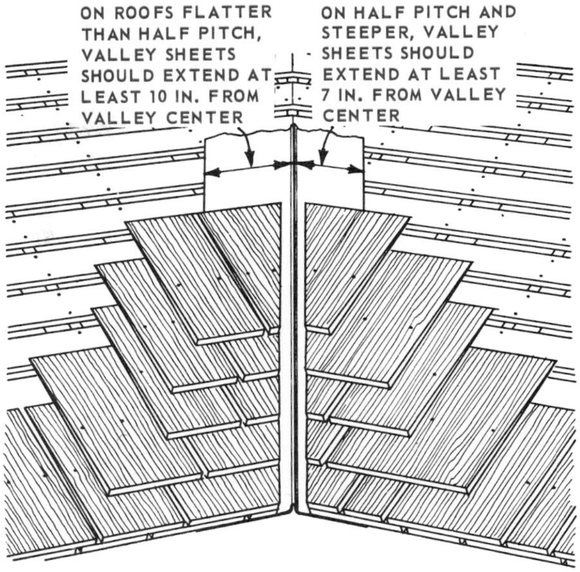

Fig. 10-31. Valley construction for wood shingles. (Red Cedar Shingle and Handsplit Shake Bureau)

Tight flashing around chimneys is also essential to a good roofing job. One method of installing base and cap or counter flashing around a chimney is shown in Fig. 10-32. In new construction, the cap flashing is laid in the joints when the chimney is built. If the chimney is laid up without flashing, the mortar joints must be chiseled out and the flashing forced in. It may be held in place by nails driven into the mortar. Finally, the joints must be filled or pointed with good mortar.

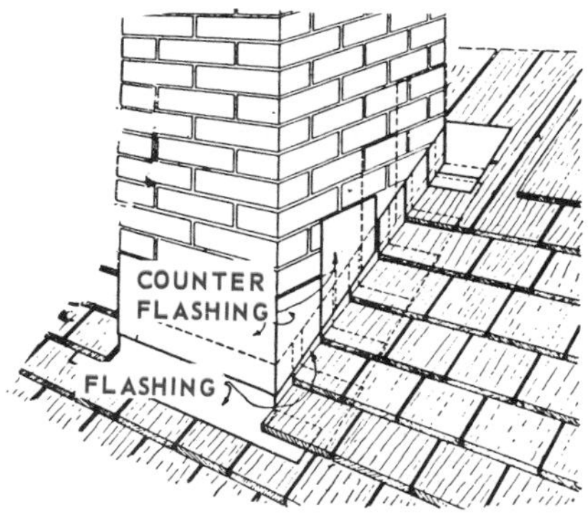

Fig. 10-32. Chimney flashing for wood shingle roofs.

Nails

In applying shingles, only rust resistant nails should be used. Hot-dipped, zinc-coated nails which have the strength of steel and the corrosion resistance of zinc, are recommended. Fig. 10-33 gives information on sizes of nails for various jobs.

Most carpenters prefer to use a shingler's or lather's hatchet, Fig. 10-34, to lay wood shingles. This has a sharpened blade and heel for splitting and trimming, and an adjustable gauge for rapidly spacing the weather exposure.

Applying Shingles

The first course of shingles at the eaves should be doubled or tripled. All shingles when laid on the roof should be spaced 1/4 in. apart to provide space for expansion when they become rain soaked. Use only two nails to attach each shingle. Of considerable importance in nailing is the proper placing of these two nails. They should be near the butt line of the shingles in the next course that is to be applied over the course being nailed, but should never be driven below this line so they will be exposed to the weather. Driving the nails 1 to 1 1/2 in. above the butt line is good practice, with 2 in. as an allowable maximum. Each nail should be placed not more than 3/4 in. from the edge of the shingle at each side. When nailed in this manner, the shingles will lie flat and give good service.

HANDY SAYS:

"Use care when nailing wood shingles. Drive the nail just flush with the surface. The wood in shingles is soft and can be easily crushed and damaged under the nail heads."

The second layer of shingles in the first course should be nailed over the first layer so the joints in each course are not less than 1 1/2 in. apart. See Fig. 10-35. A good shingler will use care in breaking the joints in successive courses, so they do not match up in three successive courses. Joints in adjacent courses should be at least 1 1/2 in. apart.

It is good practice to use a board as a straightedge to line up rows of shingles. Tack the board temporarily in place to hold the shingles until they are nailed. Two men often work together; one distributes and lays the shingles along the straightedge while another nails them in place. As the shingling progresses, check the alignment every 5 or 6 courses with a chalk line. Measure down from the ridge occasionally to be sure shingle courses are parallel to the ridge.

On a roof section where one end terminates at a valley, shingles for the valley should be carefully cut to the proper miter at the butts. Use

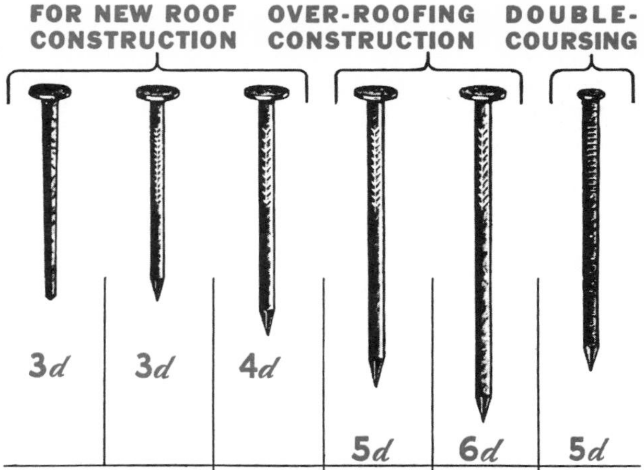

FOR 16" AND 18" SHINGLES		FOR 24" SHINGLES	FOR 16" & 18" SHINGLES	FOR 24" SHINGLES	FOR ALL SHINGLES
1 1/4" LONG	1 1/4" LONG #14 1/2 GAUGE	1 1/2" LONG #14 GAUGE	1 3/4" LONG #14 GAUGE	2" LONG #13 GAUGE	1 3/4" LONG #14 GAUGE
APPROX 376 NAILS TO LB	APPROX 515 NAILS TO LB.	APPROX 382 NAILS TO LB.	APPROX 310 NAILS TO LB.	APPROX. 220 NAILS TO LB.	APPROX. 380 NAILS TO LB

Fig. 10-33. Nails for wood shingles.

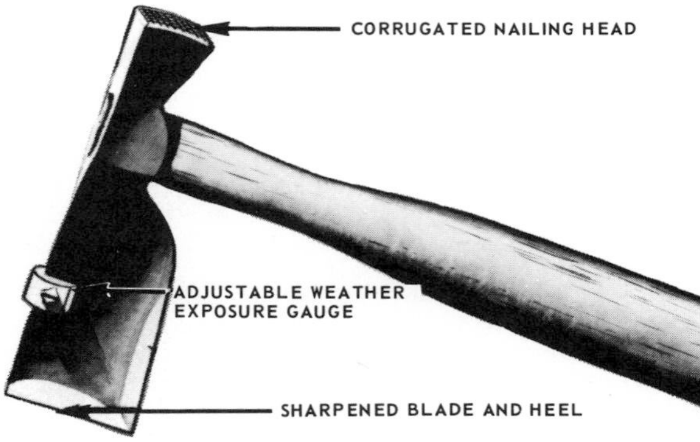

Fig. 10-34. Shingler's hatchet. (Red Cedar Shingle and Handsplit Shake Bureau)

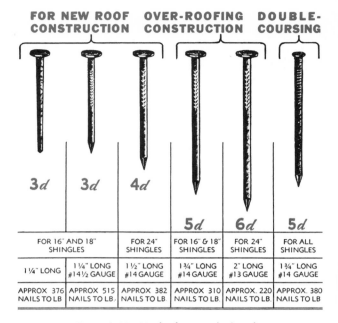

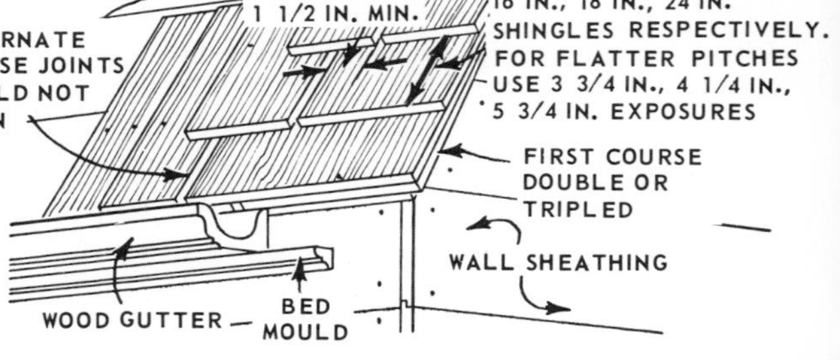

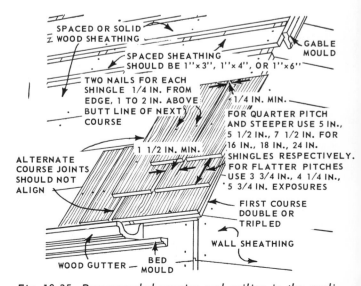

Fig. 10-35. Recommended spacing and nailing in the application of wood shingles.

Fig. 10-36. Laying wood shingles along the valley. Select wide shingles for this application.

wide shingles. Nail the shingles in place along the valley first and then lay toward the other end, Fig. 10-36.

Drip from gables may be prevented by using a piece of 6 in. bevel siding along the edge and parallel to the end rafter, Fig. 10-37.

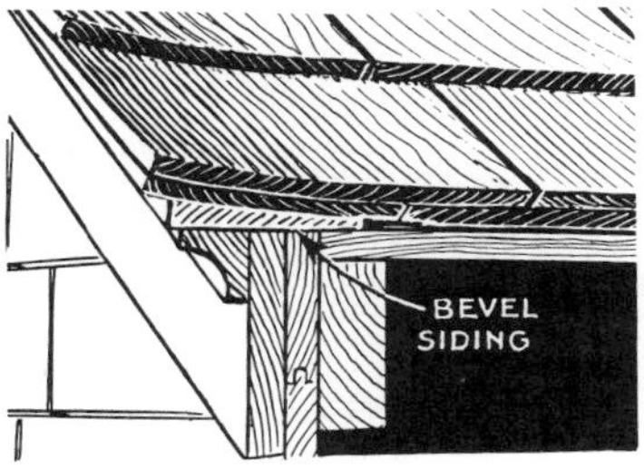

Fig. 10-37. Preventing drip from gable. A length of beveled siding gives an inward tilt to the shingles.

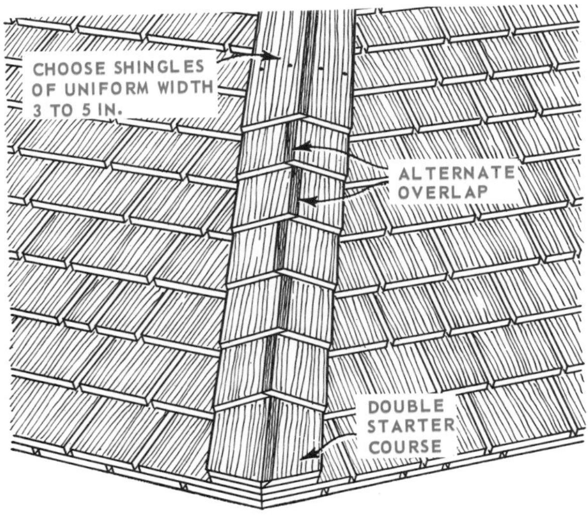

Fig. 10-38. Standard application for hips and ridges.

Shingled Hips and Ridges

Good, tight ridges and hips are required to avoid roof leakage. In the best type of hip construction, Fig. 10-38, nails are not exposed to the weather. Shingles of approximately the same width as the roof exposure are sorted out. Two lines are then marked on the shingles on the roof the correct distance back from the center line of the ridge on each side. On small houses, hips may be made narrower, the two lines being spaced closer to the center line.

Factory assembled hip and ridge units are available, Fig. 10-39. Weather exposure should be the same as that used for the regular shingles. Be sure to use longer nails that will penetrate into the sheathing.

Fig. 10-39. Prefabricated hip and ridge unit. (Red Cedar Shingle and Handsplit Shake Bureau)

Special Effects

By staggering or building up wood shingles, usually in random patterns, shadow lines and texture can be emphasized. This is sometimes a feature applied to contemporary as well as traditional architecture. Several such applications are shown in Fig. 10-40. The ocean wave effect is secured by placing a pair of shingles butt-to-butt under the regular course and at right angles to the butt line. These cross shingles should be about 6 in. wide.

The Dutch weave effect is obtained by doubling or super-imposing extra shingles, completely at random. This effect can be emphasized by using two shingles instead of one and is generally referred to as a pyramid pattern. Note that joints are always broken by at least 1 1/2 in.

Reroofing with Wood Shingles

Wood shingles may be applied to old as well as new roofs, Fig. 10-41. If the old roofing is in reasonably good shape, it need not be removed. Before applying new shingles, all warped, split, and decayed shingles should be nailed tightly or replaced. To finish the edges of the roof, the exposed portion of the first two rows of old shingles along the eaves should be cut off with a sharp hatchet and a 1 in. wood strip nailed in this space. Place the outer edge flush with the eave line. Edges along the gable ends should be treated in a similar manner.

The level of the valleys should be raised by applying wood strips, over which new flashings should be installed. Hips and ridges should be removed to provide a solid base for new shingles. New shingles should be spaced 1/4 in. apart to allow for expansion in wet weather and allowed to project 1/2 to 3/4 in. beyond the edge of the eaves.

The procedure described for new roofs should be followed, except that longer nails are required. For 16 in. and 18 in. shingles, rust-resistant or zinc clad 5d "box" nails or special over-roofing nails 1 3/4 in. long, 14 ga., should be used. The use of 6d 13 ga. rust-resistant nails is desirable for the application of 24 in. shingles.

Usually, no particular attention needs to be given to the manner in which the nails penetrate the old roof beneath--whether they strike the sheathing strips or not because, with the larger nails that are used, complete penetration is obtained through the old shingles and a sufficient number of nails to anchor all the shingles of the new roof securely will strike sheathing or nailing strips.

New flashings should be placed around chimneys without removing the old; liberal use being made of high grade non-drying mastics to obtain a watertight seal between the brick and the metal.

In reroofing houses covered with composition material, whether in the form of roll roofing or imitation shingles, it is usually best to strip off the old material. Otherwise, moisture may condense on the roof deck below and result in rapid decay of sheathing.

Fig. 10-40. Special patterns for wood shingles. Left. Ocean wave. Center. Dutch weave. Right. Pyramid.

Wood Shakes

Often called the aristocrat of roofing materials, wood shakes provide the utmost in surface texture, Fig. 10-42. They are highly durable and, if properly installed, may outlast the structure itself.

Three kinds of wood shakes generally available are: hand-split and resawn, taper split, and straight split, Fig. 10-43. Like regular wood shingles, they are available in random widths. Various lengths and thicknesses are standardized as listed in Fig. 10-44.

Shakes must be applied to roofs that have sufficient slope to insure good drainage (recommended minimum 4 in 12). Maximum weather exposure is 13 in. for 32 inch shakes, 10 in. for 24 inch shakes, and 8 1/2 in. for 18 inch shakes.

Start the application by placing a 36 in. strip of 30 lb. roofing felt along the eaves. The begin-

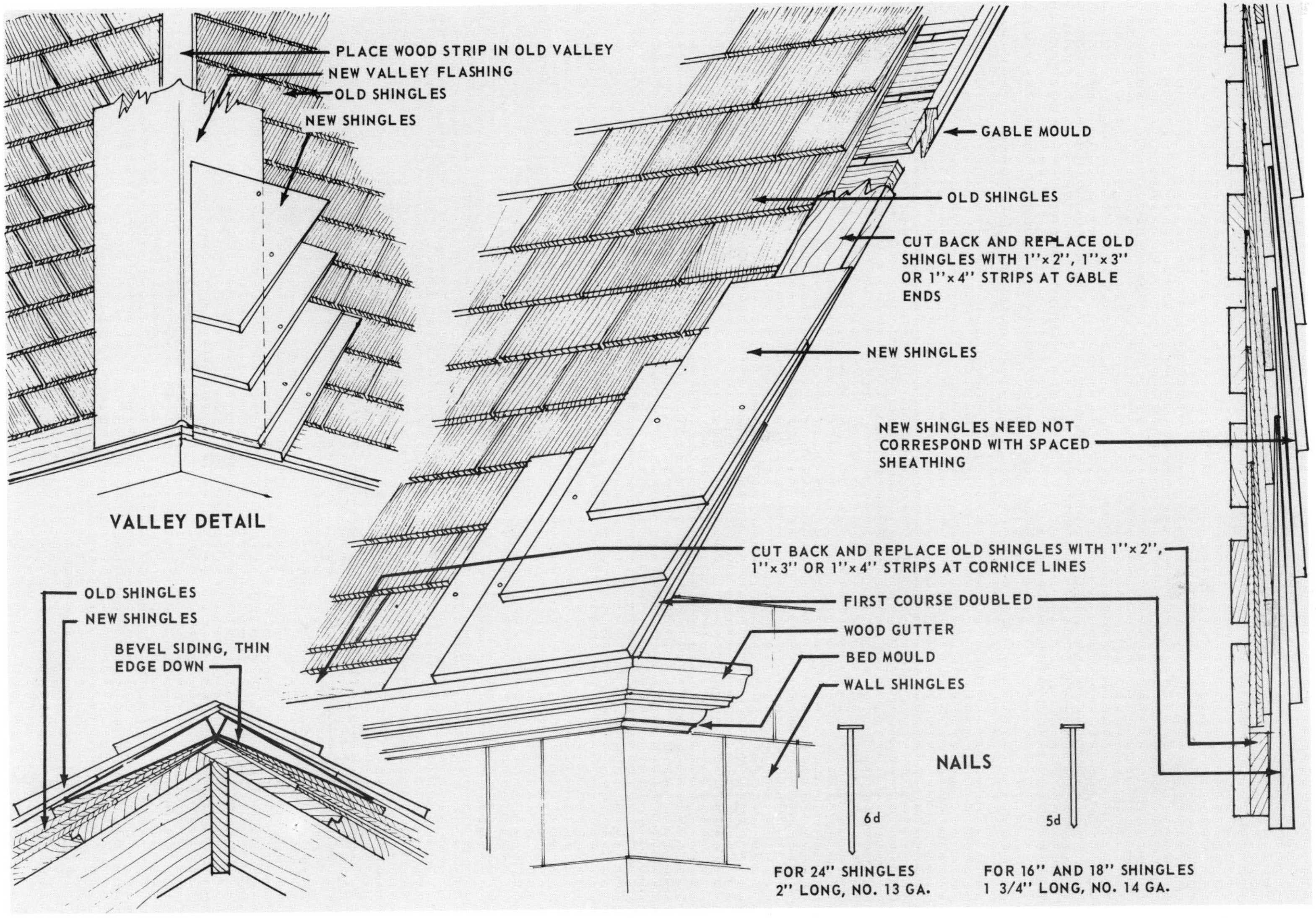

RIDGE CROSS SECTION DETAIL
CORNICE AND GABLE DETAIL
CROSS SECTION

Fig. 10-41. Application procedures and details for reroofing with wood shingles.

Fig. 10-42. Wood shakes provide an attractive and durable roof.

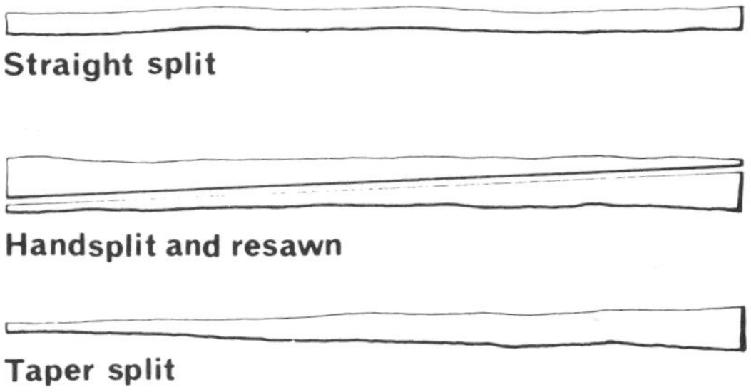

Fig. 10-43. Types of wood shakes.

Valleys are constructed by following the same procedure recommended for regular wood shingles. All valleys should be underlaid with 30 lb. roofing felt. Metal valley sheets must be at least 20 in. or wider.

Grade	Length and Thickness	20" Pack		18" Pack		Shipping Weight	Description
		# Courses Per Bdl.	# Bdls. Per Sq.	# Courses Per Bdl.	# Bdls. Per Sq.		
No. 1 HANDSP'.IT & RESAWN	15" Starter-Finish	8/8	5	9/9	5	225 lbs.	These shakes have split faces and sawn backs. Cedar logs are first cut into desired lengths. Blanks or boards of proper thickness are split and then run diagonally through a bandsaw to produce two tapered shakes from each blank.
		10/10	4				
	18" x ½" to ¾"	10/10	4	9/9	5	220 lbs.	
	18" x ¾" to 1¼"	8/8	5	9/9	5	250 lbs.	
	24" x ⅜"	10/10	4	9/9	5	225 lbs.	
	24" x ½" to ¾"	10/10	4	9/9	5	280 lbs.	
	24" x ¾" to 1¼"	8/8	5	9/9	5	350 lbs.	
	32" x ¾" to 1¼"	6/7	6			450 lbs.	
No. 1 TAPERSPLIT	24" x ½" to ⅝"	10/10	4	9/9	5	260 lbs.	Produced largely by hand, using a sharp-bladed steel froe and a wooden mallet. The natural shingle-like taper is achieved by reversing the block, end-for-end, with each split.
No. 1 STRAIGHT-SPLIT	18" x ⅜" True-Edge*	14 Straight	4			120 lbs.	Produced in the same manner as tapersplit shakes except that by splitting from the same end of the block, the shakes acquire the same thickness throughout.
	18" x ⅜"	19 Straight	5			200 lbs.	
	24" x ⅜"	16 Straight	5			260 lbs.	

Fig. 10-44. Sizes and kinds of wood shakes.

ning or starter course is doubled, just like regular wood shingles. After each course of shakes is applied, an 18 in. strip of 30 lb. felt is applied over the top portion of the shakes and extended onto the sheathing. The bottom edge of the felt is placed above the butt a distance equal to twice the exposure. For example: if 24 in. shakes are being laid at a 10 in. exposure, place the roofing felt 20 in. above the butts of the shake, Fig. 10-45.

Individual shakes should be spaced from 1/4 to 3/8 in. apart to allow for expansion. These joints should be offset at least 1 1/2 in. in adjacent courses.

Proper nailing is important. Use rust-resistant nails, preferably the hot-dipped zinc-coated type. The 6d size, which is 2 in. long, normally is adequate, but longer nails should be used if necessary because of unusual shake thickness and/or weather exposure. Nails should be long enough for adequate penetration into the sheathing boards. Two nails should be used for each shake, driven at least one inch from each edge and about one or two inches above the butt line of the following course. Do not drive nailheads into the shakes.

Fig. 10-45. Application of wood shakes. Note how the 30 lb. felt is positioned between courses. Drive the nails flush with the shingle surface but not so hard that the wood is crushed.

For the final course at the ridge line, try to select uniform size shakes and trim off the ends so they meet evenly. Carefully apply a strip of 30 lb. felt along all ridges and hips, then install shakes that have a uniform width of about 6 in. Nail these in place following the procedure described for regular wood shingles. Prefabricated hip-and-ridge units are available. Use of these will save time and provide uniformity. See Fig. 10-46.

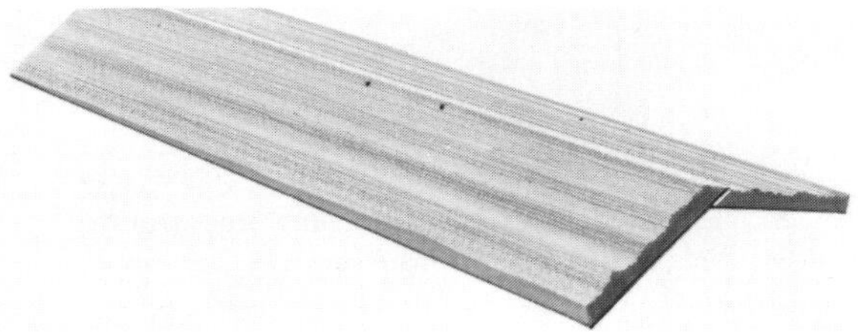

Fig. 10-46. Prefabricated hip-and-ridge units for shake roofs.

Mineral Fiber Shingles

Mineral fiber shingles (also called asbestos cement shingles) are manufactured from asbestos fiber and Portland cement. They are formed in molds under high pressure and provide a finished product that is immune to rot and decay, unharmed by exposure to salt air, unaffected by ice or snow, and fireproof because they contain nothing that will burn.

Mineral fiber shingles are available in a variety of colors and textures and may be obtained in rectangular, square, and hexagonal shapes; in single or multiple units, Fig. 10-47. They are sold by the square, and are equally well adapted for use on new buildings or over old roofs. On old roofs, the new shingles can usually be applied right over the old roof, saving the cost of removing the old shingles and giving the roof the added insulation of the old roofing.

Because the mineral fiber material is rigid and hard, nail holes are prepunched during the manufacturing process. Eave starter strips, hip and ridge shingles, and ridge rolls are prefabricated to further simplify the application. Shingles should be installed with galvanized needlepoint nails. For new roofs, use 1 1/4 in. nails. When reroofing over old shingles, use 2 in. nails.

Special equipment is required to cut the material. Dealers handling mineral fiber products will usually have several shingle cutters on hand for the use of customers, Fig. 10-48. They cut the shingles quickly, accurately and neatly; and they also have a punch for forming nail holes where extra holes are required.

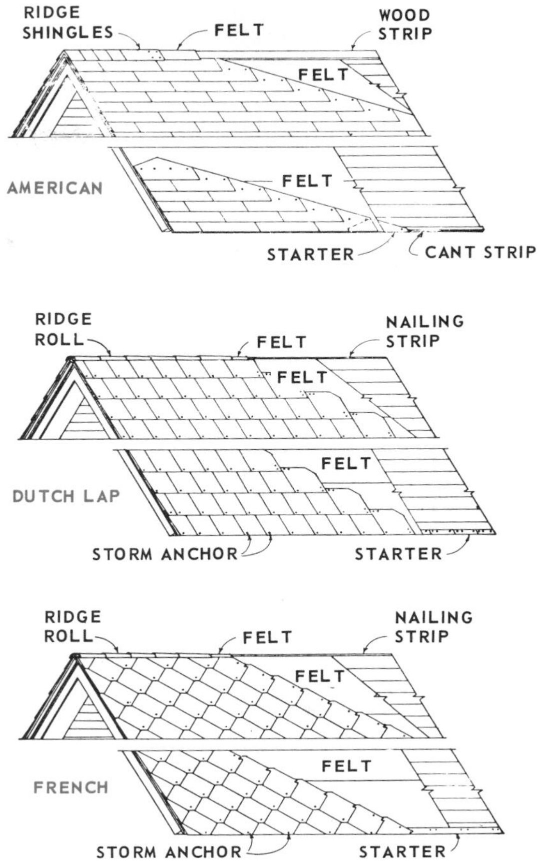

Fig. 10-47. Types of mineral fiber shingles. Minimum slope 4 in 12 for American; 5 in 12 for Dutch lap and hexagonal.

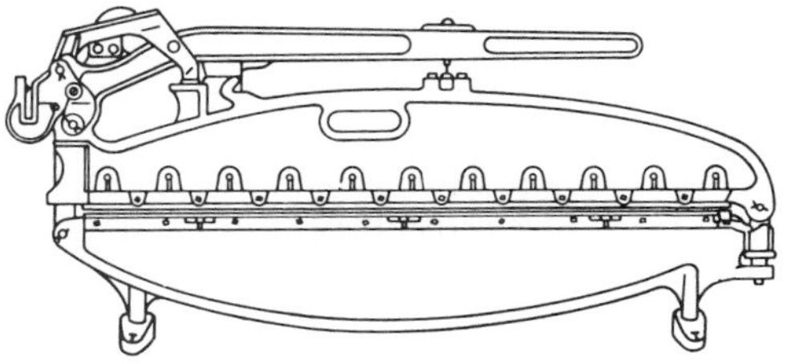

Fig. 10-48. A special cutter for mineral fiber shingles.

If a cutter is not available, asbestos shingles can be cut by hand. Use an old chisel, a drift punch, or the blade of the hatchet. Score the shingle with the tool being drawn along a straightedge. After scoring, place the shingle over a solid piece of wood and break along the scored line, Fig. 10-49. Irregular cuts or round holes are made by punching holes along the line of the cut and breaking out the piece which is to be discarded. There is a punch on the shingle cutter for punching additional holes, or they may be drilled or punched with a drift punch or other suitable pointed tool. A drift punch is recommended because it will punch a clean hole without splitting the material.

Storing Shingles

While still packed in bundles, whether in the yard or on the job, mineral fiber shingles should be kept dry. Moisture trapped between bundled shingles may cause discoloration due to efflorescence, or what ordinarily is known as "blooming." If it is necessary to use outdoor storage, stack the shingles on planks and use roofing felt, waterproof paper or a tarpaulin for cover.

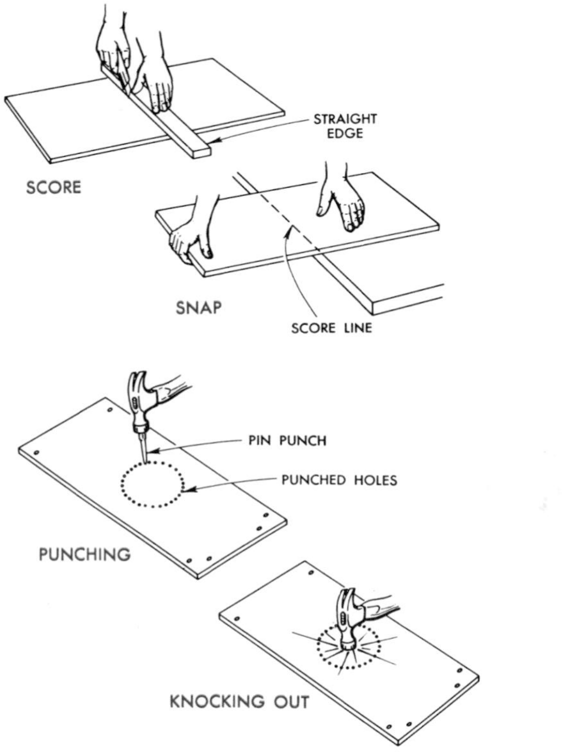

Fig. 10-49. Forming mineral fiber products by hand. Above. Score and snap method. A carbide tipped blade works best. Below. Making interior cutouts. (Mineral Fiber Products Bureau)

Application of Mineral Fiber Shingles

It is important that the roof deck be in proper condition to receive the shingles. The lumber should be well seasoned, dry and of uniform thickness. Tongue-and-groove 1 x 6 boards or plywood of adequate thickness is recommended. Nailheads must be driven down and any high spots or rough edges removed. A wood cant strip should be applied along the eaves, flush with the lower edge, to give proper pitch for the shingles.

Attach a furring strip along each side of hips and ridges to provide a nailing base for the hip and ridge covering. The roof shingles are butted against these furring strips; therefore they must be the same thickness as the shingles. The furring strips should not be over 2 in. wide, so they will be covered by the ridge units, Fig. 10-50.

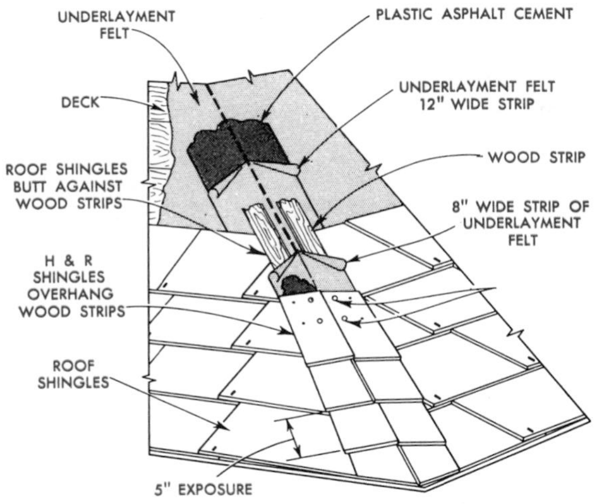

Fig. 10-50. Hip and ridge construction. (Mineral Fiber Products Bureau)

On new work or on reroofing jobs where the old materials have been removed, the roof boards should be covered with one course of waterproof sheathing paper. Horizontal laps need not be over 2 in.; end laps should be 6 in. Hips, ridges, and valleys are covered with a 12 in. lap so double thickness will be assured at these points.

Flashing for Mineral Fiber Shingles

Care should be exercised in applying valley flashings. First, a corrosion resistant metal, such as copper or stainless steel, should be used for valleys when a mineral fiber roof is being put on. Do not use aluminum. Secondly, to insure a leakproof roof, the metal valley should extend out on the roof deck well beyond the edge of the overlapping shingles and well up under the shingles on both sides. Finally, shingles should be bedded down in a layer of asphalt cement for a distance of 6 in. back from the edge where the shingles end at the valley.

Open type valleys are usually used. The edges of the metal should be turned back 1/2 in. forming a hem or water seal. The valley metal should be attached to the deck with cleats at the hem. Shingle nails should not be driven through metal valley linings.

Reroofing Over Old Shingles

Nail down all old shingles which are badly curled or warped. The use of underlayment material generally is not required under asbestos shingles when they are laid over old wood or asphalt shingles. However, missing or badly decayed wood shingles should be replaced and the surface brought to an even plane. If the top edges or corners of the new shingles do not rest on the butts of the old shingles, tilting will result. Wood strips, beveled, and equal in their greatest thickness to the butts of the old wood shingles, are satisfactory to use for leveling an old roof deck, Fig. 10-51.

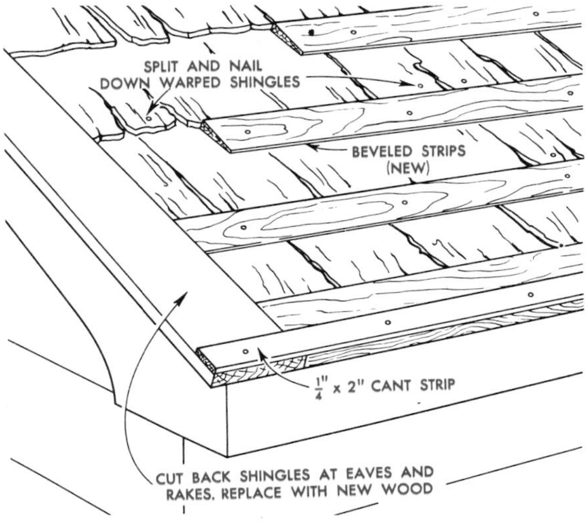

Fig. 10-51. Preparation of old roof for reroofing with mineral fiber shingles.

Sometimes at the edge of the roof, the old wood shingles are found to be in bad condition and the sheathing is not in good shape. A substantial and more attractive job can be obtained by cutting away the old shingles and laying a new board 4 to 6 in. wide by 7/8 in. thick along the edges. This provides a solid base to which the new shingles will be nailed.

Build up old valleys with wood strips of a width and thickness adequate to bring them flush with the butts of the old wood shingles. Lay waterproof felt over hips, ridges, and at valleys.

Tile Roofing

The most commonly used roofing tile are manufactured products consisting of molded, hard-burned shale or mixtures of shale and clay. Metal tile is also available.

When well made, clay tile is hard, fairly dense and durable, and may be obtained in a variety of shapes and textures. Most roofing tile of clay is unglazed, although glazed tile is sometimes used. Typical tile roof application details are shown in Fig. 10-52.

Clay tile may be used over an old roof provided the old covering is in reasonably good condition, and the roof framing is heavy enough to stand the additional weight. Additional roof framing or bracing, if required, should be added before starting with the application of the tile.

Galvanized Sheet Metal Roofing

Only galvanized sheets that are heavily coated with zinc (2.0 oz. per sq. ft.) are recommended for permanent-type construction. The sheets with lighter coatings of zinc are less durable and are likely to require painting every few years. On temporary buildings and in cases where the most economical construction is required, lighter metal can be used and will give satisfactory results if protected by paint.

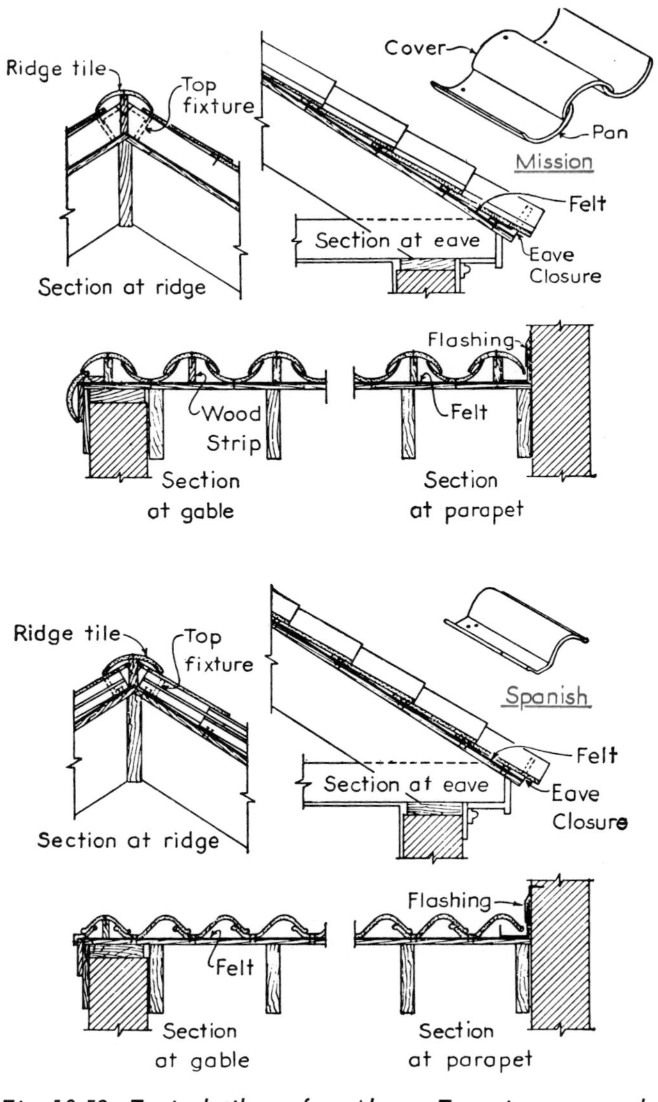

*Fig. 10-52. Typical tile roofs. **Above.** Two piece pan and cover, commonly known as Mission tile. **Below.** Spanish tile.*

Slope and Laps

Galvanized sheets may be laid on slopes as low as a 3 in. rise to the foot (1/8 pitch). If more than one sheet is required to reach the top of the roof, the ends should lap not less than 8 in. When the roof has a pitch of 1/4 or more, 4 in. end laps are usually satisfactory.

To make a tight roof, sheets should be lapped 1 1/2 corrugations at either side, Fig. 10-53. The wind is likely to drive rain water over single-corrugation lap joints. When using roofing 27 1/2 in. wide with 2 1/2 in. corrugations and 1 1/2 corrugation lap, each sheet covers a net width of 24 in. on the roof.

When 27 1/2 in. roofing is not available, sheets of 26 in. in width may be used. In laying the narrower sheets, every other one should be turned upside down. So laid, each alternate sheet laps over the two intermediate sheets. The 26 in. roofing with 2 1/2 in. corrugations cover a net width of 22 1/4 in. of roof.

Sheathing and Nails

If 26 ga. sheets are used, supports may be 24 in. apart. If 28 ga. sheets are used, supports should be not more than 12 in. apart. The heavier gauge has no particular advantage except its added strength, because the zinc coating is what gives this type roofing its durability.

For best results, galvanized sheets should be fastened with lead-headed nails or galvanized nails and lead washers. Nails properly located are driven only into tops of the corrugations. To avoid unnecessary corrosion, use nails specified by the manufacturer.

Aluminum Roofing

Corrugated aluminum roofing, if properly applied, usually makes a long lasting roof. Seacoast exposure tests reported by the Bureau of Standards indicate this material is capable of resisting corrosion in such localities unless subjected to direct contact with salt-laden spray. Where this is likely to happen, aluminum roofing is not recommended.

Aluminum alloy sheets available for roofing usually have a corrugation spacing of 1 1/4 or 2 1/2 in. Recommendations for the installation of sheet metal regarding side lap and end lap are applicable to the laying of aluminum sheets. An important precaution to observe in laying aluminum roofing is to make sure that contact with other kinds of metal is avoided. Where this is not possible, both metals should be given a heavy coating of asphalt paint wherever the surfaces are in contact.

As aluminum is soft and the sheets used for roofing are relatively thin, they should be laid on tight sheathing or on decks with openings no more than 6 in. wide. Aluminum roofing should be nailed with not less than 90 nails to a square or about one nail for each square foot. It is recommended that aluminum alloy nails be used and that nonmetallic washers be used between nail heads and the roofing.

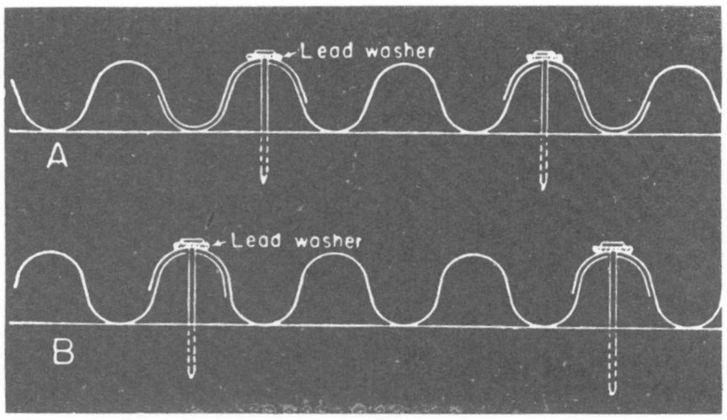

Fig. 10-53. Application of corrugated sheet metal roofing. Above. Sheets properly laid with 1 1/2 corrugation-lap. Below. Single-corrugation-lap not recommended.

If desired, the sheathing may be covered with water-resistant building paper or asphalt impregnated felt. Paper that absorbs and holds water should never be used. To avoid corrosion, aluminum sheets should be stored so that air will have free access to all sides; otherwise a white deposit will form causing them to deteriorate.

Terne Metal Roofing

Terne metal roofing is made of copper-bearing steel, heat-treated to provide the best balance between malleability and toughness. It is hot dip-coated with Terne metal, an alloy of 80 percent lead and 20 percent tin. The high weather resistance factor (notable in this type of roofing) is due primarily to the lead; tin is included because the resulting alloy makes a better bond with steel.

Grades are expressed in terms of the total weight of the coating on a given area. This area, by old trade custom, is the total area contained in a box of 112 sheets that are 20 in. x 28 in. in size, and amounts to 436 square feet. The best grade of Terne coating is 40 lb. and provides a roof surface that will last for many years.

A wide variety of sheet sizes are available, plus 50 foot seamless rolls in various widths.

This permits its use for many different types of roofs and methods of application. It is used extensively for flashing around both roof and wall openings.

For best appearance and longest wear, Terne metal roofs must be painted. A linseed oil-based iron oxide primer is recommended for a base coat, over which nearly any exterior paint and color can be used.

Gutters

Gutters or eaves troughs collect the rainwater from the edge of the roof and carry it to downspouts which in turn direct it away from the foundation or into a drainage system. The term "gutter" refers to a separate unit that is attached to the eave while the term "eaves trough" usually applies to a waterway built into the roof surface over the cornice.

Eaves troughs must be carefully designed and built since any leakage will penetrate the structure. Because of this and the extra cost of construction, they are seldom used in modern residential work.

Wood Gutters

Wood gutters of Douglas fir or red cedar are used in many parts of the country. When properly installed and maintained, the life of wood gutters is usually equal to that of the main structure.

The installation of modern wood gutters is usually made before the shingles are applied. One type is attached to the fascia board after the roof is sheathed, Fig. 10-54. Some designs are coordinated with the fascia board and are installed with this trim unit before the roof sheathing is complete. In this type of installation, the sheathing overhangs the fascia and gutter, thus minimizing the possibility of leakage. When wood gutters are specified, the architectural plans usually include installation details.

Fig. 10-54. Installing a wooden gutter before laying the finished roof. (Weyerhaeuser Co.)

Cutting, fitting and drilling is accomplished on the ground and then the various units are set in place. Most wood guttering is primed or prepainted. Always use galvanized or other types of weatherproof nails and/or brass wood screws.

Gutter ends may be sealed with blocks, returned and mitered, or butted against an extended rake frieze board, Fig. 10-55. In general, the gutter is treated like cornice molding and should present a smooth, trim appearance.

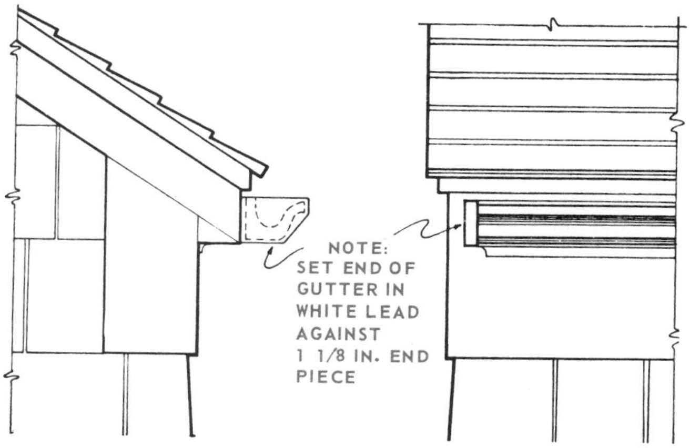
NOTE:
SET END OF GUTTER IN WHITE LEAD AGAINST 1 1/8 IN. END PIECE

Fig. 10-55. Proper position for a gutter installation.

The roof surface should extend over the inside edge of the gutter with the front top edge at approximately the height of a line extended from the top of the sheathing. For correct appearance, wood gutters are set nearly level. They will drain satisfactorily if kept clean and adequate downspouts are provided.

Downspout connections are made by cutting a hole through the gutter and inserting a sleeve of sheet metal which fits into the downspout. Locate the downspouts for best appearance and be sure to provide a sufficient number to carry away the water.

Metal Gutters

A wide variety of metal gutters are available to control roof drainage. Manufacturers have perfected gutter and downspout systems that include various component parts which can be quickly assembled and installed on the building site. Materials consist of galvanized iron and

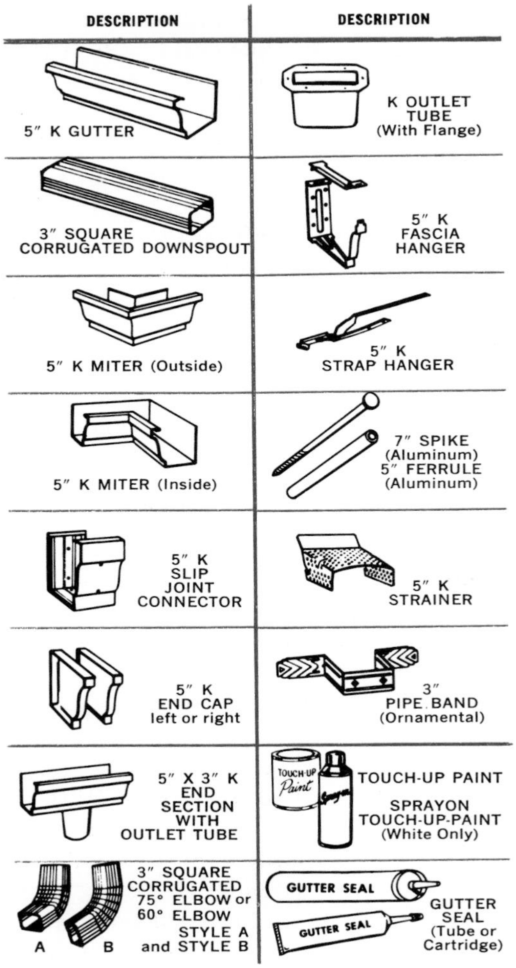

Fig. 10-56. Metal gutter system. Component parts. (Crown Aluminum Industries Corp.)

aluminum. Many systems are available in either a primed or prefinished condition to match a wide range of colors. A recent development consists of a guttering system made from molded vinyl plastic.

Gutter systems include inside and outside mitered corners, joint connectors, pipes, brackets and other items, all carefully engineered and fabricated. Parts slip together easily and are generally held with soft pop-rivets or sheet metal screws. Fig. 10-56 shows standard parts of a typical gutter and downspout system.

For best results, the size of gutters and eave troughs must correspond to the roof areas from which they receive water. For roof areas up to 750 sq. ft., a 4 in. wide trough is suitable. For areas between 750 and 1400 sq. ft., 5 in. troughs should be used. For larger areas, a 6 in. trough is recommended. Quality of gutters, like that of flashing, should correspond to the durability of the roof covering. If galvanized steel is used, it should have a heavy zinc coating.

The size of downspouts or conductor pipes required also depends on the roof area. For roofs up to 1,000 sq. ft., downspouts of 3 in. diameter have sufficient capacity if properly spaced. For larger roofs, 4 in. downspouts should be used.

Metal gutters are usually sloped 1 in. for every 12 to 16 ft. of length.

Estimating Material

The first step in estimating roofing materials is to calculate the total surface area to be covered. In new construction, the figures used to estimate the sheathing can also be used to estimate the underlayment and finished roofing materials. When these figures are not available, they can be calculated by the same methods used for roof sheathing, described in Unit 9.

Another method sometimes used to estimate roof area is to determine the total ground area of the structure, including all eave and cornice overhang. Convert the ground area to roof area by adding a percentage determined by the roof slope as follows:

Slope 3 in 12 -- add 3% of area
Slope 4 in 12 -- add 5 1/2% of area
Slope 5 in 12 -- add 8 1/2% of area
Slope 6 in 12 -- add 12% of area
Slope 8 in 12 -- add 20% of area

After the total square feet of roof surface has been estimated, divide it by 100 to find the number of squares to be covered. For example: if the total ground area plus overhang is found to be 1560 and the slope of the roof is 4 in 12, apply the following calculations.

$$\text{Roof Area} = 1560 + 1560 \times 5\ 1/2\%$$

$$= 1560 + \frac{1560 \times 11}{2 \times 100}$$

$$= 1560 + \frac{\overset{7.80}{\cancel{1560}} \times 11}{\cancel{2} \times \cancel{100}}$$

$$= 1560 + 85.80 \text{ or } 86$$

$$= 1646$$

Number of squares = 16.46 or 16 1/2

After the basic number of squares is established, additional amounts must be added. For asphalt shingles, it is generally recommended that 10 percent be added for waste. This, however, may be too much for a plain gable roof and too little for a complicated intersecting roof. Certain allowances must also be added for reduced exposure on low sloping roofs.

Usually the 10 percent waste figure can be reduced somewhat if adequate allowance is made for hips, valleys and other extras. For wood or asphalt shingles, a general procedure is to add one square for each 100 lineal feet of hips and valleys.

Quantities of starter strips, eaves flashing, valley flashing and ridge shingles must be added to the total shingle requirements. All of these are figured on linear measurements of the eaves, ridge, hips, and valleys. For a complicated structure, a plan view of the roof will be helpful in adding together these materials. All ridges and eave lines will be seen true length and can simply be scaled to find their length. Hips and valleys will not be seen as their true length and a small amount must be added. Using the percentage listed for converting ground area to roof area will usually provide sufficient accuracy.

Test Your Knowledge - Unit 10

1. The selection of roofing materials is influenced by such factors as cost, durability, appearance, application methods, and __________ of the roof.
2. In shingle application, the distance between the edges of adjacent courses measured at a right angle to the ridge is called __________.
3. Asphalt saturated felt is available in a roll that is __________ wide.
4. Three-tab square-butt shingles are ______ in. long.
5. The most commonly used underlayment for asphalt shingles is ______ lb. roofer's felt.
6. The waterway of a valley should diverge (grow wider) as it approaches the eaves at a rate of ______ in. per foot.
7. The minimum number of nails recommended for the application of each three-tab square-butt shingle unit is ______.
8. Chimney flashing consists of two parts; the cap or counter flashing and the ______ flashing.
9. Asphalt shingles can be used on roofs with a slope as low as ________ if special application procedures are followed.
10. A formed metal strip, called a ______ ______ is attached to the edge of built-up roofs.
11. In the best grade of wood shingles, the annular rings run __________ (parallel, perpendicular) to the surface.
12. When laying wood shingles, they should be spaced ______ in. apart (horizontally) to provide for expansion when they become rain soaked.
13. The vertical joints between wood shingles in adjacent courses should be spaced at least __________ in. apart.
14. A strip of 30 lb. roofing felt is placed between each course when applying _______ _________.
15. Mineral fiber roofing products are made from asbestos and ________ _________.
16. When installing corrugated steel roofing, the joints should be lapped __________ (1, 1 1/2, 2) corrugations.
17. Terne metal roofing consists of sheets of copper-bearing steel coated with an alloy of __________ and __________.
18. The vertical pipes of a gutter system are called __________ or __________.
19. Most roofing materials are figured and sold by the ________ which is equal to 100 sq. ft.

Outside Assignments

1. Make a study of the kinds and qualities of asphalt shingles used in your locality. Secure manufacturer's descriptive literature from a local builders supply store. Prepare a report including information about kinds, grades, and costs. Also include information about materials for underlayment, valley flashing, hip and ridge finish, and fasteners.
2. Prepare a report on application procedures used to install a roof on a residence in your community. Visit the building site and observe the methods used. Note the type of sheathing; special preparation of the roof deck; how valley flashing is applied; use of drip edges; type of starter courses; procedures used to align shingle courses; and how ridges are finished. ALWAYS BE SURE TO OBTAIN PERMISSION FROM THE FOREMAN OR HEAD CARPENTER WHEN VISITING A BUILDING SITE.
3. Construct a full-size visual aid showing the application of wood or asphalt shingles. Use a piece of 3/4 in. plywood about 4 ft. square to represent a lower corner of a roof deck. Apply underlayment (if required), drip edges, starter strips, and then carefully lay the shingles according to an approved pattern. By making only a partial coverage of the various layers, all of the application steps and materials used can be easily studied and observed. Present the visual aid to your class and discuss the materials and application procedures.

Construction Sequence. The residential structure shown above has been framed, sheathed, and the finished roof has been applied. Vertical siding has also been applied to the roof gable. The next step is to install the windows and exterior doors. (Weyerhaeuser Co.)

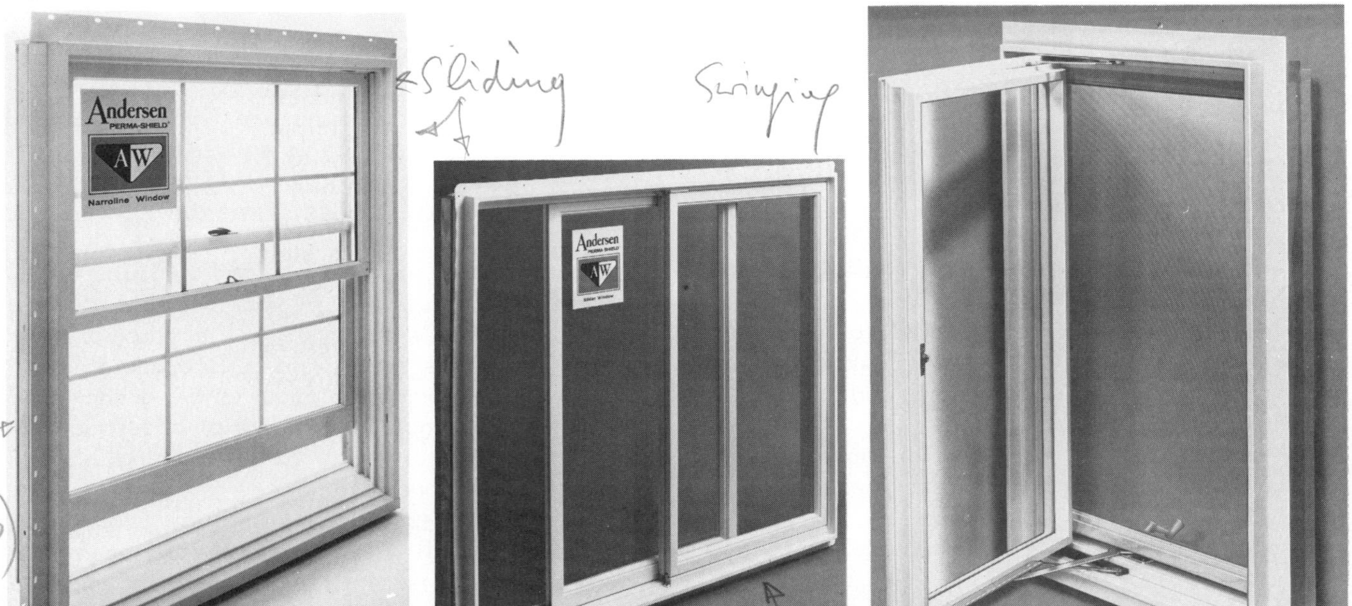

Fig. 11-1. Common types of residential windows with movable sash. The views above show outside surfaces of all the windows except the hopper type which is a view from the inside. (Andersen Corp.)

Unit 11

WINDOWS AND EXTERIOR DOORS

Because windows and doors are such an important part of a structure, the carpenter should have a basic understanding of the various types, sizes, and standards of construction and be able to recognize good quality in materials, fittings, weather stripping and finish. He must also appreciate the importance of careful installation of the various units and be an expert in this work.

Manufacture

Today windows and doors are fabricated in large manufacturing plants and then shipped to the building site as completed units ready to be installed in the openings of the structure. Woodworking factories that produce windows, doors, stairs and other finished items used in construction are often called millwork plants.

Some windows used in residences are made from aluminum and steel; however most of them are made from wood. Since wood does not transmit heat readily, there is less tendency for window frames to become cold and condense moisture vapor.

Wood will decay under certain conditions and must be treated with preservatives. Exposed surfaces must be kept painted.

Metal is stronger than wood and thus permits the use of smaller frame members around the glass. Aluminum has the added advantage of a protective film of oxide which eliminates the need for paint.

Ponderosa pine is the wood commonly used in the fabrication of windows. It is carefully selected and kiln dried to a moisture content of 6 to 12 percent.

Manufacturing Standards

The control of quality in the manufacture of windows is based on standards established by the National Woodwork Manufacturers Association. These standards cover every aspect of material and fabrication including such details as the projection of the drip cap and the slope of the sill. For example: N.W.M.A. industry standards I.S. 2-74 states that the weather stripping for a double hung window shall be effective to the point that it will prevent air leakage (infiltration) in excess of 0.50 cfm (cubic feet per minute) per linear foot of sash crack when tested at a static air pressure of 1.56 psf (pounds per square foot). The latter is equivalent to the pressure subjected by a 25 mph wind. Standards also cover the methods and procedures used in preservative treatments.

Types of Windows

In general, windows can be classified under one or a combination of three basic headings namely: (1) sliding, (2) swinging, and (3) fixed. Each of these includes a variety of designs or method of operations. Sliding windows include the double-hung and horizontal sliding. Swinging windows that are hinged on a vertical line are called casement windows while those hinged on a horizontal line can be either awning or hopper. Fig. 11-1 shows a general view of the common types of residential windows.

Double-Hung

A double-hung window consists of two sashes that slide up and down in the window frame. These are held in any given vertical position by a friction fit against the frame or by springs and various balancing devices. Double-hung windows are used extensively because of their economy, simplicity of operation, and adaptability to many architectural designs. Fig. 11-2 shows an interior view of three double-hung window units. Screens and storm sash are installed on the outside of the window to permit easy opening and closing of the sash.

Fig. 11-2. Living room window consisting of three double-hung units.

Horizontal Sliding

Horizontal sliding windows have two or more sash, at least one of which moves horizontally within the window frame. The most common design consists of two sash, both of which are movable. See Fig. 11-3. When three sash are used, the center one is usually fixed.

Fig. 11-3. Horizontal sliding windows installed above a built-in cabinet. Note that the vertical divisions do not detract greatly from the outside view.

Casement

A casement window has a sash that is hinged on the side and swings outward. Installations usually consist of two or more units, separated by mullions. Sash are operated by a cranking mechanism or a push-bar mounted on the frame. Latches are used to close and hold the sash tightly against the weather stripping.

Since the swing sash of a casement permits 100 percent opening of the window for ventilation, fixed units are often combined with operating units as shown in Fig. 11-4.

Crank operators make it easy to open and close windows located above kitchen cabinets or other built-in fixtures. Screen and storm sash are attached to inside of standard casement windows.

Fig. 11-4. A casement type window unit located above kitchen sink and counter top. (Andersen Corporation)

Awning

Awning windows have one or more sash that are hinged at the top and swing out at the bottom. They are often combined with fixed units to provide ventilation as shown in Fig. 11-5. Several operating sash can be stacked vertically in such a way that they close on themselves or on rails that separate the units.

Most awning windows have a so-called projected action where sliding friction hinges cause the top rail to move down as the bottom of the sash swings out. Crank and push-bar operators are similar to those on casement windows. Screens and storm sash are mounted on the inside. Awning windows are often installed side by side to form a "ribbon" effect as shown in Fig. 11-6. Such an installation provides privacy for bedroom areas and also permits greater flexibility in furniture arrangements along outside walls.

Consideration of outside clearance must be given to both casement and awning windows. When open, they may interfere with movement on porches, patios, or walkways that are located adjacent to outside walls.

Hopper

The hopper window has a sash that is hinged along the bottom and swings inward. It is operated by a lock-handle located in the top rail of the sash.

Fig. 11-5. Awning windows located below fixed units.

Hopper windows are easy to wash and maintain but often interfere with drapes, curtains, and the use of inside space near the window.

Multiple-Use Window

The multiple-use window is a single outswinging sash designed so it can be installed in either a horizontal or vertical position. Fig. 11-7 shows two units installed to operate similar to a casement window. These windows are simple in design and do not require complicated hardware.

Fig. 11-6. Awning windows installed in a child's bedroom.

Fig. 11-7. Multi-use windows installed in a vertical position. (Andersen Corporation)

Jalousies

A jalousie window consists of a series of horizontal glass slats that are held by a metal frame at each end. The metal frames are attached to each other by levers so the slats operate together in about the same manner as a venetian blind. Jalousie windows provide excellent ventilation. Since weather-tightness values are low in this type of window, their use in northern climates is usually limited to porches and breezeways.

Fixed Windows

The fixed window unit can be used in combination with any of the movable or ventilating units. Its main purpose is to provide daylight and a view of the outdoors. When used in this type of installation the glass is set in a fixed sash mounted in a frame that will match the regular ventilating windows. Large sheets of plate glass are often separated from other windows to form "window-walls." They are usually set in a special frame formed in the wall opening.

Window Heights

One of the important functions of a window is to provide the occupants of the structure with a view of the outdoors. The architect will be aware of the various considerations and his drawings should reflect the dimensions shown in Fig. 11-8. In residential construction the standard height from the bottom side of the window head to the finished floor is 6' - 8". When this dimension is used the heights of window and door openings will be the same. If inside and outside trim must align, 1/2 to 3/4 in. must be added to this height for thresholds and door clearances. Window manufacturers usually provide exact dimensions for their standard units. See Fig. 11-9.

Window Glass

Sheet glass used in windows is manufactured by the "float process." It has a slightly wavy surface but this is hardly noticeable in the top grades. Several standard qualities and thicknesses are available. See Fig. 11-10. Heavy sheet glass can be used for smaller "picture windows" where slight distortion is not objectionable, however polished plate glass is usually specified for such installations.

Insulating Windows

Glass areas are a major source of heat loss in winter and heat gain in summer. This transfer of heat is greatly reduced when the window consists of two layers of glass separated by an air space. A single layer of standard window glass has a U-value of 1.16; while a double pane of the same glass with a 1/2 in. or greater air space has a U-value of about .58.

Storm sash is commonly used to provide the second pane of glass. It is attached to the outside of the window frame to form the air space. Storm sash will, when properly fitted, reduce infiltration. A disadvantage is that the storm sash must be removed from operating units to provide summertime ventilation.

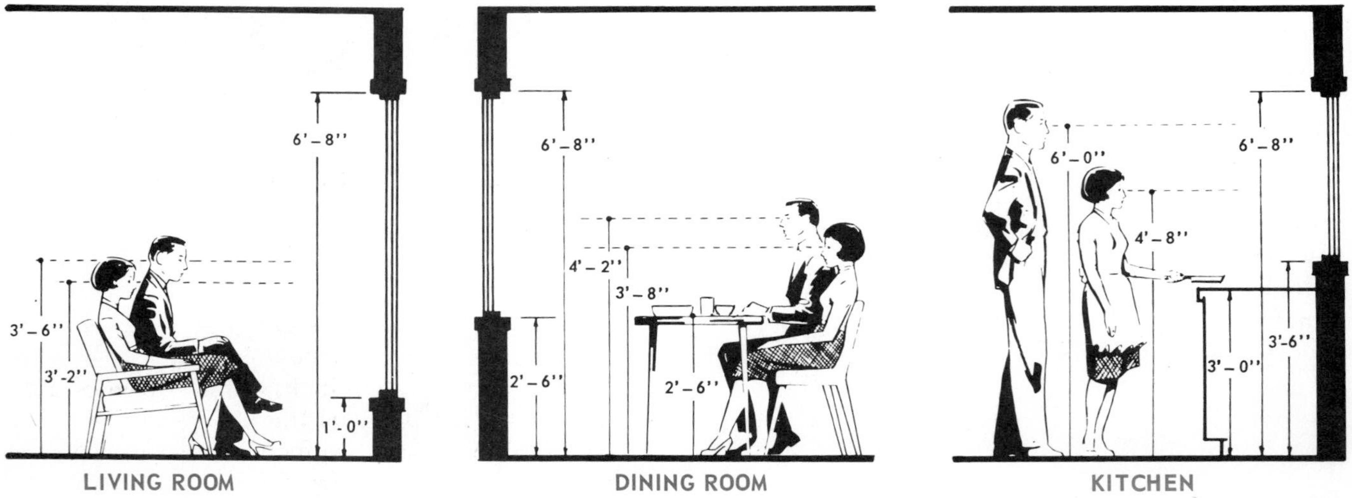

Fig. 11-8. Standard window heights: Living room, dining room, and kitchen. Horizontal window divisions should be avoided at eye-levels shown.

A newer version of the storm sash is the storm panel. It consists of a pane of glass set in a light metal frame and attached directly to the sash, usually on the inside, Fig. 11-11. The storm panel becomes a part of the sash and does not need to be removed except for cleaning. It is commonly used on horizontal sliding sash and also on casement or other hinged sash.

Sealed double glass offers many advantages. For standard movable sash, the two layers of 1/8 in. glass are fused together with a 3/16 in. air space between the layers. The double glass panel is installed in the window sash in about the same manner

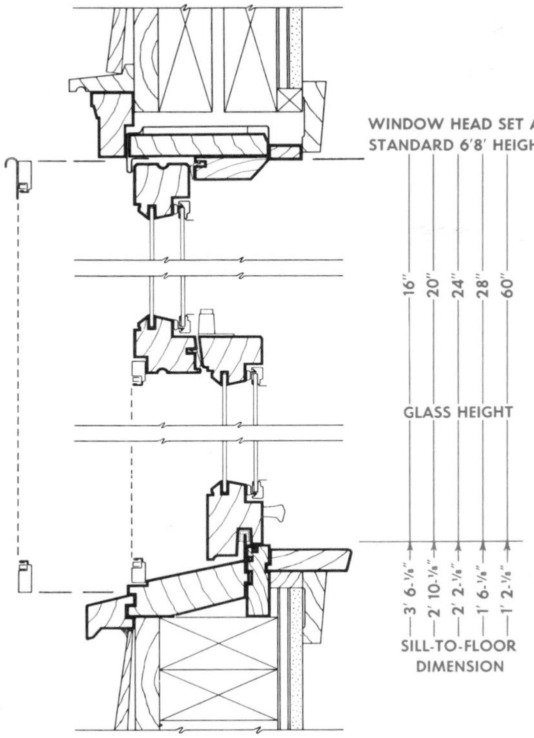

Fig. 11-9. Window head and sill heights based on a given manufacturer's product. (Rolscreen Co.)

Types	Classification	Quality	Thickness
Window	Single Strength Double Strength Double Strength	AA, A, B AA, A, B Greenhouse	.085—.100 .115—.133 .115—.133
Heavy Sheet	3/16" 7/32"	AA, A, B AA, A, B	.182—.205 .205—.230
Picture		AA, A, B AA, A, B AA, A, B	.043—.053 .058—.068 .070—.080

* All sheet glass has a characteristic surface wave, which is more apparent in larger sizes. This should be taken into consideration in the kind of glass you specify.

Fig. 11-10. Standard types of sheet glass. (Libby-Owens-Ford Glass Co.)

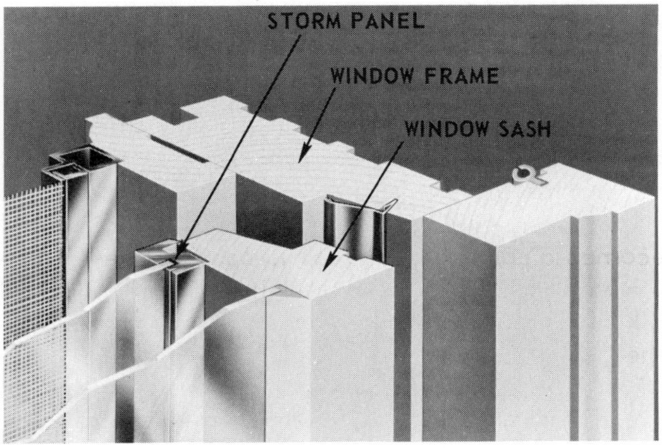

Fig. 11-11. Section through the side (jamb) of a window showing how the storm panel is mounted on the sash.

as single glass, see Fig. 11-12. Double layers of plate glass are used in large fixed units and employ special metal to glass seals. They are described later in this Unit.

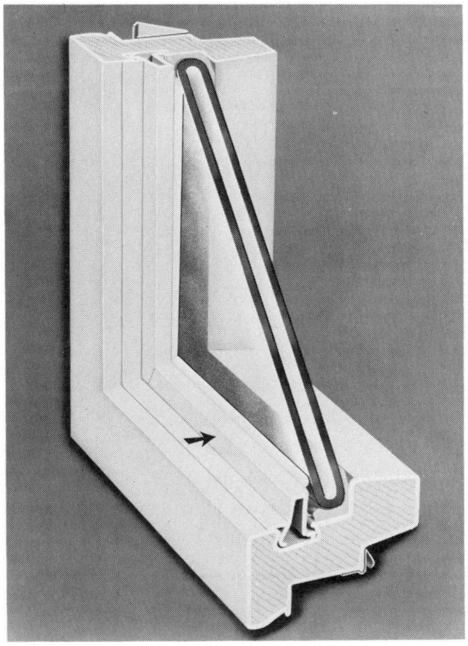

Fig. 11-12. Sealed double glass installed in a modern window sash. Rigid vinyl glazing bead (arrow) may be removed to replace glass. Regular sealed double glass is not recommended for altitudes above 4000 ft. (Andersen Corp.)

Screens

Ventilating windows require screens to keep out insects. The mesh should have a minimum of 252 openings per square inch, and be made of such noncorrosive materials as aluminum, bronze, plastic or stainless steel. Manufacturers have perfected many unique methods for mounting and storing screen panels. They are attached to the inside of the window frame on horizontal sliding and outward swinging windows. Most modern screens use a light metal frame as shown in Fig. 11-13.

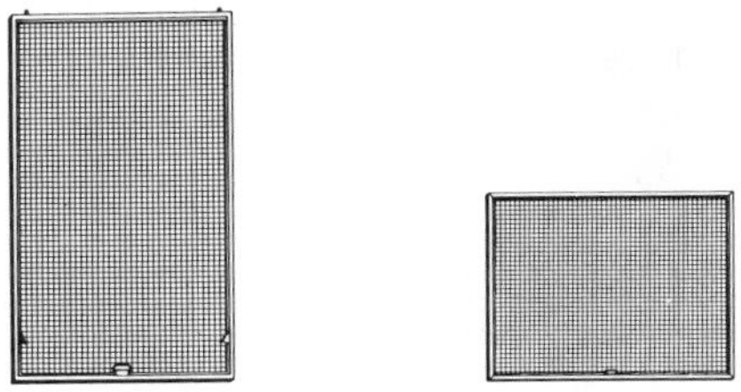

FULL SCREEN HALF-SCREEN

Fig. 11-13. Screen panels. Half-screens are used on double-hung windows and fit under the top sash and between the jambs.

Muntins

Years ago window glass was available only in small sheets. By using rabbeted strips called muntins, small panes of glass could be used to fill large openings.

Today, even though large sheets of glass are available, muntins are still used for special effects in traditional architecture. They are, how-

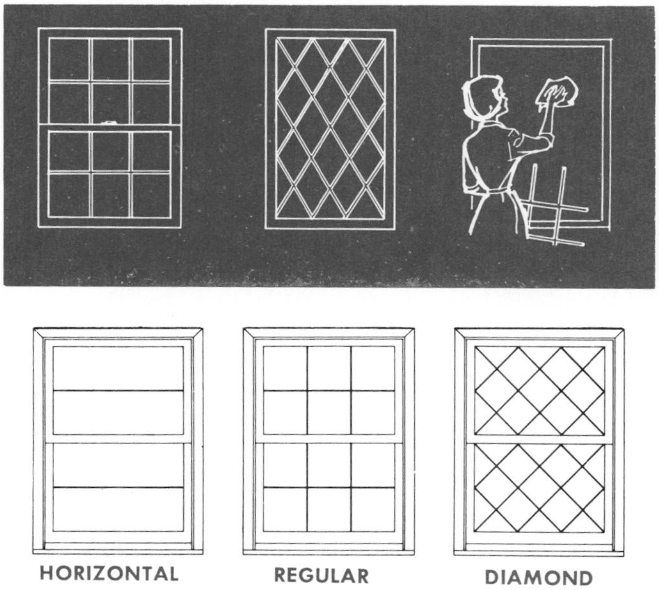

Fig. 11-14. Above. Muntin assemblies clip into place as an overlay and can be easily removed for cleaning and painting. Below. Standard muntin patterns.

ever, usually applied as an overlay and do not actually separate or support small panes of glass. Made of wood or plastic and in various patterns-- they snap in and out of the sash for easy painting and cleaning, Fig. 11-14.

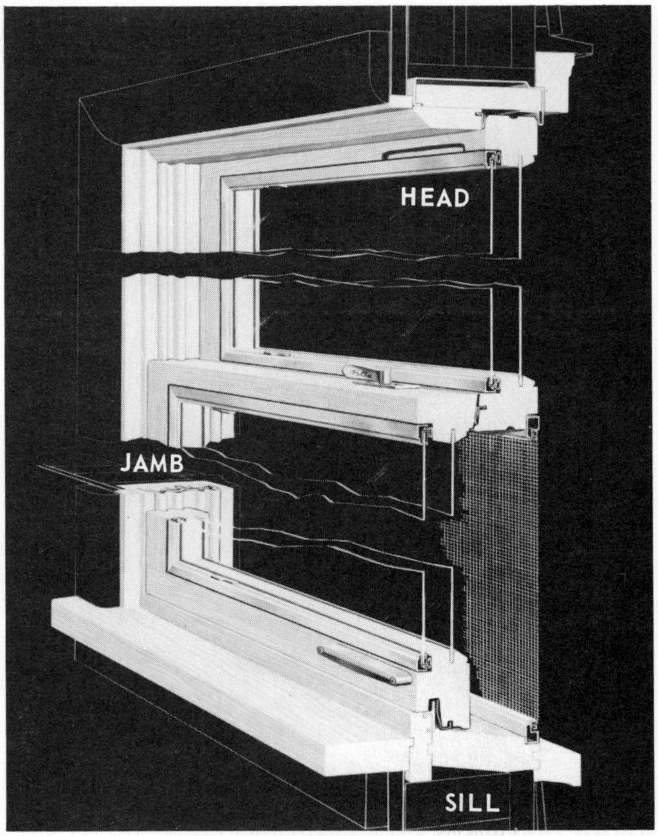

Fig. 11-15. Standard sections used to show details of all types of windows. (Rolscreen Co.)

Parts of Windows

Because much of the actual construction of a window is hidden, sectional views are used to show the parts and how they fit together. It is standard practice to use sections through the top, side, and bottom which are called the head, jamb, and sill of a window, see Fig. 11-15. These drawings include the parts of a window and also include wall framing members and surface materials. See Fig. 11-16.

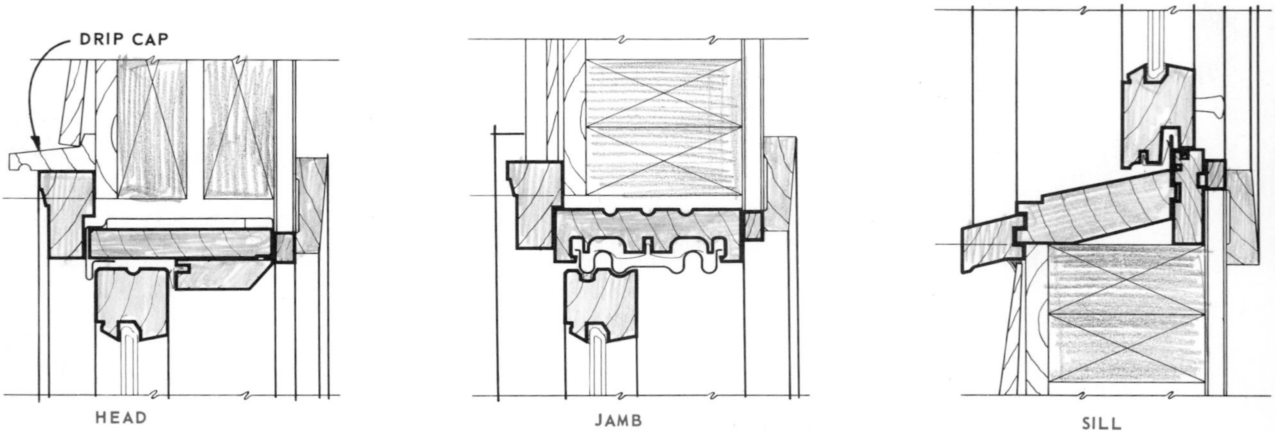

Fig. 11-16. Typical detail drawings of window sections.

A section view of a typical mullion is needed to show how window units fit together. A mullion is formed by the window jambs when two units are joined together as shown in Fig. 11-17. This figure also lists the names of the various parts of a standard double-hung unit. Other types of windows have about the same parts.

A drip cap, shown in the head section of Fig. 11-16 is designed to carry rain water out over the window casing. When the window is protected by a wide overhanging cornice this element is seldom included.

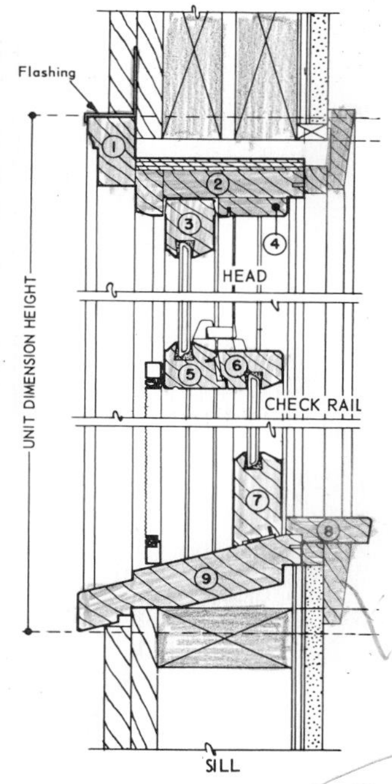

Parts Identification

1. Head Casing
2. Head Jamb
3. Upper Sash Top Rail
4. Head Stop
5. Upper Sash Check Rail
6. Lower Sash Check Rail
7. Lower Sash Bottom Rail
8. Stool
9. Sill
10. Side Casing
11. R.H. Side Jamb
12. R.H. Lower Sash Side Stile
13. L.H. Lower Sash Side Stile
14. R.H. Combination Side Stile
15. L.H. Combination Side Stile
16. Exterior Mullion Casing
17. L.H. Side Jamb
18. R.H. Upper Sash Side Stile
19. L.H. Upper Sash Side Stile
20. Side Casing

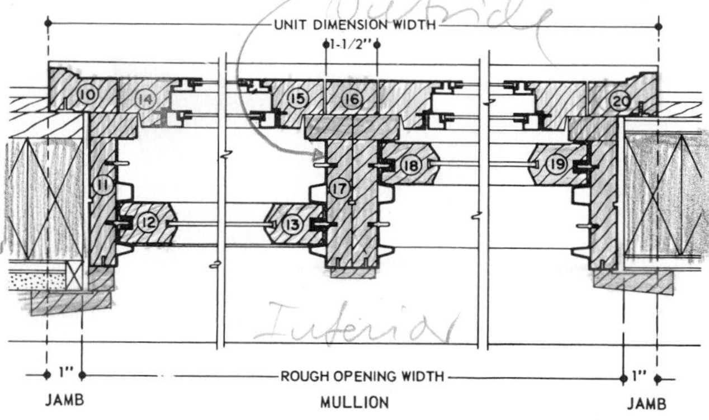

Fig. 11-17. Parts of a standard double-hung window. (Rock Island Millwork)

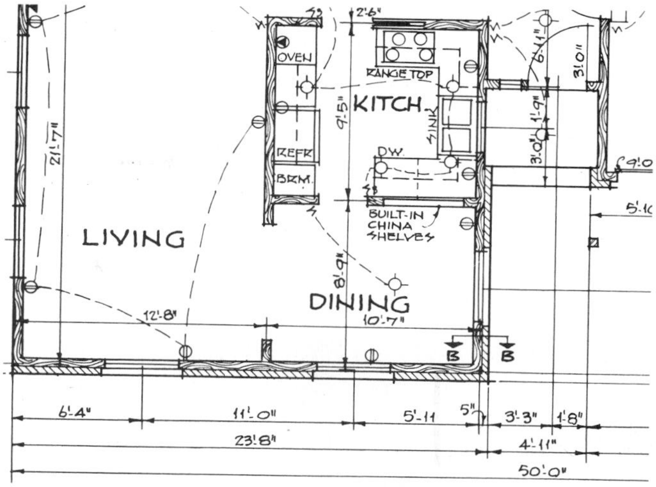

Fig. 11-18. Architectural plans show the location of windows and doors.

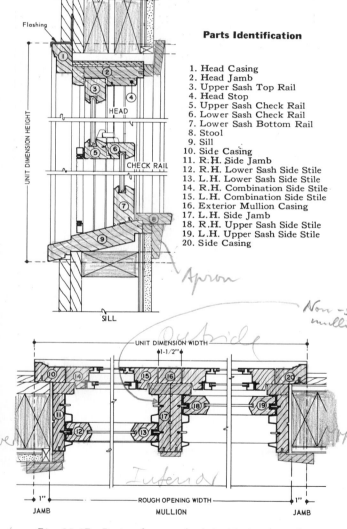

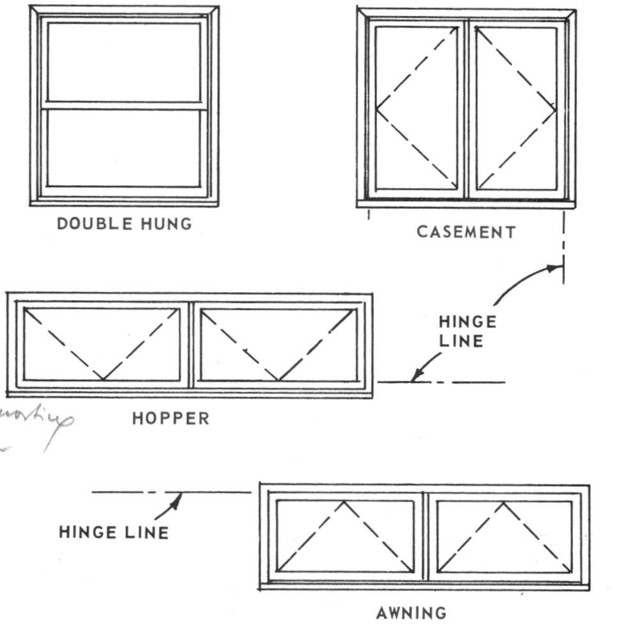

Fig. 11-19. How various types of windows appear in elevation views. Horizontal sliding windows are noted to define them from fixed units.

Windows in Plans and Elevations

The carpenter studies the plans and elevations of the working drawings to become informed about the type of windows and their location. It is common practice to locate the horizontal position of windows and exterior doors by including a dimension line to the center of the opening as shown in Fig. 11-18. In masonry construction the dimension is given to the edge of the opening. The later method is sometimes used in frame construction.

Elevations will show the type of windows, see Fig. 11-19, and may include glass size and heights. The position of the hinge line (point of dotted line) will indicate the type of swinging window. Sliding windows require a note to indicate they are not fixed units. Supporting mullions will be included in the plans and also in elevations. See Fig. 11-20.

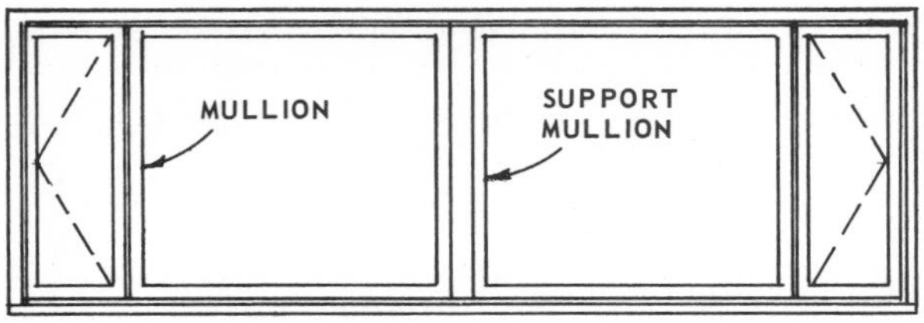

Fig. 11-20. Supporting and nonsupporting mullions.

Window Sizes

Besides the type and position of the window, the carpenter should know the size of each unit or combination of units. Window sizes may include several or all of the following: (1) glass size, (2) sash size, (3) rough frame opening, and (4) masonry or unit opening. Fig. 11-21 shows the

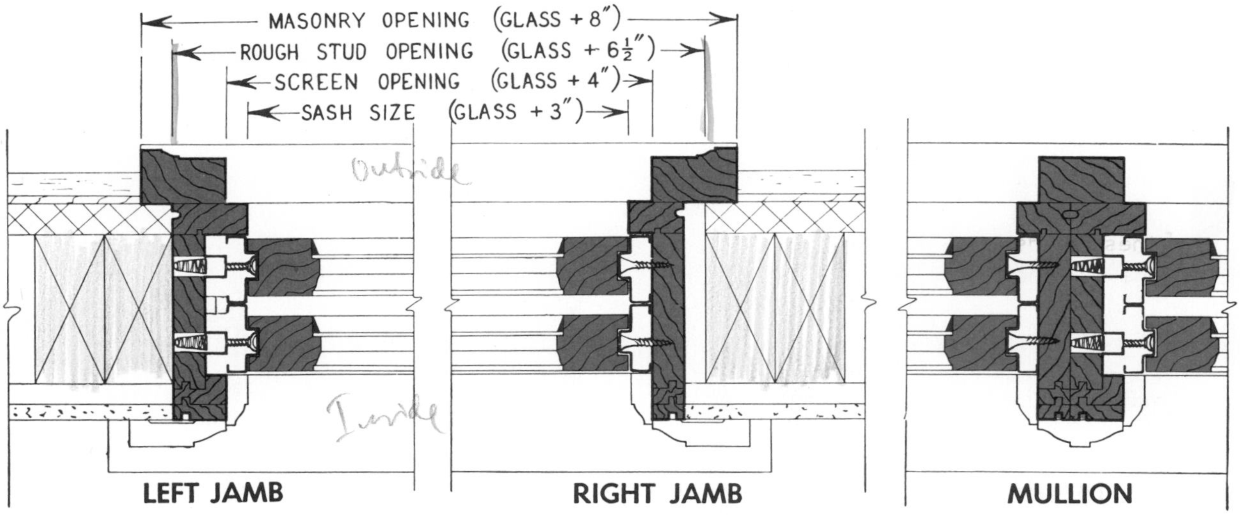

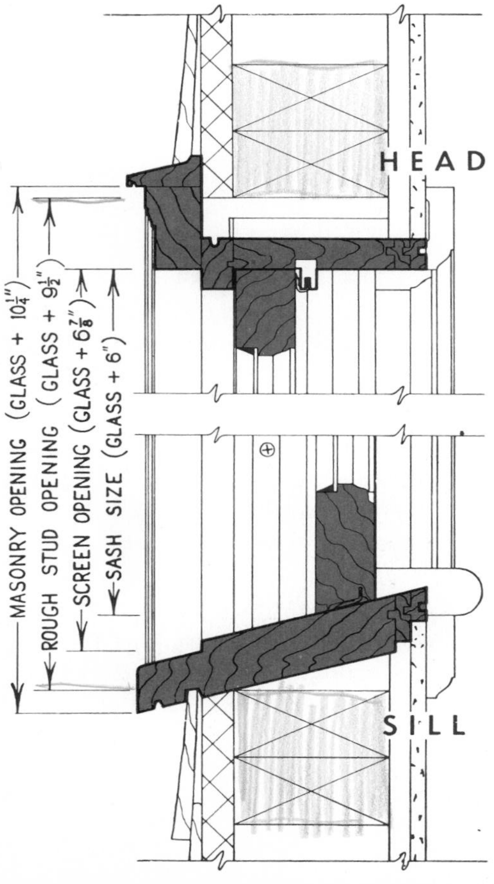

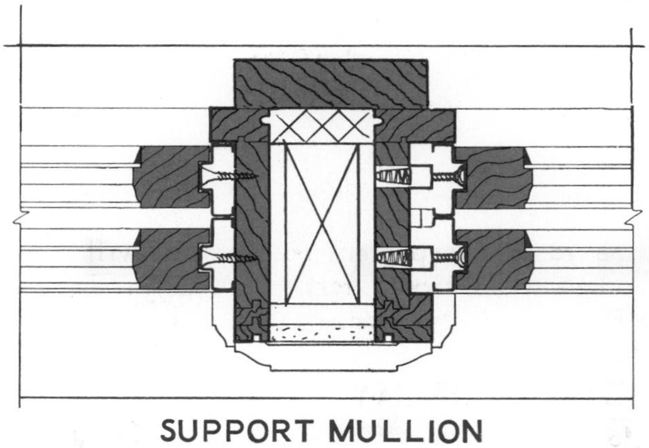

Fig. 11-21. Window sizes and location of measurements. The R. O. (rough stud opening) is larger than the overall size of the frame to permit alignment and leveling when unit is installed.

WINDOW SCHEDULE							
CODE	QUAN	NO LTS	GLASS SIZE	SASH SIZE	ROUGH OPENING	MASONRY OPENING	REMARKS
U	5	2	28" X 57 9/16"	5'-4 1/8" X 5'-3 1/16"	5'-7 1/4" X 5'-6 1/4"		+ NO. 58058 ANDERSEN GLIDING WINDOWS
V	4	1	2'-5 1/16" X HEIGHT VARIES	2'-7 13/16" X HEIGHT VARIES	2'-8 13/16" X HEIGHT VARIES		TRAPEZOIDAL UNITS (SEE DETAIL) VERIFY HEIGHTS ON JOB
W	2	1	55 1/4" X 48 1/8	4'-9 1/2" X 4'-2 3/8	5'-1 1/2" X 4'-6"		+ NO. 134 ANDERSEN PICTURE WINDOWS
X	2	2	16 1/4" X 48 11/16"	3'-1 3/4" X 4'-2 3/8"	3'-5 3/4" X 4'-6"		+ 2N4 ANDERSEN CASEMENT WINDOWS
Y	3	2	20 3/4" X 48 11/16"	3'-10 3/4" X 4'-2 3/8"	4'-2 3/4" X 4'-6"		+ W2N4 ANDERSEN CASEMENT WINDOWS
Z	1	2	20 3/4" X 36 1/2"	3'-10 3/4" X 3'-2 3/16"	4'-2 3/4" X 3'-5 7/8"		+ W2N3 ANDERSEN CASEMENT WINDOW
AA	2	1	20 3/4 X 36 1/2"	1'-10 1/2" X 3'-2 3/16	2'-2 1/2" X 3'-5 7/8"		+ W1N3 ANDERSEN CASEMENT WINDOWS
BB	2	2	20 3/4" X 48 11/16"	3'-10 3/4" X 4'-2 3/8"		4'-4 3/4" X ————	+ W2N4 ANDERSEN CASEMENT WINDOWS
CC	8	1	28" X 18"	UNIT SIZE 2'-8 1/8" X 1'-11 3/8"		2'-8 1/8" X ———— SEE DETAIL	+ 2820 ANDERSEN BASEMENT WINDOWS

* SLIDING GLASS DOORS
BY: ALWINTITE DIVISION
GENERAL BRONZE CORPORATION
GARDEN CITY, L.I., N.Y.

** PELLA DOORS
BY: ROLSCREEN COMPANY
PELLA, IOWA

\+ ANDERSEN WINDOWS
BY: ANDERSEN CORPORATION
BAYPORT, MINNESOTA

Fig. 11-22. A typical window schedule from a set of architectural plans. (Garlinghouse Plans)

position of these measurements and approximately how they are figured from the glass size. They will vary slightly from one manufacturer to another.

A complete set of architectural drawings should provide detailed information about window sizes. This information is usually listed in a table called a window and door schedule, Fig. 11-22. It includes among other things, the manufacturer's numbers and rough opening sizes for each unit or combination. An identifying letter is located at each opening on the plan and a corresponding letter is then used in the schedule to specify the required window unit and the necessary information.

When this information is not included in the architectural plans the carpenter will need to study manufacturer's catalogs and other descriptive literature. Sizes for basic units are often given in a diagramatic form as shown in Fig. 11-23. The size of the rough opening (R.O.) is of major importance to the carpenter, however he will also need other dimensions; for example, the height of the R.O. above the floor.

HANDY SAYS:

"When window sizes (also doors) are given, whether they consist of rough openings, sash size or other items, the horizontal dimension is listed first and the height is listed second. For example, the unit No. 222W shown in Fig. 11-23 requires a R.O. of 4'-8 1/2" x 2'-7"."

Rough openings and other sizes are readily secured from manufacturer's data, Fig. 11-24, when single units are used. Many windows however consist of a combination of several ventilating and fixed units and the carpenter may be required to calculate the required dimensions. Although each manufacturer will provide specific sizes and direc-

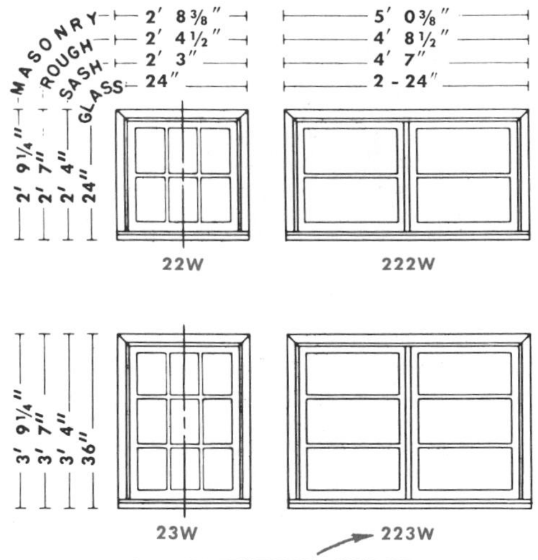

Fig. 11-23. Typical form used by window manufacturers to designate sizes of their standard units. (Rolscreen Co.)

tions for use when working with his products, the example used in Fig. 11-25 is typical of a carpenter's calculations. Study the calculations followed and then select other combinations of units and develop R.O. dimensions for them. Note that it would be incorrect to simply add together the rough opening for each unit in a given combination to secure the total rough opening.

Jamb Extensions

The thickness of window units may be adjusted to various wall structures. For example: a typical frame wall with 3/4 in. sheathing and a standard interior surface of lathe and plaster may be 5 1/8 in. compared with a single-layer dry wall construction of about 4 3/4 in. Adjustments in the window frame may be made by applying a jamb

Window Sizes

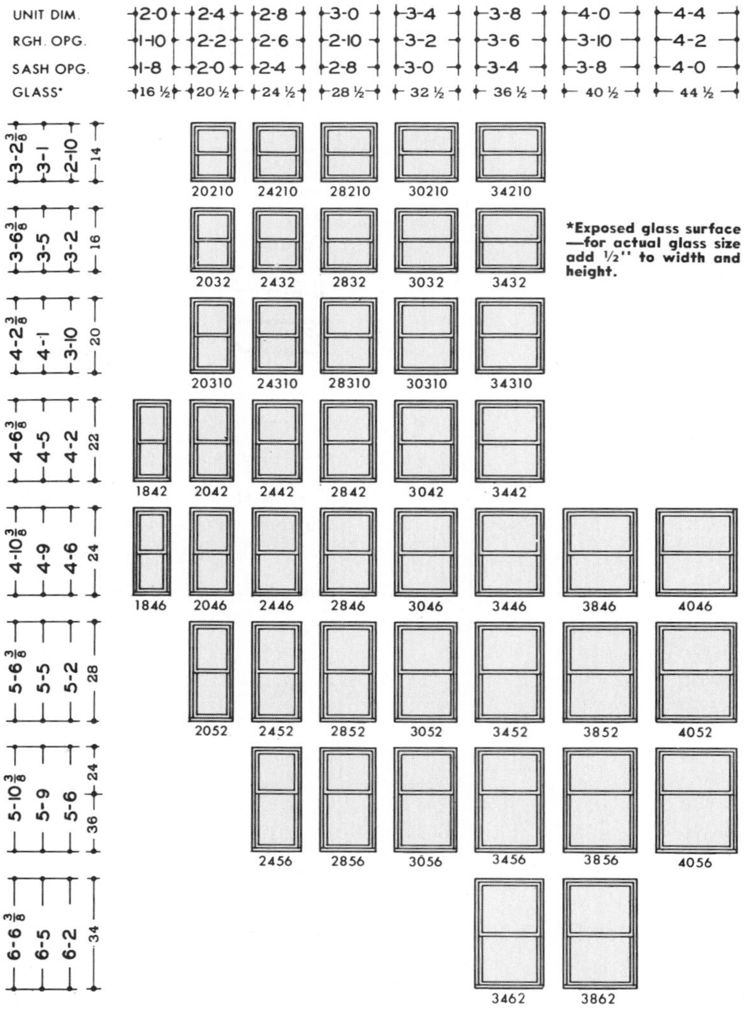

***Exposed glass surface —for actual glass size add 1/2" to width and height.**

TYPICAL COMBINATIONS

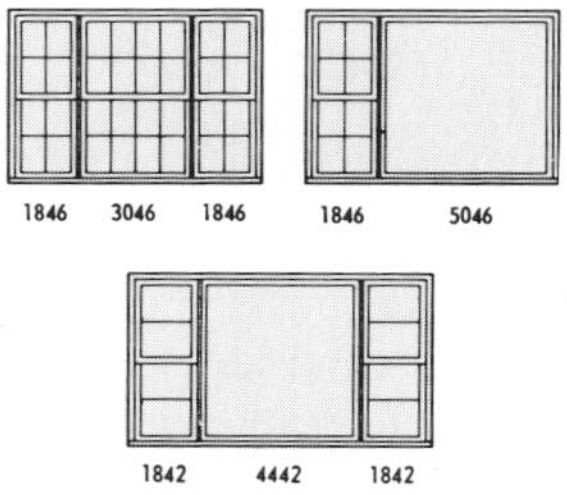

PICTURE WINDOWS

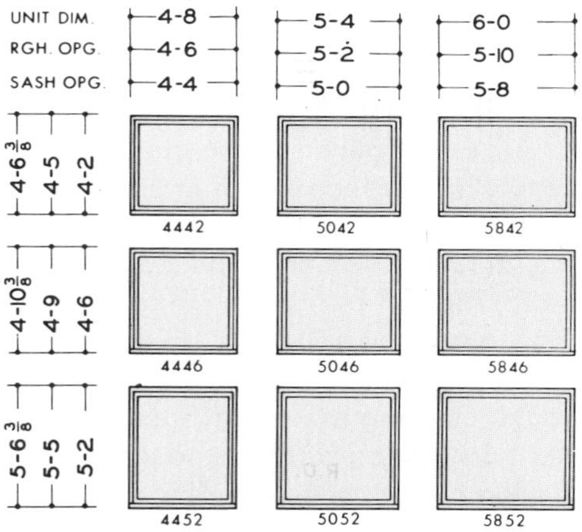

How to figure multiple openings 1-5/16" or 1 1/2" mullion post

Rough opening widths—The sum of individual sash openings plus 1 5/16" or 1 1/2" for each mullion, plus 2" for overall rough opening. **Unit dimension widths**—Add 2" to rough opening widths.

UNIT NO.	1" GLASS RABBET	1" GLASS SIZE
4442	49 x 46 1/2	48 1/2 x 46
5042	57 x 46 1/2	56 1/2 x 46 1/8
5842	65 x 46 1/2	64 1/2 x 46
4446	49 x 50 1/2	48 1/2 x 50
5046	57 x 50 1/2	56 1/2 x 50
5846	65 x 50 1/2	64 1/2 x 50
4452	49 x 58 1/2	48 1/2 x 58
5052	57 x 58 1/2	56 1/2 x 58 1/8
5852	65 x 58 1/2	64 1/2 x 58

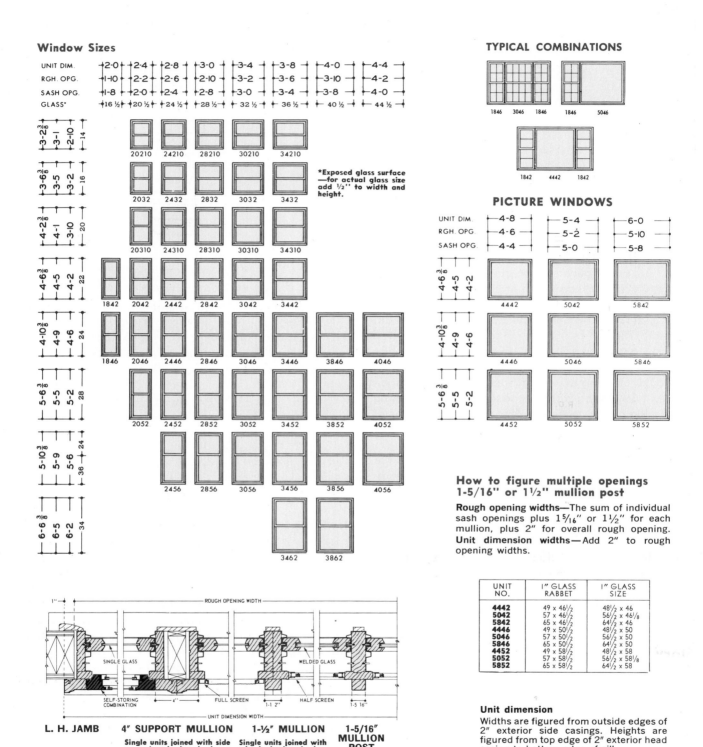

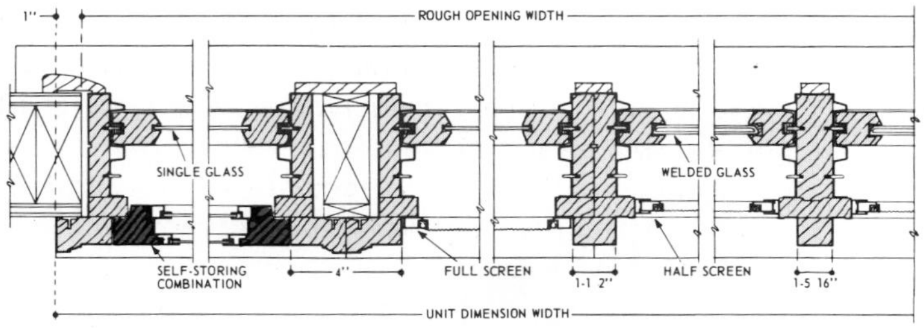

L. H. JAMB — **4" SUPPORT MULLION** Single units joined with side casings back to back. — **1-1/2" MULLION** Single units joined with jambs back to back. — **1-5/16" MULLION POST**

Unit dimension

Widths are figured from outside edges of 2" exterior side casings. Heights are figured from top edge of 2" exterior head casing to bottom edge of sill.

Fig. 11-24. Sizes of double-hung and matching fixed units available from one manufacturer. (Andersen Corporation) extension of the required size, Fig. 11-26. Some manufacturers build their frames to a basic size (4 5/8 in.) and then equip the unit with an extension as specified by the builder or architect.

Detailed Drawings

Sectional drawings showing each part of the window and how the unit is placed in a given type of wall structure are helpful to the carpenter. Fig. 11-27 shows detail drawings for a typical double-hung window. Similar drawings are available for other types of windows from other manufacturers. The architect will often include selected

SASH OPENING	2042	2'–0"
SASH OPENING	5842	5'–8"
SASH OPENING	2042	2'–0"
2–MULLIONS (1 1/2")		0'–3"
ADD FOR R.O.		0'–2"
		9'–13" = 10'–1"

TOTAL ROUGH OPENING
R.O. = 10'–1" x 4'–5"

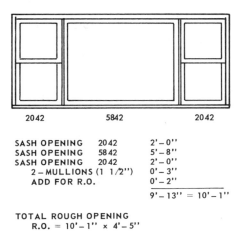

Fig. 11-25. Calculating the R. O. for a combination of units, Sizes and numbers were secured from Fig. 11-24.

detail views that apply to a specific plan. Whether included in the architectural drawings or made available through manufacturer's catalogs, detailed drawings such as these and also those in Figs. 11-28 and 11-29 will be essential in building the rough frame of the wall structure and making the proper installation of the window units.

Story Poles

A story pole or rod is a straight strip of wood upon which is laid out various vertical reference heights. A door and window story pole may be used to establish the height and size of rough openings during the construction of the building frame and is also helpful when installing doors and windows. It will help prevent mistakes in the size of openings and insure the level alignment of doors and window heads.

Fig. 11-30 shows a simple story pole layout for a given window unit, the details of which are shown in Fig. 11-31. The construction details of the window head will normally be the same throughout a building while the sill height may vary. Additional positions can be superimposed over other layouts, however, each one should be carefully labeled or indexed so the correct distance will always be applied to the proper opening. A more detailed explanation of the story pole is included in Unit 8.

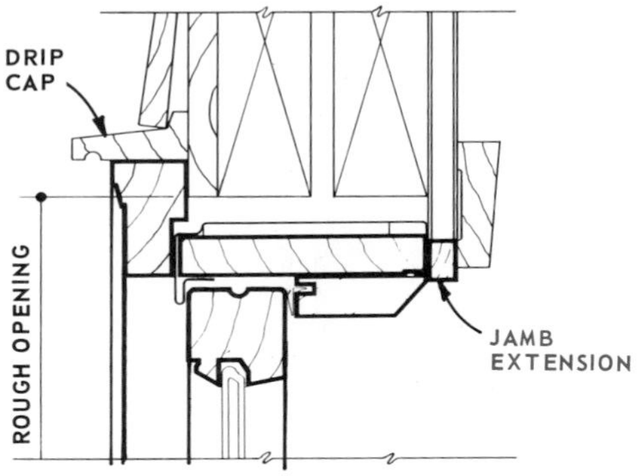

Fig. 11-26. Jamb extensions are applied to a standard window frame to adjust for various wall thickness.

Residential doors are normally 6'-8" high and the top of windows is usually held at this same height. See Fig. 11-32. An extra 1/2 in. is added to allow for thresholds under entrance doors and clearance under interior doors.

Installing Windows

If the rough openings are plumb, level, and the correct size, it is a relatively easy task to install modern windows. Manufacturers furnish directions that apply specifically to their various products and the carpenter should follow them carefully.

When windows are received on the job they should be stored in a clean, dry area. If they are not fully packaged, some type of cover should be used to prevent damage from dust and dirt. It is highly desirable that wood windows become conditioned to the prevailing humidity of the locality before they are installed.

Check carton labels and move window units (still packaged) into the various rooms and areas to which they belong.

Unpack the window and check for any shipping damage. Do not remove any diagonal braces or spacer strips until after the installation is complete. Check the rough opening to make certain it is the correct size. Most windows require at least 1/2 in. space on each side and 3/4 in. above the head for plumbing and leveling. When board sheathing is used, a strip of building paper should be tacked around the rough opening to prevent excessive infiltration.

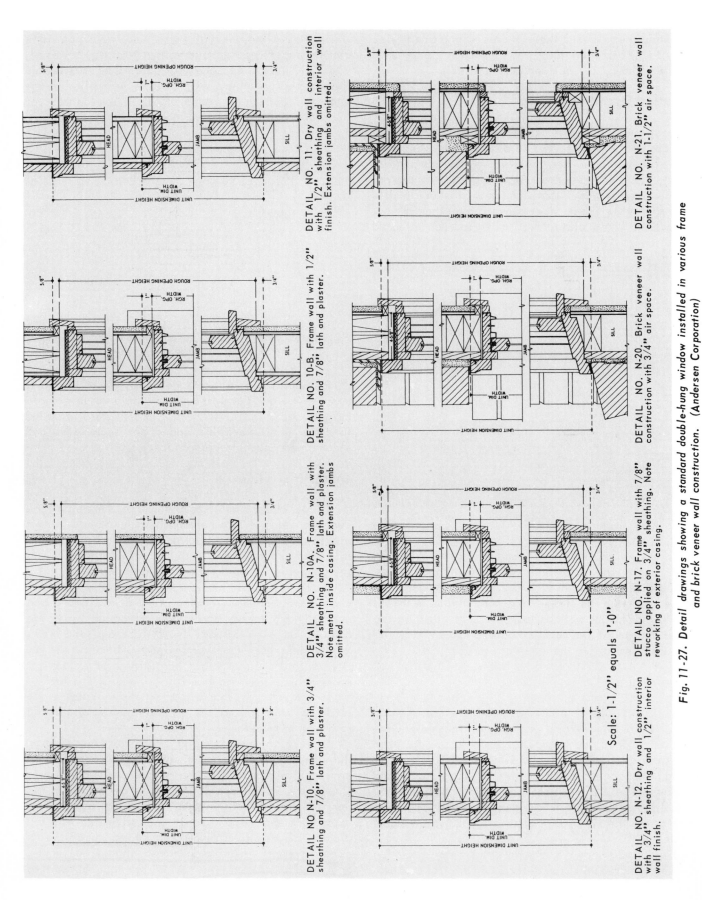

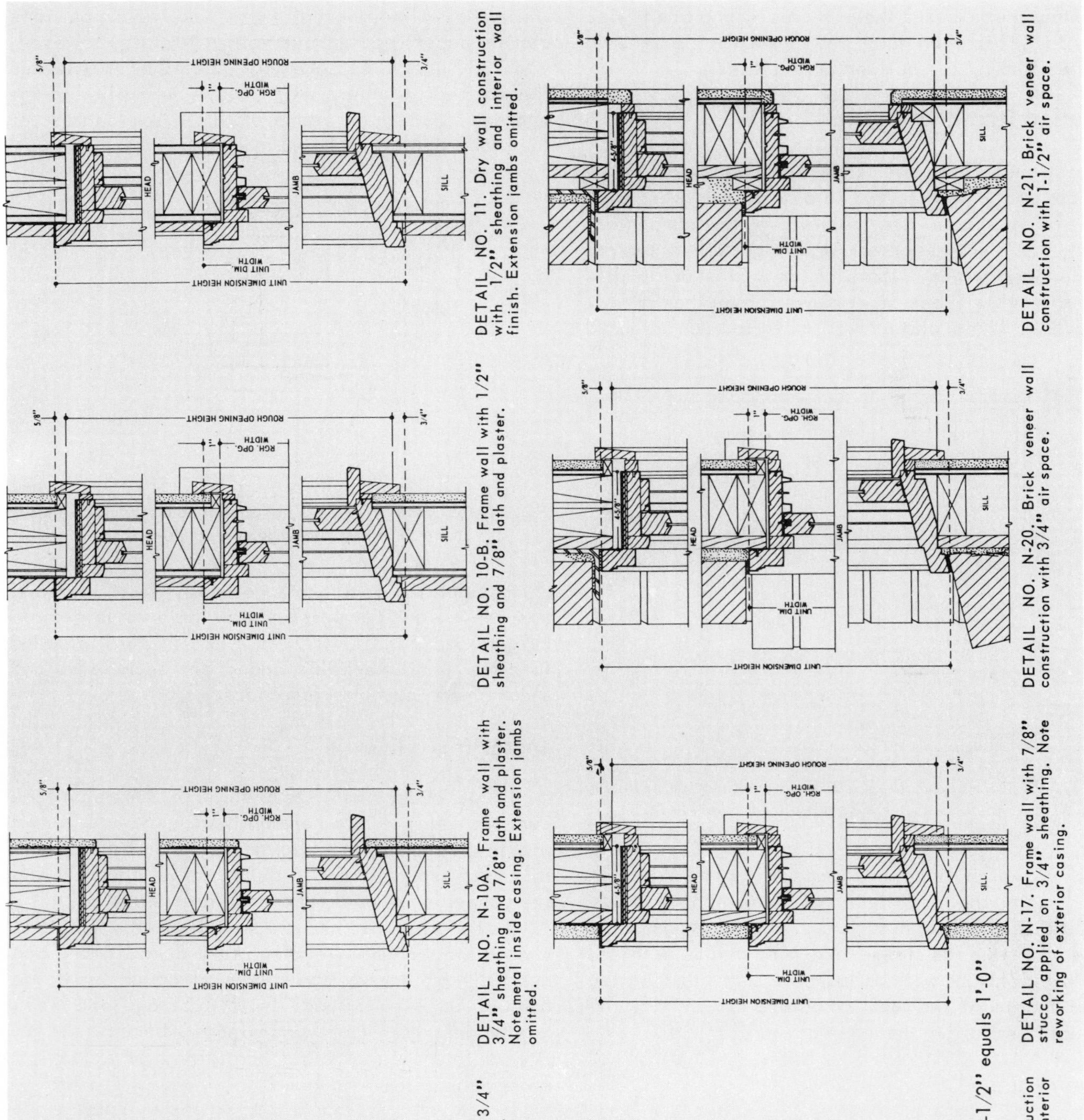

DETAIL NO N-10. Frame wall with 3/4" sheathing and 7/8" lath and plaster.

DETAIL NO. N-10A. Frame wall with 3/4" sheathing and 7/8" lath and plaster. Note metal inside casing. Extension jambs omitted.

DETAIL NO. 10-B. Frame wall with 1/2" sheathing and 7/8" lath and plaster.

DETAIL NO. 11. Dry wall construction with 1/2" sheathing and interior wall finish. Extension jambs omitted.

Scale: 1-1/2" equals 1'-0"

DETAIL NO. N-12. Dry wall construction with 3/4" sheathing and 1/2" interior wall finish.

DETAIL NO. N-17. Frame wall with 7/8" stucco applied on 3/4" sheathing. Note reworking of exterior casing.

DETAIL NO. N-20. Brick veneer wall construction with 3/4" air space.

DETAIL NO. N-21. Brick veneer wall construction with 1-1/2" air space.

Fig. 11-27. Detail drawings showing a standard double-hung window installed in various frame and brick veneer wall construction. (Andersen Corporation)

CORNER WINDOW DETAILS

The following details are suggestions for installing Andersen NARROLINE window in corners. Three different methods of framing corners are shown.

Plan section at right shows the most common way of framing corners. Note the opening dimensions are the same when using 4-2x4's. When using this type of corner the casing on both sides inside and out are the same size.

Vertical section is the same as for standard frame wall construction.

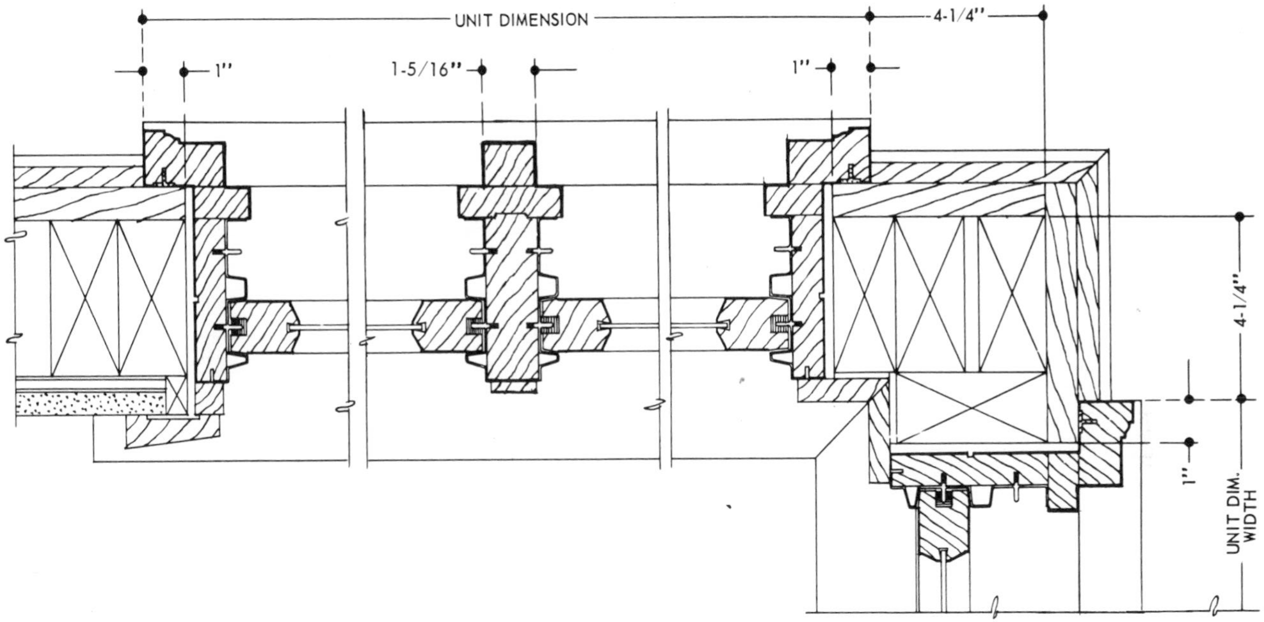

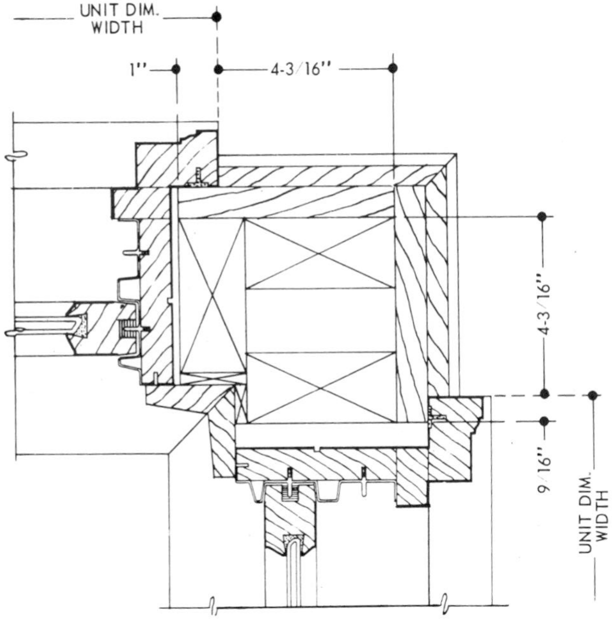

ALTERNATE CORNER SECTIONS

Detail at left shows 3-2x4's used in framing of corner. Note opening dimensions are different so inside and outside dimensions are the same.

Detail at right shows a 3" pipe column used in place of 2x4's.

Andersen Seal-Trim exterior casing, shown within dark outline; furnished when specified. All other exterior and interior casing not furnished by Andersen.

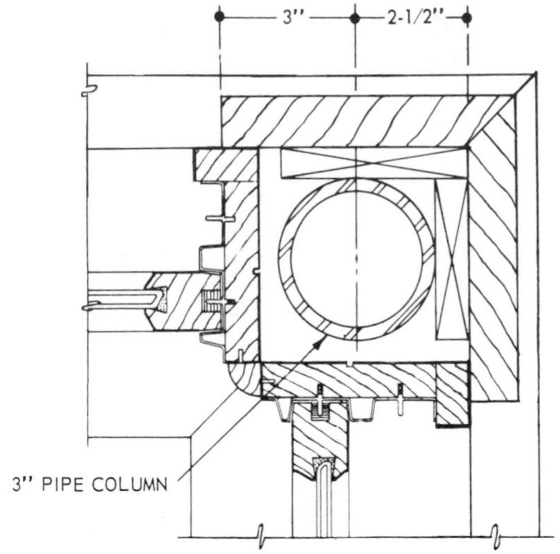

Fig. 11-28. Corner details for a double-hung window.

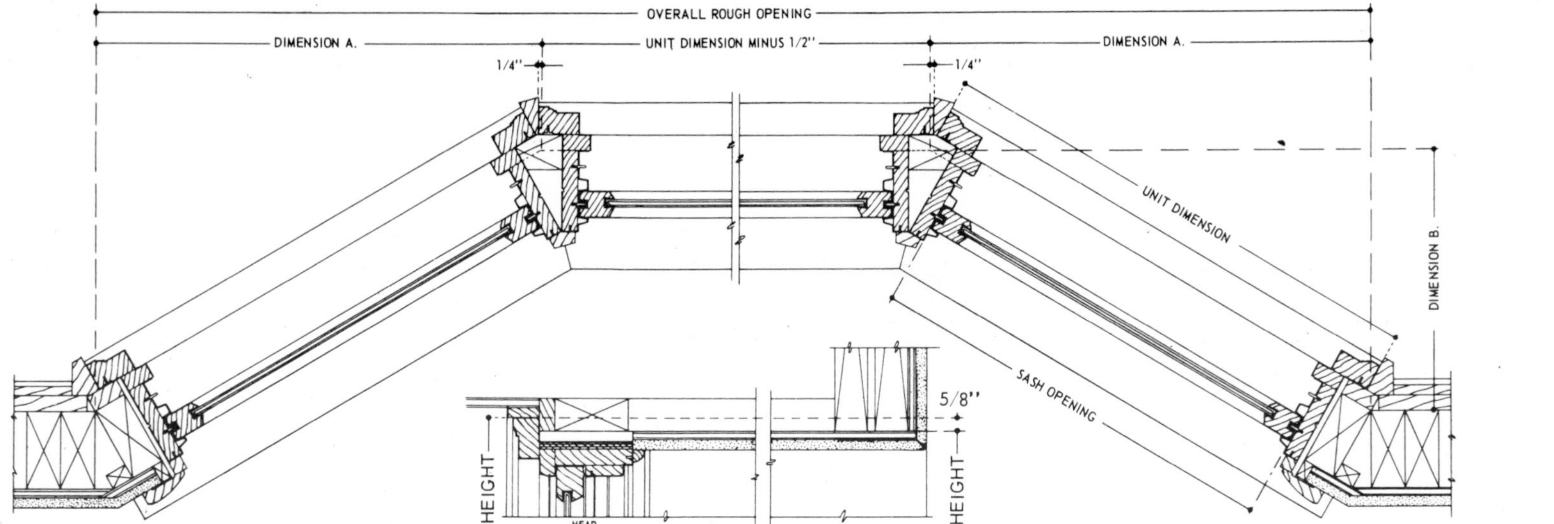

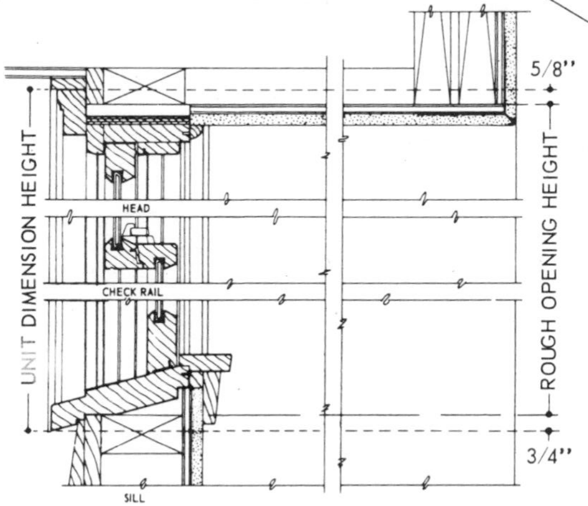

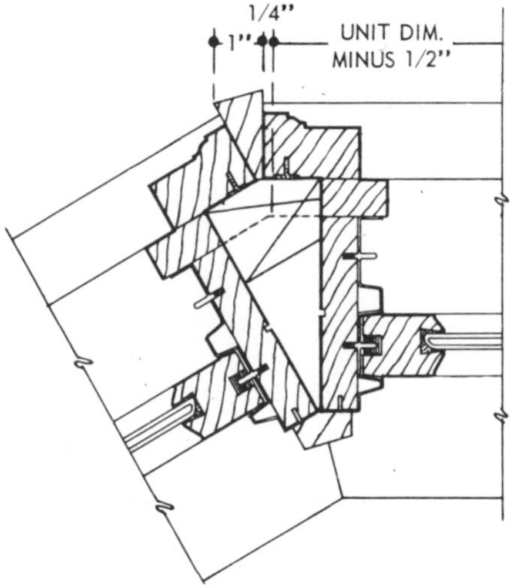

MULLION DETAIL

Details show standard units with 2" exterior casing used to form-angle bay. Exterior mullion casing and auxiliary side casing not furnished

Sash Opening	Dimension A.	Dimension B.
1-8	21-1/8	12-3/16
2-0	24-9/16	14-3/16
2-4	28	16-3/16
2-8	31-1/2	18-3/16
3-0	35	20-3/16
3-4	38-7/16	22-3/16
3-8	41-7/8	24-3/16
4-0	45-3/8	26-3/16

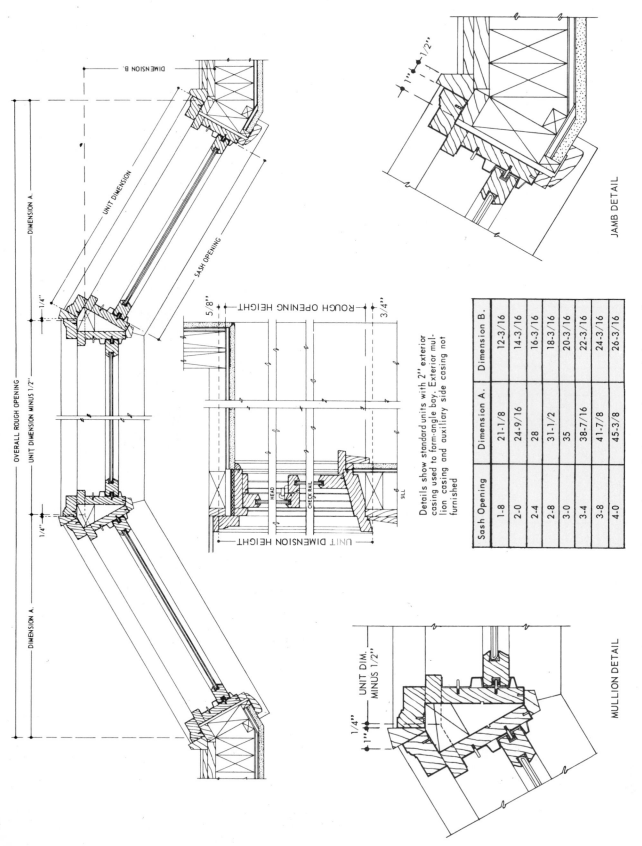

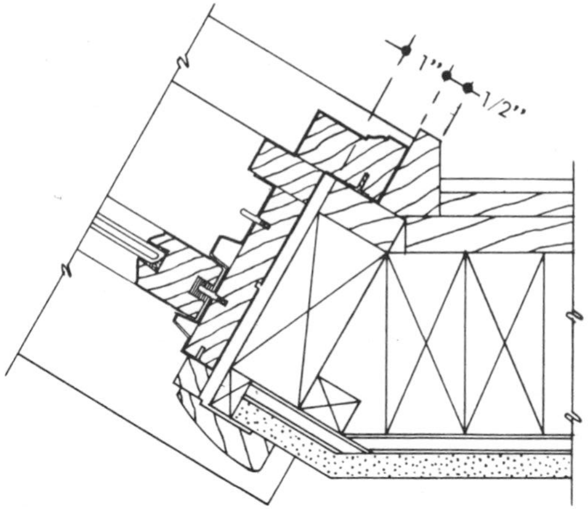

JAMB DETAIL

Fig. 11-29. Details for a 30 deg. angle bay window.

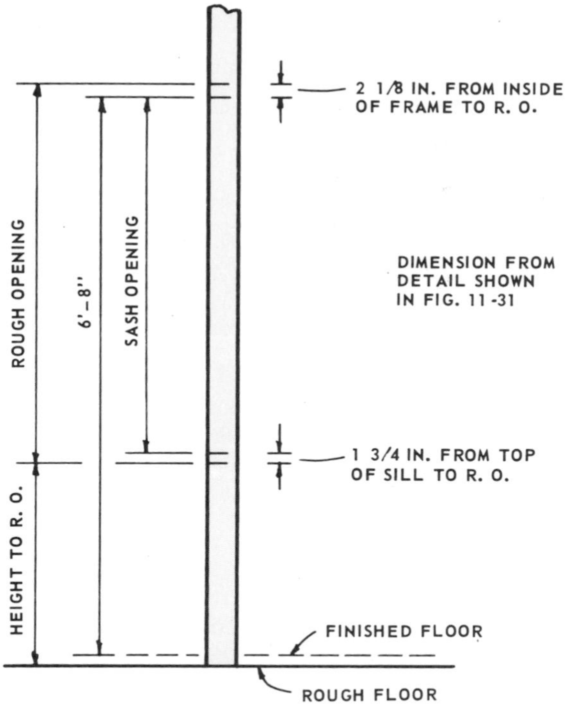

Fig. 11-30. Story pole layout for a window.

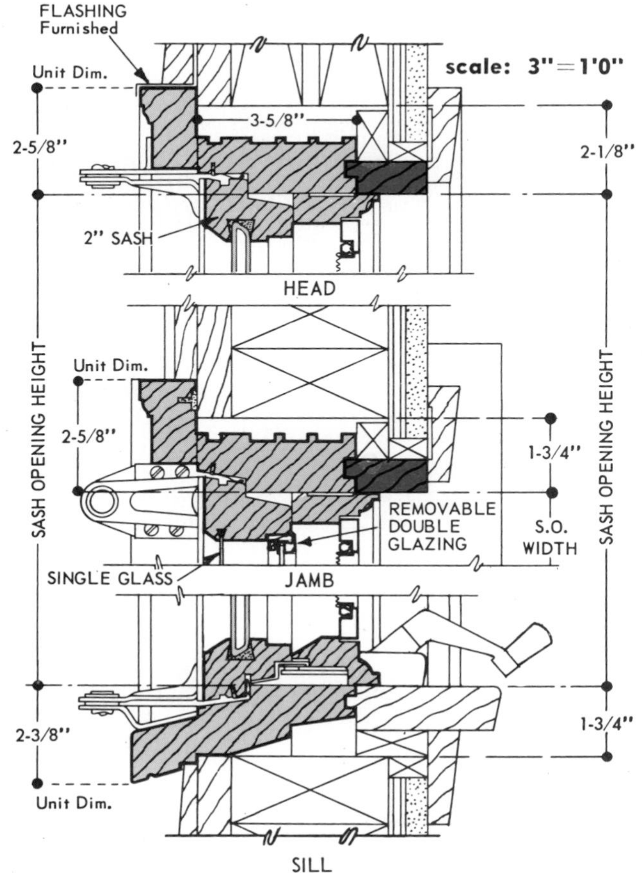

Fig. 11-31. Details of a casement window.

If the window has not been primed at the factory, this may be done before installation. Weather stripping and special channels should not be painted. Follow the manufacturer's recommendations. Although most windows will have the outside casing attached, some windows, especially those with a metal frame, are set in the opening and then the outside trim is installed.

Fig. 11-32. The top of door and window frames and trim are usually aligned as shown above. (Andersen Corporation)

Most window units and multiple unit combinations are installed from the outside. The window, if stored inside, can be easily moved to the outside through the rough opening in single story construction. Secure sufficient help to carry windows and handle them carefully as shown in Fig. 11-33. Place the window in the opening, Fig. 11-34, and secure it temporarily in place.

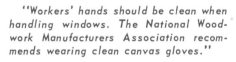

HANDY SAYS:

"Workers' hands should be clean when handling windows. The National Woodwork Manufacturers Association recommends wearing clean canvas gloves."

Place wedge blocks under the sill and raise the frame to the correct height as marked on the story pole. Also adjust the wedges so the frame is perfectly level, Fig. 11-35. Since there may be a tendency for the sill to sag on wide windows or multiple units, wedge blocks should be spaced at the ends and several places in the center area. Nail through the lower end of the side casing to secure the bottom of the frame in place.

Plumb the side jambs with the level and check the corners with a framing square. Usually the sash should be closed and locked in place. Drive nails temporarily into the top of the side casing. Now check over the entire window and see that it is

Fig. 11-33. Left. Be sure to handle windows carefully. Fig. 11-34. Right. Place the window in opening and fasten it temporarily. In windy weather, use extra props.

Fig. 11-35. Adjust shims and wedges until the frame is perfectly level. (Andersen Corporation)

square and level. Open the ventilating sash to see that they operate smoothly, Fig. 11-36. Any sag in the head or bow in the jambs should be straightened with a spacer-strip.

Finally, nail the window permanently in place with aluminum or galvanized casing nails. Space the nails about 16 in O.C. and be certain they are long enough to penetrate well into the building frame, Fig. 11-37. Window casing is made of soft wood and will dent easily, therefore, a nail set should be used for the last driving strokes.

At this point many builders prefer to cover the inside of the window with a sheet of polyethylene film to protect it during the application of inside wall surface materials.

Fig. 11-36. Check the operation of ventilating sash before final nailing of the frame. (Weyerhaeuser Co.)

Fig. 11-37. Secure the window frame by nailing through the casing and into framing members of the structure.

Installing Fixed Units

As air conditioning becomes more prevalent, the need for ventilating windows will likely be reduced. Fixed units are less expensive than those of an equal size that are movable.

When the fixed glass panel is of medium size, it is usually mounted in a sash and frame, Fig. 11-38, and then combined with matching ventilating units. The installation of such a unit is the same as that of regular windows, Fig. 11-39. They are of course larger and heavier and extra precautions should be observed in handling and making the installation. Fig. 11-40 shows an interior view of such a unit after completion.

In modern construction large fixed glass areas are usually fitted with double panes of glass (triple panes are sometimes used). These panels are called insulating glass or sealed double or triple glazing. They are composed of sheets of glass separated 1/4 to 1/2 in. and contain dehydrated-captive air, hermetically (airtight) sealed around the edges at the factory. The edge is protected by a metal channel, Fig. 11-41. Two popular trademarks are "Thermopane" and "Twindow." Insulating glass offers the following advantages:

1. Minimizes chilly downdrafts, so often the penalty of large, uninsulated glass areas.
2. Increases the surface temperature of the glass next to the room, thereby reducing body radiation to the warmer surfaces.
3. Saves on fuel bills because its insulating efficiency is much greater than ordinary single-glazed windows. It saves on air conditioning, too, since less heat gets in during the summer.
4. Offers greater "see-ability" by reducing room-side glass fogging in winter.
5. Reduces outside noise.
6. Eliminates the need for storm windows.

Fig. 11-39. Installing a "picture window" unit with casement windows on each side. (Andersen Corporation)

Fig. 11-40. Interior view of "picture window" after inside trim and cabinetwork is complete. (Andersen Corporation)

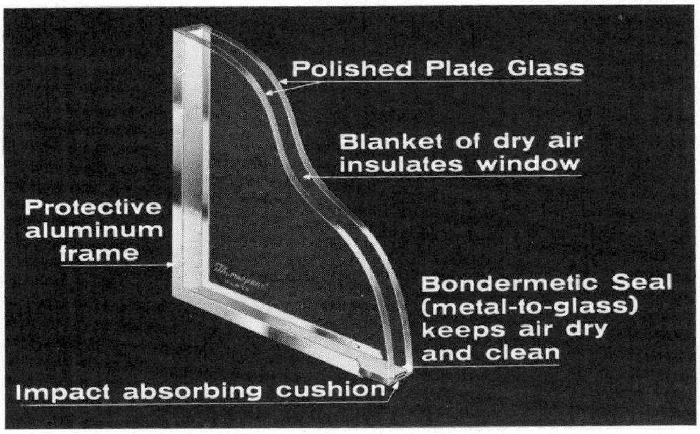

Fig. 11-41. Sealed double insulating glass. (Libby-Owens-Ford Glass Co.)

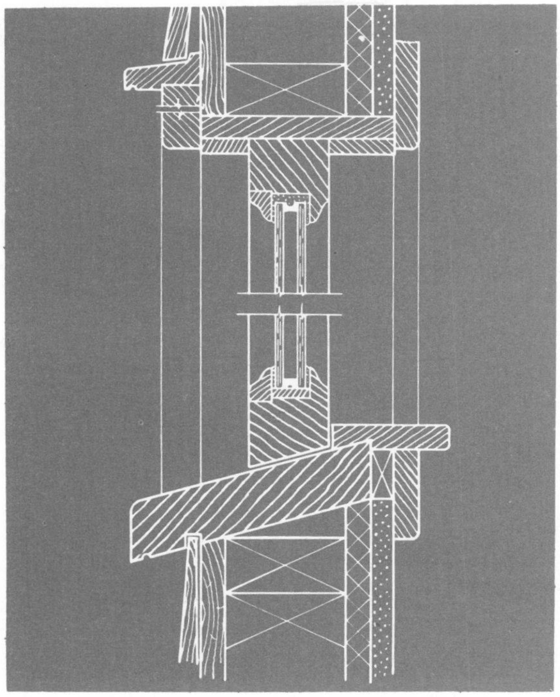

Fig. 11-38. Typical detail of 1 in. double insulating glass using a conventional sash and frame.

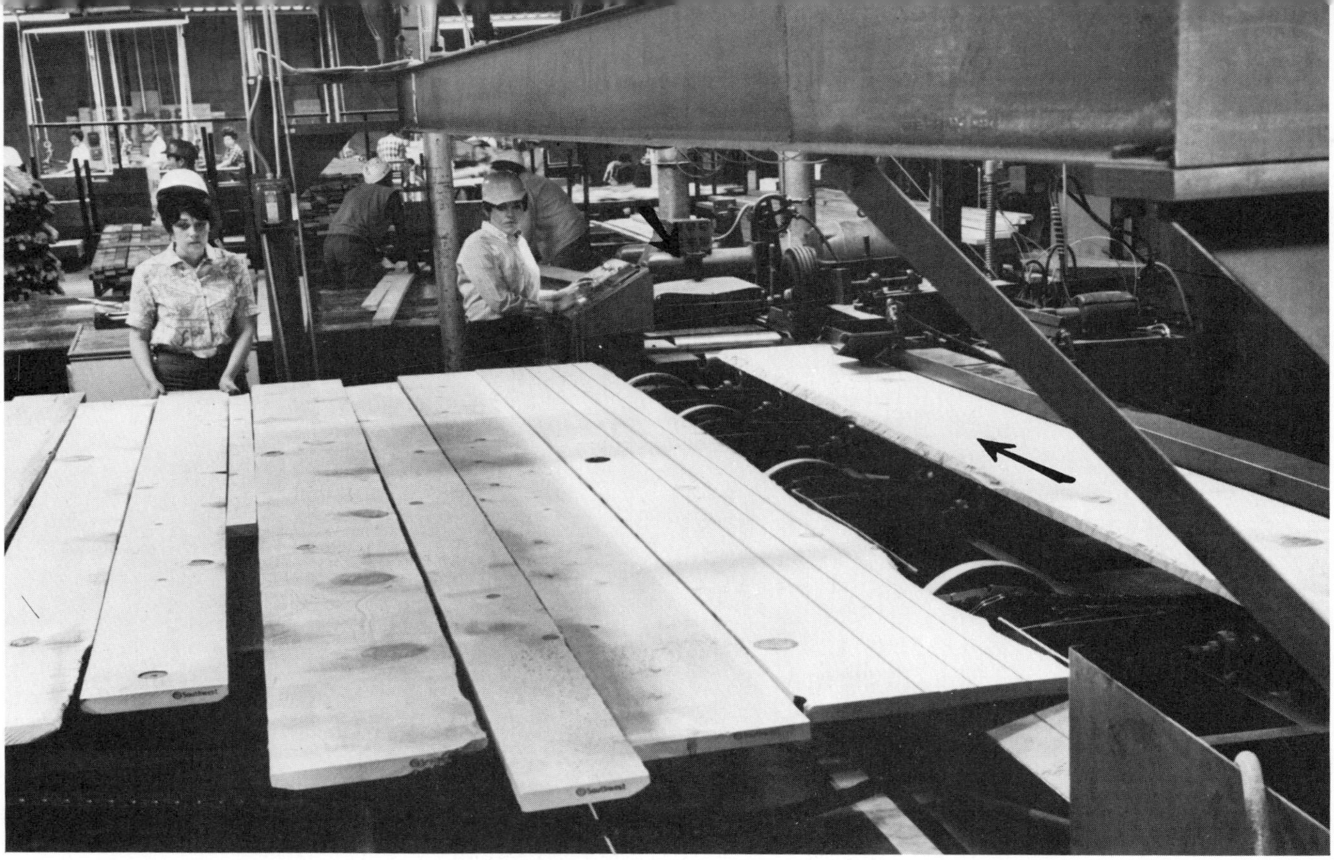

Rough mill operations in a plant that specializes in door production (above and below). Rough ripping. Operator at console adjusts shadow lines for best ripping layout. The lines are projected from an overhead box. The position of each line is recorded electronically and transferred to the gang ripsaw (arrow) where the blades are automatically spaced to make the cuts as the board passes through. (Maywood)

Rough milling. Ripped boards on right move by cut-off operators where defects are removed as pieces are cut to required lengths. Note the stops (arrow) that are used to quickly position the cut. (Maywood)

The resistance to heat flow provided by sealed double glazing is compared to single panes of glass in Fig. 11-42. The "U-value" is the amount of heat expressed in Btu's which will pass through one square foot of the window in one hour for each degree (Fahrenheit) temperature difference between the inside and outside.

	TYPE GLAZING	GLASS THICKNESS	AIR SPACE THICKNESS	"U" VALUE
	SINGLE	⅛"		1.14
	SINGLE	¼"		1.12
	DOUBLE (2 Pieces Glass) (1 Air Space)	⅛"	¼"	.63
		¼"	¼"	.61
		⅛"	½"	.57
		¼"	½"	.55

Fig. 11-42. U-values for single and double glazing. (PPG Industries, Inc.)

Large insulating units are made of 1/4 in. plate glass which results in considerable weight. They are seldom installed in window frames at the factory although they are sometimes mounted in a sash. Usually large glass units are glazed (set in opening with glazing compound) as a separate operation after the frame and/or sash have been installed.

Openings must be square, free of twists, and rugged enough to bear the weight of the glass unit. Use only high grade wood materials that are dry and free from warp. Special setting blocks and clips, Fig. 11-43, may be used to hold the glass in position with clearance on all sides. The edge should be completely surrounded with a nonhardening glazing compound (Neoprene or Vinyl) so there is no contact between the glass and frame. Fig. 11-44 shows minimum clearances recommended by one manufacturer. Since insulating glass units cannot be altered in any way on the building site, it is essential that sash and frames be carefully designed and that specified dimensions are followed.

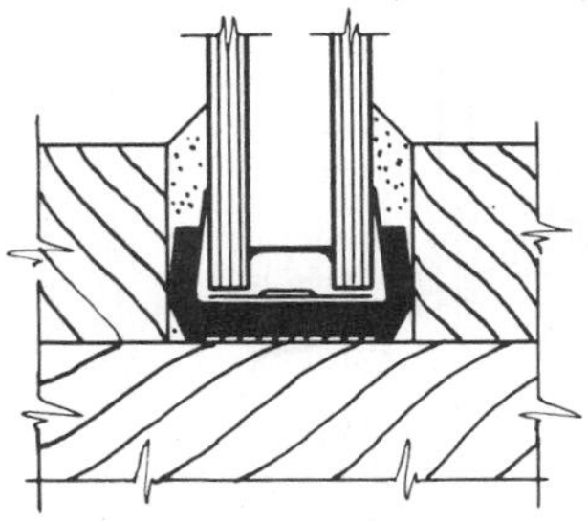
NEOPRENE SETTING BLOCK IN WOOD SASH OR FRAME

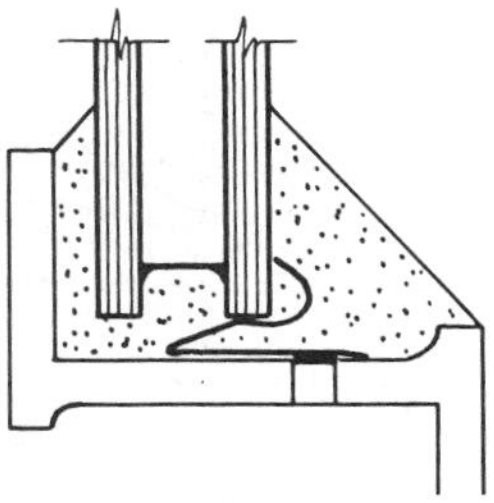
METAL GLAZING CLIP IN METAL SASH OR FRAME

Fig. 11-43. Setting blocks and clips.

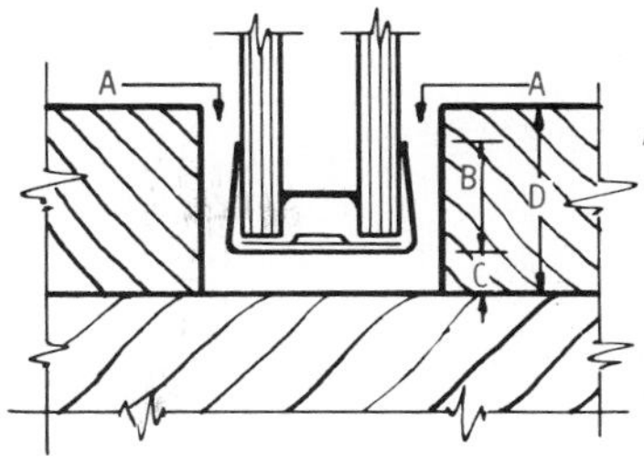

Factory-fabricated, *Thermopane* dimensions cannot be changed at point of use. Adequate clearances and rabbet depths must be provided for all units so that metal seal is not visible.

	½" and ¾" Units ⅛" Glass	⅝" and ⅞" Units $\frac{3}{16}$" Glass	¾" and 1" Units ¼" Glass
A Face Clearance	⅛"	⅛"	⅛"
B Metal Edge Depth	⅜"	⅜"	⅜"
C Edge Clearance	⅛"	¼"	¼"
D Minimum Rabbet Depth	⅝"	¾"	¾"

Fig. 11-44. Recommended clearances and rabbet depths for various glass units. (Libby-Owens-Ford Glass Co.)

A step-by-step procedure along with basic information on building a simple wood frame and installing insulating glass is given in Fig. 11-45. Standard sizes of insulating glass for fixed window units are listed in Fig. 11-46.

Suggested details for the construction of a window wall are shown in Fig. 11-47. The photo, Fig. 11-48, is an example of the dramatic effect that can be attained through the use of window walls.

Glass Blocks

Glass blocks are also used in residential construction. In outside wall openings they have insulating properties which aid in heating and air conditioning, help prevent drafts, dampen disturbing noises, cut off unpleasant views and insure privacy where it is desired.

Inside the home, partitions and screens of glass blocks add a smart touch to rooms they divide, yet provide light to show decorative schemes to good advantage.

Glass blocks are made of two formed pieces of glass fused together to leave an insulating air space between. They come in several different patterns and are usually available in three nominal sizes -- 6 x 6 in., 8 x 8 in. and 12 x 12 in. as shown in

1

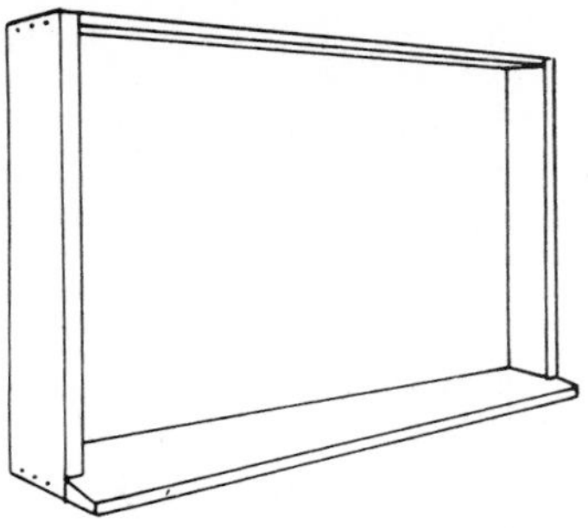

We suggest that your lumber dealer assist in selecting wood and fabrication of sill for this frame. Sill is made from a 2 x 8; sides and top from 2 x 6's. Make frame ½" larger in height and width than *Thermopane* unit.

2

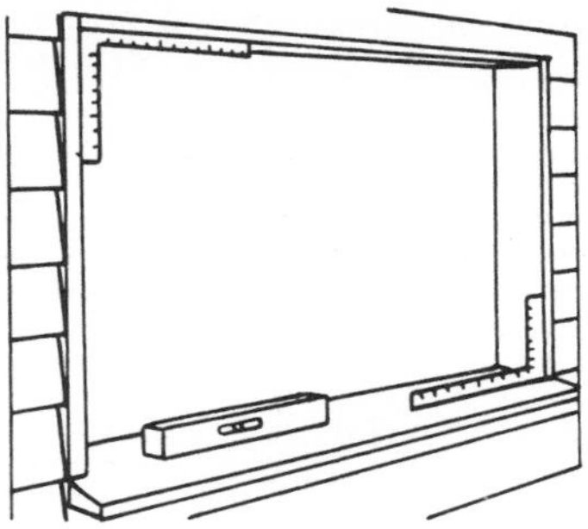

Install frame in opening and make sure sill is flat and level. Next, be sure rest of frame is square and plumb with the sill.

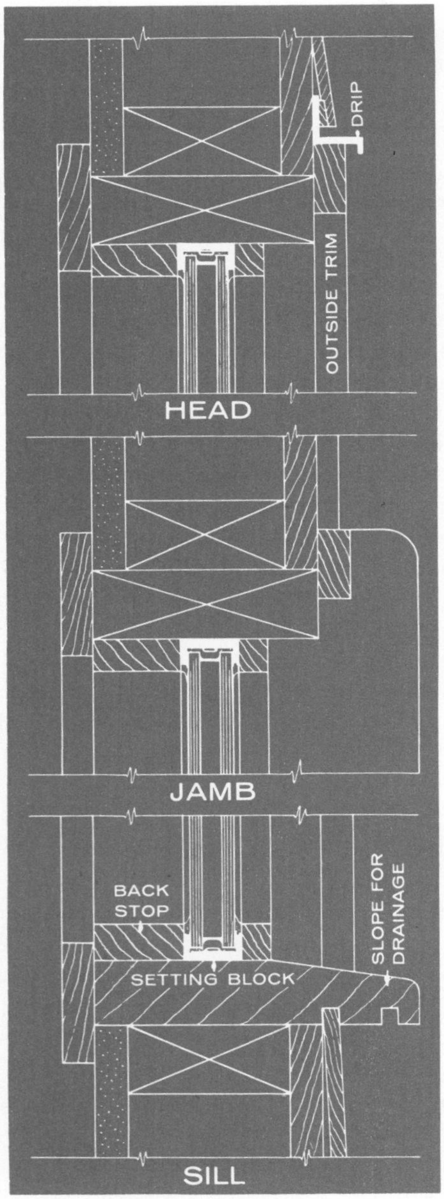

3

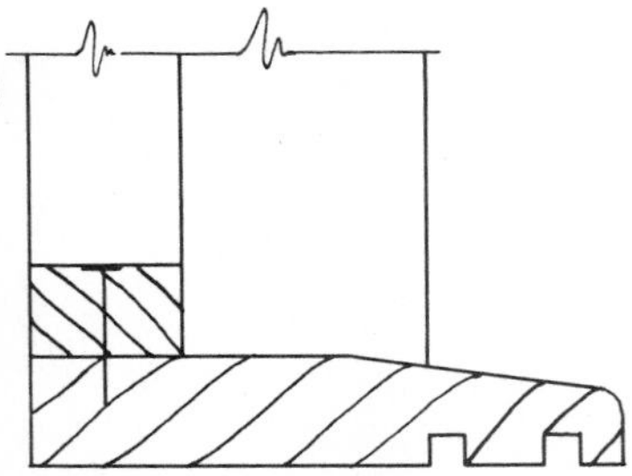

Nail back stop in place. See chart (page 5) for proper depth. Mitered corners make the best job.

4

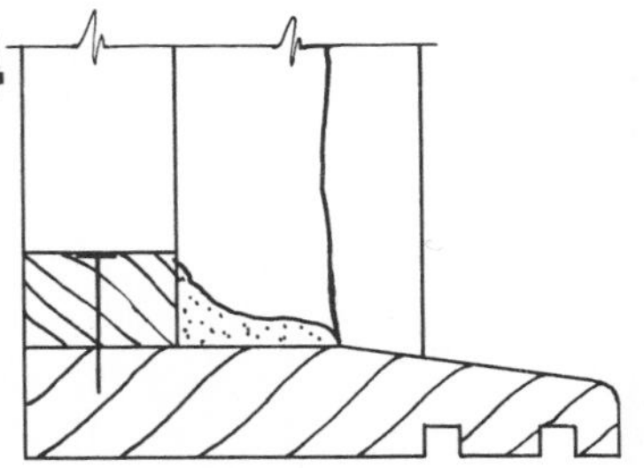

Using a good knife-grade glazing compound, apply a generous bed to back stop, a full bed on bottom of sash and enough at sides and top to make a weather-tight seal. DO NOT USE PUTTY.

5

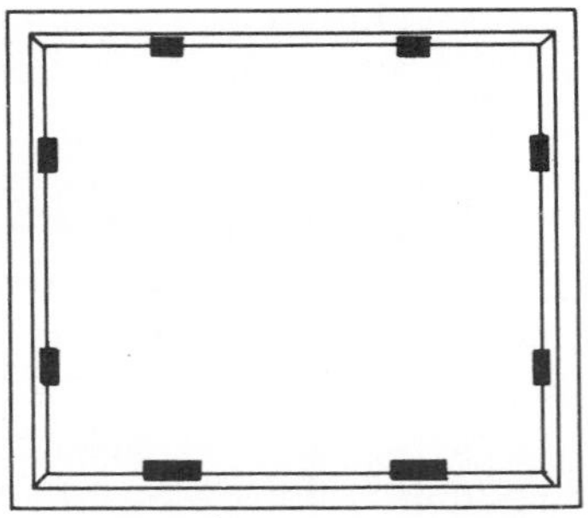

Snap two neoprene setting blocks on bottom of *Thermopane* unit one quarter of total width in from each corner. Next, snap into place sufficient neoprene spacer strips on sides and top of unit to insure uniform setting.

6

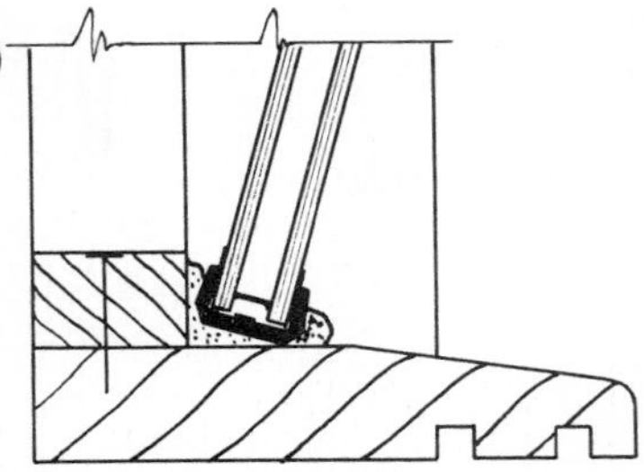

Place bottom edge of *Thermopane* in position on sill and press unit until the sides of setting blocks and spacer strips rest firmly against the back stops.

7

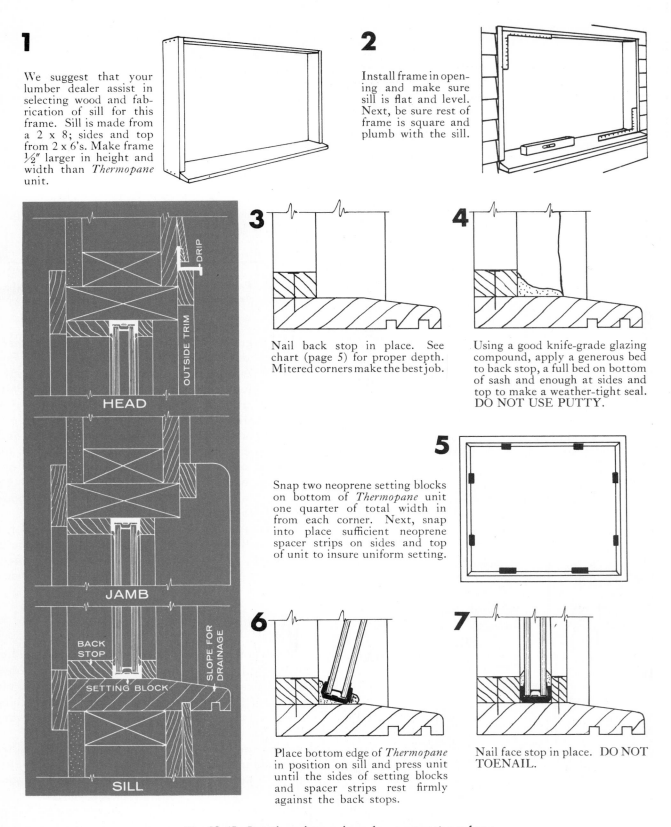

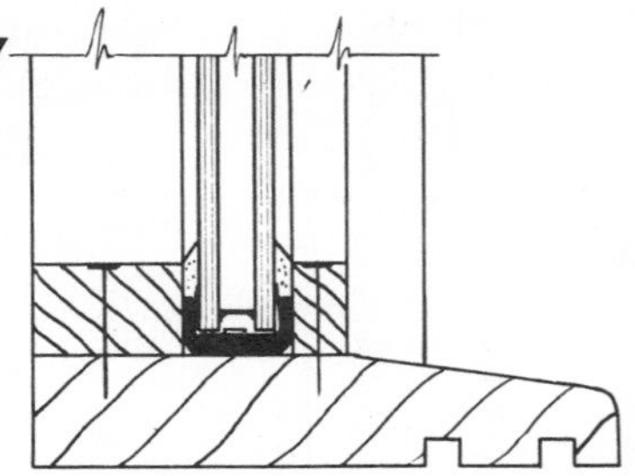

Nail face stop in place. DO NOT TOENAIL.

Fig. 11-45. Details and procedures for constructing a frame and installing insulating glass. (Libby-Owens-Ford Glass Co.)

2 panes of ¼" polished Parallel-O-Plate ½" air space. 6.5 lbs. per sq. ft.

Width	Height
33"	x 76¾"
35½"	x 36"
35½"	x 48⅛"
35½"	x 60⅜"
42"	x 66"
42"	x 72"
44½"	x 36"
44½"	x 46"
44½"	x 48⅛"
44½"	x 60⅜"
44½"	x 72¾"
45"	x 76¾"
45⅜"	x 52"
46⅛"	x 52½"
47¾"	x 50⅜"
48"	x 48"
48"	x 60"
48½"	x 42"
48½"	x 46"
48½"	x 50"
48½"	x 58"
50⅜"	x 47¾"
50⅜"	x 60⅛"
55¼"	x 36"
55¼"	x 48⅛"
55¼"	x 60⅜"
56½"	x 42"
56½"	x 46⅛"
56½"	x 50"
56½"	x 58⅛"
56½"	x 66"
57"	x 76¾"
58⅛"	x 52½"
64½"	x 46"
64½"	x 50"
64½"	x 58"
64½"	x 66"
66⅝"	x 47¾"
66⅝"	x 60⅛"
68¾"	x 36"
68¾"	x 48⅛"
68¾"	x 60⅜"
68¾"	x 72¾"
70⅛"	x 52½"
70⅛"	x 56½"
72"	x 48"
72"	x 60"
72½"	x 46"
72½"	x 50"
72½"	x 58"
72½"	x 66"
75"	x 36"
75"	x 48⅛"
75"	x 60⅜"
80½"	x 50"
80½"	x 58"
84"	x 66"
84"	x 72"
93"	x 36"
93"	x 48⅛"
93"	x 60⅜"
93"	x 72¾"
96"	x 66"
96"	x 72"
96½"	x 50"
96½"	x 58"
116½"	x 58"

2 panes of DSA window glass ¼" air space. 3.25 per sq. ft.

21¾" x 62¾"	42½" x 22½"
25¾" x 62¾"	45½" x 25½"

2 panes of 3/16" "A" heavy sheet glass with ½" air space. 5 lbs. per sq. ft.

35½" x 60⅜"	56½" x 46⅛"
48½" x 42"	56½" x 50"
48½" x 50"	56½" x 42"

Fig. 11-46. Standard sizes of insulating glass. (Libby-Owens-Ford Glass Co.)

Fig. 11-47. Framing details for insulating glass in "window wall" construction.

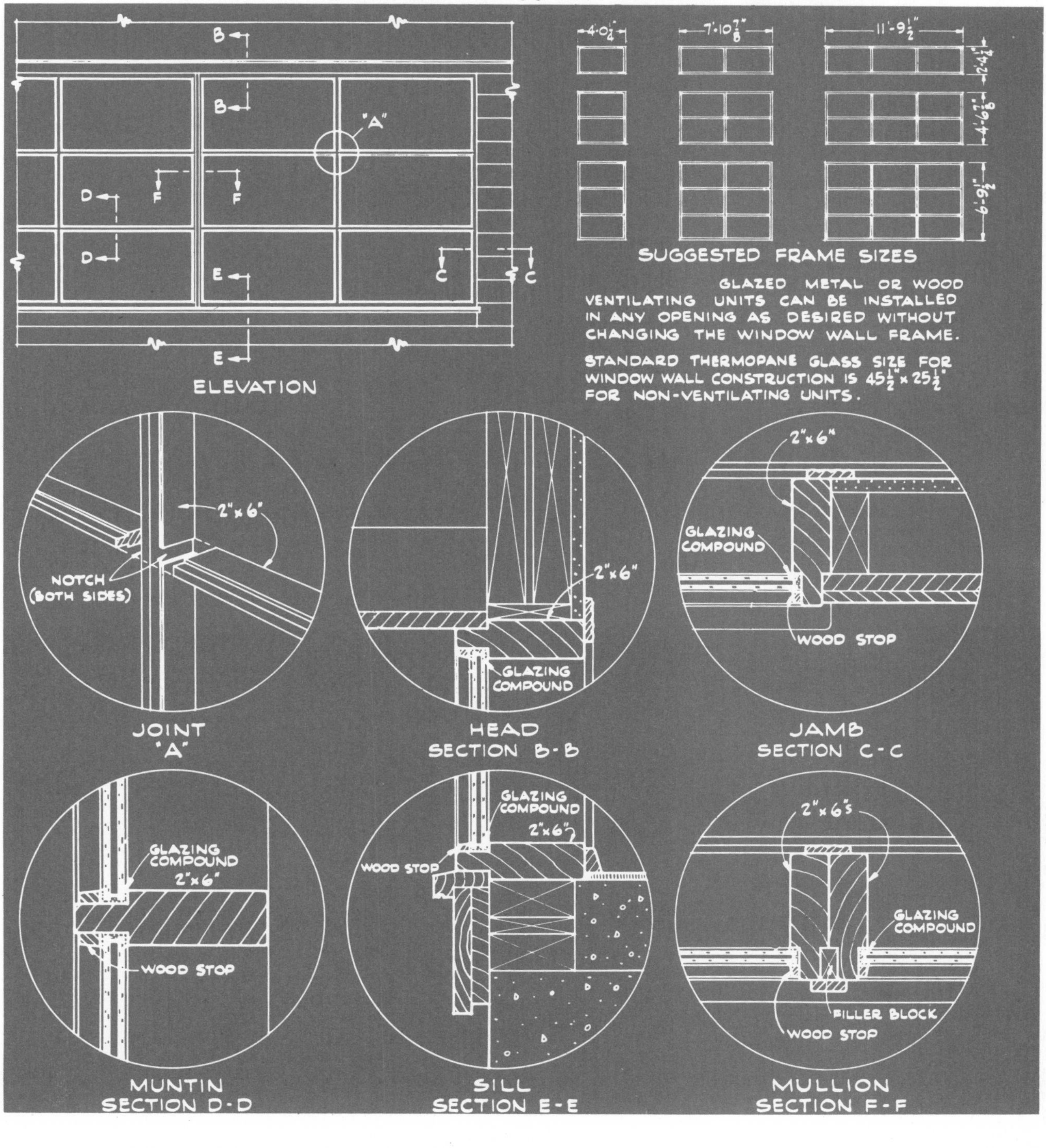

Fig. 11-48. Interior view of a window wall. (Andersen Corporation)

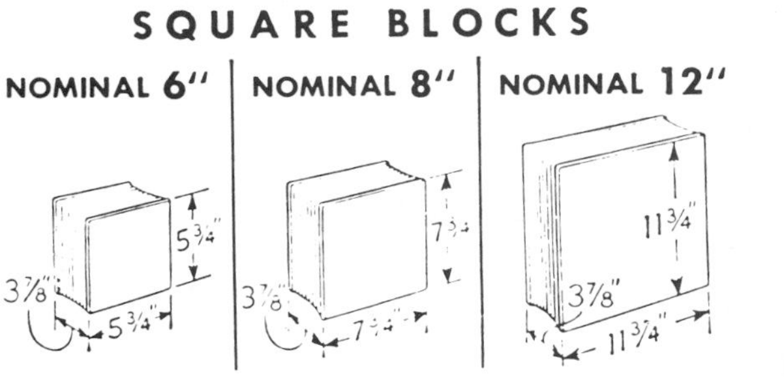

Fig. 11-49. Nominal and actual sizes of commonly used glass block.

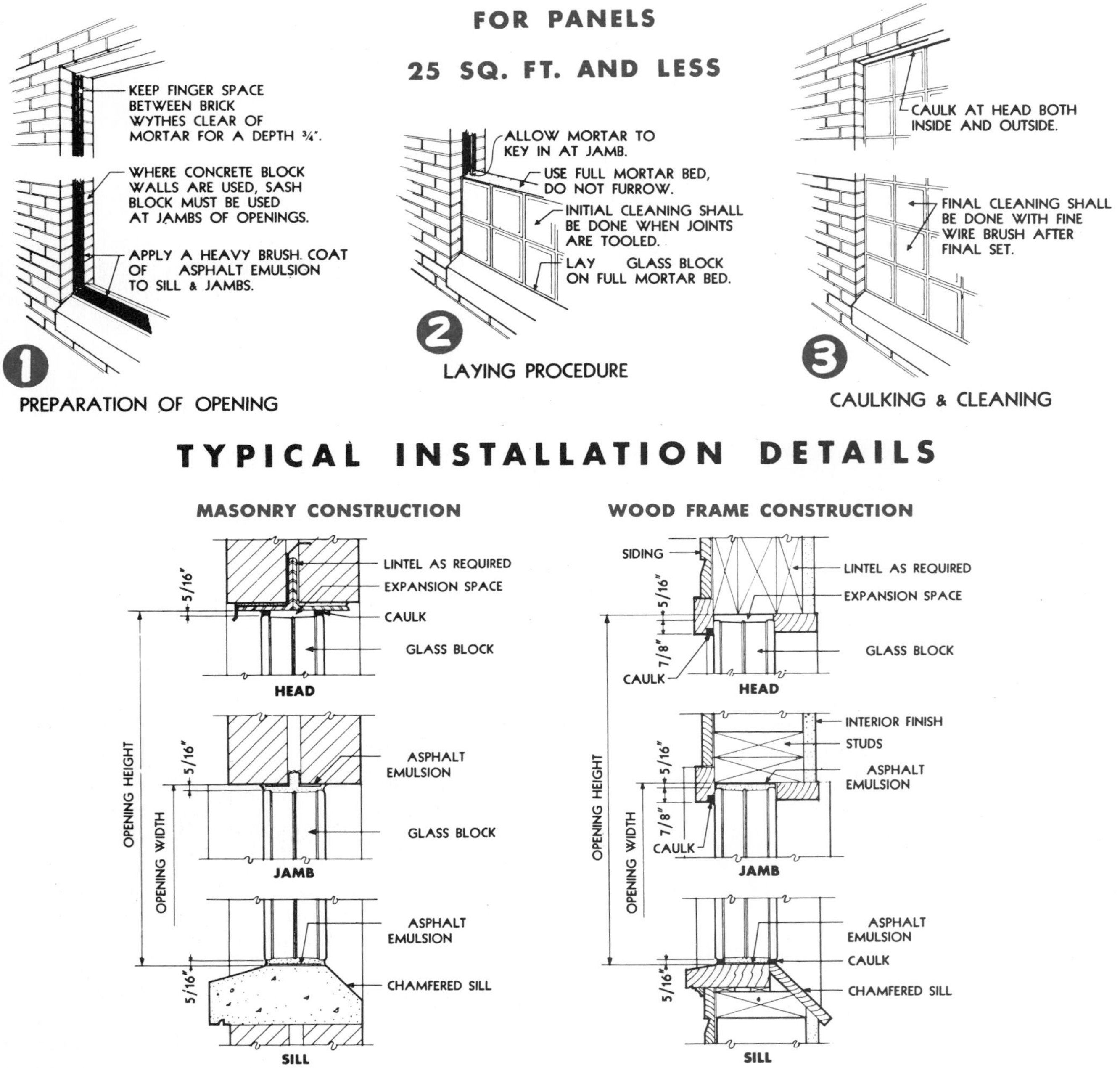

Fig. 11-50. Details and installation procedures for glass block panels of 25 sq. ft. or less. (PPG Industries, Inc.)

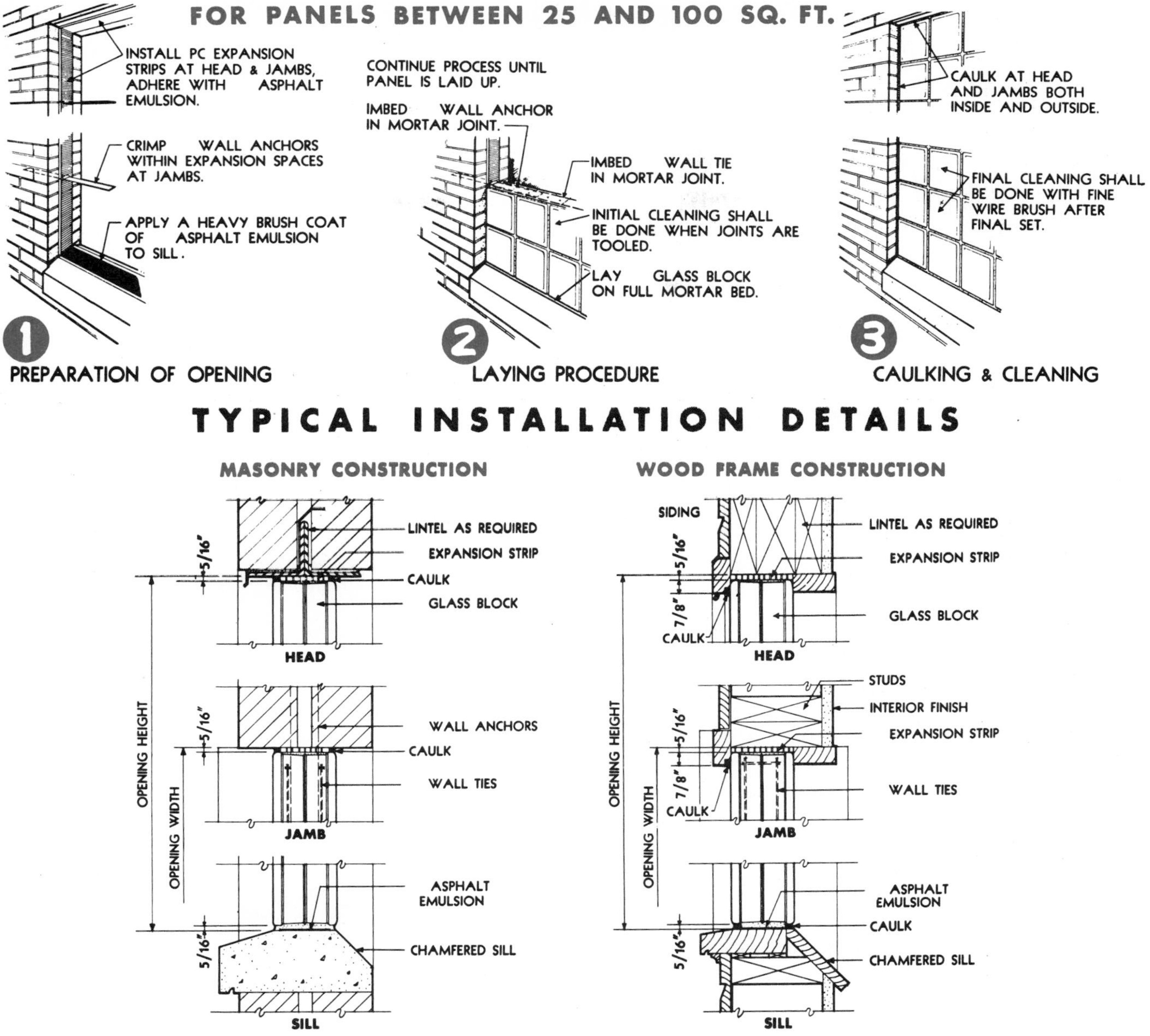
FOR PANELS BETWEEN 25 AND 100 SQ. FT.
INSTALL PC EXPANSION STRIPS AT HEAD & JAMBS, ADHERE WITH ASPHALT EMULSION.
CRIMP WALL ANCHORS WITHIN EXPANSION SPACES AT JAMBS.
APPLY A HEAVY BRUSH COAT OF ASPHALT EMULSION TO SILL.
1
PREPARATION OF OPENING
CONTINUE PROCESS UNTIL PANEL IS LAID UP.
IMBED WALL ANCHOR IN MORTAR JOINT.
IMBED WALL TIE IN MORTAR JOINT.
INITIAL CLEANING SHALL BE DONE WHEN JOINTS ARE TOOLED.
LAY GLASS BLOCK ON FULL MORTAR BED.
2
LAYING PROCEDURE
CAULK AT HEAD AND JAMBS BOTH INSIDE AND OUTSIDE.
FINAL CLEANING SHALL BE DONE WITH FINE WIRE BRUSH AFTER FINAL SET.
3
CAULKING & CLEANING
TYPICAL INSTALLATION DETAILS
MASONRY CONSTRUCTION
LINTEL AS REQUIRED
EXPANSION STRIP
CAULK
GLASS BLOCK
5/16"
HEAD
WALL ANCHORS
CAULK
WALL TIES
5/16"
JAMB
ASPHALT EMULSION
CHAMFERED SILL
5/16"
SILL
OPENING HEIGHT
OPENING WIDTH
WOOD FRAME CONSTRUCTION
SIDING
LINTEL AS REQUIRED
EXPANSION STRIP
GLASS BLOCK
5/16"
7/8"
CAULK
HEAD
STUDS
INTERIOR FINISH
EXPANSION STRIP
WALL TIES
5/16"
7/8"
CAULK
JAMB
ASPHALT EMULSION
CAULK
CHAMFERED SILL
5/16"
SILL
OPENING HEIGHT
OPENING WIDTH

Fig. 11-51. Details and installation procedures for glass block panels of 25 to 100 sq. ft.

Fig. 11-49. All are 3 7/8 in. thick. Special shapes are available for turning corners and for building curved panels. The blocks come in both light-diffusing and light-directing types.

Installing glass blocks is not difficult and regular masonry tools are used. Even though the carpenter will seldom make the actual installation, he will be required to build the framework and should have a basic understanding of the design requirements.

Installing Small Glass Block Panels

Installation details for glass block panels of 25 sq. ft. and less are given in Fig. 11-50. In such panels the height should not exceed 7 ft. and the width 5 ft.

Panels may be supported by a "mortar key" at jambs in masonry, or by wood members in frame construction (no wall anchors required). No wall ties are required in the joints and expansion space is required at the head only.

Installing Large Glass Block Panels

When panels exceed 25 square feet of area, additional requirements must be observed. See Fig. 11-51. Expansion strips (strips of resilient material) are used to partially fill expansion spaces at jambs and head of larger panel openings.

Panels should never be larger than 10 ft. wide or 10 ft. high and must be supported at jambs by use of wall anchors or wood members. A portion of each anchor is embedded in masonry and in

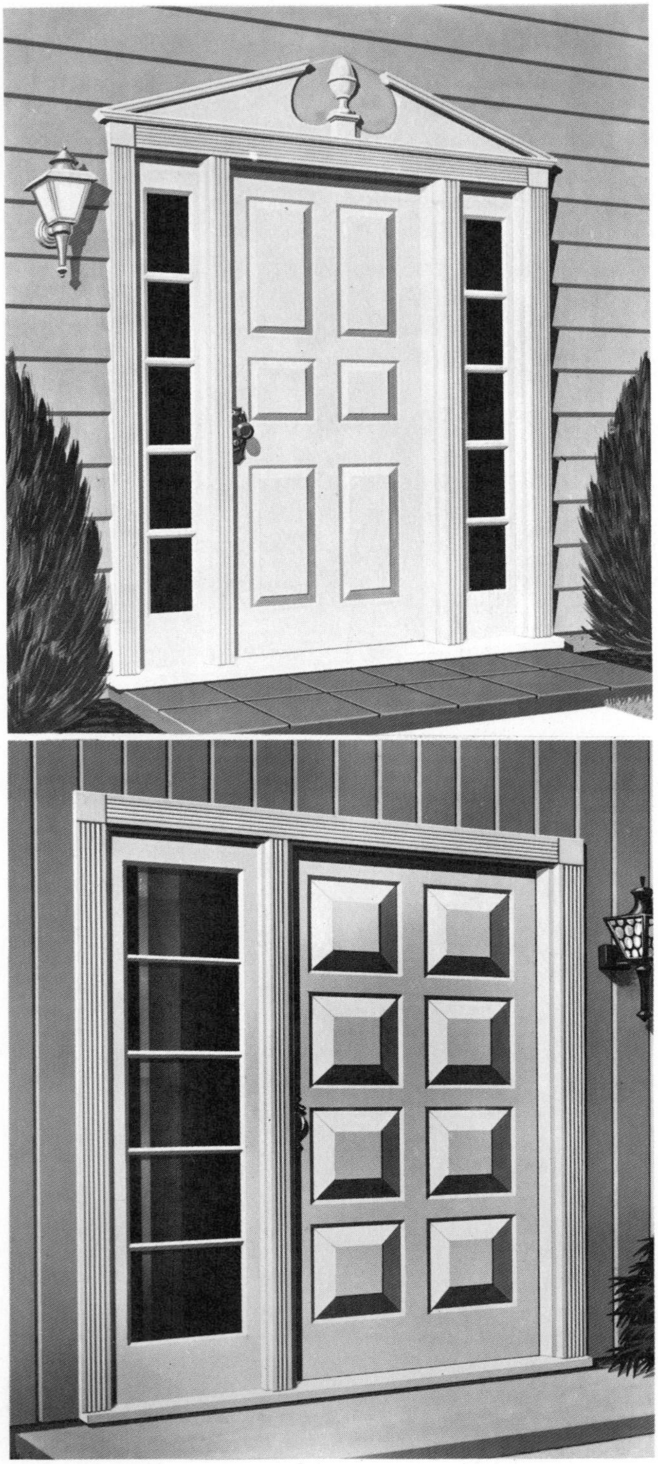

Fig. 11-52. Residential Entrances. Above. Traditional design with sidelights and reeded pilasters. Below. Contemporary styling with single sidelight. (C-E Morgan)

glass block mortar joint. They should be crimped within expansion spaces and spaced on 24 in. centers to occur in same joints as wall ties. Anchors are corrosion-resistant, 2 ft. long and 1 3/4 in. wide.

Wall ties should be installed on 24 in. centers in horizontal mortar joints of larger panels and lap not less than 6 in. whenever necessary to use more than one length of tie. Do not bridge expansion spaces. Ties are corrosion-resistant, 8 ft. long and 2 in. wide.

Openings for Glass Blocks

To determine heights or widths of openings required for panels, multiply the number of units by the nominal block size and then add 3/8 in. For example: what size opening would be required for a panel consisting of 8 in. blocks that was 4 units wide by 5 units high?

Width -- 4 x 8 + 3/8 = 32 3/8 = 2' - 8 3/8"
Height -- 5 x 8 + 3/8 = 40 3/8 = 3' - 4 3/8"

Exterior Door Frames

Outside door frames are installed at the same time as windows, following similar procedures. Secondary and service entrances usually consist of simple frames and trim members that match the windows. Main entrances however, often contain additional elements that add an important architectural feature. See Fig. 11-52.

Exterior doors in residential construction are nearly always 6'-8" high, although 7'-0" sizes are available. Main entrances usually are equipped with a single door that is 3'-0" wide, with 2'-8" and 2'-6" sizes being used for rear doors and service doors. FHA Minimum Property Standards list a minimum exterior door width of 2'-6".

Outside door frames, like windows, consist of heads, jambs, and sills. The head and jambs are made of 5/4 in. stock since they must carry not only the main door but also screen and storm doors. Door frames are shown in the elevation views of the architectural plans with the door in place, Fig. 11-53.

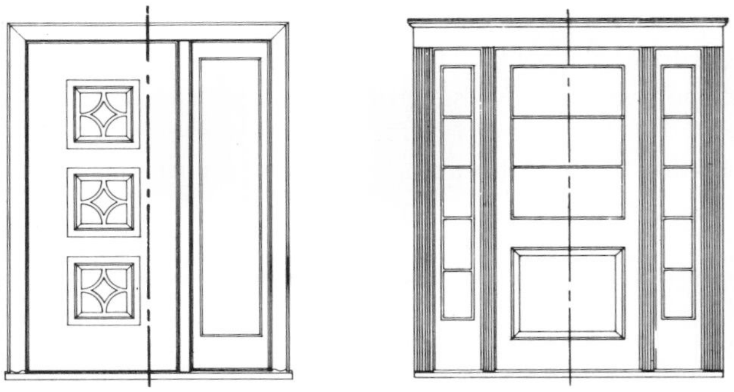

Fig. 11-53. Entrance frames and doors as they appear in the elevations of the architectural plans.

Fig. 11-54. Assembling entrance door frames at a millwork plant. (Jordan Millwork Co.)

Door frames are manufactured at a millwork plant, Fig. 11-54, and arrive at the building site, either assembled and ready to install or disassembled (knocked down, K.D.). Sometimes K.D. units are assembled in the shop of the dealer or distributor. It is relatively easy to assemble door frames "on the job" when the joints are accurately machined and the parts are carefully packaged and marked.

Details of a given frame may vary, but the general construction is the same, Fig. 11-55. The head and jambs are rabbeted, usually 1/2 in. deep, to receive the door. In residential construction, outside doors swing inward and the rabbet must be located on the inside. Stock door frames are designed for standard wall framing but can also be adapted to stone or brick veneer construction as shown in Fig. 11-56.

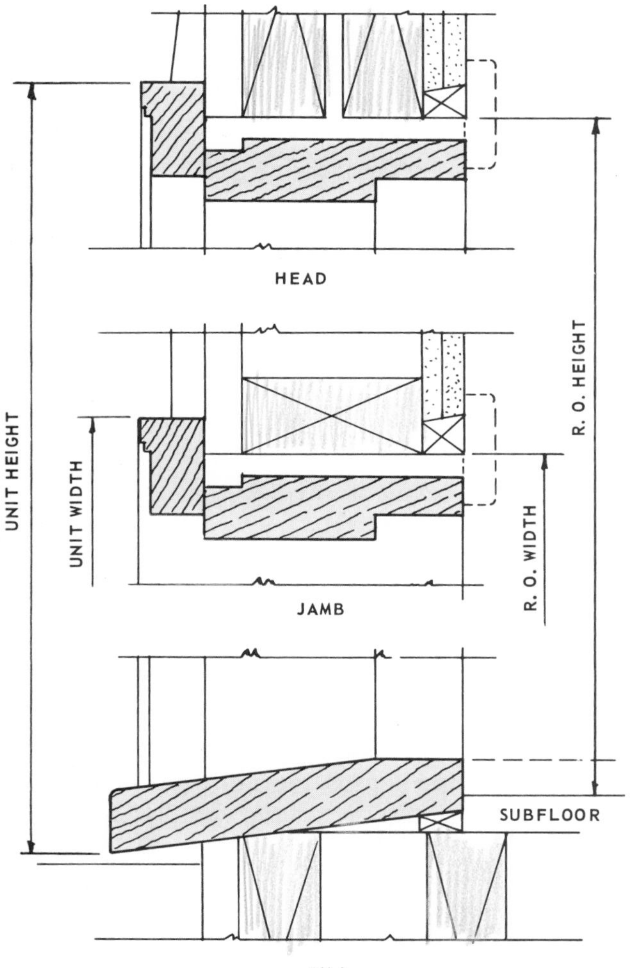

Fig. 11-55. Exterior door frame details and sizes. For most frames the following calculations can be used:
Unit Height = Door + 4 1/2 in.
Unit Width = Door + 4 in.
R.O. Height = Door + 2 1/2 in.
R.O. Width = Door + 2 1/2 in.

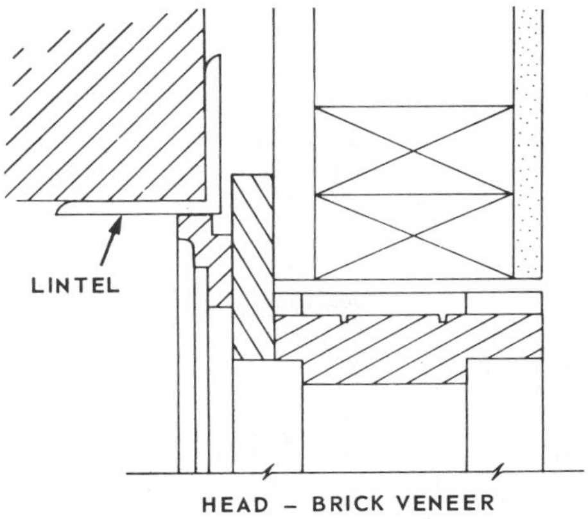

Fig. 11-56. Head detail for brick veneer construction.

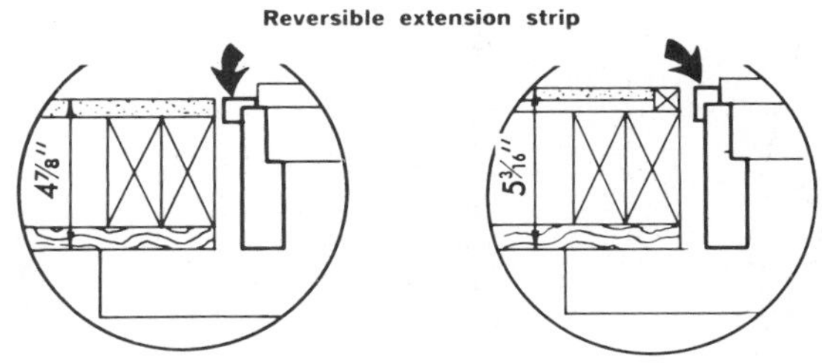

Fig. 11-57. Converting 4 5/8 in. jambs to either 4 7/8 in. or 5 3/16 in. with a reversible extension strip. (C-E Morgan)

Standard stock frames can be fitted with extension strips, Fig. 11-57, that quickly convert them to the requirements of a given wall thickness. By installing the same strip in a different position, several widths can be attained. This procedure is also used on window frames.

Sill design varies considerably, however the top is always level with the surface of the finished floor. Sills may be made of wood, metal, stone, or concrete. The outside surface at the entrance is placed just below the door sill or lowered by a standard riser height (7 1/2 in.), Fig. 11-58.

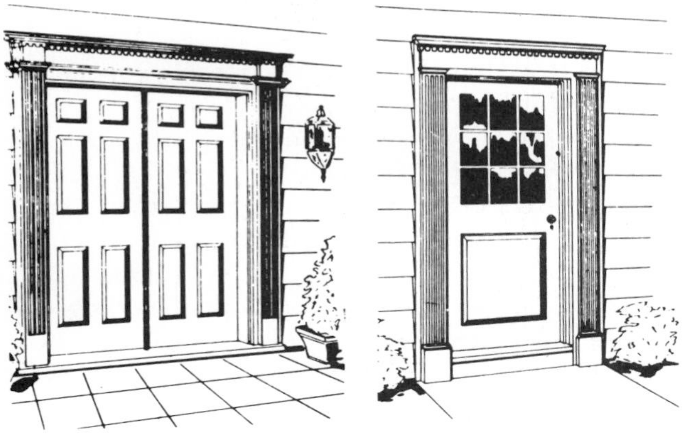

Fig. 11-58. Door sill and outside surface. Left. Located just below sill. Right. Located one standard riser height below top of sill.

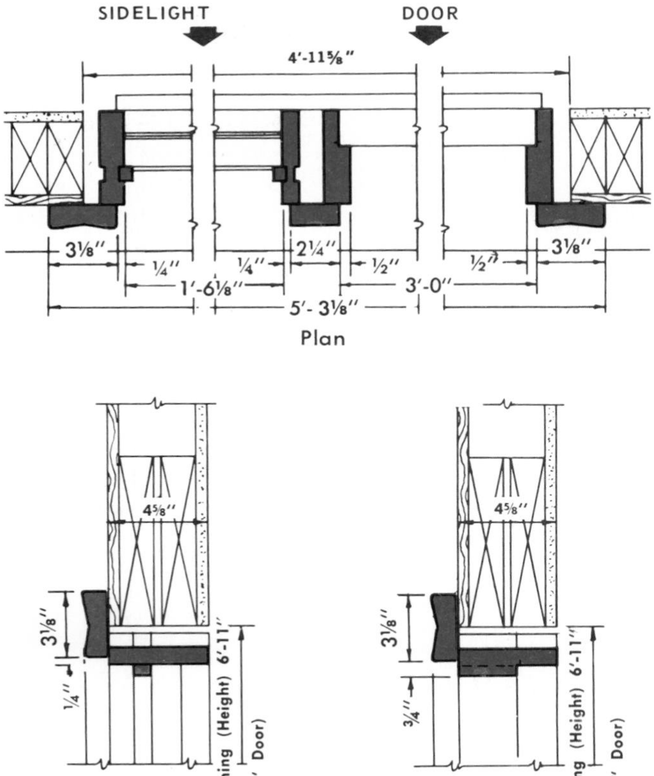

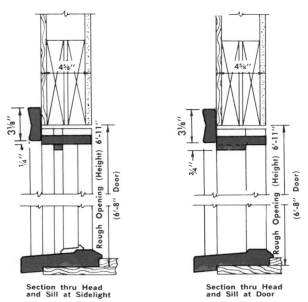

Fig. 11-59. Details of a front entrance door frame with side-light. (C-E Morgan)

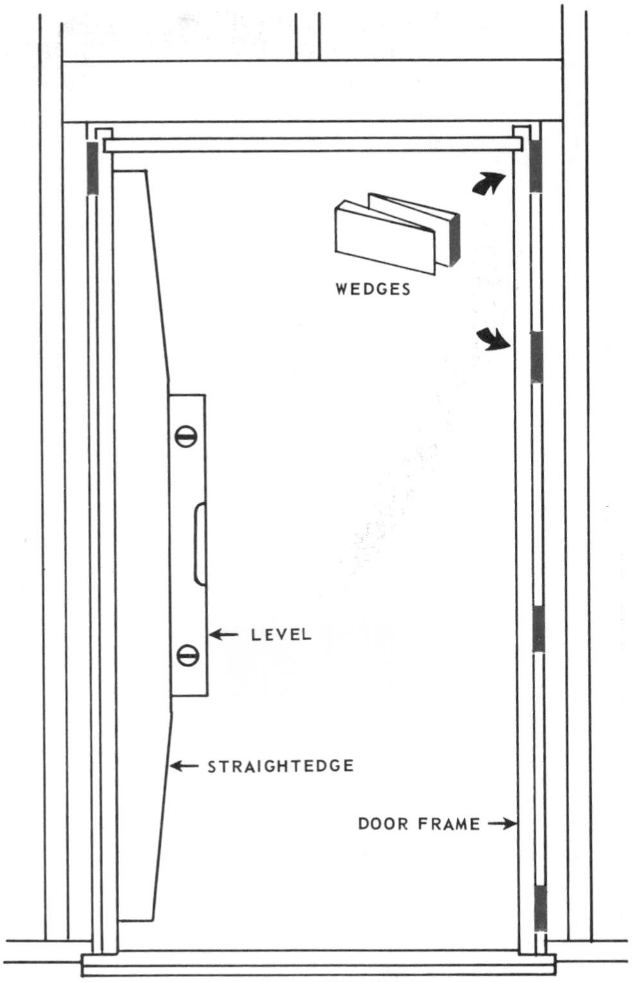

Fig. 11-60. Plumbing door jambs.

Positioning a standard door sill so it will be level with the finished floor requires the removal of a section of the rough floor and some trimming of the top edge of the floor joist. This is done at the time the frame is installed. The framing of the rough opening (R.O.), however, comes early in the construction stages before the door frame is delivered to the job and the carpenter will need to check the working drawings carefully. The size of the rough opening will usually be included in the door and window schedule. The height of the opening is shown in detail sections. For standard construction, the R.O. can be calculated by adding about 2 1/2 in. to the door width and height.

When this information is not included in the working drawings, you should consult the manufacturer's descriptive literature which will contain not only R.O. requirements but also detail drawings. This is especially important for front entrances that consist of more complicated structures and fixed sidelight window units as shown in Fig. 11-59.

Reversible extension strip for converting 4⅝" jambs to 4⅞".

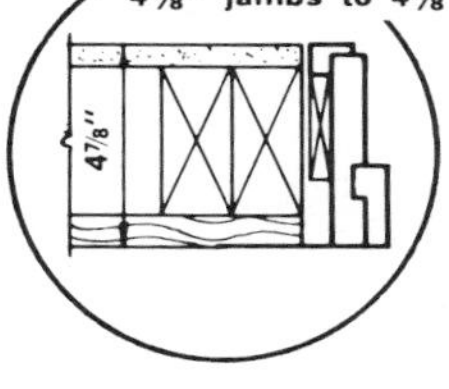

Reversible extension strip for converting 4⅝" jambs to 5$\frac{3}{16}$".

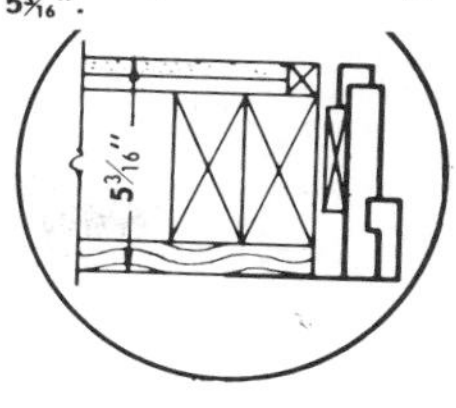

DIMENSIONS	M-48A,	M-49A
Jamb Opening Width	5'-4½"	5'-4½"
Jamb Opening Height	6'-8"	6'-8½"
Width (overall)	5'-11½"	5'-11½"
Height (overall)	7'-1½"	7'-0"
Rough Opening Width	5'-7½"	5'-7½"
Rough Opening Height	6'-11"	6'-9½"

FOR ONE JAMB OPENING 5'-4½" x 6'-8"

Elevation

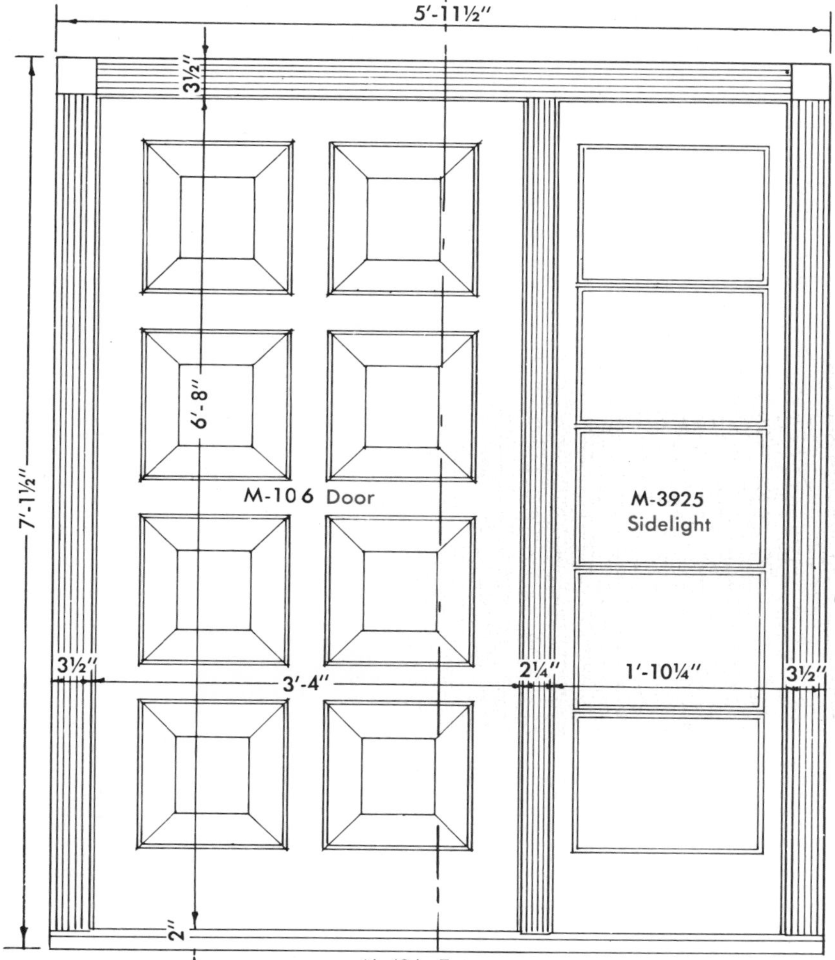

M-48A Entrance Elevation

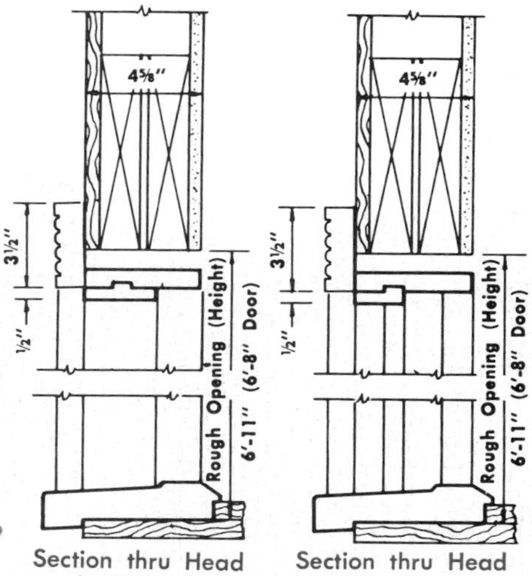

Section thru Head and Sill at Door

Section thru Head and Sill at Sidelight

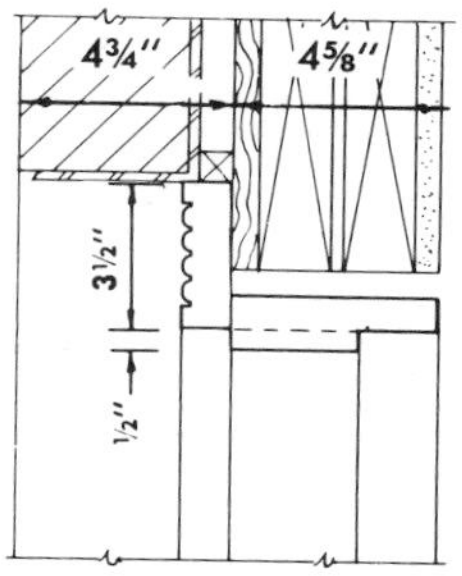

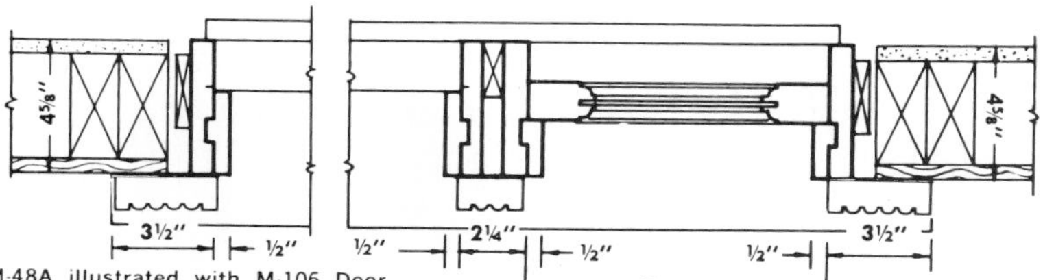

Entrance M-48A illustrated with M-106 Door and M-3925 Sidelight. Various combinations of MORGAN Door and Sidelight Designs can be used . . . see door catalog. M-48A Entrance will accommodate 3'-4" Door with 1'-10" Sidelight or 3'-0" Door with 2'-2" Sidelight and consists of basic M-48 Entrance and one M-48ML pre-assembled mullion, top and bottom spacer blocks and sidelight stops.

PLAN — M-48A, *M-49A
*WITHOUT SILL

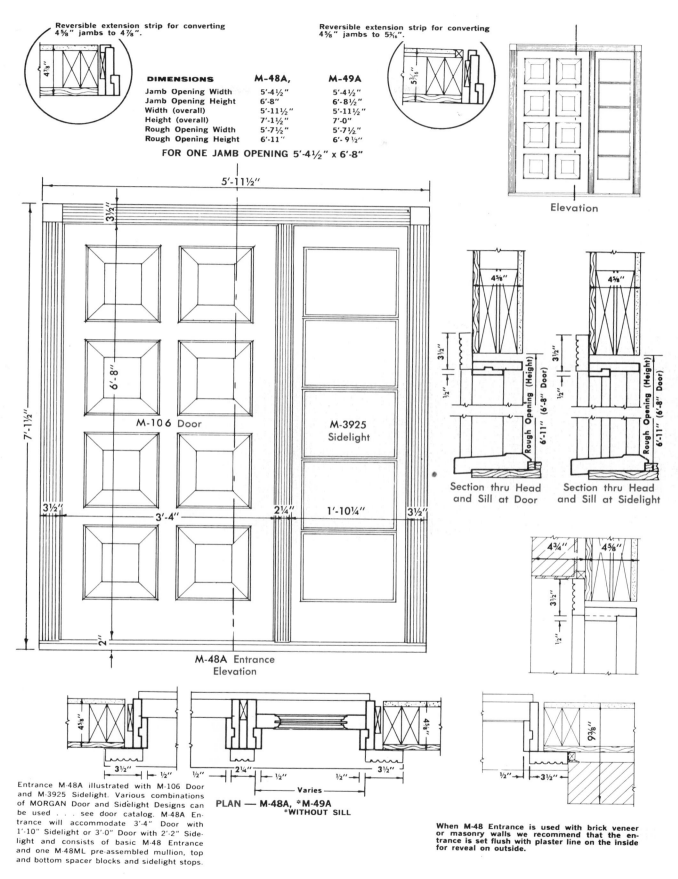

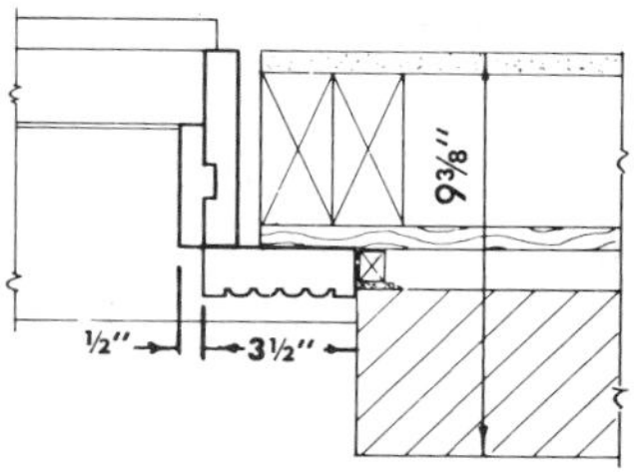

When M-48 Entrance is used with brick veneer or masonry walls we recommend that the entrance is set flush with plaster line on the inside for reveal on outside.

Fig. 11-61. Complete details of an entrance consisting of single sidelight.
(C-E Morgan)

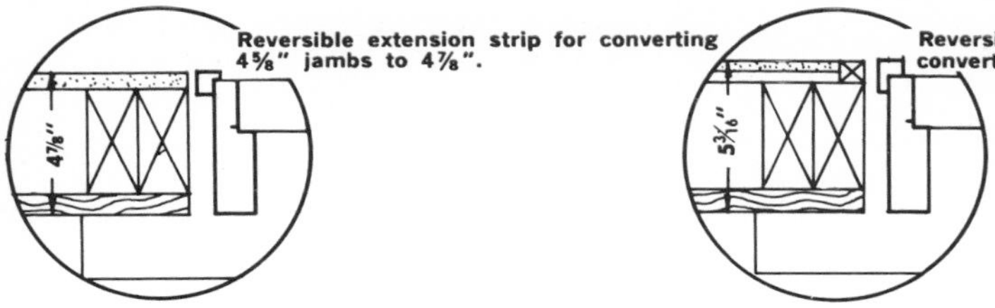

DIMENSIONS	6'-8" DOORS	7'-0" DOORS
Rough Opening	5'-4" x 6'-11"	5'-4" x 7'-3"
Width at Casing	5'-10¾"	5'-10¾"
Width (overall)	6'-2⅛"	6'-2⅛"
Height (overall)	7'-1⅟₁₆"	7'-5⅟₁₆"
Sidelight Sash	10" x 4'-4", 1⅜"	10" x 4'-4", 1⅜"

Rough opening height is taken from top of rough floor.

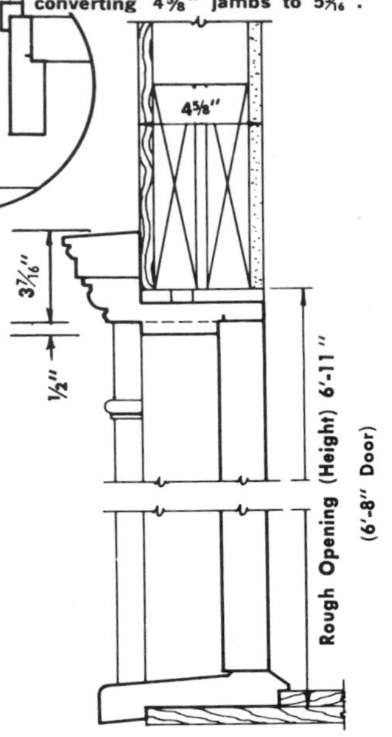

SECTION THRU HEAD & SILL

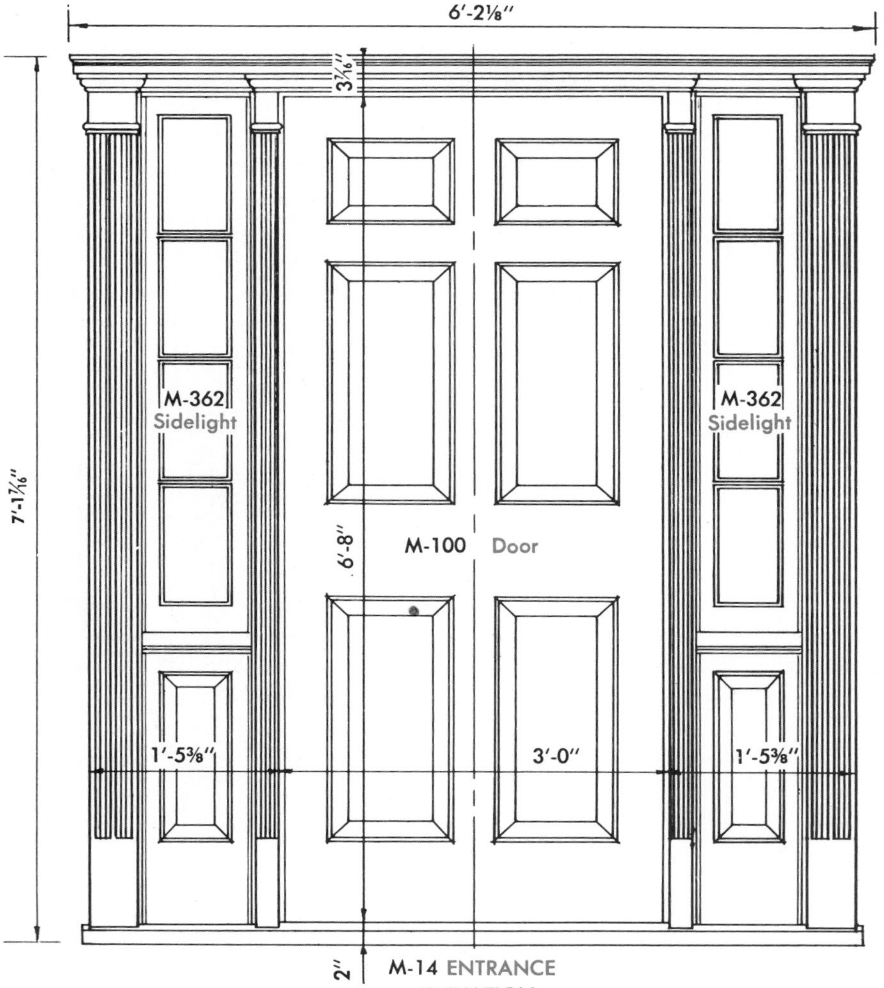

M-14 ENTRANCE ELEVATION

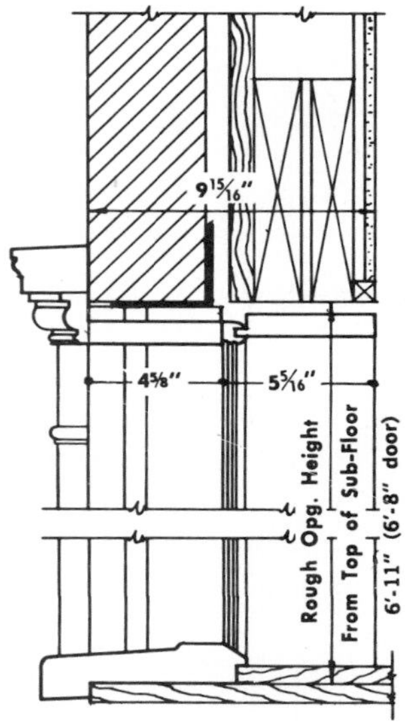

For Brick Veneer or masonry construction use M-10 Extension Jamb.

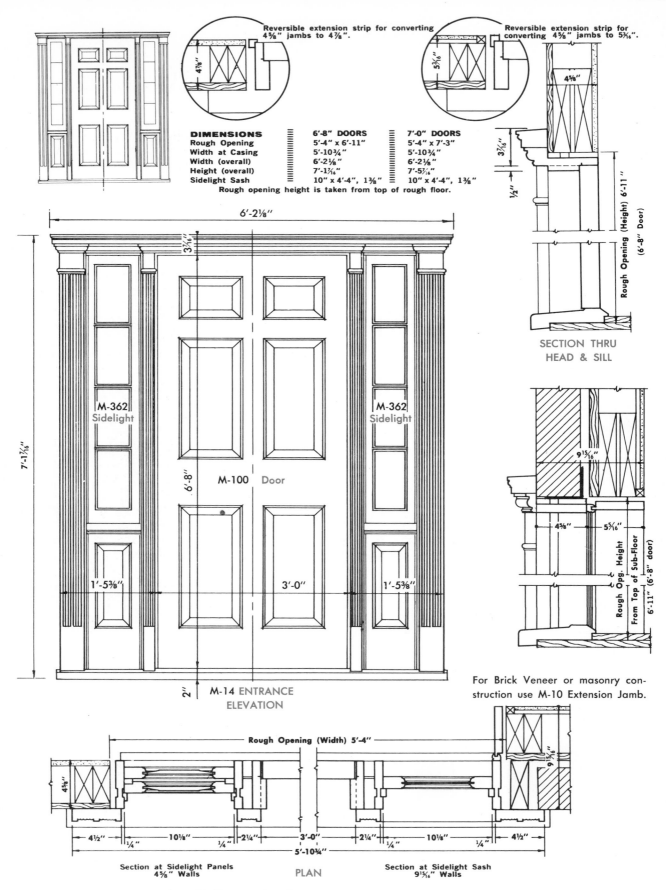

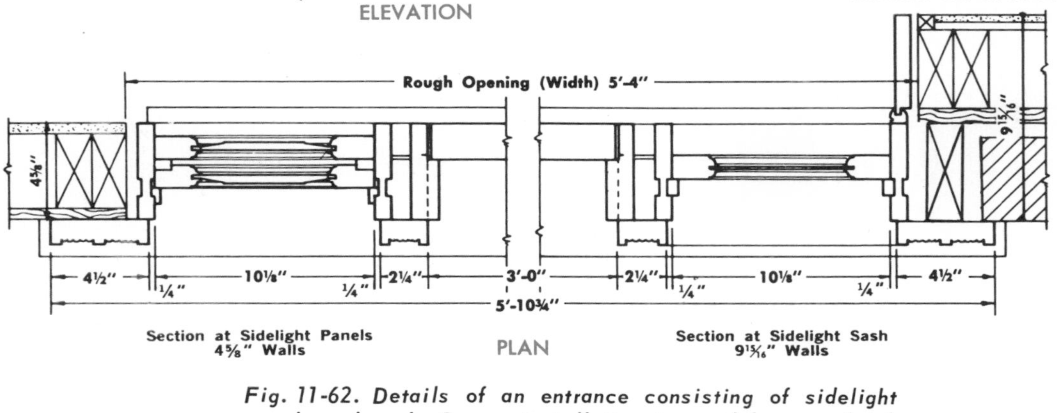

Section at Sidelight Panels 4⅝" Walls

PLAN

Section at Sidelight Sash 9¹⁵⁄₁₆" Walls

Fig. 11-62. Details of an entrance consisting of sidelight panels and sash. Covers installation in wood frame or brick veneer construction. (C-E Morgan)

Installing Door Frames

Check the size of the rough opening to make sure that proper clearances have been provided. Cut out sill area, if necessary, so the top of the sill will be the correct distance above the rough floor. In some structures it may be necessary to install flashing over the bottom of the opening.

Place the frame in the opening, center it horizontally, and secure it with a temporary brace. Using blocking and wedges, level the sill and bring it to the required height. Be sure the sill is well supported. For masonry walls and slab floors the sill is usually placed on a bed of mortar.

With the sill level, drive a nail through the casing into the wall frame at the bottom of each side. Insert blocking or wedges (wood shingles are often used) between the studs and the top of the jambs and adjust until they are plumb. Use a level and straightedge as shown in Fig. 11-60.

HANDY SAYS:

"When setting door and window frames never drive any of the nails completely into the wood until all nails are in place and a final check has been made to make sure that no adjustments are necessary."

Place additional wedges between the jambs and stud frame in the approximate location of the lock strike plate and hinges. Adjust the wedges until the side jambs are well supported and straight. Then, secure the wedges by driving a nail through the jamb, wedge, and into the stud. Finally, nail the casing in place with nails spaced 16 in. O.C. following the same precautions suggested for window frame installation.

After the installation is complete, a piece of 1/4 or 3/8 in. plywood should be lightly tacked over the sill to protect it during further construction work. At this time, many builders prefer to hang a temporary combination door in order that the interior of the structure can be secured, thus providing a place to store tools and materials.

Setting the threshold and hanging the door is a part of the interior finishing operation and will be described in Unit 17. Exterior door types, designs, and sizes are included in the same Unit. When a prehung door unit is installed the door should be removed from its hinges and carefully stored.

Figs. 11-61 and 11-62 provide examples of drawings furnished by one manufacturer showing the construction and installation details of entrance units.

Sliding Glass Doors

With todays emphasis on outdoor living, a terrace or patio door is often included in residential designs. The French or casement type door, formerly used for this purpose, has been largely replaced with the modern sliding glass door.

This type of door, riding on nylon or stainless steel rollers, is easy to operate. When equipped with quality weather stripping and insulating glass, it restricts heat loss and condensation to a level where satisfactory results are attained even in

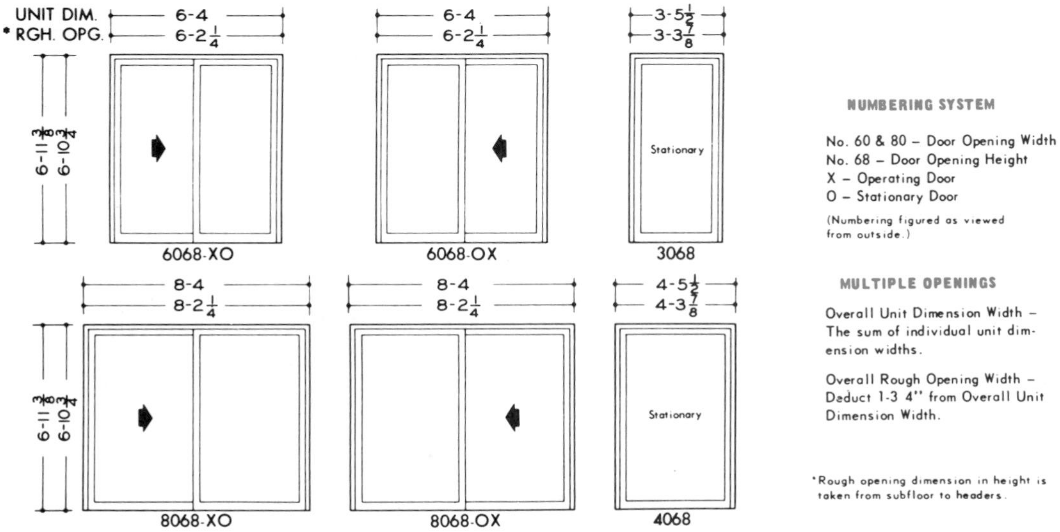

Fig. 11-63. Standard sliding glass door units and sizes. (Andersen Corporation)

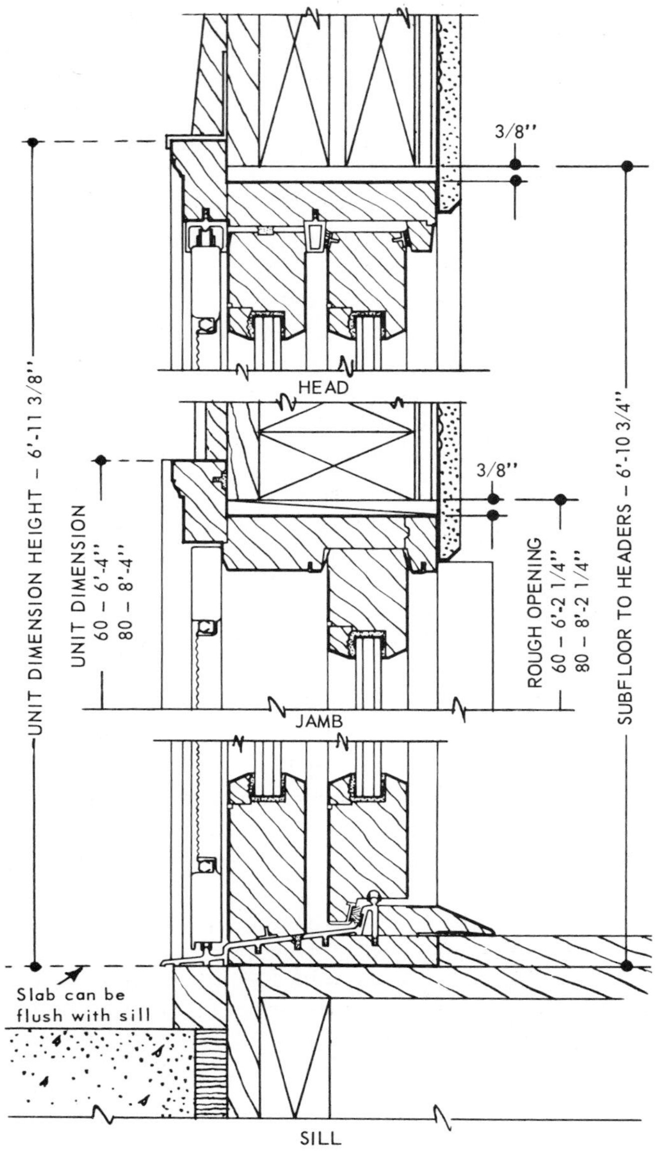

Fig. 11-64. Construction and installation details of a sliding glass door. (Andersen Corporation)

cold climates. It is available as a factory assembled frame and door unit with parts and installation details that are similar to sliding windows.

Sliding glass door units contain at least one fixed and one operating panel. Some may have three or four panels. The type of unit is commonly designated by the number and arrangement of the panels -- as viewed from the outside. The letter X denotes an operating panel and the letter O a fixed panel as shown in Fig. 11-63.

Construction details, Fig. 11-64, will vary from one manufacturer to another, as will unit sizes and rough opening requirements. If detailed drawings or R.O. sizes are not included in the architectural plans, then the carpenter should secure the information from the manufacturer's literature.

The installation of sliding door frames is similar to the procedure described for regular outside doors. Before setting the frame in place a bead of sealing compound should be laid across the opening to insure a weather-tight joint. If heavy glass doors are to slide properly the sill must be level and straight. Side jambs are plumbed and held with wedges in the same way as regular door frames. After carefully checking, complete the installation by driving weatherproof nails through side and head casings into structural frame members, spacing them not more than 16 in. O.C. See Fig. 11-65.

After the frame is installed the opening can be enclosed with plywood, or the doors may be installed and then covered with a polyethylene film or other material to protect the glass, sash, and frame from damage. Manufacturers include instructions for making the door installation and it is important to read and study this information carefully BEFORE proceeding with the work.

HANDY SAYS:

"After installing large glass units in buildings under construction, it is considered good practice to place a large X on the glass, using masking tape or washable paint. This will remind workers that the glass is in place, so they will not walk into it or damage it with tools and materials."

Garage Doors

Basically there are three types of garage doors; hinge or swinging, swing-up, and roll-up. As a result of refinement and perfection in the design of hardware and counterbalancing equipment, the two latter types have all but replaced hinged doors. See Fig. 11-66.

Today, there are a number of companies that specialize in the manufacture of garage doors. Their products are carefully engineered and made of such materials as wood, steel, aluminum and fiberglass. Many different designs are available to blend with contemporary or traditional architecture. Some sections may be fitted with glass panels to provide interior natural light.

Sizes

FHA specifies a minimum width of 8 ft. and a height of 6 ft.-4 in. for a single residential garage door, but a width of a 9 ft. is commonly

Fig. 11-65. Installing a sliding glass door frame. Using a straightedge and wedges to check and straighten jambs.

used today. A two-car or double-door is usually 16 ft. wide. Standard heights for either single or double doors are 6 ft. - 8 in. or 7 ft. Stock sizes of one manufacturer's doors are listed in Fig. 11-67.

Garage Door Frames

Frames for garage doors include side jambs and a head similar to exterior passage doors except no rabbet is required. The frame is usually included in the millwork order along with windows and doors so the outside trim will match. Although the size of the frame opening is usually the same size as the door, the manufacturer's specifications and details should be checked before placing the order.

Fig. 11-68, shows a typical jamb section for wood frame or masonry construction. Note the thickness (2 in. nominal) of the heavy inside frame to which the track and hardware will be mounted. The width of this member should be at least 4 in. wide with no projecting bolt or lag screw heads.

The rough opening width for frame construction will normally be about 3 in. greater than the door size. The height of the rough opening should be the door height plus about 1 1/2 in. as measured from the finished floor.

Another size that must be given careful consideration is the inside headroom height. A certain clear minimum distance between the top of the door and the ceiling must be provided for hardware, counterbalancing mechanisms and the door itself when in an open position. On some special low-headroom designs this distance may be as little as 6 in., however, you should always be sure to check the manufacturer's requirements for a specific door.

HANDY SAYS:

"When installing garage door frames, follow the general procedure used for regular door frames. Be certain the jambs are plumb and the head is level."

Fig. 11-66. Left. Rigid one-piece swing-up type garage door. Right. Roll-up door with hinged sections that move upward and turn to horizontal position as they move back from top of opening. Rollers attached to sections run in steel track. (Wagner Mfg. Co.)

WIDTH		HEIGHT
8'0"	x	7'0" or 6'6"
9'0"	x	7'0" or 6'6"
10'0"	x	7'0" or 6'6"
16'0"	x	7'0" or 6'6"
18'0"	x	7'0" or 6'6"

Fig. 11-67. Stock sizes of garage doors.

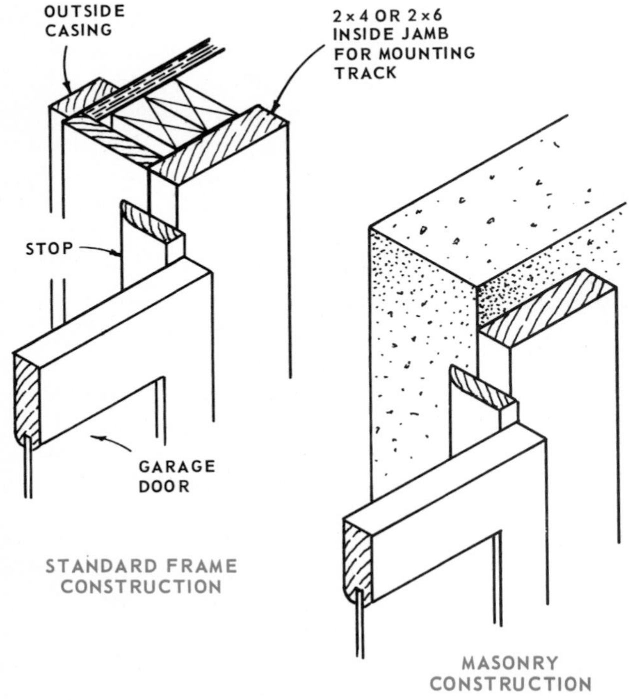

Fig. 11-68. Typical jamb construction for garage door frames. Roll-up type of door is shown.

Installation

A commonly followed procedure for a sectional roll-up door includes the following major steps: 1. Tack stops temporarily in place. 2. Assemble door sections in opening and attach hinges. 3. Attach rollers to door. 4. Place track on rollers and attach track to jamb. 5. Mount horizontal track sections. 6. Raise and prop door in open position. 7. Attach counterbalancing mechanism. 8. Open and close door and make necessary adjustments. 9. Reset stops for smooth, tight fit.

Manufacturers always supply detailed directions and procedures for the installation of their particular products and urge the carpenter to follow these printed materials carefully. They are usually well illustrated and easy to understand. See Fig. 11-69.

Hardware and Counterbalances

Garage door hardware must be well designed so the door will operate easily. Track, hinges, and bolts should be made of galvanized steel of sufficient gauge weight to last the life of the door.

To offset the weight of the door, various counterbalancing devices are used. Two of the most common types are shown in Fig. 11-70. The extension spring is commonly used on residential doors, either single or double. The torsion spring and its mechanism is somewhat more expensive but usually provides a smoother and more consistent action. It is especially recommended for wide doors and those that are heavier in construction.

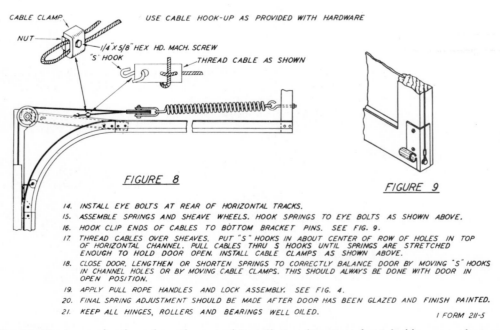

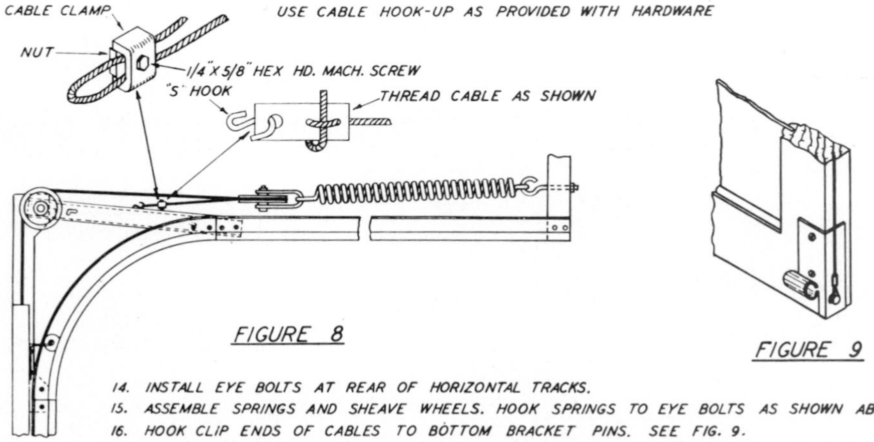

14. INSTALL EYE BOLTS AT REAR OF HORIZONTAL TRACKS.
15. ASSEMBLE SPRINGS AND SHEAVE WHEELS. HOOK SPRINGS TO EYE BOLTS AS SHOWN ABOVE.
16. HOOK CLIP ENDS OF CABLES TO BOTTOM BRACKET PINS. SEE FIG. 9.
17. THREAD CABLES OVER SHEAVES. PUT "S" HOOKS IN ABOUT CENTER OF ROW OF HOLES IN TOP OF HORIZONTAL CHANNEL. PULL CABLES THRU S HOOKS UNTIL SPRINGS ARE STRETCHED ENOUGH TO HOLD DOOR OPEN. INSTALL CABLE CLAMPS AS SHOWN ABOVE.
18. CLOSE DOOR. LENGTHEN OR SHORTEN SPRINGS TO CORRECTLY BALANCE DOOR BY MOVING "S" HOOKS IN CHANNEL HOLES OR BY MOVING CABLE CLAMPS. THIS SHOULD ALWAYS BE DONE WITH DOOR IN OPEN POSITION.
19. APPLY PULL ROPE HANDLES AND LOCK ASSEMBLY. SEE FIG. 4.
20. FINAL SPRING ADJUSTMENT SHOULD BE MADE AFTER DOOR HAS BEEN GLAZED AND FINISH PAINTED.
21. KEEP ALL HINGES, ROLLERS AND BEARINGS WELL OILED.

I FORM 211-5

Fig. 11-69. A sample sheet from the set of installation directions furnished by a manufacturer.

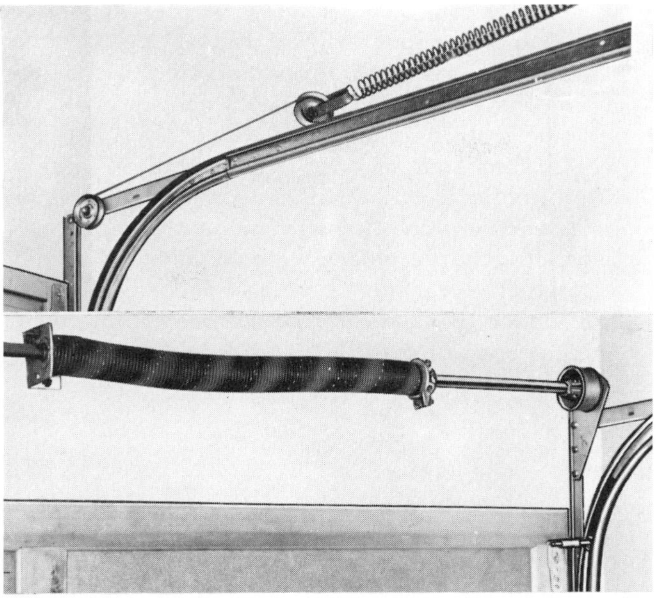

Fig. 11-70. Counterbalances for garage doors. Above. Extension spring. Below. Torsion spring.

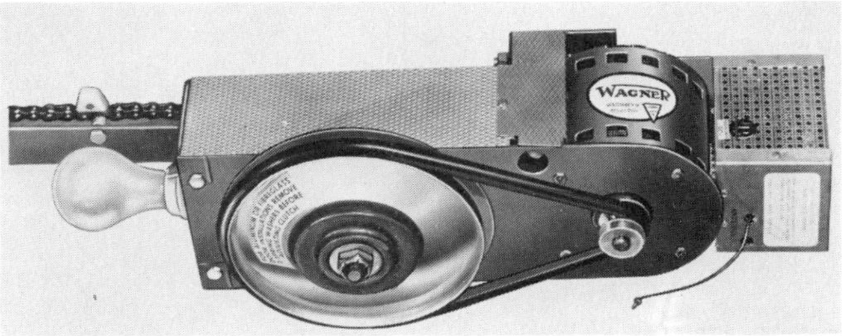

Fig. 11-71. Electric powered garage door opener.

Today, many residential plans include a two-car garage with a 16 ft. or wider door. The operation of the large door has led to the wide acceptance of electric powered door openers. Fig. 11-71 shows a model that requires no extra head room for installation. It can be equipped with push buttons for direct manual operation or with radio gear for remote control.

Test Your Knowledge - Unit 11

1. Woodworking factories that produce windows, doors, and other fabricated wood items used in building construction are often referred to as ______________ plants.
2. The kind of wood most often used to manufacture windows is __________ _________.
3. A type of window that is hinged on the side and swings outward is called a ________________.
4. The height of a standard residential door is ___________________.
5. If a single pane of glass in a window is replaced with a two layer insulating glass the heat loss will be reduced by about ________ percent.
6. The side section of a window is called the _________________.
7. When there is little or no roof overhang to protect the window, a _____________ should be installed above the head casing.
8. If the R.O. for a window is listed as 3' - 6" x 3' -5", the height of the rough opening would be ___________.
9. To adjust for various wall thickness, ________ __________ are applied to standard window frames.
10. After a window unit has been temporarily set in the rough opening the next step is to ____________ the ____________ at the correct height.
11. When installing ventilating window units it is usually best to have the sash __________. (open, closed)
12. Large insulating glass panels are made from polished plate that is ________ in. thick.
13. The thickness of standard glass block is _________ in.
14. When figuring the size of the opening for glass block panels, multiply the number of units by the __________ (actual, nominal) block size.
15. In brick veneer construction, the masonry over door and window heads is carried by a metal support called a _______________.
16. Outside wood casing for windows and doors is attached with nails spaced ________ in. O.C.
17. The two most popular types of garage doors are the swing-up and the _______________.
18. A two-car garage door is usually ________ ft. wide.
19. The two common types of spring counterbalances for garage doors are extension and _________________.

Outside Assignments

1. Prepare a written report on the manufacture of glass. Include such headings as; historical development, early production methods, modern processes, float method, drawing method, and grinding and polishing plate glass. Study encyclopedias, reference books and such booklets as Romance of Glass, available from the PPG Industries, Inc.
2. Secure a set of architectural plans for a house that does not include a window schedule. Also secure descriptive literature and data sheets

on at least one brand of windows. The latter can be obtained from lumber yards and builder's supply stores or you may want to write directly to a manufacturer. Prepare a complete window schedule for the house plan. Include unit sizes, manufacturer's numbers, rough openings, and data that might be helpful to the carpenter when making the installation.

3. Visit a home building site in your community. Try to find one where the windows have just been installed. Look over the work and ask the carpenter about the procedure he followed. See if you can identify the types and parts of the window units. Prepare a written report or make an oral report to your class. Special note: Always be sure to obtain permission to observe from the foreman or head carpenter when visiting a building site.
4. The rising cost of energy has resulted in special emphasis being placed on window design and construction. Some manufacturers are now producing triple glazed window units for homes located in northern climates. Secure information about these units that includes: U-factors, prices, special installation directions, and predicted fuel savings.

Trimming doors to size. After the edges are sawed and sanded on the first double-end machine (as shown) the door is carried into a second machine that trims the top and bottom. The doors being processed are made by bonding molded faces to frames. The faces are produced by blending wood fibers with a binder and then pressing the mixture in dies under 2700 tons of force. (Caradco)

Unit 12
EXTERIOR WALL FINISH

The term "exterior finish" includes the application of all exterior materials of a structure. It generally refers to the roofing materials, cornice trim boards, wall coverings, and trim members around doors and windows. The installation of special architectural woodwork at entrances or the application of a ceiling to a porch or breezeway area would also be included under this broad heading.

Previous units have described the application of the finished roof and the installation of the trim around windows and outside doors. This unit provides information about the construction and finish of cornice work and the materials and methods used to provide a suitable outside wall covering.

Cornice Designs

The cornice, also called an eave, is formed by roof overhang and provides a finished connection between the wall and the edge of the roof. It is an important element in the total appearance of a structure and the architectural style will determine to a large extent the design requirements. Fig. 12-1 includes a number of detailed cornice sections secured from architectural drawings of contemporary residential structures.

Diagrams of several closed or "boxed" cornice designs are illustrated in Fig. 12-2. An open cornice is sometimes used, exposing the rafters and underside of the roof sheathing. Wide overhangs are used extensively in modern buildings. These provide shade for large window areas, protect the walls, and add to attractiveness of the structure.

The rake is the part of a roof that overhangs a gable. It is usually enclosed with carefully fitted trim members.

Parts of Cornice and Rake Section

Fig. 12-3 shows the structural and trim parts of a boxed cornice. The fascia board is the main trim member along the edge of the roof. A ledger strip is nailed to the wall and carries the lookouts which are attached to each rafter. A nailing strip is sometimes attached to the back of the fascia, between each rafter. Nailing strips, along with the lookouts and ledger, provide a frame to which the plancier or soffit material can be applied. This surface material can be plywood, hardboard, solid stock, mineral fiber board or plaster.

The trim used for a boxed rake section, Fig. 12-4, requires the support of the projecting roof boards. In addition, lookouts or nailers are fastened to the side wall and the roof sheathing. As in cornice construction, these serve as a nailing base for the soffit and fascia. When the rake projects a considerable distance the sheathing does not provide adequate support and the roof framing should be extended as shown in Fig. 9-22.

The various parts of the cornice and rake structure that are exposed to view are generally called exterior trim. In typical construction these parts are cut, fitted, and nailed in place on the job. The properties desired in material used for exterior trim include; good painting and weathering characteristics; easy working qualities; and maximum freedom from warp. Decay resistance is also desirable where materials may absorb moisture. The cedars, cypress, and redwood have high decay resistance. Less durable species may be treated to make them decay resistant. Coating end joints or miters of members subjected to excessive moisture is recommended. White lead paste or special caulking compounds are commonly used for this purpose.

Cornice and Rake Construction

In most construction, the fascia boards are installed on the perimeter of the roof frame at the time the roof is sheathed. It is important that they be straight, true, and level with well fitted joints. Corners should be mitered and end joints should meet at a 45 deg. angle as illustrated in Fig. 12-5.

To frame a cornice with a horizontal soffit, first install a ledger strip along the wall. Use a carpenters' level to locate points on the wall even with the bottom edge of the fascia, then measure up from this point the required distance to the bottom of the ledger. Study the architectural drawing details to secure dimensions. Snap a chalk line between these points and nail the ledger in place.

Lookouts are usually made from 2 x 4 stock, spaced at each rafter or every other rafter de-

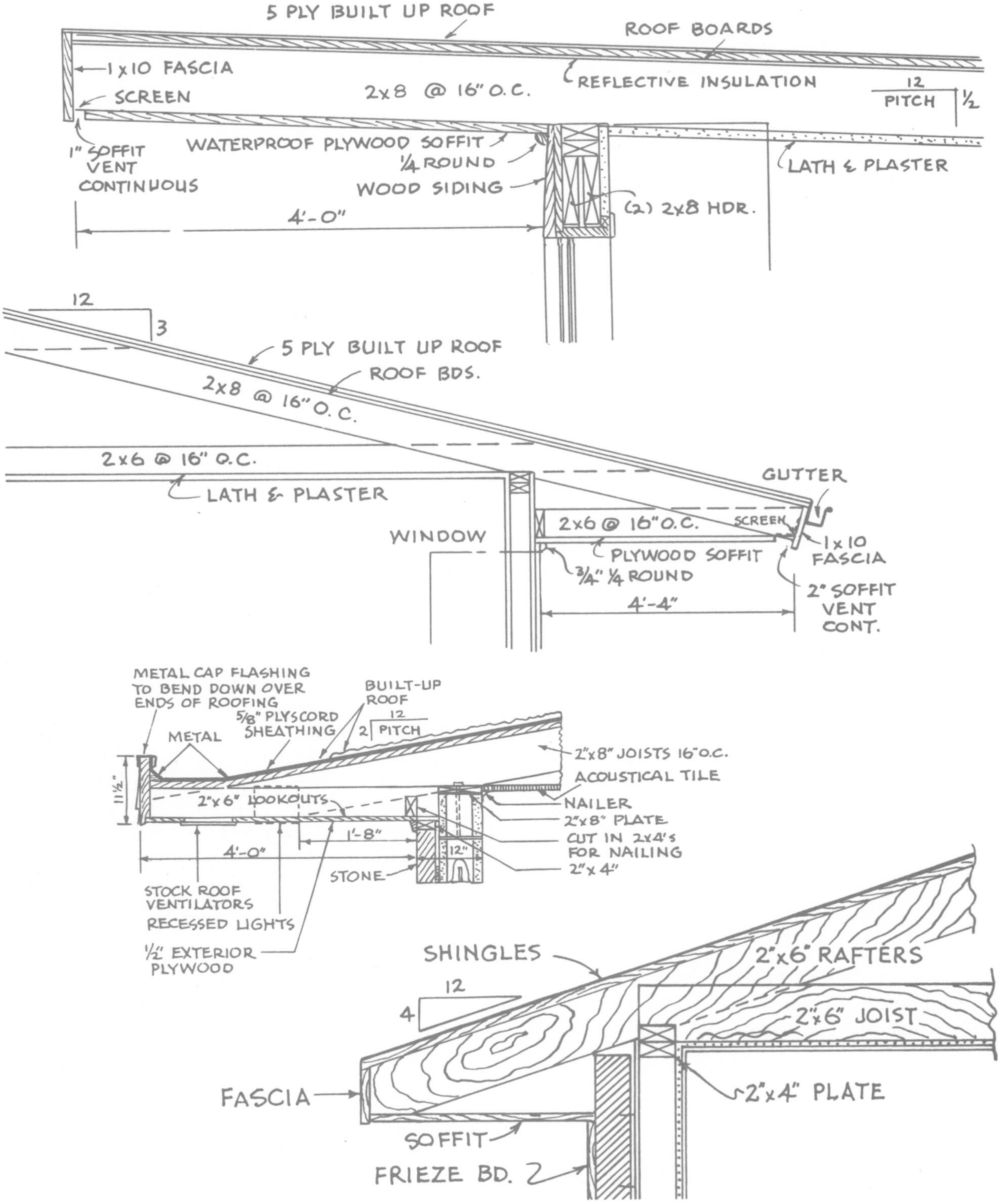

Fig. 12-1. Typical cornice details from contemporary architectural plans.

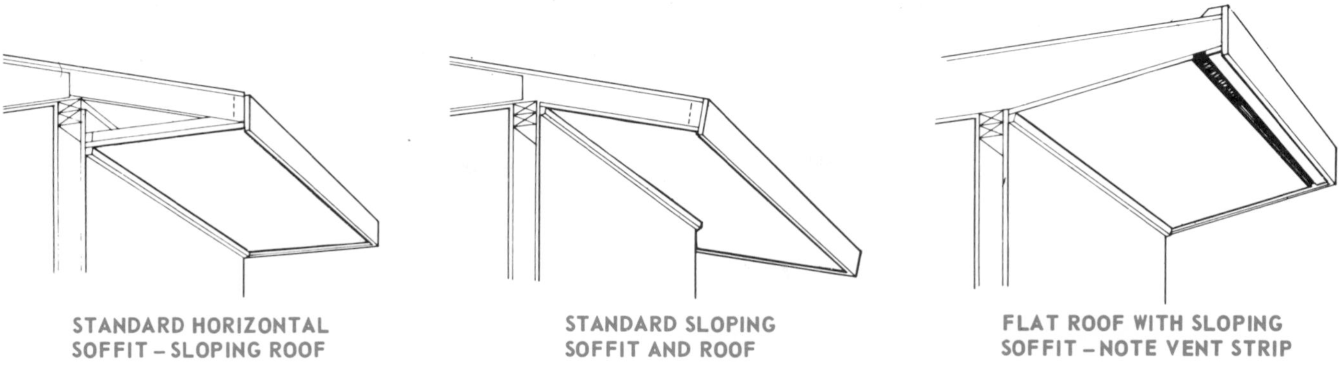

Fig. 12-2. Cornice designs.

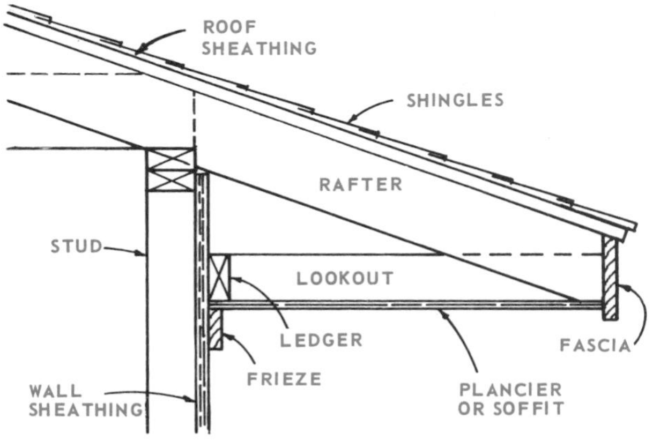

Fig. 12-3. Parts of a typical cornice construction.

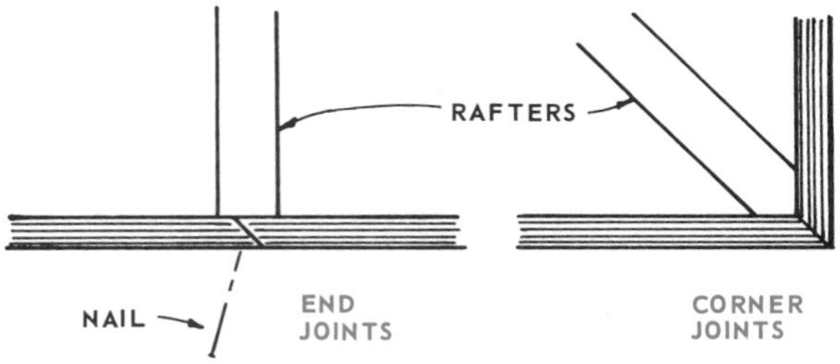

Fig. 12-5. Joints for fascia boards. (Plan view)

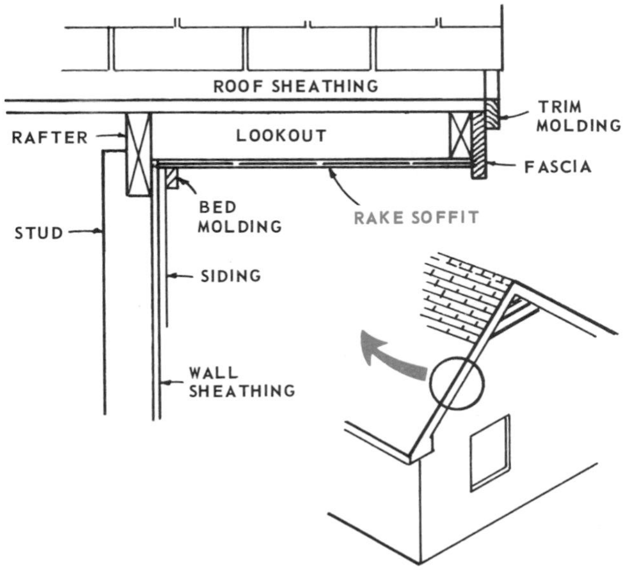

Fig. 12-4. Typical boxed rake section.

Fig. 12-6. Installing a plywood soffit in a sloping cornice section. (American Plywood Assoc.)

pending on the kind of soffit material used. Cut the lookouts to length and make the installation by toenailing to the ledger and nailing to the overhang of the rafter. When using thin material for the soffit, attach a nailing strip along the inside of the fascia to provide a nailing surface. Sometimes the back of the fascia is grooved to receive the soffit material.

Fig. 12-7. Cornice and rake complete. Vertical siding has been installed on the gable wall and the first horizontal siding board is in place.

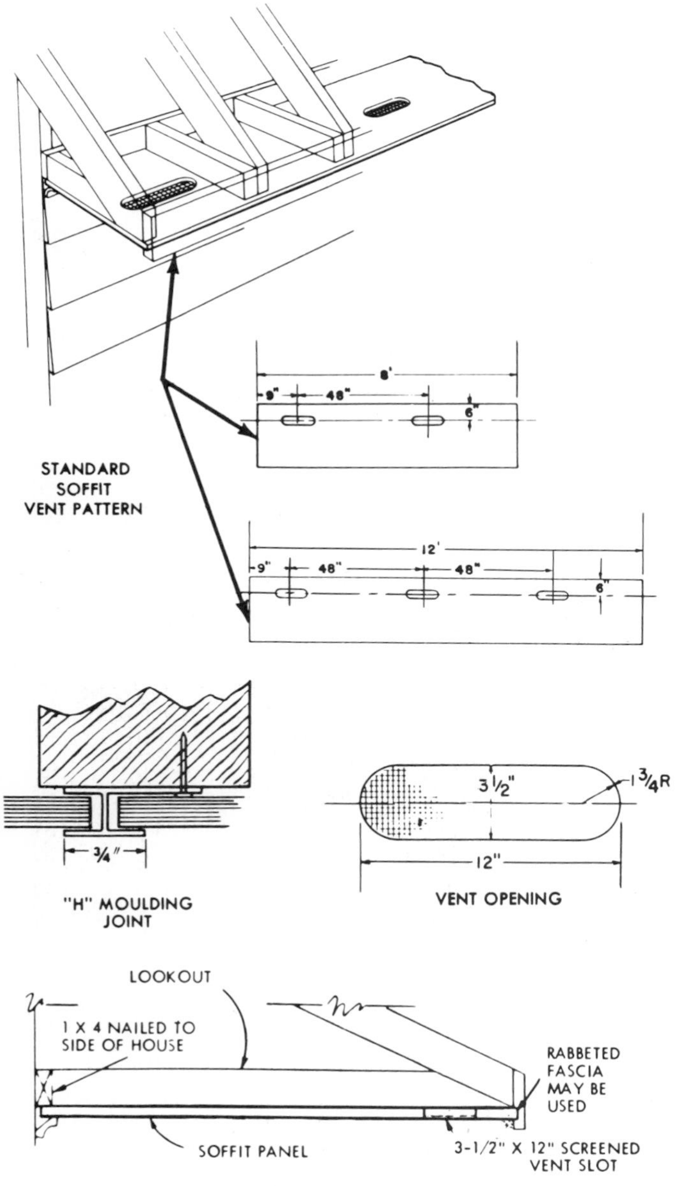

Fig. 12-8. Details of prefabricated soffit panels.

Fig. 12-9. Nailing soffit panel to lookouts. (The Upson Company)

After the frame is complete, the soffit is applied. First cut the material to the required size and then secure it in place with nails or screws that are rust-resistant. When regular casing or finish nails are used, they must be set and the holes filled with putty after the prime coat of paint. Fig. 12-6 shows the installation of 3/8 in. exterior plywood in a cornice design where soffit material is attached to the bottom edge of the rafters. Soffits should be provided with screened slots or metal ventilating units so there will be a flow of air through the enclosed cornice section and also through the attic space.

HANDY SAYS:

"Always use rust-resistant nails for outside finish work. They may be made of aluminum or galvanized (or cadmium plated) steel."

The rake section at the end of the roof over the gable wall should be constructed to match the cornice using the same general procedures. After the main trim members are installed, moldings are often set in corners to cover irregularities. In modern construction however, their use is usually held to a minimum in order to maintain a smooth trim appearance, Fig. 12-7.

Prefabricated Cornice Materials

Because cornice construction is somewhat time consuming, many builders prefer to purchase prefabricated materials that can be rapidly installed. Various systems are available that provide a neat, trim appearance. One such product consists of 3/8 in. laminated wood-fiber panels that are factory primed and available in a variety of standard widths (12 to 48 in.) and in lengths up to 12 feet. The panels can also be equipped with factory-cut vents and factory-applied screens. See Fig. 12-8.

When installing large sections of wood fiber panels, fit each panel with some clearance for expansion. Space nails (4d rust-resistant) about 6 in. apart along edges and intermediate supports. Start nailing at the end that is butted against a previously placed panel. First nail to the main supports and then along the edges. Drive the nails carefully so the underside of the head is just flush with the panel surface, Fig. 12-9.

Lookouts may be eliminated in a soffit system where special supports are attached to the upper surface of the panels. These supports are made of 20 ga. steel channels with prongs that make it easy to attach them to the back of the panels. The supports provide rigidity and the panels are secured in place only at the front and back edge.

To install the channels, place the soffit face down on a solid surface and lay the channel in position (usually located 24 in. O.C.). Insert a piece of 2 x 4 stock in the channel and drive the prongs into the soffit panel as illustrated in Fig. 12-10.

Fig. 12-10. Attaching the channel to the top side of the soffit panel.

After the channels are attached, the units are lifted into place, Fig. 12-11, and held with a special molding attached to the inside of the fascia and along a nailer attached to the wall of the structure. Special H-shaped clips are used to join the ends of adjacent panels. A view of the completed installation is shown in Fig. 12-12.

Porch and carport ceilings can be covered with factory-primed panels similar to those used for cornice soffits. Standard 4 x 8 ft. and larger units are designed for either 16 in. or 24 in. O.C. framing. Always leave a 1/8 in. spacing along all edges to

Fig. 12-11. Panel being lifted into place. (The Upson Co.)

allow for expansion. Start nailing in the center of the panels and move toward the outside. Space nails about 6 in. apart. Edges of panels are secured and joints are covered through the use of special H-strips as shown in Fig. 12-13.

Wall Finish

After the cornice and rake section of the roof are complete, the siding material is applied to the walls. When the structure includes a gable roof, Fig. 12-14, the wall surface material is usually applied to the gable end before the lower section

Fig. 12-12. Completed installation of soffit system.

is covered. This permits scaffolding to be attached directly to the wall during its application.

All exterior trim members, if not factory-primed, should be given a primer coat of paint as soon as possible after installation.

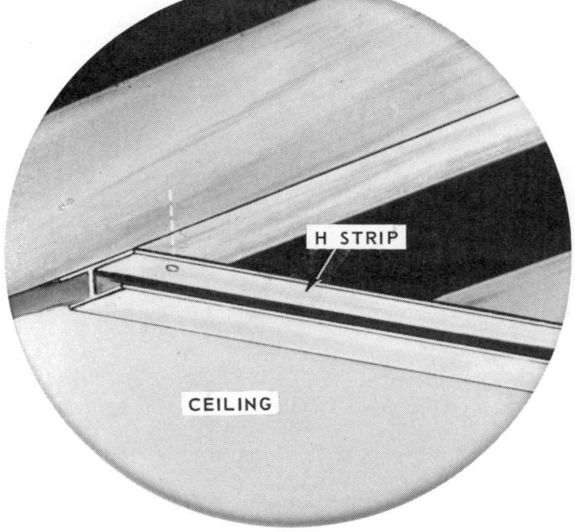

Fig. 12-13. Porch ceiling finished with factory-primed panels. Special H strips are used at joints.

Fig. 12-14. Wall surface material is applied to gable ends before lower walls are covered.

Horizontal Siding

One of the most characteristic materials used for the exterior finish of American homes is wood siding. The deep shadow cast on the wall by the butt edge, especially evident in bevel siding, emphasizes the horizontal lines that are desired by many home owners.

Siding is usually applied over sheathing, however, in mild climates or on such buildings as summer cottages, it may be applied directly to the studs. Where sheathing is omitted, or where the type of sheathing does not provide sufficient strength to resist a racking load, the wall framing should be braced as described in Unit 8.

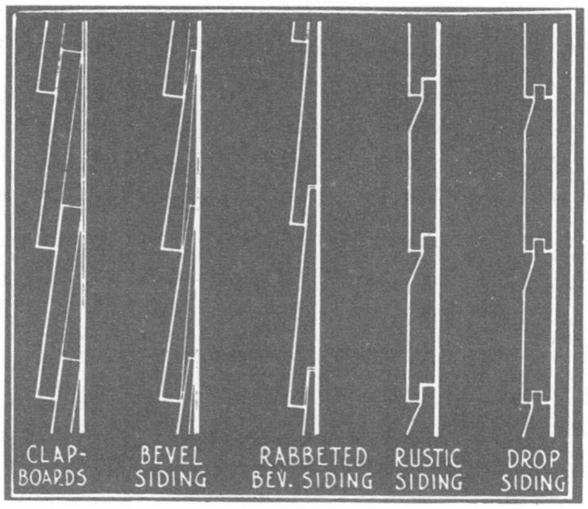

Fig. 12-15. End view of five types of horizontal siding. This shows the slight difference between drop siding and rustic siding, and makes possible quick comparison of the five types.

End views of a number of types of horizontal siding are shown in Fig. 12-15. Bevel siding which is most commonly used, is available in various widths. It is made by sawing plain surfaced boards at a diagonal to produce two wedge shaped pieces. The siding is about 3/16 in. thick at the thin edge and 1/2 to 3/4 in. thick on the other edge, depending on the width of the piece.

Wide bevel siding often has shiplapped or rabbeted joints so that the siding lies flat against the studding instead of touching it only near the joints as ordinary bevel siding does. This reduces the apparent thickness of the siding by 1/4 in. but permits the use of extra nails in wide siding and reduces the chance of warping. It is also economical, since the rabbeted joint requires less lumber than the lap joint used with plain bevel siding. The rabbet however, must be deep enough that, when the siding is applied, the width of the boards can be varied to meet window sill, head casing, and eave lines as desired. The table in Fig. 12-16 lists the more common sizes of horizontal siding and the actual size measurements.

Rustic and drop sidings are usually 3/4 in. thick, and 6 in. wide. They come in a wide variety of patterns. Drop siding usually has tongue-and-groove joints, while rustic siding has shiplap type joints.

Drop siding is heavier, has more structural strength and tighter joints than bevel siding.

SIDING	NOMINAL SIZE WIDTH INCHES	DRESSED DIMENSIONS STANDARD THICKNESS INCHES	DRESSED DIMENSIONS STANDARD FACE WIDTH INCHES
Bevel	4	** 7/16 by 3/16	3 1/2
	5	10/16 by 3/16	4 1/2
	6	. . . .	5 1/2
Wide Beveled	8	** 7/16 by 3/16	7 1/4
	10	9/16 by 3/16	9 1/4
	12	11/16 by 3/16	11 1/4
Rustic and Drop. . . . . (Shiplapped)	4	9/16	3 1/8
	5	3/4	4 1/8
	6	. . . .	5 1/16
	8	. . . .	6 7/8
Rustic and Drop. . . . . (Dressed and Matched)	4	9/16	3 1/4
	5	3/4	4 1/4
	6	. . . .	5 3/16
	8	. . . .	7

In patterned siding, 11/16, 3/4, 1, 1 1/4 and 1 1/2 inches thick, board measure, the tongue shall be 1/4 in. wide in tongued-and-grooved lumber, and the lap 3/8 in. wide in shiplapped lumber, with the overall widths 1/4 in. and 3/8 in. wider, respectively, than the face widths shown above.

** Minimum thicknesses.

Fig. 12-16. Table listing sizes of horizontal siding. The nominal sizes, which are used in computing the footage of lumber, are based upon the sizes of boards which are cut from the logs. These rough, green boards shrink somewhat in width and thickness as they dry, and their size is further reduced by machining to pattern.

Because of this, it is often used on garages and other buildings where sheathing is not included in the wall structure.

Wood used for exterior siding should be of a select grade that is free of knots, pitch pockets and other defects. Edge-grain material is less likely to warp than flat-grained surfaces. The moisture content at the time of application should be equal to that which it will attain in service. This is approximately 12 percent, except for the Southwestern States, where the moisture content should average about 9 percent.

Siding should be handled carefully when it is delivered to the building site. The wood from which siding is made is usually quite soft and the surface can be easily damaged. It is desirable to store siding within the structure or keep it covered with a weatherproof material, until it is applied.

Wall Sheathing and Flashing

Horizontal siding can be applied over various sheathing materials. When the sheathing consists of solid wood, plywood or nail-base fiberboard, the siding is nailed directly to the material at about 24 in. intervals. End joints in the siding may occur between framing members. Gypsum board and regular fiberboard sheathing cannot be used as a nailing base and the siding should be attached by nailing through the sheathing and into the frame. Certain special application methods may require wood strips to form a nailing base. Additional information about wall sheathing is included in Unit 8.

HANDY SAYS:

"Be sure the structural frame and sheathing are dry before applying the siding material. If excessive moisture is present during the application, later drying and the resulting shrinkage will likely cause the siding to buckle."

Sheathing paper is applied directly to the wall frame to resist the infiltration of air and moisture when sheathing is not used. It is also used to cover common board and shiplap sheathing. Sheathing paper is not necessary over plywood, fiberboard, or treated gypsum sheathing. Sheathing paper (also called building paper) may be an asphalt-saturated felt which has a low vapor resistance. Materials such as coated felts or laminated waterproof papers that have a high moisture vapor resistance will act as a vapor barrier and should not be used. See Unit 13.

Before the application of siding, flashing should be installed where it is required around openings. Metal flashing is usually included over the drip caps of doors and windows, Fig. 12-17. In areas not subjected to wind driven rain, head flashing may be omitted when the vertical height between the top of the finished trim of the opening and the soffit of the cornice is equal to or less than one-fourth of the horizontal width of the overhang.

For structures with unsheathed walls, jambs of doors and windows should be flashed with a 6 in. wide strip consisting of either metal, 3 oz. copper-coated paper, or 6 mil polyethylene film.

Installation Procedures

Wood siding is precision-manufactured to standard sizes and is easily cut and fitted into place. Plain beveled siding is lapped so it will shed water and provide a windproof and dustproof covering. A minimum lap of 1 in. is used for 6 in. widths, while 8 and 10 in. siding should lap about 1 1/2 inch.

To install horizontal siding, first prepare a story pole by laying out the distance from the soffit to about 1 in. below the top of the foundation, Fig. 12-18. Divide this distance into spaces equal to the width of the siding minus the lap. Adjust the lap allowance (maintain minimum requirements) so the spaces are equal. When possible, adjust the spacing so single pieces of siding will run continuously above and below windows or other wall openings without notching, Fig. 12-19. When the layout is complete mark the position of the top of each siding board on the story pole.

Now hold the story pole in position at each inside and outside corner of the structure and transfer the layout to the wall, Fig. 12-20. Also make the layout along window and door casings. Some carpenters prefer to set nails at these layout points, since lines can be quickly attached to them and used to align the siding stock. They can also be used to hold the chalkline if guidelines are laid out by this method.

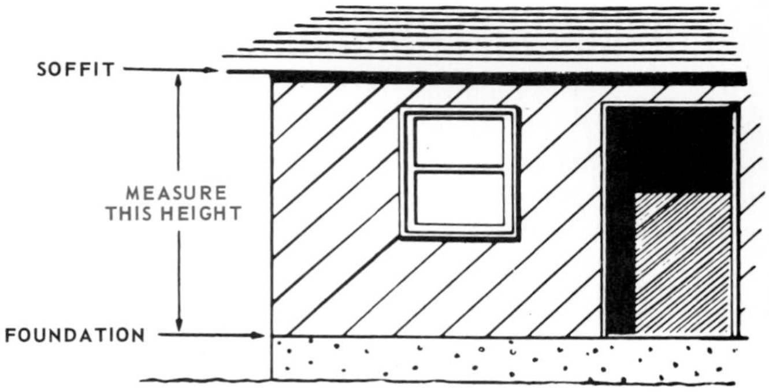

Fig. 12-18. Height used in story pole layout.

Fig. 12-19. Coordinate the siding courses with the bottom and top of windows.

Fig. 12-20. Transfer story pole layout to all inside and outside corners and also to door or window casings.

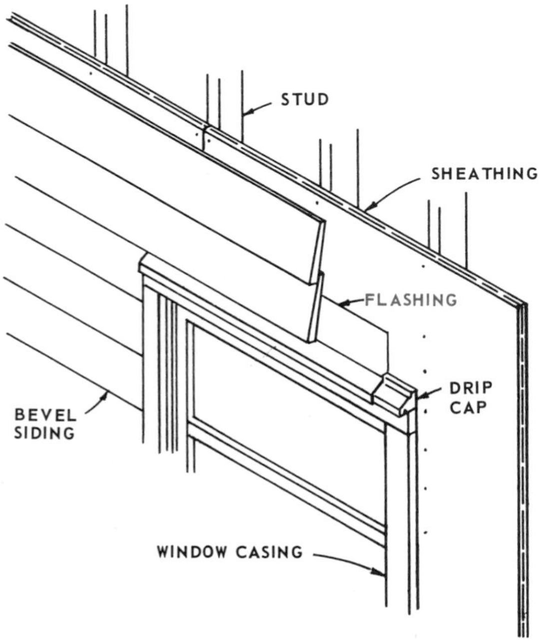

Fig. 12-17. Metal flashing is applied over the drip caps of doors and windows that are not protected by roof overhang.

HANDY SAYS:

"When snapping a chalk line stretched over a long distance, it is a good idea to hold it against the surface at the mid-point, then snap it on each side."

After the layout has been made and carefully checked, start the application of bevel siding by first nailing a strip along the foundation line equal to the thin edge of the siding, Fig. 12-21. This will provide the proper slant for the first course. Now apply the first piece, allowing the butt edge to extend below the strip to form a drip edge.

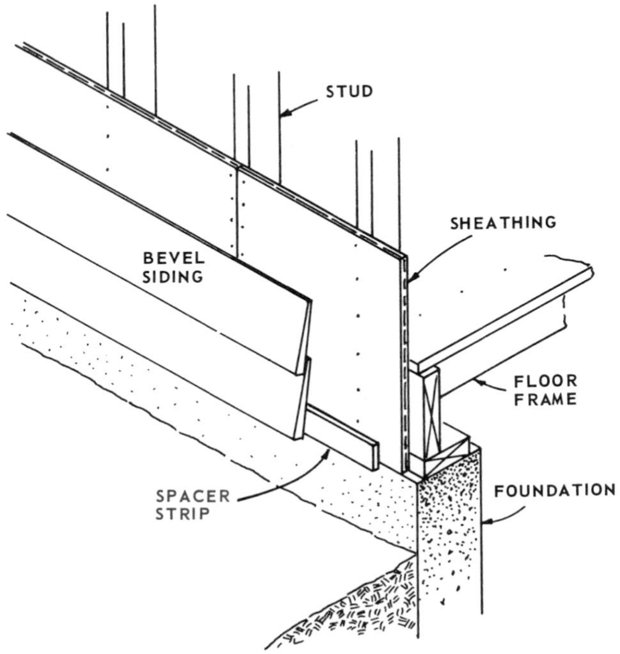

Fig. 12-21. Bevel siding requires a spacer strip under the first course.

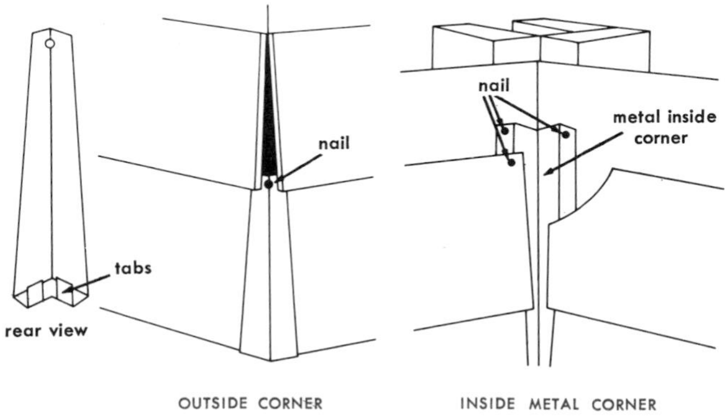

Fig. 12-22. Metal corners for horizontal siding.

Inside corners can be formed with a square piece of wood, or, metal corners can be used as shown in Fig. 12-22. Although outside corners could be lapped or mitered, in modern construction, metal corners are used almost universally. They can be installed quickly and provide a neat, trim appearance, Fig. 12-23.

Fig. 12-23. Installing an outside metal corner. (Masonite Corp.)

Corner boards, Fig. 12-24, are sometimes used for siding installation. They may be formed by assembling two pieces of lumber (thickness depends on siding) and then attaching the assembly to the structure. Corner boards may be plain or molded depending on the architectural treatment required. After the corner boards are in place, the siding is fitted tightly against them.

Cut and fit horizontal wood siding tightly against window and door casings, corner boards, and adjoining boards. For quality work, the carpenter first cuts the lengths with a fine tooth saw and then smoothes the ends with a few strokes of a block plane. Square butt joints are used between adjacent pieces of siding and should be staggered as widely as possible in successive courses.

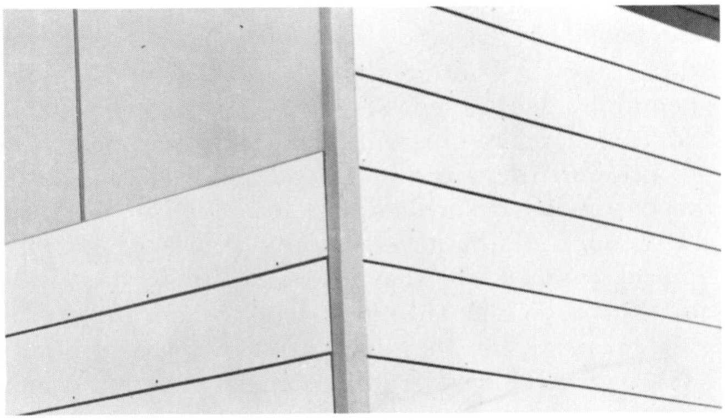

Fig. 12-24. Outside corner finished with a corner board. (Armstrong Cork Co.)

Wood siding can be given a coat of water-repellent preservative before it is installed, or the water-repellent can be brushed on after the installation. Preservatives are sold by lumber dealers and paint stores. They contain waxes, resins, and oils. In addition to this treatment, joints in siding may be bedded in a special caulking compound to insure watertightness.

Nailing

To fasten siding in place, zinc-coated steel, aluminum, or other noncorrosive nails are recommended. Using plain steel-wire nails, especially the large-headed ones designed for flush driving, often results in unsightly rust spots on most paints. Even small-headed plain steel nails, countersunk and puttied are likely to cause trouble.

Fig. 12-25. Installing horizontal siding using a pneumatic powered nailer loaded with special noncorrosive nails. (Senco Products, Inc.)

Horizontal siding should be face-nailed to each stud, Fig. 12-25. For 1/2 in. siding over wood or plywood sheathing, use sixpenny nails; and over fiberboard or gypsum sheathing, use eightpenny nails. For 3/4 in. siding over wood or plywood sheathing, use sevenpenny nails; and over fiberboard or gypsum sheathing, use ninepenny nails.

The nail is generally placed about 1/2 in. above the butt edge, in which case it passes through the upper edge of the lower course of siding. Another nailing method for bevel siding is to drive the nails through the siding just above the lap so the nail misses the thin edge of the piece of siding underneath. This permits expansion and contraction of the siding board with seasonal changes in moisture content. This eliminates the tendency for the siding to cup or split where both edges are

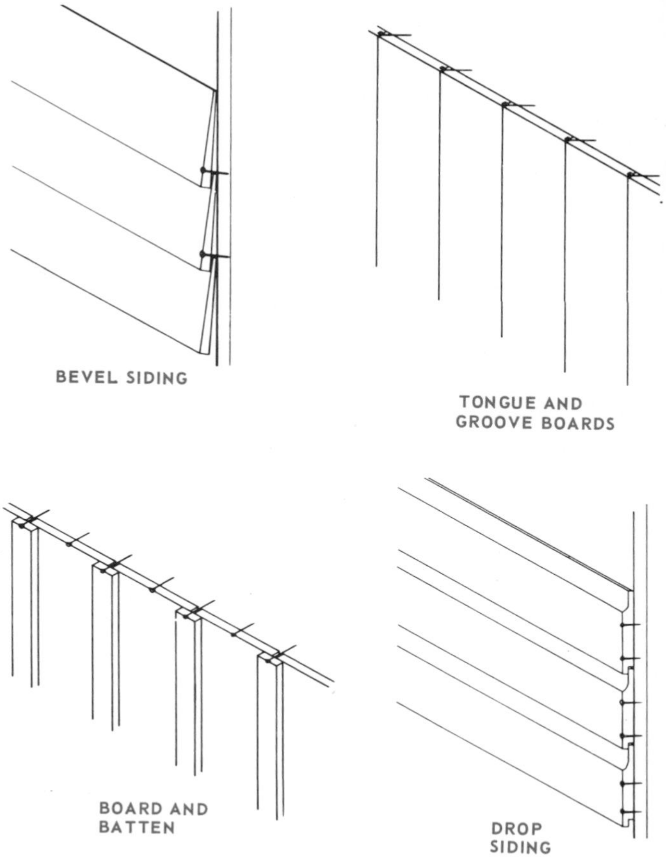

Fig. 12-26. Nailing patterns for wood siding.

nailed. Since the amount of swelling and shrinking is proportional to the width of the material, nailing above the lap is more important in wide siding than in narrow siding.

If there is a possibility the material will split when nailing end joints, holes should be drilled for the nails. Fig. 12-26 shows some wood siding nailing patterns.

Although the usual procedure for the installation of horizontal siding is to proceed upward along the wall, some carpenters prefer to start at the top and work down, Fig. 12-27. This is adaptable for multi-story or split level structures where scaffolds are attached to wall. Another advantage -- the siding is less likely to be damaged after it is applied. When following such procedure, chalk lines are set at the butt edge of the siding instead of the top edge.

Painting and Maintenance

When wood siding deteriorates, it is from decay or weathering. Neither will occur if simple precautions are taken. Decay is the disintegration of wood caused by the growth of fungi. These fungi grow in wood when the moisture content is exces-

Fig. 12-27. Applying horizontal siding from the top, downward.

sive. If the structure is built on a foundation which has been carried well above the ground and the construction is such that water runs off instead of into the walls, decay should never be a problem.

The priming coat should be put on as soon as possible after the siding has been applied. If an unexpected rain should wet unprimed wood siding, the first coat of paint should not be applied until the excessive moisture has evaporated.

Estimating Amount of Siding

In figuring the quantity of siding required for a structure, it is necessary to increase the footage by an amount sufficient to compensate for the difference between nominal and finished sizes. Additional amounts must also be added for the machining of joints and the overlap in beveled siding. The table in Fig. 12-28 provides approximate percentages that should be added to the net square footage of the wall surface to be covered. Below is an example showing how they may be applied to a typical problem.

Example: 1 x 10 Bevel Siding - 1 1/2 in. lap
Wall Height = 8 ft.
Wall Perimeter = 160 ft.
Door & Window Area = 240 sq. ft.

Total Area to be covered = (8 x 160) - 240
= 1280 - 240
= 1040 sq. ft.

Siding Estimate = 1040 + (1040 x 29%)
= 1040 + 302
= 1342 bd. ft.

Area for gable ends can be calculated by multiplying the height above the eaves by the width and dividing by two. Considerable waste occurs in covering triangular areas and at least 10 percent should be added to this calculation. When the structure includes an excessive number of corners due to projections and recesses in the wall line, an additional allowance of at least 5 percent should be added to the percentages shown in Fig. 12-28.

			ADD
Bevel Siding	1x4 – 3/4	lap	45%
	* 1x5 – 7/8	lap	38
	1x6 – 1	lap	33
	1x8 – 1 1/4	lap	33
	1x10 – 1 1/2	lap	29
	1x12 – 1 1/2	lap	23
		ADD	
Rustic and Drop Siding (Shiplapped)	1x4	28%	
	* 1x5	21	
	1x6	19	
	1x8	16	
Rustic and Drop Siding (Dressed and Matched)	1x4	23	
	* 1x5	18	
	1x6	16	
	1x8	14	

* Unusual Sizes.

Fig. 12-28. Percentage that must be added to net wall surface when estimating amounts of horizontal wood siding.

Vertical Siding

Vertical siding, Fig. 12-29, is commonly used to feature entrances or gable ends. It is also often used for the main wall areas. Vertical siding may be plain-surfaced matched boards, patterned

Fig. 12-29. Vertical siding installed at the main entrance of a modern home. (Masonite Corp.)

Fig. 12-30. Battens over rough boards present a rustic appearance and provide for expansion and contraction. (Western Wood Products)

matched boards, or square-edge boards covered at the joint with a batten strip.

Matched vertical siding, made from solid lumber, should preferably be not more than 8 in. wide. It should be installed with two 8d nails not more than 4 ft. apart. Backer blocks should be placed between studs to provide a good nailing base. The bottom of the boards are usually undercut to form a water drip.

Batten-type applications are nearly always used with wide square-edged boards, Fig. 12-30. which, because of their width, are subjected to considerable expansion and contraction. The batten strips used to cover the joints should be nailed to only one siding board so the adjacent board can swell and shrink without splitting the boards or the batten strip.

Board and batten effects may be obtained by first covering the wall with large vertical sheets of plywood or composition material and then attaching vertical strips over the joints and at several intervening positions between the joints. Fig. 12-31 shows an application of this type with solid wood strips being applied to the surface of exterior plywood sheets.

Fig. 12-31. Applying batten strips over sheets of exterior plywood.

Wood Shingles

Wood shingles are often used for wall covering, and a large selection of types is available. Some are especially designed for sidewall application with a grooved surface and factory applied paint or stain. Shingles are very durable and can be applied in various ways to provide a variety of architectural effects. Handsplit shingles are occasionally used, however they are more expensive and difficult to install.

Most shingles are made in random widths, varying in the No. 1 grade from 3 in. to 14 in. with only a small proportion of the narrow width permitted. Shingles of a uniform width known as dimension shingles are also obtainable. For side wall applications, maximum exposure to the weather recommended for 16 in. shingles is 7 1/2 in.; for 18 in. shingles 8 1/2 in.; for 24 in. shingles 11 1/2 in. Shingles on side walls are frequently laid in what is called "double-coursing." This is done by using a lower grade shingle under the shingle exposed to the weather. The exposed shingle butt extends about 1/2 in. below the butt of the under course. When butt nailing is used, a greater weather exposure is possible, frequently as much as 12 in. for 16 in. shingles, 14 in. for 18 in. shingles and 16 in. for 24 in. shingles. Fig. 12-32 lists the sizes of a standard sidewall shingle with grooved surface. Approximate coverage for various lengths and exposures is included.

In mild climates, sheathing spaced apart on centers equal to the shingle exposure and shingled with a high grade product provides a satisfactory wall. Tar paper should always be used with such construction; either between the shingles and sheathing or between the sheathing and the studding. Spaced sheathing is also satisfactory on implement sheds, garages, and other structures where protection from the elements is the principal consideration.

To obtain the best effect and to avoid unnecessary cutting of shingles, butt-lines should be even with the upper lines of window openings, and also, wherever possible, with the lower lines of such openings. It is better to tack a temporary strip to the wall to use as a guide for placing the butts of the shingles squarely, rather than to attempt to shingle to a chalk line.

Grade	Length	Thickness (at Butt)	No. of Courses Per Bdl/Carton	Bdls/Cartons Per Square	Shipping Weight	Description
No. 1	16" (Fivex) 18" (Perfections) 24" (Royals)	.40" .45" .50"	33/33 28/28 13/14	1 carton 1 carton 4 bdls.	60* lbs. 60* lbs. 192 lbs.	Same specifications as rebutted-rejointed shingles, except that shingle face has been given grain-like grooves. Natural color, or variety of factory-applied colors. Also in 4-ft. and 8-ft. panels.

NOTE: * 70 lbs. when factory finished.

LENGTH AND THICKNESS	Approximate coverage of one square (4 bundles) of shingles based on following weather exposures												
	3½"	4"	4½"	5"	5½"	6"	6½"	7"	7½"	8"	8½"	9"	9½"
16" x 5/2"	70	80	90	100*	110	120	130	140	150Ɪ	160	170	180	190
18" x 5/2¼"	....	72½	81½	90½	100*	109	118	127	136	145½	154½Ɪ	163½	172½
24" x 4/2"	....	....	....	....	....	80	86½	93	100*	106½	113	120	126½

LENGTH AND THICKNESS	10"	10½"	11"	11½"	12"	12½"	13"	13½"	14"	14½"	15"	15½"	16"
16" x 5/2"	200	210	220	230	240†	....	....	....	....	....	....	....	....
18" x 5/2¼"	181½	191	200	209	218	227	236	245½	254½†	....	....	....	....
24" x 4/2"	133	140	146½	153Ɪ	160	166½	173	180	186½	193	200	206½	213†

NOTES: * Maximum exposure recommended for roofs. Ɪ Maximum exposure recommended for single-coursing on sidewalls. † Maximum exposure recommended for double-coursing on sidewalls.

Fig. 12-32. Sizes and approximate coverage of sidewall shingles. The thickness is based on the number of butts required to equal a given measurement. For example; a 5/2" means that 5 butts will measure 2 inches.

Single Coursing of Side Walls

The single-coursing method for sidewall application is similar to roof application, the major difference being in the exposures employed. In roof construction, maximum permissible exposures are slightly less than one-third of the shingle length. This produces a three-ply covering. Vertical surfaces of side walls present less weather-resistance problems than do roofs and accordingly a two-ply covering of shingles is usually adequate.

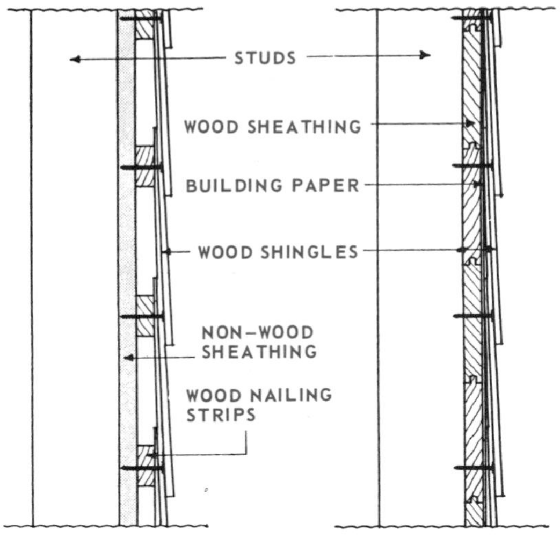

NON-WOOD SHEATHING (NAILING STRIPS REQUIRED)   WOOD SHEATHING (NO NAILING STRIPS)

Fig. 12-33. Single-course method of applying shingles to sidewalls. Solid backing and nailing base is provided by the use of wood sheathing or nailing strips over non-wood sheathing.

In single-coursed side walls, weather exposure of shingles should never be greater than half the length of the shingle, minus 1/2 in. so two layers of wood will be found at every point in the wall. For example, when 16 in. shingles are used, the maximum exposure should be 1/2 in. less than 8 in. or 7 1/2 in.

Single-course side walls, Fig. 12-33, should have concealed nailing. This means that the nails must be driven approximately one inch above the butt line of the succeeding course, so the shingles of this course will adequately cover them. Two nails should be driven in each shingle up to 8 in. in width, and each nail placed about 3/4 in. from the edge of the shingle. On shingles wider than 8 in. a third nail should be driven in the center of the shingle, at the same distance above the butt line as the other nails. Rust-resistant nails should be used, 3d size, 1 1/2 in. in length.

Estimating Quantities

In estimating the quantity of shingles required for side walls, actual areas to be shingled should be calculated in terms of square feet. Window and door areas should be deducted. By consulting the table, Fig. 12-32, the coverage of one square (4 bundles) at the exposure to be used (for example, 150 in the case of 16 in. shingles at 7 1/2 in. exposure) should be divided into the wall area to be covered. The figure obtained will be the number of squares needed. To this figure, add about 5 percent to allow for waste in cutting and fitting around openings, and for the double starter course.

Double Coursing of Side Walls

The attractive appearance of side walls when shingles are given extra wide exposure, combined with deep shadow lines in the sun, has gained wide acceptance for the double-coursing method of applying shingles.

In double coursing, a low-cost shingle is generally used for the inner layer and this is covered with a No. 1 grade shingle on the outside. A variety of types of shingles are available for the outer course. Pre-stained shingles, which are available in attractive colors, are particularly

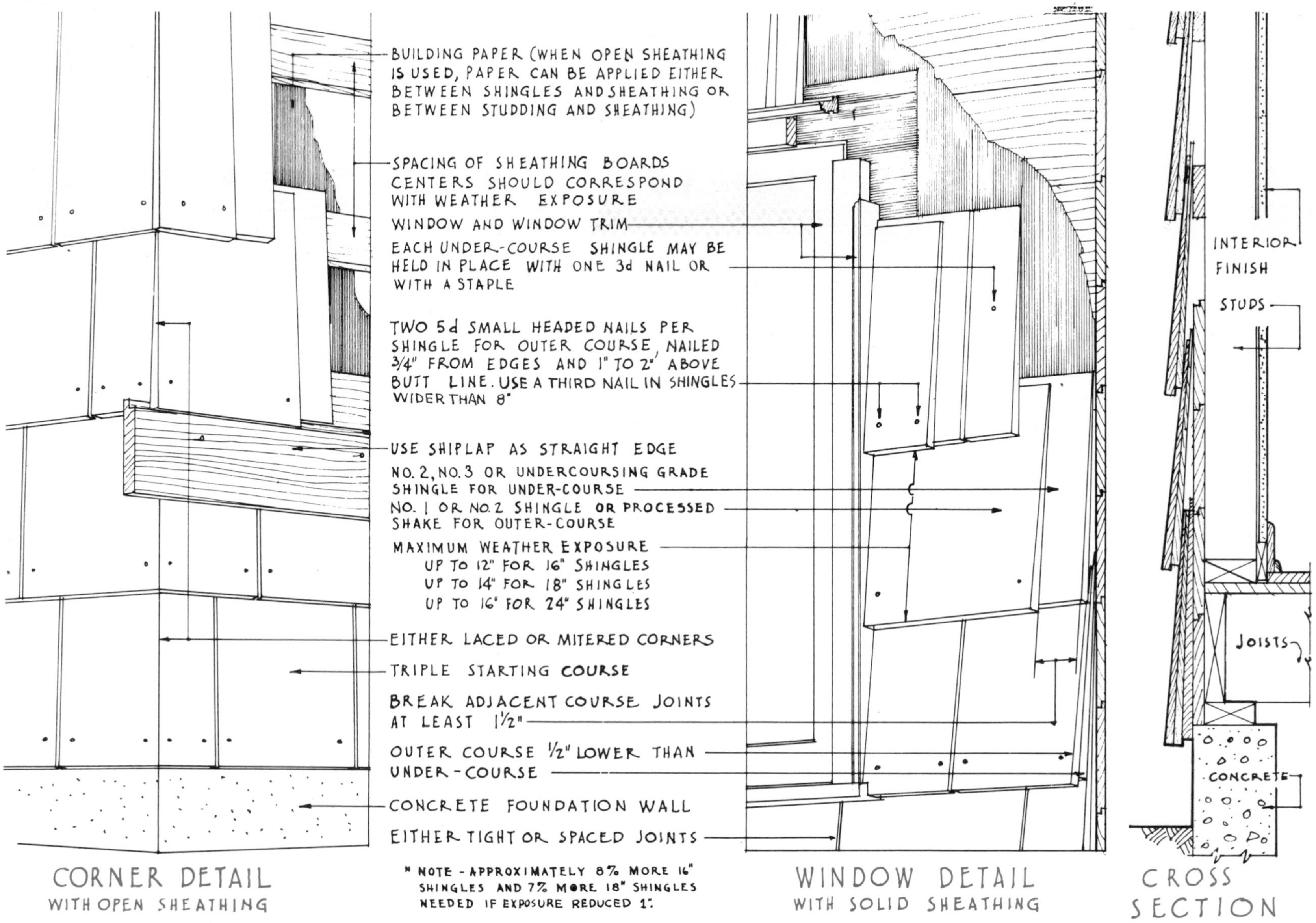
BUILDING PAPER (WHEN OPEN SHEATHING IS USED, PAPER CAN BE APPLIED EITHER BETWEEN SHINGLES AND SHEATHING OR BETWEEN STUDDING AND SHEATHING)
SPACING OF SHEATHING BOARDS CENTERS SHOULD CORRESPOND WITH WEATHER EXPOSURE
WINDOW AND WINDOW TRIM
EACH UNDER-COURSE SHINGLE MAY BE HELD IN PLACE WITH ONE 3d NAIL OR WITH A STAPLE
TWO 5d SMALL HEADED NAILS PER SHINGLE FOR OUTER COURSE, NAILED 3/4" FROM EDGES AND 1" TO 2" ABOVE BUTT LINE. USE A THIRD NAIL IN SHINGLES WIDER THAN 8"
USE SHIPLAP AS STRAIGHT EDGE
NO. 2, NO. 3 OR UNDERCOURSING GRADE SHINGLE FOR UNDER-COURSE
NO. 1 OR NO. 2 SHINGLE OR PROCESSED SHAKE FOR OUTER-COURSE
MAXIMUM WEATHER EXPOSURE
UP TO 12" FOR 16" SHINGLES
UP TO 14" FOR 18" SHINGLES
UP TO 16" FOR 24" SHINGLES
EITHER LACED OR MITERED CORNERS
TRIPLE STARTING COURSE
BREAK ADJACENT COURSE JOINTS AT LEAST 1 1/2"
OUTER COURSE 1/2" LOWER THAN UNDER-COURSE
CONCRETE FOUNDATION WALL
EITHER TIGHT OR SPACED JOINTS
INTERIOR FINISH
STUDS
JOISTS
CONCRETE
CORNER DETAIL
WITH OPEN SHEATHING
* NOTE - APPROXIMATELY 8% MORE 16" SHINGLES AND 7% MORE 18" SHINGLES NEEDED IF EXPOSURE REDUCED 1".
WINDOW DETAIL
WITH SOLID SHEATHING
CROSS SECTION

Fig. 12-34. Double-coursed method for shingle application.

suitable. Although wide exposures imply the use of long shingles, this effect is obtained in double coursing by the application of doubled layers of regular 16 in. or 18 in. shingles. The maximum exposure to the weather of 16 in. shingles double coursed is 12 in. For 18 in. shingles it is 14 in.

The application of shingles on a double-course side wall is illustrated in Fig. 12-34. Most procedures used for regular siding can be followed. When the application is over composition or spaced sheathing, include the position of the nailing strips on the story pole when it is laid out.

Shingle and Shake Panels

Shingles and shakes for side wall application are available in panel form. The panels consist of individual shingles (usually Western red cedar) permanently bonded to a backing. Standard size panels include lengths of 4 ft. and 8 ft., Fig. 12-35. These are available in various textures, either unstained or factory-finished in a variety of colors.

A special shingle panel, Fig. 12-36, is fabricated by mounting machine grooved shingles on an insulated backer board, forming a shiplap edge as shown. The panel is factory-finished and designed for exposures up to 14 in. Matching metal corners provide a tight, weatherproof fit, Fig. 12-37.

Shingle panels are applied by following about the same precautions and procedures described for regular shingles. Installation time however, is greatly reduced. Additional on-site labor is also saved when factory-primed or factory-finished units are used. When applying the latter, the installation should be made with nails of matching color, supplied by the manufacturer. Fig. 12-38 shows a partial view of a completed residential structure where shingle panels were used for the outside wall covering.

HANDY SAYS:

"When you are making an application of a specialized or prefabricated product such as shingle panels, be sure to follow the recommendations furnished by the manufacturer."

Re-siding with Wood Shingles

Shingles or shakes can be applied over old siding or other wall coverings that are sound and will hold nailing strips. First apply building paper over the old wall, then attach nailing strips as previously described. Usually it is necessary to

Fig. 12-35. Shingle panels for sidewall application. Above. Natural pattern with staggered butt line. Below. Brushed surface to emphasize grain pattern. (Shakertown Corp.)

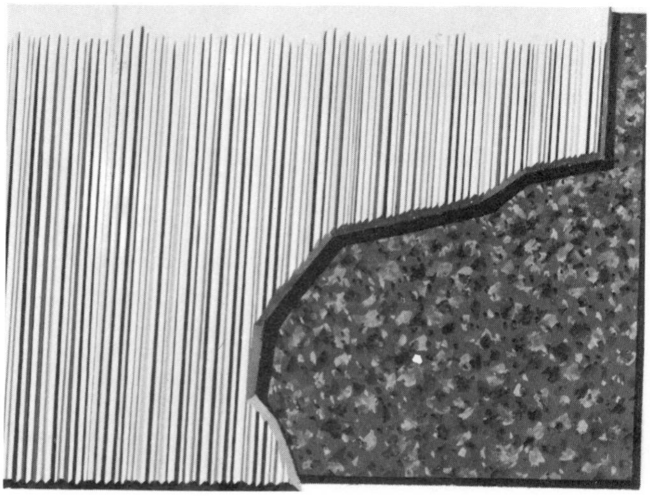

Fig. 12-36. Special shingle panel. Cutaway view shows insulated backing board. Designed for an exposure up to 14 in.

Fig. 12-37. Installing matching metal corners on shingle panels.

Fig. 12-38. Completed installation of shingle panels.

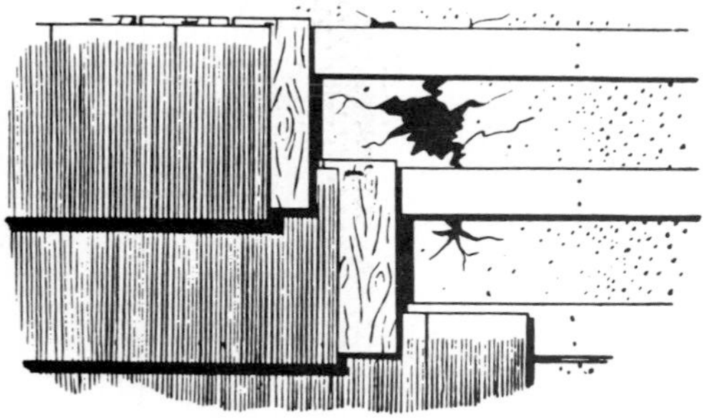

Fig. 12-39. Re-siding a stucco wall with wood shingles.

add new molding strips around the edge of window and door casings to trim the edge of the shingles. Fig. 12-39 illustrates how nailing strips are applied over an old stucco surface. Fig. 12-40 shows a shingle panel being applied over drop siding.

Mineral Fiber Siding

This product, also called asbestos cement siding, is produced from the same materials as mineral fiber roof shingles, (See Unit 10). Siding of this type is usually prefinished with modern long lasting factory baked coatings and is available in a variety of textures and colors. Both textured and smooth surfaced siding may be obtained with either straight or wavy exposed butt lines.

Mineral fiber siding usually comes in units (called shingles) 24 in. wide and 12 in. deep. When applied, the shingles are lapped 1 1/2 in. at the head, leaving an exposed area of 10 1/2 by 24 in. Some units, however, are as large as 48 in. wide. The siding is sold in squares; a square being sufficient to cover 100 square feet of wall surface. Ordinarily there are three bundles per square, each containing 57 pieces of the 12 by 24 in. material. Approximate shipping weight per square is 187 lbs.

Follow the same general procedures and methods for storing, handling, and cutting as listed in Unit 10 for mineral fiber shingles. The layout of a

Fig. 12-40. Using shingle panels for a re-siding application.

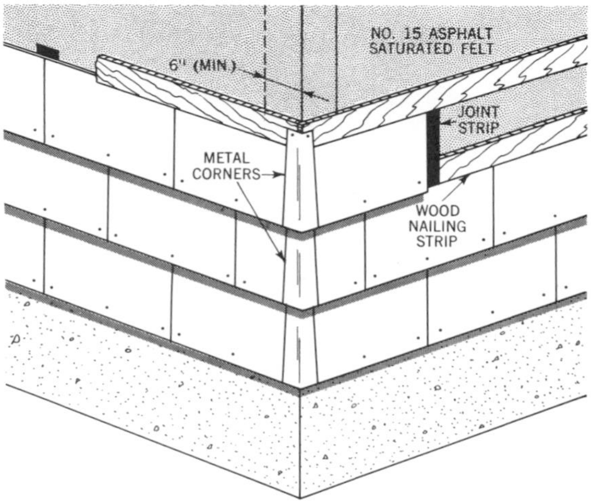

Fig. 12-41. Mineral fiber shingle application. Use double layer of felt around corner. (Mineral Fiber Products Bureau)

wall surface with a story pole and treatment of corners, Fig. 12-41, is basically the same as for other siding materials. Fig. 12-42 illustrates an approved method of installing flashing above openings.

The application of standard siding units over composition sheathing, which will not hold nails, is illustrated in Fig. 12-43. The nailing strips are placed to overlay the top of the lower course by about 3/4 in. The ends of the nailing strips must be located over studs. Vertical joints between shingle units must not occur over nailing strip joints. Note that the mineral fiber siding unit is applied with the bottom edge overlapping the wood strip 1/4 in. to provide a drip edge.

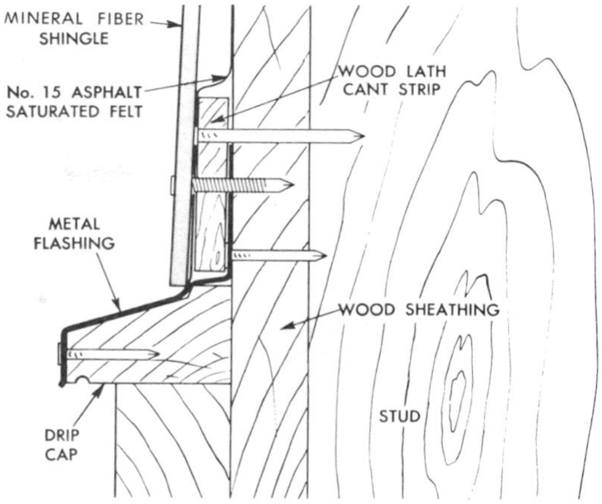

Fig. 12-42. Flashing above doors and windows.

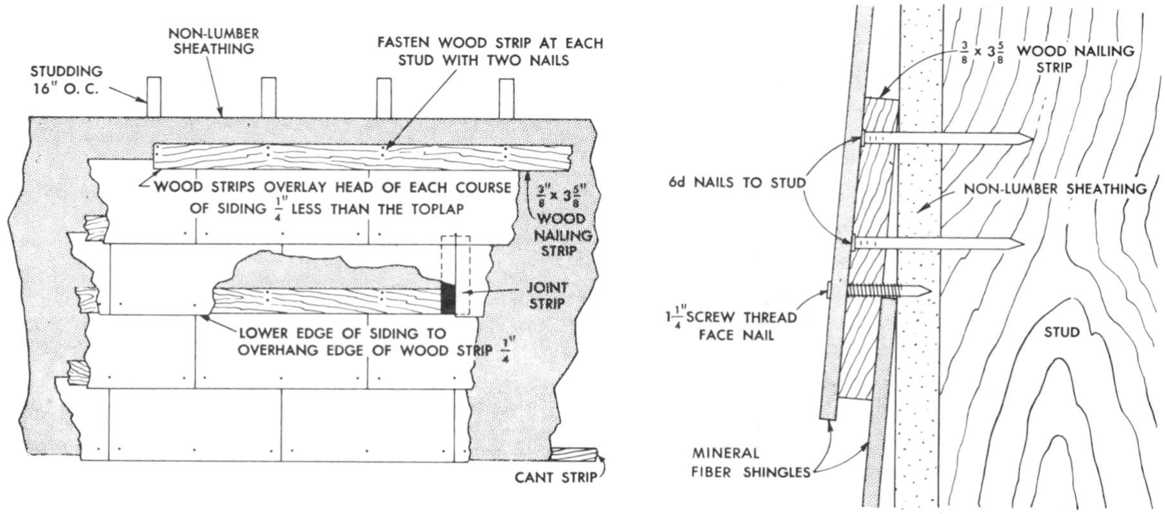

Fig. 12-43. Typical application of siding units over composition sheathing.

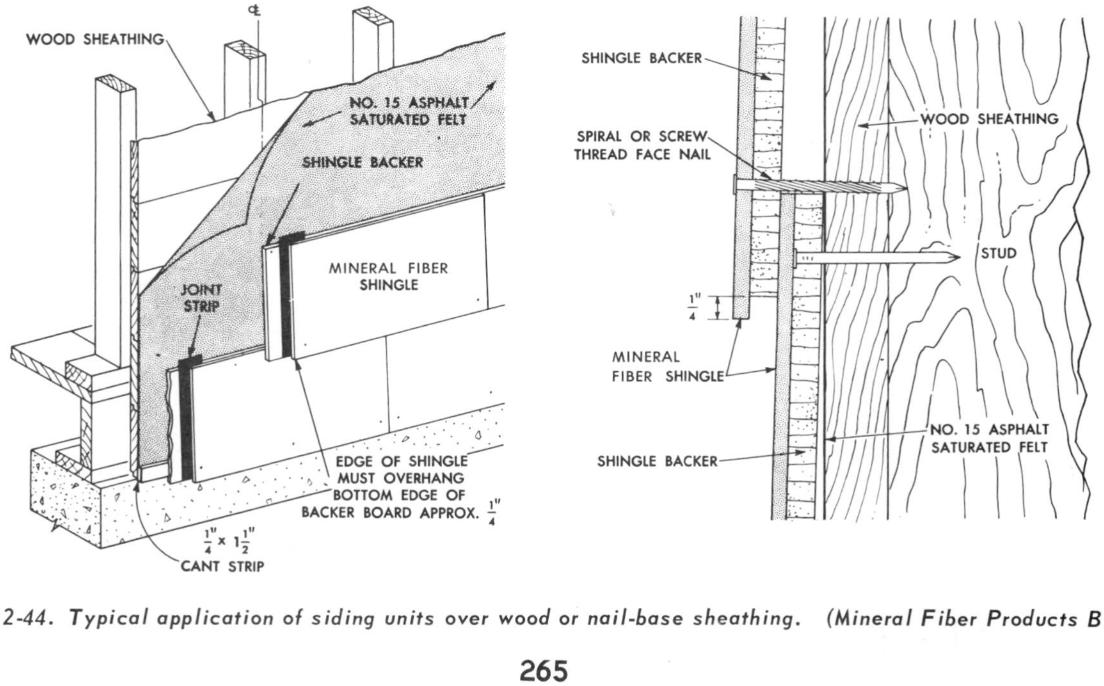

Fig. 12-44. Typical application of siding units over wood or nail-base sheathing. (Mineral Fiber Products Bureau)

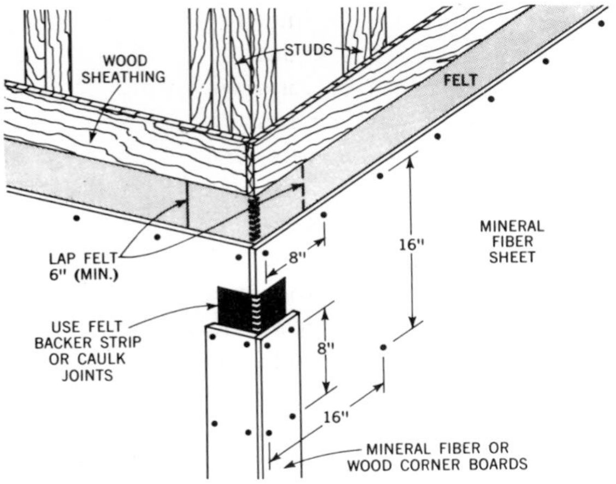

Fig. 12-45. Corner construction for mineral fiber sheet applications. (Mineral Fiber Products Bureau)

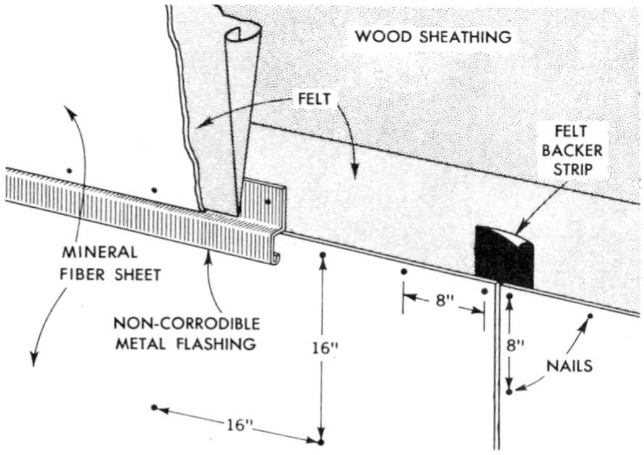

Fig. 12-46. Horizontal butt joint with metal flashing strip. Use backer strips behind all vertical joints.

If the sheathing consists of lumber, plywood or other materials that will serve as an adequate nailing base, the application can be made as shown in Fig. 12-44. The shingle backer produces an attractive, heavy shadow line along the lower edge of each course of siding. Be sure the end joints of the backer and shingle units do not occur at the same location.

When the shingle backer method is employed, first nail the backer in place with 1 1/4 in. galvanized nails spaced about 3 in. from the top edge. If this is carefully positioned along the chalkline layout, the siding units can be quickly lined up along the top edge as shown. Joint flashing strips are not required when the backer board consists of a material that is water repellent or nonstaining.

Flat Sheets

Mineral fiber in flat sheets is usually available in 4 x 4 ft. and 4 x 8 ft. sizes. In some areas larger sizes can be obtained. Thicknesses ranging from 1/8 to 1/2 in. permit the use of this product on a wide range of structural designs. Like siding shingle units, the sheets come unfinished, preprimed and prefinished in a variety of colors.

Follow the same general requirements in making an installation of flat sheets as for shingle units. Vertical joints must occur over a firm nailing base and a backer strip must be included. Sheets should be butted tightly together. Sometimes joints are covered with wood or mineral fiber batten strips. For decorative effects, additional battens may be added at various locations between the joints. Corners may be finished as shown in Fig. 12-45.

Horizontal joints may be butted or lapped. When butted it is best to install a metal flashing strip as illustrated in Fig. 12-46. If metal flashing is not used, the joints may be covered with a batten strip carefully bedded in a recommended caulking compound.

HANDY SAYS:

"Before making an installation of mineral fiber products, secure full information from manufacturers."

Asphalt Siding

Asphalt materials especially designed for siding applications are available in two forms; roll products and shingles. Roll materials are mineral surfaced. Many are embossed to simulate brick or stone. Shingles may be either the strip type or individual type. When remodeling, these products may be applied on either new wall surfaces or over old materials.

The same general procedures described for the application of roof shingles can be applied to side wall installation. Fig. 12-47 illustrates a method commonly used for individual shingles. The exposed nails can be eliminated if the tabs are sealed down with plastic cement. Be sure to use both horizontal and vertical chalklines to align the work.

At each outside corner a metal corner bead manufactured for the purpose is applied. A neat appearance will result if the width of the finishing shingle portion in each course is equal to the width of the starting shingle portion in the same course on the adjacent intersecting wall. At inside corners use a 12 in. wide flashing strip of No. 30 asphalt saturated felt.

If the trim around wall openings is flush with the old siding or backer-board, either a narrow wood molding or a special metal stop molding should be applied 1 in. in from the outside trim edge. The shingles are then butted against the strip. If desired the edges of the shingles may be laid in a narrow bead of plastic cement or caulking compound. When the trim projects beyond the face of the prepared wall surface, the end shingles are cut to fit against the edge of the trim and under the window sills. Asphalt plastic cement should be used under the joint between the shingles and trim.

Plywood Siding

The use of plywood as an exterior wall covering permits a wide range of application methods and decorative treatments. Fig. 12-48 illustrates and describes briefly several of the commonly used designs and surface textures. Plywood may be used to compliment other siding materials by using it for gable ends; flat panels above or below windows; or for a continuous band at various levels along an entire wall.

All plywood siding must be made from exterior type plywood, in which the layers of veneer are bonded together with a waterproof glue. Douglas fir is the most commonly used species, however cedar and redwood are also available. Panels come in either a sanded condition or with factory applied sealer or stain. Plywood siding with a special long lasting surface or coating is also available. Information on grading standards may be obtained from Unit 4.

Panel sizes are 48 in. wide by 8, 9, and 10 feet long. A 3/8 in. thickness is normally used for direct-to-stud applications while a 5/16 in. thickness may be used when installed over an approved sheathing. Thicker panels are required when the texture treatment consists of deep cuts.

Applications consisting of large sheets are generally made with the panels in a vertical position, thus eliminating the need for horizontal joints. Fig. 12-49 shows the installation of a vertical panel being made over a sheathed wall. For unsheathed walls, Fig. 12-50, the thickness of the plywood should be not less than 3/8 in. for 16 in. stud spacing, 1/2 in. for 20 in. stud spacing, and 5/8 in. for 24 in. stud spacing. Vertical joints must occur over studs and horizontal joints must be over solid blocking. Standard application requirements are given in Fig. 12-51.

Plywood lapped sidings can provide the same appearance as regular beveled siding. Heavy shadow lines are secured by using spacer strips at the lapped edges. Application requirements are given in Fig. 12-52. A bevel of at least 30 deg. is recommended. The lap should be at least 1 1/2 in. Vertical joints should be butted over a shingle and centered over a stud unless wood sheathing at least 3/4 in. thick is used. Nail siding to each of the studs along the bottom edge and not more than 4 in. O.C. at vertical joints. Nails should penetrate studs or wood sheathing at least 1 in. If plywood lap siding is wider than 12 in. a wood taper strip should be used at all studs with nailing at alternate studs. Outside corners should butt against corner molding or be covered.

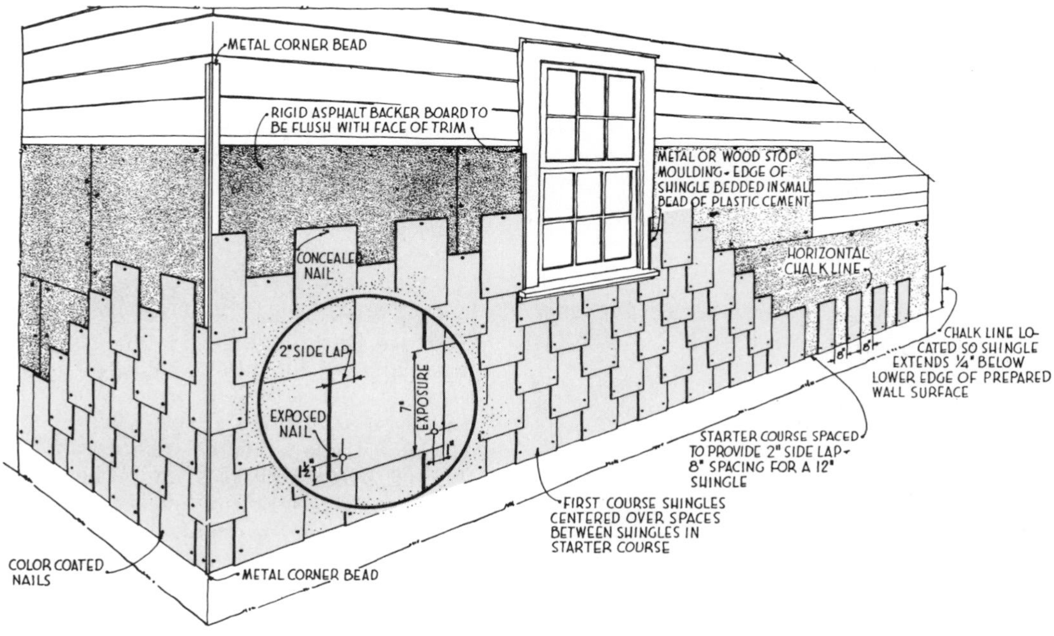

Fig. 12-47. Typical application when re-siding with asphalt shingles.

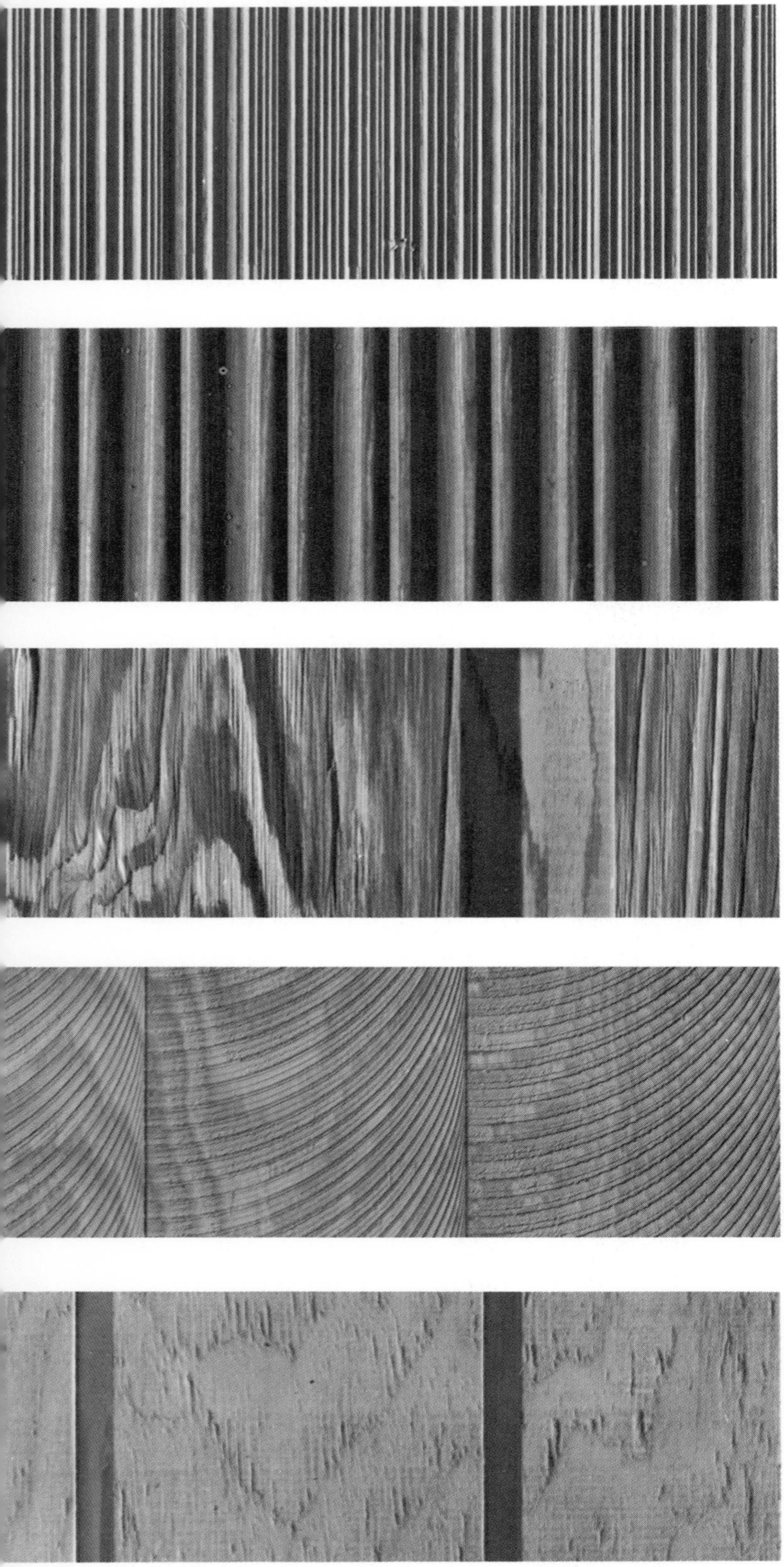

Striated
Random width, closely spaced grooves form vertical patterns in surface. The striations conceal nail heads, checking and grain raise, and eliminate unsightly joints.

Finish with exterior paint or pigmented stain.

Corrugated Surface
Closely spaced corrugations fabricated into surface veneer. Available with irregularly spaced grooves of uniform depth or with irregularly spaced grooves of varying depth.

Finish with exterior paint or pigmented stain.

Textured Reverse-Boards & Batten
Deep, wide grooves cut into brushed or rough textured cedar faces to provide sharp shadow line. Grooves 1/4" deep, 1-1/2" wide, spaced 12" o.c., panel thickness 5/8".

Finish with exterior pigmented stains or leave natural, without finish, for weathered rustic effect.

Plank-Textured
Rough sawn sections create a lap effect with distinct shadow lines every 8 inches. Long edges are shiplapped for continuous patterns.

Especially suitable for exterior pigmented stains.

Texture-One-Eleven
Deep grooves cut into face for sharp shadow lines. Grooves 1/4" deep, 3/8" wide, 4" or 2" o.c., panel thickness 5/8". Other groove spacings on order. Standard type available with unsanded face containing small knotholes and other natural wood characteristics for rustic effect. Also available with Medium Density overlaid, sanded, brushed or striated faces on order.

Fig. 12-48. Surface textures and designs generally available in plywood siding. (American Plywood Assoc.)

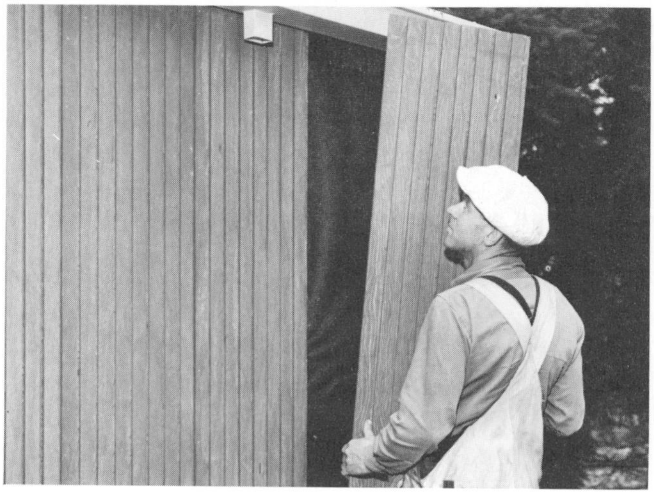

Fig. 12-49. *Textured plywood siding panels being installed over insulation board sheathing.*

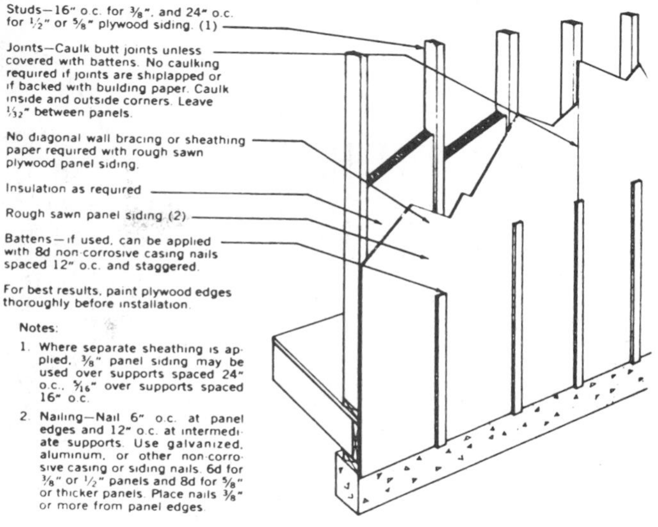

Studs—16" o.c. for 3/8", and 24" o.c. for 1/2" or 5/8" plywood siding. (1)
Joints—Caulk butt joints unless covered with battens. No caulking required if joints are shiplapped or if backed with building paper. Caulk inside and outside corners. Leave 1/32" between panels.
No diagonal wall bracing or sheathing paper required with rough sawn plywood panel siding.
Insulation as required
Install diagonal bracing and sheathing paper when lap or bevel siding is installed directly in studs.
Rough sawn panel siding (2)
Battens—if used, can be applied with 8d non-corrosive casing nails spaced 12" o.c. and staggered.
For best results, paint plywood edges thoroughly before installation.
Notes:
1. Use same nail schedule for installing lap siding direct to studs or over sheathing. When installed over sheathing, siding joints may occur away from studs.
1. Where separate sheathing is applied, 3/8" panel siding may be used over supports spaced 24" o.c., 5/16" over supports spaced 16" o.c.
2. Nailing—Nail 6" o.c. at panel edges and 12" o.c. at intermediate supports. Use galvanized, aluminum, or other non-corrosive casing or siding nails. 6d for 3/8" or 1/2" panels and 8d for 5/8" or thicker panels. Place nails 3/8" or more from panel edges.

Fig. 12-51. *Standard application requirements for vertical plywood siding panels. (American Plywood Assoc.)*

Fig. 12-50. *Installing plywood siding panels directly to studs. Panel thickness is 5/8 in.*

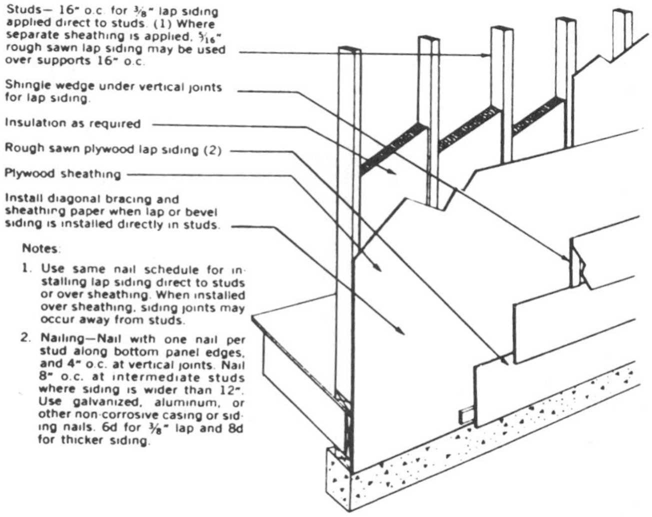

Fig. 12-52. *Standard application requirements for lapped plywood siding.*

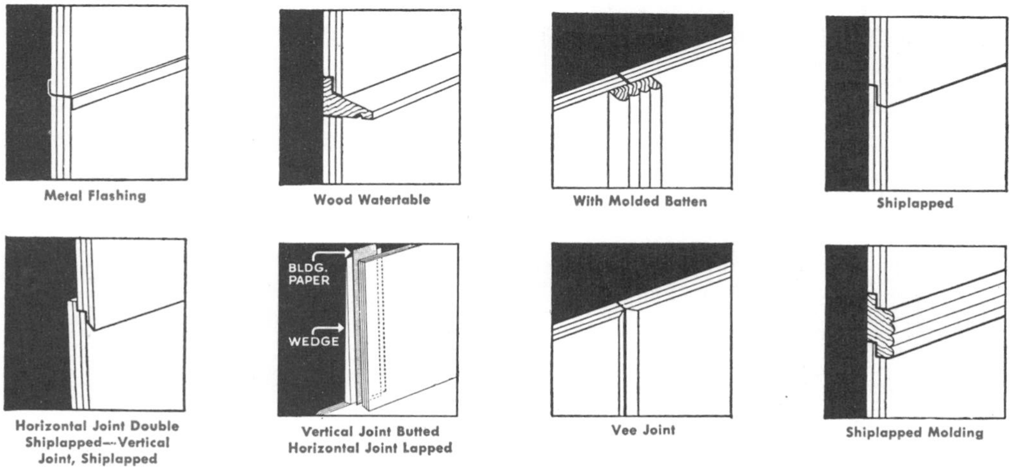

Wood Watertable
With Molded Batten

Vee Joint

Fig. 12-53. *Joint details for plywood siding. All edges of plywood should be sealed with paint or special caulking compound. (American Plywood Assoc.)*

Fig. 12-54. Lap siding made of hardboard. Textured to resemble natural wood. (Masonite Corp.)

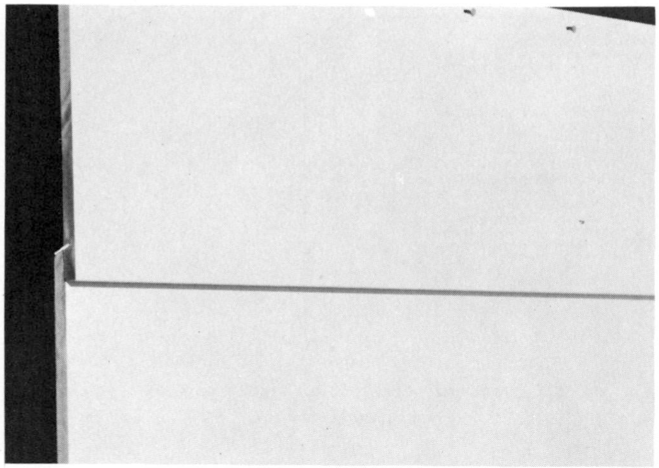

Fig. 12-55. View of lap siding with a special mounting strip that hooks over the lower course.

Fig. 12-53 shows several ways to handle joints between plywood panels. All edges of plywood siding--whether butted, V-shaped, lapped, covered or exposed--should be sealed with a heavy application of high grade exterior primer, aluminum paint, or heavy lead and oil paint. Special caulking compounds are also recommended.

HANDY SAYS:

"Because large sheets of plywood and hardboard siding provide tight, draft-free wall construction, it is important to have an effective vapor barrier near the warm surface of the wall."

Hardboard Siding

Improved techniques in the manufacture of hardboard have resulted in the production of siding materials that are durable, easy to apply, and adaptable to various architectural effects. Installation methods are similar to those normally used and previously described for plywood sidings. Hardboard sidings do not usually exhibit dimensional stability factors equal to plywood and special precautions should be observed in the application. Manufacturers usually emphasize the need to leave a slight space where hardboard siding butts against adjacent pieces or trim members.

Hardboard siding panels are available in standard widths of 4 ft. and lengths of 8, 9, and 10 ft. Lap siding units are usually 12 in. wide by 16 ft. long, however narrower widths can be purchased. The most common thickness is 7/16 in. Like plywood, hardboard sidings are furnished in a wide range of textures and surface treatments. See Fig. 12-54. Most panels have a factory-applied primer coat. Prefinished units with matching batten strips and trim members are available.

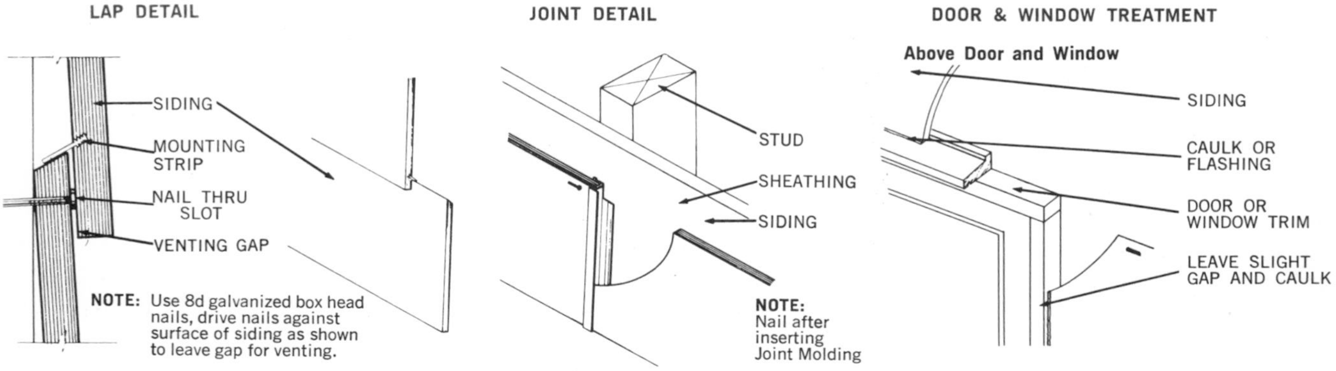

Fig. 12-56. Application details of siding system.

When applying hardboard siding, follow standard installation procedures as described for regular siding materials. Studs should not be spaced greater than 16 in. O.C. A firm and adequate nailing base is essential. Use a fine-tooth handsaw or power saw equipped with a combination blade to cut the panels. Nails must be galvanized or otherwise weatherproofed. Wood trim and corner boards should be at least 1 1/8 in. thick. Join units lightly together and space nails at least 1/2 in. in from edges and ends. Use an approved caulking compound at joints.

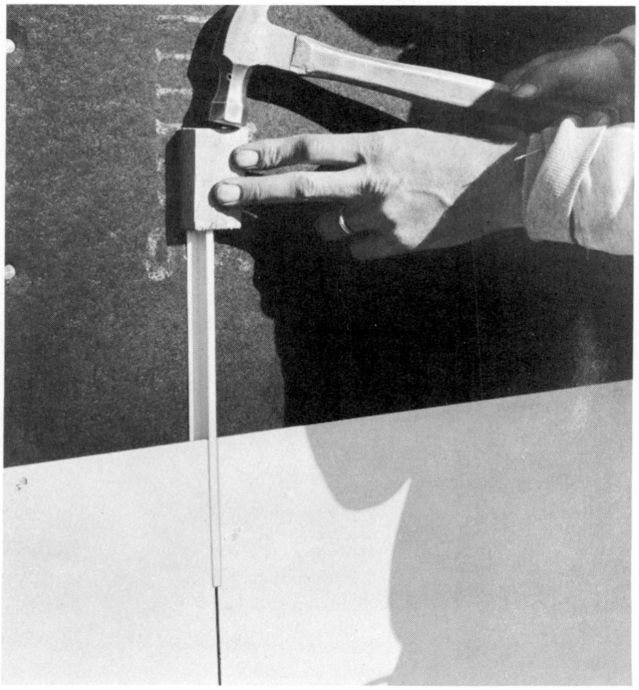

Fig. 12-57. Driving joint molding into place.

Siding Systems

Many manufacturers produce siding products that include special designs or patented devices which simplify the methods of application.

Fig. 12-55 shows a lap siding system where the units are equipped with a mounting strip. This strip hooks over the top of the previously applied course and the unit is nailed in place through prepunched slots along the top edge. The slots permit the siding to expand or contract slightly under the nail head, thus eliminating possibility of distortion. Application details are illustrated in Fig. 12-56.

A special joint molding made of metal is furnished with the siding system. This is driven in place as illustrated in Fig. 12-57. On unsheathed walls the joints must occur at studs. The siding is prefinished in a variety of colors with matching metal corners, joint molding, and trim. Colored caulking is also available.

Vertical siding is available in the same system with special matching trim and colored nails. On some installations, batten strips are used to cover hardboard batten backers. If desired, special snap-on batten strips can be furnished to cover hardboard batten backers, thus eliminating any exposed nailing. In addition to covering all vertical joints, intermediate batten strips may be installed to fulfill design requirements.

Another system is based on a patented aluminum fastener that eliminates face nailing and also provides venting to minimize moisture traps behind the siding. A metal starter strip is attached to the bottom of the sheathing or sill plate to carry the first course. The top edge of the siding is secured by fasteners, nailed in place at 16 in. intervals, Fig. 12-58. After the top edge is attached, the bottom edge is driven into the prongs of the fastener in the lower course by placing a softwood block over the position and striking it with two or three blows of a hammer. Fig. 12-59 shows a completed installation.

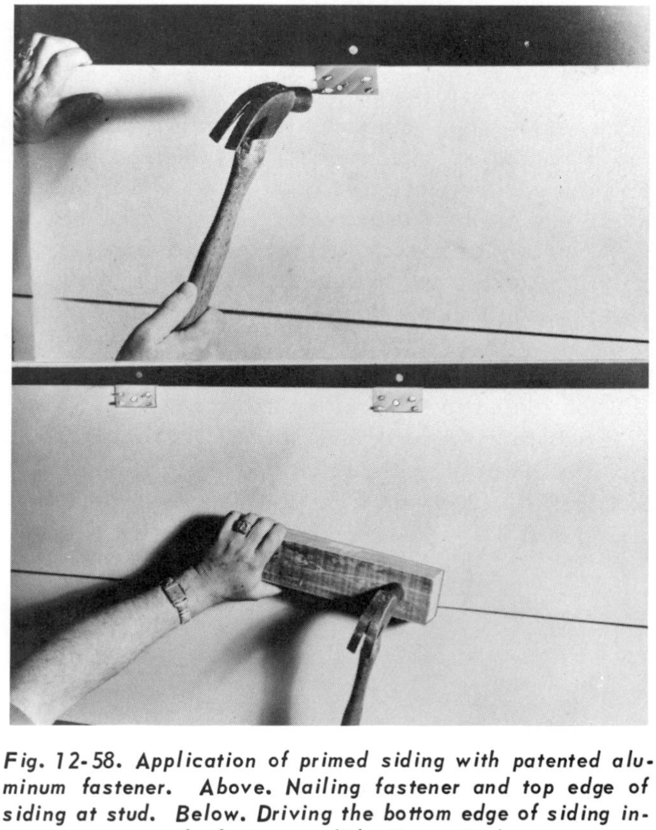

Fig. 12-58. Application of primed siding with patented aluminum fastener. Above. Nailing fastener and top edge of siding at stud. Below. Driving the bottom edge of siding into the fastener. (The Upson Co.)

Fig. 12-59. Completed siding installation with special fastener -- no exposed face nailing necessary.

Aluminum Siding

This type of siding material offers low maintenance costs. It is factory finished with baked-on enamel, and provides an appearance that closely resembles painted wood siding. Aluminum siding is designed for use on new or existing construction and can be applied over wood, stucco, concrete block, and other surfaces that are structurally sound. Basic specifications (alloy and gauge) for aluminum siding are established by FHA and the Aluminum Siding Association.

A variety of horizontal and vertical panel styles in both smooth and textured designs are produced with varying shadow lines and size of face exposed to the weather. An insulated panel is also produced that consists of an impregnated fiberboard material laminated to the back surface.

Each manufacturer supplies directions for the installation of his particular product. These should be carefully followed. Fig. 12-60 shows a horizontal siding unit being fastened in place. Panels are fabricated with prepunched nail and vent holes and special interlocking design. Special trim strips, Fig. 12-61, which are easily attached around windows and doors provide a weathertight joint. Corners, similar to those used on regular lap siding are installed as shown in Fig. 12-62.

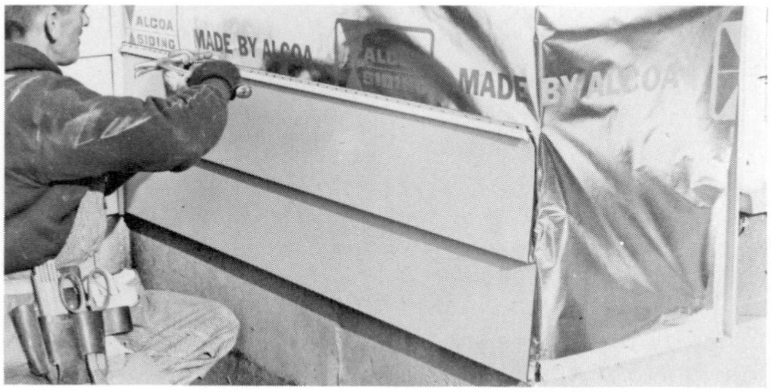

Fig. 12-60. Installing an aluminum siding panel. Lower edge interlocks with previously applied course. (Alcoa Building Products)

As a precaution against faulty electrical wiring or appliances that might energize aluminum siding and create an electrical hazard, grounding should be included. The Aluminum Siding Association recommends that a No. 8 or larger wire be connected to any convenient point on the siding and to the cold water service or the electrical service ground. Connectors should be UL (Underwriters Laboratories) approved.

Fig. 12-61. Applying special aluminum trim strip under window.

Fig. 12-62. Installing corners that match the aluminum siding.

Vinyl Siding

Recent developments in the chemical industries have resulted in economical production of a rigid polyvinyl chloride compound that is tough and durable. This material, commonly called vinyl, is extruded into siding units (either horizontal or vertical) and accessories, Fig. 12-63. Panel thickness of about 1/20 in. is available in various widths up to 8 in.

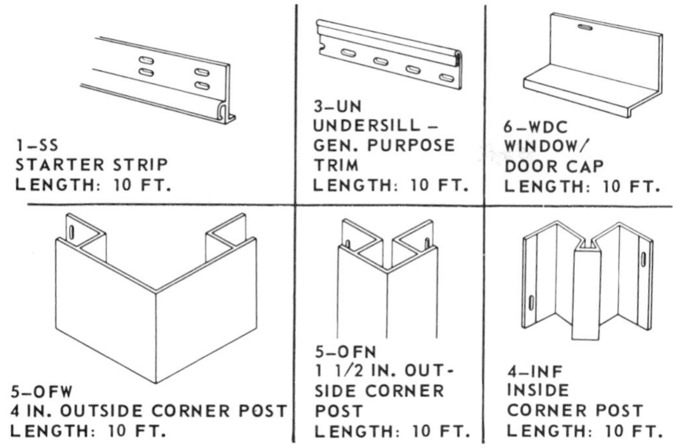

Fig. 12-63. Vinyl siding accessories. (Bird and Son, Inc.)

Like aluminum siding, vinyl siding is usually installed with a backer board or insulation board behind each sheet. This backer adds rigidity and strength as well as insulation. Panels are designed with interlocking joints that are moisture proof. Since the siding must be allowed to expand and contract slightly with temperature changes, the nail holes are slotted to permit the resulting movement.

HANDY SAYS:

"When making an installation of vinyl siding, be sure to read and follow the directions furnished by the manufacturer."

Stucco

When properly applied, stucco makes a satisfactory exterior wall finish. The finish coat may be tinted by adding coloring or the surface may be painted with a suitable material. Where stucco is used on houses more than one story high, the use of balloon framing for the outside walls is desirable. If platform framing is used, shrinkage of joists may cause distortion or cracks in the stucco.

The base for stucco consists of wood sheathing, sheathing paper, and metal lath, Fig. 12-64. The metal lath should be heavily galvanized and spaced at least 1/4 in. away from the sheathing so the base coat (called scratch coat) can be easily forced through, thoroughly embedding the lath. Metal or wood molding with a groove that "keys" the stucco is applied at edges and around openings. Galvanized furring nails, metal furring strips, and self-furring lath are available. Nails should penetrate the sheathing at least 3/4 in. When fiberboard or gypsum sheathing is used, nailing with adequate penetration should occur over studs.

Brick or Stone Veneer

A veneer wall is usually not referred to as a masonry wall but as a frame wall with masonry being used in place of the siding. In this type of wall, the masonry (which may be brick, concrete units, or stone) supports its own weight only.

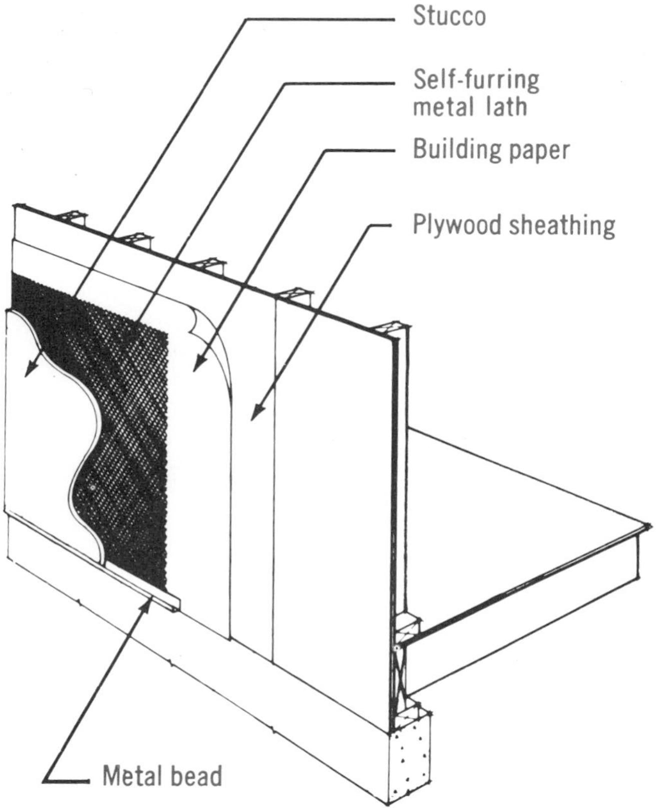

Fig. 12-64. General construction details for a stucco finish. Building paper (also called sheathing paper) is recommended for both interior and exterior plywood sheathing. (American Plywood Assoc.)

The foundation must be sufficiently wide to provide a base for the masonry, Fig. 12-65. A base flashing of noncorrosive metal should extend from the outside face of the wall, over the top of the ledge, and at least 6 in. up behind the sheathing as shown. When the construction consists of plywood sheathing, and an air space of 1 in. is included, sheathing paper is not required.

Corrosion-resistant metal ties are used to secure the veneer to the framework. These are usually spaced 32 in. apart horizontally and 15 in. vertically. Where other than wood sheathing is used, the ties should be secured to the studs. Weep holes (small openings in the bottom course) permit the escape of any water or moisture that may penetrate the wall. They are spaced about 4 ft. apart.

Select a type of brick suitable for exposure to the weather. Such brick is hard, and low in water absorption. Sandstone and limestone are most commonly used for stone veneer. These materials vary widely in quality. Be sure to select materials known locally to be durable.

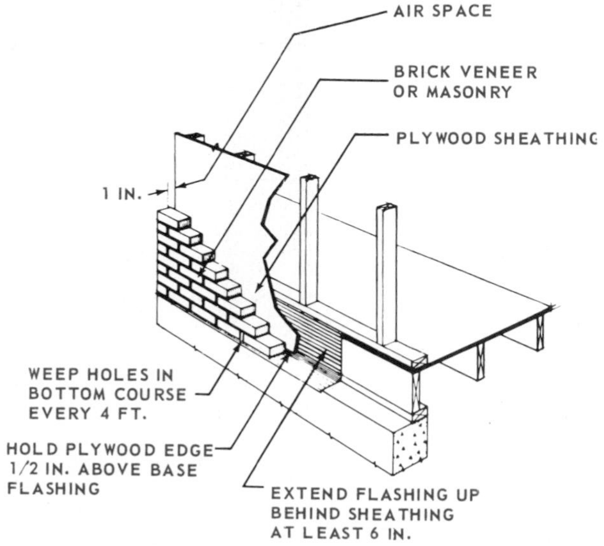

Fig. 12-65. General construction details for brick veneer construction using plywood sheathing. (American Plywood Assoc.)

Blinds and Shutters

Some architectural designs may require the installation of blinds and shutters at the sides of window units. These consist of frame assemblies with solid panels or louvers, Fig. 12-66. In the

Fig. 12-66. Shutters are sometimes used to provide a decorative effect. (Western Wood Products)

Fig. 12-67. In this home shutters provide an architectural feature and emphasize the horizontal lines.

early days of our country they served an important function since they could be closed over the window, thus protecting the glass which was very expensive. Closing and locking the shutters also provided some security to the inhabitants.

Today, shutters and blinds are used mainly for decorative effects. They tend to extend the width of the window unit, emphasizing the horizontal lines of the structure, Fig. 12-67. Hinging devices are seldom employed and they are usually affixed to the exterior wall with screws or other fasteners so they can be easily removed for painting and maintenance. Stock sizes include various heights to fit standard window units; with widths ranging from 14 in. to 20 in. by 2 in. increments (unit widths are increased 2 in. at a time -- 14 to 16, etc.).

Test Your Knowledge - Unit 12

1. In cornice construction the strip nailed to the wall to support the lookouts and soffit is called a ______________.
2. Some prefabricated soffit systems utilize steel channels to provide rigidity to the soffit material, thus eliminating the need for______.
3. Of the various types of horizontal wood siding, ______________siding usually has the most strength and tightest joints.
4. The best grade of solid wood siding is made from ___________(edge-grain, flat-grain) material.
5. Head flashing can usually be omitted over an opening if the vertical distance to the soffit is less than_____the width of the overhang.
6. Plain beveled siding in a 10 in. width should be lapped about_____in.
7. Corner boards are installed ________(before, after) the horizontal siding units are fastened in place.
8. When plain square-edged boards are used for vertical siding the joint is usually covered with a ___________strip.

Vertically grooved plywood siding provides a practical exterior wall finish for these apartments on a university campus. Each apartment unit (called a module) was prefabricated in an industrial plant and then assembled on the building site. (American Plywood Assoc.)

9. Standard wood shingle lengths for side wall coverage are 16 in., 18 in., and _____ inches.
10. In single-coursed side walls, the weather exposure of wood shingles should not be greater than _____ their length, minus 1/2 in.
11. The most commonly used mineral fiber siding shingle is _____ in. wide.
12. When using plywood siding over an unsheathed wall, the thickness should not be less than _______ in. when studs are spaced 16 in. O.C.
13. Large sheets of plywood siding usually result in tight construction, therefore it is highly important that a vapor barrier be included on the _________ (warm, cold) side of the wall.
14. The most common thickness for hardboard siding material is ______ in.
15. As a protection against faulty electrical wiring or appliances, aluminum siding should be _____________.
16. Small openings located in the bottom course of veneer construction that permits moisture to escape are called ____________.

Outside Assignments

1. Observe and study the cornice work on your home or some other residential structure in in your neighborhood. Prepare a detail drawing (use a scale of 1" = 1' - 0") showing a typical cross section. Since most of the construction will likely be hidden, you will need to develop your own structural design and select sizes for some of the parts. Study the details shown in various architectural plans and those in Fig. 12-1 of this Unit. Make your drawing similar to these and be sure to include the sizes of all materials.
2. Visit a builders' supply store or lumber yard in your locality. Study the various types of siding carried in stock. Secure descriptive folders about various prefinished products and special siding systems. Also secure approximate costs. Prepare a written report that includes a general description of the materials and summarizes the installation procedures. List some advantages and disadvantages of the various types.
3. Visit a building site in your neighborhood where the exterior wall finish is being applied. Be sure to secure permission from the builder or head carpenter. Observe the methods and materials being used. Make notes about: type of framing and stud spacing; type of sheathing; type of flashing; type, size, and quality of siding material; layout methods; how units are cut and fitted; nails and nailing pattern; and special joint treatment. Carefully organize your notes and make an oral report to your class.

Exterior side walls. Above. Using cedar shakes -- single course with concealed nailing. Nails are driven about 1 in. above butt line of succeeding course, so shingles will cover them. Left. A board and batten plywood siding. Right. Reverse board and batten siding. The plywood siding comes in panels 48 x 96, 108, or 120 in.

Unit 13

THERMAL AND SOUND INSULATION

Today, a high level of comfort is obtainable in our homes and other buildings, through the use of modern insulating materials. The successful operation of heating and air conditioning systems depends to a considerable extent on carefully constructed floors, walls, and ceilings where adequate consideration has been given to thermal insulation and vapor barriers. Insulation is essential in comfort-conditioned buildings -- whether they are located in northern climates, or in the warm southern regions.

Economical operation, another basic objective, can be achieved through a well-insulated structure. In cold climates, fuel will be saved. When summer heat is the predominate factor, insulation will reduce the size requirements of the air conditioning equipment and also reduce the cost of operation.

Fig. 13-1. Modern insulation materials are packaged to insure convenient handling and storage.

A wide range of insulation materials is available to fill the requirements of modern construction. They are "engineered" for efficient installation and come in conveniently sized packages that are easy to handle and store, Fig. 13-1.

Accoustical treatments and sound control employ many of the same materials used for thermal insulation and is therefore included in this Unit.

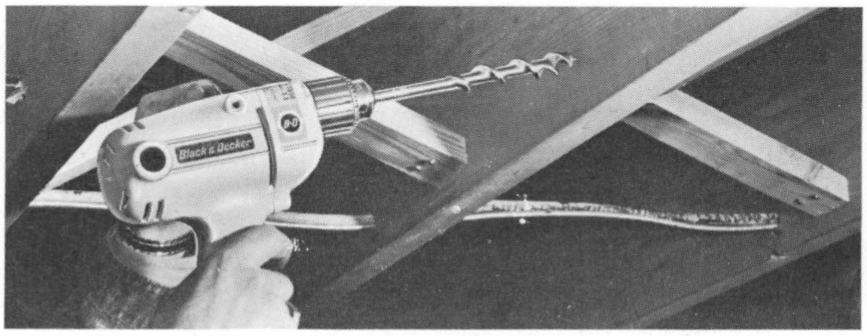

Fig. 13-2. Installing electrical wiring.

Building Sequence

While carpenters are busy completing the exterior finish of the structure, other tradesmen work on the inside. Duct work is installed in the floors, walls, and ceiling for heating and air conditioning. The electrician strings rough wiring, Fig. 13-2. He also installs other types of conduit in the framework and sets the metal boxes for convenience outlets, switches, and lighting fixtures that must be installed at this time.

Fig. 13-3. The plumber must install pipes before insulation is placed.

The plumber installs the drains, and vent stacks and also the pipes that carry the water service, Fig. 13-3. Built-in plumbing fixtures (bath tubs and shower drains) are installed and then carefully covered with building paper or other materials to protect them during the inside finish operations.

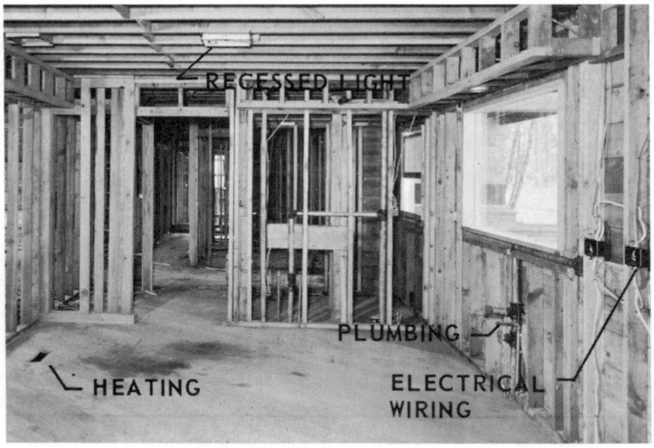

Fig. 13-4. Interior view of a residential structure with electrical, plumbing, and heating rough-in complete and ready for insulation materials.

During the installation of the heating and plumbing "rough-in", the tradesman often needs to cut through structural framework. The carpenter should check this work. Framing members should be reinforced wherever necessary. Fig. 13-4 shows an interior view of a house with the rough-in of heating, plumbing, and electrical work completed and the structure ready for insulation and interior wall finish.

How Heat is Transmitted

Although the carpenter is seldom required to design buildings or figure heat losses, he should have some knowledge and understanding of the theory and factors that are involved. Thus, he will more fully appreciate how important it is to carefully select and install thermal insulation materials.

Heat seeks a balance with surrounding areas. When the inside temperature is controlled within a given comfort range there will be some flow of heat; from the inside to the outside in winter and from the outside to the inside during hot summer weather.

Heat is transferred through walls, floors, ceilings, windows, and doors at a rate directly related to the difference in temperature, and the resistance to heat flow provided by intervening materials. The transfer of heat takes place by one or more of three methods -- conduction, convection, and radiation. Fig. 13-5.

CONDUCTION. Transmission of heat from one molecule to another within a given material or from one material to another when they are held in direct contact. Dense materials such as metal or stone conduct heat more rapidly than porous materials such as wood and fiber products. Any material will conduct some heat when a temperature difference exists between its surfaces.

CONVECTION. This term applies to the transfer of heat by another agent, usually air. In large spaces, the molecules of air can carry heat from warm surfaces to cold surfaces. When air is heated it becomes lighter and rises, thus causing a flow of air (called convection currents) within the space. Air is a good insulator when confined to a small space or cavity where convection flow is limited or nonexistent. In walls and ceilings, air spaces are considered restrictive to convection currents and will reduce the flow of heat.

RADIATION. Heat can be transmitted by wave motion in about the same manner as light. This process is called radiation because it represents radiant energy. Heat obtained from the sun is radiant heat. The waves do not heat the space through which they move, but when they come in contact with a colder surface, a part of the energy is absorbed while some may be reflected. Effective resistance to radiation is attained through reflec-

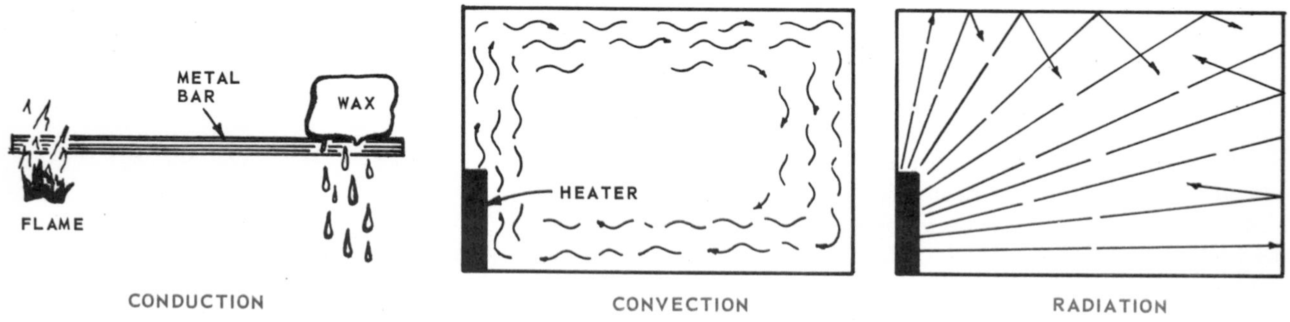

Fig. 13-5. How heat is transferred.

tion. Shiny surfaces, such as aluminum foil, are often used to provide this type of insulation.

Actually, heat transmission through walls, ceilings, and floors will be a result of all three of the methods, operating in various degrees. In addition to this -- some heat is lost through cracks around doors, windows, and other openings.

Thermal Insulation

All building materials resist the flow of heat (mainly conduction) to some degree, depending on their porosity or density. As previously stated, air is an excellent insulator when confined to the tiny spaces or cells inside a porous material. Dense material such as masonry or glass contain few if any air spaces and are poor insulators. Fibrous materials are generally good insulators not only because of the porosity in the fibers themselves but also because of the thin film of air that surround each individual fiber.

Commercial insulation materials are made of glass fibers, glass foam, mineral fibers, organic fibers, and foamed plastic. A good insulation material should be fireproof, vermin-proof, moisture proof, and resistant to any physical change that would reduce its effectiveness against heat flow.

Selection must be based on initial cost, effectiveness, durability, and the adaptation of its form to that of the construction and installation methods.

Heat Loss Coefficients

The thermal properties of common building materials and insulation materials are known or can be accurately measured. The amount of heat flow (transmission) through any combination of these materials can be calculated. First it is necessary to know and understand certain terms.

Btu -- The abbreviation for British thermal unit and is the amount of heat needed to raise the temperature of 1 lb. of water 1 deg. F.

"k" -- The amount of heat (in Btu's) transferred in one hour through 1 square foot of a given material that is 1 in. thick and has a temperature difference between its surfaces of 1 deg. F. It is also called the coefficient of thermal conductivity.

"C" -- Represents the conductance of a material (regardless of its thickness) and shows the amount of heat (Btu's) that will flow through the material in one hour per square foot of surface with 1 deg. temperature difference. Example: the "C" value for an average hollow concrete block is .53.

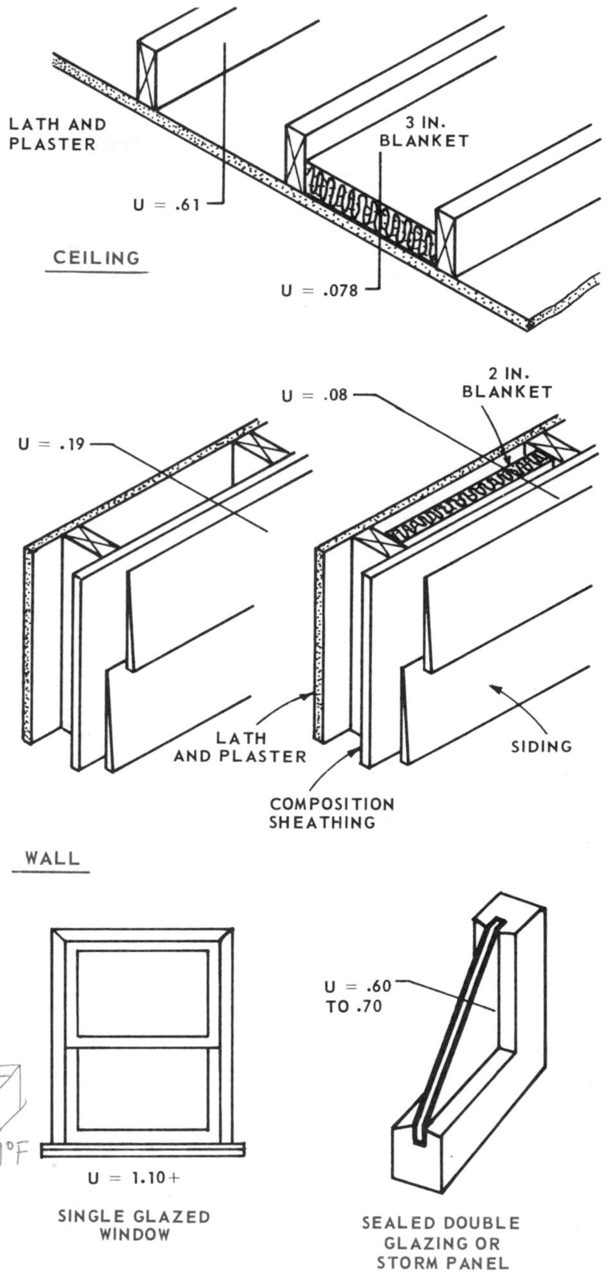

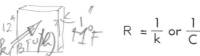

Fig. 13-6. How insulation reduces the flow of heat (U) in various constructions.

"R" -- Represents the resistivity or resistance which is the reciprocal of conductivity or conductance. A good insulation material will have a high "R" value.

$$R = \frac{1}{k} \text{ or } \frac{1}{C}$$

"U" -- Represents the total heat transmission (Btu's) per square foot, per hour with 1 deg. temperature difference for a structure (wall, ceiling, floor) which may consist of several

"U" VALUES OF COMMON WALL CONSTRUCTION — 3/8" GYPSUM LATH AND 1/2" PLASTER INTERIOR

TYPE OF WALL*	Construction Indicated	WITH INSULATION BETWEEN FRAMING (gypsum lath and plaster interior) 1" Batt or Blanket	2" Batt or Blanket	3" Batt or Blanket	3⅝" Wool Loose Fill	Reflective Insulation E = 0.05 1 Air Space	2 Air Spaces	3 Air Spaces
FRAME — Wood siding or wood shingle								
1. With wood sheathing	0.24	0.13	0.087	0.065	0.065	0.17	0.12	0.079
2. With ½" reg. density insulation board sheathing	0.23	0.12	0.086	0.065	0.065	0.17	0.11	0.078
3. With 25/32" insulation board sheathing	0.19	0.11	0.080	0.061	0.061	0.14	0.10	0.071
4. With shingle backer, wood shingles and 25/32" insulation board sheathing	0.18	0.11	0.078	0.060	0.060	0.14	0.10	0.069
VENEER — Brick on Frame								
1. With wood sheathing	0.27	0.14	0.090	0.067	0.067	0.19	0.12	0.083
2. With ½" reg. density insulation board sheathing	0.25	0.13	0.088	0.066	0.066	0.18	0.12	0.081
3. With 25/32" insulation board sheathing	0.21	0.12	0.083	0.063	0.063	0.16	0.11	0.075
8" CINDER BLOCK	0.25	0.13	0.088	0.066	0.066	0.18	0.12	0.081
8" SOLID BRICK (Face and Common)	0.29	0.14	0.092	0.069	0.069	0.20	0.13	0.086

"U" VALUES OF COMMON FRAME CEILINGS

Heat Flow Up

TYPE OF CEILING*	INSULATION BETWEEN OR ON TOP OF JOISTS (No Flooring Above) None	Regular Density Insulation Board on top of joists		Blanket or Batt insulation between joists			Perlite or Vermiculite between joists		Reflective Insulation E = 0.05		
		½"	1"	1"	2"	3"	2"	4"	1 Air Space	2 Air Spaces	3 Air Spaces
No Ceiling		0.38	0.24								
Gypsum Board (⅜")	0.65	0.26	0.19	0.19	0.11	0.079	0.18	0.11	0.38	0.21	0.14
Gypsum Lath (⅜") Plastered	0.61	0.26	0.19	0.19	0.11	0.078	0.17	0.10	0.37	0.21	0.14
Plywood (⅜")	0.59	0.25	0.18	0.18	0.11	0.078	0.17	0.10	0.36	0.20	0.14
Insulation Board (½" Regular Density)	0.38	0.20	0.16	0.16	0.10	0.074	0.15	0.09	0.26	0.17	0.12

"U" VALUES OF COMMON FRAME FLOORS

Heat Flow Down

TYPE OF FLOOR*	INSULATION BETWEEN OR ON BOTTOM OF JOISTS None	Regular Density Insulation Board on bottom of joists		Blanket or Batt insulation between joists			Reflective Insulation E=0.05 1 Air Space	2 Air Spaces	3 Air Spaces
		½"	1"	1"	2"	3"	Heat Flow Down		
Single Wood	0.35	0.18	0.14	0.13	0.087	0.066	0.096	0.052	0.040
Double Wood	0.28	0.16	0.13	0.12	0.082	0.063	0.089	0.050	0.039

*Refer to ASHRAE Guide and Data Book for reference to these and other roof and floor assemblies.

Fig. 13-7. "U" values for various constructions commonly used in residential building.

materials or spaces. A standard frame wall with composition sheathing, gypsum lath and plaster, with a 1 in. blanket insulation will have a "U" factor of about .11. To calculate the "U" factor for a given wall or building section where the R values are known, apply the following formula:

$$U = \frac{1}{R_1 + R_2 + R_3 \text{ ---}}$$

The "U" value for a framed wall construction, using conventional materials is given in Fig. 13-6. When a 2 in. batt or blanket of insulation is added, the "U" value is reduced from .19 to .08. Gypsum lath and plaster are dense materials and offer little resistance to heat flow. If you refer again to Fig. 13-6, you will note that a 3 in. blanket or batt will reduce the "U" value of the ceiling to .078. Complete tabulations for all building materials and types of constructions can be found in heating and cooling handbooks. The original source of most data on this subject is the American Society of Heating, Refrigerating, and Air Conditioning Engineers Guide and Data Book (ASHRAE). See also Figs. 13-7 and 13-8.

How Much Insulation

Comfort and health in a dwelling during cold weather depend not only on the air temperature of the heated space but also on the temperatures of

Basic wall construction*		Plain wall no plaster	Wall direct (½-in. plaster on)	⅜-in. plaster-board (½-in. plaster on ¾-in. furring with)	½-in. rigid insulation (½-in. plaster on ¾-in. furring with)
Concrete masonry (cores not filled)	8-in. sand and gravel or limestone	0.53	0.49	0.31	0.22
	8-in. cinder	0.37	0.35	0.25	0.19
	8-in. expanded slag, clay or shale	0.33	0.32	0.23	0.18
	12-in. sand and gravel or limestone	0.49	0.45	0.30	0.22
	12-in. cinder	0.35	0.33	0.24	0.18
	12-in. expanded slag, clay or shale	0.32	0.31	0.23	0.18
Concrete masonry (cores filled with insulation)**	8-in. sand and gravel or limestone	0.39	0.37	0.26	0.19
	8-in. cinder	0.20	0.19	0.16	0.13
	8-in. expanded slag, clay or shale	0.17	0.17	0.14	0.12
	12-in. sand and gravel or limestone	0.34	0.32	0.24	0.18
	12-in. cinder	0.20	0.19	0.15	0.13
	12-in. expanded slag, clay or shale	0.15	0.14	0.12	0.11
4-in. face brick plus:	4-in. sand and gravel or limestone unit	0.53	0.49	0.31	0.23
	4-in. cinder, expanded slag, clay or shale unit	0.44	0.42	0.28	0.21
	4-in. common brick	0.50	0.46	0.30	0.22
	8-in. sand and gravel or limestone unit	0.44	0.41	0.28	0.21
	8-in. cinder, expanded slag, clay or shale unit	0.31	0.30	0.22	0.17
	8-in. common brick	0.36	0.34	0.24	0.19
	1-in. wood sheathing, paper, 2x4 studs, wood lath and plaster	—	0.27	0.27	0.20

*All concrete masonry shown in this table are hollow units All concrete masonry wall surfaces exposed to the weather have two coats of portland cement base paint. Surfaces of all walls exposed to the weather subject to a wind velocity of 15 miles per hour.
**Values based on dry insulation. The use of vapor barriers or other precautions must be considered to keep insulation dry.

Fig. 13-8. "U" values for various masonry walls used in residential and commercial buildings. (Portland Cement Association)

ALL-WEATHER COMFORT STANDARD

	U-VALUE	INSULATION "R" NUMBER
Ceilings	0.05	R19
Walls	0.07	R11
Floors Over Unheated Spaces	0.07	R13

MODERATE COMFORT AND ECONOMY STANDARD

	U-VALUE	INSULATION "R" NUMBER
Ceilings	0.07	R13
Walls	0.09	R8
Floors Over Unheated Spaces	0.09	R9

MINIMUM COMFORT STANDARD

	U-VALUE	INSULATION "R" NUMBER
Ceilings	0.10	R9
Walls	0.11	R7
Floors Over Unheated Spaces	0.11	R7

Fig. 13-9. "U" values and comfort standards. (National Forest Products Assoc.)

the room surfaces. It is quite possible to maintain a satisfactory level of air temperature with extra heat and still have cold walls and floors. The human body will lose heat by radiation (or conduction if in contact) to these colder surfaces. Minimizing the difference in temperature between inclosed surfaces and room air can be achieved by adding insulating materials that will reduce the loss of heat through these surfaces.

Heating engineers have conducted numerous studies that have resulted in the establishment of standards and recommendations concerning the amount of insulation. These recommendations are usually based on the geographical location of the structure.

One of the widely used standards proposes three levels of comfort which may be attained by varying amounts of insulation in the ceiling, walls, and floors, Fig. 13-9. The "All Weather Comfort Standard" is especially developed to apply to houses which are air conditioned or electrically heated. Both "R" numbers and "U" values are listed and will serve as a guide in selecting the kind and amount of insulating materials.

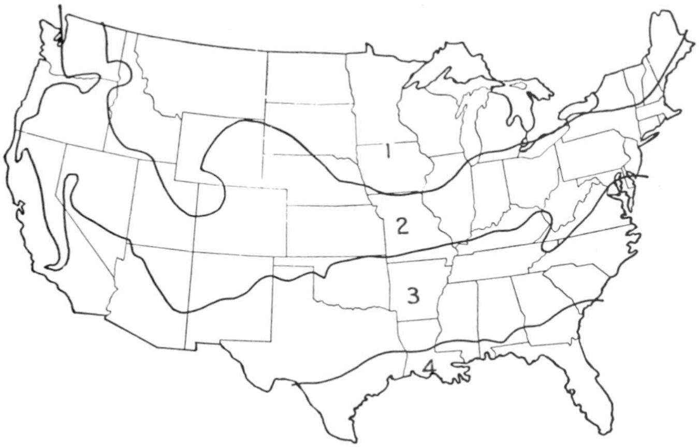

Fig. 13-10. U. S. Department of Agriculture climate zones.

The comfort standards may be related generally to climate zones throughout the country. See Fig. 13-10. Zone 1, which covers the northern part of the country will have cold weather with temperatures as low as 15 degrees below zero and colder. The highest standards listed in the table should be met since the "degree days per year" will normally exceed 6000.

HANDY SAYS:

"A degree day is the product of one day and the number of degrees F. the mean temperature is below 65 F. Figures are usually quoted for a full year and are used by the heating engineer to determine the design and size of the heating system."

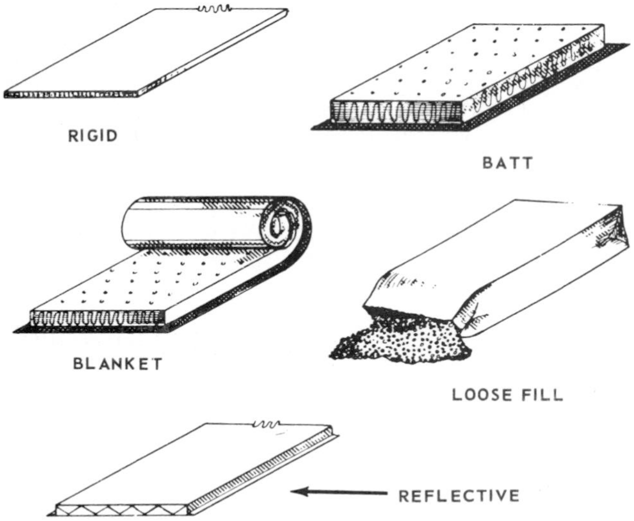

Fig. 13-11. Basic forms of insulation.

In the warm climates of the lower belt of Zone 3, minimum amounts of insulation are needed to provide comfort during the heating season. None is required in Zone 4. Air conditioned homes in these regions however, will need well insulated walls and ceilings to assure economy in the operation of cooling equipment.

As you study tables listing "U" values, it will be obvious that heat transmission decreases as the insulation thickness increases, but not in a direct relationship to the increase. The first inch reduces the "U" value more (proportionally) than the second inch. For example, in a frame wall with a "U" value of .24, the addition of 1 in. of insulation will reduce the value by 46 percent to .13; the second inch of insulation will reduce it only 16 percent more to .09. Still further thickness will continue to lower the "U" value but at a still lower percentage.

Types of Insulation

Insulation is made in a variety of forms and types, Fig. 13-11, and may be grouped into four broad classifications; 1. Flexible. 2. Loose fill. 3. Rigid. 4. Reflective.

Flexible insulation is manufactured in two types: Blanket or quilt, and batt. Blanket insulation is generally furnished in rolls or strips of convenient length and in various widths suited to standard stud and joist spacing. It comes in thicknesses of 1 to 3 1/2 inches, Fig. 13-12.

Fig. 13-12. A roll of blanket insulation, 15 in. wide. to fit between studs 16 in. O.C. The thermal resistance to heat flow "R" varies for floor, wall, or ceiling installation. Note the arrows indicating the direction of heat flow and "R" value.

The body of the blanket is made of loosely felted mats of mineral or vegetable fibers, such as rock, slag, glass wool, wood fiber, and cotton. Organic fiber mats are usually treated chemically to make them resistant to fire, decay, insects, and vermin. Blanket insulation is usually enclosed in paper covers with tabs on the side for attachment, Fig. 13-13. The covering sheet on one side may be treated to serve as a vapor barrier. In some cases the covering sheet is surfaced with aluminum foil or other reflective insulation.

Batt insulation is made of the same fibrous material as blankets. Thickness can be greater in this form and may range from 2 in. to 6 in. They are generally available in widths of 15 and 23 in. and in 24 and 48 in. lengths. Batts are available with a single flanged cover or with both sides covered as shown in Fig. 13-14.

Loose fill insulation is composed of various materials used in bulk form and supplied in bags or bales, Fig. 13-15. It may be poured, blown, or placed by hand and is commonly used to fill spaces between studs or to build up any desired thickness on a horizontal surface.

Loose fill insulation is made from such materials as rock, glass, slag wool, wood fibers, shredded redwood bark, granulated cork, ground or macerated wood pulp products, vermiculite, perlite, powdered gypsum, sawdust and wood shavings. One of the chief advantages of this type is that when insulating an older structure only a few boards need to be removed in order to blow the material into the walls.

Rigid insulation as ordinarily used in residential construction, is made by reducing wood, cane, or other fiber to a pulp and then assembling the pulp into lightweight or low-density boards that combine strength with heat and acoustical insulating properties.

It is available in a wide range of sizes, from tile 8 in. square, to sheets 4 ft. wide and 10 ft. or more long. Insulating boards are usually 1/2 in. to 1 in. in thickness. Boards of greater thickness are made by laminating together boards of standard thickness.

"C" or "K" factors for individual materials are not directly applicable to heat loss calculations. Instead they must first be converted into a resistance value "R". Engineering handbooks will list "R" values for a wide range of specific items. Fig. 13-16 lists an accepted general value for some commonly used materials.

Insulating boards are used for many purposes including roof and wall sheathing; subflooring; interior surface of walls and ceilings; base for plaster; and insulation strips for foundation walls and slab floors.

Although somewhat expensive, foamed glass or cork board makes an excellent rigid form of insulation. A recent development that is widely used today is foamed plastic (polystyrene). Because of

Fig. 13-13. Standard blanket insulation. Note flange along the lower edge used to attach blanket to joists and studs. (Owens-Corning Fiberglas Corp.)

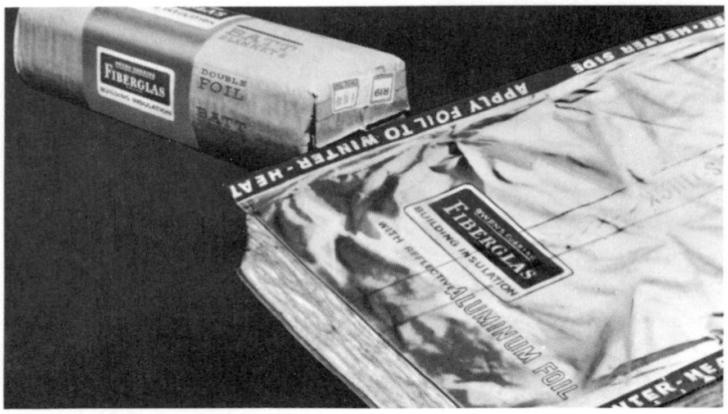

Fig. 13-14. Batt with aluminum foil cover. Opposite side has a porous foil breather paper.

Fig. 13-15. Loose fill insulation. Each bag covers a nominal 25 square feet to a depth of 4 inches.

MATERIAL	THICKNESS	R
Brick Veneer	4 IN.	.44
Concrete Block	8 IN.	.50
Hardboard	1/4 IN.	.18
Plywood	3/8 IN.	.59
Insulation Board	25/32 IN.	2.06
Gypsum Sheathing	1/2 IN.	.45
Wood	25/32 IN.	.98
Gypsum Lath and Plaster	7/8 IN.	.41
Mineral Wool Blanket	1 IN.	3.70
Plastic (Foamed)	1 IN.	3.45
Corkboard	1 IN	3.70
Vermiculite (Loose Fill)	1 IN.	2.08
Air Space (Vertical)	3/4 IN. OR MORE	.97

Fig. 13-16. "R" values for common construction and insulation materials.

its water-resistant quality it is especially adaptable to masonry work.

Insulating boards should not be confused with ordinary wallboard, which is more tightly compressed and has less insulating value. Insulating sheathing board ordinarily comes in two thicknesses, 1/2 and 25/32 in. They are available in 2 x 8 ft. sheets for horizontal application, and 4 x 8 sheets or longer for vertical application.

Reflective Insulation

Reflective insulation, which is usually a metal foil, or foil-surfaced material, differs from other insulating materials in that the number of reflecting surfaces, not the thickness of the material, determines its insulating value. In order to be effective, the metal foil must be exposed to an air space, preferably 3/4 in. or more in depth.

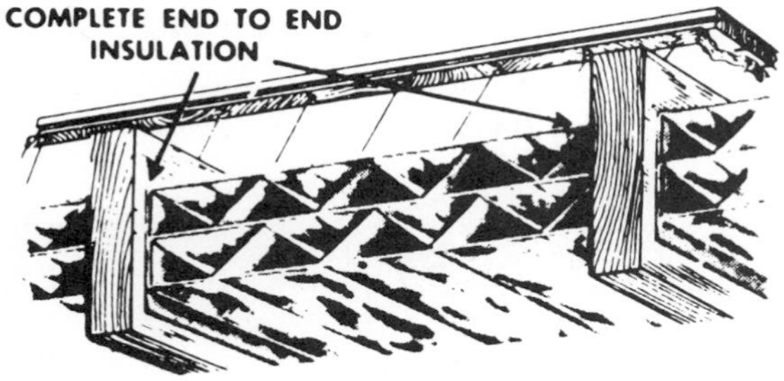

Fig. 13-17. Reflective accordion type insulation. It comes to the building site in a flat condition and then is "opened up" to form spaces between the surfaces when installed.

Aluminum foil is available in sheets or corrugations supported on paper. It is often mounted on the back of gypsum lath. One of the most effective forms of reflective insulation is multiple spaced sheets (accordion insulation) as shown in Fig. 13-17.

Other Types of Insulation

There are available on the market today many insulations which do not fit the classifications covered previously in this section. Some examples include the confetti like material mixed with adhesive and sprayed on the surface to be insulated; multiple layers of corrugated paper; and lightweight aggregates like vermiculite and perlite used in plaster to reduce heat transmission. Lightweight aggregates made from blast furnace slag, burned clay products, and cinders are commonly used in concrete and concrete blocks to improve the insulation qualities of these materials.

Where to Insulate

Heated areas, especially in cold climates, should be surrounded with insulation by placing it in the walls, ceiling, and floors, Fig. 13-18. It is best to have it as close to the heated space as possible. For example: if an attic is unused, the insulation should be placed in the attic floor rather than in the roof structure.

If however, it is desirable to heat attic space or certain portions of the area, walls and ceilings should be insulated. Where the insulation is placed in the roof frame, see Fig. 13-19, be sure to allow space between the insulation and the sheathing for free air circulation. The floors of rooms above unheated garages or porches require insulation if maximum comfort is to be maintained.

Where a basement space is to be remodeled into a room, insulation of the walls is recommended. In addition to saving heat and attaining greater comfort, better acoustical qualities will also be acquired.

When rooms are added to a house or porches are enclosed to provide additional space, insulation should be installed with the same care as in new construction.

Insufficient insulation or ventilation directly under low pitched roofs used in modern construction may cause a special problem, Fig. 13-20. During winter weather, excessive heat escaping from rooms below may cause snow on the roof to melt and water to run down the roof. In reaching the overhang, the water may freeze again causing a ledge or dam of ice to build up. Additional water reaching this dam may back under the shingles and leak into the building.

Basementless Structures

Floors over unheated space directly above the ground require the same degree of insulation as walls in the same climate zone. This space (called

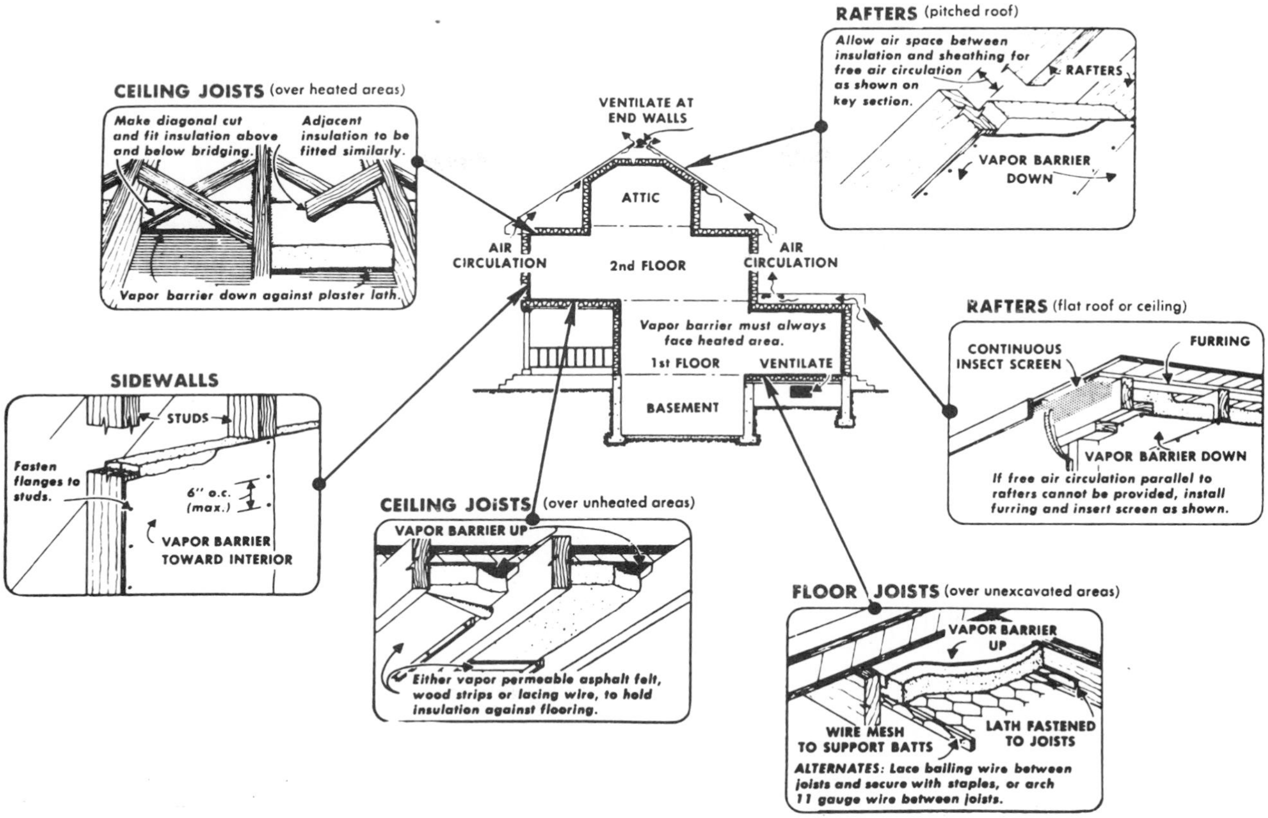

Fig. 13-18. Where and how to insulate in residential construction. (Acoustical and Board Products Assoc.)

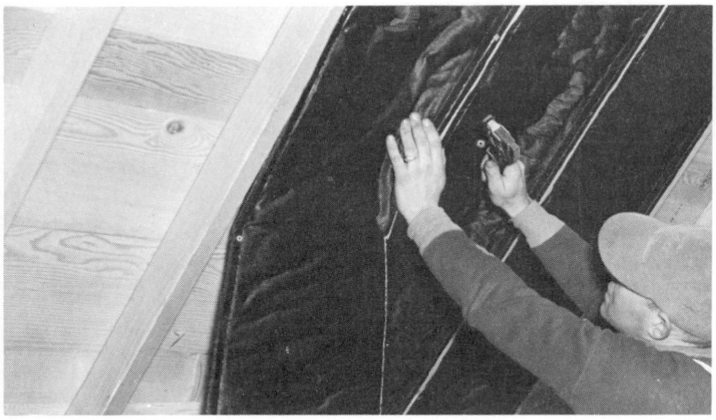

Fig. 13-19. Installing blanket insulation between the rafters in an attic area.

a crawl space) if enclosed by foundation walls, must be ventilated and will therefore approach the outside temperature.

Fig. 13-21 shows the essential requirements for crawl space insulation and ventilation. A vapor barrier should be positioned either on top of the insulation as shown or between the rough and finish floor. To prevent the insulation from sagging or tearing loose, it should be supported by wire mesh or wooden strips.

Moisture coming up through the ground can be controlled by covering it with 4 mil (.004) polyethylene plastic film or roll roofing weighing at least 55 lbs. per square. The material should be laid over the surface of the soil with edges lapping at least 4 in. If the ground is rough, it is advisable

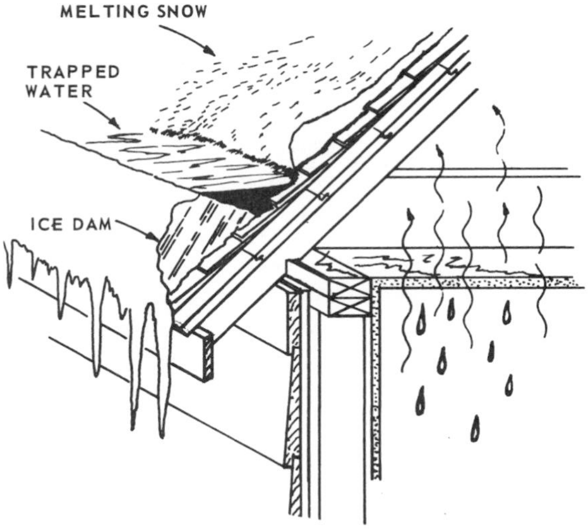

Fig. 13-20. Lack of insulation in ceiling can result in an ice dam and trapped water.

to apply a layer of sand or fine gravel before laying the vapor barrier. Covers of this kind greatly restrict the evaporation of water and somewhat less ventilation is needed than when no covers are used. The soil surface below the building should be above the outside grade if there is a chance that water might get inside the foundation wall. The soil cover is especially valuable where the water table is continually near the surface, or the soil has high capillarity. Be sure the covering is carried well up along the foundation wall.

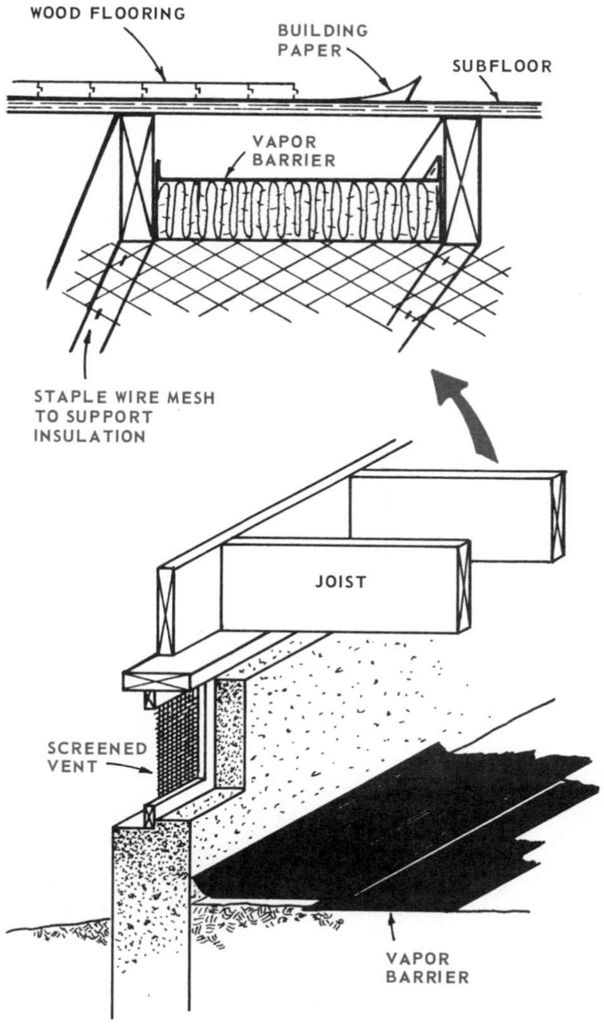

Fig. 13-21. Essential construction to provide insulation and ventilation for a crawl space. Use roll roofing or polyethylene film for the ground cover.

Crawl spaces are sometimes closed and used as a plenum (large heating duct) to distribute warm air throughout the structure. A detail of this construction is shown in Unit 7.

Today many homes and other structures are built on concrete slab floors. Such floors should

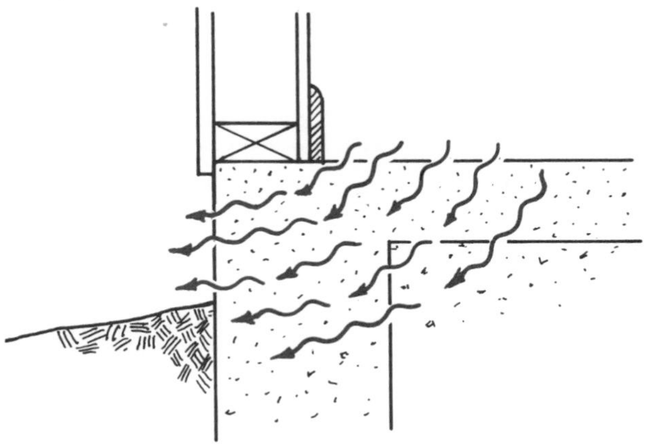

Fig. 13-22. How heat is lost from a concrete slab floor.

contain insulation and a vapor barrier. Fig. 13-22 shows how the heat flows for a slab floor. The loss is concentrated along the perimeter. Very little heat is lost into the ground under the central part of the floor. Although the vapor barrier must be continuous under the entire floor only the perimeter needs to be insulated. The insulation can be installed horizontally (about 2 ft.) under the floor or vertically along the foundation walls as shown in Fig. 13-23.

Condensation

Moisture in the form of water vapor is always present in the air. It acts like a gas and penetrates wood, stone, concrete and most other building materials. Warm moisture-laden air within a heated building forms a vapor pressure which constantly seeks to escape and mix with the colder, drier air outside.

Water vapor results from many sources within a living space. It is generated by cooking, bathing, clothes washing, and drying, or by humidifiers which are often used to maintain a comfortable level of humidity.

When warm air is cooled, some of its moisture will be released as condensation. The temperature at which this occurs for a given sample of air is called the dew point. If you live in a colder climate -- any region where the January temperature is 25 deg. F. or less -- the dew point can occur within the wall structure or even within the insulation itself. The resulting condensation will reduce the efficiency of the insulation and may cause permanent damage to the structural members over a period of time.

Moisture that collects within a wall during the winter months usually finds its way to the exterior finish in the spring and summer, causing deterioration of siding and/or paint peeling. The siding is usually a porous material and will hold a considerable amount of moisture. The paint is nonporous and as a result the moisture gathers under the paint film, causing blisters and separation from the wood surface. Recent developments in paint manufacturing have provided products that are somewhat porous and permit moisture to pass through, thus minimizing this problem in structures without adequate vapor barriers.

During warm weather, condensation may occur in basement areas or on concrete slab floors in contact with the ground. When warm humid air comes in contact with cool masonry walls and floors, some of the moisture will condense causing wet surfaces. Covering these surfaces with insulation will minimize condensation. Operating a dehumidifier in areas surrounded by cool surfaces will prove helpful during humid weather.

HANDY SAYS:

"The "dew point" is the temperature at which the air is completely saturated with moisture. Any lowering of the air temperature will cause condensation to occur."

Vapor Barriers

A properly installed vapor barrier (membrane through which water vapor cannot readily pass) will protect ceilings, walls, and floors from moisture originating within a heated space. See Fig. 13-24. Within an insulated wall or construction, the temperature will be warm on the inside and cool on the outside. The vapor barrier must be located on the WARM SIDE to prevent moisture from moving through the insulation to the cool side and condensing.

Many of the insulation materials produced today have a vapor barrier applied to the inside surface. Also, many interior wall surface materials are backed with vapor barriers. When these materials are properly applied they usually provide satisfactory resistance to moisture penetration.

If the insulating materials do not include a satisfactory vapor barrier, then one should be installed as a separate element. Vapor barriers in wide continuous rolls include: asphalt-coated paper; aluminum foil, Fig. 13-25; polyethylene films; and various combinations of these materials. To prevent accidental puncturing, it should be installed after heat ducts, plumbing, and electrical wiring are in place. Cut and carefully fit the barrier around such openings as outlet boxes.

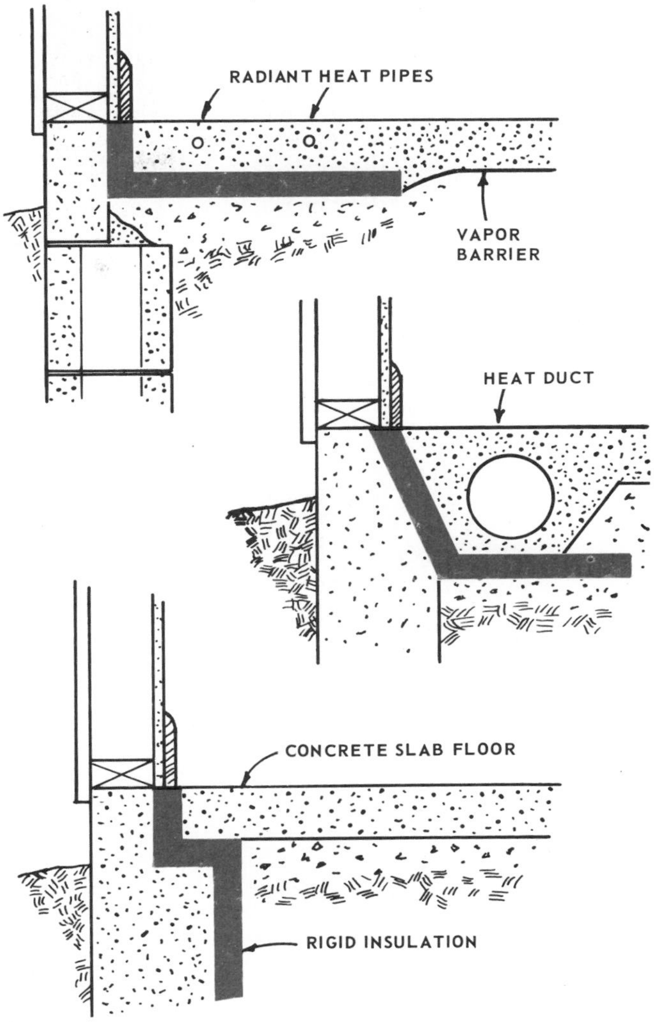

Fig. 13-23. Typical methods of installing perimeter insulation in a slab floor. Vapor barriers should be continuous under the entire floor.

Some architects specify the use of 2 or 4 mil polyethylene film, Fig. 13-26, making a continuous cover over walls, ceilings, and windows just before the plaster base or wallboard is applied. The covering protects the window unit during plastering operations; permits light to enter; and can easily be trimmed out of the opening just before the finished wood trim is installed.

Ventilation

In addition to the proper placement of vapor barriers, adequate ventilation of the constructions will help to insure against moisture problems. The cold side (outside) of walls should be weathertight but still permit the wall to "breathe." Building paper can be used over wood sheathing to reduce infiltration if it is not waterproof and will permit moisture within the wall to escape.

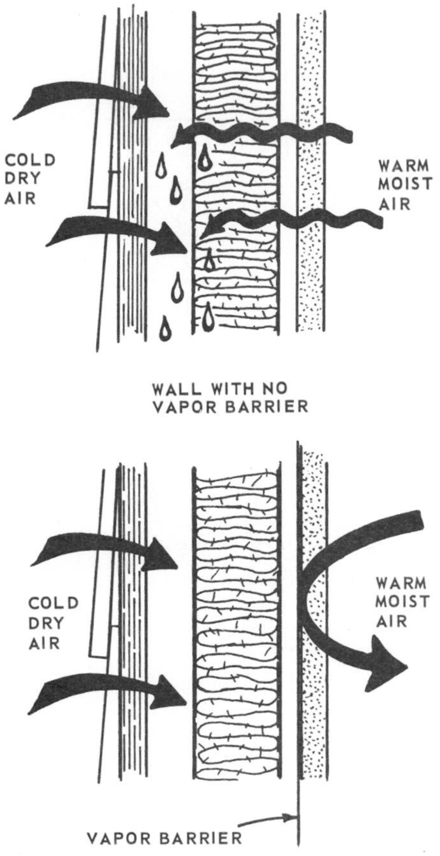

Fig. 13-24. How a vapor barrier protects a wall from condensation.

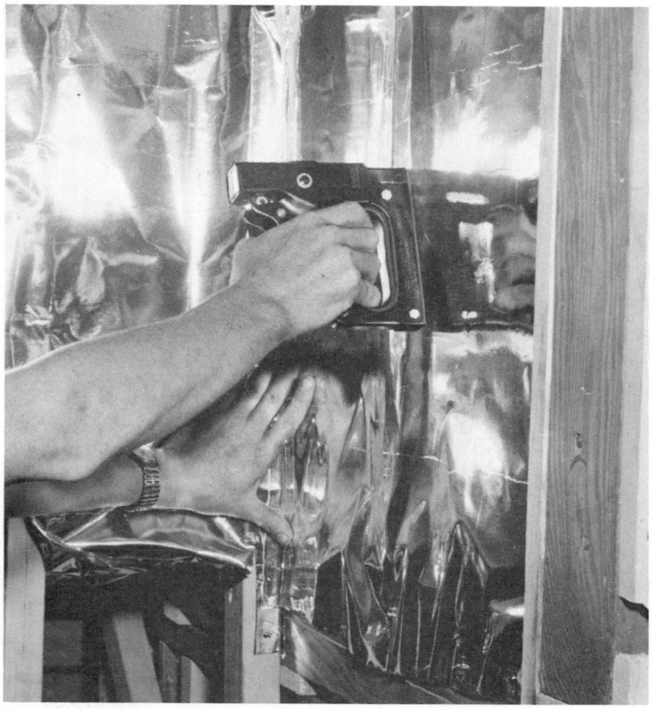

Fig. 13-25. Aluminum foil provides a good vapor barrier when joints are carefully lapped.

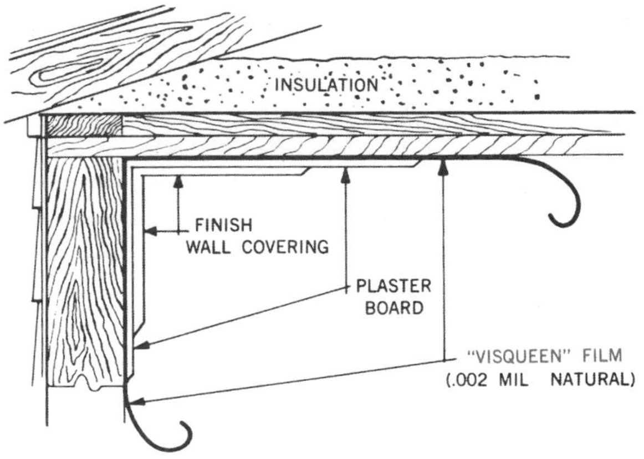

Fig. 13-26. Using polyethylene film as a vapor barrier. Lap all joints at least 3 in. and make tight applications around electrical outlets, doors, windows, and other openings. (Ethyl Corp.)

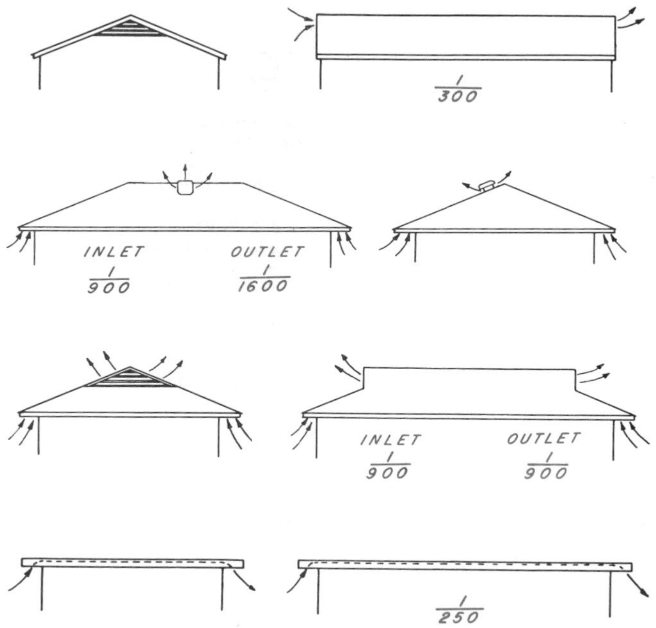

Fig. 13-27. Ventilation requirements for various types of roofs. The figures indicate the fractional part of the total ceiling area to be used in the vents.

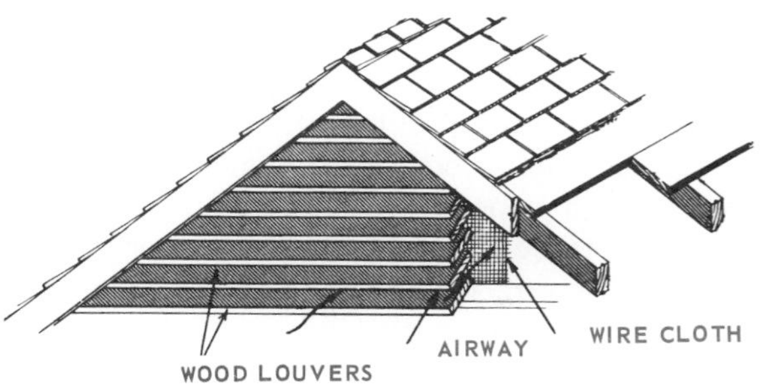

Fig. 13-28. Louver-type ventilator for a gable or Dutch hip roof. Vent opening should be designed to keep snow and rain from entering and should also be screened against insects.

It is especially important to provide ventilation for an unheated attic or space directly under a low pitched or flat roof. Fig. 13-27 shows the most common systems for ventilation and the recommended size of the vent area. For a standard gable roof, vents like the one shown in Fig. 13-28, are located in each end; providing a total ventilating area equal to 1/300 of the ceiling area.

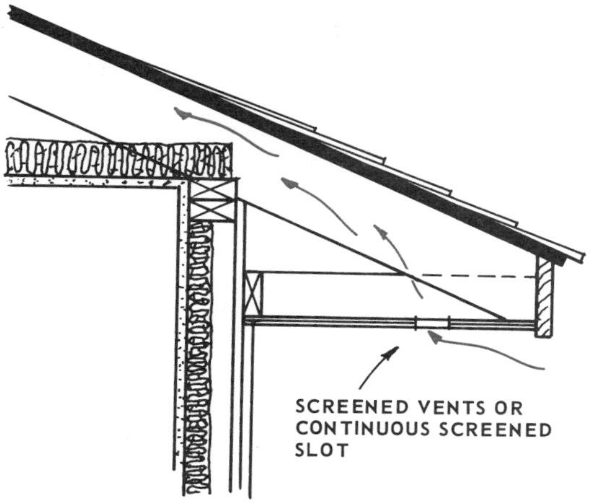

Fig. 13-29. Vents located in the building cornice. Be sure there is free access of the air to the attic area.

Fig. 13-29 shows a method of ventilating a cornice section. A wide range of prefabricated ventilators for roofs, gable ends, and cornices are available. One type of roof ventilator is shown in Fig. 13-30. Fig. 13-31 describes one model that is manufactured in many sizes.

Fig. 13-30. Ridge ventilator provides uniform airflow from soffit to ridge.

Ventilators that are located in the roof surface may result in leaks if not properly installed. Whenever possible these ventilators should be installed on a section of the roof that slopes to the rear. Under-roof ventilation is especially important in low-sloped roofs to prevent excessive accumulation of hot air during summer months. Sometimes it may be possible to utilize a false flue or a section of a chimney for attic ventilation, Fig. 13-32.

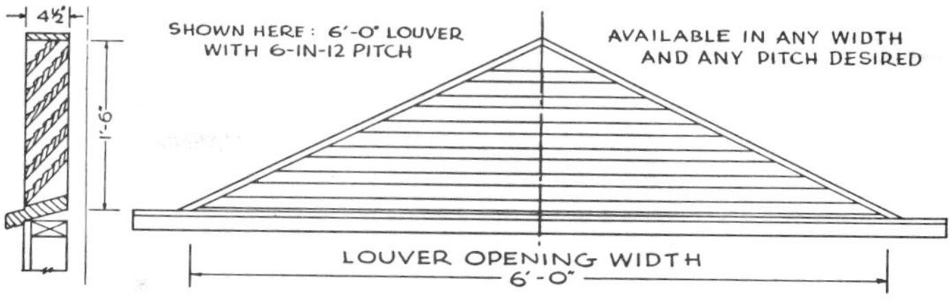

Fig. 13-31. Prefabricated wood ventilator furnished completely assembled and ready to install. (Ideal Co.)

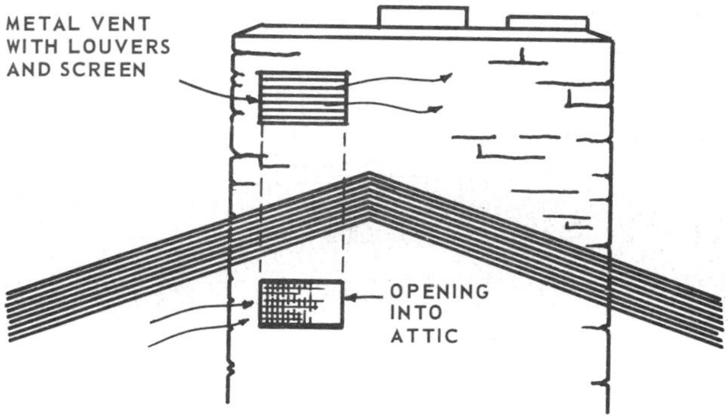

Fig. 13-32. Using a section of a chimney for attic ventilation.

Wall structures of existing frame buildings that indicate inadequate "cold side" ventilation can be corrected by installing patented ventilators as shown in Fig. 13-33. Most of these units consist of a metal tube with a cover that includes tiny louvers.

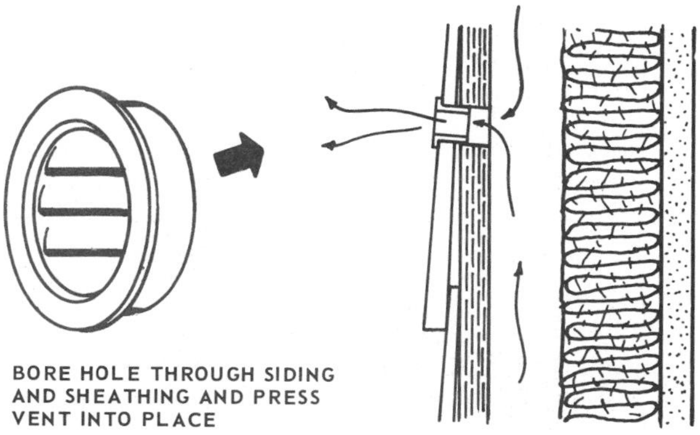

Fig. 13-33. Miniature vents installed in existing structures. 1 in. size is usually sufficient for most wall structures. Sizes range from 1 to 4 in.

They are installed by simply pressing them into a hole bored through the siding and sheathing -- in each stud space. For maximum ventilation, install one at the bottom of the wall and one at the top.

Installing Batts and Blankets

Insulation materials must be properly installed if they are to perform efficiently. Even the best insulation will not provide its rated resistance to heat flow if the manufacturer's instructions are not followed or if it is torn or otherwise damaged during or after installation.

Blankets or batts can be cut with a fine-tooth, crosscut hand saw; shears or a large knife. Measure the space and then cut the insulation 2 to 3 in. longer. Remove a portion of the batt from each end so that you will have a flange of the backing or vapor barrier to staple to the framing.

When working with blanket insulation, it is usually best to mark the required length on the floor, and then unroll the blanket, align it with the marks and cut the required number of pieces. For wall installation, first staple the top end to the plate and then staple down along the studs, aligning the blanket carefully. See Fig. 13-34. Finally secure the bottom edge to the sole plate.

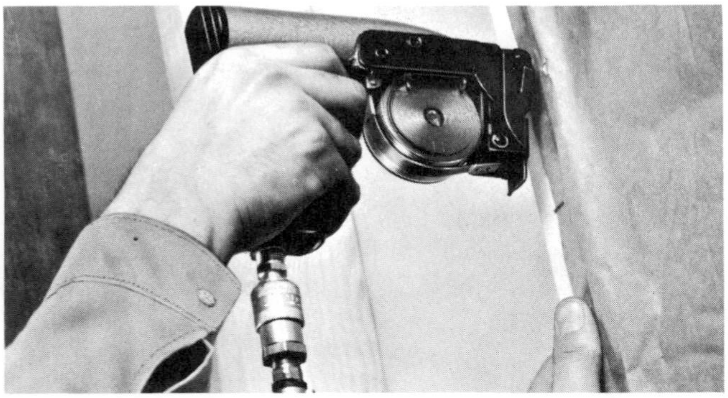

Fig. 13-34. Installing blanket insulation using an air driven stapler. (Spotnails Inc.)

To install batts in a wall section, place the unit at the bottom of the stud space and press it into place. Start the second batt at the top with it tight against the plate. Sections can be joined at the mid-point by butting them together. The vapor barrier should be overlapped at least 1 in. unless a separate one is installed. Some batts are designed without covers or flanges and are held in place by friction.

Flanges, common to most blankets or batts, are stapled to the face or side of the framing members, Fig. 13-35. Pull the flange smooth and space the staples no more than 12 in. apart. The

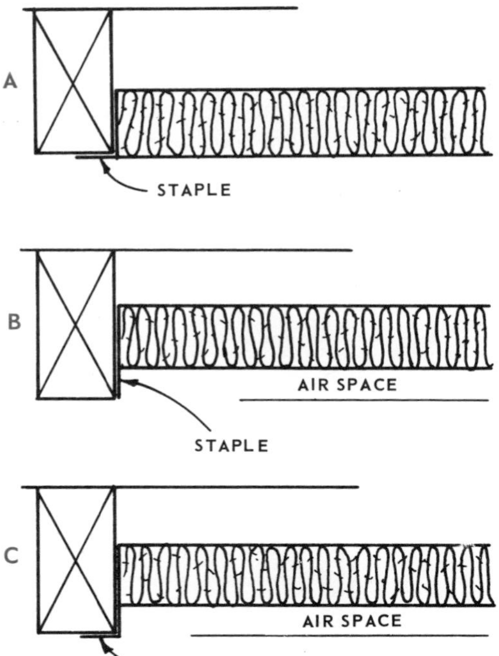

Fig. 13-35. Methods of installing blankets and batts. A–Flush with inside surface of stud. B–Flange stapled along side of stud to form air space. C–Special flange.

interior finish, when applied, will serve to further seal the flange in place when it is fastened to the stud face. Some blankets have special folded flanges which enable them to be fastened to the face of the framing and also form an air space as shown in C, Fig. 13-35.

In dry wall construction, specifications may require that the faces of the framing be left uncovered. When so applied, the flanges of the insulation should fit smoothly to the sides of the framing and the staples should be spaced 6 in. or closer. Be sure that all gaps or "fish mouths" (wrinkles) are eliminated. When it is necessary to secure maximum vapor protection, a separate vapor barrier should be applied over the entire wall or ceiling area. Avoid any perforations and lap joints fully.

Thread the insulation carefully behind any drain pipes that might be located in the wall as shown in Fig. 13-36. If water service pipes (in cold climates they should never be located in outside walls) are present, it is well to add a separate vapor barrier between the pipe and interior surface to prevent moisture condensation on the cold pipe.

Ceiling insulation can be installed from below or from above if attic space is accessible. When batts are used they are usually installed from below as shown in Fig. 13-37, following about the same general procedure as recommended for walls. They must be butted snugly together at the ends

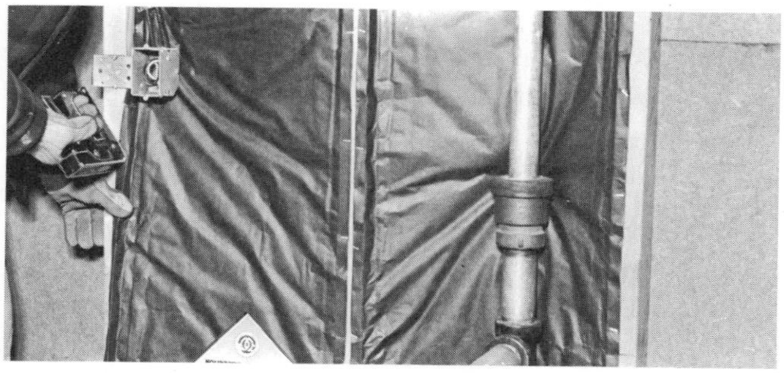

Fig. 13-36. Insulation should be fitted carefully around plumbing pipes and electrical outlet boxes. (Acoustical and Board Products Assoc.)

Fig. 13-37. Installing insulation batts between ceiling joists, with vapor barrier down.

and carried out over the outside wall plates, Fig. 13-38. In multi-story construction, especially in cold climates, the perimeter of floor frames should be insulated as shown in Fig. 13-39. Note position of insulation between joists.

HANDY SAYS:

"When installing ceiling insulation from below, wear tight fitting clothing and a cap or hat with a wide brim. Goggles may also be helpful. This is especially important when working with insulation made of tiny glass fibers which may irritate your skin or eyes."

After the main large areas of walls and ceilings are complete, spaces above and below windows should be insulated. Small cuttings and short pieces remaining from the main sections can be used. Cut the sections to fit as shown in Fig. 13-40. Take the time necessary to carefully apply the insulation around electrical outlets and other wall openings. BE CERTAIN THAT YOU DO NOT COVER OUTLET BOXES OR THEY MAY BE MISSED WHEN THE WALL SURFACE IS APPLIED.

A thorough insulation job will require that all spaces be filled. For example: the space between window and door frames, and the rough framing as shown in Fig. 13-41. Cuttings left over from larger spaces can be utilized. Pack the insulation into the small space with a stick or screwdriver and then cover the area with a vapor barrier.

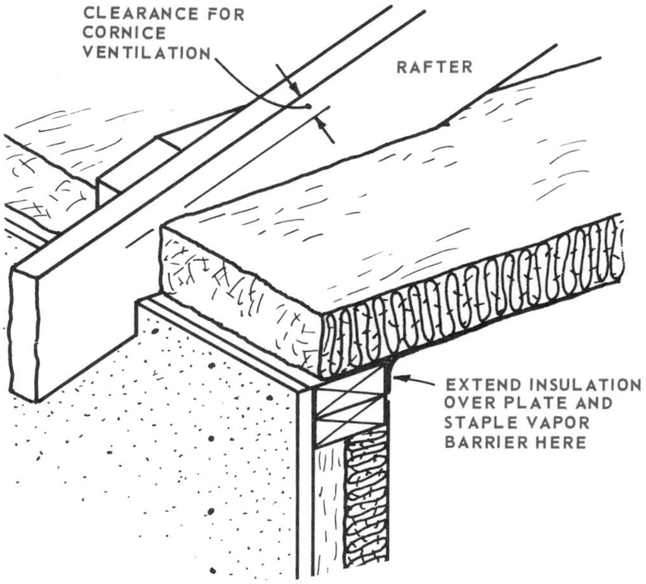

Fig. 13-38. Placement of ceiling insulation at outside wall.

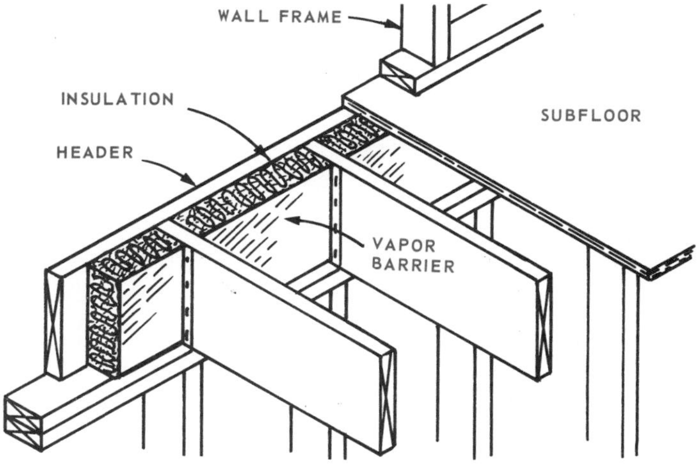

Fig. 13-39. Insulating perimeter of floor frame.

Installing Reflective Insulation

Reflective insulation is available in single sheet form or in blankets that open like an accordion to form multiple reflecting surfaces. They are installed in about the same manner as regular fibrous materials by stapling them over or between framing members.

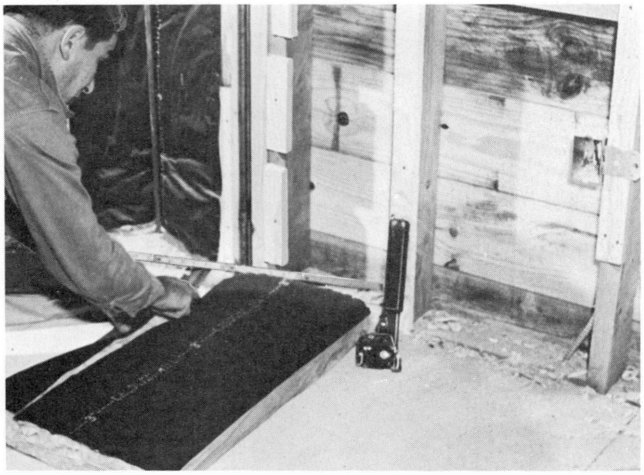

Fig. 13-40. Cutting insulation to fit odd size spaces. (National Forest Products Assoc.)

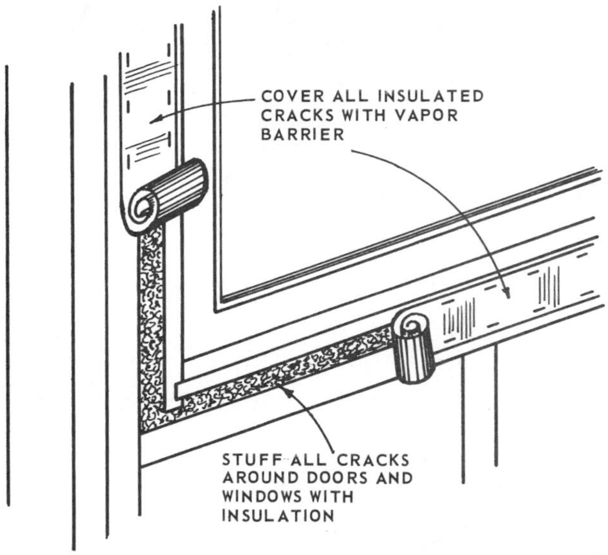
COVER ALL INSULATED CRACKS WITH VAPOR BARRIER
STUFF ALL CRACKS AROUND DOORS AND WINDOWS WITH INSULATION

Fig. 13-41. Insulate around window and door frames.

Proper application is very important with this type of insulation because its effectiveness depends on air spaces next to the reflective surfaces. The spaces in vertical sections should be 3/4 in. or more. In ceilings and floors, an air space of 1 1/2 in. is usually recommended. (Always study the manufacturer's recommendations.) Fig. 13-42 shows reflective insulation being stapled to ceiling joists.

Installing Fill Insulation

Fill insulation is poured or blown into place. It is especially adaptable to existing structures where "blown-in" fill may be used, without resorting to extensive alteration in the wall surfaces.

Ceilings are easily insulated with fill materials. It is poured directly from bags into the joist spaces and then leveled with a straight-edge as shown in Fig. 13-43. A vapor barrier should be installed on the underside of the joists before the ceiling finish is applied. It will control the flow of moisture and also prevent fine particles (present in some forms of fill insulation) from sifting through cracks that might develop in the ceiling.

Fill insulation is often poured into the core of block walls, see Fig. 13-44, or in the space in masonry cavity walls. Thermal resistance is greatly increased. For example, the "U" factor of a standard concrete block (.53) is reduced to .36 when the cores are filled with insulation. A lightweight 8 in. block will be reduced from .33 to .17.

Installing Rigid Insulation

The installation of slab or block insulation varies with the type of product. The manufacturers' specifications should always be studied and carefully followed.

Insulating board is used extensively for exterior walls and its application is covered in Unit 8. Sometimes it is used as the sheathing material for roofs or simply as an insulating material installed over the roof deck. A number of products are especially designed to insulate concrete slab floors. See Fig. 13-45. Waterproof materials such as glass fibers or foamed plastic provide desirable characteristics.

Fig. 13-46 shows a plastic foam insulation board applied to a masonry wall. It is bonded to the basic wall surface with a special mastic and provides a permanent insulation and vapor barrier. After the boards are installed, conventional plaster coats

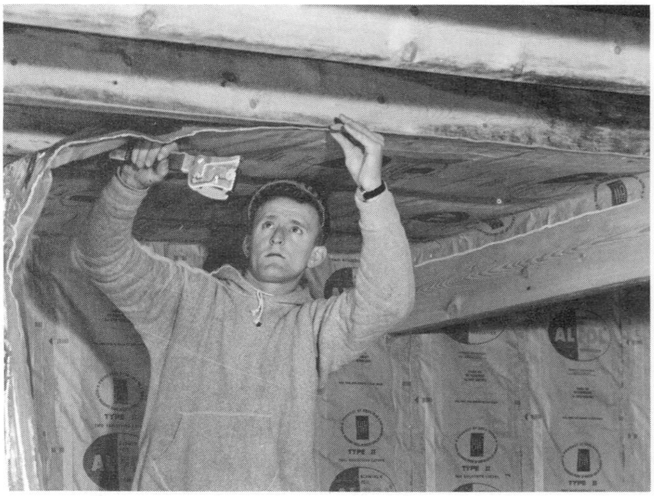

Fig. 13-42. Installing reflective insulation. (Bostitch, Inc.)

Fig. 13-43. Leveling fill insulation over ceiling joists. (Vermiculite Institute)

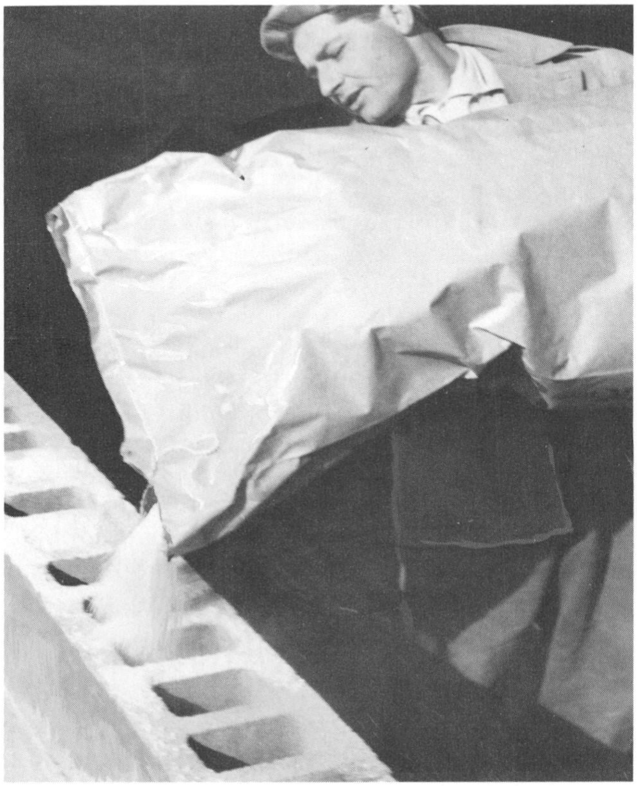

Fig. 13-44. Filling cores of concrete block walls with insulation. (Perlite Institute, Inc.)

Fig. 13-45. Installing a rigid form of insulation in the perimeter of a slab floor. (Owens-Corning Fiberglas Corp.)

can be applied to the surface. Plastic foam (Styrofoam) insulation has gained wide acceptance as a rigid insulating material and has been successfully applied to a wide variety of constructions.

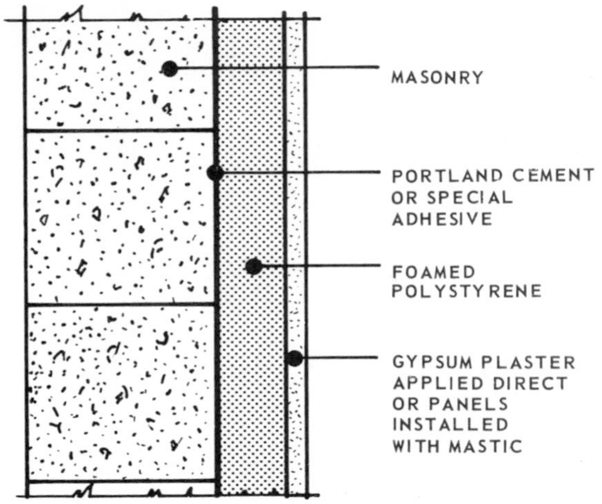

Fig. 13-46. Insulating a masonry wall with rigid insulation board made from foamed polystyrene "Styrofoam."

Insulating for Electric Heat

Shortages of fuel oil and natural gas have brought about a tremendous increase in the use of electricity for residential heating. Today, electricity is expensive and when it is used for heating, the structure must be especially well insulated if located in a cold climate. Recommended minimum amounts of insulation include:

Concrete slab floors: 2 in. of rigid insulation carried 24 in. down on the foundation wall or in 24 in. around the perimeter of the floor.

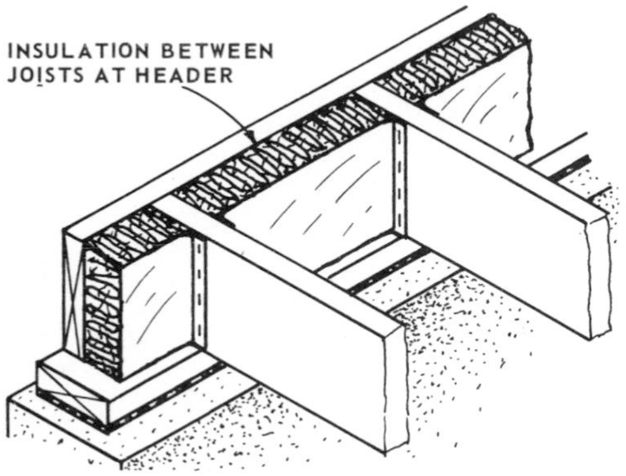

Fig. 13-47. For electrically heated structures include insulation around the perimeter of the floor.

Floor over crawl space: 4 in. of insulation over space ventilated with 1 sq. ft. of vent area for every 1500 sq. ft. of floor area.

Walls: 3 1/2 in. of insulation. Add insulation between joists as shown in Fig. 13-47.

Ceilings: 6 in. of insulation. When heating cables are installed in the ceiling, 8 in. of insulation should be used.

Insulating Existing Structures

Special precautions must be taken when insulating an existing structure where no vapor barrier can be installed. Inside humidities should be controlled. "Cold side" ventilation is essential. A fairly satisfactory vapor barrier can be secured by applying: two coats of either lead and oil paint, rubber emulsion paint, or aluminum paint; or a vapor barrier wall paper. The application should be carefully made on all surfaces of outside walls and ceiling.

Windows and Doors

Storm windows and storm doors, when properly installed, form dead air spaces, reducing the flow of heat to the outside and preventing infiltration of cold air.

Heat loss through the window frames as well as the glass must also be considered. Wooden frames and sash are generally considered to be better in this respect than metal. In addition to the heat loss, the relatively high conductivity of metal can cause condensation on interior surfaces.

Even with the best selection of windows that are double-glazed or equipped with storm units, heat losses through them will be 8 to 10 times as great as through a well insulated wall.

Infiltration

In addition to the heat loss through floors, walls, ceilings, doors and windows, there is still another consideration. This is the air leakage around openings and cracks, called infiltration. It can be minimized through the use of weather-stripping made from such materials as metal, rubber, felt, and plastic, or a combination of these materials.

Modern residential windows are built in a factory and equipped with appropriate weather stripping during fabrication. Outside doors, however, are often fitted to the door frame on the job and the weather stripping is installed by the carpenter. The various types will require different methods of installation. You should follow the manufacturer's instructions. Fig. 13-48 shows a standard type of metal weather stripping.

Sometimes you may think that a window is leaking air when you are near it and feel a slight draft. This is usually due to the air in contact with the cold glass, becoming colder and heavier than the rest of the room air. It moves downward to the floor and across the room. Heating registers and convectors are positioned to offset or minimize these "down-drafts."

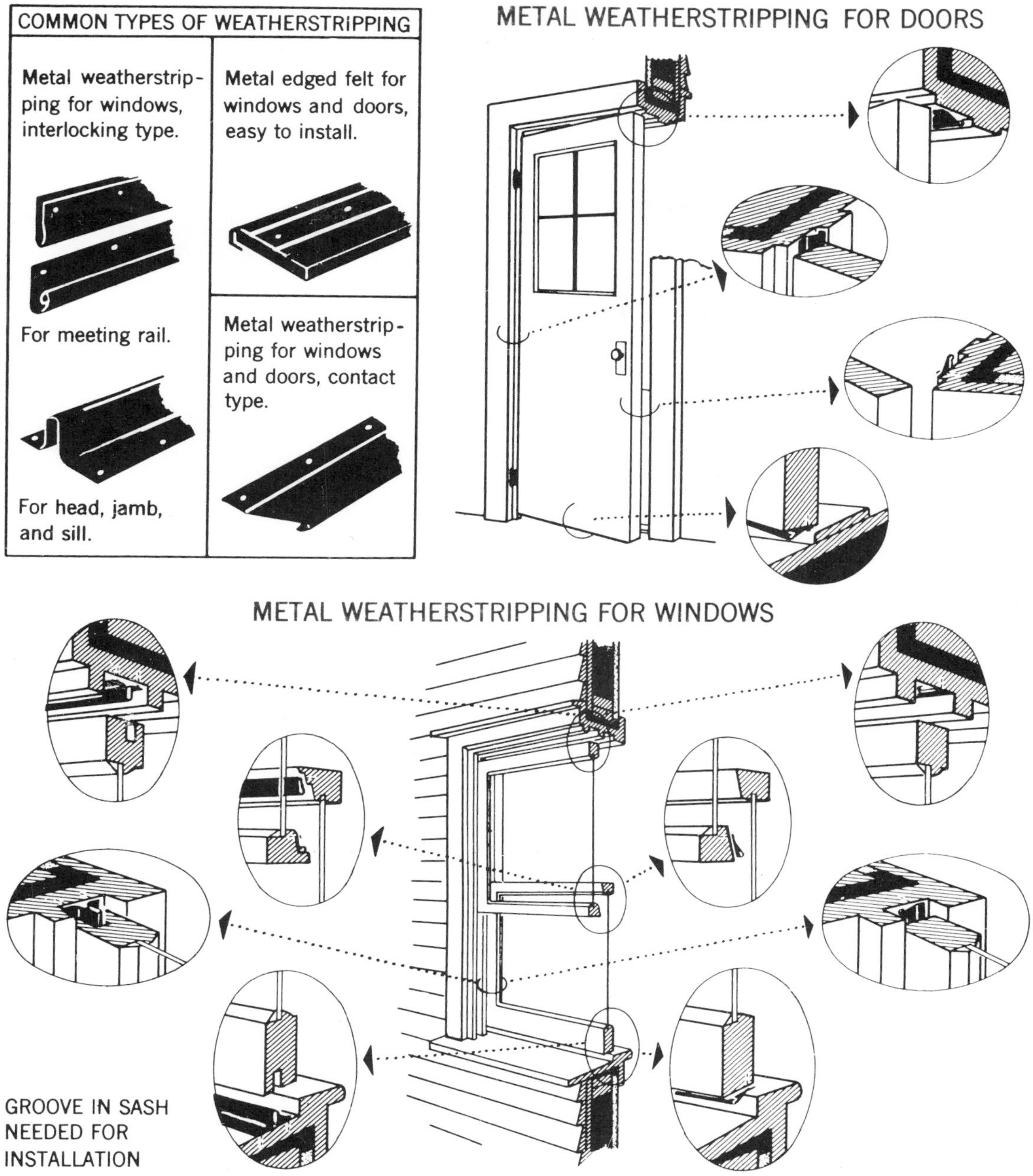

Fig. 13-48. Forms of weather stripping. In modern window construction, rigid polyvinyl chloride (PVC) strips are widely used.

Estimating Materials

The amounts of insulating materials are figured on the basis of area (square feet) and the thickness is then specified as separate data. The size of packages vary considerably depending on the type and thickness. For example: one manufacturer packages 1 1/2 in. blankets in rolls of 140 sq. ft. A 3 in. blanket in the same width comes in rolls of 70 sq. ft.

To determine the amount of insulation for exterior walls, first add up the total perimeter of the structure, then multiply this by the ceiling height. From this total, deduct the area for doors and windows. Many carpenters will deduct only large windows or window-walls and disregard doors and smaller openings. This extra allowance will make up for loss in cutting and fitting, and provide for additional material needed around plumbing pipes, recessed lighting boxes, and other items. Here is an example:

Perimeter	= 30 + 40 + 36 + 20 + 6 + 20
	= 152
Area	= 152 x 8
	= 1216
Window Wall	= 12 x 8
	= 96
Net Area	= 1216 - 96
	= 1120 sq. ft.

Batts are furnished in packages (sometimes called tubes) that contain as much as 100 sq. ft. A 6 in. batt is usually packaged in units containing 50 sq. ft.

When estimating the amount for floors and ceilings, use the same figures that were listed in calculating the subfloor area. Stair wells and openings for large fireplaces could be deducted. However, here again these amounts may provide the extra need for waste and special packing around fixtures.

The same procedure can be used to figure reflective insulation. A greater allowance for cutting and waste should be made, especially when using accordion type which cannot be spliced effectively. Rolls of reflective insulation contain from 250 to 500 sq. ft.

Rigid insulation is also figured on the basis of area and can be calculated from dimensions shown on the working drawings. For a perimeter insulation strip in a concrete slab floor, multiply the perimeter of the building by the width of the strip. There is little waste on such an installation since even small pieces can be utilized.

Fill insulation comes in bags that usually contain 3 or 4 cu. ft. The cubic feet required can be calculated as follows:

Area	= 1200 sq. ft.
Thickness	= 4 in. = 1/3 ft.
Cu. ft. required	= 1200 x 1/3 = 400
*Less 10%	= 400 - 40
Net Amount	= 360
Number of bags (4 cu. ft.)	= 90

* Allowance for joists 16" O.C.

Manufacturers information and specifications will usually list tables that provide a direct reading of the number of bags required for a certain thickness of application. These are especially helpful in estimating amounts needed for such items as filling the core of concrete blocks. See Fig. 13-49.

Acoustics and Sound Control

Noise (unwanted sound), a by-product of our modern world, has reached a magnitude that demands consideration in every home and building. It is unpleasant, reduces human efficiency, and can cause undue fatigue.

Houses, especially those built in northern climates, will have wall and roof structures heavy enough to repel average outside noises; therefore noise or sound control will apply mainly to interior partitions, floors, and surface finishes.

Sounds in an average home will be generated by conversation, television, radios, record players, typewriters, and musical instruments. Vacuum cleaners, washing machines, food mixers and garbage disposers are examples of mechanical equipment that creates considerable noise. Plumbing, heating and air conditioning systems may be a source of excessive noise if poorly designed and installed. The activities in play rooms and workshops may create sounds that will be undesirable if transmitted to relaxing and sleeping areas.

The solution to problems of sound and noise control can be divided into three parts: 1. Reducing the source; 2. Controlling sound within a given area or room; 3. Controlling sound transmission to other rooms or areas. The carpenter should have some understanding of how the latter two can be accomplished and be familiar with sound conditioning material and constructions. Just as in the case of thermal insulations, he must appreciate the importance of careful work and proper installation methods if the required efficiency in sound control is attained.

APPROXIMATE COVERAGE

NUMBER OF BAGS (4 CU. FT.) REQUIRED

WALL AREA	CORE FILL BLOCK SIZE			CAVITY FILL CAVITY WIDTH			
(SQ. FT.)	6 IN.	8 IN.	12 IN.	1 IN.	2 IN.	2 1/2 IN.	3 IN.
100	5	7	12	2	4	5	6
500	23	33	58	10	21	26	31
1000	46	65	118	21	42	52	62
2000	91	130	236	42	84	104	125
3000	137	195	354	62	124	155	187

Fig. 13-49. Table providing estimated amounts of fill insulation for masonry walls. (Perlite Institute, Inc.)

Acoustical Terms

Sound - A vibration or wave motion that can be heard. It usually reaches the ear through air. The air itself does not move but vibrates back and forth in TINY molecular motions of high and low pressure.

Decibel - The unit of measurement used to indicate the loudness or intensity of sound. Comparable to the "degree" as a measurement of heat or cold. See Fig. 13-50.

Reverberation Sounds - These are airborne sounds

	Decibels	Threshold of Feeling
	120	
Deafening	110	Thunder, Artillery Nearby Riveter Elevated Train Boiler Factory
	100	
Very Loud	90	Loud Street Noise Noisy Factory Truck Unmuffled Police Whistle
	80	
Loud	70	Noisy Office Average Street Noise Average Radio Average Factory
	60	
Moderate	50	Noisy Home Average Office Average Conversation Quiet Radio
	40	
Faint	30	Quiet Home or Private Office Average Auditorium Quiet Conversation
	20	
Very Faint	10	Rustle of Leaves Whisper Sound Proof Room Threshold of Audibility
	0	

Fig. 13-50. Levels of common sounds and noises.

which continue after the actual source has ceased; caused by reflections from floors, walls and ceiling.

Frequency - Sound frequency is the rate at which sound energized air molecules vibrate. The higher the rate, the more cycles per second (cps). Examples are the low frequency of a bass drum and the high frequency of a piccolo.

Impact Sounds - These are the sounds that are carried through a building by the vibrations of the structural materials themselves. Footsteps heard through the floors of a structure are an example of impact sounds.

Masking Sounds - These are the normal sounds within habitable rooms which tend to "mask" some of the external sounds entering the room.

Decibels Reduction - An expression used to indicate the sound insulating properties of a wall or floor panel.

Sound Transmission Loss (TL) - Sound insulating efficiency of wall or floor construction, measured in decibels. Transmission loss is number of decibels which sound loses when transmitted through a wall or floor. Transmission loss of any wall or floor depends on the materials and techniques used in construction.

Sound Absorption - The capacity of a material or object to reduce sound waves by absorption. Acoustical materials, such as acoustical ceiling tile, are designed to absorb sounds within a given area; sounds that otherwise would be reflected and cause excessive reverberation and build-up of intensity within that area.

Noise Reduction Coefficient (NRC) - The sound absorption of acoustical materials is expressed as the average percentage absorption at the four frequencies which are representative of most household noises, namely 250, 500, 1000 and 2000 cycles per second.

Sound Transmission Class (STC) - The STC is a single number which represents the minimum performance of a wall or floor at all frequencies. The higher the STC number, the more efficient the wall or floor will be in reducing sound transmission.

Sound Intensity

The number of decibels indicates the loudness or intensity of the sound. Roughly speaking, the decibel unit is about the smallest change in sound that is audible to the human ear. Actually, the decibel has the same relation to a scale of loudness as the degree has to a thermometer. Reference to Fig. 13-50 will show that the rustle of leaves or a low whisper is on the threshold of audibility, that is, the sound is barely discernible to the ear. The top of the scale is a painfully loud sound of approximately 120 decibels, called the "threshold of feeling." Between these two limits lies the range of ordinary sound.

One sound level is 10 decibels greater than another if its intensity is 10 times greater than the other. There is a logarithmic relation, on the decibel scale, to the amount of sound energy involved. If the sounds differ by 20 decibels, the ratio of their intensities is 10^2 or 100 times greater. If by 30 decibels, the ratio is 10^3 or 1,000 times greater, and so on up the scale.

Transmission

When sound is generated within a room, the waves strike the walls, floor and ceiling. A large portion of this sound energy is reflected back into the room; the rest is "absorbed" by the surfaces. If there are cracks or holes through the wall (no matter how minute), a part of these sound waves will travel through as airborne sounds. The sound waves striking the wall will cause it to vibrate as a diaphragm, reproducing these waves on the other side of the wall.

Sound transmission through theoretically airtight partitions is the result of such diaphragm action. The sound insulation values of such substances is therefore almost entirely a matter of their relative weight, thickness, and area. In partitions of normal dimensions, this value depends mostly upon weight.

As the sound moves through any type of wall or barrier its intensity will be reduced. This reduction is called Sound Transmission Loss (TL) and is expressed in decibels. A wall with a TL of 30 db. will reduce the loudness level of sound passing through it from 70 db. to 40 db. See Fig. 13-51. The transmission loss of any floor or wall will be determined by the materials, design, and quality of constructional techniques.

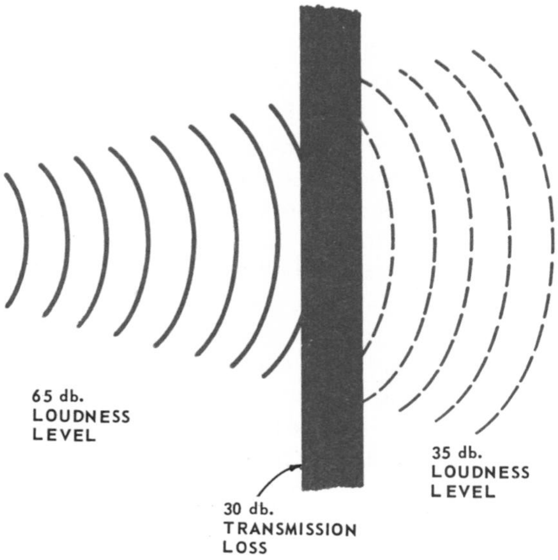

Fig. 13-51. Sound transmission loss through a wall. Values will vary somewhat depending upon the frequency of the sound waves.

Although transmission loss rated in decibels is still used, another system of rating sound-blocking efficiency is widely accepted. It is called the Sound Transmission Class (STC) system. Standards have been established through extensive research by such associations as the Insulation Board Institute, and the National Bureau of Standards.

STC numbers have been adopted by acoustical engineers as a measure of the resistance to sound transmission of a building element. Like the resistance in thermal insulation (R), the higher the number, the better the sound barrier. Fig. 13-52 shows how a composite STC rating is applied to a given wall construction which has various TL values in decibels through a specified range of frequencies. Fig. 13-53 further describes these STC ratings by making a simple application to a wall separating two apartments.

Actually there is another factor present which should be considered when designing any sound insulating panel. These are usually referred to as MASKING SOUNDS. In theory, an inaudible sound of zero (0) on the decibel scale is for a perfectly quiet room. Since there are noises in every habitable room which tend to mask the sound entering, it is only necessary to reduce sound to the level of the more or less maintained sound level within the space to be insulated. Assume that there is a radio playing soft music in the listening room (about 30 decibels). Thus, sound of less than 30 decibels entering the room would be completely masked.

Wall Construction

How high an STC rating must be for a given wall will depend largely on the type of areas it separates. For example: partitions between bedrooms in an average home usually will not require special soundproofing while those between bedrooms and activity or living rooms should have a high STC rating. Partitions surrounding a bathroom should also have a high STC number; with extra attention given to the placement of insulation around pipes and the installation of air cushions in water service lines.

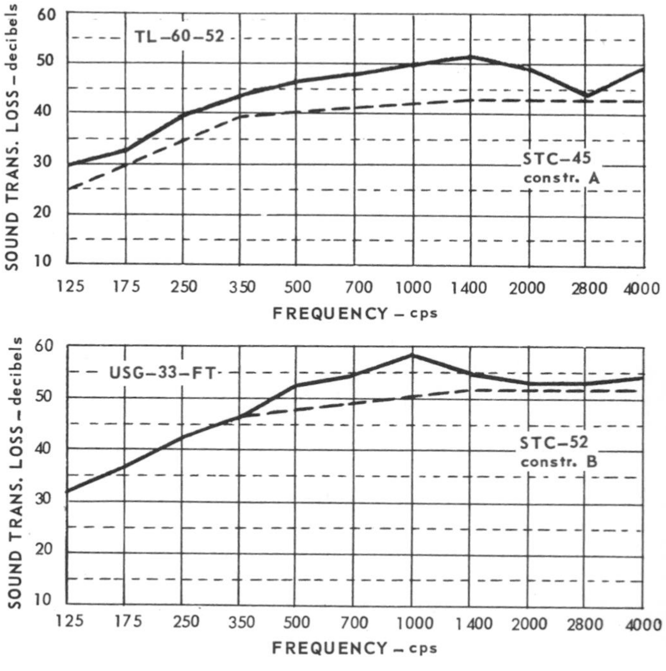

Fig. 13-52. Graphs showing the TL values in decibels at various frequencies for two STC rated constructions.

Fig. 13-54 shows a number of practical constructions for partitions and their STC ratings. Some of them include sound deadening board; (structural insulation board product designed es-

30 STC
LOUD SPEECH CAN BE UNDERSTOOD
FAIRLY WELL

35 STC
LOUD SPEECH AUDIBLE BUT NOT
INTELLIGIBLE

42 STC
LOUD SPEECH AUDIBLE
AS A MURMUR

50 STC
LOUD SPEECH NOT AUDIBLE

Fig. 13-53. How various STC ratings apply to a partition between two apartments.

pecially for use in sound control systems). This is made principally from wood and cane fibers in a nominal 1/2 in. thickness. Standard sizes of sound deadening board are the same as regular insulation board materials. It is usually identified with the words; IBI - RATED SOUND DEADENING BOARD on each sheet or package to distinguish it from other insulation board products.

In order to secure the high STC ratings in a wall structure when using a sound deadening board, application details supplied by the manufacturer should be carefully followed. Where nails are used, the size, type and application patterns are critical. In dry wall construction, the joints should be taped and finished and the entire perimeter sealed. Openings in the wall for convenience outlets and medicine cabinets require special consideration. For example, electrical outlets on opposite faces of the partition should not be located in the same stud space.

Double Walls

Partitions between apartments are often constructed to form two separate walls as shown in Fig. 13-55. Standard blanket insulation is installed in about the same manner as for thermal insulation purposes. It should be stapled to only one row of the framing members.

For economical and space saving construction, a slit stud design can be used in nonload-bearing partitions. Cutting should be done before erecting. Lumber with large knots should be avoided. The cut must be centered in the wide face of the stud and extended to a point within about 3 in. from each end, Fig. 13-56.

Floors and Ceilings

In general, the considerations for soundproofing that were applied to walls can also be applied to floors and ceilings. Floors are subjected to impact sounds (walking, moving furniture, operating vacuum cleaners and other vibrating equipment) and sound control is somewhat more difficult. Often an impact sound may cause more annoyance in the room below than it does in the room where it is generated. The addition of carpeting or similar material to a regular hardwood floor will be effective in minimizing impact sounds.

Sound deadening board, properly installed will increase the STC rating. See Fig. 13-57. The use of patented metal clips to attach the ceiling material is a practical solution. Various suspended ceiling systems also provide high levels of sound control.

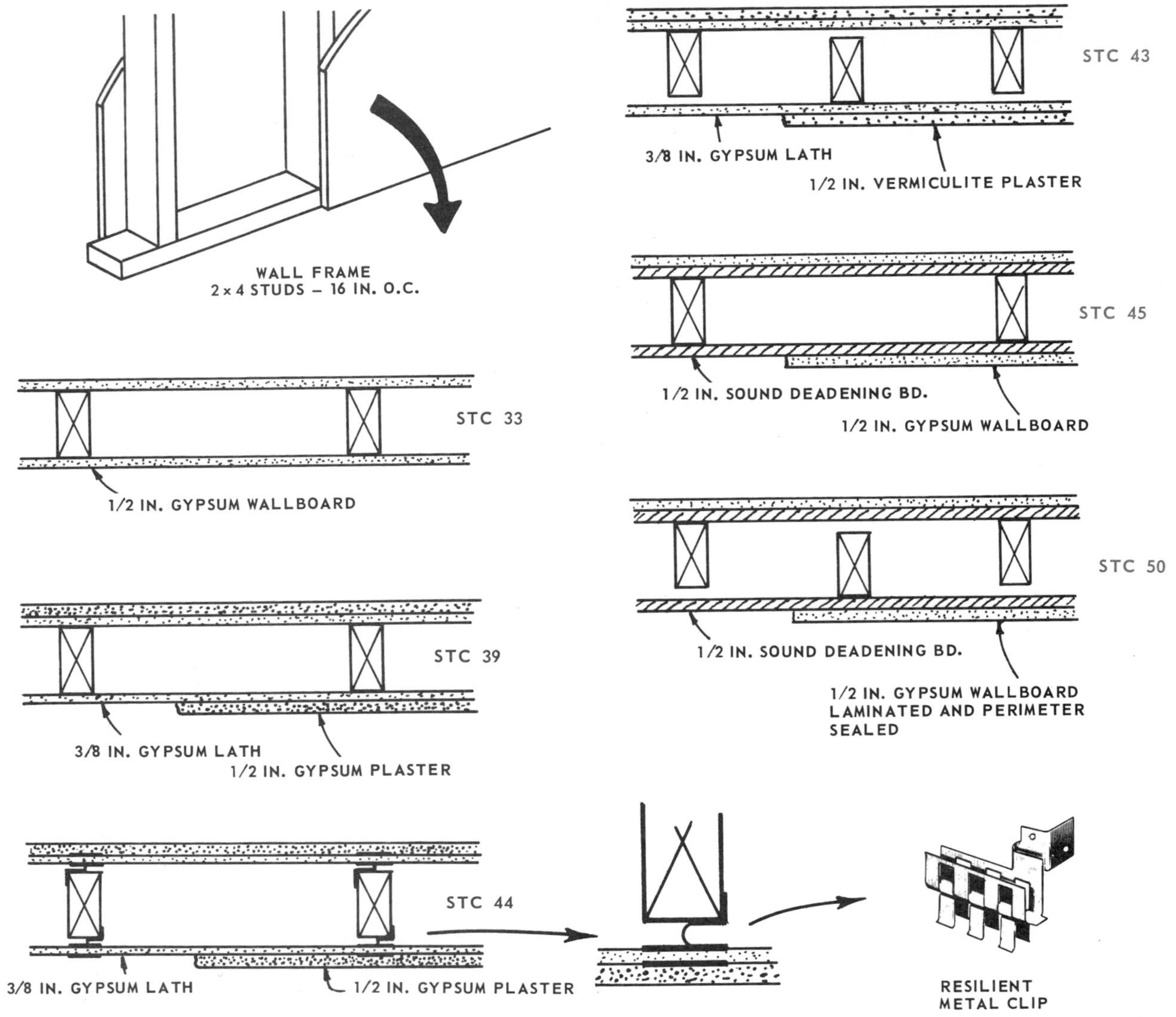

Fig. 13-54. Interior partitions with sound transmission class (STC) ratings. Including an insulation blanket within the structures will usually increase the STC rating.

Fig. 13-55. Double wall construction provides a high STC rating for a partition between apartments. Stapling insulation blanket to top plate. (Owens-Corning Fiberglas Corp.)

Fig. 13-56. Slit stud partition design. Wallboard joints and edges must be sealed.

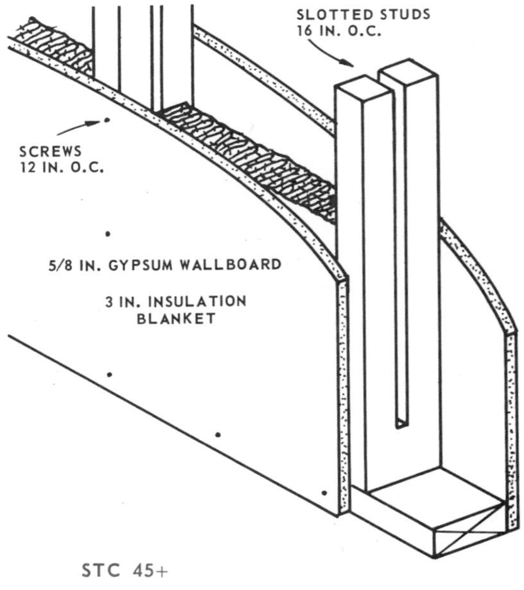

An existing floor can be soundproofed through a modification as illustrated in Fig. 13-58. Sleepers of 2 x 3 material are laid in place over a glass wool blanket but are not nailed to the old floor. if sound insulation is to be achieved. Doors between rooms are probably the greatest factor in this category. A 1/4 in. crack around a wood door (2 in. thick) would admit four times as

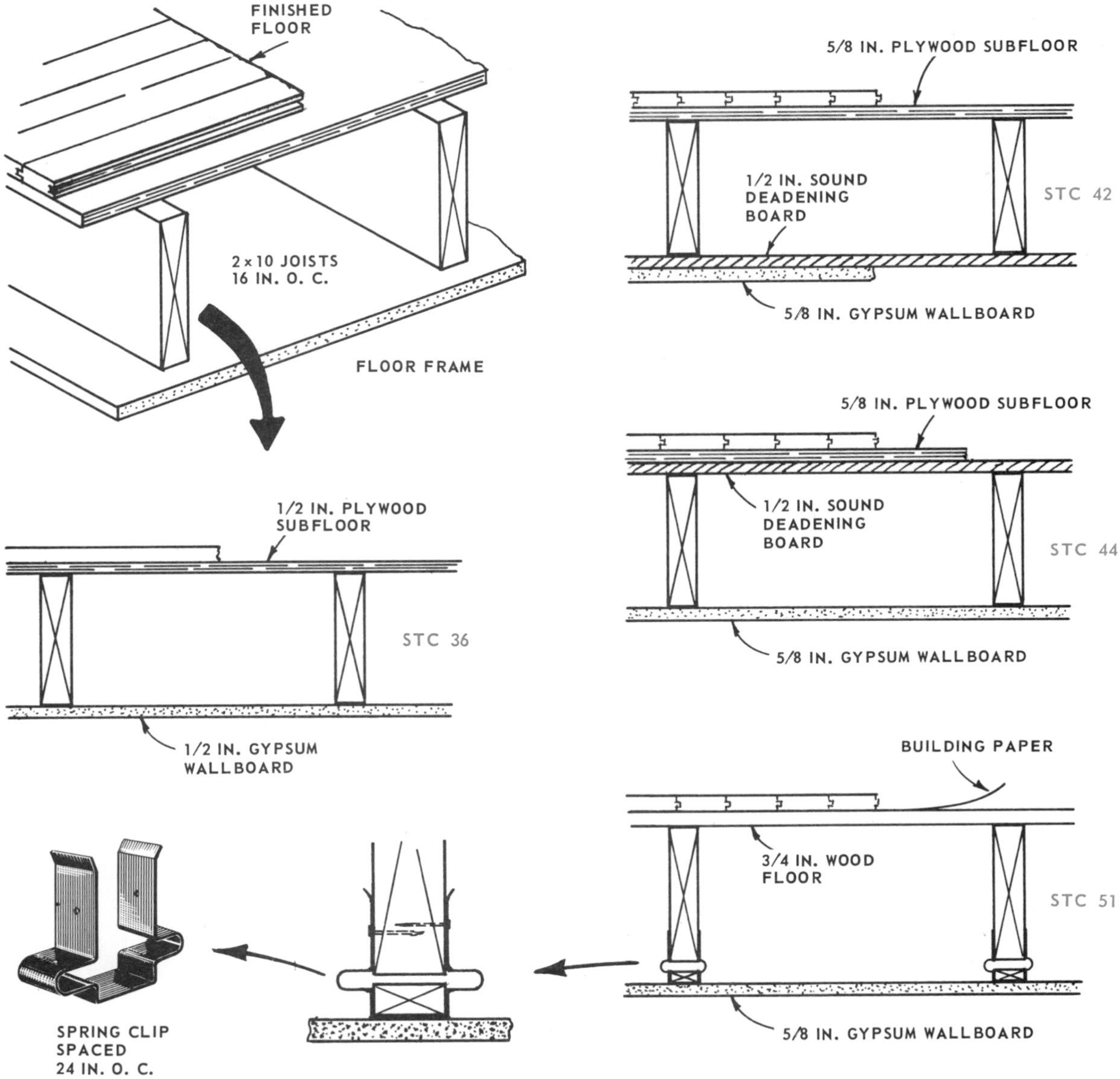

Fig. 13-57. Floor and ceiling constructions with various STC ratings.

When the new floor is laid, be sure the nails do not go all the way through the sleepers. The only contact between the new and old floor is the glass wool blanket which will compress to about 1/4 in. under the sleepers.

Doors and Windows

Sound has a tendency to spread out after passing through an opening. Thus cracks and holes should be avoided in every type of construction

much sound of medium intensity as the door itself. Felt, rubber or metal strips around the jambs and head are desirable. Conditions can be further improved by some form of "draft excluder" at the sill, such as a threshold or felt weatherstop.

Hollow-core interior doors that are well fitted will have a sound reduction value from 20 to 25 decibels. Similar double doors, hung with at least 6 in. air space between, will have a sound reduction factor as high as 40 decibels. This factor can be increased considerably by using felt or rubber strips around the stops.

For special installations, "soundproof" doors are available and can be built to suit almost any condition. Special hardware is used on this type of door to prevent sound transmission through the doorknobs.

Similar precautions should be observed with glazed openings as with doors. Cracks around these openings may counteract other precautions to eliminate sound transmission. Windows or glazed openings should be as airtight as feasible. Double glazing adds considerably to the sound reduction effect of the opening. Glass blocks have a sound reduction factor of about 40 decibels; being effective where transparent glazing is not needed.

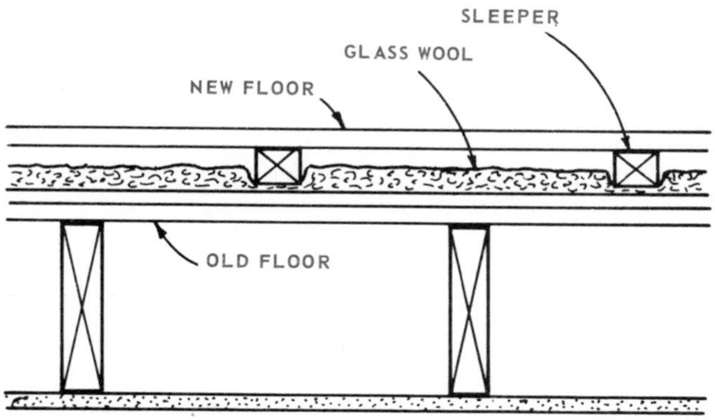

Fig. 13-58. Soundproofing an existing floor.

Noise Reduction Within a Space

While it is important to design walls and floors that will reduce sound transmission between spaces, it is also advisable to treat the enclosure so that sound will be trapped or reduced at its source. Reducing the noise level within the room will not only have an appreciable effect on the transmission of sound, but will make living conditions within the room more desirable. Areas in homes where noise reduction is most important include kitchens, utility rooms, family rooms, and hall ways.

There are a number of different types of acoustical material available to the builder. These come in a wide range of sizes -- from a 12 by 12 in. tile to boards 4 by 16 ft. Those with the best acoustical properties will absorb up to 70 percent of the sound that strikes them. Some are distinctly an applicator or "acoustical engineer" product. The latter requires either special equipment or trained men to install. The most common types and probably the most used methods are perforated or porous fiberboard units, perforated metal pan units, cork acoustical material and acoustical plaster. All of these materials provide high absorption qualities and except for the acoustical plaster, have a factory applied finish. Since the sound absorbing properties of any of these materials depends on its sponge-like quality, they are relatively light and do not require any building reinforcement or structural changes.

How Acoustical Materials Work

The sound absorbing value of most materials is dependent to a greater or lesser degree on its porous surface. Sound waves entering these pores, or holes, get "lost" and are said to be dissipated into heat energy. See Fig. 13-59. Other materials depend on a similar absorption action to reduce sound. The material used has a vibration point which approaches zero. Heavy draperies or hangings, hair felt, and other soft flexible materials function in this manner.

The efficiency of an acoustical unit or product is measured by its ability to absorb sound waves. Since noise is a mixture of confused sounds of many frequencies, this efficiency is measured by the noise-reduction coefficient (NRC) of a material for the average middle range of sounds. For most installations, this figure can be used to compare the values of one material over another.

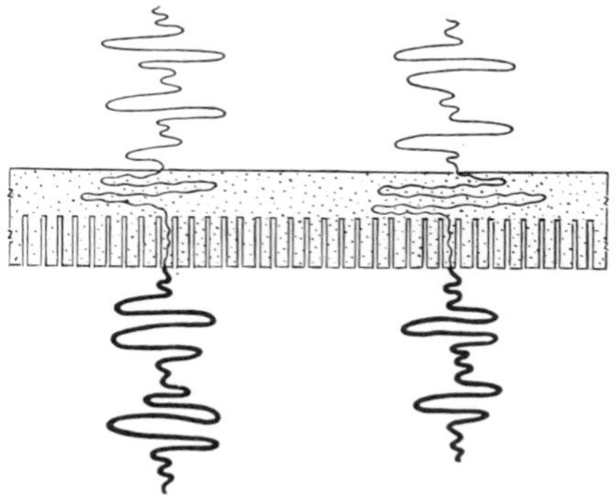

Fig. 13-59. Sound waves are absorbed by porous materials. The sound-absorption coefficient (also called a noise-reduction coefficient) of a material is defined as the fractional part of the energy of a sound wave that is absorbed at each reflection.

However, some materials are designed to do a better job for high or low frequency sounds and for special cases such as music studios, auditoriums and theaters, an acoustical engineer should be consulted. Most manufactuers of acoustical materials furnish this service.

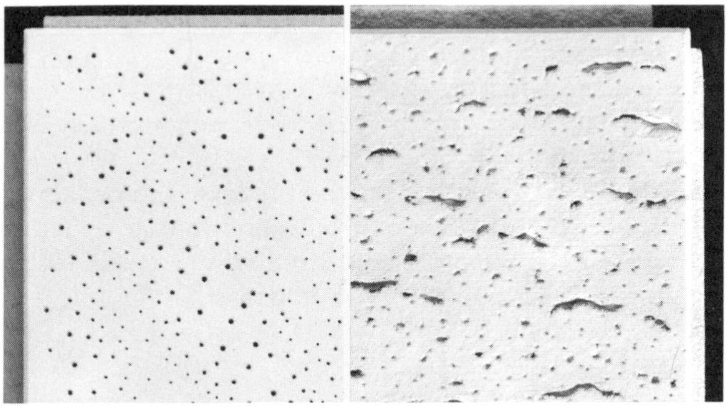

Fig. 13-60. Ceiling tiles designed to provide high sound-absorption factors. Left. Drilled holes. Right. Special "fractured" surface. (Wood Conversion Co.)

The perforated fiberboard acoustical materials are made in tile shape of a low density fibrous composition. Holes of various sizes are drilled almost through the tile. The sounds which strike these units are trapped in the holes. The walls of the holes, being relatively soft and fibrous in nature, form tiny pockets which absorb the sound. See Fig. 13-60. There are also some fibrous products which are not drilled, but depend on the porosity of the surface and its low vibration point to absorb the sound. Most of these products have a relatively smooth surface, providing a high degree of light reflection without glare.

They are usually installed on the ceiling or upper wall surface by gluing or stapling. Where the surface is in poor condition, furring strips can be used and the material nailed or glued to them. Each manufacturer has specifications on installation.

The perforated metal-pan type performs in somewhat the same manner. The sound enters through the holes in the surface of the pans and is trapped by the backing material which is of a soft resilient nature. It is commonly used in institutional and commercial buildings where ease of maintenance and fire resistance are important factors.

Fig. 13-61. Suspended ceiling system uses acoustical panels and provides space for recessed lighting. (Armstrong Cork Co.)

Suspended Ceilings

These factory finished units are installed on suspended metal runners which are designed to allow the large panels to simply be dropped into place, Fig. 13-61. They are used where suspended ceilings are advisable to conceal pipes, electrical wiring, and structural beams. The entire ceiling or any part of it may be removed and relocated without damage to the material. This factor alone might influence the installation of this type of acoustical treatment where it had not been considered before because of the necessity of access to furred spaces.

Units are made from various materials. Some are made from ground cork or glass fibers pressed into acoustical panels of various sizes and thickness. Because of the resistance of both these materials to moisture, they are ideal for use in indoor swimming pools, commercial kitchens or any place where humidity is a problem. Additional information on suspended ceilings is included in Unit 14.

Acoustical Plaster

There are a number of lightweight or fibrous materials used in the preparation of acoustical plaster. To secure the best appearance and highest sound-absorption qualities, the plaster should be sprayed on the surface, Fig. 13-62. Acoustical plasters are generally used where an unlined or plain surface wall or ceiling is desired or where curved or intricate planes make the use of unit types impractical.

To install acoustical plaster instead of plain plaster, omit the finish plaster coat and instead apply two coats of the acoustical plaster.

Installation of Materials

Manufacturers' recommendations as to the manner of installation should be carefully followed on all acoustical materials. Many of these materials exhibit large variations in their properties when the method of installation has been changed from that recommended. In many cases, the amount of air space back of the material is a factor in its sound absorption qualities.

Fig. 13-62. Spraying acoustical plaster on a ceiling. (Vermiculite Institute)

Because of the relatively soft nature of most acoustical materials, it is the usual practice to install them on the ceiling or upper portion of sidewalls. For sounds originating in the average room, the ceiling offers a sufficient area for sound absorption materials. Directions concerning the methods and procedures for installing ceiling tile are included in Unit 14.

Maintenance

Painting of perforated boards usually does not lower the efficiency, if properly done. The dirt clogging the pores of the material should be removed first. This may often be done with a vacuum cleaner or by brushing with a soft-hair brush. Some acoustical material can be cleaned by washing.

Spray painting is usually preferable to brush painting as a thinner mixture has less tendency to clog the pores of the material. Improperly applied paint would soon fill the pores of the material and destroy its efficiency.

HANDY SAYS:

"Manufacturers of acoustical tiles and other products have prepared detailed instructions for installation and maintenance. Be sure to follow them carefully."

Test Your Knowledge - Unit 13

1. When heat moves trom one molecule to another within a given material, the method of heat transmission is called __________.
2. Heat can be transmitted by wave motion. This method is referred to as __________.
3. The selection of an insulation material should be based on its cost, effectiveness, __________, and the adaptation to the construction.
4. The coefficient of thermal conductivity is represented by the letter______.
5. The "U" factor for a standard frame wall; covered with composition sheathing and gypsum lath and plaster, and including a one inch blanket of insulation; is about __________.
6. A standard insulation batt is usually______ or ______ inches long.
7. The temperature at which condensation occurs for a given sample of air is called the _______ _______.
8. The vapor barrier in a wall structure should be located on the __________ (cold, warm) side of the insulation.
9. When stapling blanket insulation between studs start at the __________. (top, bottom)
10. Fill insulation can be poured or__________ into place.
11. Ceilings of residential structures, heated by electricity, should contain at least_____ inches of insulation.
12. Heat loss through a double-glazed window may be as much as ______ times the loss through a well insulated wall.
13. Air leakage around windows and doors is called __________.
14. The unit of measure used to indicate the loudness or intensity of a sound is called the __________.
15. A wall with a TL of 40 db. will reduce a loudness level passing through it from 70 db. to _______ db.
16. A method of rating the sound-blocking efficiency of a wall, floor, or ceiling structure is called the _____ ______ ______ system.
17. Sound deadening board is made largely from wood and _________ fibers.
18. To insure a high STC rating, electrical outlets on opposite faces of a partition should not be located in the same ________ ________.
19. In slit-stud construction, the cut should extend to within about ______ inches of the ends of the studs.
20. The efficiency of an acoustical ceiling unit is expressed in a NRC rating which is an abbreviation for _______ _______ ________.

Outside Assignments

1. Secure samples of various thermal insulating materials. Include glass, mineral, and organic fibers and also foamed plastic materials. Also

include loose fill insulations made from such material as vermiculite. Enclose the fibrous and granular materials in small envelopes made of polyethylene plastic film. Mount the samples on a display board with descriptive titles that include k and R factors.

2. Prepare a brief study of the most common types of heating and cooling systems used in your locality. Make several drawings showing how they operate and how they are controlled. Study local building codes and outline the basic requirements that must be followed in the installation. Visit with a heating contractor and secure basic information concerning the heating and cooling load-calculations for residential structures. Also obtain information about manufacturers, approximate costs, and installation procedures. Organize your information by preparing an outline and make an oral report to your class.
3. From a study of reference books and trade magazines, prepare a report on radiant heating systems. Include information about panels that are heated with electricity as well as those heated with hot water. Place special emphasis on the methods of installation since structural design may need to be modified when these systems are employed. Also secure information concerning types and amounts of insulation materials; including special application procedures that may be required.
4. Obtain the use of a tape recorder and carefully record the various sounds produced by equipment and devices found in a modern home. Include laundry equipment, dishwashers, food mixers, garbage disposers, plumbing fixtures, and vacuum cleaners. Also include such situations as walking on hard surface floors, and closing passage doors or cabinet doors. Play the recording for your class and then lead a discussion on how the various sounds can be minimized or controlled through proper design and the use of special materials and construction.

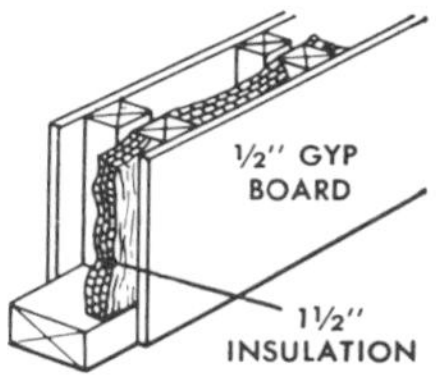

Slit-stud wall with 1½" blanket insulation hung from top plate offers greater resistance to noise transmission.
STC-43 **N-63**

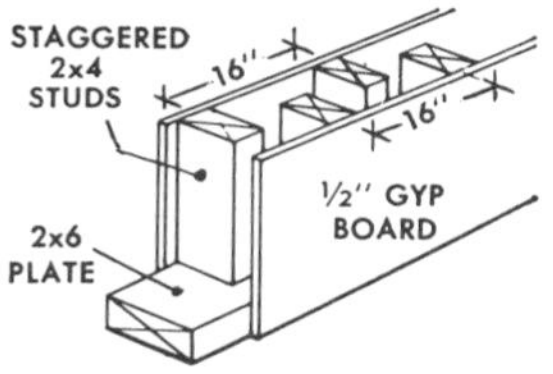

Staggered studs provide complete separation between wall faces. Staggering 2x3s on 2x4 plate is almost as effective.
STC-41 **N-60**

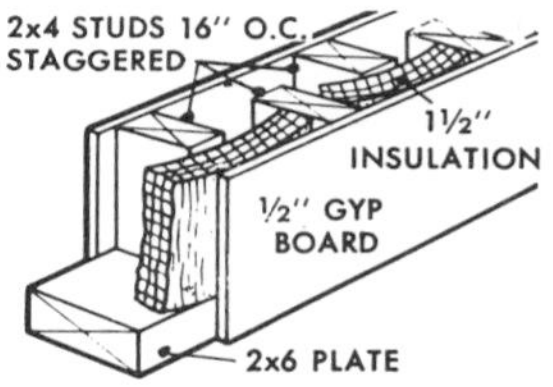

Blanket insulation between staggered studs performs as well as weaving continuous insulation behind studs.
STC-49 **N-65**

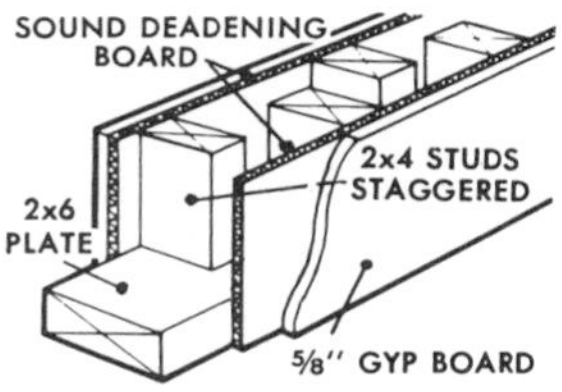

Sound deadening board absorbs noise, dampens wall vibration, adds surprisingly to effectiveness of staggered-stud wall.
STC-52 **N-70**

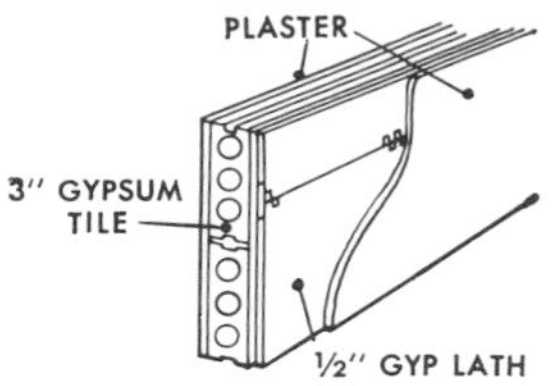

Fireproof wall of 3" gypsum tile has been considered a good acoustical wall, yet its rating is surprisingly low.
STC-38 **N-54**

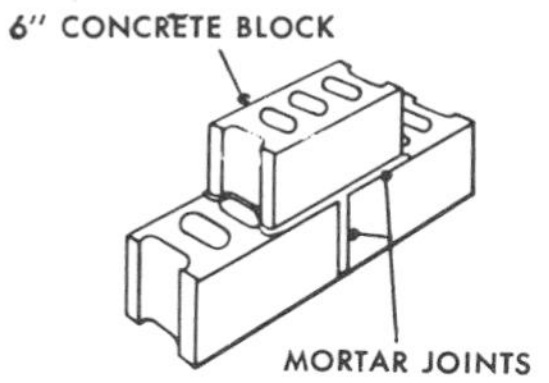

Concrete block wall becomes more effective when pores are sealed with two coats of paint.
Unpainted: STC-43 **N-59**
Painted: STC-44 **N-61**

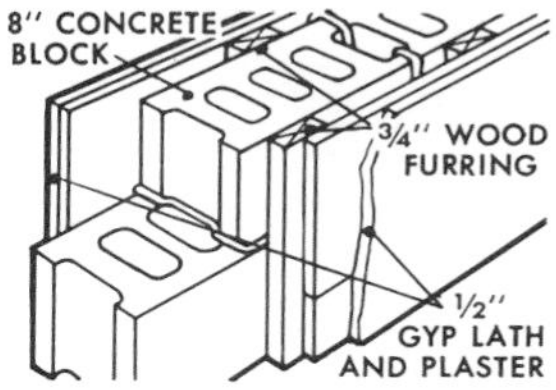

Concrete block with furred gypsum lath and plaster both sides has excellent properties in mid and high frequencies.
STC-49 **N-66**

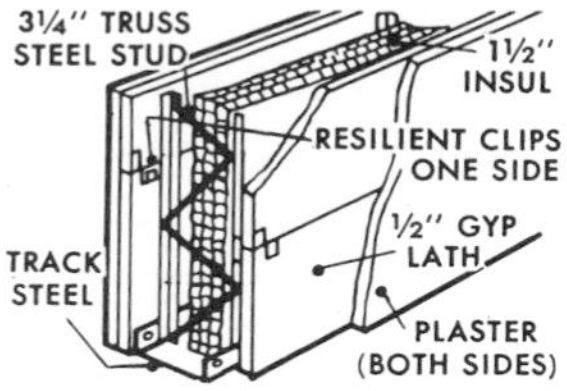

Steel truss studs, 3¼" wide with gypsum lath and plaster both sides, require insulation to be effective.
STC-41 **N-59**

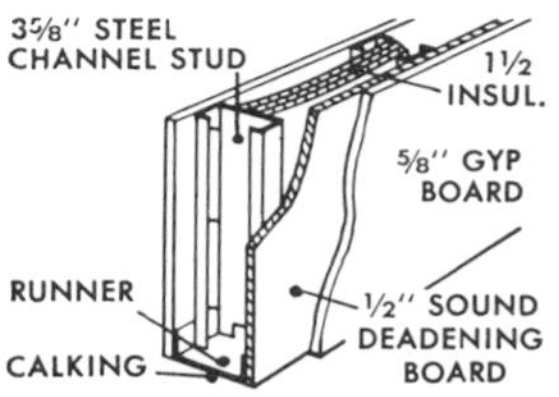

Steel channel studs with insulation plus sound deadening board on one side rate high. Calking seals gaps.
STC-52 **N-70**

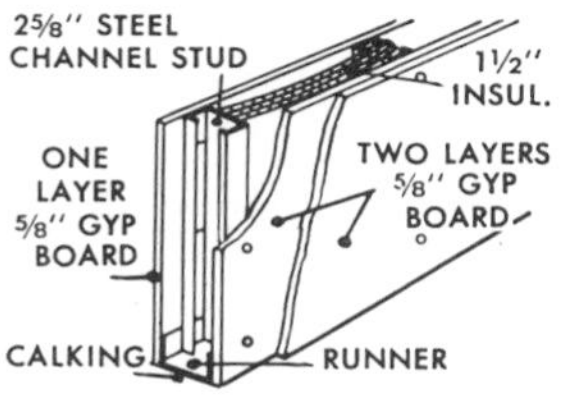

Extra layer of ⅝" gypsum board on one side of 2⅝" channel studs gives fair performance with insulation.
STC-46 **N-69**

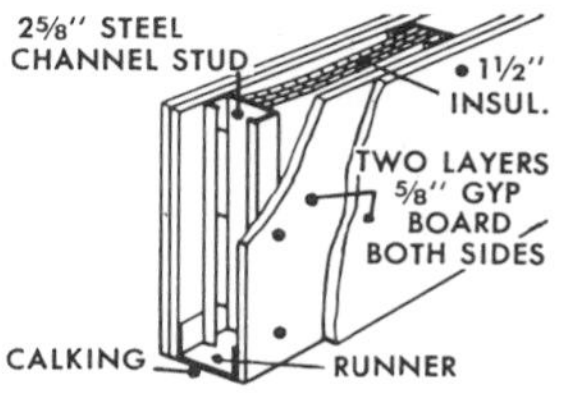

Laminated ⅝" gypsum board on both sides of 2⅝" steel channel studs plus insulation gets very high rating.
STC-55 **N-74**

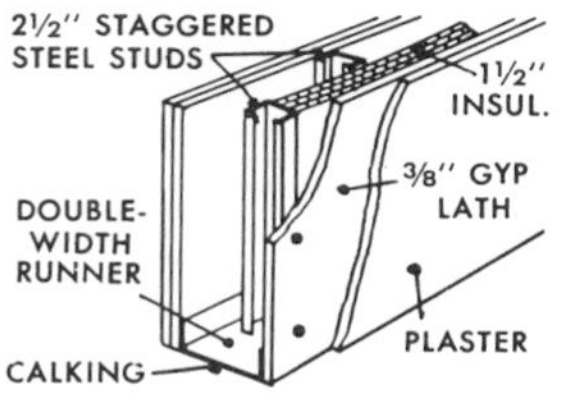

Staggered steel studs with ⅜" gypsum lath and plaster compares favorably with staggered wood studs and insulation.
STC-49 **N-69**

Various wall designs and sound transmission factors. The speech rating number (N) meets a desirable standard at about 60, when conservation cannot be understood in the next room.

Wet end of gypsum board machine. Unit on left pours streams of gypsum mixture on backing paper carried on moving belt. Blade (arrow) spreads and levels mixture to correct thickness. Paper for the front surface is then applied and rolled around the edges. The gypsum hardens rapidly as it passes through heated rollers and platens and is then cut to length. (Georgia-Pacific)

Installing gypsum wallboard in an apartment building. Note that a vapor barrier is not being used since the floor above is heated space. The purpose of the insulation is to reduce sound transmission. (National Gypsum Co.)

Unit 14

INTERIOR WALL AND CEILING FINISH

This Unit on interior wall and ceiling finish is a study of the materials and methods used to cover the framed areas inside the structure. Many types of materials are employed for this purpose; the two principle ones being plaster and dry wall construction.

The term dry wall construction is commonly applied to a wall finish consisting of large gypsum boards that are carefully applied. Nail holes and joints are smoothly covered and finished so that the completed surface appears much like regular plaster.

Plaster

Through the years, gypsum plaster, Fig. 14-1, has provided qualities that are desired in a wall and ceiling finish--beauty, durability, economy, fire protection, structural rigidity and resistance to sound transmission. It also has a high factor of adaptability since it can be readily applied to curved or irregularly shaped surfaces, as well as those that are flat.

Plaster is made from gypsum, one of the common minerals found in the earth. It is known chemically as hydrous calcium sulphate, which is calcium sulphate in combination with two molecules of water ($CaSO_4 - 2\ H_2O$). Gypsum is a highly fireproof material. When exposed to fire, the chemically combined water, called water of crystallization, must first be vaporized and this dissipates some of the heat. As the water leaves the gypsum mass, a white chalky material remains on the surface to further insulate against the flame. Fire protection engineers, recognizing the fire resistance qualities of plaster, have developed accurate ratings which are listed by the National Bureau of Standards and the National Bureau of Fire Underwriters. These ratings are used as a basis for establishing requirements in various building codes.

When plaster is used for the interior wall and ceiling surface, the carpenter usually applies the plaster base and installs the grounds that serve as guides for the plasterer. In addition to a thorough knowledge of the requirements and methods of making this installation, he should also have a general knowledge of how the plaster coats are applied.

Fig. 14-1. Applying the first coat (base coat) of plaster over perforated gypsum lath. (Gypsum Association)

Plaster Base

A plaster finish requires some type of base upon which the plaster can be spread. For many years, wood lath was used for this purpose, however in modern construction, sheet materials and metal lath have eliminated its use. Plaster is sometimes applied directly to masonry surfaces or special gypsum block units.

Today, a commonly used plaster base is gypsum lath, Fig. 14-2. A standard size panel measures 16 in. by 48 in. and is applied horizontally to the framing members of the structure. It consists of a rigid gypsum board with a special paper cover.

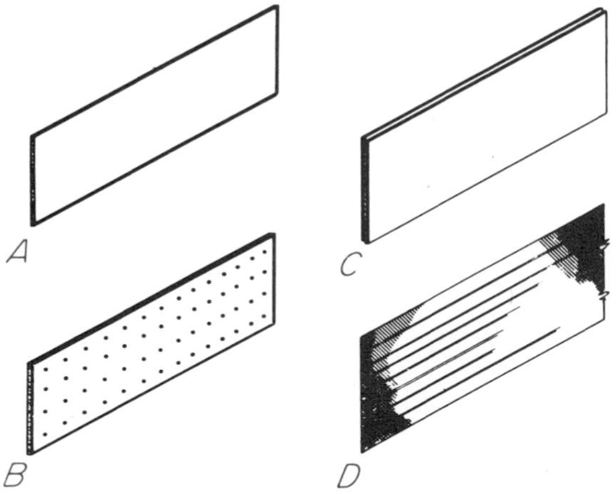

Fig. 14-2. Plaster base materials. A–Gypsum lath. B–Perforated gypsum lath. C–Insulating fiberboard lath. D–Metal lath.

For a stud or ceiling joist spacing of 16 in. on center, a 3/8 in. thickness is used. For a 24 in. spacing a 1/2 in. thickness is required. Gypsum lath can be obtained with an aluminum foil back that serves as a vapor barrier; and provides some insulating value if it faces an air space. It is also available with perforations which improve the plaster bond and thus lengthens the time the wall surface will remain intact when exposed to fire. Some building codes specify this type.

Insulating fiberboard lath is also used as a plaster base. It comes in a 3/8 in. and 1/2 in. thickness with a shiplap edge. Fiberboard lath has considerable insulation value and is often used on ceilings or walls adjoining exterior or unheated areas. Fig. 14-3 shows sizes available.

Expanded metal lath consists of a copper alloy sheet steel, slit and expanded to form a multitude of openings for keying the plaster. The two most common types are diamond mesh and flat rib. Standard size pieces are 27 in. wide by 96 in. long. Metal lath is usually dipped in black asphaltum paint, or is galvanized, to make it rust-resistant.

TYPE	THICKNESS (Inches)	WIDTH (Inches)	LENGTH (Inches)
PLAIN	3/8	16	48 or 96
	1/2	16	48
PERFORATED	3/8	16	48 or 96
	1/2	16	48
INSULATING	3/8	16	48 or 96
	3/8 or 1/2	24	as requested to 12 ft.
LONG LENGTH	1/2	24	as requested to 12 ft.

Fig. 14-3. Gypsum and insulating lath sizes.

Applying Lath

Before making the application, the framing should be inspected for proper spacing and alignment. Check to see that corners and openings have nailing members to support the ends of the lath.

Units should be applied with the long dimension perpendicular to the framing and with the end joints staggered between adjacent courses, Fig. 14-4. Turn the folded or lapped paper edges of gypsum lath toward the framing. Edges and ends of lath should be in moderate contact.

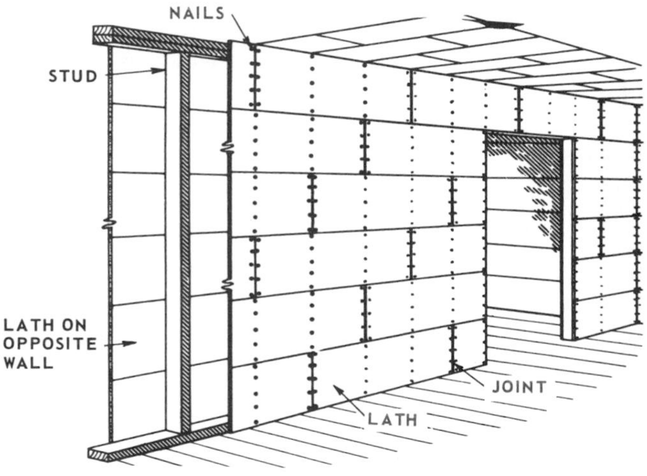

Fig. 14-4. Application pattern for standard gypsum lath.

THICKNESS OF LATH AND DESCRIPTION OF FASTENER	SPACING
NAILS 3/8" thick lath: 13 gauge, 1 1/8" long, 19/64" flat head, blued	16" wide lath, 4 nails per lath per framing support, spaced 5", and placed not less than 3/8" from ends or edges. (Note: Some fire resistive constructions require 5 nails per lath per support.) For 24" wide lath, use 5 nails per lath per framing support.
1/2" thick lath: Same as above, except 1 1/4" long	Same as above except that where placed on supports 24" o.c., 5 nails, spaced at 4" o.c. per lath per support, shall be used. For 24" wide lath, use 6 nails per lath per support.
STAPLES 3/8" thick lath: Minimum recommended staple: 16 U.S. Standard gauge flattened galvanized wire, 7/16" wide crown (outside measure), 7/8" long legs with divergent points.	Staples shall be driven with the crown parallel to the long dimension of the framing member, using not less than 4 staples per framing member and not more than 5" o.c., in such manner that the crown bears tightly against the lath but does not cut into the face paper. The legs of the staples shall be no less than 3/8" from the ends and edges of the lath.
1/2" thick lath: Same as above, but staples 1" long.	Same as above, except that where placed on supports 24" o.c., 5 staples, spaced at 4" o.c. per lath per support, shall be used. For 24" wide lath, use 6 staples per lath per support.

Fig. 14-5. Nailing requirements for gypsum lath.

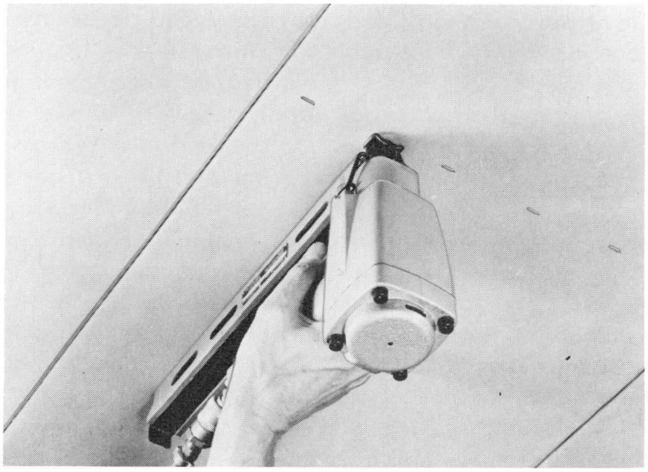

Fig. 14-6. Using a pneumatic powered stapler to attach 3/8 in. gypsum lath to ceiling joists. The staples have a 7/16 in. crown. (Spotnails Inc.)

Gypsum lath (3/8 in. thickness) is usually applied with 13 ga. gypsum-lathing nails 1 1/8 in. long. Nail and staple sizes for various installations are given in Fig. 14-5. The lath must be nailed at each stud or joist crossing. Insulating lath is installed in about the same manner as gypsum lath, except that 13 ga. 1 1/4 in. blued nails should be used. In modern construction, staples are often used to attach the lath and are driven with pneumatic powered staplers as shown in Fig. 14-6.

Gypsum lath is easily formed to size by scoring one or both sides with a pointed or edge tool and then breaking along the line. Be sure to make neatly fitted cut-outs for plumbing pipes and electrical outlets.

HANDY SAYS:

"When applying gypsum lath, drive the nails "home" so the head will be just below the paper surface. Avoid additional hammer blows that will crush the gypsum core."

Metal Lath

In commercial construction, metal lath is commonly used throughout the structure, while in residential work it may be employed only in shower stalls or bathroom areas where the wall surface is subjected to excessive moisture and/or extra strength is required. The metal lath, embedded in plaster, provides a rigid wall surface when the construction is properly designed.

Metal lath must be of the proper type and weight for the support spacing. Sides and ends are lapped and corners are returned (overlapped). Before application of the lath in bathroom areas the studs are usually covered with a 15 lb. asphalt-saturated felt, Fig. 14-7. Portland cement plaster is often used as the first coat when the surface will be finished with ceramic tile.

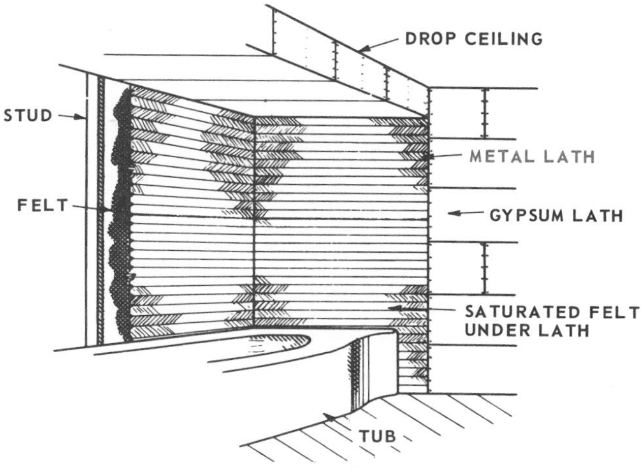

Fig. 14-7. Standard installation of metal lath in tub alcove.

Reinforcing

Since some drying will nearly always occur in wood frame structures, shrinkage can be expected. This will likely cause plaster cracks to develop around openings, in corners, or wherever there is a concentration of cross-grain wood.

To minimize or eliminate this cracking, expanded metal lath is often used in key positions over the plaster base. Strips, approximately 8 in. wide, are applied diagonally at the corners of doors and windows as shown in Fig. 14-8. Tack or staple these lightly in place so they become a part of the plaster base. If nailed securely to the framing; warping, shrinking, and twisting of the frame will be transmitted into the plaster coats and cause cracks. Metal lath should be used under or around wood beams that will be covered with plaster. Be

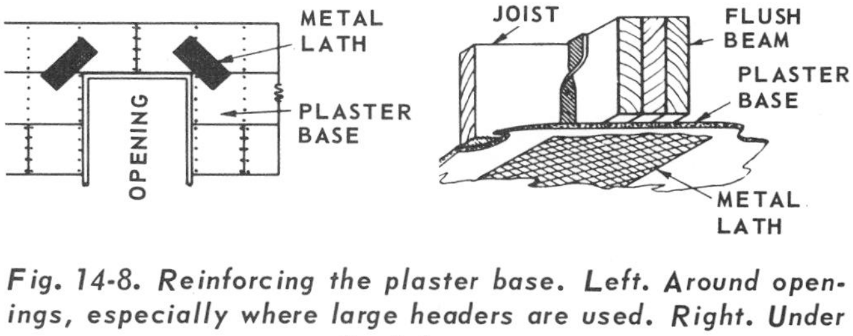

Fig. 14-8. Reinforcing the plaster base. Left. Around openings, especially where large headers are used. Right. Under flush beams.

sure to extend the edges of the reinforcing well beyond the structural element being covered.

Inside corners may be reinforced with a special formed metal lath or wire fabric called Cornerite, Fig. 14-9. Minimum widths should be 5 in. thus forming a 2 1/2 in. width on each surface of the internal angle. In some plaster base systems that

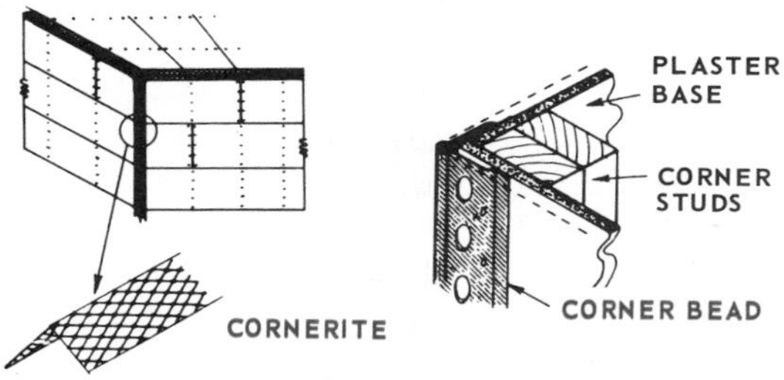

Fig. 14-9. Reinforcing inside and outside corners.

employ a special attachment clip, Cornerite is not recommended. Outside corners, where no wood trim will be applied, are reinforced with metal corner beads. They must be carefully applied-- plumb or level -- since they serve not only as a reinforcement for the plaster but as a ground (guide) for its application. When applying corner bead, use a straightedge and level or plumb line. Also use a spacer block to check the distance between the corner of the bead and the surface of the plaster base. This distance must be equal to the thickness of the plaster coats.

Plaster Grounds

Plaster grounds are usually wood strips, the same thickness as the total thickness of the plaster base and plaster (3/8 in. plus 1/2 in. for average residential construction). They are installed before the plaster is applied and serve as a guide or

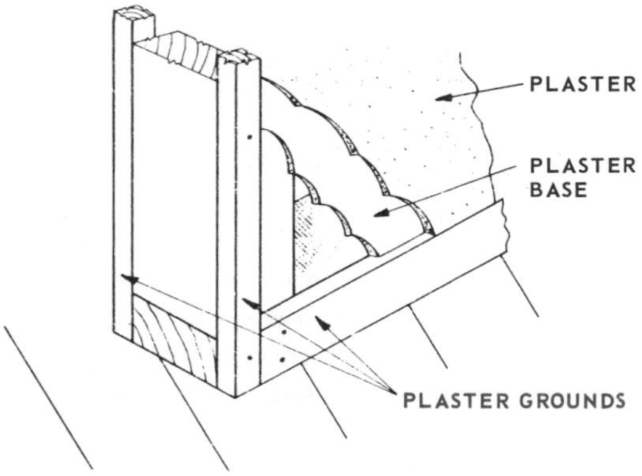

Fig. 14-10. Plaster grounds installed at an opening and along the floor line.

leveling surface, and may be used later as a nailing base for attaching trim members. Grounds are used around doors, windows, and other openings; and sometimes at the bottom of walls along the floor line, Fig. 14-10. In some wall systems, especially large commercial and institutional buildings, metal edges and strips serve as grounds.

Today, modern windows are usually equipped with jamb extensions that are adjusted for the various thicknesses of materials used in the wall structure. These jamb extensions serve as plaster grounds and the carpenter seldom needs to make any alterations.

Grounds around some openings are removed after the plastering is complete. Those used at door openings must be carefully set (plumbed) and conform to the width of the door jamb to insure a good fit of the casing. The width of standard interior door jambs is usually 5 1/4 in.

The carpenter often constructs a jig or frame that is temporarily attached to the door opening. He then can quickly nail the grounds in place along the straightedges of the jig. Instead of using two strips, some carpenters prefer to use a single piece of 3/8 in. exterior plywood that is ripped to the same width as the door jamb. Such grounds, if carefully removed, can be reused on future jobs with the same requirements, Fig. 14-11.

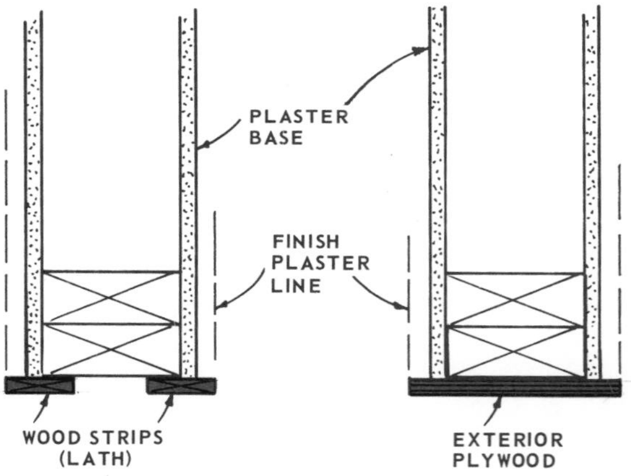

Fig. 14-11. Removable grounds installed at door openings.

HANDY SAYS:

"When the plaster base is complete, mark lines on the subfloor at the center line of each stud. After the plaster has been applied and is dry, these marks are transferred to the wall and are helpful when installing baseboards, cabinets, and fixtures."

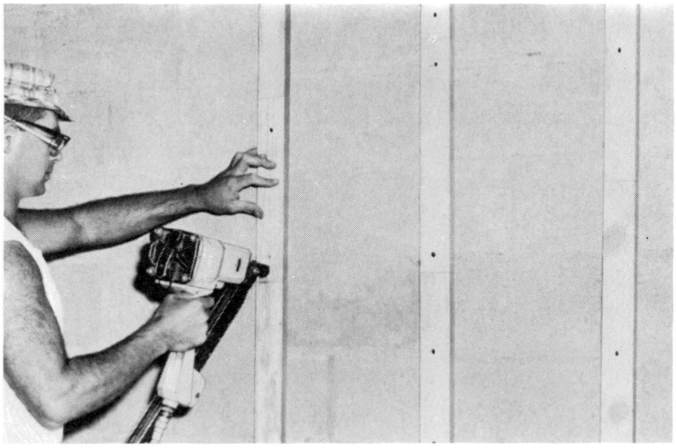

Fig. 14-12. Attaching furring strips to a concrete block wall with a pneumatic nailer. Special 6d round head nails are being used. (Duo-Fast Fastener Corp.)

Plaster Base on Masonry Walls

Gypsum plaster can be applied directly to most masonry surfaces. To prevent excessive heat loss through outside walls, furring strips are usually installed which carry the plaster base materials and provide an air space. Strips are attached by using case hardened nails driven into the mortar joints, metal or wood nailing plugs placed in the wall during construction, or newly developed adhesive materials and patented fasteners. Fig. 14-12 shows strips being attached with a pneumatic nailer.

When heat loss needs to be further reduced, blanket insulation can be installed as shown in Fig. 14-13. Some constructions employ the use of

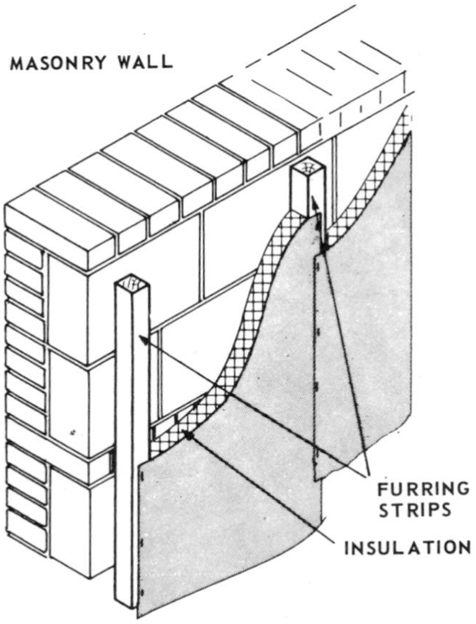

Fig. 14-13. Method of insulating an exterior masonry wall.

plastic foam panels that are attached directly to the wall with adhesives. The plastic foam serves as a plaster base. Follow manufacturers' recommendations when making an installation of this kind.

Plastering Materials and Methods

Plaster is applied in two or three coats. The base coats are prepared by mixing gypsum with an aggregate, either at the gypsum plant or on the job. These aggregates may consist of wood fibers, sand, perlite or vermiculite -- sand is the most commonly used.

Never use a batch of plaster that has started to set. All such plaster should be discarded. Wash out the mixer with clean water after each batch has been prepared and keep tools and equipment clean.

In three-coat work, the first coat is called the scratch coat and is applied directly to the plaster base, Fig. 14-14. It is cross raked or scratched after it has "taken up" (stiffened) and is then allowed to set and partially dry. The second coat -- brown coat -- is then applied and leveled with the grounds and screeds using a darby (a long flat tool used by plasterers when working on ceilings) and rod (straightedge). After the brown coat has set and somewhat dry the third or finish coat is applied.

In two-coat work the scratch coat and brown coat are applied almost simultaneously. The cross raking of the scratch coat is omitted and the brown coat of plaster is usually applied (doubled-back) within a few minutes. This application method is the one most frequently used over gypsum or insulating lath plaster bases commonly used in residential construction.

Minimum plaster thickness (all coats) should not be less than 1/2 in. when applied to regular gypsum or insulating lath bases. A 5/8 in. thickness is usually required over brick, tile or masonry. When plaster is applied to the metal lath it should measure 3/4 in. in thickness from the back side of the lath.

A plaster job should be inspected constantly for basecoat thickness. Unless grounds and screeds are used on ceilings or large wall areas it is extremely difficult to keep the thickness uniform. Should the thickness be reduced, the possibility of checks and cracks is much greater. A 1/2 in. thickness of plaster possesses almost twice the resistance to bending and breaking as a 3/8 in. thickness.

Recent developments in the plastering trade center around the plastering machine, Fig. 14-15.

Fig. 14-14. Applying a base coat of plaster. Note opening for electrical outlet at lower left.

Fig. 14-15. Using a plastering machine to apply a base coat of plaster. (Gypsum Association)

It not only saves a great deal of labor but also improves the quality of the plaster application since the lapsed time between mixing and application is shortened. The machine also makes possible the control of the densification or compaction of the plaster coat.

The final or finish coat (about 1/16 in. thick) consists of two general types; the sand-float or textured surface and the putty or smooth finish. In the sand-float finish, special sand is mixed with gypsum or lime and cement. After the plaster is applied to the surface it is smoothed with a float to produce various effects depending on the floating method and the coarseness of the sand.

A smooth finish is produced by applying a putty-like material consisting of lime and gypsum or cement, Fig. 14-16. It is troweled perfectly smooth, somewhat like concrete.

Space in this book does not permit more than a very brief explanation of the methods and procedures of the plastering industry. For a more complete study - secure a copy of the manual entitled, Lathing and Plastering, prepared by the Gypsum Association, 1603 Orrington Ave., Evanston, Ill. 60201.

Fig. 14-16. Applying the finish coat of plaster.

Fig. 14-17. Installing Gypsum wallboard. (National Gypsum Co.)

Veneer Plaster

Veneer plaster (also called "thin-coat") is a high-strength plaster for application as a very thin coat, about 1/8 in. thick. It requires a base especially designed for this purpose. The joints of the plaster base are treated with a fiber glass tape. Special corner beads must be used. Because of the composition and thinness of the plaster coat -- it dries very rapidly -- trim and decoration work may proceed after a minimum drying time of about 24 hours.

Veneer plaster coats can be applied with a smooth, troweled finish or with a textured surface similar to regular plaster work.

TYPE	THICKNESS	EDGES
Regular	1/4" 3/8" 1/2" 5/8"	Tapered Square Square Tapered Bevel
Type "X"	1/2" 5/8"	Square Tapered Bevel
Insulating (Aluminum Foil on Back Surface)	3/8" 1/2" 5/8"	Square T & G Tapered Round
Backing Board	3/8" 1/2" 5/8"	Round Square Square T & G Round
Type "X" Backing Board	1/2" 5/8"	Round T & G Square
Coreboard (Homogeneous or Laminated)	1"	Square T & G Ship Lap
Predecorated	3/8" 1/2" 5/8"	Bevel Round Square Bevel Square

Fig. 14-18. Types of Gypsum wallboard.

Dry Wall Construction

Gypsum Wallboard

The use of dry wall materials, such as gypsum wallboard, has steadily increased in modern construction. See Fig. 14-17. Many builders prefer to use this because of the time saving factor. Regular plaster requires considerable drying time. It may be difficult to keep carpenters and other tradesmen effectively employed during the drying period. Both types of finish present some advantages and disadvantages. Dry wall construction, for example, requires that studs and ceiling joists be perfectly straight and true -- otherwise the wall surface will be uneven. The wood framing material must also have a moisture content very near to that which it will eventually attain in service to prevent "nail pops" and joint cracks.

Gypsum wallboard is a sheet of material that has a noncombustible core. It is basically a gypsum product, surfaced with a treated paper or other sheet materials. A variety of thicknesses, edge joints, and types are available, Fig. 14-18. Some boards have a predecorated surface while others have a surface suitable for the application of finish after installation. Several types have a surface especially designed for use as a base (backer board) in multilayer construction.

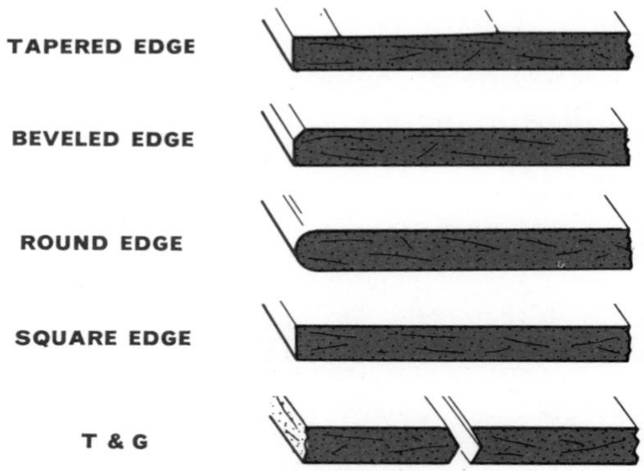

Fig. 14-19. Standard edge shapes.

Standard wallboard sizes include widths of 4 ft. and lengths as long as 16 ft. Fig. 14-19 shows standard edge shapes. The tapered edge forms a shallow channel between adjacent boards. This channel permits the use of tape and filler to form a smooth, continuous ceiling or wall surface.

Single Layer Construction

A single layer of wallboard can be applied directly to the wood framework. Use a thickness of 1/2 in. or 5/8 in. Ceilings are applied first and then the sidewalls. Horizontal application -- long edges at right angles to studs and joists -- is preferred since it results in few joints and stronger construction. Sidewall joints, which may run vertically, must be located at studs, Fig. 14-20.

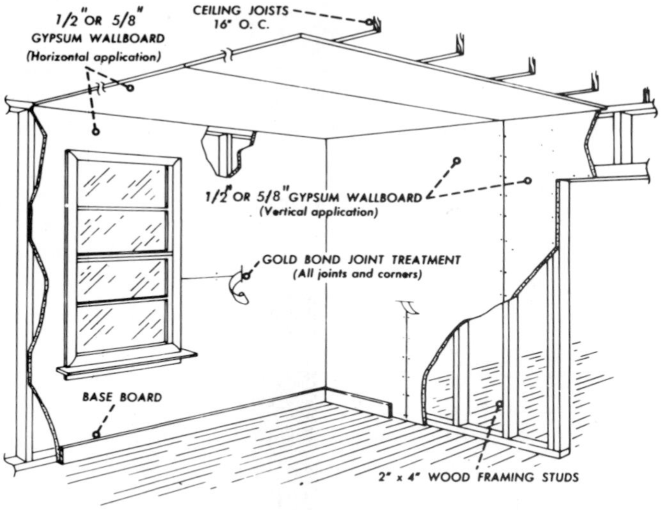

Fig. 14-20. Single layer dry wall construction.

Measuring and Cutting

All measurements should be carefully taken from the exact location where the wallboard will be installed. Usually it is best to make two readings, one on each side of the panel. Following this procedure will eliminate errors and reveal openings and framework that are not plumb or square. Use a 12 foot steel tape rule.

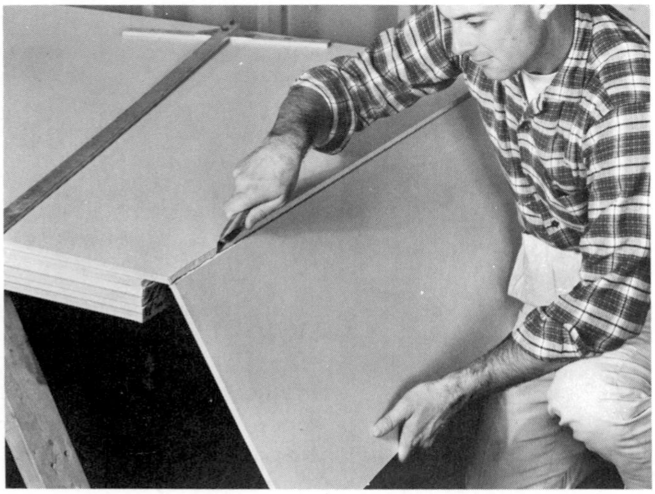

Fig. 14-21. Cutting the paper on the back side of the board after the break. Note the large steel T-square that was used to score the face side.

Straight cuts across the width or length of a board are made by first scoring the face with a knife pulled along a straightedge. With the main section of the board supported close to the scored line, snap the core by pressing downward on the overhang. Finally cut the back paper as shown in Fig. 14-21. When necessary the cut can be smoothed with a file or coarse sandpaper mounted on a block of wood.

HANDY SAYS:

"When scoring wallboard, always use a sharp knife that will make a "clean" cut through the paper face and penetrate slightly into the core."

Irregular shapes and curves can be cut with a coping saw, compass saw, or electric sabre saw. Fig. 14-22 shows an opening for a convenience outlet being made with a portable power saw.

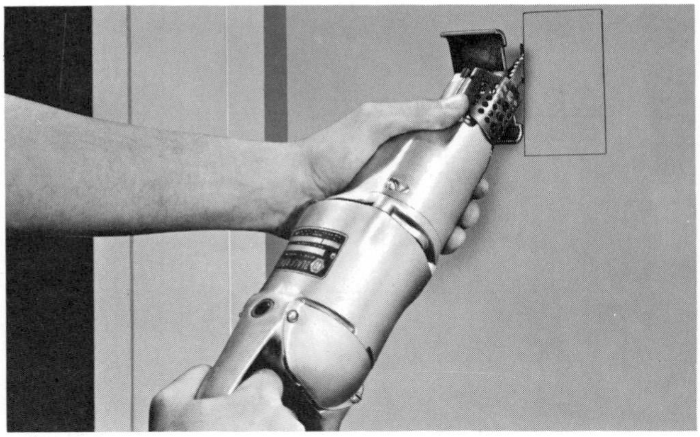

Fig. 14-22. Cutting Gypsum wallboard with a portable power saw.

Nails and Screws

Nail spacing will vary somewhat depending on the materials being used. For single layer construction they are spaced not to exceed 7 in. on ceilings and 8 in. on walls. Annular ring nails with a 1/4 in. head and 1 1/4 in. long are generally recommended.

All wallboard must be drawn tightly against the framing so there can be no movement of the board on the nail shank. In the area adjacent to the point of nailing, the board must be held firmly against the framing while the nail is being driven. Fig. 14-23 shows how large panels can be held with braces during the application on a ceiling.

Fig. 14-23. Using braces (also called shores) to hold a ceiling panel in place while nailing. (The Flintkote Co.)

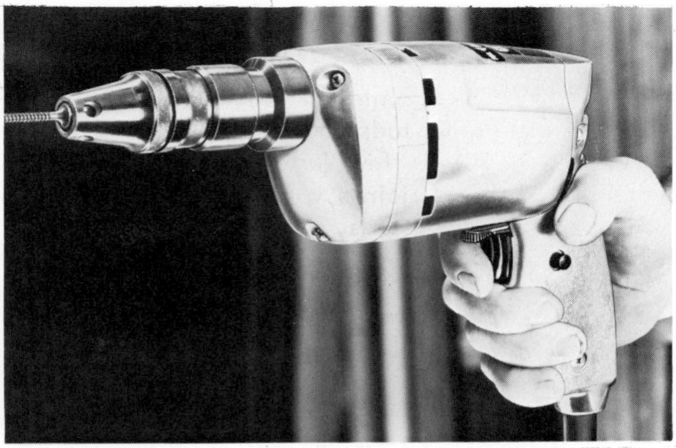

Fig. 14-25. Using an electric drill equipped with a special depth-adjusting clutch to drive wallboard screws. Clutch disengages when nose surfaces on panel. (Stanley Power Tools)

Drive nails straight and true. Use extra care during the final strokes so the nailhead rests in a slight dimple formed by the crowned head of the hammer. Be careful not to break the paper face of the board.

The double nailing method of attachment provides additional insurance for firm contact with framing. Panels are applied as required for conventional nailing, except that nails in the field of the board should be spaced 12 in. on center. After the panel is secured, another nail is spaced approximately 2 in. from the first; driven and dimpled. If necessary, the first nail should receive another blow to assure snug contact.

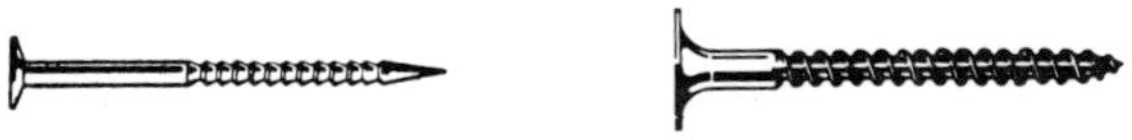

Fig. 14-24. Wallboard fasteners. Left. Annular ring nail. Right. Phillips head screw.

Gypsum wallboard screws, Fig. 14-24, provide firm, tight attachment to wood framing, Fig. 14-25. Special self-tapping screws are used for metal-framed wallboard systems. Fasteners of this type must be driven so the screwhead rests in a slight dimple formed by the driving tool. The paper face of the wallboard should not be cut, neither should the gypsum core be fractured.

Since screws hold the wallboard more securely than nails -- ceiling spacing can be extended to 12 in. and side walls to 16 in.

Adhesive

The application of wallboard with a special adhesive largely eliminates the need for nails or screws and the resulting concealment of the heads. It also produces a sturdier wall that is somewhat more resistant to impact sounds.

Fig. 14-26. Applying a bead of adhesive to studs.

Apply a continuous bead of adhesive to the center of all studs, joists or furring, Fig. 14-26. Where two pieces of wallboard join on a framing member, use a zig-zag form of bead. The size of the adhesive bead should be approximately 1/4 to 3/8 in. so that when the board is in place it will be held by a band that is at least 1 in. wide and 1/16 in. thick. Use temporary nailing or bracing to insure full contact of the wallboard until the adhesive develops the required bonding strength.

Joint and Fastener Concealment

To conceal joints, first apply a bedding coat of compound into the depression formed by the tapered edges of the board and over all butt joints. Use a 5 or 6 in. joint knife especially designed for making this application. Center the reinforcing tape over the joint and smooth out to avoid wrinkling or buckling. Press tape into com-

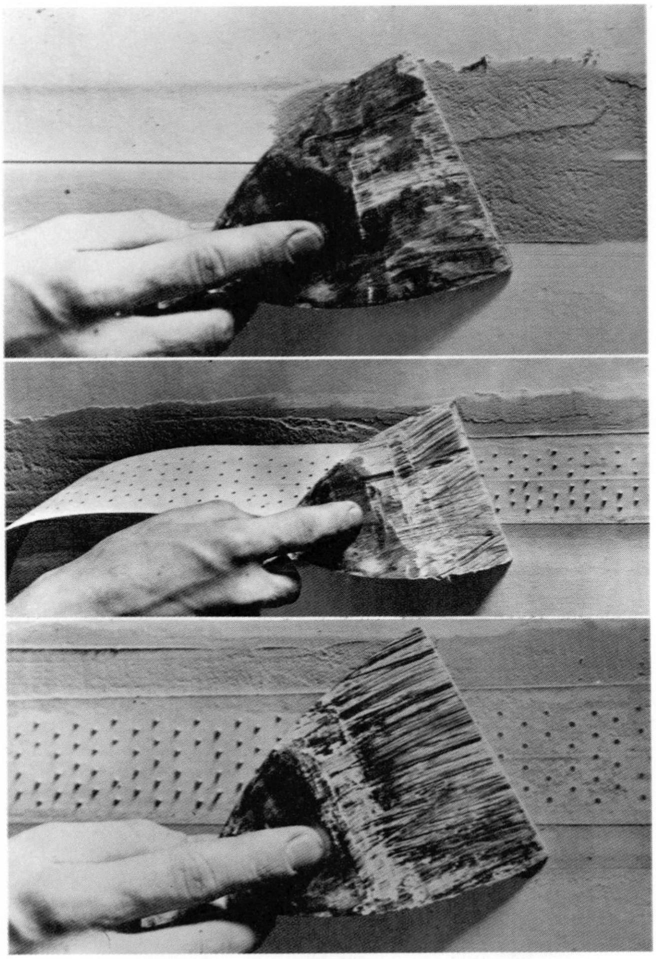

Fig. 14-27. Taping wallboard joint. Top. Applying compound into the channel formed by the tapered edges of the wallboard. Center. Embedding tape centered over joint. Bottom. Applying skim coat over tape. (National Gypsum Co.)

pound by drawing knife along the joint with enough pressure to remove any excess compound and then apply a skim coat. See Fig. 14-27.

After the embedding coat is completely dry, a second coat is applied over the tape with the edges feathered out slightly beyond the edges of the first coat as shown in Fig. 14-28. When this coat is completely dry, a third coat is applied with the edges feathered out about 2 in. beyond the second coat. After the last coat is dry, sand lightly if necessary. Fasteners are also concealed with compound - each coat being applied at the same time the joints are covered.

The use of a pressure sensitive fiberglass reinforcing tape reduces the time required to finish flat joints. It is simply pressed into place with hand pressure. Plain fiberglass tape is also available and attached to the board surface with staples, Fig. 14-28. Two coats of fast-setting joint compound are applied over the tape and sanded smooth. When the final surface is spray textured, the joints are usually finished with a single coat of compound.

Corners

Exterior corners are reinforced with a metal corner bead, fastened to the framing through the wallboard. One type of bead is made of metal with paper flanges, Fig. 14-29. After installation they are concealed with joint compound in about the same manner as regular joints.

At internal corners, both horizontal and vertical, reinforcing tape is used. First apply a bedding coat of joint compound to both sides of the corner, and then fold the tape along the centerline and smooth it into place. Excess compound is removed and the surfaces are subsequently finished along with the other joints.

A wide variety of metal channel trim is available to finish and reinforce edges around doors, windows, and other openings. Fig. 14-30 shows several shapes and sizes.

Double Layer Construction

Double layer (also called two-ply) wallboard applications over wood framing insures a strong wall surface and improves fire protection and sound insulation qualities. This method is adaptable

Fig. 14-28. Using fiberglass reinforcing tape. Left. Stapling tape over joint. Center. Applying joint compound over pressure sensitive tape. Right. Spray texturing surface over a single coat of joint compound. (National Gypsum Co.)

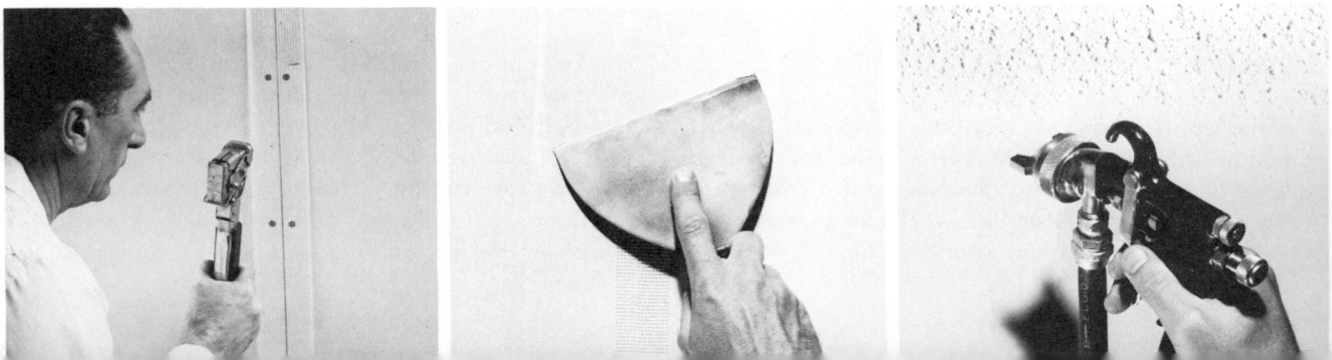

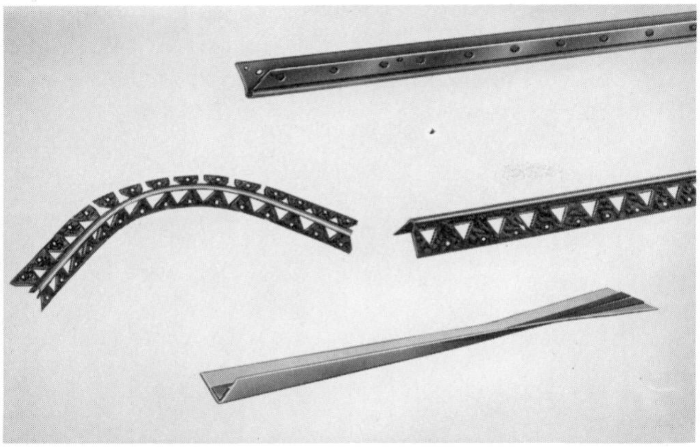

Fig. 14-29. Metal corner beads. Bead located at the bottom of the view has paper flanges.

to either the use of predecorated panels or standard beveled wallboard with treated joints.

The base layer, Fig. 14-31, may be regular gypsum wallboard or backer board. Sound deadening board is also used for the base layer and is often specified when high STC ratings are desired. The various materials are applied with power-driven screws or staples or nailed to the framing in the conventional manner.

An adhesive is used to laminate the finish layer to the base layer. The finish layer is applied so that the joints are offset at least 10 in. from those of the base layer. In some applications the layers are placed in opposing horizontal and/or vertical positions to each other. Usually the adhesive is applied to the entire surface, however, strip lamination (spaced ribbons of adhesive) may be used for certain installations.

Many methods of adhesive applications are available, from manually operated trowels to powered devices. Regardless of the method used, spacing and size of the beads of adhesive must provide the required spread when the panels are pressed into position. Fig. 14-32 shows the use of a mechanical spreader.

Temporary bracing may be used to maintain the position of the board until bonding has taken place. When nails are used, they should provide a minimum penetration of 3/4 in. into the wood framing members.

HANDY SAYS:

"In the application of any dry wall system, be sure to study and follow the recommendations of the manufacturer of each of the products (board, fasteners, compounds, adhesives) used."

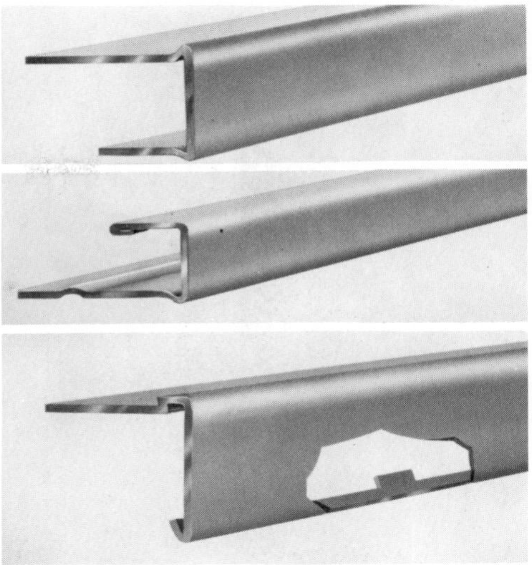

Fig. 14-30. Metal channel trim for edge protection around openings and where molding is required.

Predecorated Wallboard

A variety of predecorated gypsum wallboard is available in various designs and simulated wood grain patterns. This is usually applied vertically because of the difficulty involved in successfully

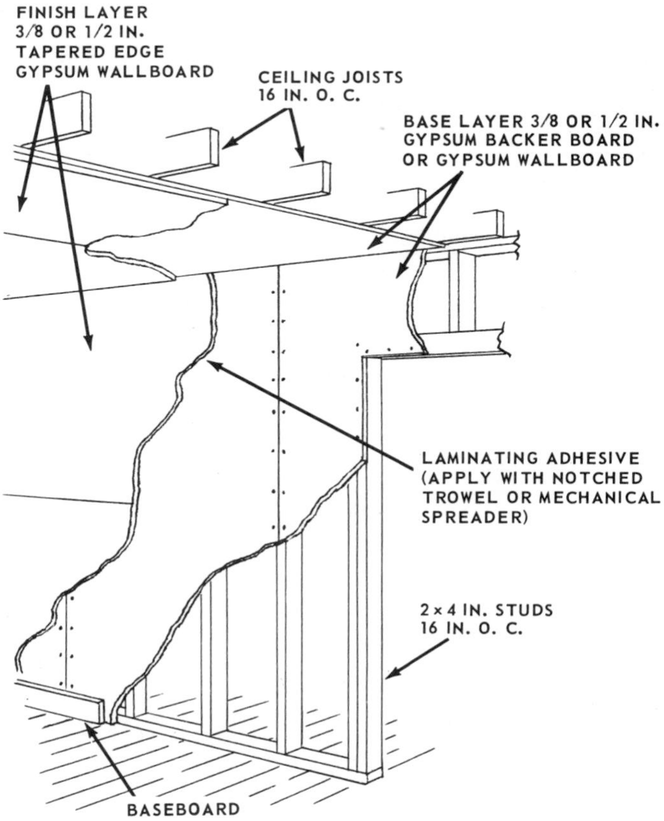

Fig. 14-31. Double layer construction over wood framing.

Skilled operator inspects prefinished hardwood panels as they move by on conveyor line. By simply moving a control he can remove a defective panel and route it to a repair station. (Georgia-Pacific)

Partial view of assembly area in a woodworking plant that specializes in the manufacturing of bi-fold door units. Workers in right foreground load parts into special air-operated clamping frame. (Maywood)

Fig. 14-32. Using a mechanical adhesive spreader. Left. Working from ceiling to midpoint of wall. Center. Completing application from floor to midpoint. Right. Ceiling application.

matching and finishing butt joints. The use of an adhesive to bond the panels to a base layer is common practice, however, matching colored nails are available. Be sure to drive colored nails with a plastic-headed hammer, a rawhide mallet, or a special cover placed over the face of a regular hammer. Nails should never be spaced closer than 3/8 in. to the ends or edges of the wallboard.

To trim edges and joints of predecorated panels, aluminum moldings are produced that match the finished surface of the wallboard, Fig. 14-33. These are cut to length with a hack saw and nailed in place with flat-head wire nails spaced 8 in. to 10 in. apart. When attaching divider strips, first place the molding on one panel that is carefully aligned, nail the exposed flange in place, then insert the adjacent panel.

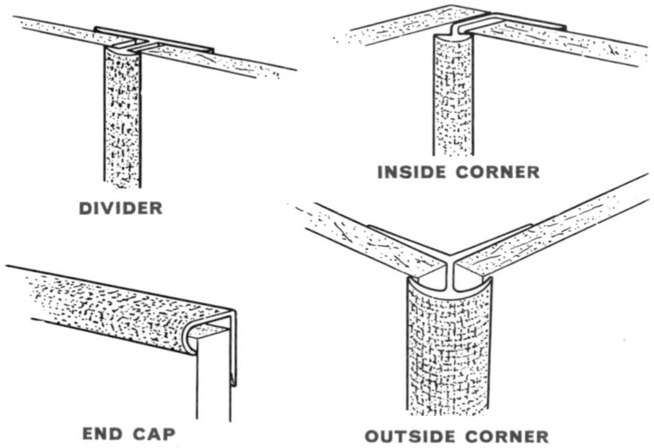

Fig. 14-33. Moldings for predecorated wallboard.

Masonry Walls

Gypsum wallboard can be installed over metal or wood furring strips attached to a masonry wall. Where the structure consists of an interior wall that is straight and true, the panels can be laminated directly to the masonry surface with a special adhesive.

Exterior walls must be thoroughly waterproofed and insulation should be included if the structure is located in a cold climate. Fig. 14-34 shows a masonry wall application made with metal furring channels and insulated with plastic foam. When wood furring strips are used, they should be of a nominal 2 in. width and of a thickness 1/32 in. greater than the insulation. The plastic foam slabs are usually bonded to the masonry surface with adhesive. Wallboard joints and nail holes are concealed, following the treatment previously described.

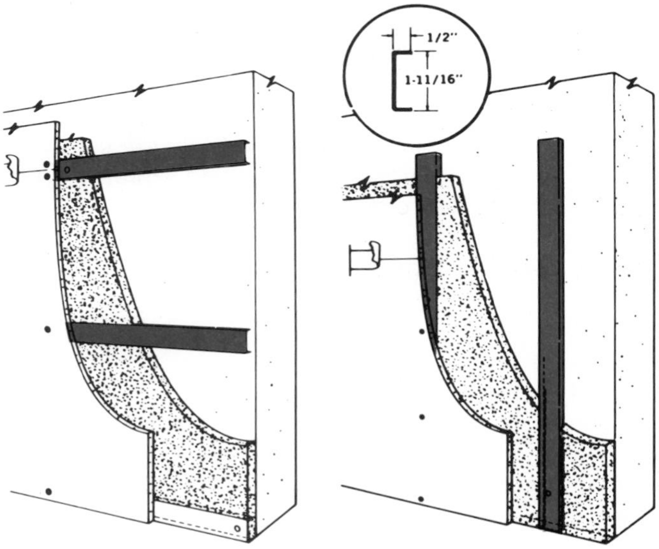

Fig. 14-34. Wallboard attached to a masonry wall with metal furring channels. Plastic foam is used for insulation.

Installations of wallboard on exterior masonry walls, especially those below grade, must be carefully designed. Always follow recommendations furnished by the manufacturers of the various products.

Waterproofing Gypsum Wallboard

Wallboard base surfaces can be used in shower and tub areas if properly applied and waterproofed. Use full size sheets to eliminate butt joints and run the sheets horizontally. Be sure the paperbound edge extends down over the fixture rim with a clearance of 1/4 in. All fasteners and joints should be treated with a minimum of two coats of joint compound.

The surface of the wallboard is waterproofed with a 1/16 in. thickness (skim coat) of tile adhesive, applied according to the manufacturer's specifications and allowed to thoroughly dry -- usually 24 hours. Other special sealers are also available. The bedding coat of adhesive is then

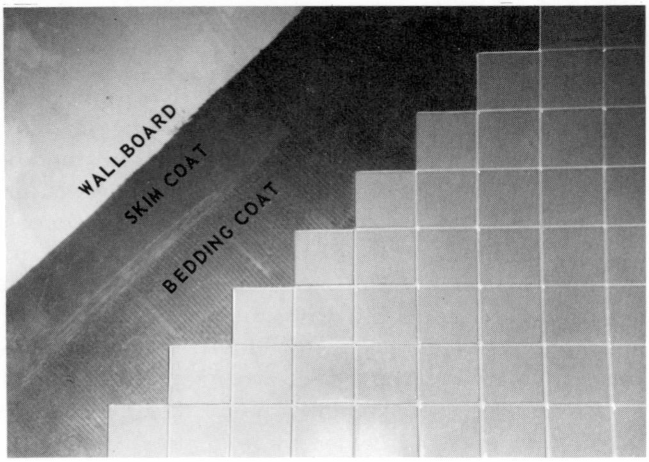

Fig. 14-35. Sequence of coats used in preparing base for tile. (Gypsum Association)

Fig. 14-37. Installing solid wood paneling. Boards are edge and end matched. (Forest Products Laboratory)

applied and the tile surface is installed. Fig. 14-35 shows the sequence of coats. Special waterproofed backer wallboards can be obtained which do not require the sealer coat.

Solid Lumber Paneling

Solid wood paneling makes a durable and attractive interior wall surface and may be appropriately used in nearly any type of room. A number of different species of hardwood and softwood are available. Sometimes, grades that contain numerous knots are used to secure a special appearance. Defects, such as the deep fissures in pecky cypress, can provide a dramatic effect.

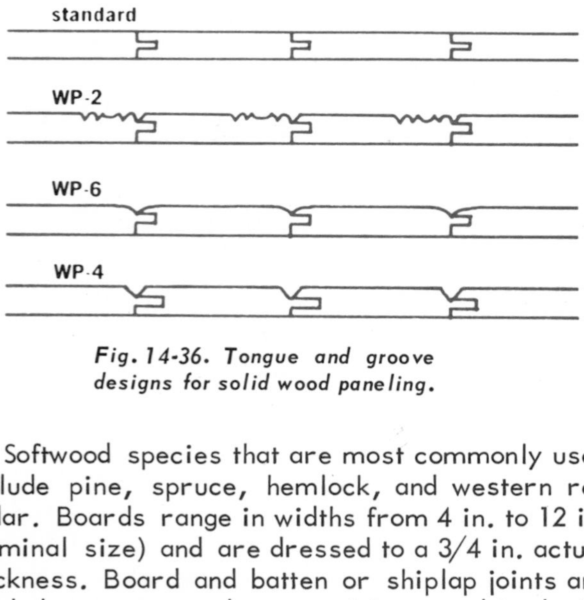

Fig. 14-36. Tongue and groove designs for solid wood paneling.

Softwood species that are most commonly used include pine, spruce, hemlock, and western red cedar. Boards range in widths from 4 in. to 12 in. (nominal size) and are dressed to a 3/4 in. actual thickness. Board and batten or shiplap joints are used, but tongue and groove joints combined with shaped edges and surfaces are more popular in most localities, Fig. 14-36.

When solid wood paneling is applied horizontally, furring strips are not required and the boards are nailed directly to the studs. Inside corners are formed by butting the paneling units flush with the other walls. If random widths are used, boards on adjacent walls must match and be accurately aligned. Vertical installations require furring strips at the top and bottom of the wall and at various intermediate spaces, depending on the type of material, Fig. 14-37. Sometimes 2 x 4 in. pieces are installed between the studs to serve as a nailing base. Even when heavy tongue and groove boards are used, these nailing members should not be spaced more than 48 in. apart.

Narrow widths (4 to 6 in.) of tongue and groove (T & G) paneling are blind nailed (nailed so heads of nails do not appear on finished surface). This eliminates the need for countersinking and filling nail holes and provides a smooth blemish-free surface -- especially important when clear finishes are used. Use a 6d finish nail and drive it at a 45 deg. angle into the base of the tongue and on into the bearing point.

Exterior wall constructions, where the interior surface consists of solid wood paneling, should include a tight application of building paper located close to the back side of the boards. This will prevent the infiltration of wind and dust through the joints. In cold climates, using insulation and vapor barriers is important.

HANDY SAYS:

"Interior solid wood paneling may cause problems resulting from expansion and shrinkage. Be sure the material used has a moisture content about equal to that which it will attain in service. This should be about 8 to 10 percent for most parts of the United States."

Plywood

A tremendous variety of plywoods -- both hardwood and softwood -- are available for interior wall finish. These are available in plain or processed surfaces that simulate driftwood and many other decorative effects, Fig. 14-38. Some sheets are scored to imitate solid wood paneling.

Today, most of the plywood used for interior walls has a factory-applied finish that is tough and durable. Manufacturers can furnish matching trim and molding that is also prefinished and easy to apply. Color-coordinated putty sticks are used to conceal traces of the application.

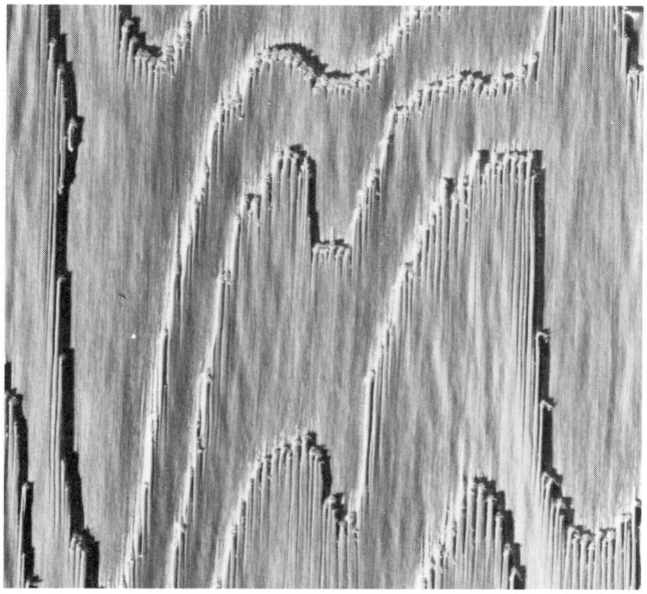

Fig. 14-38. Surface of a softwood panel that has been "brushed" to accentuate the grain pattern and texture. (American Plywood Assoc.)

Joints between plywood sheets can be treated in a number of ways. Some panels are fabricated with special edges that permit almost perfect joint concealment. Usually it is easier to accentuate the joints with grooves or use various shaped battens and strips, Fig. 14-39.

Before making an installation of plywood, the panels should become adjusted (conditioned) to the temperature and humidity of the space or room. Prefinished plywood should be removed from cartons and carefully stacked horizontally, with 1 in. spacer strips between each pair of face-to-face panels. This should be done at least 48 hours before the application is to be made.

Plan the layout carefully to minimize the amount of cutting and reduce the number of joints. Fig. 14-40 shows several application designs and emphasizes the importance of aligning panels with openings whenever possible. When a clear finish is to be applied, it is well to stand the panels around the room and position them to secure the most pleasing effect in color and grain patterns. To avoid errors, number the panels in sequence after their position has been established.

Plywood can be attached directly to the wall studs with nails or special adhesives. Use 3/8 in. plywood for this type of installation. When studs are poorly aligned or when the installation is made over an existing surface that is in poor condition, it is usually advisable to use furring. Nail 1 x 2 in. furring strips horizontally to the studs starting at the floor line and continue up the wall, spacing the strips 16 in. on center. Nail vertical strips every four feet to support panel edges. Level uneven areas by placing shims behind the furring strips. All four edges of each panel must be supported.

Begin installing panels at a corner. Scribe and trim the edge of the first panel so it is plumb. Fasten it in place and then fit and install the next panel. Allow approximately 1/4 in. clearance at the top and bottom. After all panels are in place, molding is used to cover this space along the ceiling. Baseboards provide finish at the floor line.

On some jobs 1/4 in. plywood is installed over a base layer of 1/2 in. gypsum wallboard. This backing tends to bring the studs into alignment; provides a rigid finished surface; and improves the fire resistant qualities. The plywood is bonded to the gypsum board by using adhesives and following about the same procedure as described in double-layer dry wall construction.

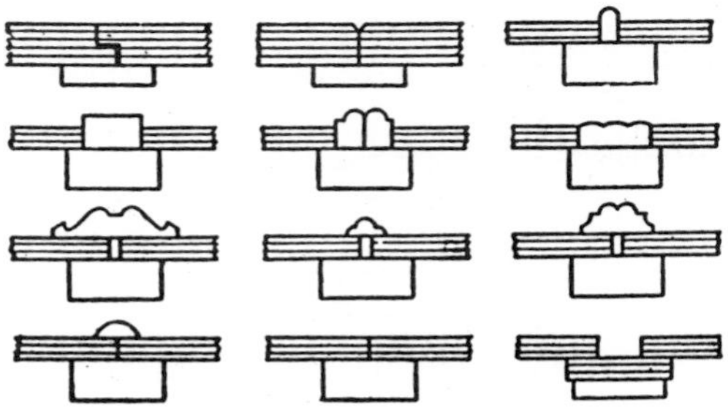

Fig. 14-39. Suggested joint treatment for interior plywood walls.

Hardboard

Hardboard is manufactured in a broad range of face patterns and textures, and is widely used for finished interior wall surfaces. See Fig. 14-41. A popular product has factory applied wood grain

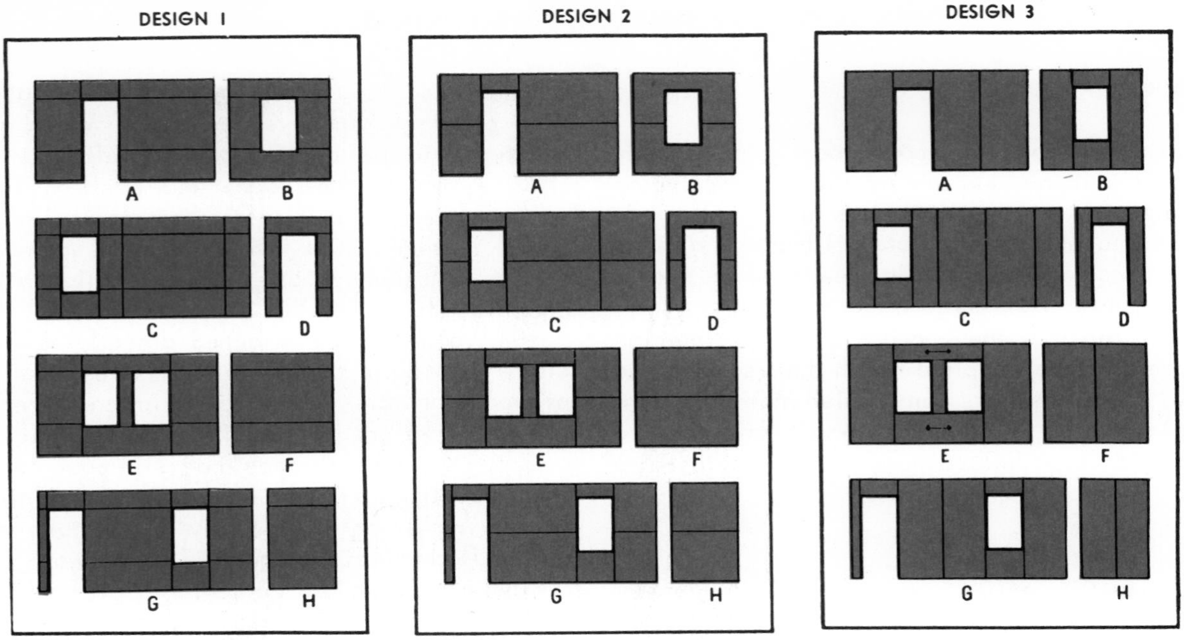

Shown above are three plywood wall paneling design suggestions for each of eight typical wall elements. The basic rule followed in each is to "work from the openings," i.e., first line up vertical joints above doors, above and below windows, then divide the plain wall space into an orderly pattern.

DESIGN I is a horizontal pattern obtained by placing panels in three pieces so that continuous horizontal joints are at 1.) window and door head level, and 2.) window sill or other level in proportion to the upper joint. If the height of window and door tops are not the same, choose the lower to establish joint line.

Note that the panel arrangement is keyed by vertical joints placed on either side of the top of the door and above and below windows as shown in Elevations A, C, E and G. Where panel length exceeds wall-element width, as shown in Elevations B and D, vertical joints above openings may be omitted.

DESIGN 2 is a simple two-panel horizontal arrangement. The single continuous horizontal joint is placed midway between door and ceiling. Vertical joints at openings (elevations A, C, E and G) again key panel design, except that they may be omitted where panel length exceeds wall element width as in Elevations B and D.

DESIGN 3, a vertical panel arrangement is another illustration of the basic principle of initiating panel design by lining up vertical joints with wall openings. The plain wall space is then divided vertically in widths proportionate to that of openings. In a vertical panel arrangement where width of a door or window opening exceeds panel width, panels may be placed horizontally as shown by arrows in Elevation E. Such combinations of vertical and horizontal arrangements may be used in the same room with pleasing effect.

Fig. 14-40. Plywood panel arrangements for interior walls.

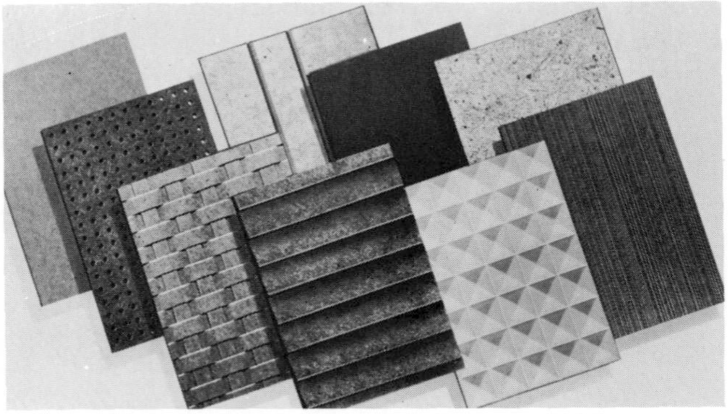

Fig. 14-41. Hardboard patterns and textures. (American Hardboard Assoc.)

finish with random grooves that can hardly be distinguished from actual wood panels. Various embossed panels may be used to customize walls in activity rooms, offices, or display areas. Through special processing, hardboard can be fabricated with a very low moisture absorption rate. This type is often scored to form a tile pattern and is used in bathrooms and kitchens. Panels for wall application are usually 1/4 in. thick.

Since hardboard is made from wood fibers, the panels will expand and contract slightly with changes in humidity. These should be installed when they are at their maximum size otherwise there will be a tendency for them to buckle between the studs or attachment points. Manufacturers of prefinished hardboard panels recommend that they be unwrapped and then placed separately around the room for at least 48 hours before application.

Procedures and attachment methods are similar to those previously described for plywood. Special adhesives are available and also metal or plastic moldings in matching colors. Drilling nail holes is advisable for the harder types.

HANDY SAYS:

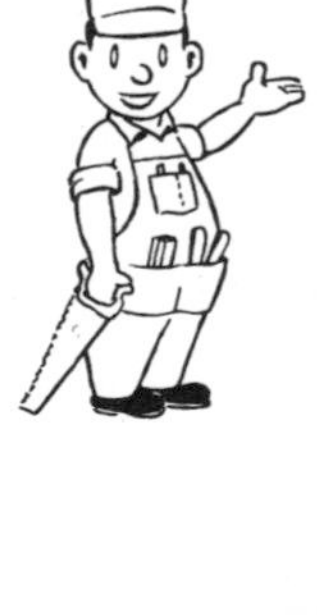

"When making an application of any of the various factory finished wallboard, plywood, or hardboard materials, always follow the recommendations furnished by the manufacturer."

Plastic Laminates

Plastic laminates are sheets of synthetic material that is hard, smooth and highly resistant to scratching and wear. Although basically designed for table and counter tops, they are also used for wainscotting and wall paneling in homes and commercial buildings.

Fig. 14-42. Installing wall panels manufactured from particle board with a 1/32 in. plastic laminate face. (Formica Corp.)

Since the material is thin (1/32 to 1/16 in.), it must be bonded to other supporting panels. Contact bond cement is commonly used to make this application. In recent years, manufacturers have developed prefabricated panels where the plastic laminate is already bonded to a base or backer material. Fig. 14-42 shows the installation of a prefabricated panel consisting of a 1/32 in. plastic laminate mounted on 3/8 in. particle board. Edges are tongue and grooved so that units can be blind-nailed into place. Various matching corner and trim moldings are available. Fig. 14-43 shows a completed installation.

Ceiling Tile

Ceiling tiles are suitable for both old and new construction. They can be installed over furring strips, solid plaster, plaster board or any smooth, continuous surface.

Fig. 14-43. Completed installation of a plastic laminate wall surface. (Formica Corp.)

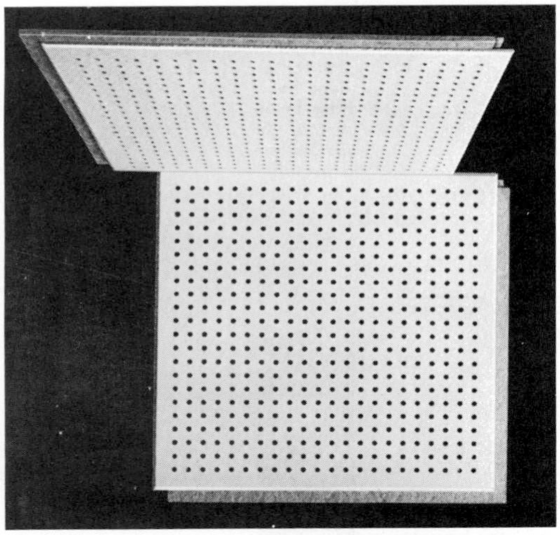

Fig. 14-44. Regular 12 x 12 ceiling tile made from cellulose fibers. Available in 1/2 and 3/4 in. thickness. (Wood Conversion Co.)

There is a considerable range in the types of material used to fabricate ceiling tile. Generally available are: fiberboard tile, mineral tile, perforated metal tile, glass-fiber tile, and perforated asbestos tile. When selecting a product, consideration should be given to appearance, light reflection, fire resistance, sound absorption, maintenance, cost, and ease of installation.

A standard size tile is 12 x 12 in., Fig. 14-44. However, the tiles are available in larger sizes; for example, 24 x 24 in. and 16 x 32 in. A wide range of surface patterns and textures are manufactured. Fig. 14-45 shows a typical fiberboard

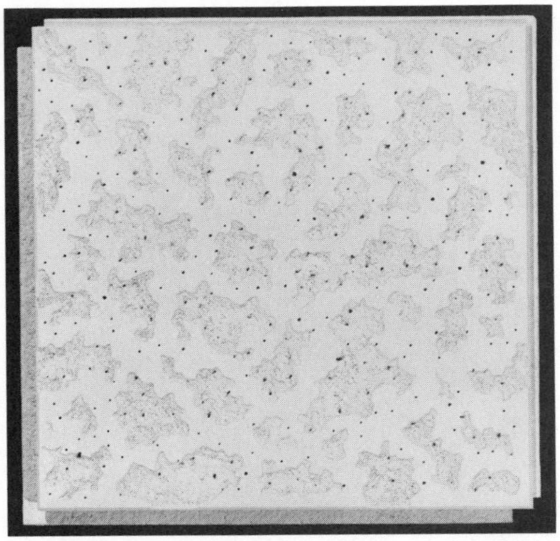

Fig. 14-45. Ceiling tile with a sculptured design that provides the effect of a continuous surface after installation.

tile design that would be appropriate for a residential installation.

Fig. 14-46 illustrates the unique type of overlapping, tongue and groove edge that provides a wide flange to receive staples. This type joint also permits efficient installation when an adhesive is used to hold the tile in place.

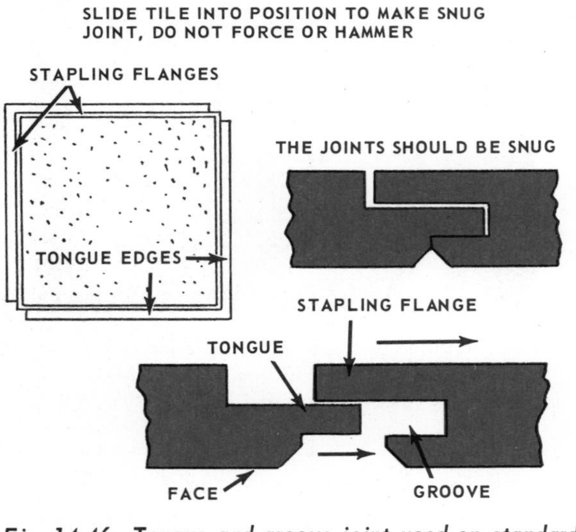

Fig. 14-46. Tongue and groove joint used on standard ceiling tile.

Layout Procedure

First, measure the two short walls and locate the midpoint of each one; then snap a chalk line, Fig. 14-47, to establish the center line. In the middle of this line establish a chalk line that is at right angles to the long center line. All tiles are installed with their edges parallel to these lines.

For a symmetrical appearance, the border courses along opposite walls should be equal. For example, in a room width of 10 ft. 8 in. use 9 full tile and a border tile trimmed to 10 in. on each side. After determining the width of the border tile, snap chalklines parallel to the center lines that will provide a guide for the installation of these border tile or the furring strips that will support them.

HANDY SAYS:

"Most manufacturers recommend that fiberboard tile be unpacked in the area where they will be installed, at least 24 hours before application. This will allow the tile to adjust to room temperature and humidity."

Installing Furring

Furring (1 x 3 wood strips) should be attached to the ceiling joists. If applied over an old ceiling surface, be sure to accurately locate the joists before nailing the strips in place. Locate the first furring strip flush against the wall at right angles to the joists and nail it in place with two 8d nails at each joist crossing, Fig. 14-48.

Placement of the next strip will be determined by the width of the border tile. Nail this strip in place so that it will be centered over the edge of the border, using the line previously laid out. All other strips are then installed 12 in. O.C. (for 12 x 12 in. tile). Fig. 14-49 shows furring complete. Some carpenters prefer to start from the center and proceed on each side, toward the walls. If the furring strips are accurately sized, a spacer jig, Fig. 14-50, may expedite the application. Double check the position of the furring strips from time to time to make sure they will be centered over the tile joints.

It is essential that the lower face of the furring strips be level with each other. Check alignment with a carpenters' level or straightedge. To bring strips into alignment, drive tapered shims between the strips and the joist as shown in Fig. 14-51. If only one or two joists extend below the plane of the others, it may be best to notch the joists before installing the furring strips.

If pipes or electrical conduit are to be located below the ceiling joists, it will be necessary to install a double layer of furring strips, Fig. 14-52. The first course is spaced 24 to 32 in. on center and attached directly to the joists. The second course is then applied perpendicular to the first -

spaced as previously described. Large pipes or ducts that project below the ceiling joists should be boxed in with furring strips before the tile is installed. Wood or metal trim can be used to finish corners and edges.

An alternate furring method, which could be called "sheet furring", consists of installing gypsum wallboard to the joists or an existing ceiling surface. The tile is then attached to the wallboard either with adhesive or special staples.

Fig. 14-47. Using a chalk line to mark the center of the ceiling layout. 14-48. Installing the first furring strip. Use two 8d nails at each joist. 14-49. Completed installation of furring. Note closer spacing along wall to accommodate for the width of border tile. 14-50. Using a spacer jig to locate furring strips. 14-51. Level the furring by driving tapered shims between the joist and strip. 14-52. Double layer of furring provides space for conduit and pipes located below joists. (Armstrong Cork Co.)

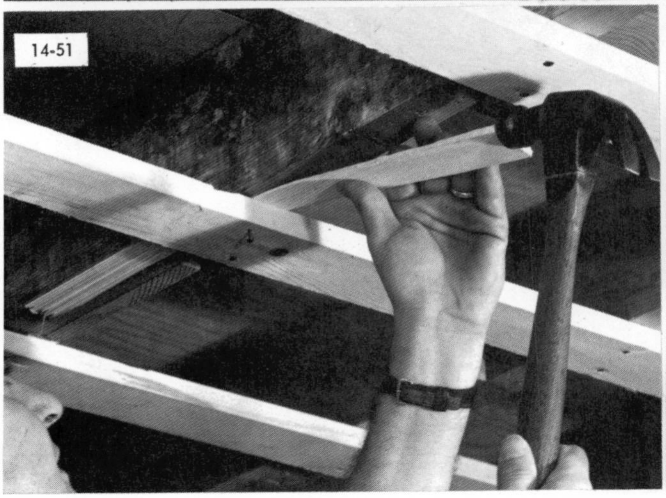

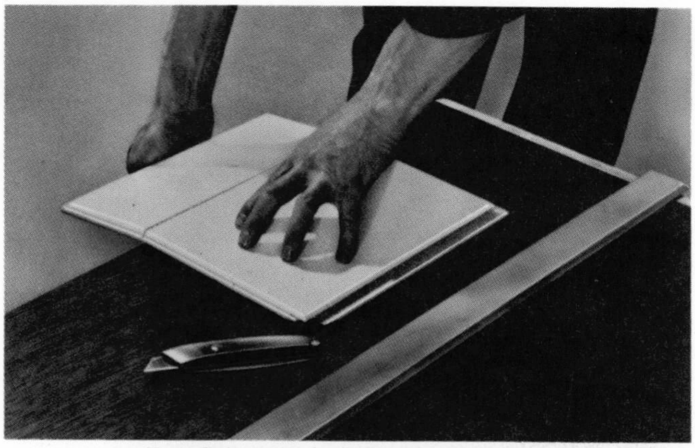

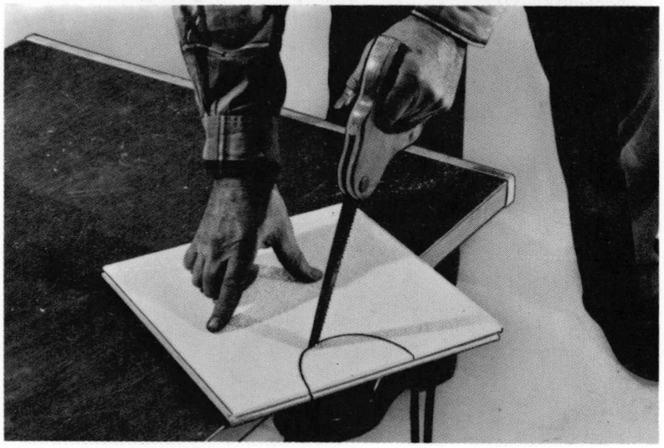

Fig. 14-53. Above. Cutting border tile. Remove the tongue edge, leaving the wide flange for staples. Below. Using a compass saw to cut a circular opening. (Flintkote Co.)

Installing Tile

After the furring is in place, check it over carefully and then snap chalklines on the strips to provide a guide for setting the border tile. Do this along each wall and be certain the lines are correct and in accord with the initial layout previously described. Make a double check to see that the chalk lines form 90 deg. angles at the corners.

To cut the border tile, first score the face deeply with a knife drawn along a straightedge, and then break along this line by placing it over a sharp edge as illustrated in Fig. 14-53. Irregular cuts around light fixtures or other projections may be made with a coping or compass saw. Power tools can also be used. Some tiles are made of mineral fibers which rapidly dulls regular cutting edges.

Start the installation with a corner tile and then set border tile out in each direction. Fill in full size tile. When you reach the opposite wall the border tile will have the stapling flanges removed, and will need to be face-nailed. Locate these nails close to the wall so they will be covered by the trim molding.

Fig. 14-54 shows tile being attached with a stapler. Be sure the tile is correctly aligned and then hold it firmly in place while setting the staples. For 12 x 12 tile use three staples along each flanged edge; four staples for 16 x 16 tile. Staples should be at least 9/16 in. long.

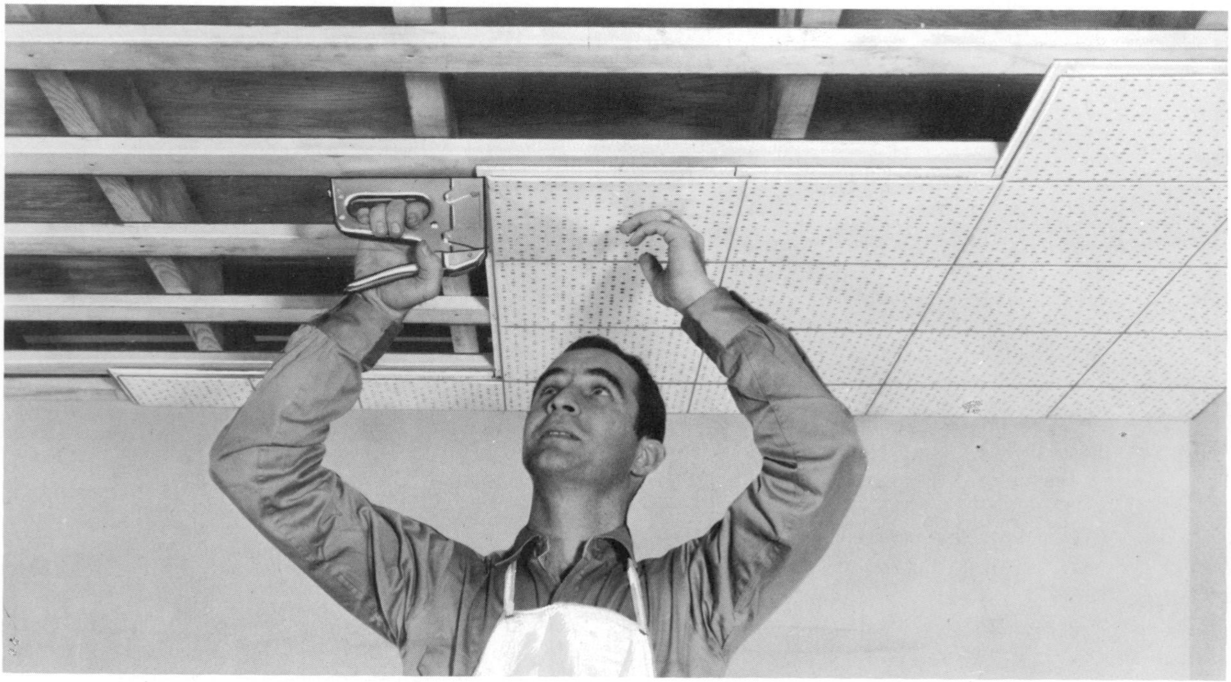

Fig. 14-54. Stapling tile to furring strips.

Fig. 14-55. Installing tile with a putty type adhesive.

There are several types of adhesive designed especially for installing ceiling tile. The thick putty type is applied in daubs about the size of a walnut, Fig. 14-55. Apply adhesive to each corner (12 x 12 tile) and about 1 1/2 in. away from each edge. Now place the tile in position, slide it back about 1/2 in. and then slide it forward into position against the other tile as shown in Fig. 14-56. This motion along with firm pressure will spread the adhesive so the layer will be about 1/8 in. thick.

Some adhesives are thinner and are applied with a brush. In remodeling work be sure the old ceiling is clean and that any paint or wallpaper is adhering well. Always follow the recommendations and directions provided by the manufacturer of the products being used.

HANDY SAYS:

"Be sure to keep your hands clean while handling ceiling tile."

Suspended Ceilings

When heating ducts and plumbing lines interfere with the application of a finished surface, a suspended ceiling provides a practical solution. Modern installations consists of a metal framework especially designed to support a given type or size of tile or panel, Fig. 14-57.

The height of the ceiling must first be determined and then a molding is attached to the perimeter of the room. Use a level chalkline as shown in Fig. 14-58.

Carefully calculate room dimensions, lay out positions of main runners and then install screw eyes (4 ft. O.C.) in existing ceiling structure as shown in Fig. 14-59, left. Panels next to the wall (border panels) may need to be reduced in width to provide a symmetrical arrangement. Plan the layout in about the same way as previously described for regular ceiling tile.

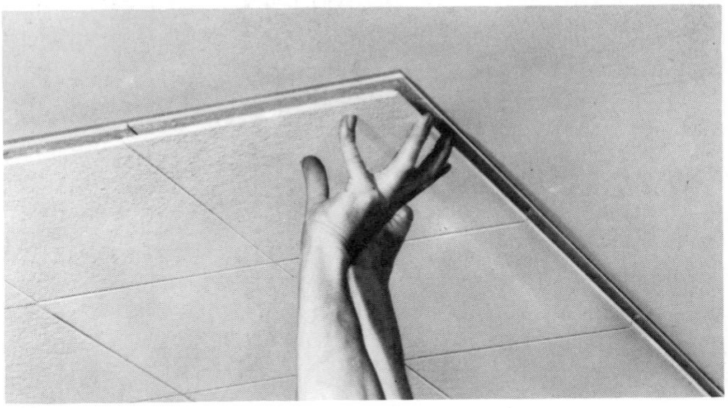

Fig. 14-56. Tile in position -- carefully aligned with previously set tile.

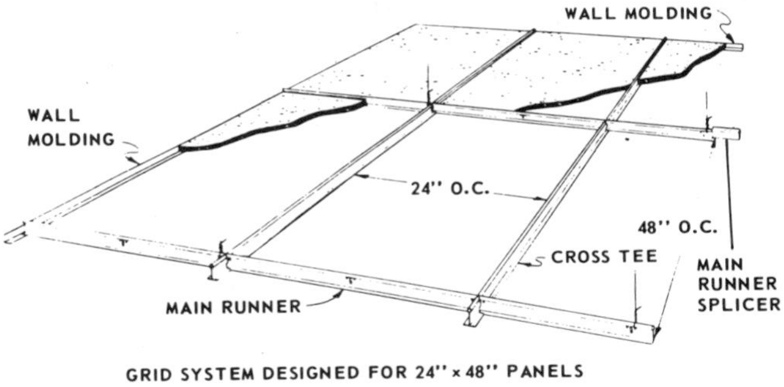

Fig. 14-57. A typical metal framework used for suspended ceilings.

Fig. 14-58. Attaching molding to wall -- first step in installing a suspended ceiling.

Fig. 14-59. Left. Installing screw eyes in joist to support metal runners. Right. Tying runner to screw eye with wire.

Install the main runners by resting them on the wall molding and attaching to the wires tied to the screw eyes, Fig. 14-59, right. Use chalk lines or string, stretched between the wall molding to insure a level assembly. Sections of runners are easily spliced and odd lengths can be cut with a fine-tooth hack saw or aviation snips.

After all the main runners are in place, carefully leveled and securely attached to the supporting wires, install the cross tees. See Fig. 14-60. Check the required spacing for the tiles or panels and simply insert the end tab of the cross tee into the runner slot.

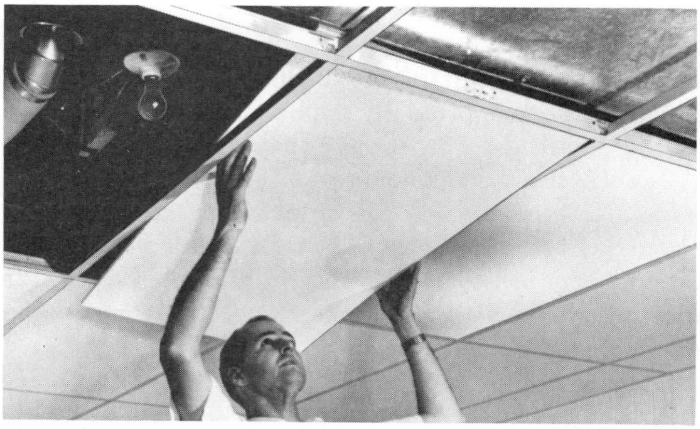

Fig. 14-61. Ceiling panels are raised above the framework and then lowered into place. (Armstrong Cork Co.)

Fig. 14-60. Installing cross tees between main runners.

With the suspension framework complete, panels can be installed. Each panel is tilted upward and turned slightly on edge so it will "thread" through the opening as illustrated in Fig. 14-61. After the entire panel is above the framework, turn it to a horizontal position and lower it onto the grid flanges.

Another design of a suspension system is shown in Fig. 14-62. This provides an installation where the metal hangers and runners are concealed. The tongue and groove joint is similar to the joint on

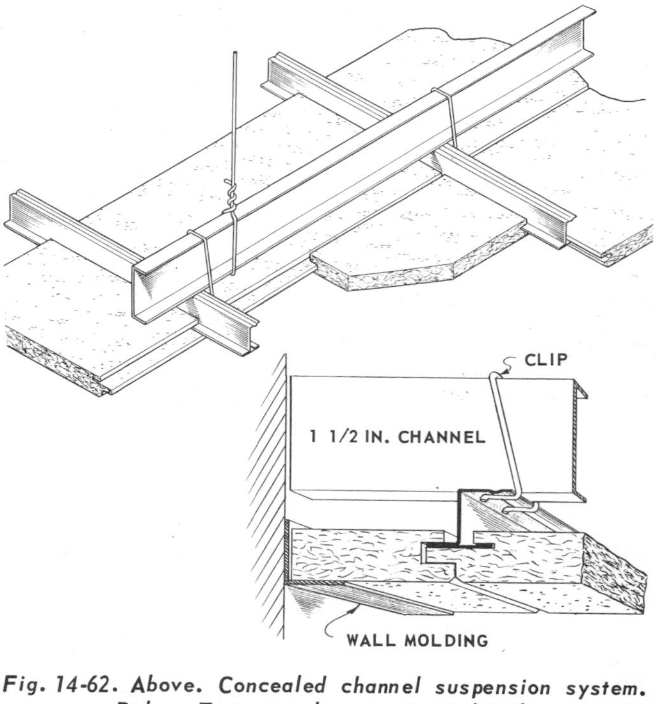

Fig. 14-62. Above. Concealed channel suspension system. Below. Tongue-and-groove joint detail.

regular ceiling tile. Joints are interlocked with the flange of the special runner when the installation is made, Fig. 14-63.

One of the advantages of suspended ceilings is their adaptability to modern recessed lighting. Luminous panels can be inserted in the framework, located directly below flourescent lighting fixtures. Luminous ceilings are especially popular in kitchens and bathrooms where large quantities of well diffused light is desirable. Fig. 14-64 shows this type of lighting employed in a game or activity room.

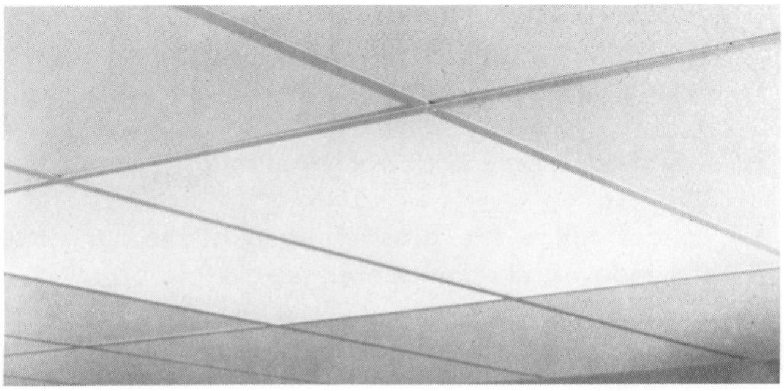

Fig. 14-64. Suspended ceiling system with recessed lighting units. (Armstrong Cork Co.)

Estimating Materials

The amount of standard gypsum lath required when this is used as a plaster base is estimated by first calculating the total number of square feet of wall and ceiling area. Regular size door and window openings are disregarded -- these provide allowance for waste. However, large window walls and picture windows may be subtracted.

Fig. 14-63. Ceiling panel joints interlock with supporting flange. (Wood Conversion Co.)

Usually the ceiling area can be considered equal to the floor area. Add to this the length of all interior wall surfaces, multiplied by the wall height. For example:

Floor area = ceiling area = 1250 sq. ft.
Total length of walls = 14 + 12 + 14 + 12 + 10 ---
= 400
Wall area = 400 x 8
= 3200 sq. ft.
Total Area = 3200 + 1250
= 4450 sq. ft.
Std. Lath Bundle = 64 sq. ft.
Bundle Estimate = 70

Plasterers usually base their prices and estimates on the number of square yards. The total square footage can be converted to square yards by dividing by 9 (1 sq. yd. = 9 sq. ft.).

Basically, the same procedure can be followed in developing a rough estimate for dry wall construction. When ordering wallboard, however, lengths must be specified. Always plan to use the longest practical length so that butt joints will be held to a minimum or eliminated. Each room must be considered separately with dimensions taken directly from the wall structures or secured by carefully scaling the plan. Estimated amounts of joint compound, adhesive, and nails can be obtained from tables and charts prepared by manufacturers.

The amount of solid wood paneling is based on its nominal or unfinished size. Seasoning and planing bring it to its dressed size while forming joints may further reduce the actual face widths. For example: a 1 x 6 tongue and groove board has a face width of 5 1/8 in. First calculate the square footage of the wall to be covered and then multiply this by various factors obtained from lumber tables (for 1 x 6 tongue and groove boards use 1.17; for 1 x 8 use 1.16). When making a standard vertical application, add about 5 percent for waste when required lengths can be selected or the lumber is end-matched.

Although plywood and hardboard panels are priced by the square foot, full sheets (4 x 8 standard size) must be purchased. Since most ceilings are 8 feet high it is only necessary to calculate the number of widths required when making a standard vertical application. When working with expensive hardwood panels it is usually advisable to make a scaled layout similar to the diagrams in Fig. 14-40.

Quantities of ceiling tile are estimated by figuring the square footage of the area to be covered. Round out any fractional parts of a foot in width and length to the next larger full unit when making this calculation. Also add extra units when it is necessary to balance the installation pattern with border tile along each wall. When using 12 x 12 tile the number required will equal the square footage plus the extra allowance described. Standard 12 x 12 ceiling tile are packaged 64 sq. ft. to each carton.

Test Your Knowledge - Unit 14

1. Gypsum is known chemically as hydrous ______ ______.
2. A standard size panel of gypsum lath measures 3/8 x 16 x ______.
3. To reinforce the plaster base at the corners of window and door openings, a strip of ______ ______ is used.
4. Strips of wood or metal installed at the edge of openings to provide a guide for the plasterer are called ______.
5. The most common aggregate used in plaster mixes is ______.
6. In standard three-coat plaster applications, the brown coat is applied ______. (first, second, last)
7. When applied to regular gypsum lath, the total thickness of the plaster coats should be not less than ______ in.
8. In single layer dry wall construction, the gypsum wallboard thickness should be ______ or ______ in.
9. When nailing wallboard in place, the nails are spaced closer together on the ______. (walls, ceilings)
10. In addition to the skim coat applied over the tape, wallboard joints usually require ______ (1, 2, 3,) coats of joint compound.
11. When making a double layer wallboard application, joints running parallel to each other should be offset at least ______ in.
12. Solid wood wall paneling should have a moisture content of about ______ to ______ percent for most areas of the United States.
13. Prefinished plywood wall paneling should be "room conditioned" for at least ______ hours before application.
14. Using a base layer of gypsum wallboard under plywood paneling improves alignment and rigidity, and also makes the construction more ______ ______.
15. Standard ceiling tile have a wide stapling or nailing flange on ______ (1, 2, 3, 4) edges.
16. For a symmetrical ceiling tile pattern in a room 11 ft. - 6 in. wide, use 10 full (12 x 12) tile and a ______ in. border tile.
17. The metal framework of a modern suspended ceiling is supported mainly by ______ tied to the building structure directly above.
18. It is general practice for plasterers to base their cost estimates on the number of ______ ______ of ceiling and wall areas.

Outside Assignments

1. Secure and study literature from companies that process and fabricate gypsum products. Learn about the history and development of plastering and modern processes used in making plaster, gypsum lath and gypsum wallboard. Obtain samples of the products commonly used in your locality, along with approximate prices and costs of installation. Prepare a written report or outline the information carefully and make an oral report to your class.
2. Visit with a local dry wall contractor and learn about some of the special problems and remedies typical to this type of wall finish. Secure such informations as: care and handling materials; repairing and adjusting warped studs and framework; repairing damaged boards and surfaces; cause and remedies of tape blisters; and definition and prevention of nail pops. Organize the information you secure under appropriate headings and make an oral presentation to your class.
3. From various reference books, including manuals on architectural standards; learn about methods and construction details for ceiling systems that include radiant heating. Cover both hot water and electrical systems. Prepare scaled drawings (larger than actual size) of sections through various ceiling constructions, showing the heating pipes and elements. Include notes concerning material specifications and critical temperatures for the various constructions.
4. Suspended ceilings in institutional and commercial buildings often include a ventilation system that provides both heating and cooling. The space above the ceiling serves as a plenum and air enters the room through small slots or holes either in the tile or at special joints between the tile. Make a study of this method of air distribution and prepare a written report with drawings and other illustrations.

Unit 15
FINISH FLOORING

The term "finish flooring" applies to the material used as the final wearing surface of a floor construction. A wide selection of materials are fabricated for this purpose. Hardwoods and softwoods are available as strip flooring in a variety of widths and thicknesses, and as random width planks or unit-blocks. Many new and improved materials are being used in the production of composition flooring (called resilient flooring). Commonly used items of this type include asphalt tile, vinyl-asbestos tile, vinyl tile, rubber tile and linoleum tile; the last three of which are also available in sheet form.

In addition to the above, flagstone, slate, brick, and ceramic tile are frequently selected for special areas -- such as entrances, bathrooms or multipurpose rooms. Floor structures usually need to be specifically designed to carry this type of finish surface.

When the finish flooring consists of wood, it is usually laid after the wall and ceiling surfaces are complete, and before interior door frames and other trim are applied. The surface should be covered to protect it while other inside finish work is in progress. Sanding and the application of finish to the floor surface becomes the last major operation as the interior is completed. Where prefinished wood flooring is used, this too must be carefully covered and protected. Likewise the installation of resilient flooring and prefinished wood flooring must be among the last in the sequence of interior finishing steps.

Care should be exercised in the selection and installation of the finish flooring. The material must be attractive, however above all, it must be strong and long lasting. Since it is a fixed and relatively permanent part of the structure it should be of high quality so that replacement and repair will seldom be necessary.

Wood Flooring

Through the years, wood has been a universally accepted material for flooring in residential structures. Wood flooring, especially hardwood, has the strength and durability to withstand wear and also provides a highly attractive appearance.

Fig. 15-1. Strip flooring. Plain-sawed oak.

Oak is a widely used species, however maple, birch, beech and other hardwoods also have desirable qualities. Softwood flooring includes such species as pine and fir. This is fairly durable when produced with an edge-grain surface.

Types of Wood Flooring

There are three general types of wood flooring used in residential structures; strip, plank, and block. As the name implies, strip flooring consists of pieces cut into narrow strips. It is laid in a random pattern of end joints, Fig. 15-1. Most strip flooring is tongue and grooved, both on the sides and ends. This design is also referred to as side-and-end matched. Another feature of modern strip flooring is the undercut. See Fig. 15-2. This con-

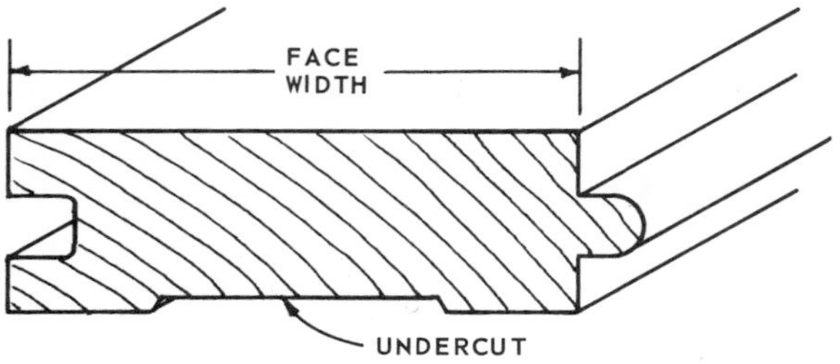

Fig. 15-2. Typical section of strip flooring. Ends are also tongue-and-grooved.

sists of a wide groove on the bottom of each piece which enables it to lay flat and stable even though the subfloor surface may contain slight irregularities.

Plank flooring, Fig. 15-3, provides an informal atmosphere. It is particularly appropriate for Colonial and Ranch style homes. Plank floors are usually laid in random widths. The pieces are bored and plugged to simulate the wooden pegs originally used to fasten them in place. Today this type of floor has tongue and groove edges and is laid in about the same manner as regular strip floors.

Block flooring presents an installed appearance similar to conventional parquetry -- an elaborate design formed by small wood blocks. Modern unit blocks consist of short lengths of flooring, held together with glue, metal splines or other fasteners. Square and rectangular units are produced. Generally, each block is laid with its grain turned at a right angle with the surrounding units, Fig. 15-3. Blocks, called laminated units, are produced by gluing together several layers of wood.

Sizes and Grades

Today hardwood flooring is generally available in widths ranging from 1 1/2 to 3 1/4 in. Standard thicknesses include 3/8, 1/2, and 25/32 in.

Solid planks are usually 25/32 in. thick, however, greater thicknesses are available. Widths range from 3 to 9 in. in multiples of 1 in. Unit-blocks are also commonly produced in a 25/32 thickness. Dimensions (width & length) are in multiples of the widths of the strips from which they are fabricated. For example, squares assembled from 2 1/4 in. strips will be 6 3/4 x 6 3/4 in., 9 x 9 in. or 11 1/4 x 11 1/4 in.

Uniform grading rules are established by manufacturers in cooperation with the U.S. Bureau of Standards and such organizations as the National Oak Flooring Manufacturers' Association or the Maple Flooring Manufacturers Association. Grading is based largely on appearance -- with consideration given to knots, streaks, color, pinworm holes, and sapwood. The percentage of long and short pieces is also a consideration.

Oak, for example, is classified into two grades of quarter-sawed stock and five grades of plain-sawed. In descending order plain-sawed grades are clear, select, No. 1 common, No. 2 common and 1 1/4 foot shorts. White and red oak species are ordinarily separated in the highest grades. In addition to other specifications the No. 1 common grade, as an example, requires that each bundle (24 board feet) contain no more than 10 pieces of lengths less than 18 in. long and that the average length be 32 in. and over.

Fig. 15-3. Left. Room floored with plank flooring. Right. Unit-block flooring. (Forest Products Lab.)

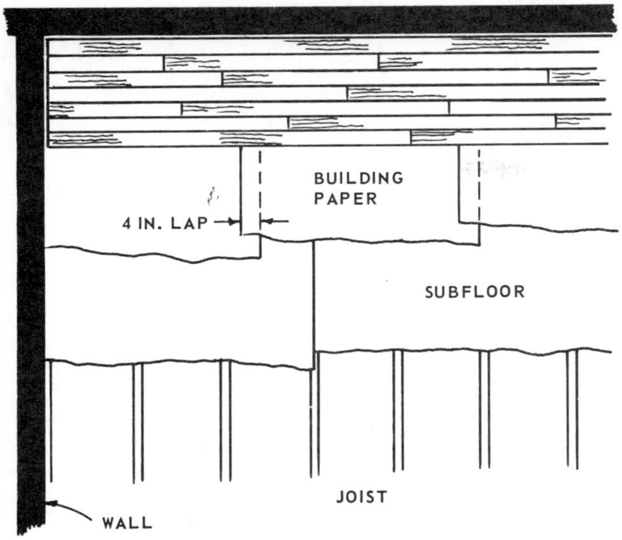

Fig. 15-4. Cutaway view of conventional floor structure for strip flooring.

Manufacturers recommend that wood flooring be delivered four or five days before installation. It should be piled loosely throughout the structure and an inside temperature of at least 70 deg. should be maintained. This period of "conditioning" permits the wood to equalize its MC (moisture content) with that in the building. Extra precautions of this nature will minimize excessive expansion or contraction that may result in cracks and distorted surfaces after the installation is complete.

Subfloors

In conventional joist constructions, most building codes specify a sound subfloor. It adds considerable strength to the structure and serves as a base for attaching the finish flooring.

For regular strip or plank flooring, it is generally recommended that the subflooring consist of good quality boards about 1 in. thick and not more than 6 in. wide. Space the boards about 1/4 in. apart and face nail them solidly at every bearing point. Inadequate or improper nailing of the subfloor almost invariably results in squeaky floors.

In modern construction, plywood is commonly used for the subfloor. It must be installed according to recognized standards. See Unit 7 for sizes and installation methods.

Installing Wood Strip Flooring

Check over the subfloor to make certain it is clean and that nailing patterns are complete. Lay a good quality building paper, extending from wall to wall, with a 4 in. lap as shown in Fig. 15-4. The location of the joists should be chalk lined on the paper -- especially when plywood is used as the subfloor.

Where floor areas are located directly over the heating plant or hot air ducts, it is advisable to use a double-weight building paper. Also, insulation may be attached directly to the underside of the subfloor. This extra precaution will prevent excessive heat from reaching the finish flooring; causing cracks and open joints.

Strip flooring should be laid at a right angle to the floor joist. It will usually present the best appearance when this direction aligns with the longest dimension of a rectangular room. Since floor joists will normally span the shortest dimension of the living room, this will establish the direction of the floor in other rooms.

HANDY SAYS:

"MC usually recommended for flooring at the time of installation is 6 percent for the dry Southwestern States; 10 percent for the more humid Southern States; and about 7 or 8 percent for the remainder of the country."

Nailing

Floor squeaks or creaks are caused by the movement of one board against the other -- either in the subfloor or finished floor. The use of an adequate number and proper size of nails is an important means of minimizing these undesirable noises. When possible, the nails used for the finish

Flooring Dimen., In.	Type, Size, Spacing of Nails
Tongued-and-grooved flooring. To be blind-nailed.	
25/32 x 3-1/4 25/32 x 2-1/4 25/32 x 1-1/2	7d or 8d screw type or cut steel nail.* 10-12 in. apart.
1/2 x 2 1/2 x 1-1/2	5d screw type or cut steel or wire nail. 10 in. apart.
3/8 x 2 3/8 x 1-1/2	Bright casing nail—4d wire or cut nail, or screw nail. 8 in. apart. Wood subfloor required.
Square-edge flooring. Face-nailed.	
5/16 x 2 5/16 x 1-1/2	1-in. 15 gauge fully barbed flooring brad, preferably cement coated. 2 nails every 7 in. Wood subfloor required.

** If steel wire flooring nail is used, it should be 8d, preferably cement coated. Newly developed machine-driven barbed fasteners of the size recommended by the manufacturer are acceptable.*

Fig. 15-5. Nails for strip flooring.

floor should extend through the subfloor and into the joist. This is especially important when plywood is used for the subfloor. Various sizes of nails are specified for the different types of floors as listed in Fig. 15-5.

Start the installation in a single room by laying the first strip along either sidewall. Select long pieces and if the wall is not perfectly straight and true, set the first course with a chalkline. Turn the groove edge next to the wall and leave an expansion space of not less than 1/2 in. This space will be covered later by the baseboard and baseshoe.

Make sure the first strip is perfectly aligned and then face nail it in place as illustrated in Fig. 15-6. In face nailing the nail head must be set and the resultant hole filled.

Succeeding strips are blind nailed with the nail penetrating the flooring where the tongue joins the shoulder. The nail is driven at an angle of about 50 deg. Use a nail set to finish the driving so that the edge of the strips will not be damaged. Each strip should fit tightly against the preceeding strip. When it is necessary to drive the strip into position, use a piece of scrap flooring for a driving block.

Cut and fit a number of pieces and lay them ahead of the installed strips as shown in Fig. 15-7. Use different lengths, matching them so they will extend from wall to wall with about 1/2 in. clearance at each end. Do not permit end joints in successive courses to come closer than 6 in. to each other. Try to arrange the pieces so the joints are well distributed with as much blending of color and grain pattern as practical. Pieces cut from the end of a course are carried back to the opposite wall to start the next course.

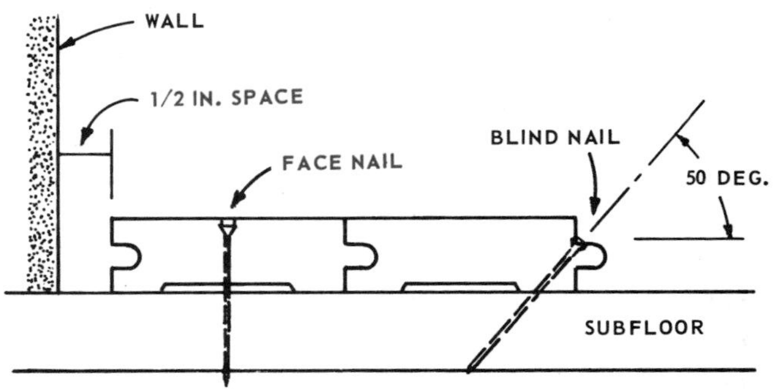

Fig. 15-6. Nailing starter strip.

Laying Around Projections

Flooring strips should run continuously through door ways and into adjoining rooms. When there is a projection into the room, such as a wall or partition, follow the procedure described in Fig. 15-8.

Fig. 15-7. Using a portable nailer to install strip floor. (Rockwell Mfg. Co.)

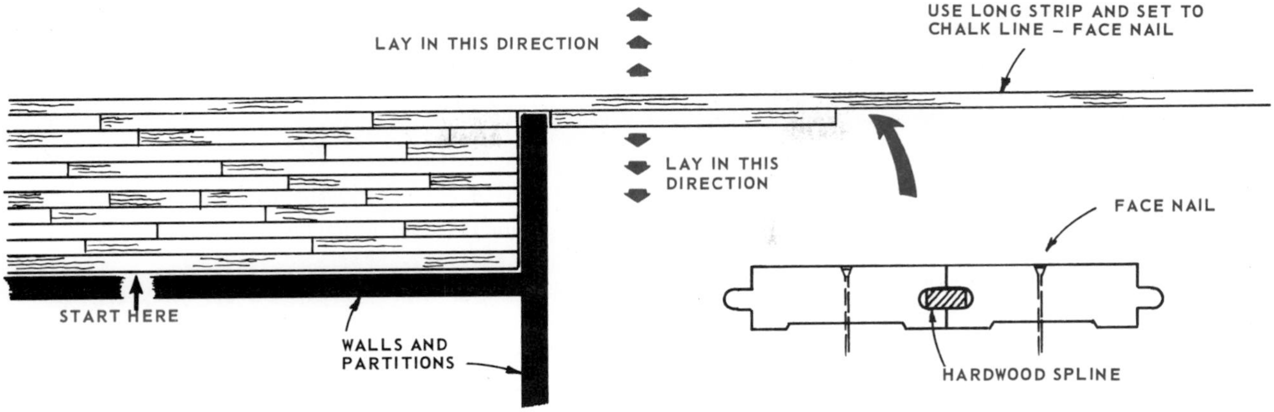

Fig. 15-8. Recommended procedure when laying strip floors around a projecting wall or partition.

First lay the main area to a point even with the projection and then extend the next course all the way across the room. Set the extended strip to a chalkline and then face nail it in place. Form a tongue on the grooved edge by inserting a hardwood spline and then install the flooring in both directions from this strip as shown.

If a large area is to be laid with strip or plank flooring, it is sometimes advantageous to set up a starter strip at or near the center. Use a spline in the groove of the starter strip as previously described. Be sure to measure and accurately align the starter strip with the walls on each side of the room.

When the floor has been brought to within 2 or 3 feet of the far wall, the room should be checked again to find out if the strips are parallel to the wall. If not, the grooved edges can be dressed off slightly at one end until the strips have been adjusted to run parallel. This is necessary in order that the last piece may show an even alignment with the baseboard.

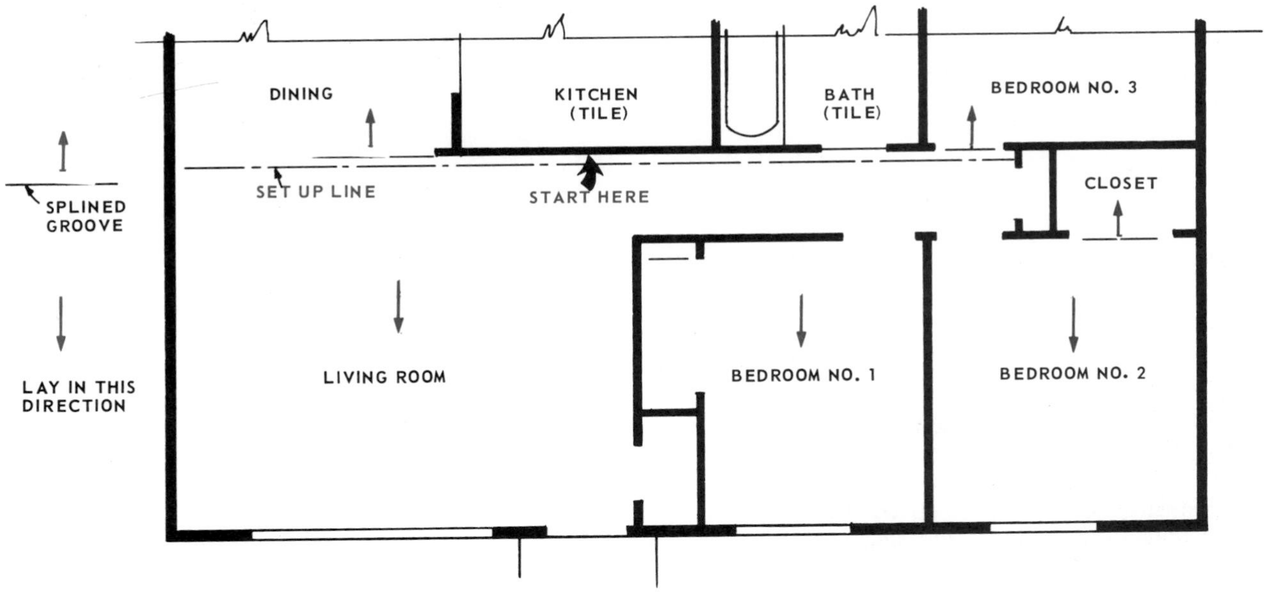

Fig. 15-9. Suggested procedure in laying strip floors throughout a number of rooms in a typical residential plan.

Multi-Room Layout

When flooring installation is carried throughout the major part of a building, make a thorough study of the floor plan to determine the most efficient procedure to follow.

Fig. 15-9 shows a typical residential floor plan with a suggested procedure for the installation of strip flooring. A setup line (chalkline) is first laid out two flooring widths (plus 1/2 in.) from the partition as shown. The starter courses are aligned with the setup line and face nailed in place.

Laying is continued across the living room, across the hall, and carried into and through bedrooms No. 1 and No. 2.

A spline is set in the groove of the starter strip and the floor is laid from this point into the dining room and bedroom No. 3. In bedroom No. 2 a splined groove is also used to lay the floor back into the closet.

In small closet areas such as shown in bedroom No. 1, it is usually impractical to reverse the direction of laying with a splined groove and the pieces are simply face nailed in place.

The last strip laid along the wall of a room will usually need to be ripped so it will fit into place with the required clearance (1/2 in. minimum). Also the last several strips must be face nailed. Do not use ripped strips where they might detract from the appearance. It is recommended that full length strips always be used around entrances and across doorways.

In general, plank flooring is installed by following the same procedures used for strip flooring. In addition to the regular blind nailing, screws are set and concealed in the face of wide boards.

Estimating

To determine the number of board feet of strip flooring needed to cover a given area, first calculate the area in square feet and then add the percentage listed for the particular size being used. See Fig. 15-10. The figures listed are based on laying flooring straight across the room. They provide an allowance for side-matching, plus about 5 percent for end matching. Where there are many breaks and projections in the wall line -- additional amounts should be added.

Example: Total Area	= 900 sq. ft.
Flooring size	= 25/32 x 2 1/4
Bd. ft. of flooring	= 900 + 38 1/3% of 900
	= 900 + 345
	= 1245
Number of bundles	= 52 (24 Bd. Ft. to bundle)

Wood Flooring Over Concrete

Finish wood flooring systems can be successfully installed over a concrete slab. When the concrete floor is suspended, with an air space below, a moisture barrier is usually not required. The installation consists of sleepers (wood strips) attached to or embedded in the concrete surface. The sleepers then serve as a nailing base for the flooring material.

55% for	25/32" x 1½"
42½% for	25/32" x 2"
38-1/3% for	25/32" x 2¼"
29% for	25/32" x 3¼"
38-1/3% for	⅜" x 1½"
30% for	⅜" x 2"
38-1/3% for	½" x 1½"
30% for	½" x 2"

Fig. 15-10. Percentage added to total area (square feet) to calculate number of board feet of flooring required.

If the concrete is placed directly on the earth, either at or below ground level, an approved membrane moisture barrier must be incorporated. Fig. 15-11 illustrates in a general way a system of waterproofing and sleeper installation that will provide a base for 25/32 strip flooring. A coat

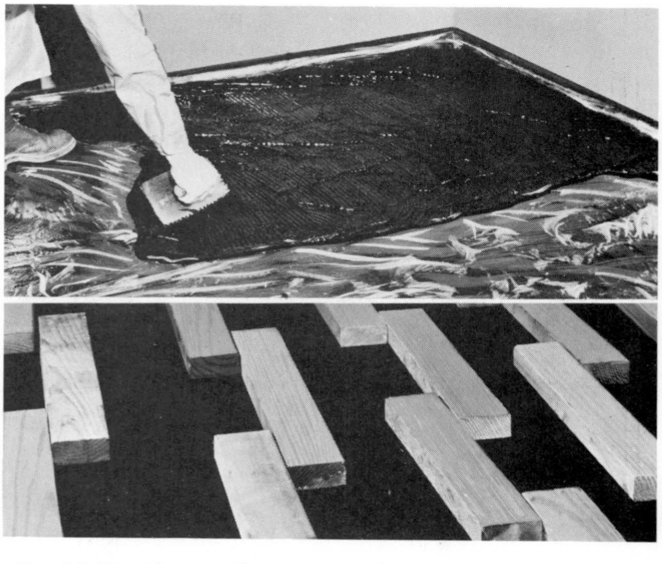

Fig. 15-11. Above. Moisture-proofing a concrete slab-on-grade with asphalt mastic and polyethylene film. Below. 2 x 4 sleepers set in asphalt mastic provide nailing base for wood flooring. (E. L. Bruce Co.)

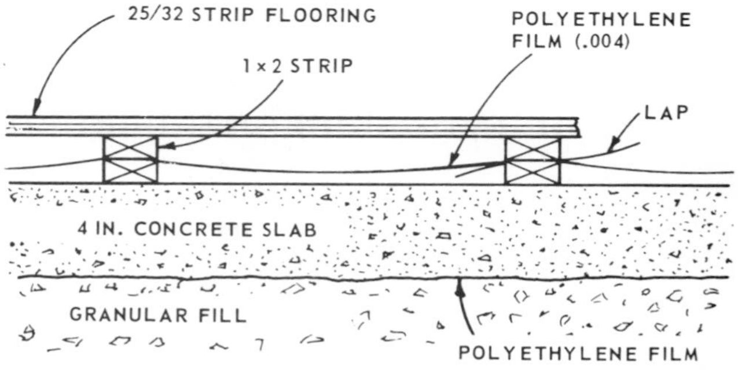

Fig. 15-12. Strip floor system over concrete slab. Sleepers consist of doubled 1 x 2 in. strips.

of asphalt is first evenly applied over the concrete surface and then covered with polyethylene film. Another coat of a special asphalt mastic is then applied as shown and wood sleepers are set in

Fig. 15-13. Installing strip hardwood flooring over concrete slab. A–Applying rivers of mastic along layout lines. B–Nailing bottom strips in place. C–Spreading polyethylene film over bottom strips. D–Nailing top strips to bottom strips. E–Nailing strip flooring to sleepers. (National Oak Flooring Assoc.)

place. The sleepers consist of 2 x 4 lumber about 30 in. long, laid flat side down, and running at a right angle to the proposed direction of the flooring. They must be lapped at least 3 in. for 2 1/4 in. and 4 in. for 3 1/4 in. flooring. Spacing between centers is usually 12 to 16 in.

A newer system of laying strip floors over a concrete slab-on-grade offers savings over older methods. It consists of a double layer of 1 x 2 wood sleepers nailed together, with a moisture barrier of 4 mil polyethylene film placed between the two pieces, Fig. 15-12. This is accepted by FHA.

The sequence of photographs in Fig. 15-13 illustrates the basic steps to follow in making the installation. The floor is first cleaned and primed,

Fig. 15-14. Spreading a coat of mastic over a plywood base — first step in laying a block floor.

and then chalklines are snapped 16 in. apart. Bands (called rivers) of a special adhesive for bonding wood to concrete are then applied along layout lines.

Treated wood sleepers (1 x 2 in. random lengths) are embedded in the adhesive and secured in place with 1 1/2 in. concrete nails, spaced approximately 24 in. apart. These strips could also be attached with explosion-actuated equipment. After all bottom sleepers are in place a layer of polyethylene film is laid over the strips. Sheets are joined by forming a full lap over a sleeper.

After the vapor barrier is in place, the second layer of strips (no treatment necessary) are nailed to the bottom courses with 4d nails spaced about 16 in. apart. Strip or plank flooring is then installed in about the same manner as previously described. An additional requirement is that no two adjoining flooring strips should break joints in the same space between sleepers.

HANDY SAYS:

*"Installation of wood flooring over concrete must be carefully designed and installed. Secure detailed specifications from manufacturers of the products to be used, or such organizations as the National Oak Flooring Manufacturers' Association."**

*National Oak Flooring Association. 814 Sterick Building, Memphis, Tenn. 38103

Fig. 15-15. Laying laminated wood blocks in special mastic. The same method can be used to lay 3/8 in. planks made of hardwood veneers. The planks are usually 3, 5, and 7 in. wide with tongue and groove on sides and ends. (E. L. Bruce Co.)

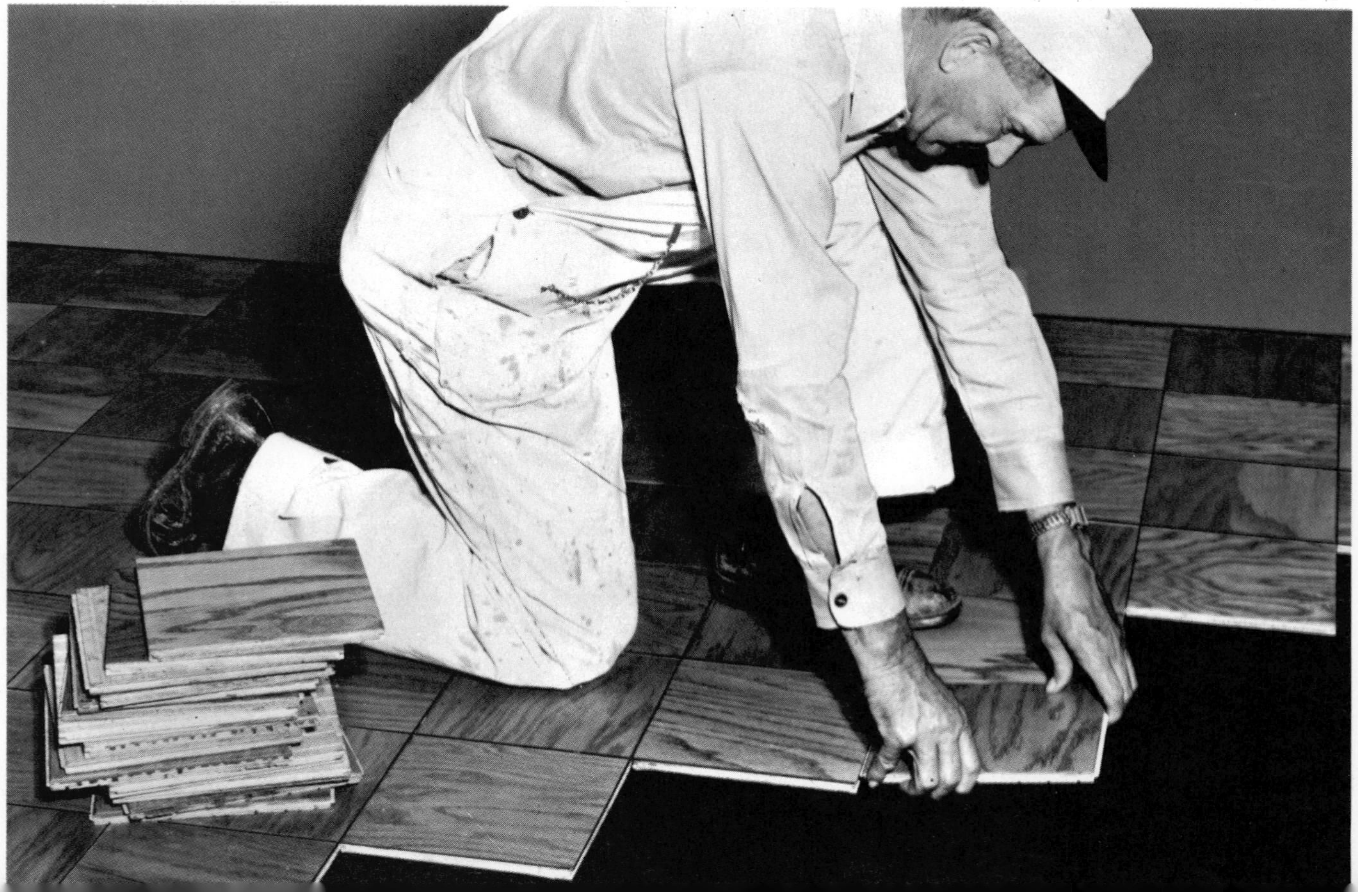

Wood Block Flooring

Block flooring is available in two main types; unit blocks and laminated blocks. The unit blocks, composed of several short lengths of flooring, have tongue edges at right angles to each other. Laminated blocks are made of three plies of hardwood bonded together with moisture resistant glue. The tongues are located on opposite edges.

Installation procedures for both types are basically the same. Unit blocks, however, require an allowance for expansion. Rubber strips are sometimes used for this purpose. Generally, it is recommended that unit block installations be made with a 1 in. space at walls or other vertical surfaces.

Either of the blocks can be nailed in place or laid in a special mastic. Nailing is done through the tongue edges in about the same manner as strip or plank floors. When laid in mastic a thin coat is first applied over the base which must be smooth, level, and dry. See Fig. 15-14. A layer of 30 lb. asphalt saturated felt is laid in place and another coat of mastic (about 3/32 in. thick) is spread. The blocks are then laid in this top coat as illustrated in Fig. 15-15.

Installation patterns can be either square or diagonal. Chalklines are first snapped equidistant from the sidewalls if laying is to proceed from a center point. Layout procedures are similar to those described for resilient flooring materials included later in this Unit.

Since block flooring and adhesive materials vary somewhat with different manufacturers, detailed information concerning installation procedures should always be secured from the manufacturer of the specific products being used.

Prefinished Wood Flooring

Most of the types of flooring previously described are available with a factory applied finish. Because the application of this finish is nearly always made under controlled conditions, using modern methods, equipment, and materials -- it is generally considered to be superior to that which can normally be applied on-the-job. Another advantage of using prefinished flooring is that it speeds completion of the structure; floors are ready for service immediately after application.

Disadvantages arise from the fact that the installation must be made with special care and accuracy. Although face nailing can be covered with special filler materials -- it must be held to a minimum when working with prefinished materials. Hammer marks caused by careless blind nailing are extremely difficult to repair.

Since the installation of a prefinished floor must necessarily be the last step in the interior finish sequence, other trim must be set with proper allowances. For example, interior door jambs and casing will require that spacer blocks (equal to the floor thickness) be placed under the units while they are being installed. Unless prefinished baseboards are installed after the floor is laid, they must also be placed at a level to allow necessary clearance. Other special provisions will likely need to be made around built in cabinets, and at stairways and entrances.

Underlayment

Flooring materials such as asphalt, vinyl, linoleum, and rubber will usually reveal rough or irregular surfaces in the flooring structure upon which they are laid. Conventional subflooring does not provide a satisfactory surface and an underlayment of plywood or hardboard is required. On concrete floors a special mastic material is sometimes used when the surface condition does not meet the requirements of the finish flooring.

In addition to providing a smooth surface the underlayment prevents the finish flooring materials from checking or cracking when slight movements take place in a wood subfloor due to moisture and temperature changes.

Underlayment, where used for carpeting and resilient materials, is usually installed as soon as wall and ceiling surfaces are complete, and at about the same time unfinished hardwood flooring is laid.

Hardboard

Hardboard products (material made from wood fibers) are fabricated in a special way to meet the requirements of an underlayment board. They are available in a standard thickness of about 1/4 in. (.215) with a panel size of 3 x 4 and 4 x 4 feet.

This type of underlayment material will bridge small cups, gaps, and cracks, but where larger irregularities exist they should be repaired, before the application is made. High spots should be sanded down and low areas should be filled. Hardboard underlayment, being a wood product, is subject to dimensional change under varying levels of humidity. Panels should be unwrapped and placed separately around the room for a period of at least 24 hrs. before they are installed.

To make the application, start in one corner of the room and fasten each panel securely before laying the next. Some manufacturers print a nailing pattern on the "up" side of the panel. See Fig. 15-16.

Fig. 15-16. Installing a 4 x 4 ft. hardboard panel. Allow a 1/8 in. space next to the wall.

Allow at least a 1/8 in. space along an edge next to a wall or any other vertical surface.

Stagger the joints of the underlayment panel. See Fig. 15-17. The direction of the continuous joints should be perpendicular to those in the subfloor. Be especially careful to avoid alignment of any joints in the underlayment with those in the subfloor. Fit the panels together accurately with a 1/32 in. space at each joint.

Fig. 15-17. Stagger joints and leave a 1/32 in. space between panels.

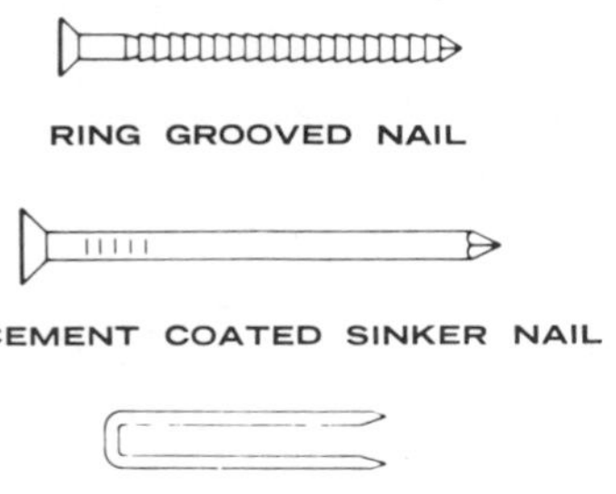

Fig. 15-18. Fasteners used for underlayment.

Underlayment should be attached to the subfloor with approved fasteners. See Fig. 15-18. They should be spaced not over 6 in. on center throughout the entire area of the panel. Space nails or staples in 3/8 in. from the edge, Fig. 15-19. Be sure to drive nail heads flush with the surface. When fastening underlayment with staples, use a type that is etched or galvanized and at least 7/8 in. long. Space staples not over 4 in. apart along panel edges.

Fig. 15-19. Hardboard should be nailed every 4 in. over the entire area of the panel. Nailing patterns are printed on some products. (Armstrong Cork Co.)

Special adhesives have been developed for bonding underlayment to subfloors. This eliminates the possibility of nail-popping under resilient floors which is caused by movement (expansion and contraction) in the subfloor and frame.

Plywood

Plywood fills the requirements for an underlayment material and is preferred by many carpenters. It is dimensionally stable and spacing between joints is not critical. Since a range of thicknesses are available, alignment of the surfaces of various finish flooring materials is easily accomplished, Fig. 15-20.

In the application of plywood underlayment, follow the same general procedures previously described for hardboard. Turn the grain of the face plys to run at a right angle to the framing supports and stagger end joints. Nailing patterns may be spaced somewhat greater for plywood but should not exceed a field spacing of 10 in. (8 in. for 1/4 and 3/8 in. thicknesses) and an edge spacing of

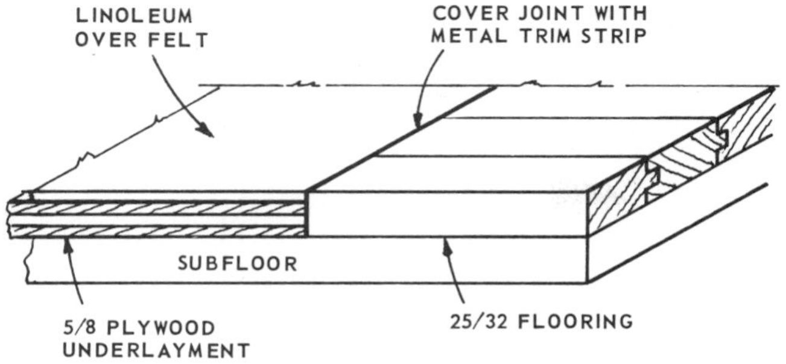

Fig. 15-20. Alignment of linoleum with strip flooring using plywood underlayment.

6 in. on center. Use ring groove or cement coated nails of an appropriate size -- depending on the thickness of the plywood and type of subfloor. Plywood can also be stapled. Using a pneumatic powered tool, Fig. 15-21, is desirable.

HANDY SAYS:

"When preparing a base for resilient flooring materials by applying an underlayment over a solid board subfloor – do not drive long nails through the underlayment, subfloor and on into the joists. Subsequent shrinkage of the subfloor and frame (especially in new construction) will likely cause these nails to rise above the surface of the underlayment and form tiny "blisters" in the floor surface."

Installing Resilient Floor Tile

After the underlayment is securely fastened and before applying the floor covering material, sweep and vacuum the surface carefully. Check to see that surfaces are smooth and joints level. Rough edges should be eliminated with sandpaper or a block plane.

The smoothness of the surface is extremely important, especially when the more pliable materials (vinyl, rubber, linoleum) are to be installed. Over a period of time these materials will "telegraph" (show on the surface) even the slightest irregularities and/or rough surfaces. Linoleum is especially susceptible and for this reason a base layer of felt is often applied over the underlayment when this material, either in tile or sheet form, is to be installed.

Today there is a tremendous range of resilient flooring materials. They vary in types, grades, thicknesses, and application requirements. Adhesives used for their installation are also available with a wide range of characteristics to fit many different requirements. Special attention should be given to their selection and should be based on the type of flooring material, underlayment or base, methods of application, and conditions under which the installation is made.

Because of the complexity of all these factors, it is essential that each application be made according to the recommendations and instructions furnished by the manufacturer of the product. Space does not permit more than a general description of methods and application procedures for the installation of tile floors. Sheet materials are not included, since their installation requires highly specialized procedures - normally handled by the professional flooring mechanic.

Start the layout of a floor tile installation by locating the center of the end walls of the room. Disregard any breaks or irregularities in the contour. Establish a main centerline by snapping a chalkline between the two points, Fig. 15-22. When snapping long lines, remember to hold the line at various intervals and snap only short sections as shown.

The next step consists of laying out another centerline at right angles to the main one. Use a

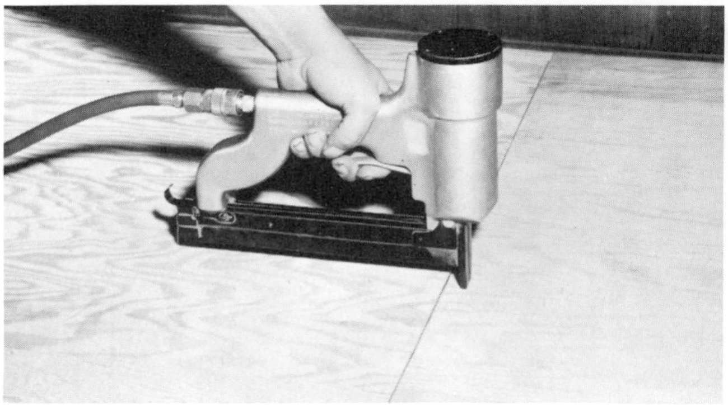

Fig. 15-21. Installing a plywood underlayment using a pneumatic powered nailer-stapler. Model shown drives a staple up to 1 1/8 in. long. (Senco Products, Inc.)

Fig. 15-22. Snapping a chalk line to establish the main center line for a floor tile installation.

Modern prehung door machine produces up to 25 units per hour. Door moves first into saw (arrow) where lock edge is bevel-trimmed and sanded. At the center station, door is bored for hardware, jambs are set in place and automatically guided routers cut gains for hinges. Air-driven screwdrivers are then used to install hinges. At the last station, casing is attached to the jambs and the unit is prepared for shipment. See page 374 for prehung door installation procedures. (*Frank Paxton Lumber Co.*)

Fig. 15-23. Using a carpenters' square to lay out a center line at a right angle.

carpenters' square as illustrated in Fig. 15-23 or set up a right triangle (base 4 ft., altitude 3 ft., hypotenuse 5 ft.). A chalkline can be used or draw the line along a straightedge.

With the centerlines established, make a trial layout of the tile along the centerlines as shown in Fig. 15-24. Measure the distance between the wall and last tile. If the distance is less than 2 in.

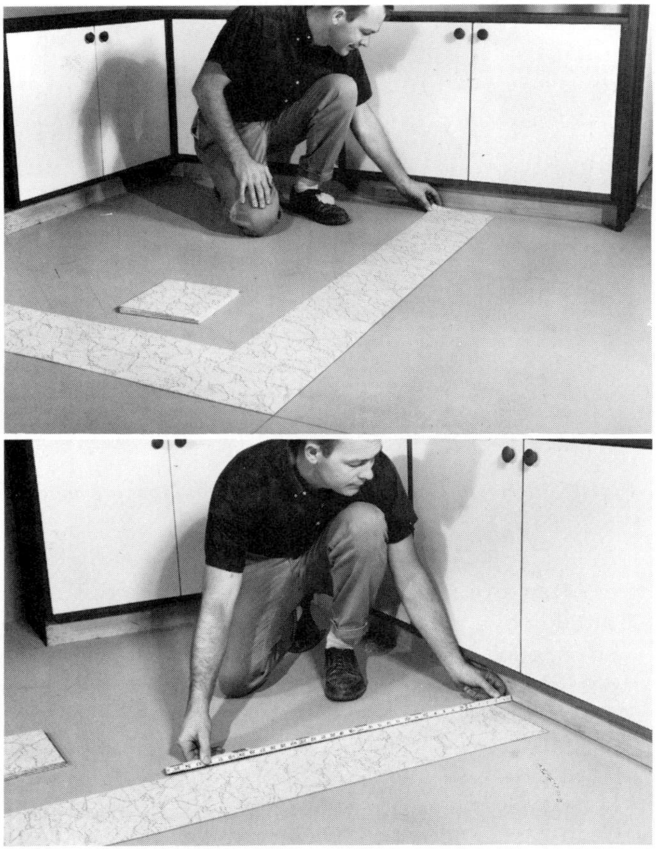

Fig. 15-24. Above. Making a trial layout of tile. Below. Measuring for border tile.

or more than 8 in., move the center line 4 1/2 in. (for 9 x 9 tile) closer to the wall. This adjustment will eliminate the need to install border tiles that are too narrow to fit easily into place. Check the layout along the other center line in the same way. Since the original center line is moved exactly one-half of the tile size, the border tile will remain uniform on opposite sides of the room.

Spreading Adhesive

Remove the loose tile, check to see that the floor surface is perfectly clean and then spread the adhesive over one-quarter of the total area, Fig. 15-25. Make the spread even with the chalkline but do not cover it. Be sure to use the type of spreader (trowel or brush) recommended by the manufacturer of the adhesive.

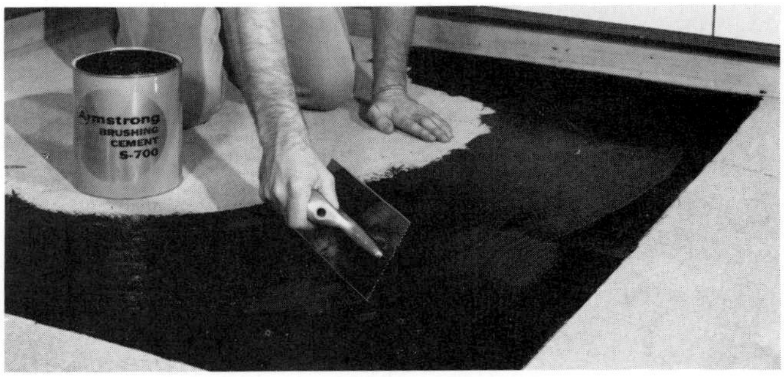

Fig. 15-25. Spreading adhesive with trowel.

The spread of adhesive is very important: if it is too thin, the tile will not adhere properly; if too heavy, it will creep up between the joints.

The adhesive must be allowed to take an initial set before a single tile is laid. The time required will vary from about 15 minutes to a much longer time depending on the type of adhesive used. Test the surface with your thumb. It should feel slightly tacky but should not stick to your thumb.

Laying Tile

Start laying the tile at the center of the room. Make sure the edges of the tile align with the chalkline. If the chalkline was partially covered with the adhesive, snap a new one or tack a thin, straight strip of wood in place to act as a guide in placing the tile.

Butt each tile squarely to the adjoining tile, with the corners in line, Fig. 15-26. Be careful to lay each tile in place--do not slide. Sliding may cause the adhesive to ooze up between the joints and prevents a tight fit. Take sufficient

Fig. 15-26. Laying the tile. Align the joining edges first and then lower the rest of the tile into place.

time so each tile will be positioned correctly. There is usually no hurry since most adhesives can be "worked" over a period of several hours.

Asphalt and vinyl-asbestos tile do not need to be rolled. Rubber, vinly and linoleum are usually rolled after a section of the floor is laid. Be sure to follow the manufacturer's recommendations.

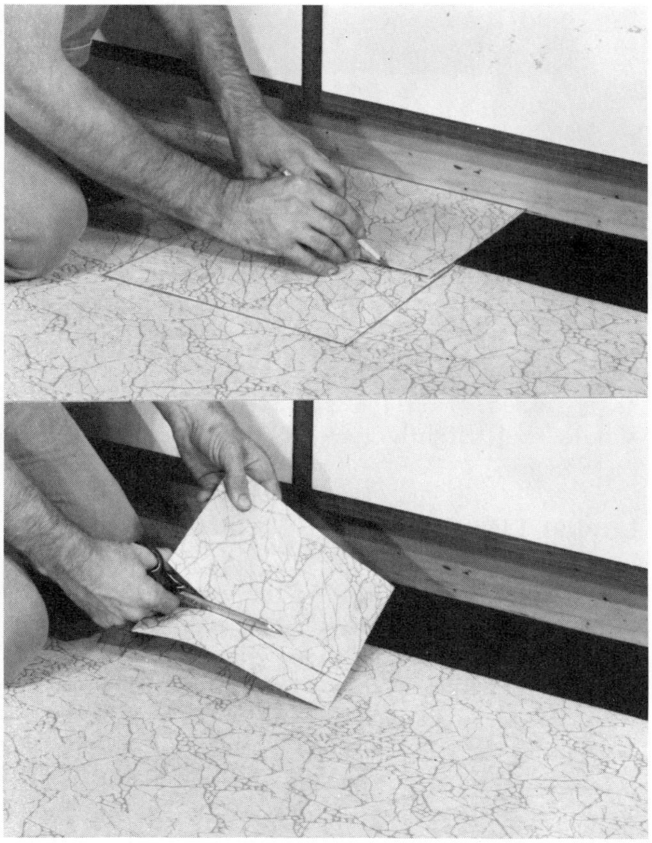

Fig. 15-27. Above. Marking border tile. Below. Cutting tile, (vinyl-asbestos type).

After the main area is complete, set the border tile as a separate operation. To lay out a border tile, place a loose tile (the one that will be cut and used) over the last tile in the outside row. Now take another tile and place it in position against the wall and mark a sharp pencil line on the first tile, Fig. 15-27. Cut the tile along the marked line, using heavy-duty household shears or tin snips. Some types of tile require a special cutter or they may be scribed and broken. Asphalt tile can be readily cut with snips if it is first heated.

Various trim and feature strips are available to customize a tile installation. They are laid by following the general procedures previously described for the regular tile, Fig. 15-28.

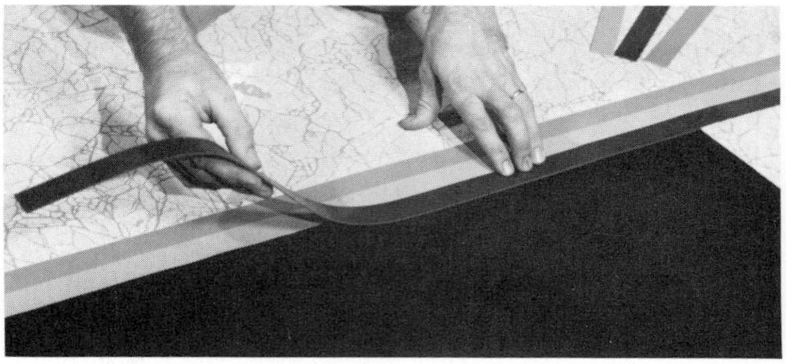

Fig. 15-28. Setting feature strips to create a custom effect.

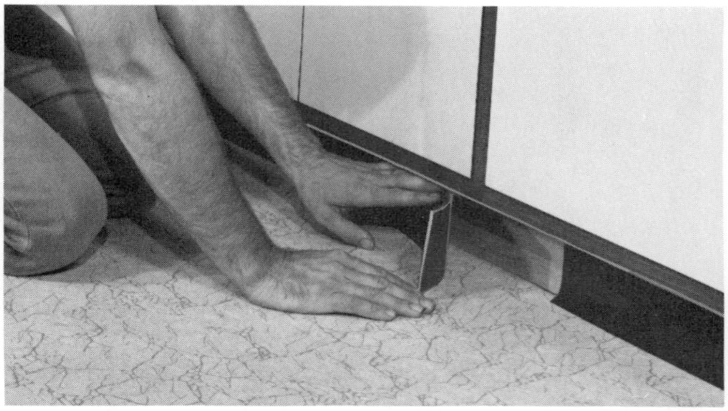

Fig. 15-29. Installing cove base. Apply a special adhesive to the material.

After all sections of the floor have been completed, cove base can be installed along the wall and around fixtures as shown in Fig. 15-29. A special adhesive is available for this operation. After cutting the proper lengths and making a trial fit -- apply the adhesive to the cove base and press it into place.

Check over the completed installation carefully. Remove any spots of adhesive with a great deal of care using cleaners and procedures approved by the manufacturer. Fig. 15-30 shows a general view

Fig. 15-30. Completed vinyl-asbestos tile applied to a concrete floor below grade. Linoleum and many of the other resilient tiles should not be laid on concrete unless it has been fully waterproofed with an approved vapor barrier. (Armstrong Cork Co.)

of a basement area made into attractive living space. The carefully designed and applied vinyl-asbestos floor tile is an outstanding feature.

Test Your Knowledge - Unit 15

1. The species of wood most commonly used for finish flooring in residential structures is __________.
2. Standard thicknesses of hardwood flooring include 3/8, _______ and 25/32 in.
3. When laying strip floors that start against a wall, lay the first course with the __________ (tongue edge, groove edge) turned toward the wall.
4. When the starter strip is located in the center of an area, install a hardwood _________ in the grooved side to permit laying in both directions.
5. How many board feet are there in a standard bundle of hardwood flooring?
6. The two types of block flooring most commonly used in residential installations are unit-blocks and _________ blocks.
7. Hardboard panels used for underlayment are laid with a space of ______ in. at each joint.
8. Approved metal fasteners for underlayment include ring-groove nails, divergent staples, and _________ nails.
9. A resilient flooring material that is most likely to show small irregularities in the base is __________.
10. Before cutting asphalt tile, they should be _________.

Outside Assignments

1. From a study of reference books located in the library or pamphlets secured from a local building supply firm, prepare a written report describing the procedures and processes used in the manufacturing of hardwood flooring. Include kiln drying requirements and moisture content standards. Also include grading rules that are applied to the species and qualities commonly used in your geographical area.
2. Prepare an oral report on resilient flooring materials. Possibly you can secure samples from a floor covering contractor or home furnishing store. Include information about thicknesses, colors, tile sizes and approximate costs. Also include a list of the advantages and disadvantages of the various types. If time permits, include information about the composition (basic materials used in manufacture) of each type and general requirements for installation.

TYPE OF FLOORING	ASPHALT		SEMI-FLEXIBLE ASBESTOS	VINYL				FLEXIBLE HOMO-GENEOUS	RUBBER		LINOLEUM		CORK
				BACKED TYPES									
	REGULAR	GREASE-PROOF		REGULAR		ALKALI RESISTANT							
FORM AVAILABLE	TILE	TILE	TILE	SHEET	TILE	SHEET	TILE	TILE	SHEET	TILE	SHEET	TILE	TILE
DIMENSIONS*													
Size or Width, Inches	9 x 9	9 x 9	9 x 9	72	9 x 9	45,72	9 x 9	9 x 9	36	9 x 9	72	9 x 9	9 x 9
Thickness, Inches	0.125 0.1875	0.125 0.1875	0.0625 0.080 0.09375 0.125	0.0625 0.070	0.0625	0.0625 0.080	0.0625 0.080	0.080 0.09375 0.125	0.080 0.125 0.1875	0.080 0.125 0.1875	0.070 0.090 0.125	0.065 0.090 0.125	0.09375 0.125 0.1875 0.3125
Approximate For Thickness Indicated	0.125″ 0.1875″	0.125″ 0.1875″	0.0625″ 0.125″	0.0625″ 0.070″	0.0625″	0.0625″ 0.080″	0.0625″ 0.080″	0.080″ 0.09375″ 0.125″	0.080″ 0.125″ 0.1875″	0.080″ 0.125″ 0.1875″	0.070″ 0.090″ 0.125″	0.065″ 0.090″ 0.125″	0.09375″ 0.125″ 0.1875″ 0.3125″
USE LEVEL**													
Suspended	Yes	Yes	Yes	Yes	Yes	Yes	Yes	Yes	Yes	Yes	Yes	Yes	Yes
On Grade	Yes	Yes	Yes	No	Yes	Yes	Yes	Yes	Yes	Yes	No	No	Yes
Below Grade	Yes	Yes	Yes	No	No	Yes	Yes	Yes	Yes	Yes	No	No	No
PHYSICAL CHARACTERISTICS													
Thermal Conductivity Btu/Hr./Eq. Ft./°F/In.	3.1	3.1	3.1	1.4	1.4	1.2-3.3	1.2-3.3	5.3	5.3	5.3	1.5	1.5	0.5
Relative Maximum Static Load Without Permanent Indentation Expressed in Lbs./Sq. In.	25(Fair)	25(Fair)	25(Fair)	75(Good)	75(Good)	75(Good)	75(Good)	200 (Excellent)	200 (Excellent)	200 (Excellent)	75(Good)	75(Good)	75(Good)
Under Foot Comfort	Fair	Fair	Good	Good	Good	Good	Good	Excellent	Excellent	Excellent	Good	Good	Excellent
Apparent Warmth to Touch	Fair	Fair	Good	Good	Good	Good	Good	Good	Good	Good	Good	Good	Excellent
Quietness (Noise Level)	Fair	Fair	Good	Good	Good	Good	Good	Very Good	Very Good	Very Good	Good	Good	Excellent
Surface Alkali Resistance	Excellent	Excellent	Excellent	Excellent	Excellent	Excellent	Excellent	Excellent	Good	Good	Fair	Fair	Fair
Grease Resistance	Poor	Very Good	Excellent	Excellent	Excellent	Excellent	Excellent	Excellent	Good	Good	Excellent	Excellent	Fair
Ease of Maintenance	Fair	Good	Good	Very Good	Very Good	Very Good	Very Good	Very Good	Good	Good	Very Good	Very Good	Fair

Data on resilient flooring materials.

Unit 16
STAIR CONSTRUCTION

A stairs is a series of steps leading from one level of a structure to another. When the series is a continuous section without breaks formed by landings or other constructions, the term "run of stairs" or "flight of stairs" is sometimes used. Other terms that can be properly substituted for stairs include "stairway" and "staircase."

In residential buildings, the popularity of the one-story structure has, for many years, minimized requirements for stair construction. Regular carpenters could usually handle the relatively simple task of constructing the service stairs leading from the first floor to the basement level. Recently, however, the trends in traditional styling along with split-level and multi-level designs have placed a new emphasis on this important aspect of carpentry.

The construction of a principal or main stairs -- often constituting an architectural feature in an entrance hallway or other area -- requires a high degree of skill. See Fig. 16-1. The quality of the workmanship should compare favorably with that found in fine cabinetwork.

Today, the parts for main stairways are usually fabricated in millwork plants and then assembled on the job. Even when this procedure is followed, the assembly work must be performed by a skillful carpenter who understands the basic principles involved in stair design and has a thorough knowledge of layout and construction procedures.

Main stairways are usually not constructed or installed until after interior wall surfaces are complete and finish flooring or underlayment has been laid. Basement stairs should not be installed until the concrete floor has been placed. Carpenters build temporary stairs from framing lumber to provide easy access to the various levels of the structure until the permanent stairs are installed. These are usually designed as a detachable unit so they can be moved from one project to another. Sometimes carriages are installed during the rough framing and temporary treads are attached. Later, during the interior finishing operations, these treads are replaced and other finished parts are installed.

Fig. 16-1. A main stairs can provide an attractive architectural feature. Its design and installation has long been considered one of the highest forms of joinery. (Morgan Co.)

Types of Stairs

There are a number of types and classifications of stairs. Basically stairs are divided into service stairs and main stairs. Either of these may be closed, open or a combination of open and closed,

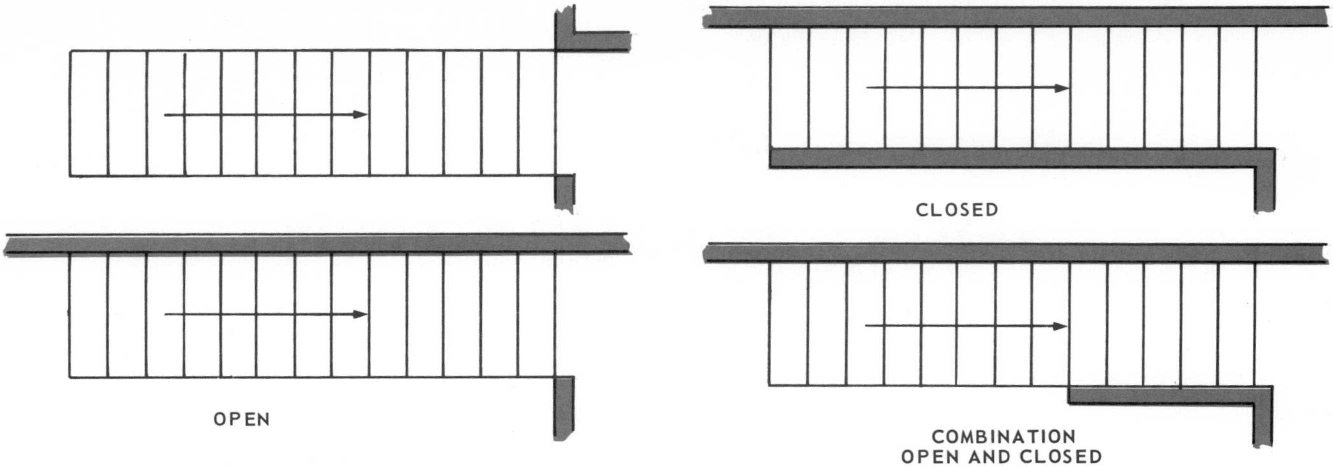

Fig. 16-2. Open and closed stairs. The term "open stairs" applies even though a wall is located on one side.

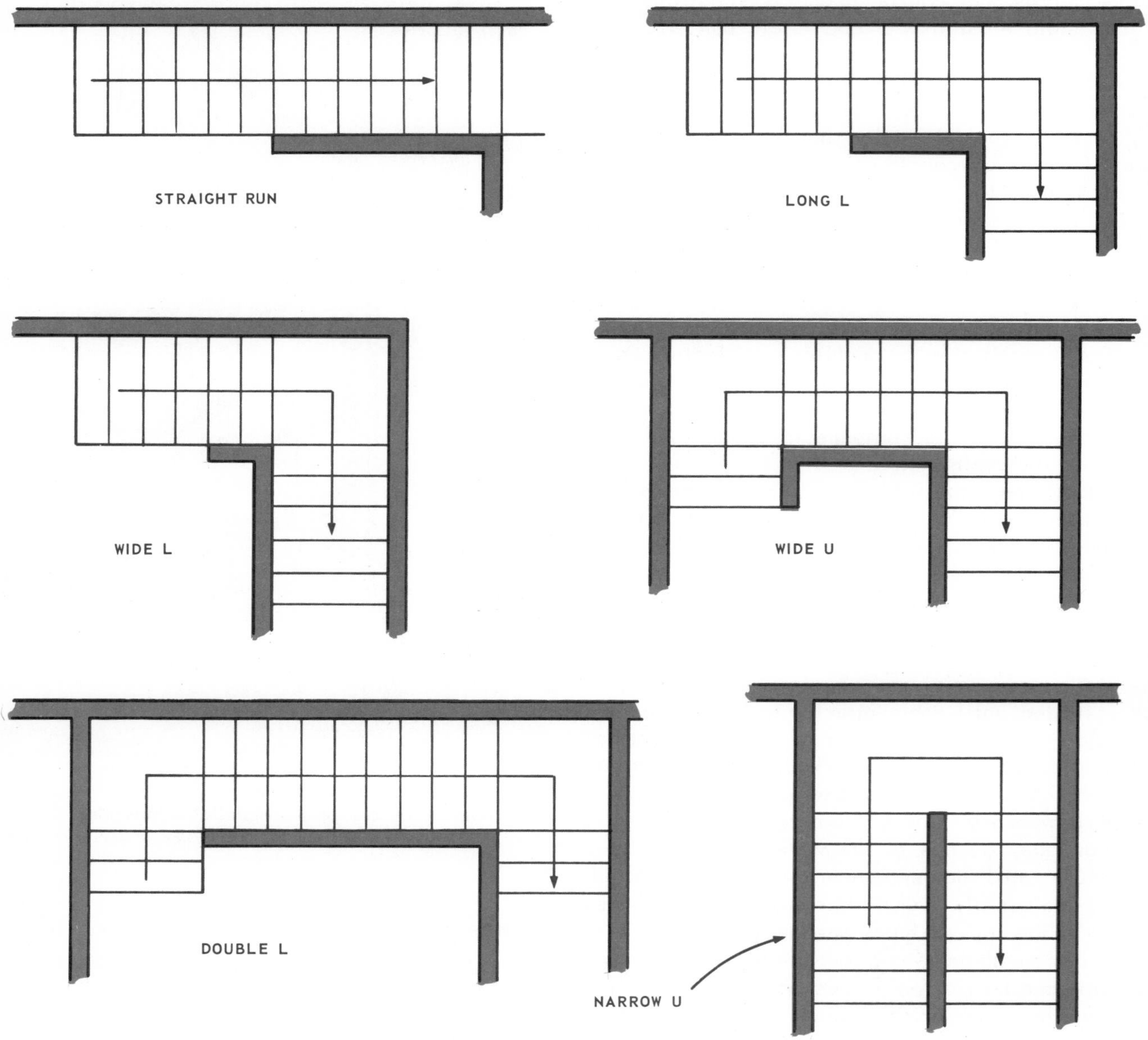

Fig. 16-3. Descriptive terms commonly used to define stair types.

Fig. 16-2. From the standpoint of design the types usually listed are straight run, platform, and winding. The platform type includes landings where the direction of the stair runs is usually changed. Such descriptive terms as L type (long L and wide L), double L type and U type are commonly used. See Fig. 16-3.

The straight run stairs is continuous from one floor level to another without landings or turns and is therefore the easiest to build. In standard multi-story designs they require a long stairwell which often presents a problem in smaller structures. A long run, consisting of 12 to 16 steps, also has the disadvantage of being tiring because it offers no opportunity for a pause in ascent.

In modern split-level designs the runs are short. These are generally defined as straight runs since they connect directly to the next floor level. Usually stair runs in this type of design are located in such a manner that headroom is automatically provided by the section directly above, Fig. 16-4.

Winding stairs also called "geometrical stairs" are circular or elliptical and gradually change directions as they ascend from one level to another. These often require curved wall surfaces that are difficult to build. Due to the complex nature of circular stair construction and the accompanying cost, such stairs are seldom used in a residential structure.

Parts and Terms

Stairs consist primarily of risers and treads supported by stringers. The height of the riser is called the unit rise and the width of tread (nosing excluded) is called the unit run. The sum of all the risers is the total rise and the sum of all the tread is the total run, Fig. 16-5.

Space above the stair is called headroom and is the vertical distance from a line along the front edges of the tread to the enclosed surface or header above. This distance is usually specified in local building codes. FHA requires a minimum headroom of 6 ft. - 8 in. for main stairs and 6 ft. - 4 in. for basement or service stairs.

Stairwell Framing

Stair building procedures vary widely from one locality to another. A carriage consisting of stringers may be cut and installed during the wall and floor framing, Fig. 16-6, or all aspects of the stairwork may be postponed until the interior finishing stages are in progress.

Regardless of procedures followed, the rough opening in the floors that will form the stairwell must be carefully laid out and constructed. If the architectural drawings do not include dimensions and details of the stair installation then the carpenter will need to calculate the sizes -- following recognized standards and local code restrictions.

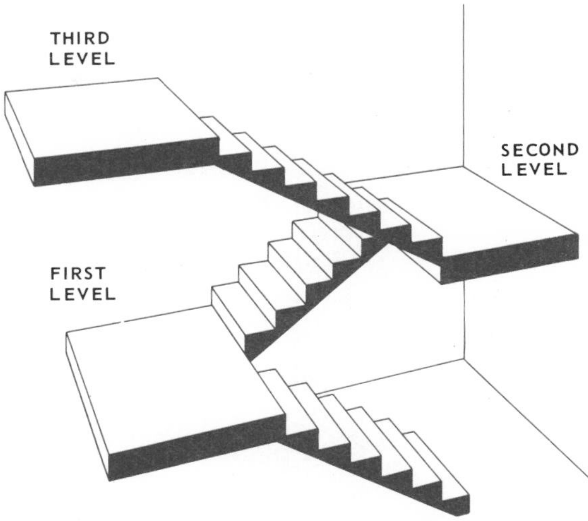

Fig. 16-4. Stair runs in split-level designs.

Trimmers and headers in the rough framing should be doubled, especially when the span is greater than 4 feet. Headers more than 6 feet in length should be installed with framing anchors

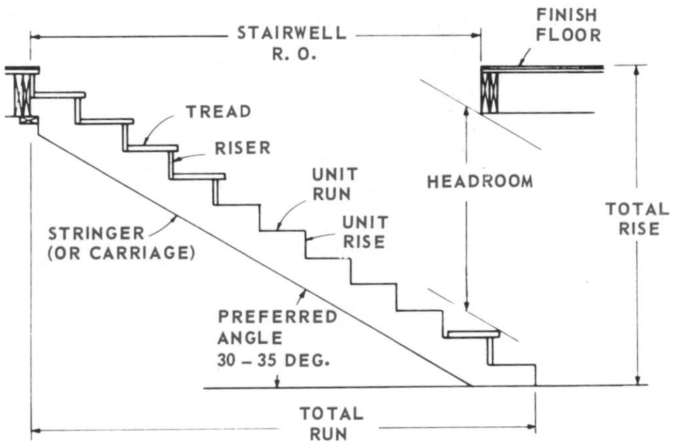

Fig. 16-5. Basic stair parts and terms. Note that the total number of risers is always one greater than the total number of treads.

unless supported by a beam, post or partition. Tail joists over 12 feet long should also be supported by framing anchors or a ledger strip. See Unit 7 for procedures in framing rough openings.

Providing adequate headroom often constitutes a problem, especially in smaller structures.

Stair carriage consisting of three stringers cut and installed during rough framing operations. Note the 2 x 4 spacer (arrow) placed on each side to provide clearance for the application of wall finish.

Wall finish of gypsum wall board and veneer plaster has been applied. Flat joints were reinforced with pressure sensitive fiberglass tape. The bottom side of the stairs has been sealed with a heavy sheet of gypsum board according to local code requirements.

Newel has been installed by cutting through the subfloor and anchoring the bottom end securely to the floor joists. Closed stringer (arrow) is in place. Note how open stringer is attached to supporting stringer with blocking. Plywood underlayment has been installed on the floor.

Skilled carpenter installing the first two risers. Top edge of riser must be perfectly aligned with supporting members so no crack will show when treads are installed.

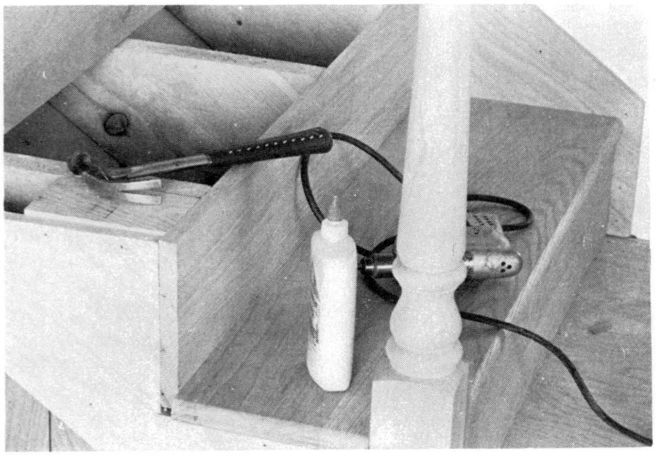

First tread installed. Since the stair parts (secured from a millwork plunt) are oak it is necessary to drill nail holes. Glue is also applied to each joint. In addition to face nailing of risers and treads, several nails are driven through the back, lower edge of the riser into the tread.

Balustrade assembled. The handrail is temporarily installed and then the position of the balusters are laid out on the tread and projected (plumbed) to the rail. Holes for the balusters are bored in the handrail and tread. After final assembly of the balustrade, trim pieces called brackets (see Fig. 16-25) are installed along the open stringer to cover the end grain of the risers.

Fig. 16-6. Typical sequence followed in the construction of a main stairs.

Installing an auxiliary header close to the main header, Fig. 16-8, will permit a slight extension in the floor area above a stairway. When a closet is located directly above, the closet floor is sometimes raised so additional headroom is gained.

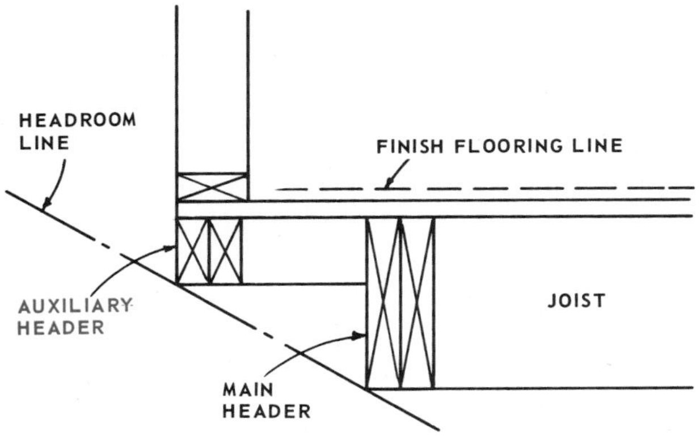

Fig. 16-8. Extending floor area with an auxiliary header. Partition above must be nonsupporting.

Stair Design

The most vital factor in stair design is the relationship between the rise (riser) and run (tread minus nosing). Well established rules governing this relationship should always be followed. When the rise-run combination is incorrect, the stair will be tiring and cause extra strain on the leg muscles -- also the toe of the foot may kick the riser when it is placed on the tread.

A unit rise of 7 to 7 5/8 in. high with an appropriate tread will combine both comfort and safety. Main or principal stairs are usually planned to have a rise within this range. Service stairs are often steeper but should not exceed a riser height of more than 8 in. As the height of the rise is increased the run must be decreased. See Fig. 16-9. Three generally accepted rules for calculating the rise-run (riser-tread) ratio are:

Rule No. 1 - The sum of 2 risers and 1 tread should equal 25 in.

Rule No. 2 - The sum of 1 riser and 1 tread should equal 17 to 18 in.

Rule No. 3 - The product obtained by multiplying the height of the riser by the width of the tread should equal approximately 75 in.

A riser of 7 1/2 in. would, according to Rule No. 1, require a tread of 10 in. -- a 6 1/2 in. riser would require a 12 in. tread. In residential structures, treads (excluding nosing) are seldom less than 9 in. or more than 12 in. wide. In a given run of stairs it is extremely important that all of the treads and all risers be the same size. A person tends to measure (subconsciously) the first few risers and will probably trip on subsequent risers that are misaligned.

HANDY SAYS:

"In a given run of stairs, be sure to make all of the risers the same height and all of the treads the same width. An unequal riser, especially one that is too high, may cause a fall."

The width of a main stair should allow two people to pass without contact and also provide sufficient space so furniture can be moved up or down. A minimum width of 3 ft. is generally

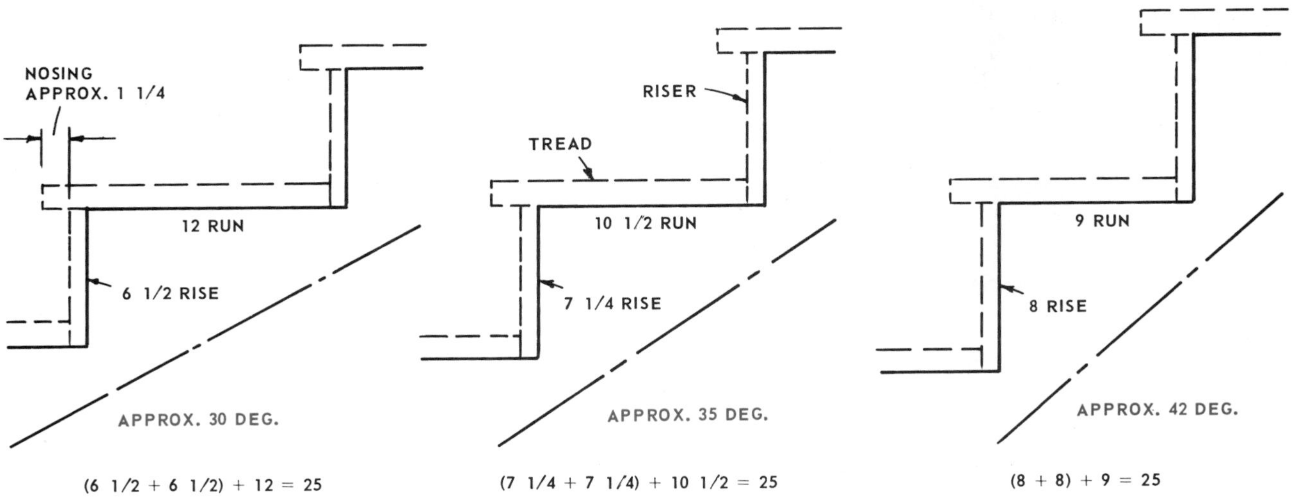

Fig. 16-9. Rise-run relationships in stair design. Rule No. 1 has been applied in the calculations.

recommended, Fig. 16-10. FHA permits a minimum of width, measured clear of the hand rail, of 2 ft. - 8 in. On service stairs, the requirement is reduced to 2 ft. - 6 in. Furniture moving is an important consideration and extra clearance should be provided in closed stairs of the L and U type; especially those that include winders.

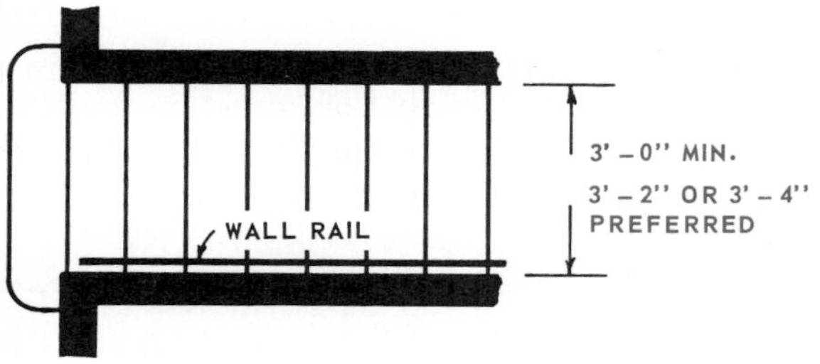

Fig. 16-10. Recommended width for a main stair.

When ascending or descending a stairway, a person must have the opportunity for support furnished through a continuous rail along the side. A handrail (also called a stair rail) is used on open stairways that are constructed with a low

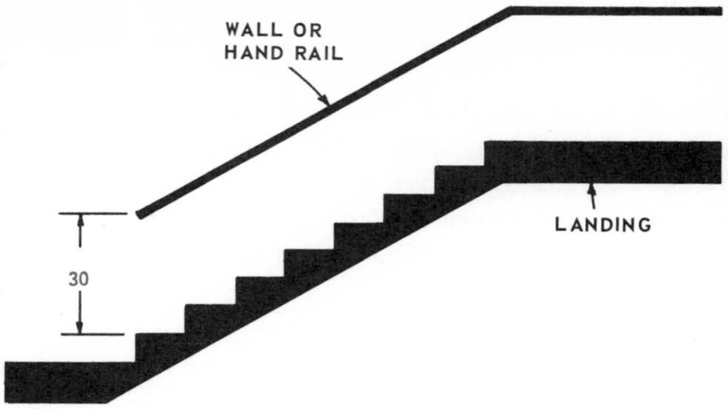

Fig. 16-11. Recommended rail heights along a run of stairs and at landings.

partition or banister. In closed stair designs the rail is called a wall rail and is attached with special metal brackets. Except for very wide stairs, a rail on only one side is sufficient. Fig. 16-11 illustrates the correct height for making a rail installation.

A complete set of architectural plans should include detail drawings of main stairs, especially when the design includes an unusual feature. For example, the stair layout in Fig. 16-12 shows a split-level entrance with open-riser stairs leading to upper and lower floors. An exact description of

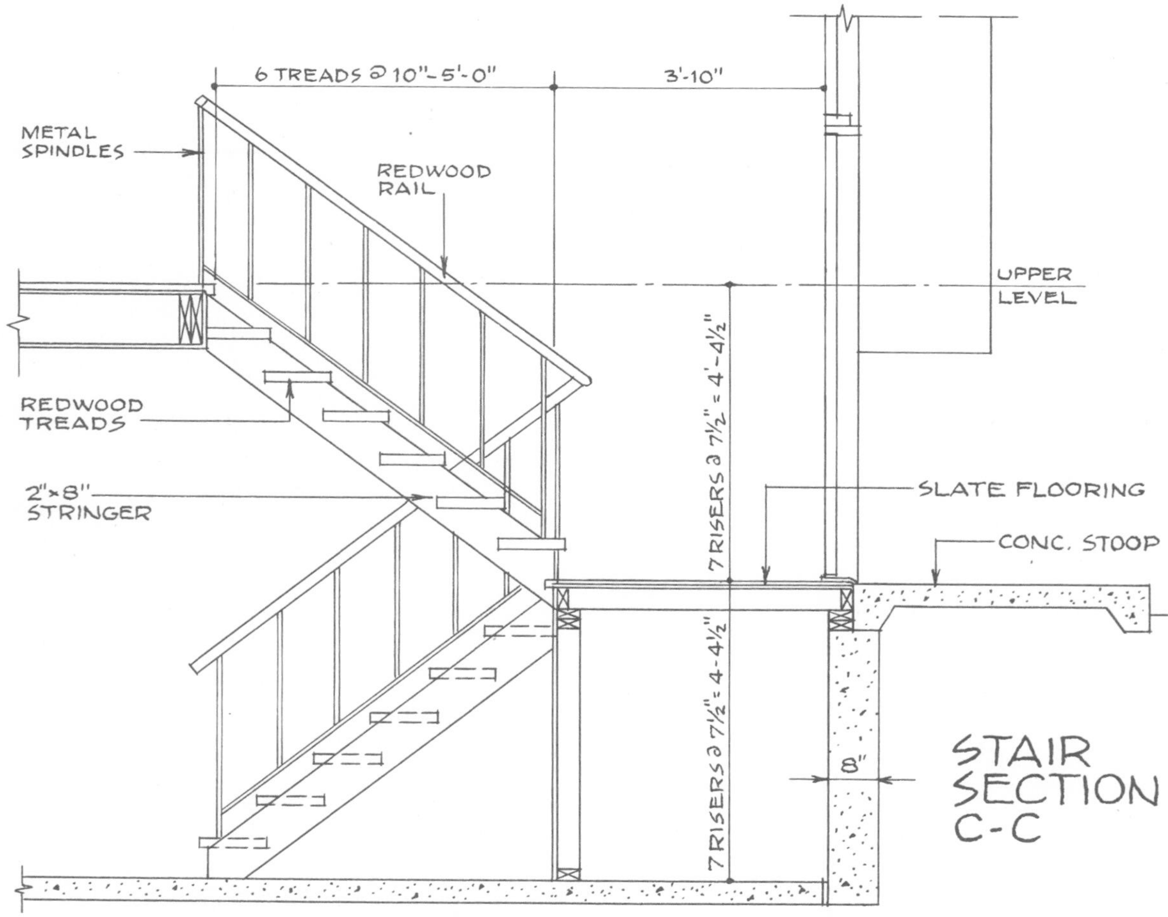

Fig. 16-12. Typical stair layout as shown in a set of architectural drawings.

tread mountings, overlap or nosing requirements, and height of the handrail is not included. These items of construction become the responsibility of the carpenter who must have a thorough understanding of basic stair design and how to lay out and make the installation.

All stairs, whether main or service, will be shown on floor plans. When details of the stair design are not included in the complete set of plans, the architect will usually specify on the plan view the number and width of the tread for each stair run. Sometimes the number of risers and the riser height is also included.

Stair Calculations

To calculate the number and size of risers and treads (less nosing) for a given stair run, first divide the total rise by 7. For example: if the total rise for a basement stairway is 7' - 10" or 94", the answer will be 13.43. Since there must be a whole number of risers, select the one closest to this figure (13.43) and divide it into the total rise;

94" ÷ 13 = 7.23 or 7 1/4"
Number of risers = 13
Riser height = 7 1/4"

In a given stair run the number of treads will be one less than the number of risers. A 10 1/2 in. tread will be correct for the example and the total run would be calculated as follows:

Number of treads = 12
Total run = 10 1/2" x 12
= 126"
= 10' - 6"

The stairs in the example will have 13 risers 7 1/4 in. high, 12 treads 10 1/2 in. wide and a total run of 10 ft. - 6 in.

Since the example was assumed to be a basement stairs, the total run could be shortened by using a steeper angle (decrease the number of risers and shorten the treads). If the decision was made to reduce the number of risers by one, the calculations would be as follows:

94" ÷ 12 = 7.83 = 7 5/6"
Number of risers = 12
Height of risers = 7 5/6"
Tread width selected = 9"
Number of treads = 11
Total run = 9" x 11
= 99"
= 8' - 3"

Stairwell Length

The length of the stairwell opening must be known and applied during the rough framing operations. If not included in the architectural drawings it can be calculated from the size of the risers and treads.

It is first necessary to know the headroom required. Add to this the thickness of the floor structure and divide this total vertical distance by the riser height which will give the number of risers in the opening length. When counting down from the top to the tread from which the headroom is measured, there will be the same number of treads as risers. Therefore to find the total length of the rough opening, multiply the tread width by the number of risers previously determined. Some carpenters prefer to make a scaled drawing (elevation) of the stairs and floor section to check the calculations.

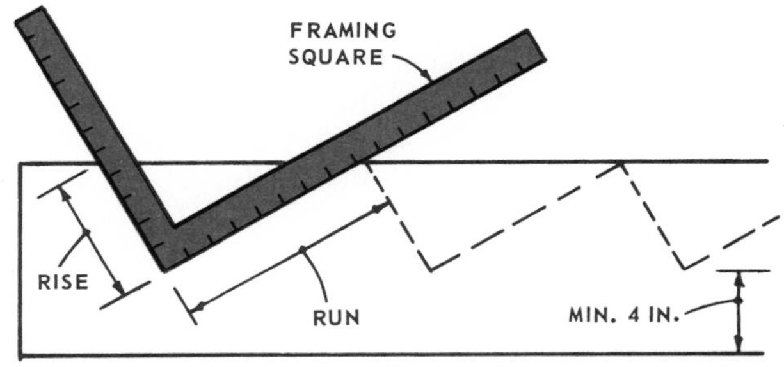

Fig. 16-13. Using framing square to lay out a stringer.

Stringer Layout

In the actual laying out of the stair stringer, it is first necessary to determine the riser height. Set a story pole (straight strip of 1 x 4 lumber) in a vertical position on the finished floor below and mark the location of the top surface of the finished floor above.

Set a pair of dividers to the calculated riser height and step off the distances. There will likely be a slight error in the first layout so adjust the setting and try again. Continue adjusting the dividers and stepping off the distance on the story pole until the last space is equal to all the others. Measure the setting of the dividers which will be the exact riser height to be used in laying out the stringers.

For a simple basement stairs, select a straight piece of 2 x 10 or 2 x 12 stock of sufficient length, and place it on sawhorses to make the layout. Begin at the end that will be the top and hold the framing square in the position shown in Fig. 16-13. Draw a line along the outside edge of

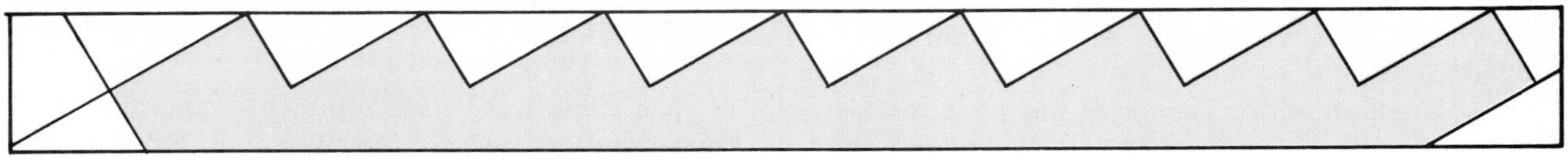

Fig. 16-14. Stringer layout completed.

the blade and tongue. Now move the square to the next position and repeat. The procedure is similar to that described for rafter layout in Unit 9. Accuracy can be assured in this layout by using framing square clips or clamping a strip of wood to the blade and tongue.

HANDY SAYS:

"Extreme accuracy is required in laying out the stringer. Be sure to use a sharp pencil or knife and make the lines meet on the edge of the stock."

Continue until the required number of risers and unit treads have been drawn, Fig. 16-14. The stair begins with a riser at the bottom so extend the last tread line to the back edge of the stringer as shown. At the top extend the last tread and riser line to the back edge.

One other adjustment must be made before the stringer is cut. When the bottom tread is installed the height of the first riser will be increased by the thickness of the tread material -- the top riser will be shortened by the same amount. To compensate for this, the bottom riser must be reduced by an amount equal to the tread thickness, Fig. 16-15.

Treads and Risers

The thickness of a main stair tread is generally 1 1/16 or 1 1/8 in. Hardwood or softwood may be used. FHA requires that stair treads shall be hardwood, vertical grain softwood or flat-grain softwood covered with a suitable finish flooring material.

Material for risers is usually 3/4 in. thick and should match the tread material, especially when the stairs are not covered. In most construction, the riser extends behind the tread, making it possible to reinforce the joint with nails or screws driven from the back side of the stairs. Where the top edge of the riser meets the tread, glue blocks are sometimes used, or a rabbeted edge of the riser may fit into a groove in the tread. A rabbet and groove joint may also be used where the back edge of the tread meets the riser.

Stair treads must include a nosing. This is the part of the tread that overhangs the one directly below. Nosing serves the same purpose as toe space along the floor line of kitchen cabinets . . . . provides space for the toe of the foot. The width of the nosing may vary from about 1 1/8 to 1 1/2 in. and should seldom be greater than 1 3/4 in. In

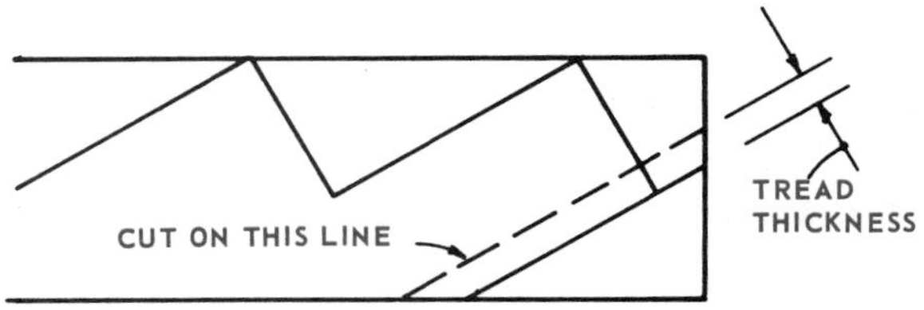

Fig. 16-15. Trimming the bottom end of a stringer to adjust for tread thickness.

general, as the tread width is increased, the nosing can be decreased. Fig. 16-16 illustrates a number of nosing forms. Cove molding may be used to close the joint between riser and tread and also cover nails used to attach the riser to the stringer or carriage.

Basement stairs are often constructed with an open riser (no riser board installed). Sometimes an open riser design is incorporated into a main stair to provide a special effect. Various methods of support or suspension may be used, often

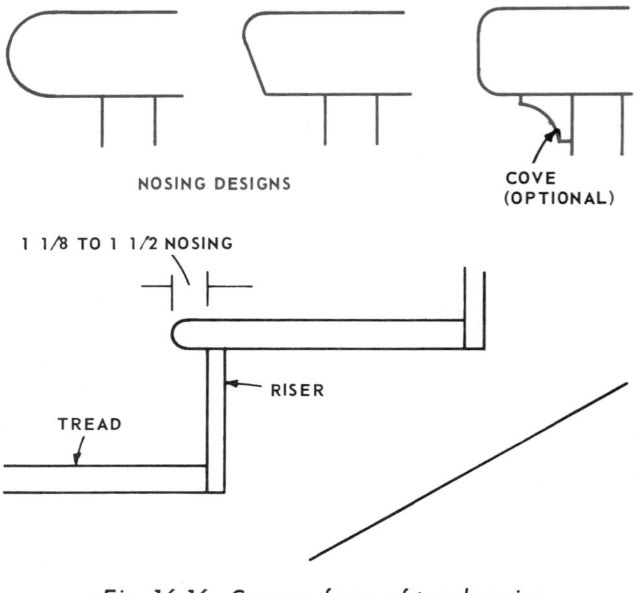

Fig. 16-16. Common forms of tread nosing.

requiring custom-made metal brackets or other devices. Fig. 16-17 shows basic types of riser designs. A slanted riser is sometimes used in concrete steps since it provides an easy way to form a nosing.

A popular type of stair construction consists of a stringer with tapered grooves into which the treads and risers fit. It is commonly called housed construction. Wedges, with an application of glue, are driven into the grooves under the

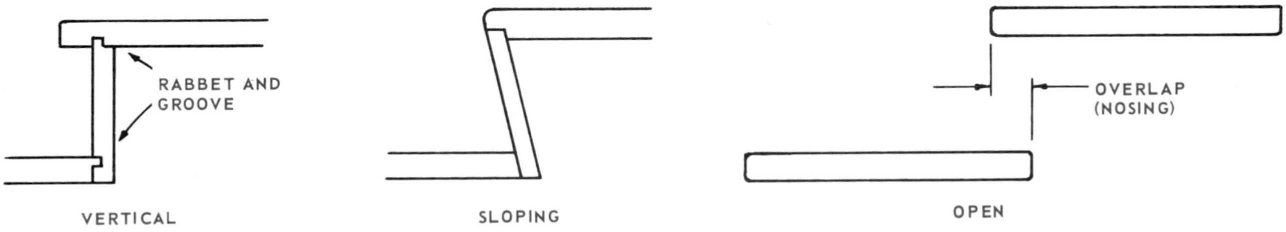

Fig. 16-17. Basic types of stair risers. For the open riser the tread should overlap at least 2 in.

Types of Stringers

Treads and risers are supported by stringers or carriages that are solidly fixed to the wall or framework of the building. For wide stairs, a third stringer is installed in the middle to add the necessary support.

The simplest type of stringer is formed by attaching cleats on which the tread can rest. Another method consists of cutting dados into which the tread will fit, Fig. 16-18. This type is often used for basement stairs where no riser enclosure is specified.

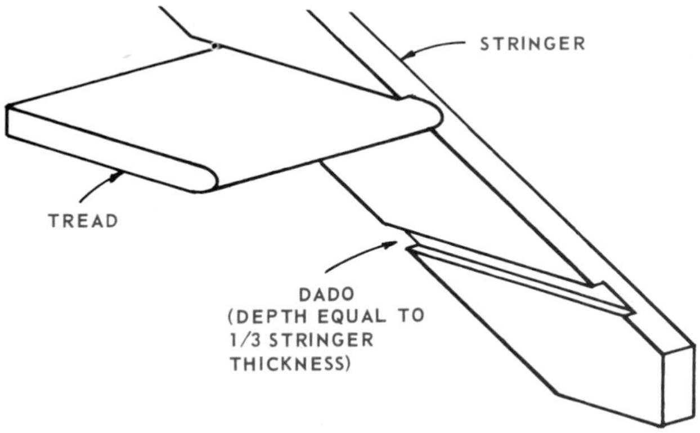

Fig. 16-18. Open riser stairs with treads set into dados cut in the stringer.

Standard cutout stringers (type used in layout description) are commonly used for either main or service stairs. Prefabricated treads and risers are often applied to this type of support. An adaptation of the cutout stringer, called semihoused construction, is illustrated in Fig. 16-19. The cutout stringer and backing stringer may be assembled and then installed as a unit or each part may be installed separately.

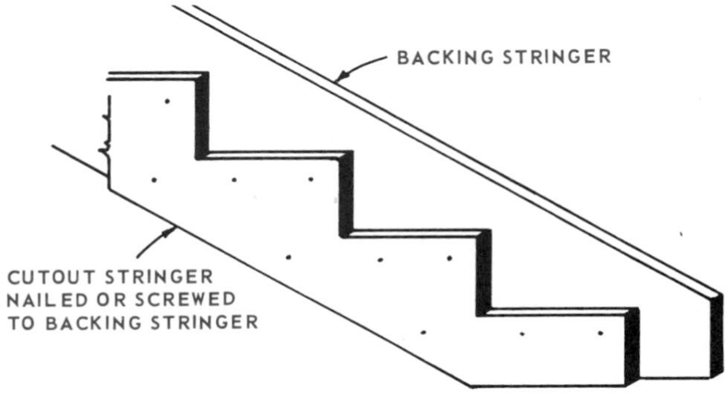

Fig. 16-19. Semihoused stringer design.

tread and behind the riser, Fig. 16-20. The treads and risers are joined together with rabbeted edges and grooves or glue blocks. This type of construction produces a stairs that is strong and dust tight, and one that will seldom develop squeaks.

Housed stringers can be purchased completely cut and ready to install or they can be cut on the

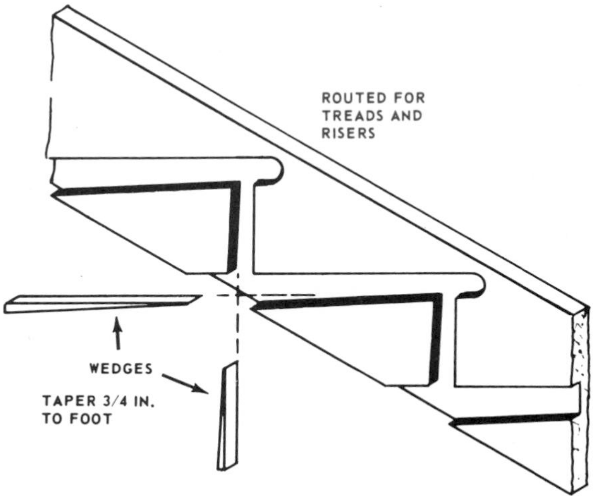

Fig. 16-20. Housed stringer design.

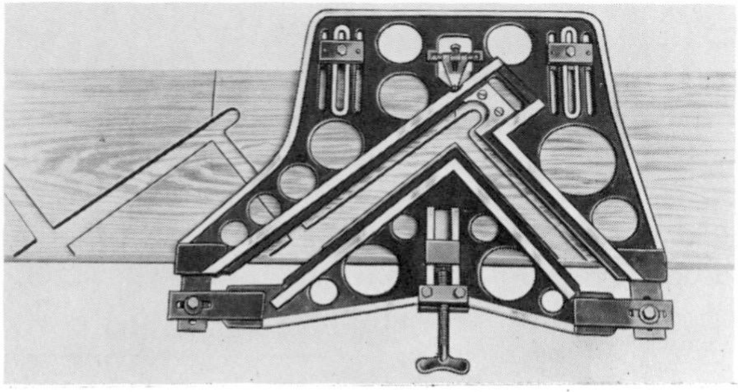

Fig. 16-21. Metal template for routing housed stringers. Fully adjustable for various tread and riser sizes and material thickness. (Rockwell Mfg. Co.)

job, using an electric router and template. Fig. 16-21 shows a metal template unit that is fully adjustable not only for various widths and heights of treads and risers but also for different thicknesses of materials.

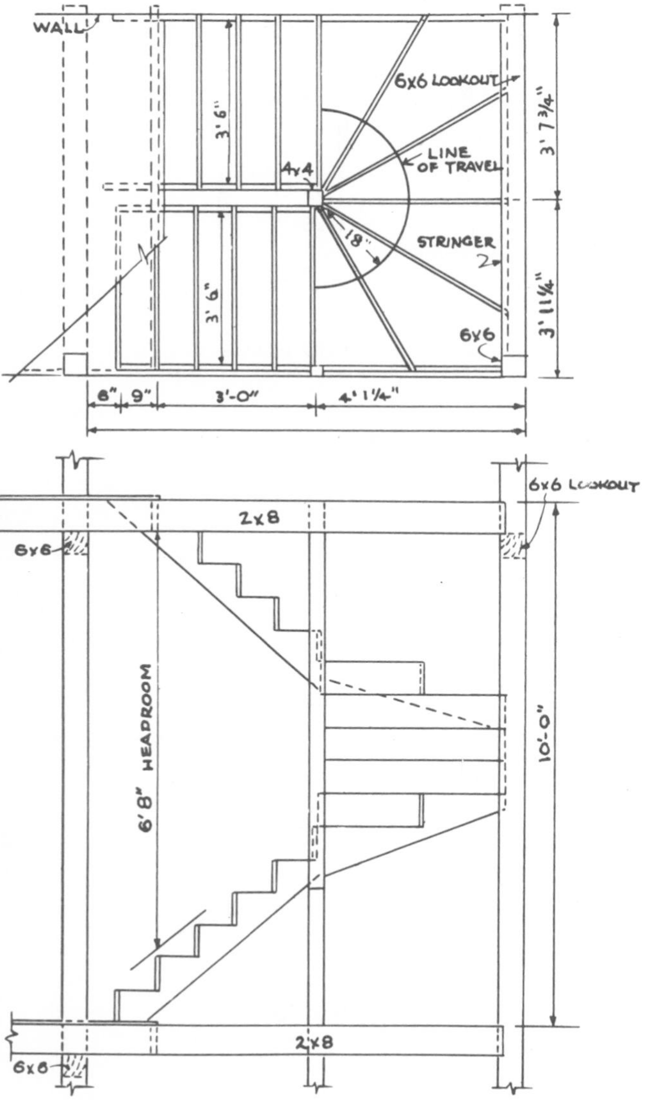

Fig. 16-22. Typical framing for a winder stairs. Tread width on winder should be the same at line of travel as tread in straight run.

To assemble the stairs, the housed stringer is spiked to the wall surface and into the wall frame. The treads and risers are then set into place, working from the top downward, using wedges and glue. This type of stringer shows above the profiles of the treads and risers and provides a finish strip along the wall. The design should permit a smooth joint where it meets the baseboard of the upper and lower level.

Winder Stairs

Winder stairs, Fig. 16-22, present stair conditions that are frequently regarded as undesirable. Their use, however, may sometimes be necessary where space is limited. The chief consideration in their design is to maintain a winder-tread width along the line of travel that is equal to the tread width in the straight run.

An adaptation of the standard winder layout is illustrated in Fig. 16-23. Here the risers converge at a point outside the stairs, thus providing some tread width at the corner. Before starting the construction of this type of stairs, the carpenter should make a full-size or carefully scaled layout (plan view) so that the best radius for the line of travel can be determined.

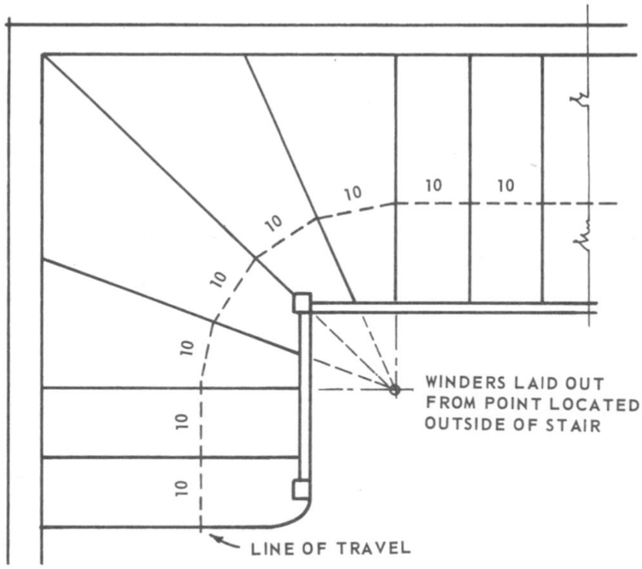

Fig. 16-23. Winder stairs with center of convergence located outside the stair construction. When winders are used, it is best to locate them near the bottom of the run.

Open Stairs

Main stairs that are open on one or both sides require some type of decorative enclosure and support for a handrail. Typical designs consist of

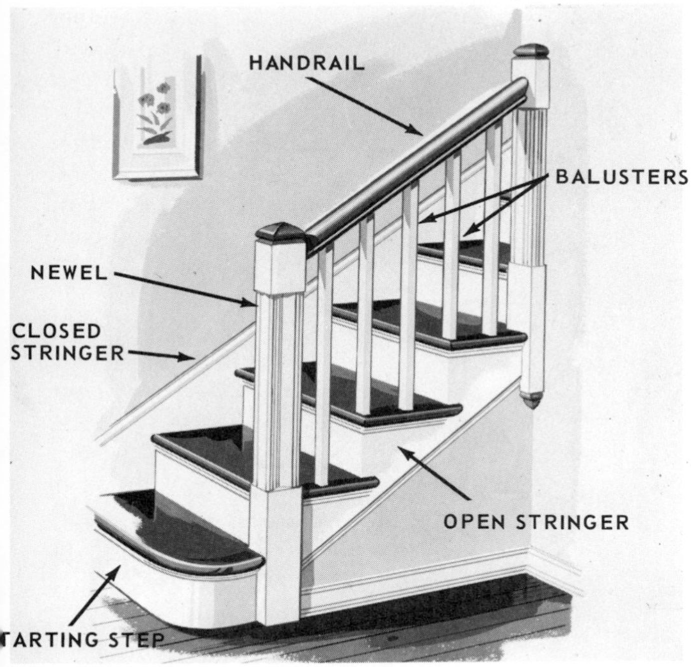

Fig. 16-24. Parts of an open stairs. The assembly of newels, balusters, and rail is called the balustrade. (Morgan Co.)

an assembly of parts called a balustrade, Fig. 16-24. The principal members of a balustrade are newels, balusters, and rails. They are usually factory-produced and are assembled on the job by the carpenter.

The starting newel must be securely anchored either to the starter step or carried down through the floor and attached to the building frame. Balusters are joined to the stair treads using either a round or square mortise, with a spacing of two or three units on each tread.

Using Stock Stair Parts

While many parts of a main staircase could be cut and shaped on the job, the usual practice in modern residential construction is to use factory-made parts. These are available in a wide range of stock sizes and can be selected to fill requirements for most standard stair designs. See Figs. 16-25 and 16-26. Stair parts are ordered through lumber and millwork dealers and are shipped to the building site in heavy, protective cartons along with directions for fitting and assembly.

Fig. 16-27 shows some suggested assemblies of balustrades, using stock parts. Hardware, especially designed for stairwork, is illustrated in Fig. 16-28.

Disappearing Stair Units

Where attics are used primarily for storage and where space for a fixed stairway is not available, hinged or disappearing stairs are often used. Such stairways may be purchased ready to install. They operate through an opening in the

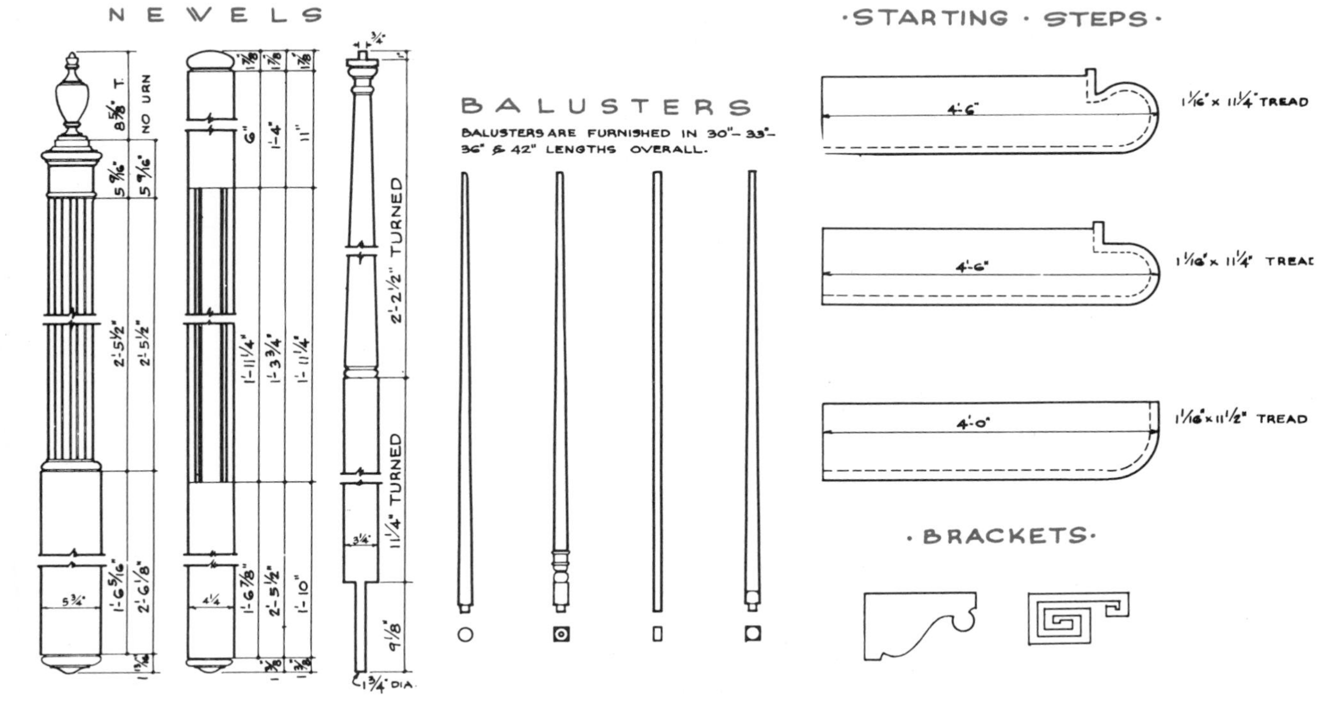

Fig. 16-25. Typical stock parts commonly available for stair construction.

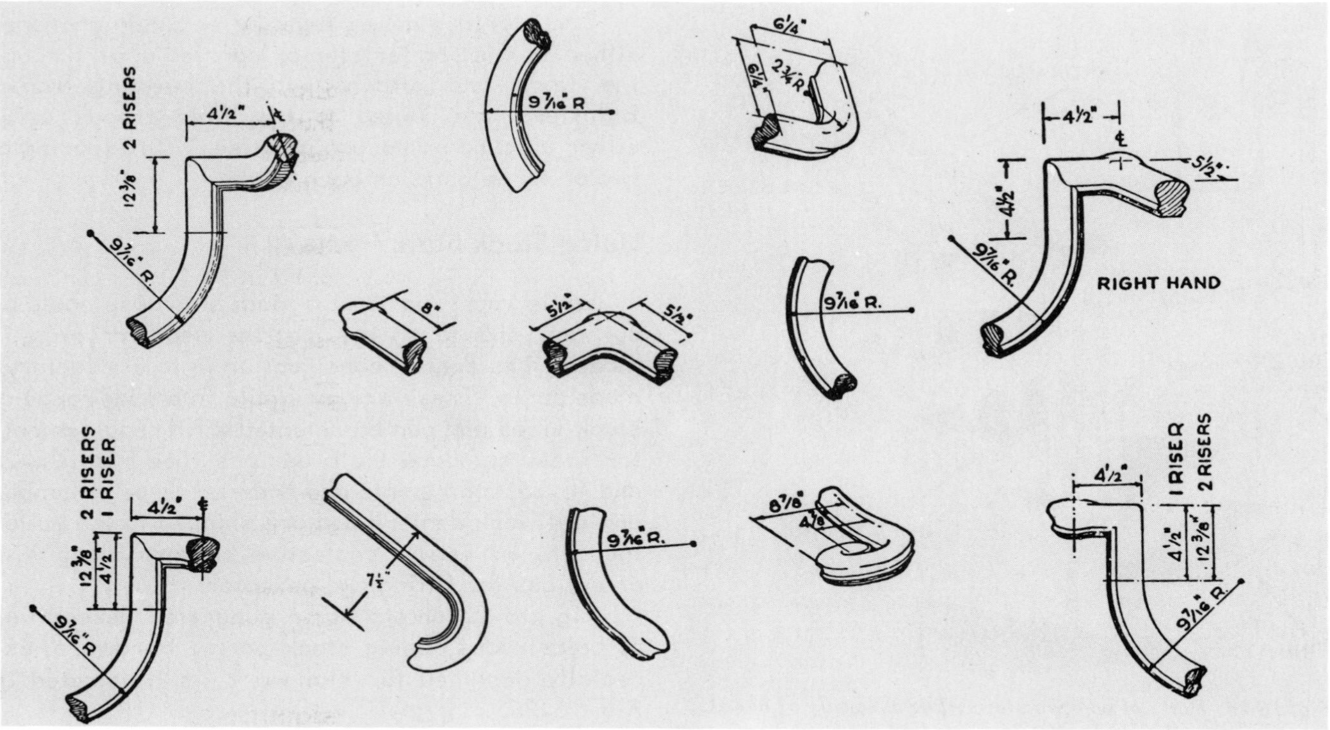

Fig. 16-26. Stock parts for handrails.

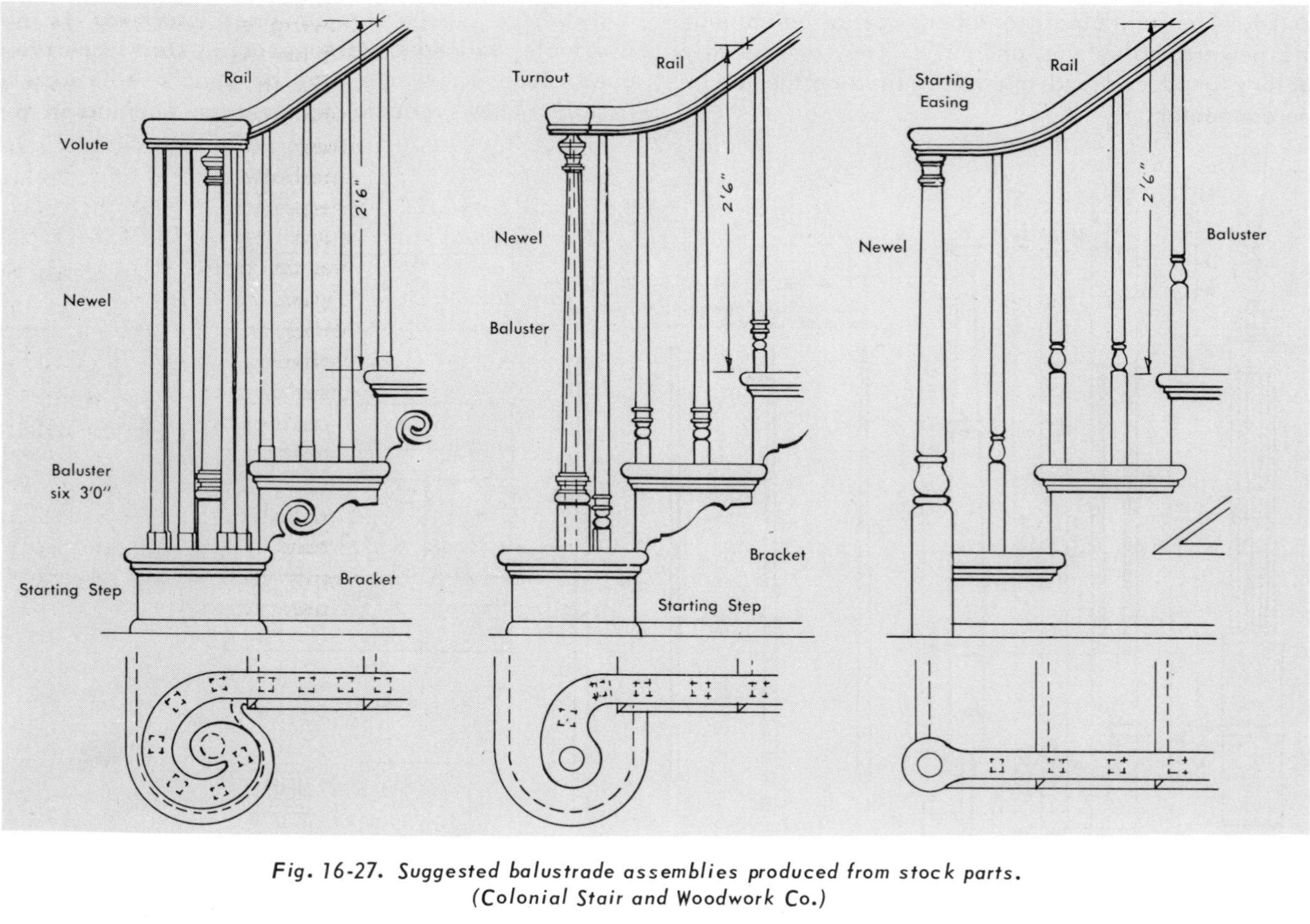

Fig. 16-27. Suggested balustrade assemblies produced from stock parts. (Colonial Stair and Woodwork Co.)

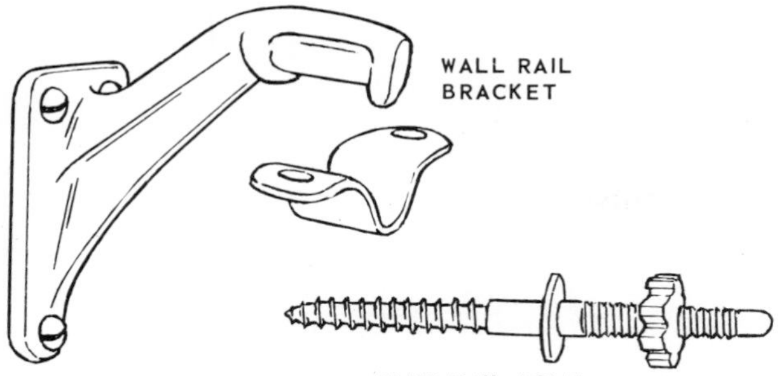

Fig. 16-28. Rail hardware. The rail bolt is consealed in the center of a joint and the nut is engaged and tightened through a hole bored in the underside. The nut is turned by striking the edges with a nail set.

ceiling and swing up into the attic space when not in use, Fig. 16-29. Where such stairs are to be provided, the attic floor should be designed for regular floor loading and the rough opening should be constructed at the time the ceiling is framed.

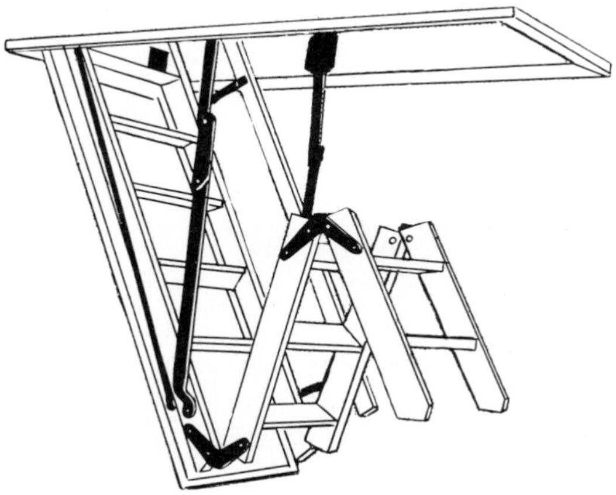

Fig. 16-29. Disappearing Stair Unit. (Rock Island Millwork)

Test Your Knowledge - Unit 16

1. The platform type of stairway includes______ where the direction of the stair runs is usually changed.
2. The minimum headroom for a main stairway as specified by FHA is_________.
3. The riser height of a service stairs should not exceed______in.
4. One of the rules used to calculate riser-tread relationship states that the sum of two risers and one tread should equal_______.
5. The front edge of the tread that overhangs the riser is called the _________.
6. A stairs in a split-level home has 6 risers with a tread width of 11 in. The total run of the stairs is _________.
7. The three basic types of risers are vertical, sloping and _________.
8. A semi-housed stair stringer is formed by attaching a_________stringer to a backing stringer.
9. Wedges used to assemble risers and treads in housed stringers should have a taper of___ in. to the foot.
10. The three principal members of a balustrade are called newels, rails, and __________.

Outside Assignments

1. Secure a set of architectural plans where the main or service stairway is not drawn in detail. Study the stair requirements carefully and then prepare a detail drawing somewhat like the diagram in Fig. 16-5 or the drawing in Fig. 16-12. Use a scale of 1/2 in. equals 1 ft. Select and calculate the riser-tread ratio carefully and be sure the number and size of risers is correct for the distance between the two levels. Check the headroom requirements against your local building codes and determine the stairwell sizes. Submit the completed drawing and size specifications to your instructor.
2. Study a millwork catalog and become familiar with the stock parts shown for a main stairway. Working from a set of architectural plans or a stair detail that you may have drawn, prepare a list of all the stair parts you would need to construct the stairway. Include the number of each part needed and also its size, quality, kind of wood, and catalog number. Take your list to a building supply dealer and secure a cost estimate for the materials.

Wood moldings being produced in a millwork plant. Blanks are fed into a machine (called a moulder or sticker) equipped with four high speed cutterheads that shape all sides of the blank in a single pass. In some setups a single blank is formed into two or more strips of molding. Note the vertical pipes that carry away the shavings. (Western Wood Moulding and Millwork Producers)

Fig. 17-1. Moldings. The curved and sloping surfaces produce shadows and highlights. (E. L. Bruce Co.)

Unit 17
DOORS AND INTERIOR TRIM

Installing door frames, hanging doors, and fitting trim around openings and at the intersection of wall, floor, and ceiling surfaces is an important part of the interior finish. This requires a high level of craftsmanship. The accuracy with which mating parts fit together can greatly enhance or detract from the final appearance of the trim.

Moldings

Moldings are specially shaped strips of wood which add to the decorative effect, and also serve essential functions. For example: window and door casings cover the space between the jamb and the wall surface and add rigidity to the construction. Stops are an important part of a window unit; as are glass beads, brick mold, and drip caps.

Moldings are used on the exterior as well as the interior of a structure. Architectural style must be considered in molding selection. As an example, for authentic colonial design, the use of appropriate moldings for trimming eaves, soffits, and roof rake is highly essential. In addition to the usual interior trim work, moldings may be applied to the surfaces of flush doors or cabinets to provide a decorative effect.

A wide range of types, patterns, and sizes are available. See Fig. 17-1. Several commonly used shapes are illustrated in Fig. 17-2. In addition to those shown, a complete list would also include; crown and cove molding, cap and brick molding, battens, glass beads, drip caps, corner guards, picture molding and screen mold. Information on molding patterns along with a numbering system and grading rules is included in a manual which is available from Western Wood Products Association, Portland, Oregon.

Interior Door Frames

The door frame is the part of the interior doorway which forms the lining of the opening and covers the edges of the partition. This consists of two side jambs and a head jamb, Fig. 17-3.

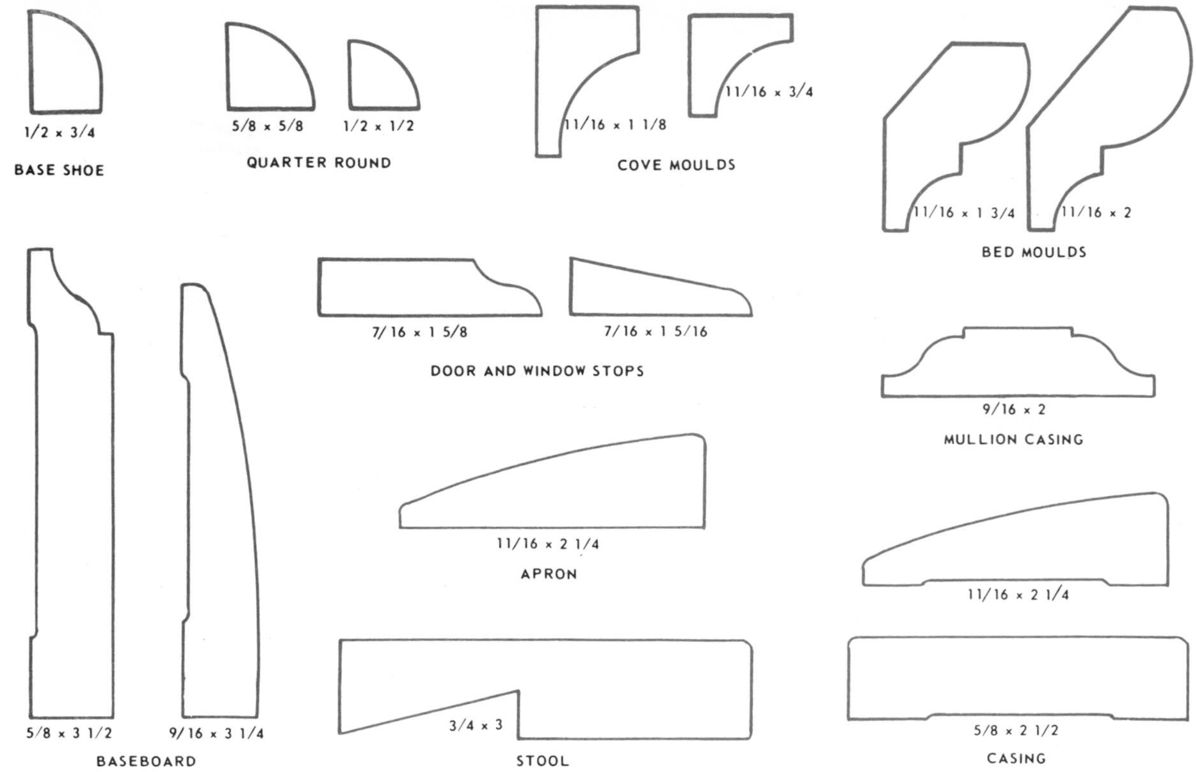

Fig. 17-2. Typical molding patterns used for interior trim. Note the hollow back of the baseboard and casing which provides clearance for irregularities in the wall surface.

Interior frames are simpler than exterior frames since the jambs are not rabbeted and no sill is included. See Unit 11.

Standard jambs for regular 2 x 4 stud partitions, made from nominal 1 in. material for plastered walls, are 5 1/4 in. wide. The back side is usually kerfed to minimize the tendency toward cupping (warping). The edge of the jamb is beveled slightly so the casing will bear against it with no visible crack.

Side jambs are dadoed to receive the head jamb. The side jambs for residential doorways are made 6 ft. - 8 1/2 in. long (clear under head jamb), thus providing 1/2 in. clearance at the bottom of the door.

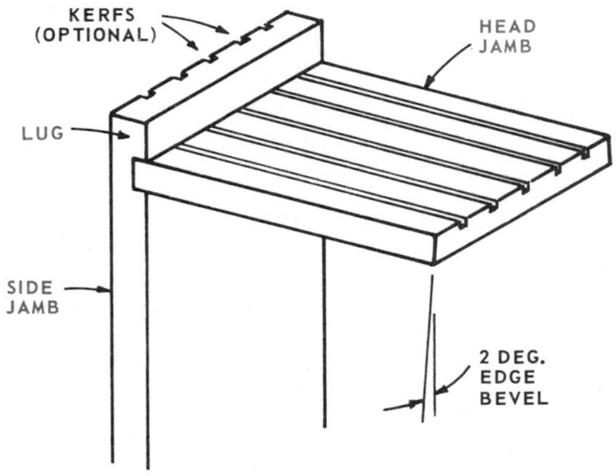

Fig. 17-3. Parts of an interior door frame.

Modern door frames are usually cut, sanded, and fitted in millwork plants. This makes possible quick assembly on the job. Door frames should receive the same care in storage and handling as other finished woodwork.

Installing Door Frames

Before assembling the door frame, check the length of the head and side jamb to determine if they are correct for the opening. Nail the jambs together using 8d casing or finish nails. If there is not sufficient vertical clearance in the rough opening, trim away part of the lug.

Place the frame in the opening with the side jambs resting on the finish flooring or spacer blocks of the required thickness, if the final flooring surface has not been laid. Check to see if the head jamb is level. If not, trim the bottom of the side jamb that is high. Now place a spreader (1 x 6 with length equal to the head jamb, minus depth of dados) between the side jambs at the floor level, Fig. 17-4.

On each side jamb, gauge a light pencil line in from the edge of the door side, a distance equal to the door thickness plus 7/8 in. All nailing of the jambs is done along this line since it will later be covered by the door stop.

Center the frame in the opening and secure it in place with double-shingle wedges at the top and bottom on each side. Plumb the jambs, using a straightedge and level, or a long carpenters' level as shown in Fig. 17-5. Make adjustments in the double-shingle blocking until each side is correct, then fasten the top and bottom of each side jamb with an 8d casing nail.

Complete the blocking by placing more double-shingle wedges back of each jamb as illustrated in Fig. 17-4. On the hinge jamb, locate one block 11 in. up from the bottom and one 7 in. down from the top and a third block equidistant between these two. Continue to check the jamb with a straightedge while adjusting the wedges, then nail through the blocking into the studs. Generally it is best to use two 8d nails, and stagger them about 1/2 in. on either side of the nailing line.

HANDY SAYS:

"When setting a door frame, do not drive any of the nails "home" until all blocking has been adjusted and the jambs are straight and plumb."

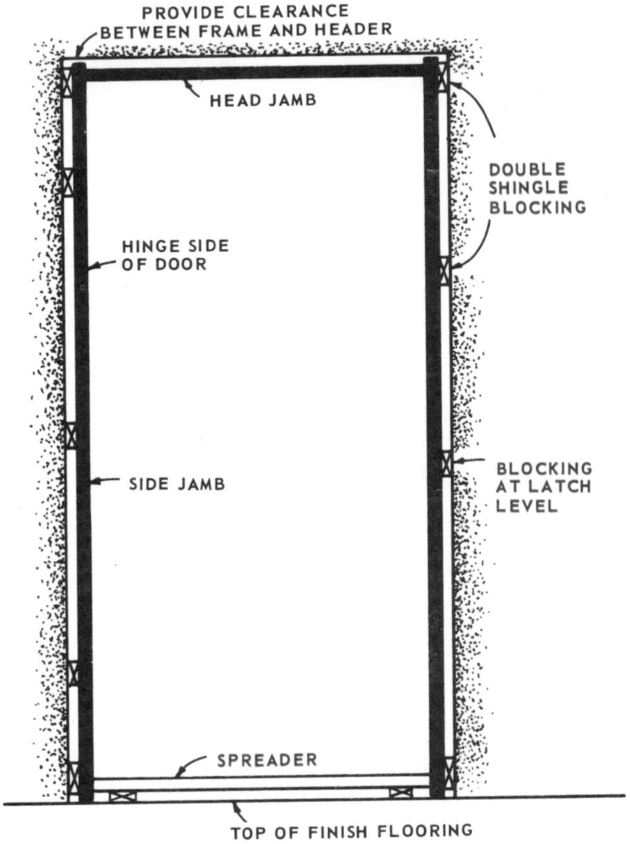

Fig. 17-4. Door frame in position with blocking and spreader.

Fig. 17-5. Checking side jamb to make sure it is plumb and straight.

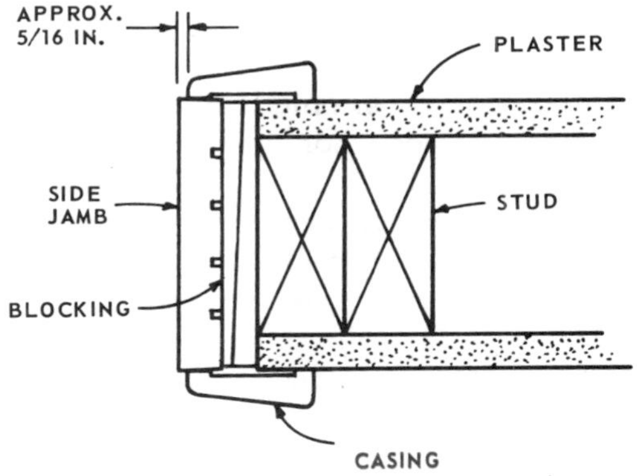

Fig. 17-6. Section through door jamb showing position of casing.

Door Casing

Door casing is applied to each side of the door frame to cover the space between the jambs and the wall surface. This secures the frame to the wall structure and makes the jambs rigid so they will carry the door. Fig. 17-6 shows a section view through a door jamb and illustrates how the casing covers the blocking and is attached to the jamb and wall surface.

To apply the casing, first select the necessary pieces and place them near the opening. Some carpenters prefer to gauge a light pencil line on the edge of the jamb 1/4 in. back from the face. Check the bottom end of the side casing to see that it is square and will rest tightly on the finished floor.

With the side pieces held in place, mark the position of the miter joint at the top end. Use a miter box and/or a wood trimmer, Fig. 17-7, to make an accurate cut. Nail the side casings temporarily in place with casing or finish nails, and mark, cut, and fit the head casing. If the miters do not fit properly, trim with a block plane.

Fig. 17-7. A wood trimmer makes smooth, accurate surfaces for miter joints. Millwork plants can usually furnish casing with miter joints already cut.

Finally, drive nails home and complete the nailing pattern. Use 4d or 6d nails along the jamb edge and drive 8d nails through the other edge into the studs. Each pair should be spaced about 16 in. O.C. Fig. 17-8 shows the completed casing at the miter joint after the nails have been set. When using hardwood casing, it is advisable to drill nail holes.

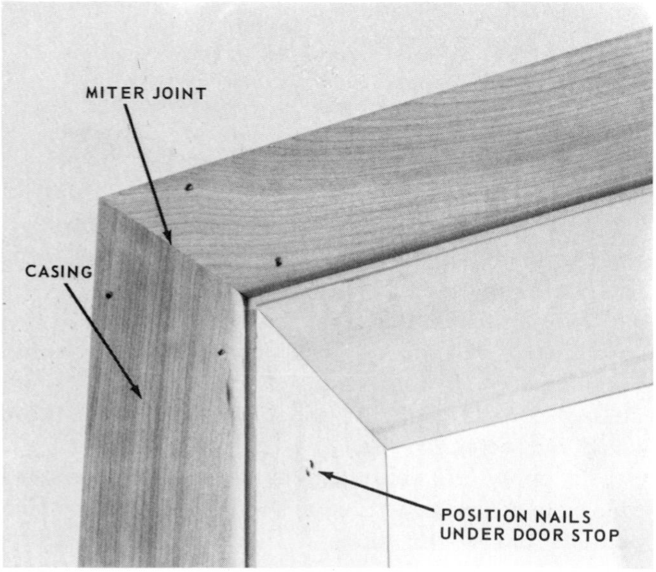

Fig. 17-8. Application of casing at the upper left-hand corner of a doorway. Note tight fit of miter joint.

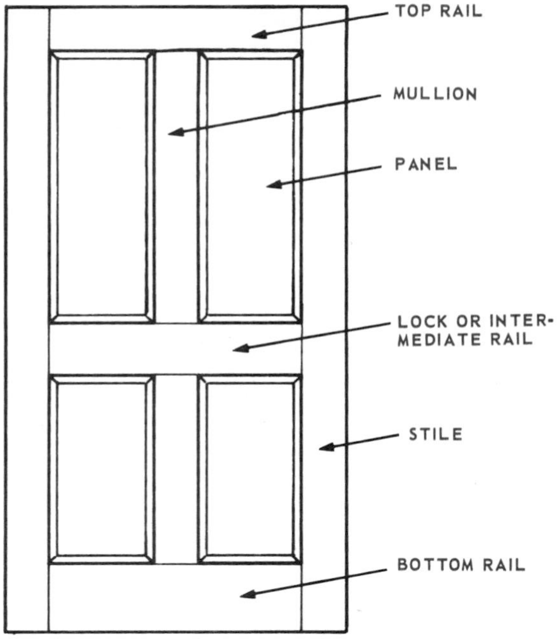

Fig. 17-9. Parts of panel door.

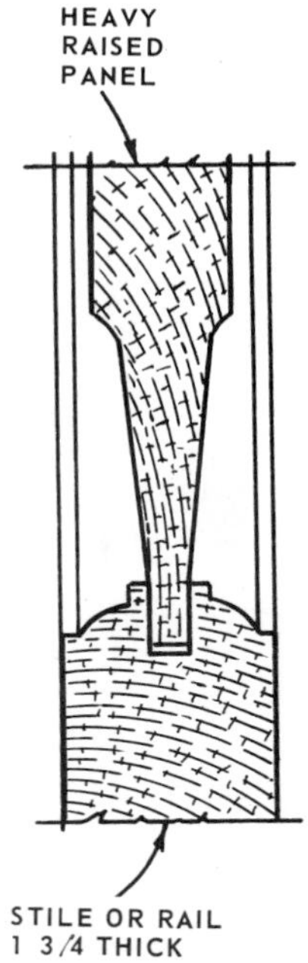

Fig. 17-10. Left. Photo of panel door used for an entrance. Note added lines and surfaces created by the raised panels. Right. Section through stile shows shape of raised panel. (Ideal Co.)

Panel Doors

There are two general types of doors; panel and flush. The panel door is also referred to as a stile-and-rail door and describes the type of construction used in sash, louver, storm, screen, and combination doors. Sash doors are similar to panel doors in appearance and construction but have one or more glass lights which replace the wood panels.

A panel door, Fig. 17-9, consists of stiles and rails with panels of plywood, hardboard or solid stock. The rails and stiles are usually made of solid material, however some consist of veneer over a lumber core. A variety of designs are formed by varying the number, size, and shape of the panels.

Special effects are secured by installing raised panels which add line and texture, Fig. 17-10. This kind of a panel is formed of thick material which is reduced around the edges where it fits into the grooves in the stiles and rails.

Flush Doors

A flush door consists of a wood frame with relatively thin sheets of material applied to both faces. Improvements in the manufacture of plywood and adhesives have made it possible to produce a flush door that is strong and durable. Today, flush doors account for a high percentage of the wood doors used in all areas of building construction.

Face panels (also called skins) are commonly made of 1/8 in. plywood, however, hardboard, plastic laminates and metal are also used. Fig. 17-11 shows a flush door constructed with metal face panels.

Doors of this type are made with either solid cores of wood or various composition materials; or with hollow core construction, Fig. 17-12. The basic frame is usually made of softwood that

Fig. 17-11. Flush type entrance door constructed of wood stiles and rails and 24 gauge steel face panels. Core consists of foamed-in-place polyurethane (K factor .13). Decorative moldings are factory applied. (Stanley Door Systems)

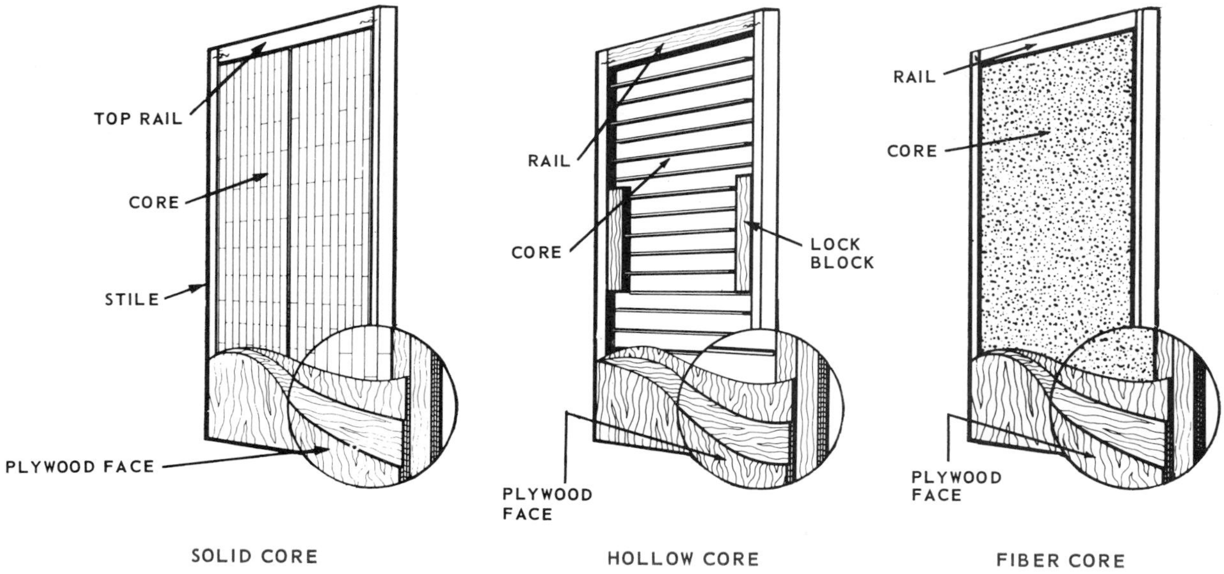

Fig. 17-12. Parts and basic types of cores used in modern flush door construction. (Mohawk Flush Doors, Inc.)

Fig. 17-13. Forming solid core for flush door with wood blocks bonded together in an electronic clamping and curing machine. (Andersen Corp.)

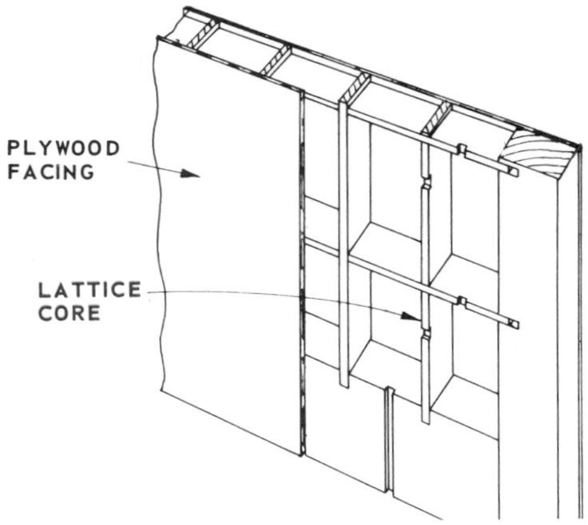

Fig. 17-14. Lattice core construction. Note notches in strips and lower rail that permit interior cavities to adjust readily to changes in atmospheric pressure. (Forest Products Lab.)

Fig. 17-15. Exterior door designs (selected group) produced by a manufacturer of door systems.

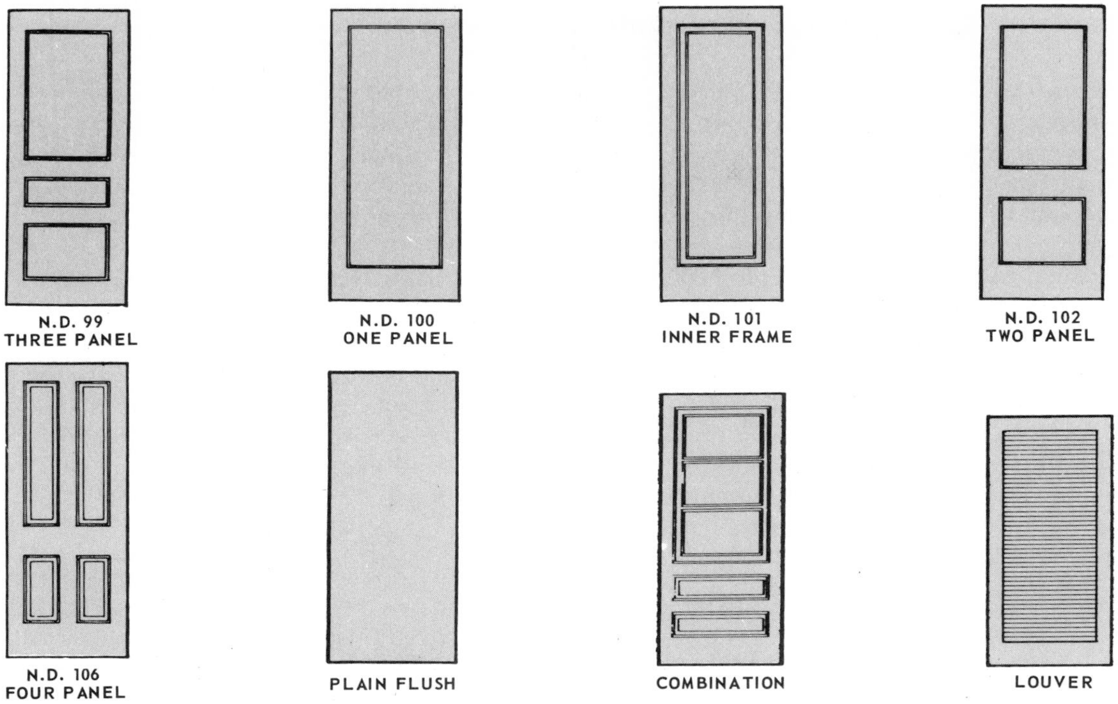

CONSTRUCTION DETAILS

Design No.	Stiles	Top Rail	Cross Rail	Lock Rail	Intermediate Rails	Mullions or Muntins	Bottom Rail	Panels
N.D. 99	4¾"	4¾"	4⅝"	4⅝"	......	......	9⅝"	Flat
N.D. 100	4¾"	4¾"	......	......	......	......	9⅝"	Flat
N.D. 101	4¼" Face	4¼" Face	......	......	......	......	9¼" or 9½" Face	Flat
N.D. 102	4¾"	4¾"	......	8"	......	......	9⅝"	Flat
N.D. 106	4¾"	4¾"	......	8"	......	4⅝"	9⅝"	Raised
N.D. 107	4¾"	4¾"	......	......	4⅝"	......	9⅝"	Raised
N.D. 108 (1)(2)(3)	4¾"	4¾"	......	8"	3⅞" or 4⅝"	3⅞" or 4⅝"	9⅝"	Raised
N.D. 111 (1)(2)(3)	4¾"	4¾"	......	8"	3⅞" or 4⅝"	3⅞" or 4⅝"	9⅝"	Raised

NOTES: (1) Height of Top Panels Overall 7⅛". (2) Doors 1'-6" are made 1 Panel Wide.
(3) Bottom and lock rails may be reversed when specified.
*Also for exterior use.

POPULAR SIZES

1'-6" x 6'-6"	2'-0" x 6'-8"	2'-4" x 6'-8"	2'-6" x 6'-8"	2'-8" x 6'-8"
1'-6" x 6'-8"	2'-0" x 7'-0"	2'-4" x 7'-0"	2'-6" x 7'-0"	2'-8" x 7'-0"
2'-0" x 6'-0"	2'-4" x 6'-0"	2'-6" x 6'-0"	2'-8" x 6'-0"	3'-0" x 6'-8"
2'-0" x 6'-6"	2'-4" x 6-6"	2'-6" x 6'-6"	2'-8" x 6'-6"	3'-0" x 7'-0"

(National Woodwork Manufacturers Association)

Fig. 17-16. Sizes and patterns commonly used for interior doors. Table lists construction details of various parts.

matches the color of the face veneers. The most common type of solid (also called slab) core construction consists of wood blocks bonded together with the end joints staggered. See Fig. 17-13.

Flush doors with hollow cores are widely used for interior doors, and may also be used for exterior doors if the construction is bonded with waterproof adhesives. Flush doors do not ordinarily provide as much thermal (heat) and sound insulation as solid core doors, and usually have a lower fire-resistance rating. The basic hollow core types include lattice (also called mesh or grid), ladder, and implanted blanks.

A lattice type core construction, Fig. 17-14, has strips spaced approximately 4 in. O.C. An implanted core consists of rectangular blanks, tubular columns, or other shapes of wood or composition material placed throughout the cavity to support the faces of the door.

Sizes and Grades

The standard thickness for exterior doors is 1 3/4 in.; interior passage doors are 1 3/8 in. Widths of 2 ft.-8 in. and 3 ft.-0 in. are most commonly used for exterior doors, Fig. 17-15.

Residential doors, both interior and exterior, have a standard height of 6 ft. - 8 in. but 7 ft. doors are sometimes used for entrances or special interior installations. A 7 ft. height is usually considered standard for commercial buildings.

Interior door widths vary with the installation. FHA specifies a minimum size of 2 ft. - 6 in. for bedrooms and a 2 ft. width for bathrooms. Closet doors may also be 2 ft., however, single doors are seldom used in modern construction since various sliding and folding doors have been perfected for this type of installation. Interior door sizes and patterns are illustrated in Fig. 17-16.

Grades and manufacturing requirements for doors are listed in Industry Standards developed by the National Woodwork Manufacturing Association (NWMA) and the Fir and Hemlock Door Association (FHDA). The purpose of these standards is to establish nationally recognized dimensions, designs, and quality specifications for materials and workmanship.

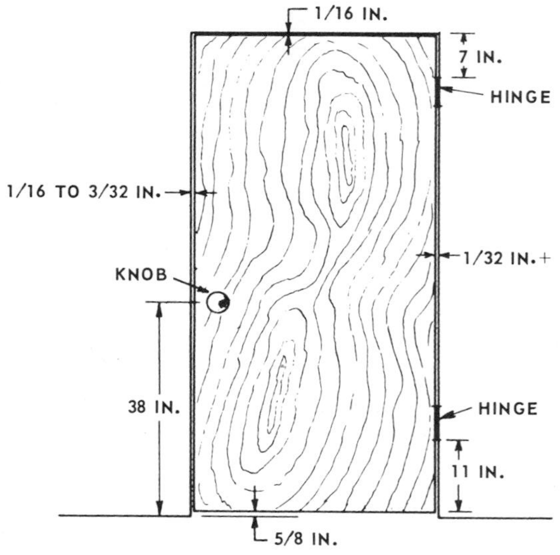

Fig. 17-17. Recommended clearance around an interior door. Some carpenters use a quarter (25 cent piece) to check clearance at top and on lock side. In quality construction, a third hinge is located mid-way between the two shown.

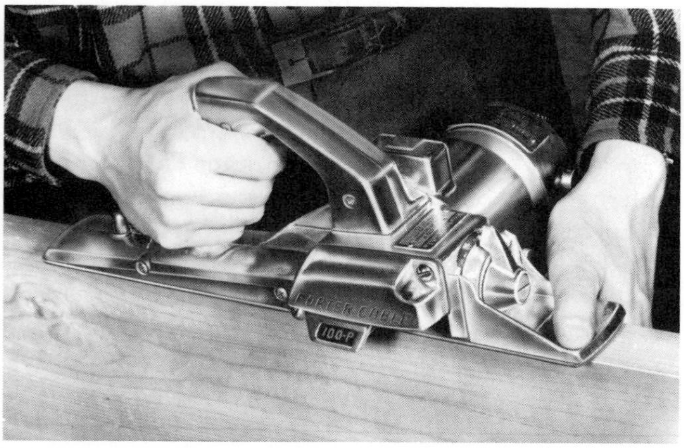

Fig. 17-18. Trimming edge of door with power plane.

Door Installation

Check the architectural drawings and door schedule to determine the correct type of door for the opening and the direction in which it will swing. Mark the door jamb that will receive the hinges and the edge on which they will be mounted.

HANDY SAYS:

"Handle doors carefully. Be careful not to soil unfinished doors. If doors must be stored for more than a few days, stack them horizontally on a clean flat surface and keep them covered.

Doors should be conditioned for several days so they will reach the average prevailing humidity before hanging or finishing."

Doors may be trimmed for fitting but should not be cut to fit smaller openings. If a large amount of the perimeter is removed, the structural balance of the door may be disturbed and warping may result. Cutouts for glass inserts in flush doors should never be more than 40 percent of the face area and the opening should not be closer than 5 in. to the edge.

First trim the door to fit the opening. Today, most doors are carefully sized at the millwork plant, leaving only a slight amount of on-the-job fitting and adjustment. The amount of planing necessary can be laid out with a rule.

Clearances should be 1/16 to 3/32 on the lock side and 1/32 on the hinge side, Fig. 17-17. A clearance of 1/16 at the top and 5/8 at the bottom is generally satisfactory. If the door is to swing across heavy carpeting, the bottom clearance should be increased. Thresholds are used under exterior doors. Where weatherstripping is used around exterior door openings to reduce infiltration, additional clearance is needed to make the installation -- about 1/8 in. on each side and on the top edge. The threshold for exterior doors may be installed before or after the door is hung.

Door trimming can be done with a hand plane but the modern carpenter generally uses a powered plane. See Fig. 17-18. While planing, the door may be held securely on edge either by clamping to saw horses or by using a special door holder, Fig. 17-19.

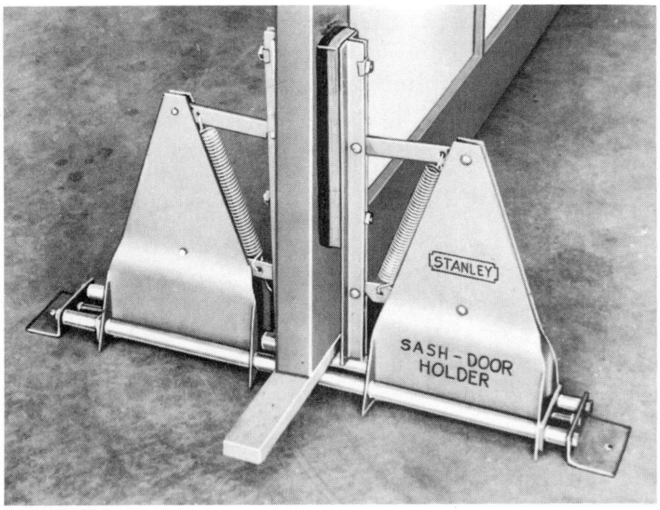

Fig. 17-19. Door holder with clamps which are cushioned to protect surface of the door.

After the door is brought to the correct size, a bevel must be planed on the lock side to provide clearance for the edge when it swings open. This bevel should be about 1/8 in 2 in. or approximately 3 1/2 deg. Narrow doors require a greater bevel than wide doors since the arc of swing is smaller. The type of hinge and the position of the pins should be considered in determining the exact bevel required.

After the bevel is cut and the fit of the door is checked, use a block plane to soften (round) corners on all edges of the door and then smooth with sandpaper.

Millwork plants can furnish prefitted doors that are machined to the size specified with the lock edge beveled and corners slightly rounded. Doors can also be furnished with gains (inset) cut for hinges, and/or holes bored for lock installation.

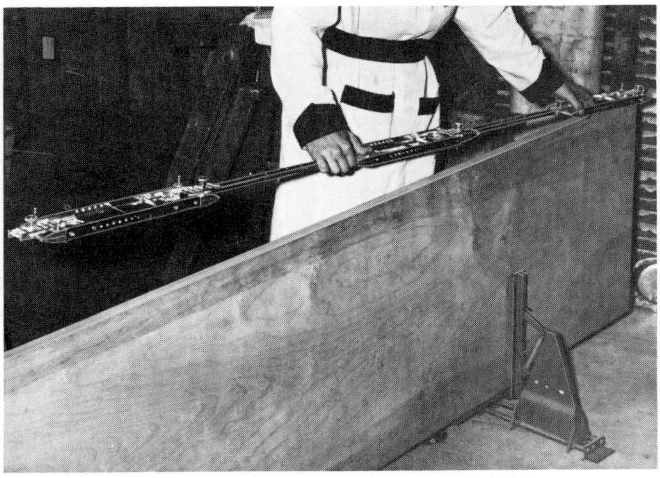

Fig. 17-20. Setting routing template in position on hinge edge of door. (Stanley Tools)

Installing Hinges

In modern construction gains for hinges are usually cut with an electric router. A door-and-jamb template is generally used to save time and insure accuracy.

The template is positioned on the door, Fig. 17-20. After the gains are routed, it is attached to the door jamb and matching cuts are made. Adjustments can be made for various door thicknesses and heights, also for different size butts (hinges). See Fig. 17-21. Most templates are designed so it is impractical to mount them on the wrong side of the door or jamb.

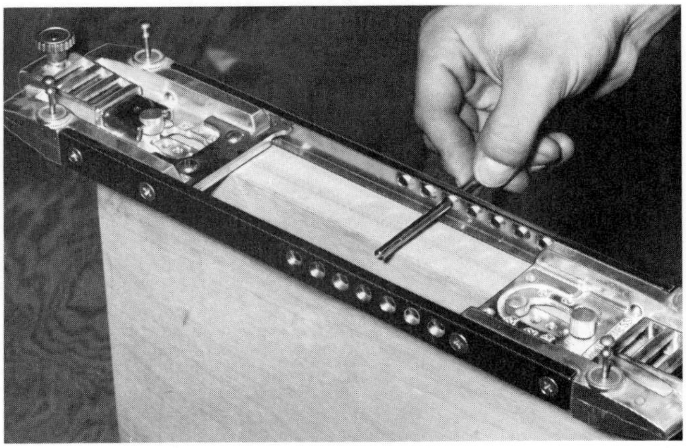

Fig. 17-21. Adjusting template.

Fig 17-22 shows a router in place on the template and the cut being made. For this type of equipment hinges with rounded corners may be used thus saving time required to square the corner with a wood chisel.

Place the hinge in the gain so the loose pin will be up when the door is in operating position. Drive the first screw in slightly toward the back edge to draw the leaf of the hinge tightly into the gain. To speed up the work, use an electric drill with a special screw-driving attachment, Fig. 17-23. Follow the same procedure and attach the free leaf of the hinge to the jamb. Some carpenters prefer to set only one or two screws in each hinge leaf, and check the fit before installing the remaining screws.

After all hinges are installed, hang the door in place and check the clearance on all edges. Required corrections should be made by planing of the door edges or adjusting the depth of the hinge gains.

Minor adjustments may be made by applying cardboard shims back of the hinge leaf, Fig. 17-24. As the illustration shows, the center of the hinge

Fig. 17-22. Above. Using router to cut a gain with round corners for a hinge. Below. Checking completed cut. Note the rounded corners on the hinge leaf.

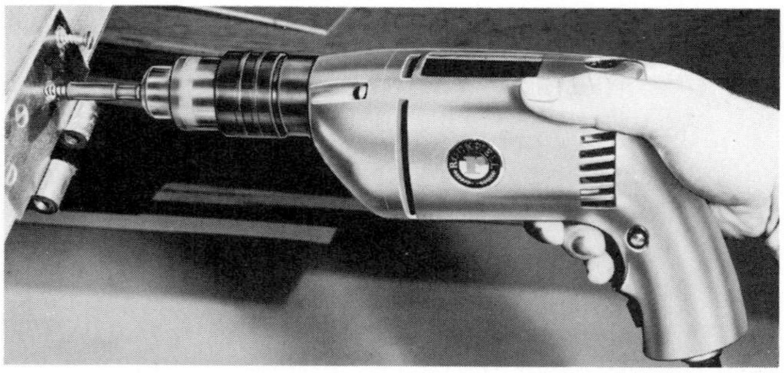

Fig. 17-23. Using electric drill fitted with special chuck for driving screws.

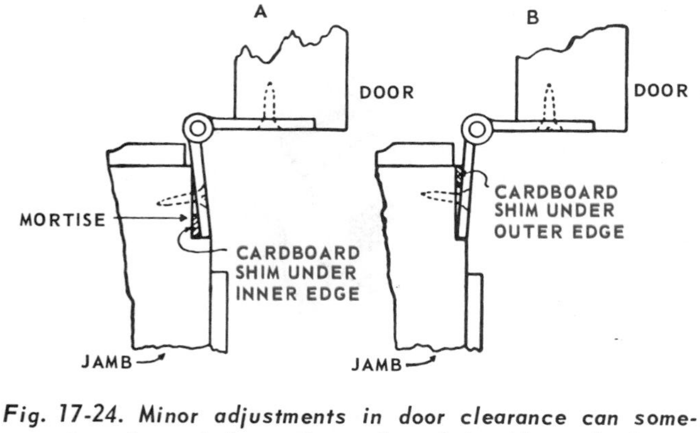

Fig. 17-24. Minor adjustments in door clearance can sometimes be made by applying shims behind hinge leaf.

can be shifted so more clearance is provided along the lock jamb (View A) or the space can be closed with the shim applied on the outside edge (View B). When adjusting a door that has been in service it may be necessary to use longer screws.

Door Stops

The door stops are usually the final trim members to be installed. Many carpenters cut and tack the stops in place before installing the lock, then secure them permanently after the lock installation has been completed.

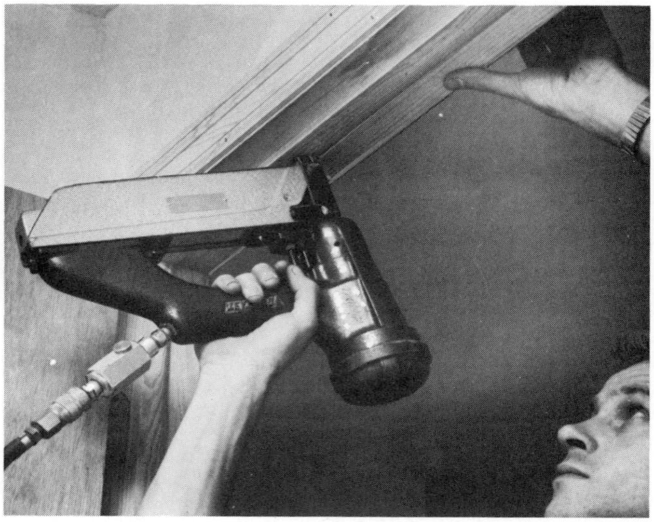

Fig. 17-25. Applying door stop to head jamb with a pneumatic powered nailer. (Fastener Corp.)

With the door in a closed position, set the stop on the hinge jamb with a clearance of 1/16 in. The stop on the lock side is set tight against the door. Set the stop on the head jamb so it aligns with the stops on the side jambs. Use miter joints and attach with 4d nails spaced 16 in. O.C. Using power operated nailers, Fig. 17-25, will save time. Fig. 17-26 shows a completed installation.

Door Locks

Four types of passage door locks are illustrated in Fig. 17-27. Cylindrical and tubular locks are used extensively in modern residential work because they can be installed easily and quickly. Unit locks are installed in an open cutout in the edge of the door and need not be disassembled when installed. Such locks are commonly used on entrance doors for apartments and some commercial buildings where locks must be changed from time to time.

Fig. 17-26. Door stop on hinge jamb is set with about 1/16 in. clearance so door will not rub when opened.

Cylindrical locks permit a sturdy, heavy-duty mechanism that provides security for exterior doors, Fig. 17-28. They require boring a large hole in the door face, a smaller hole in the edge and a shallow mortise for the front plate. The tubular lock is similar but requires a smaller hole in the door face.

When ordering door locks it is often necessary to describe the way in which the door swings.

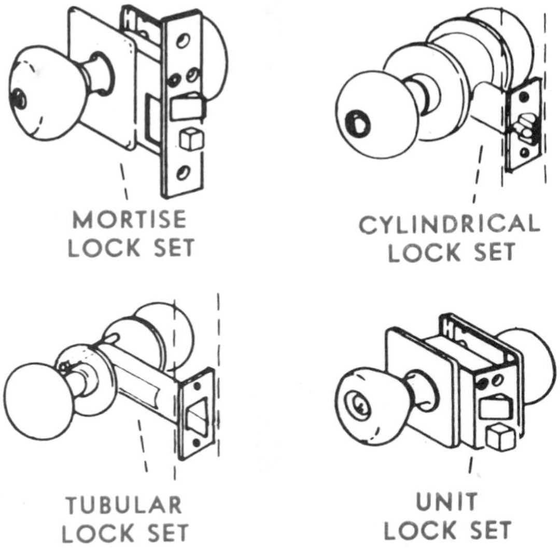

Fig. 17-27. Some basic types of door locks.

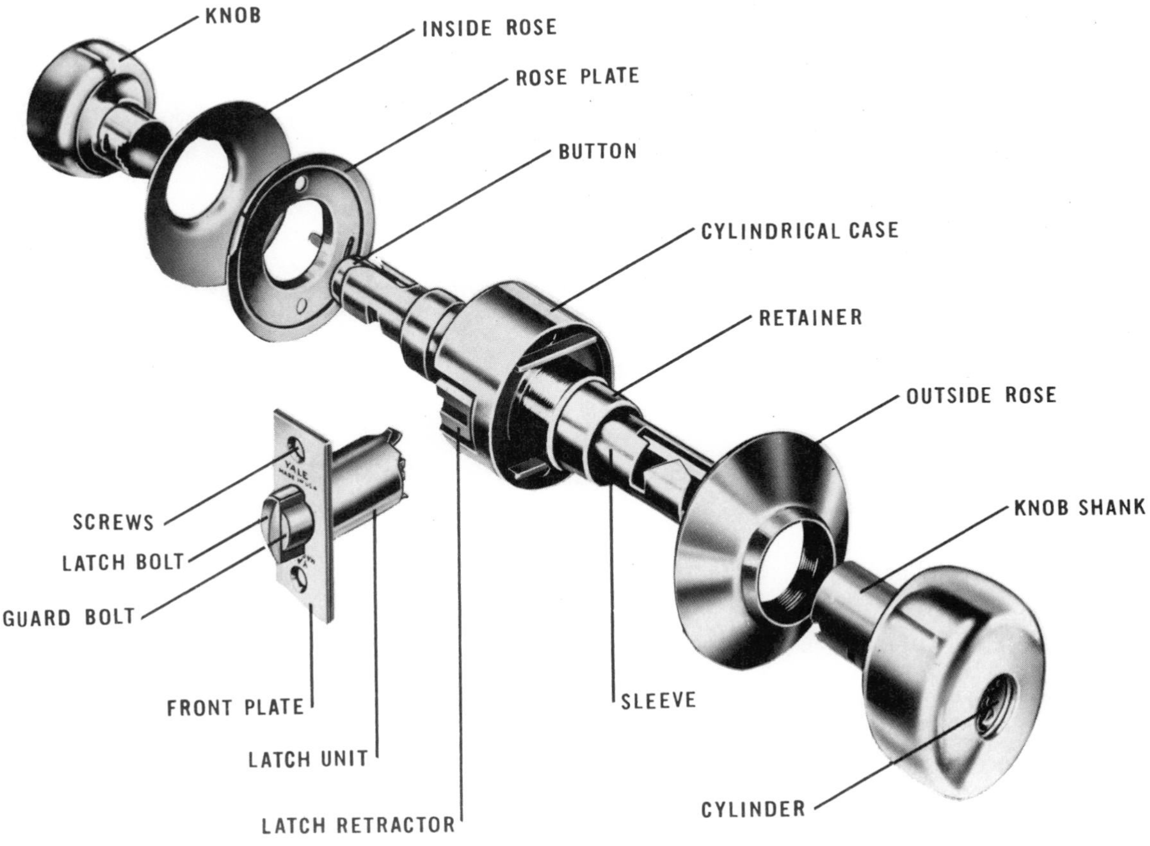

Fig. 17-28. Parts of a cylindrical lock set. (Eaton Yale & Towne Inc.)

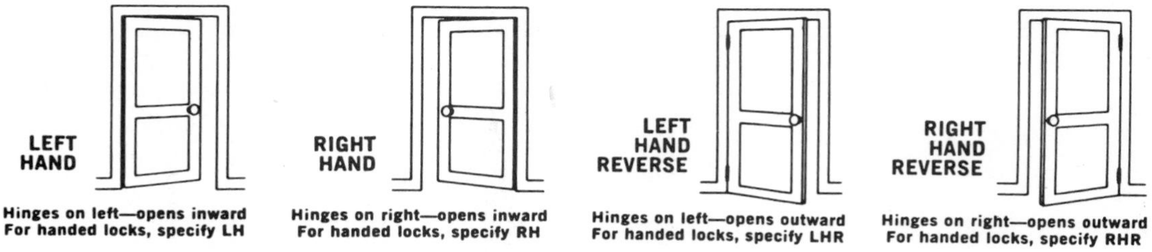

Fig. 17-29. How to determine the Hand of a Door. Use abbreviations shown.

1 MARK DOOR

Mark height line across edge of door. 38" is the usual height above the floor.

Fold template over edge of door, centering on height line.

Mark centers of $\frac{7}{8}$" and $2\frac{1}{8}$" holes.

2 INSTALL LATCH UNIT

Bore $2\frac{1}{8}$" hole through door, and $\frac{7}{8}$" hole into edge of door at points marked on template.

Cut out for latch front and install latch unit.

3 INSTALL STRIKE

Mark height line for strike on jamb. Mark vertical centerline on jamb. This centerline MUST BE same distance from stop molding as latch case centerline is from edge of door which will hit stop molding.

Cut mortise in jamb for strike and box.

Insert box and strike, and tighten screws securely.

4 ADJUST LOCK

To adjust this lock for a $1\frac{3}{8}$" door, unscrew outside rose plate $\frac{5}{16}$" from case cutout. To adjust for a $1\frac{3}{4}$" door, unscrew outside rose plate to provide $\frac{1}{2}$" between rose plate and case cutout. To adjust for any thickness between $1\frac{3}{8}$" and $1\frac{3}{4}$", set the rose plate at a suitable intermediate position.

FOR $1\frac{3}{8}$" DOOR

CASE CUTOUT

OUTSIDE ROSE PLATE

FOR $1\frac{3}{4}$" DOOR

CASE

CASE CUTOUT

5 INSTALL LOCK

With latch case in place, insert lock assembly into $2\frac{1}{8}$" hole, making sure that lock case hooks retainer legs and retractor hooks bolt tails. **Do not force.**

RETRACTOR

RETAINER LEG

BOLT TAIL

LOCK CASE

6 INSTALL ROSE PLATE

Slip on rose plate and locate screw holes on vertical centerline with "Top" up. Insert machine screws and tighten alternately to obtain secure attachment.

SEE INSTRUCTIONS ABOVE

VERTICAL ₵

7 INSTALL INSIDE ROSE

Place inside rose over rose plate with notch in rose over spring retainer and snap rose down so rose is flush with door.

SPRING

NOTCH

8 INSTALL INSIDE KNOB

Align lug on knob with narrow slot on side of spindle and push knob all the way in until retainer clicks into slot on knob.

9 IMPORTANT

This lock is set for a right hand door. When properly installed, the name YALE must be above the key slot. If not, the hand must be changed.

RIGHT WAY WRONG WAY

TO CHANGE LOCK HAND

knob must be in unlocked position. Turn outside knob counterclockwise approximately 45° and insert small nail in hole of trim cap. Depress retainer and slide knob off. Turn knob 180° (till name YALE is up), and replace knob.

YALE & TOWNE

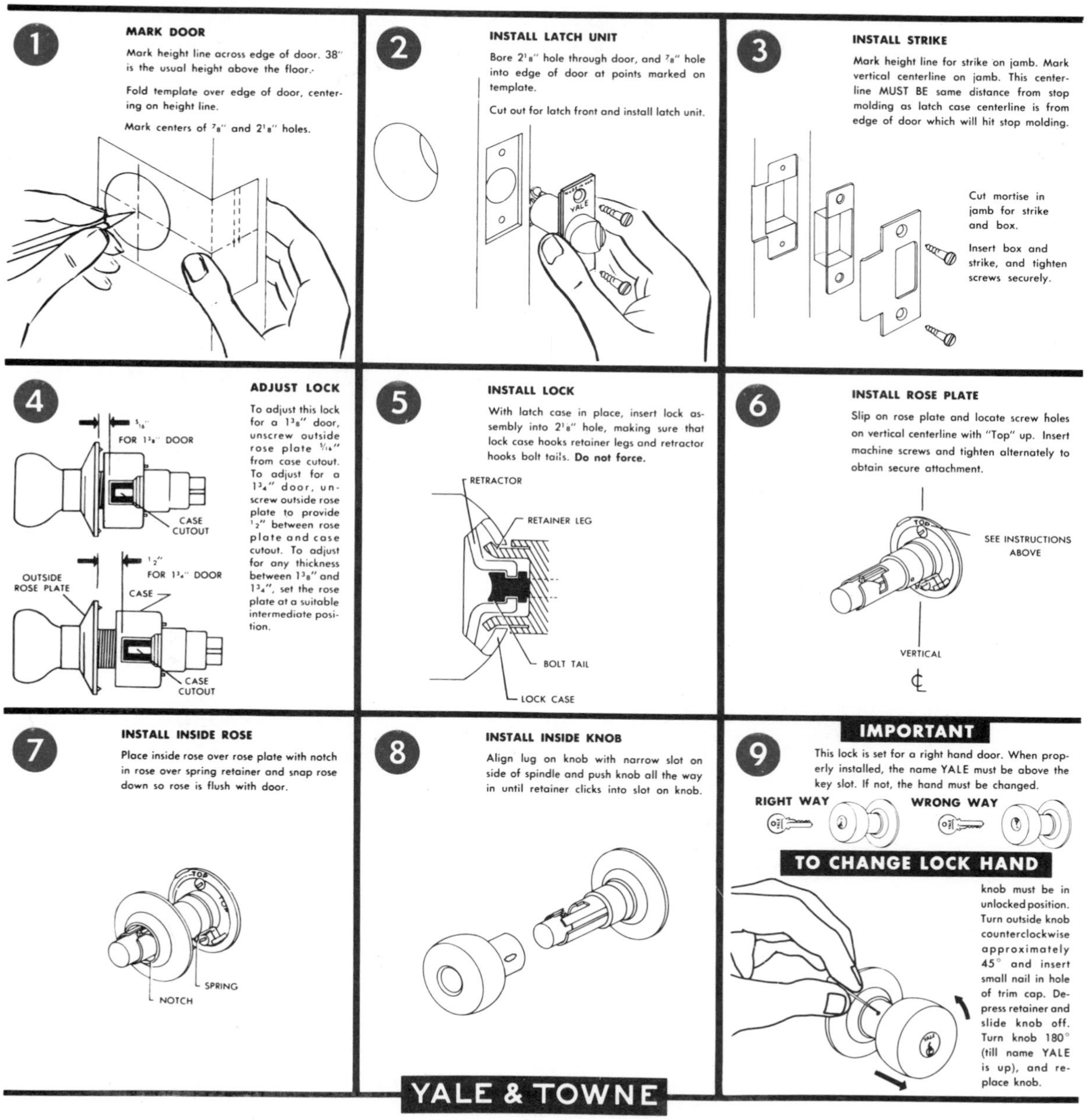

Fig. 17-30. Instructions furnished by hardware manufacturer for the installation of cylindrical lock set.

This is referred to as the "Hand of the Door" and is determined by facing the outside of the door. The outside is the street side of an entrance door and the corridor side of an interior door. Fig. 17-29 illustrates standardized procedure in determining this specification.

Lock Installation

Open the package that contains the lock set and check the content. Instructions furnished by the manufacturer should be carefully followed.

Open the door to a convenient position and block it in place with wedges placed underneath. Measure up from the floor a distance of 38 in. (36 in. is sometimes used) and mark a light horizontal line which will be on a level with the center of the lock. Position the template (furnished with the lock set) on the face and edge of the door and lay out the centers of the holes, Fig. 17-30. Continue to follow the instructions included with the lock set. Procedures for all cylindrical and tubular lock sets will be similar however, the installation details will vary slightly.

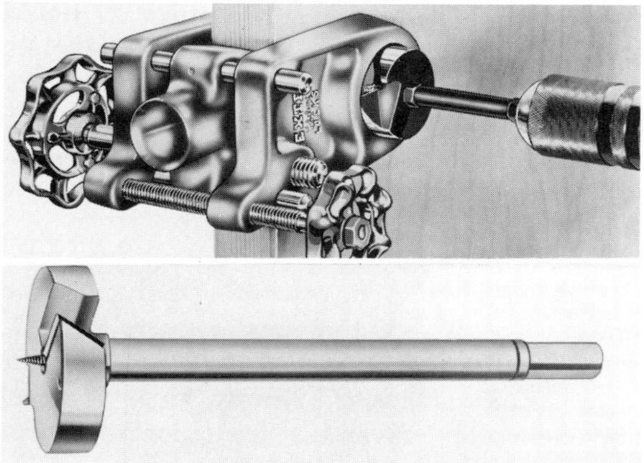

Fig. 17-31. Above. Boring jig for door hardware saves time and insures accuracy. Below. Boring bit. (Dexter Industries Inc.)

Boring jigs, Fig. 17-31, assure accurate work. When using a boring jig, a template layout is not required since the jig is designed to make holes in the correct positions. Either hand operated or power driven bits can be used to bore the holes.

The shallow mortise on the edge of the door can be laid out and cut with standard wood chisels. Speed and accuracy can be gained however, through the use of a faceplate mortise marker (also called a marking chisel), Fig. 17-32. After the perimeter is cut with this device the wood inside can be quickly removed with a standard wood chisel of appropriate width.

Fig. 17-32. Using faceplate mortise marker.

Thresholds and Door Bottoms

Exterior doors require a trim unit called a threshold to seal the space between the bottom of the door and the door sill. For many years oak or other hardwoods have been machined to a special shape for this purpose. Modern wooden thresholds are usually equipped with rubber or vinyl sealing strips.

Today, a wide range of threshold designs are made from aluminum extrusions. These are available in a clear anodized finish or gold color. Special vinyl strips are inserted into the threshold, Fig. 17-33, or to matching units attached to the door. Thresholds of this type are effective in providing an under-the-door seal. Manufacturers furnish detailed instructions for making the installation.

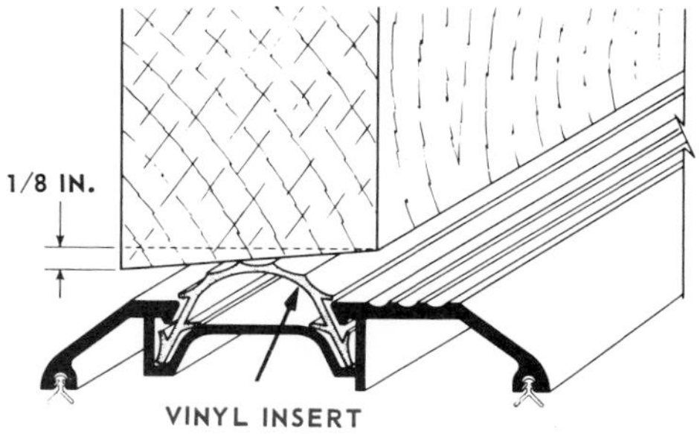

Fig. 17-33. Aluminum threshold with vinyl sealing strip. Note suggested bevel on bottom edge of door. (Pemko Mfg. Co.)

Interior doors do not require a threshold but may be equipped with various sealing strips such as the automatic door bottom shown in Fig. 17-34. This prevents air movement, and reduces sound transmission. The unit has a movable strip which drops to the floor when the door is closed and lifts when the door is opened.

Prehung Door Unit

A prehung door unit consists of a door frame and a door already installed. The door frame includes both sides of casing -- fitted and attached. Lock hardware may or may not be installed, however machining for its installation has usually been completed. Quite often the door is prefinished. The entire unit is carefully packaged and shipped to the job.

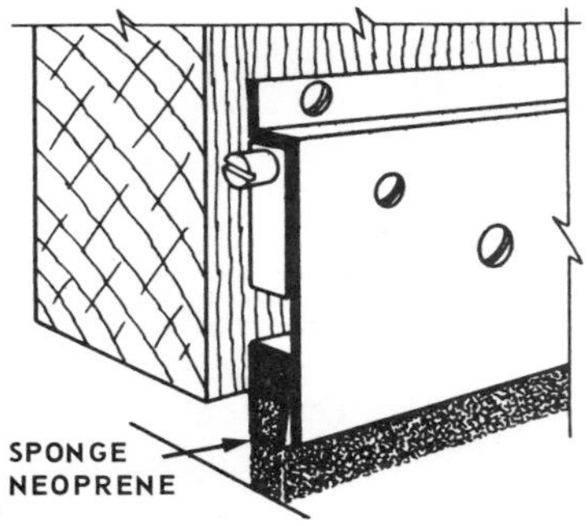

Fig. 17-34. Automatic door bottom. Sealing strip moves upward when door is opened.

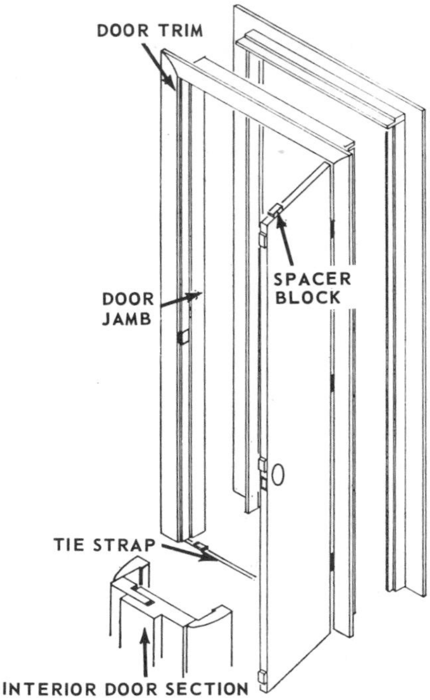

Fig. 17-35. Prehung door unit with split jamb which is adjustable to wall thickness.

Several frame designs for prehung doors are available. One with a split jamb is shown in Fig. 17-35. This is adjustable for varying wall thicknesses and can be installed in a short time. Fig. 17-36 shows a general outline of the procedure followed in making an installation of a prehung door unit with a split jamb.

Sliding Doors (Pocket-Type)

This type of door offers a space-saving feature since it is opened by simply sliding it into an opening in the partition. The door frame consists of a split side-jamb attached to a framework built into the wall. The rough opening in the structural frame must be large enough to include the finished door opening and the pocket, Fig. 17-37. The pocket framework and track is installed during the rough framing stages. Pocket-frame units are available from millwork plants in a number of standard sizes.

Several manufacturers that specialize in builders hardware have developed steel pocket-door frames, Fig. 17-38, that are easy to install and provides a firm base for wall surface materials.

Fig. 17-39 shows a cutaway view of a typical track and roller assembly for a pocket door. The hanger (wheel assembly) snaps into the plate

Fig. 17-36. General procedure for installing split-jamb type door unit. 1–Carefully remove crating and separate two sections. 2–Set section of frame that includes door into opening. 3–Plumb door frame and nail casing to wall structure. Be sure all spacer blocks are in place between jambs and door. 4–Move to other side of wall and install shims behind side jambs. Shims should be located at points where spacer blocks make contact with door. 5–Take remaining half of door frame and insert it into grooved section already installed. Nail casing to wall structure. 6–Nail through stop into jambs, remove spacer blocks and check operation of door. (Frank Paxton Lumber Co.)

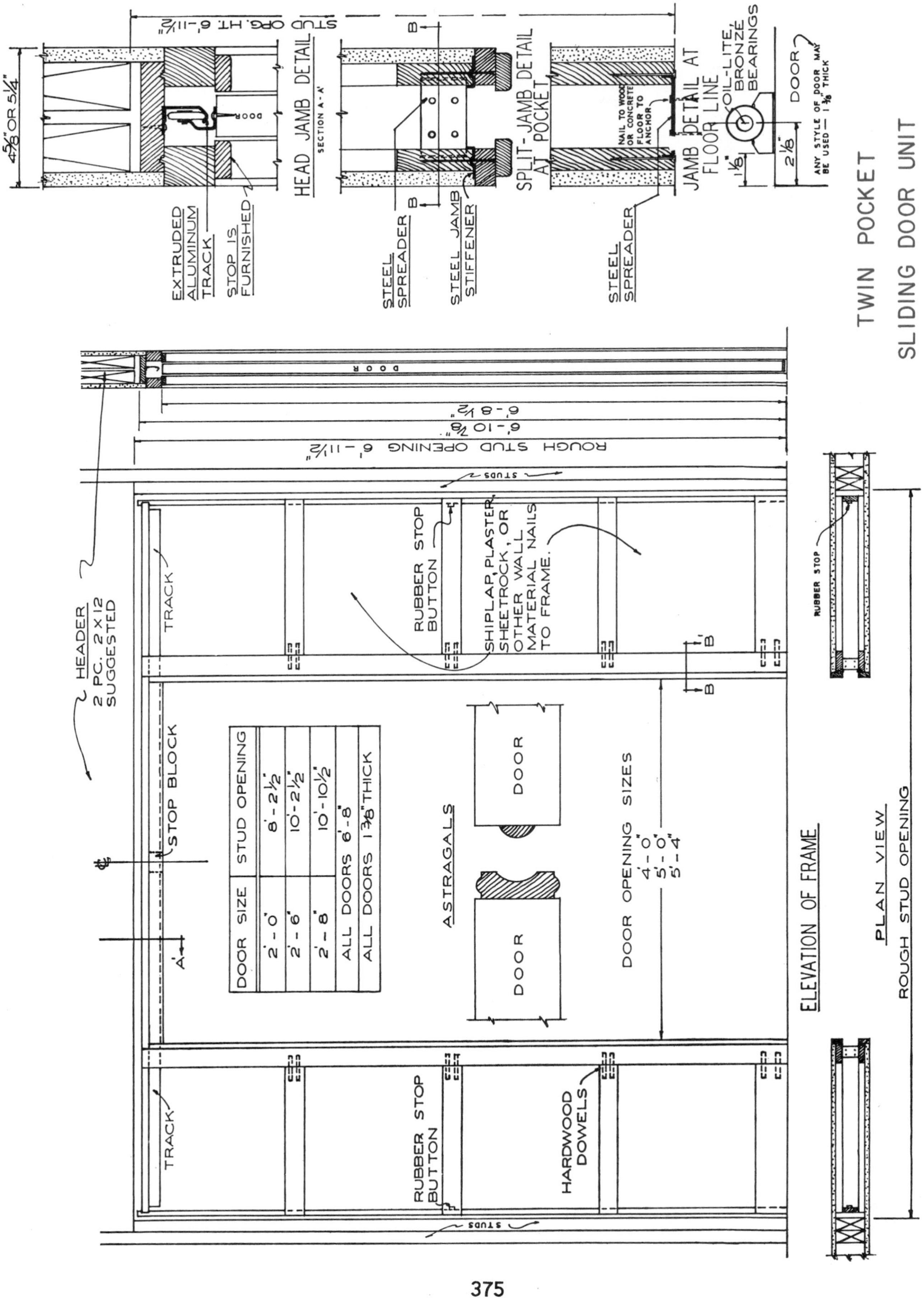
TWIN POCKET
SLIDING DOOR UNIT
HEAD JAMB DETAIL
SECTION A-A'
SPLIT-JAMB DETAIL AT POCKET
JAMB DETAIL AT FLOOR LINE
STUD OPG. HT. 6'-11½"
4⅞" OR 5¼"
EXTRUDED ALUMINUM TRACK
STOP IS FURNISHED
STEEL SPREADER
STEEL JAMB STIFFENER
STEEL SPREADER
NAIL TO WOOD OR CONCRETE FLOOR TO ANCHOR
OIL-LITE, BRONZE BEARINGS
DOOR
1⅛"
2⅛"
ANY STYLE OF DOOR MAY BE USED — 1⅜" THICK
HEADER 2 PC. 2 × 12 SUGGESTED
ROUGH STUD OPENING 6'-11½"
6'-10⅞"
6'-8½"
STUDS
TRACK
STOP BLOCK
RUBBER STOP BUTTON
SHIPLAP, PLASTER, SHEETROCK, OR OTHER WALL MATERIAL NAILS TO FRAME.
HARDWOOD DOWELS
DOOR SIZE | STUD OPENING
2'-0" | 8'-2½"
2'-6" | 10'-2½"
2'-8" | 10'-10½"
ALL DOORS 6'-8"
ALL DOORS 1⅜" THICK
ASTRAGALS
DOOR
DOOR OPENING SIZES 4'-0" 5'-0" 5'-4"
ELEVATION OF FRAME
PLAN VIEW
ROUGH STUD OPENING
RUBBER STOP
A'
B
B

Fig. 17-37. Structural details of a pocket type sliding door unit. (Ideal Company)

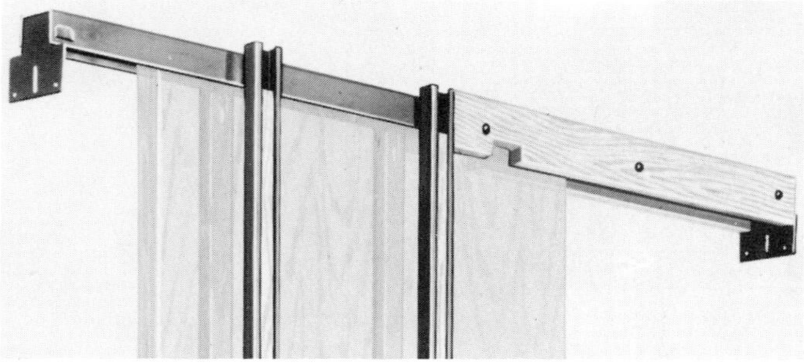

Fig. 17-38. Top section of steel framework unit for a pocket door. The package includes track, hangers, and guides. (Ekco Building Products Co.)

attached to the top of the door and can be easily adjusted up or down to plumb the door in the opening. Wheels are made of nylon which insures quiet operation.

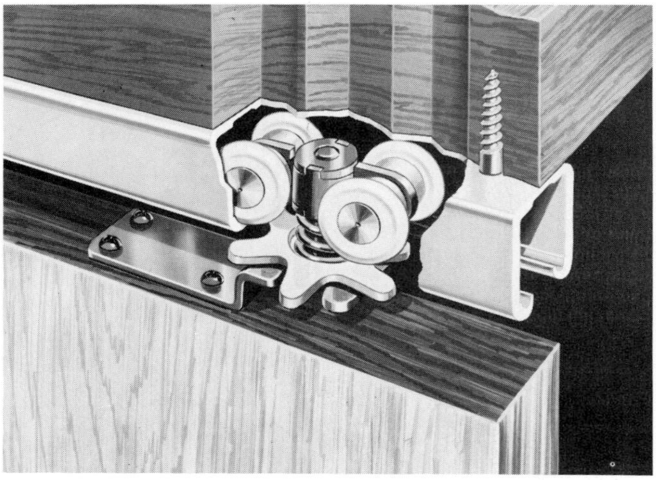

Fig. 17-39. Cutaway view of track and roller assembly for pocket door.

Sliding Doors (Bypass Type)

Standard interior door frames can be used for a bypass sliding door installation. When the track is mounted below the head jamb the height of a standard door must be reduced and a trim strip installed to conceal the hardware, Fig. 17-40. Head jamb units are available with a recessed track that permits the doors to ride flush with the underside of the jamb.

Cutaway views of standard bypass track and hangers are shown in Fig. 17-41. Each type can be used with either 3/4 or 1 3/8 in. doors. Note the hanger adjustment that raises or lowers the door for alignment after installation.

Sliding door hardware for bypass doors is packaged complete with track, hangers (rollers), floor guide, screws, and instructions for making the installation.

A disadvantage of bypass sliding doors is that access to the total opening at one time is not possible. They are, however, easy to install and are practical for wardrobes and many other installations.

Folding Doors

In modern construction, a popular folding door unit consists of pairs of doors hinged together. Folding action is guided by an overhead track. The unit, commonly called a "bi-fold" unit, may consist of a single pair of doors, or two or more pairs of doors. See Fig. 17-42. Folding door units are well suited to wardrobes, closets, and certain openings between rooms.

The opening for bi-fold units is trimmed with standard jambs and casing. Fig. 17-43 illustrates an installation of hardware. Pivot brackets and

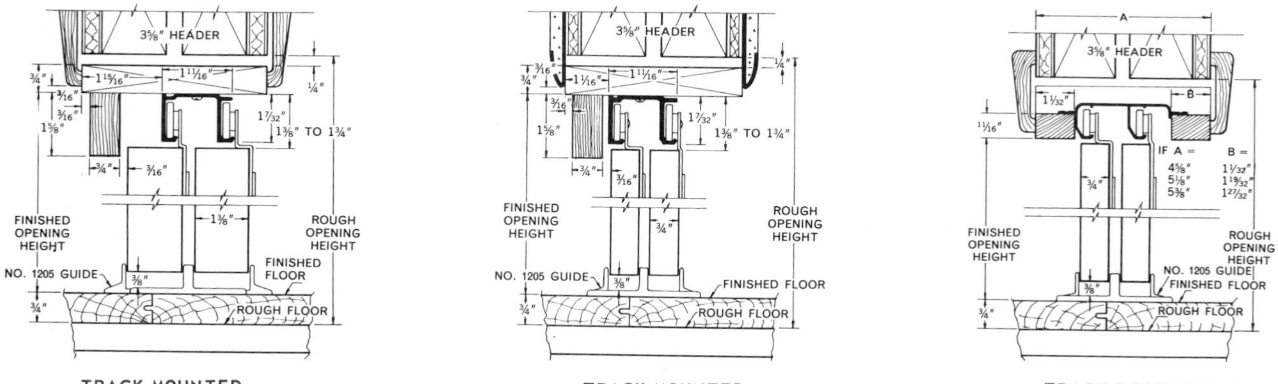

TRACK MOUNTED UNDER HEAD JAMB 1 3/8 IN. DOORS

TRACK MOUNTED UNDER HEAD JAMB 3/4 IN. DOORS

TRACK RECESSED IN HEAD JAMB 3/4 IN. DOORS

Fig. 17-40. Drawings of typical bypass sliding door installations.

Fig. 17-41. Cutaway views of bypass sliding door installations. Above. Track mounted below head jamb. Note trim member needed to conceal track. Below. Track recessed into head jamb. Hangers include vertical adjustment. (Ekco Building Products Co.)

Fig. 17-42. Folding door (bi-fold) installed in linen closet.

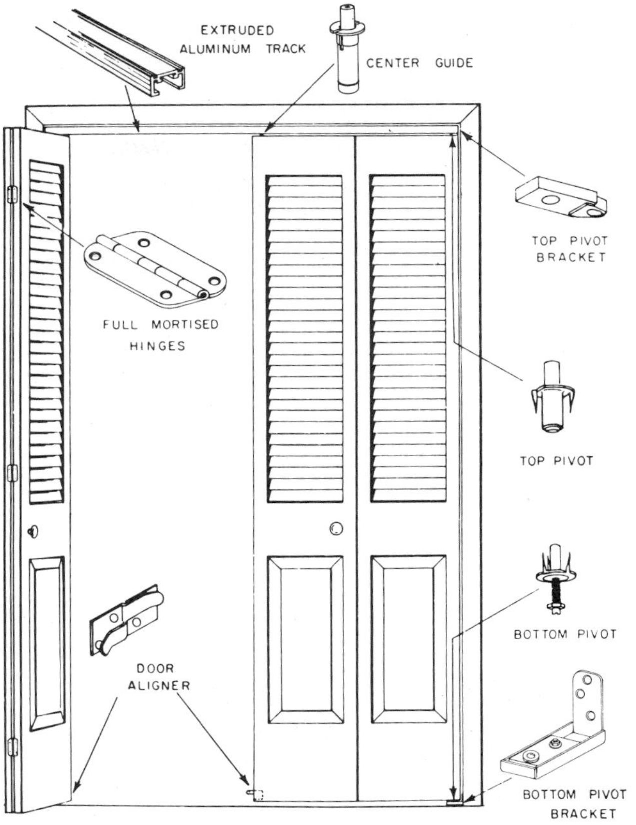

Fig. 17-43. Typical hardware used for folding door installation. The hinged units are supported by brackets. No weight is carried by center guide. (Ideal Company)

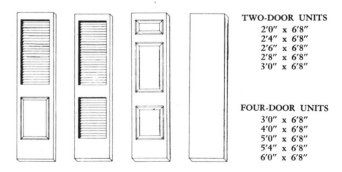

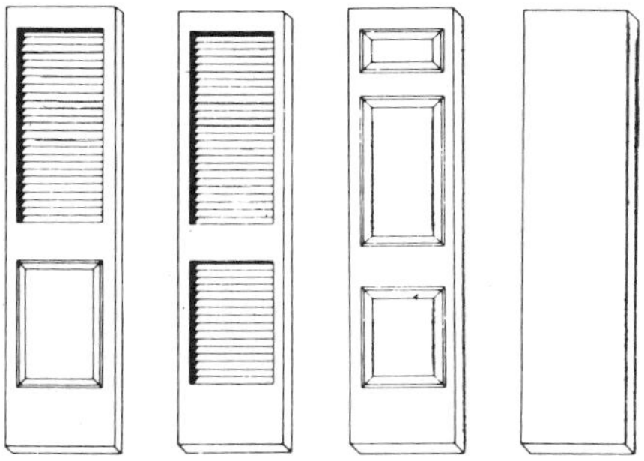

TWO-DOOR UNITS
2'0" x 6'8"
2'4" x 6'8"
2'6" x 6'8"
2'8" x 6'8"
3'0" x 6'8"

FOUR-DOOR UNITS
3'0" x 6'8"
4'0" x 6'8"
5'0" x 6'8"
5'4" x 6'8"
6'0" x 6'8"

Fig. 17-44. Typical door designs and sizes commonly available in packaged units.

center guides have self-lubricating nylon bushings that insure smooth, quiet operation. The weight of the doors is supported by the pivot brackets and hinges between the doors -- not by the overhead track and guide.

A variety of standard door designs is shown in Fig. 17-44. Two-door units generally range from 2 to 3 ft. wide while four-door units are available in total widths of from 3 to 6 ft.

When the total opening for a four-door unit is greater than 6 ft. (two-door unit over 3 ft.) it is usually necessary to install heavier hardware and use a supporting roller-hanger instead of a regular center guide. See Fig. 17-45. This type of heavy-duty hardware is also used to carry multi-folding door units that serve as room dividers, Fig. 17-46.

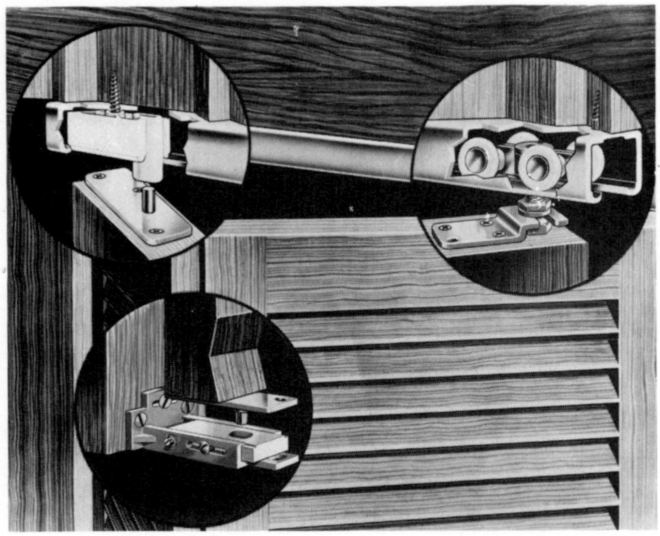

Fig. 17-45. Typical hardware used for installation where center roller guide provides support for folding unit.

Fig. 17-46. Multi-folding door units are carried by concealed overhead track. (Ekco Building Products Co.)

Folding door hardware is supplied in a package that includes hinges, pivots, guides, bumpers, aligners, nails, and screws. Instructions for making the installation are also included. Millwork plants can supply matching doors with prefitted hardware.

Some folding doors are fabricated from narrow wood panels, held together with concealed springs that serve as hinges. These are carried on an overhead track and form a "stack" when in an open position. A similar type, called an accordion-fold door, consists of fabrics and vinyl coated materials attached to a metal framework.

Window Trim

A complete side of interior window trim consists of casing, stool, apron, and stops, Fig. 17-47. Millwork companies select and package the proper length trim members to finish a given unit or combination of units.

The stool is the horizontal member that laps over the sill and extends beyond the side casing. It is cut and applied first. Hold the stool level with the sill and mark the position of the side jambs. Also mark a line on the face where it will fit against the wall surface. For a standard double-hung window this line will usually be directly above the square edge of the stool rabbet.

Carefully cut out the corners and check the fit. Be sure to allow about 1/16 in. between the front edge of the stool and the window sash. Set a piece of side casing in place and measure beyond it about 3/4 in. and make the cutoff line for the ends of the stool. Cut the ends of the stool, sand the surface and nail it into place. Some installations require that the stool be bedded in caulking compound or white lead.

Side casing is applied next. Set a piece in position on the stool, Fig. 17-48 and mark the position of the miter. Some carpenters prefer to set the casing back from the jamb about 1/8 in. following about the same procedure as previously described for door casing. Cut the miter. Drill nail holes if the casing is made of hardwood. After the side casing is nailed in place, cut and apply the head casing.

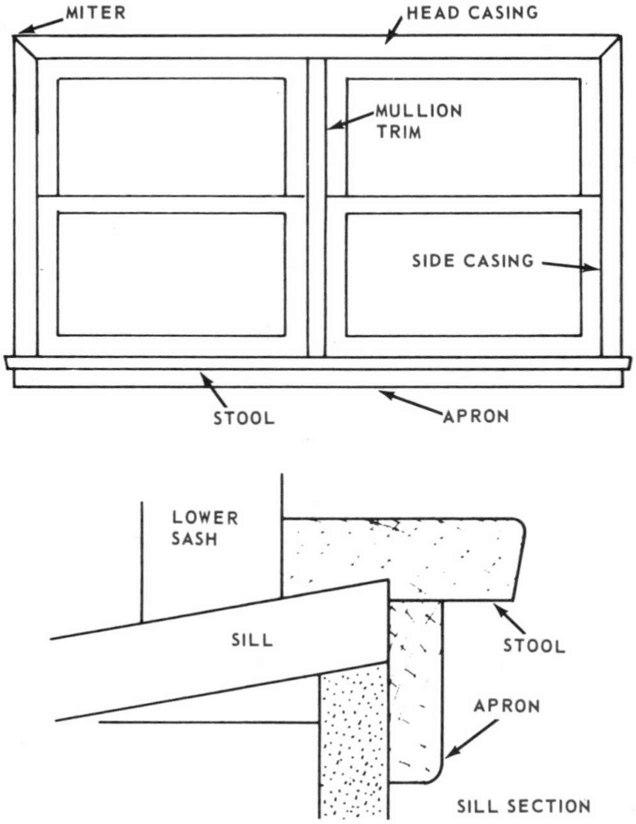

Fig. 17-47. Trim members used for standard double-hung window.

Fig. 17-48. Measuring and fitting side casing. Bottom end is squared and rests on stool.

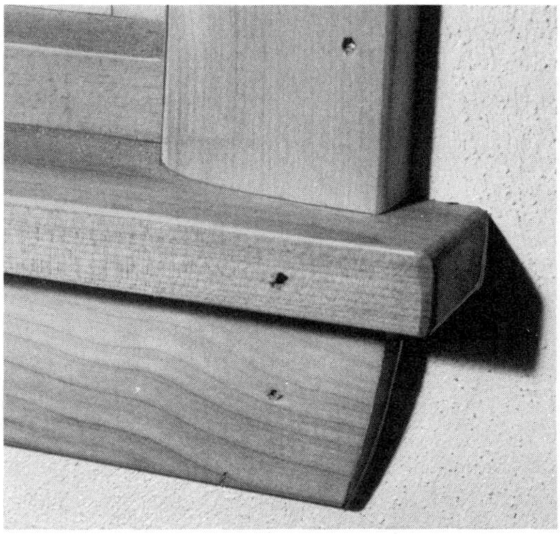

Fig. 17-50. View of lower right-hand corner of window showing trim members complete.

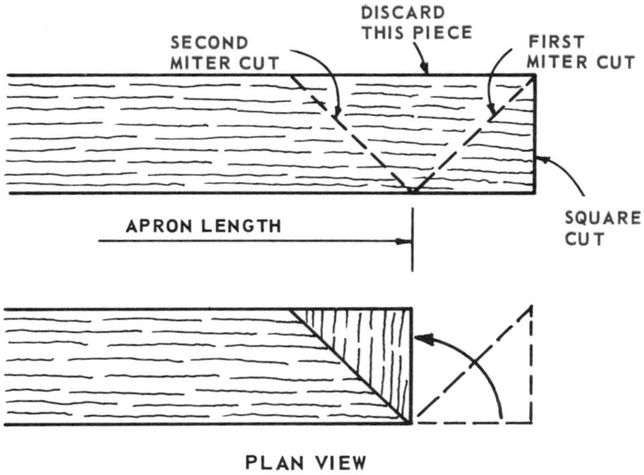

Fig. 17-49. Procedure for cutting returned end. Use glue to attach end piece.

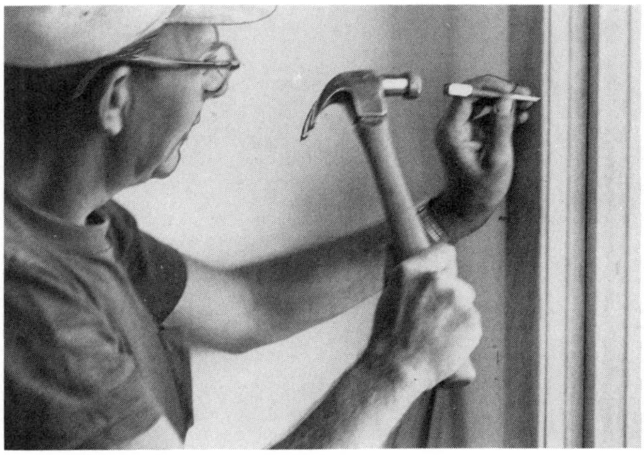

Fig. 17-51. Setting nails after all trim members are in place.

The apron is the last member to be applied. Its length should be equal to the distance between the outside edges of the side casing. Sand all saw cuts before nailing any trim in place, especially ends.

When the profile of the apron is curved the ends should be returned or coped so that the shape of ends will be the same as the sides. The returned end is commonly used and formed with miter cuts as illustrated in Fig. 17-49. A view of the completed trim at the lower right-hand corner of a window is shown in Fig. 17-50. Note the spacing of the various members and the end contour of the apron which has been returned.

In modern construction the stool and apron is sometimes eliminated and a piece of beveled sill liner is installed along with a standard piece of casing. This is commonly known as "picture frame" trimming.

After all trim members are in place, complete the nailing pattern and set the nails, Fig. 17-51. The nail holes are filled later as a step in the application of surface finishes.

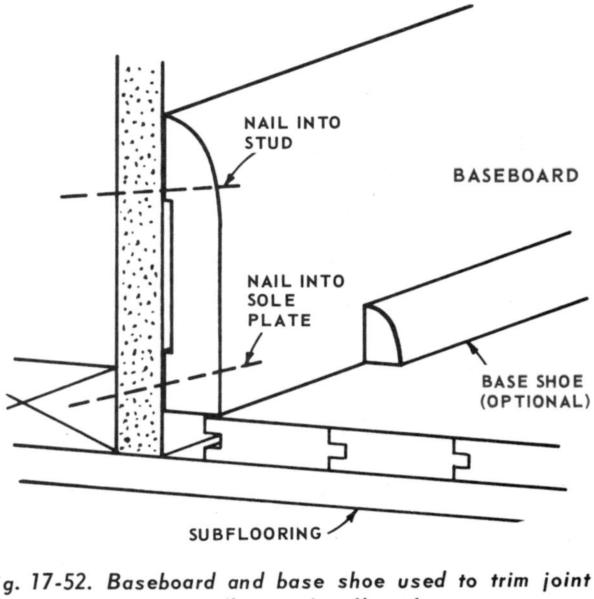

Fig. 17-52. Baseboard and base shoe used to trim joint between floor and wall surface.

Baseboard and Base Shoe

The baseboard covers the joint between the wall surface and the finish flooring. It is among the last of the interior trim members to be installed since it must be fitted to the door casings and cabinetwork. Base shoe is used to seal the joint between the baseboard and the finished floor, Fig. 17-52. It is usually fitted at the time the baseboard is installed but is not nailed in place until after surface finishes (lacquer, varnish, paint) have been applied. Base shoe is often used to cover the edge of resilient tile or carpet.

Baseboards run continuously around the room between door openings, cabinets, and built-ins. The joints at internal corners should be coped while those at external corners are mitered. See Fig. 17-53.

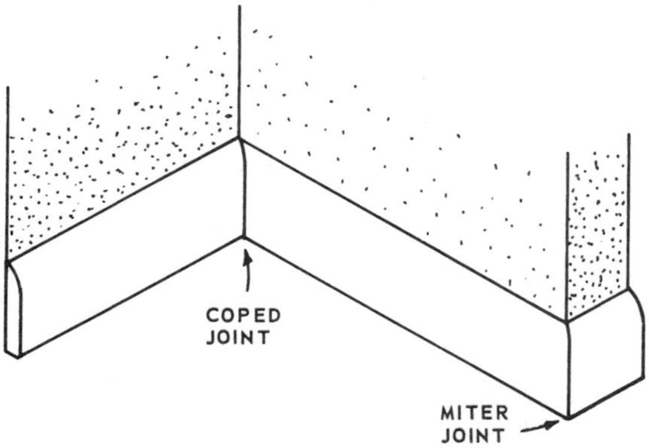

Fig. 17-53. Joints used in baseboard and base shoe installations.

Select and place the baseboard material around the sides of the room. Sort the pieces so there will be a minimum amount of cutting and waste. Where a straight run of baseboard must be joined, use a mitered-lap joint (also called a scarf joint). Be sure to locate the joint so it can be nailed at a stud.

HANDY SAYS:

"Baseboard installation is easy if stud locations have been recorded; first on the rough floor ahead of plaster and then on the wall surface before the finish flooring or underlayment is applied."

If the studs have not been marked, tap along the wall with a hammer until a solid sound is heard. Drive nails into the wall to locate the exact position of the stud, then mark the location of others by measuring the stud spacing (usually 16 in. O.C.).

Cut and fit the first piece of baseboard so it makes a tight butt joint with the intersecting wall surface. The next piece, running from an internal corner, is joined to the first with a coped joint, Fig. 17-54. To form this type of joint, first cut an

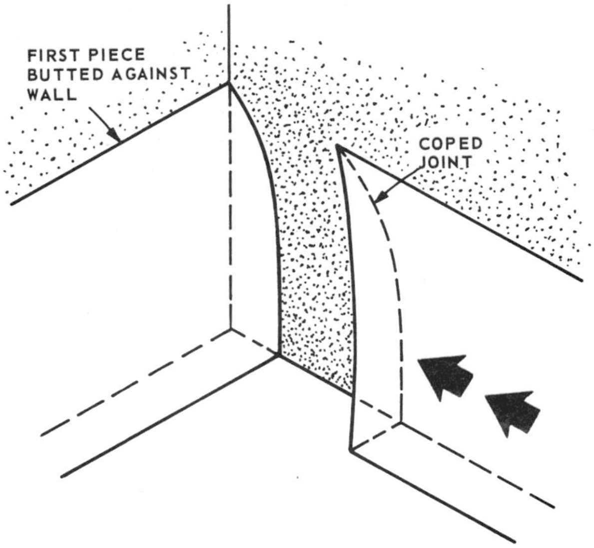

Fig. 17-54. How a coped joint fits together.

inside miter on the end of the baseboard. Then, using a coping saw, cut along the line where the sawed surface of the miter joins the contoured surface of the baseboard, Fig. 17-55. Make the cut perpendicular to the back face, thus forming an end profile that will match the face of the baseboard. Some carpenters undercut the coped end slightly to insure a tight fit at the front edge.

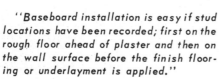

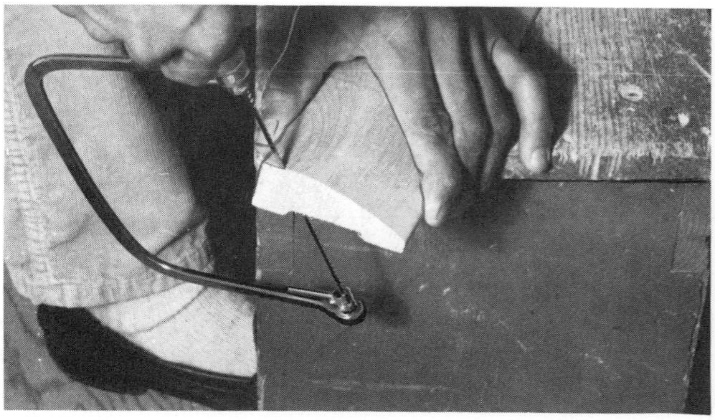

Fig. 17-55. Forming a coped joint by making a perpendicular cut along the front edge of an inside miter joint.

The coped joint takes a little longer to make than a plain miter but results in a better joint at an inside corner. It will not open when the baseboard is nailed into place. Neither will a noticeable crack appear if the wood shrinks after installation.

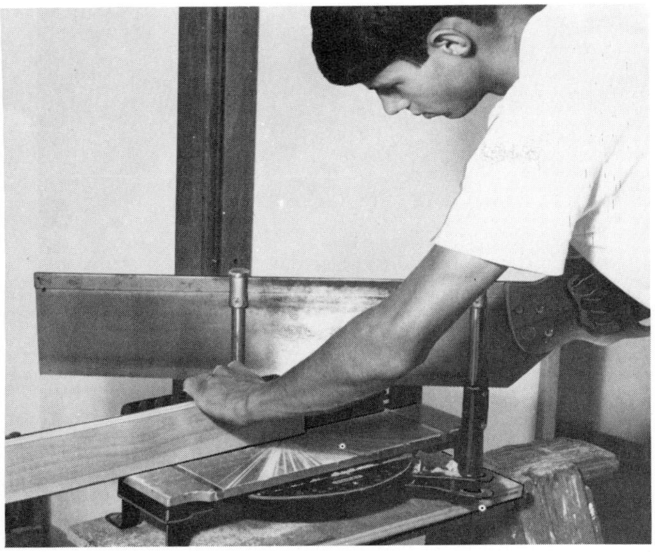

Fig. 17-56. Using miter box to cut miter joint at an outside corner.

All outside or external corners are joined with a miter joint. Hold the baseboard in position and mark at the back edge. Make the cut using a miter box, Fig. 17-56.

Before installing a section of baseboard, check both ends to make sure the cut and fit is correct. To install, hold the board tightly against the floor and nail it in place with finishing nails that are long enough to penetrate well into the studs. The

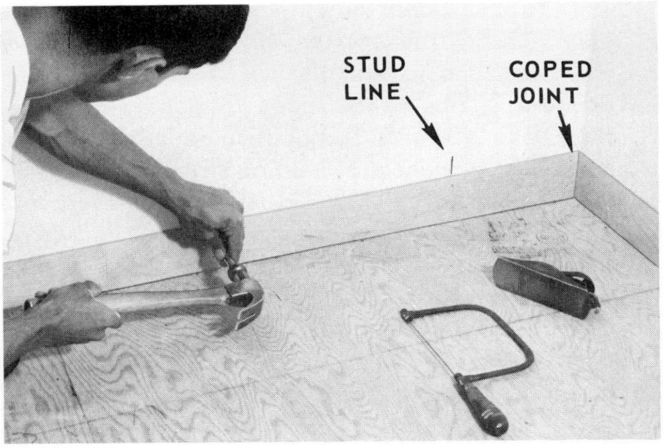

Fig. 17-57. Setting nails after installing section of baseboard. Right-handed workers usually prefer to make installation by proceeding around room in counterclockwise direction.

lower nail is angled slightly down so it is easier to drive and will enter the sole plate. Set the nail heads as shown in Fig. 17-57.

In modern construction, baseboards are normally butted against the door casing as illustrated in Fig. 17-58. The edge of the casing is designed with sufficient surface to accomodate the slightly thinner dimension of the baseboard. Base shoe, if used, is terminated at the casing with a miter cut as shown.

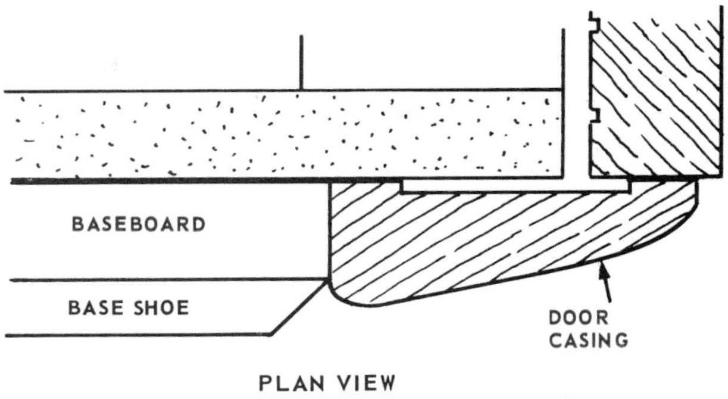

Fig. 17-58. Baseboard is butted against door casing.

How Moldings are Produced

Moldings (baseboard, casing, base shoe, window stool) are usually produced on a machine called a moulder or sticker. Strips of wood are first cut to size on a straight-line rip saw and are then fed into the moulder. All four sides of the piece are formed in a single pass through the machine.

Fig. 17-59. Six-knife cutter head used to produce standard casing pattern. The milled-to-pattern knives are sharpened by grinding a single bevel. (Jordan Millwork Co.)

As the stock enters the moulder it passes under a cutter and then moves by cutterheads located at the sides. Finally it passes over a cutter located in the table which smoothes the back and cuts out the hollow required in window and door casing and baseboard material. Fig. 17-59 shows a cutterhead set up to produce a standard casing design.

Test Your Knowledge - Unit 17

1. Interior door jambs for a standard framed wall with a plastered finish should be ______ in. wide.
2. If a door frame is slightly high for the opening, part of the ________ should be trimmed.
3. When nailing the door jambs, locate the nails so they will be covered by the ________.
4. To cover the space between the wall surface and the door jambs, door________ is applied to each side of the frame.
5. A panel door consists of rails, panels, and ________.
6. The face panels of flush doors are usually made of 1/8 in. plywood and are commonly called door________.
7. Three basic types of hollow core door constructions include lattice, ________, and implanted blanks.
8. A standard size bedroom door is ______ wide and ______ high.
9. When setting the door stop on the hinge jamb, provide a clearance of ______ with the door.
10. If you are facing the outside of an entrance door and it swings toward you with the hinges on your left, it is called a ________ (LH, RH, LHR, RHR).
11. The two most commonly used types of residential door locks are cylindrical and ________.
12. The type of sliding door that operates by moving it into an opening in the wall or partition is called a ________.
13. After the stool has been set, the next step in trimming a single window unit is to install the ________.
14. The last trim member applied to a window unit is the ________.
15. When installing baseboards, it is recommended that a ________ joint be used at internal corners.

Outside Assignments

1. Visit a residential building site in your community where the inside finish work is in progress. Note the type of trim being applied and the procedures being followed. Possibly one of the carpenters would discuss with you the advantages and disadvantages of prehung door units. Prepare carefully organized notes of your observations and make an oral report to the class. ALWAYS BE SURE TO OBTAIN PERMISSION FROM THE FOREMAN OR HEAD CARPENTER BEFORE MAKING THE VISIT.
2. Develop a door schedule for a preliminary or presentation drawing of a residence. Sizes of doors will usually not be shown and you will need to make the selection. A door schedule should include the location, width, height, thickness, type and design, kind of material, and quality requirements. Sometimes a manufacturer's number is included. To extend the assignment, secure estimated costs. Refer to a builders supply catalogue or visit a local lumber dealer.
3. Build a full-size sectional mock-up of a partition with a doorway. Include all parts -- studs, wall finish, door jamb, casing, baseboard and door. Carry the section up about 16 in. above the floor and extend the partition out from the door only about the same amount. A 3/4 in. thickness of particleboard could serve as the base and represent the finish flooring.

   It will be best to glue most of the parts together since the structure may not withstand much nailing. Laminate several layers of plywood to form the door section or "band" a door cutting. Work carefully and try to show the correct position and clearance for all the parts and trim pieces.

Fig. 18-1. Modern kitchen cabinets are attractive and provide efficiency and convenience. (International Paper Co.)

Unit 18
CABINETMAKING

Cabinetwork, as used in interior finish, refers to kitchen and bathroom cabinets, and in a general way to such work as closet shelving, wardrobe fittings, desks, and dressing tables. The term "built-in" emphasizes that the cabinet or unit is located within or attached to the structure. The use of built-in cabinets and storage units is an important development in modern architecture and design.

Modern kitchen cabinets, Fig. 18-1, present an attractive appearance and help to increase kitchen efficiency. Here, and in other areas of the home, storage units should be designed with attention given to the items that will be stored. Space must be carefully allocated and drawers, shelves, and other elements should be proportioned to satisfy specific needs.

Three types of cabinetwork are: that which is built-on-the-job by the carpenter; custom-built units constructed in local cabinet shops or millwork plants; and mass-produced cabinets available from factories that specialize in this area of manufacturing. Except in giant "mass-housing" projects, combinations of these types of cabinetwork are usually found on most jobs. Even when a large part of the cabinets are factory produced, the carpenter is responsible for the installation, a task that requires skill and careful attention to detail.

Drawings for Cabinetwork

Architectural plans usually include details of built-in cabinetwork. The floor plan shows cabinetwork location, while elevations, usually drawn to a larger scale, provide detailed dimensions. A typical drawing of a base cabinet, desk, and room divider is illustrated in Fig. 18-2. Drawings of this type are scaled and thus the need for extensive dimensioning is eliminated. The drawings serve as a guide to the carpenter in his construction work, or are followed in the selection of factory-built components. When built on the job, the detail of joints and structure becomes the responsibility of the carpenter, who should be a skilled craftsman in the area of cabinetmaking.

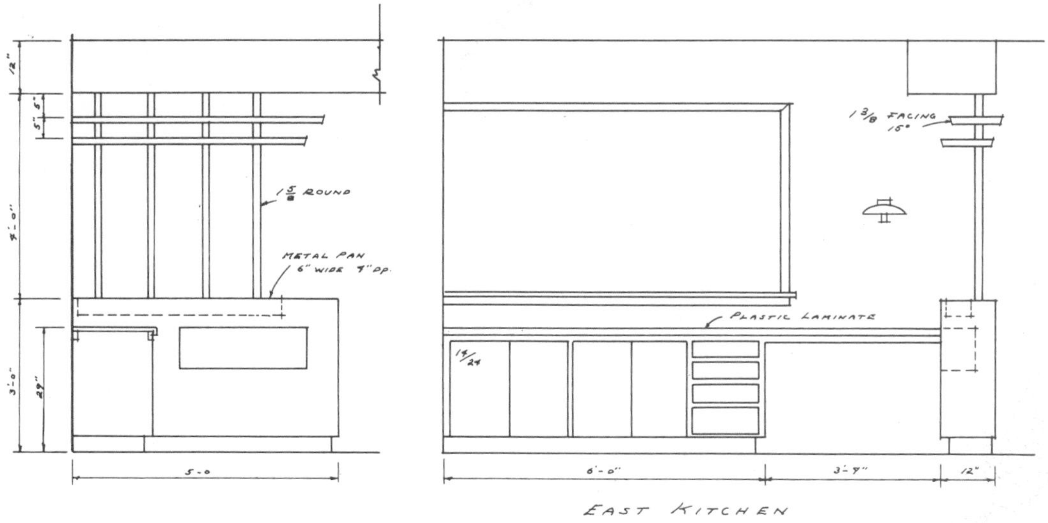

Fig. 18-2. Above. Typical built-in cabinetwork detail. Since the drawings are carefully scaled, many dimensions have been eliminated. Below. Completed cabinetwork. Drawers include a recess under the front edge that serves as a finger-pull.

Architectural drawings may include more specific details concerning the cabinetwork. This is justified on large commercial contracts where a number of individuals will be associated with the project -- and when the cabinetwork contains special materials and constructions. Fig. 18-3 shows kitchen cabinets designed for an apartment complex. Written specifications will further define the type of joinery and quality of materials.

Standard Sizes

Although the overall heights and other sizes of built-in units are usually included in the architectural plans, the carpenter should be familiar with basic design requirements in this area.

Base cabinets for kitchens are usually 36 in. high and 24 in. deep with the countertop extending about 1 inch, Fig. 18-4. The vertical distance between the base unit and the wall unit may vary from 15 to 18 in. over regular work surfaces. FHA specifies a minimum height of 24 in. when the wall cabinet is located over a cooking unit or sink.

Cabinets with built-in lavatories in bathrooms or dressing rooms are normally 31 in. high. The depth varies depending on the type of fixture. Knee-room is especially important in this type of cabinet.

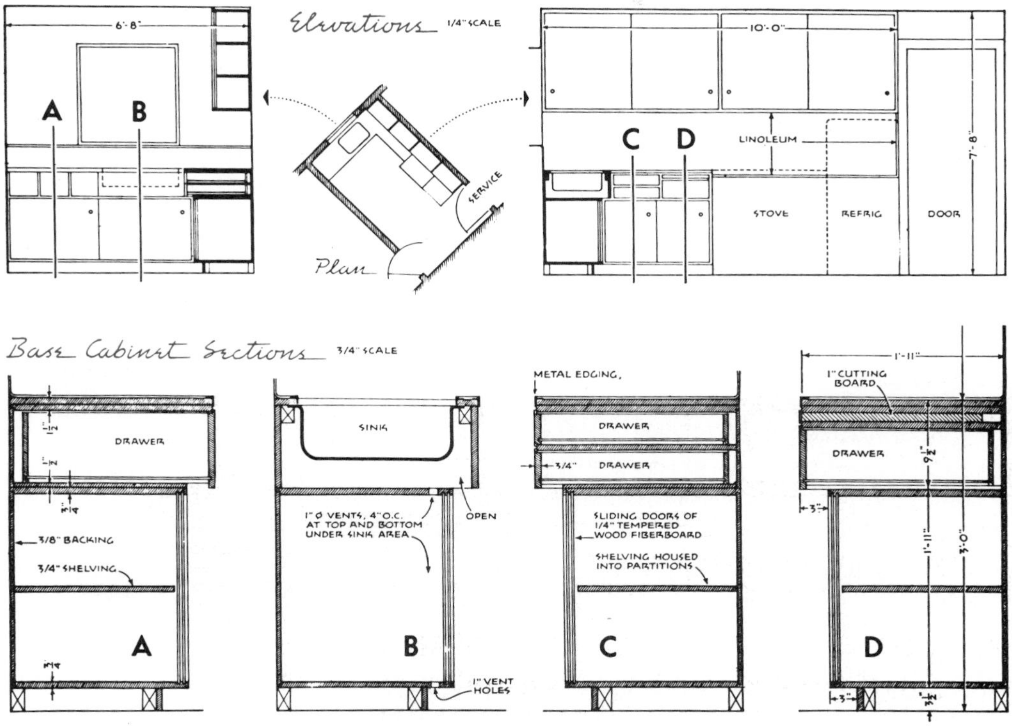

Fig. 18-3. Drawings for cabinetwork in a large apartment building. (Architectural Woodwork Institute)

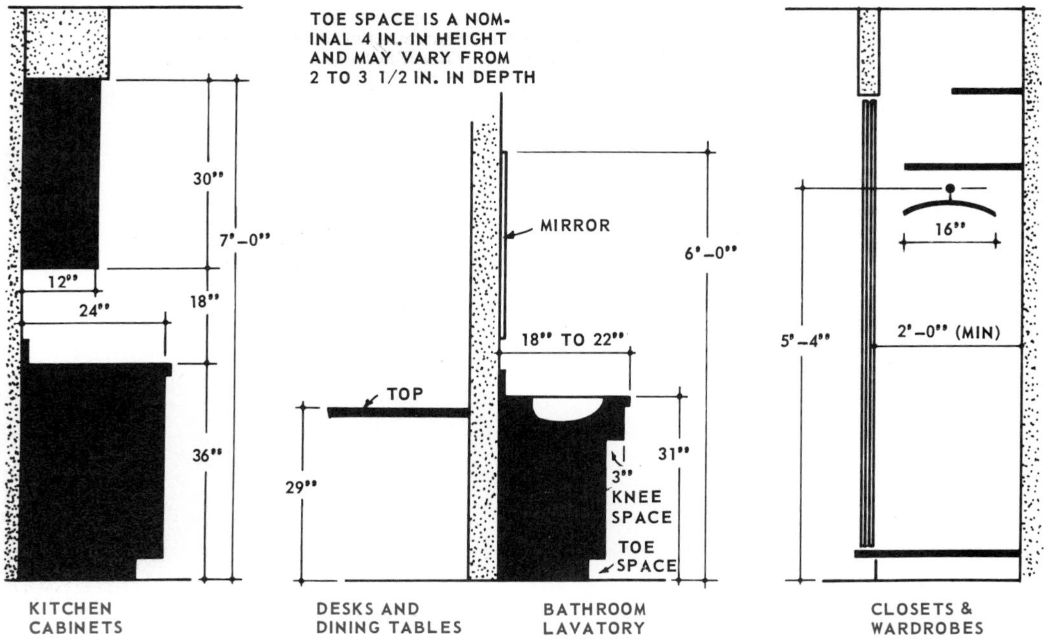

Fig. 18-4. Standard sizes for cabinetwork.

Standards for closets and wardrobes vary widely depending on items to be stored. The minimum clear depth for clothing on hangers is 24 in. When hooks are mounted on doors or the rear wall, this distance must be extended. Refer to architectural standards manuals for additional information.

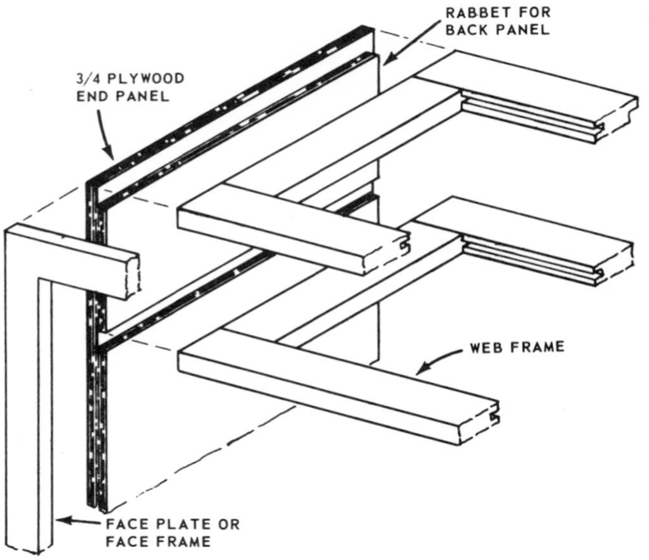

Fig. 18-5. Typical frame construction of cabinets built as a separate unit and then placed in position. Terminology as recommended by Architectural Woodworking Institute.

Construction Procedures

Two procedures are commonly used when building cabinets on the job. The first consists of cutting the parts and assembling them "in place"; attaching each piece to the floor, wall, or other members. After the basic structure is assembled, facing strips and other fittings are marked, cut to size, and attached.

In the second procedure, the entire unit is first assembled and then placed in position on the floor or attached to the wall. The structure is formed with end panels, partitions, and backs; joined with horizontal frames. See Fig. 18-5. Basically, this is the method used in custom cabinet shops or for the mass-production of cabinets in factories. When units of this type are built on the job, they can be much larger (longer sections) than is practical to construct in shops or factories since the latter type must be of a size easily transported to the building site.

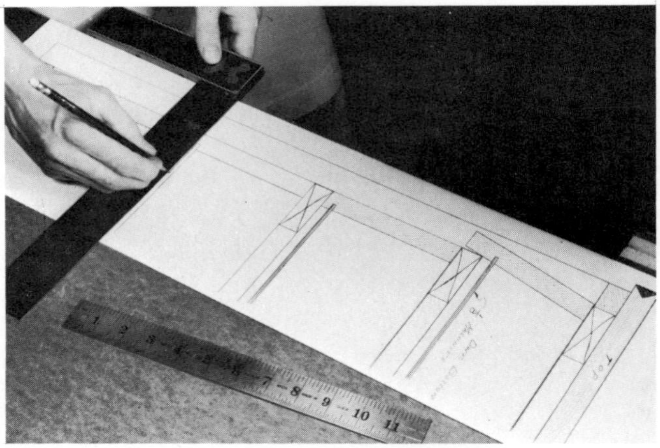

Fig. 18-6. Making a full-size master layout of a base cabinet.

Master Layouts

Before cutting out the parts for a cabinet, especially when following the second procedure described, it will be helpful to prepare a master layout on plywood or cardboard, Fig. 18-6. Several layouts may be required. These should be drawn as section views, where the structure is complicated with drawers, pull-out boards, and special shelving.

Follow the overall dimensions provided in the architectural plans and details. Draw each member full size and show various clearances that may be required. This master layout will be valuable when cutting side or end panels to size and locating joints. It can also be referred to for exact sizes and locations of drawer parts and other detailed dimensions that are not included in the regular drawings.

Fig. 18-7. Constructing the base. Note the layout lines marked on the wall for an end panel and partition.

Basic Framing

When following the assembled-in-place procedure, some of the basic layout can be made directly on the floor and/or wall surface. Each end panel and partition is represented with two lines. Be sure these lines are plumb since they can be used to line up the panels when they are installed.

The base is constructed first, as illustrated in Fig. 18-7. Use straight 2 x 4s and nail them to the floor and to a strip attached to the wall. If the floor is not level, place shims under the various members of the base. Exposed parts can be faced with a finished material or the front edge can be made of a finished piece.

After the base is complete, cut and install the end panels, Fig. 18-8. Attach a strip along the wall between the end panels and level with the top edge. Be sure the strip is level throughout its length and that it is nailed securely to the wall studs.

Next, cut the bottom panels and nail them in place on the base. Follow this with the installation of the partitions which are notched at the back corner of the top edge so they will fit over the wall strip. Plumb the front edge of the partitions and end panels and secure them in this position with temporary strips nailed along the top as shown on the left-hand cabinet in Fig. 18-9.

HANDY SAYS:

"The carpenter may prefer to install the wall cabinets before the base units. This makes it easier to work on the wall cabinets since he can work directly below them."

Wall units are constructed about the same way as the base units. Make lay out lines directly on the wall and ceiling surfaces and attach mounting strips by nailing through the wall surface into the studs. At an internal corner, end panels can be attached directly to the wall, Fig. 18-10.

Fig. 18-8. Marking an end panel while it is held in place. Typical of the method used when constructing cabinets by the assembled-in-place procedure.

Fig. 18-9. Base cabinet partially complete. Lazy Susan (also called carousel shelving) unit must be installed at this stage.

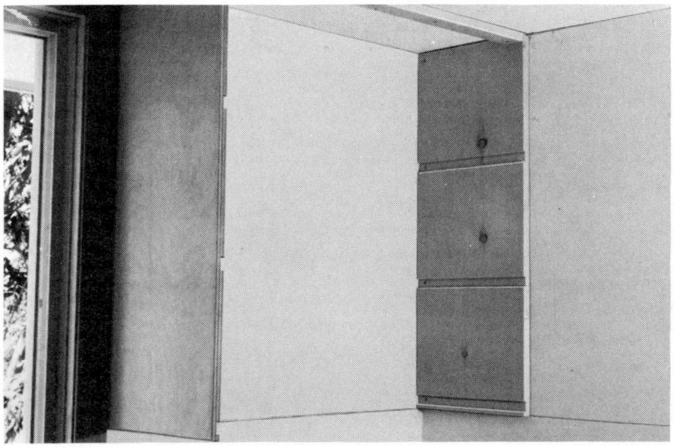

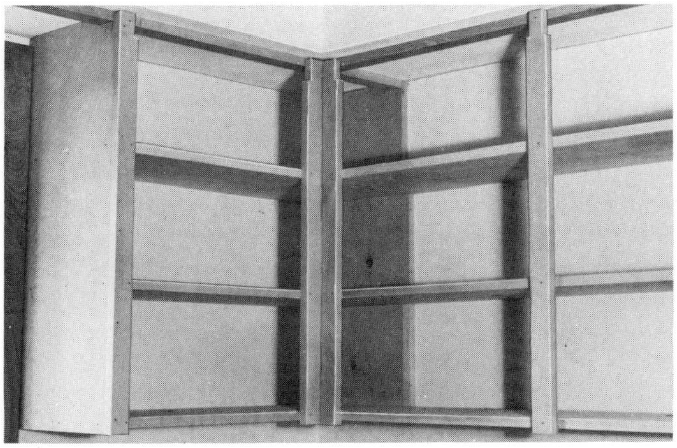

Fig. 18-10. Left. End panels of a wall cabinet in place. Right. Completed framing with facing partially applied.

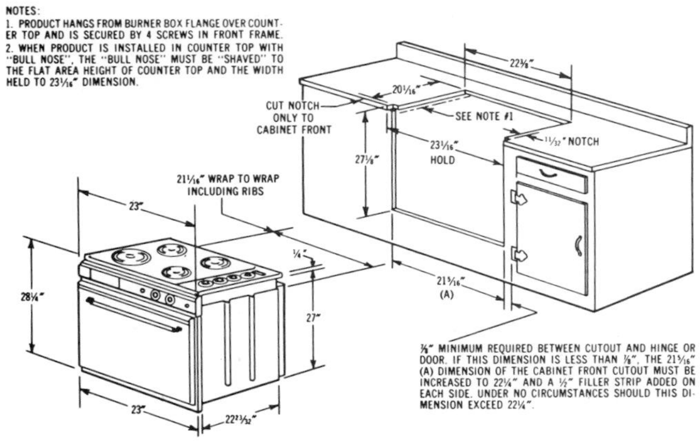

(Rough-in dimensions only. See installation instructions for exact details.)

Fig. 18-11. Typical drawing showing rough-in dimensions for a built-in range unit. (Whirlpool Corp.)

During the basic framing operations, care must be exercised in constructing the openings for built-in appliances so they will be the correct size. Secure the rough-in drawings and specifications, Fig. 18-11, furnished by the manufacturer of the appliances that have been selected, and follow them carefully. See Fig. 18-12 which shows a complete cabinet with a built-in range.

Fig. 18-12. Completed installation of a built-in range. (I-XL Furniture Co.)

Facing

Finished facing strips are applied to the front of the basic frame of the cabinet. In factories and cabinet shops, these strips are often assembled into a framework (called a face plate or face frame) before they are attached to the basic cabinet structure. The vertical members are called stiles and the horizontal members are called rails.

For assembled-in-place cabinets, each piece is installed separately. The size is laid out by holding the facing stock in position as shown in

Fig. 18-13. Lengths of facing strips are marked while holding them in place against the cabinet frame.

Fig. 18-13; then the finished cuts are made. This part can be used to lay out duplicate pieces.

Generally the stiles are installed first and then the rails are inserted, Fig. 18-14. Sometimes an end stile is attached plumb and rails are used to space the position of the adjacent stile. The parts are glued and nailed in place with finishing nails. When nailing hardwoods, drill nail holes at the ends of piece where splitting is likely to occur.

A variety of joints can be used to join the stiles and rails. A practical design is illustrated in Fig. 18-15. The depth of the gain or dado is maintained at about 3/8 in. so it will be covered by the lip of the door or drawer. Lap joints and dowel joints are also commonly used to make this attachment.

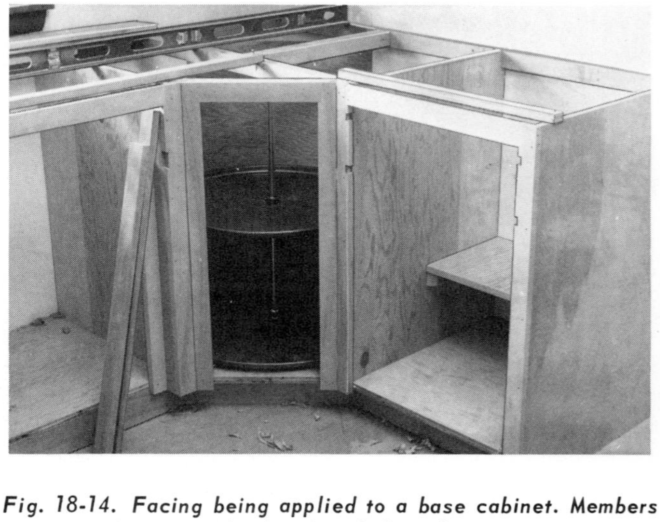

Fig. 18-14. Facing being applied to a base cabinet. Members are glued and nailed in place.

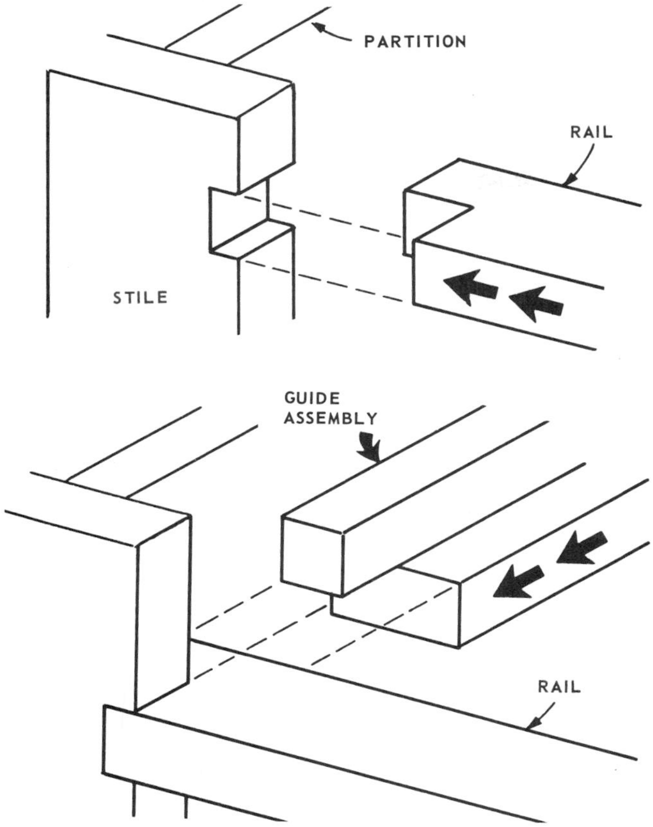

Fig. 18-15. Practical joinery for assembled-in-place cabinet facing. Above. Stile and rail joint. Below. Drawer guide assembly fits behind facing and rests on rail.

Drawer Guides

Three common types of drawer guides include: corner guides, center guides, and side guides, Fig. 18-16. The corner guide may be formed in the cabinet by the side panel and frame. It may be necessary to add a spacer strip to hold the drawer in alignment with the front facing. An adaptation of the corner guide is used in the cabinet, Fig. 18-17. Note that the guide unit located in the center is designed to carry a drawer on each side.

Center guides are often used in cabinetwork and consist of a strip or runner fastened between the front and back rails. A guide which is attached to the underside and back of the drawer, rides on this runner. The runner is attached to the frame or rails with screws, and can be adjusted so the clearance on each side is equal and the face of the drawer aligns with the front of the structure.

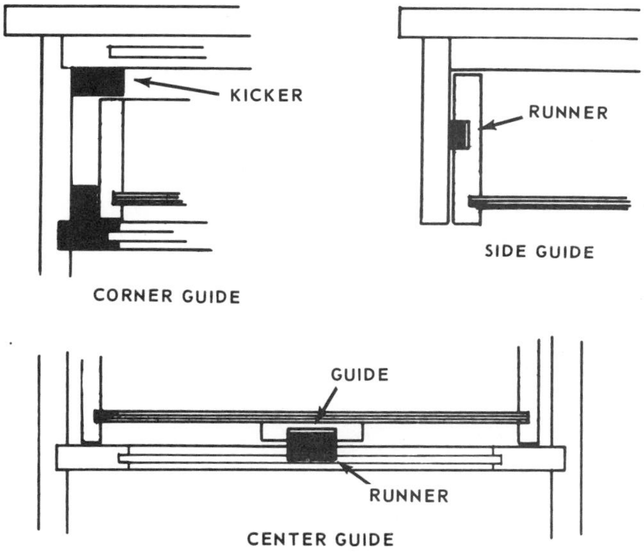

Fig. 18-16. Basic types of drawer guides.

Fig. 18-17. Lavatory cabinet in a bathroom area. Left. General view of framing with facing applied. Right. Close-up view showing drawer guides. The back of the drawer will slide along the kicker which prevents the drawer from tilting downward when opened.

In drawer openings where there is no lower frame, a side guide may be used. Grooves are cut in the drawer side before it is assembled and matching strips are fastened to the side panels. This type of guide can be used for shallow drawers or trays in wardrobe units where framing between each drawer opening is unnecessary and would waste space.

The drawer carrier arrangement may require a "kicker" to keep the drawer from tilting downward when it is pulled open. The kicker may be located over the drawer side or centered to hold down the drawer back. Wooden drawer guides should be carefully fitted and the parts given a sealer coat. If the sealer on the moving parts is sanded lightly and waxed, the drawer will work smoothly.

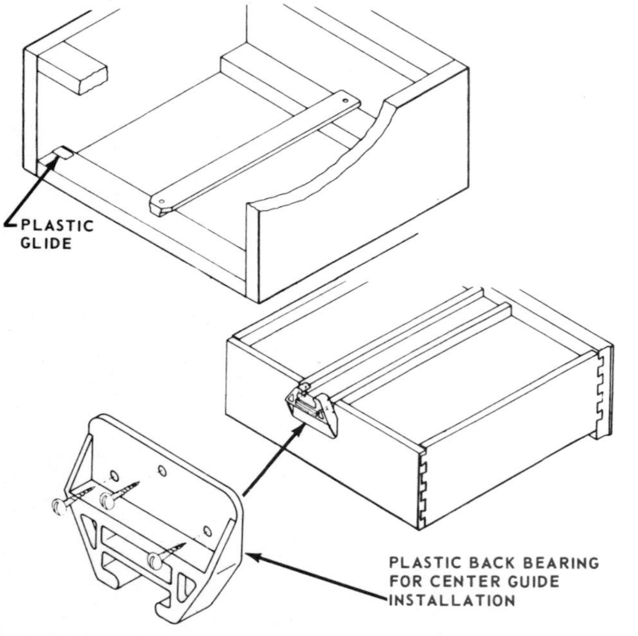

Fig. 18-18. A center guide installation using a patented back bearing. (Ronthor Reiss Corp.)

Fig. 18-18 shows a commercially made back bearing for a center slide and how it is installed. For large drawers, or those that will carry considerable weight, a special 3-roller drawer slide like the one shown in Fig. 18-19 may be used. This supports a weight up to 50 lbs. and is available in several sizes.

Drawers

There are two general types of drawers: lip and flush. Flush drawers, which must be carefully fitted, are commonly used in furniture. Lip drawers have a rabbet along the top and sides of

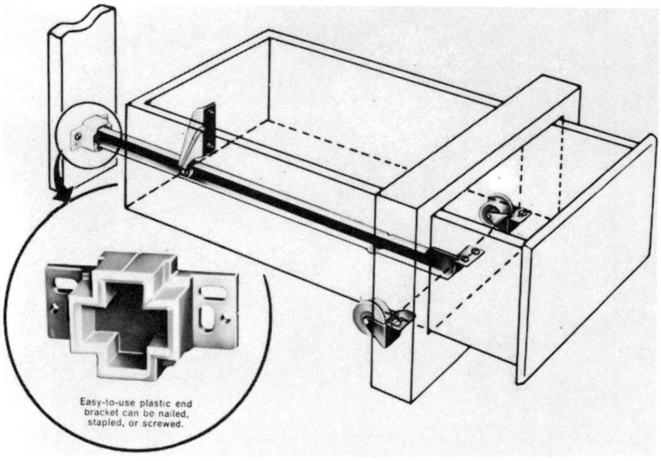

Fig. 18-19. Three-roller drawer slide assembly with flanged plastic rollers. (Amerock Corp.)

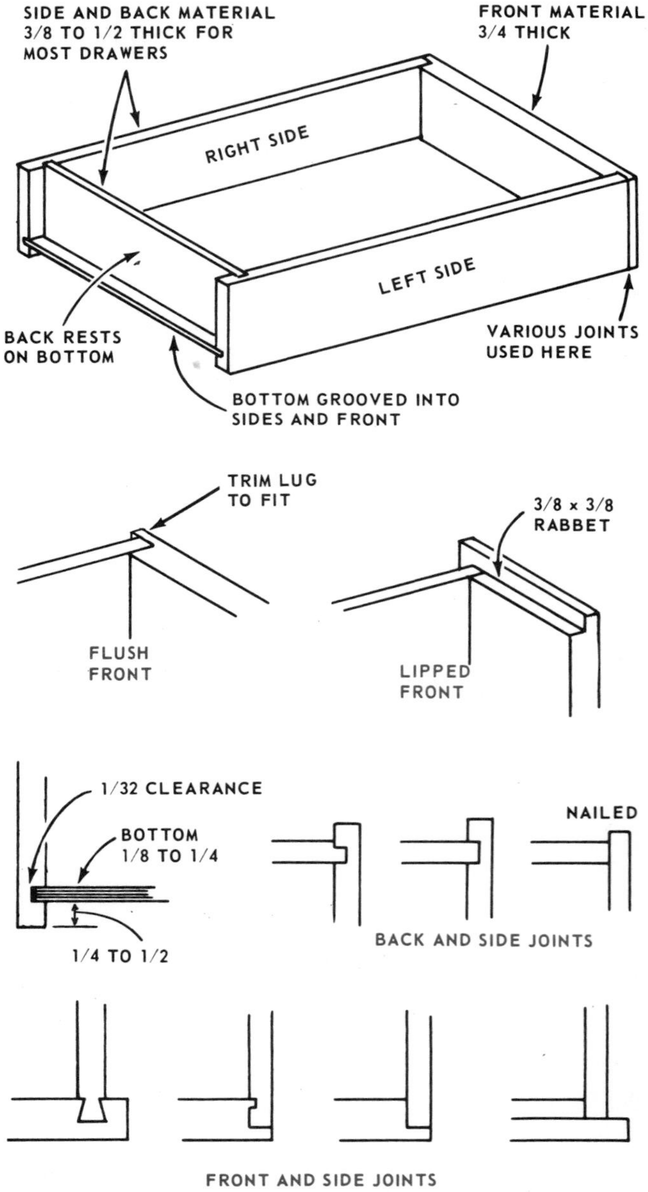

Fig. 18-20. Standard drawer construction.

the front which covers the clearance and therefore permits greater freedom in the fit. They are used mainly in built-in cabinetwork.

Sizes and designs in drawer construction vary widely. Fig. 18-20 shows standard construction with several types of joints commonly used. The joint between the drawer front and the drawer side receives the greatest strain and should be carefully designed and fitted. Often in high quality work, the corners are dovetailed and the bottom grooved into the back as well as the front and sides. Also the top edge of the drawer sides is rounded. Flush drawer fronts may be constructed with the sides set about 1/16 to 1/8 in. deeper into the fronts so that a slight lug is formed. After the drawer is assembled, this lug is trimmed to form a good fit.

Fig. 18-21. Prefabricated drawer fronts are available in many designs. Above. Raised panel effect. Below. Ogee edge. (Frank Paxton Lumber Co.)

Drawers are usually not constructed until after the cabinetwork has been assembled. The following procedure is generally recommended:

1. Select the material for the fronts. Grain patterns should match or blend with each other. Solid stock or plywood may be used. Today a variety of prefabricated drawer fronts are available in many standard sizes. See Fig. 18-21.

2. Cut the drawer fronts to the size of the opening (if flush type) allowing a 1/16 in. clearance on each side and on the top. This clearance will vary depending on the depth of the drawer and the kind of material. Deep drawers with solid fronts will require greater vertical clearance. For lip drawers, add the depth of the rabbets to the dimensions.

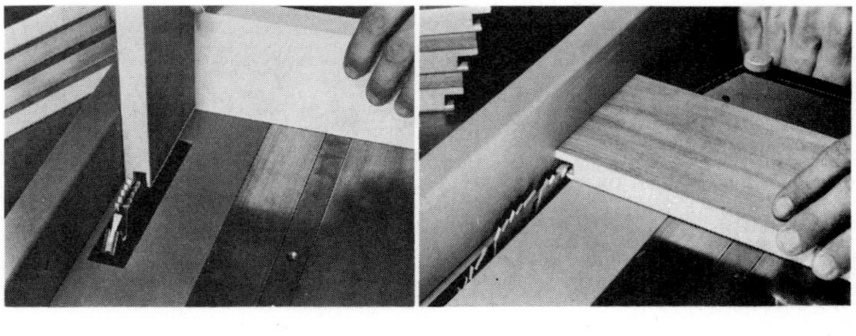

Fig. 18-22. Cutting lock joint in drawer front. Left. Cutting groove. Right. Trimming tongue to length.

3. Select and prepare the stock for the sides and back. A less expensive hardwood or a softwood can be used. Plywood is usually not satisfactory for these parts. The surfaces should be sanded either before or after cutting to length.

4. Select material for the bottom. Hardboard or plywood should be used. The bottoms can be trimmed to final size after the joints for the other drawer members are cut.

5. Cut groove for the bottom in the front and sides.

6. Cut joints in the drawer fronts that will hold the drawer sides. Fig. 18-22 shows a sequence for cutting a locked joint. For lip drawers, cut the rabbet first, then the required joint.

7. Cut the matching joint in the drawer sides. BE SURE TO CUT A LEFT AND RIGHT SIDE FOR EACH DRAWER.

Fig. 18-23. Above. Trial assembly of drawer parts. Below. Parts laid out and ready to be glued.

8. Cut the required joints for the drawer sides and backs.

9. Trim the bottom to the correct size and make a trial assembly as shown in Fig. 18-23. Make any adjustments in the fit that may be required. The parts should fit smoothly and not be so tight that they will be difficult to assemble after the glue is applied.

10. Disassemble and sand all parts. The top edge of the sides are often rounded between points about 1 in. from each end.

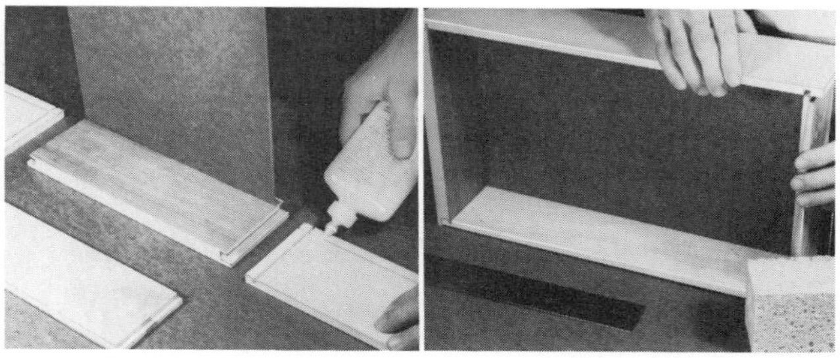

Fig. 18-24. Gluing drawer parts together. Left. Bottom in place and glue being applied to a front corner. Right. Placing the second side in position after drawer has been turned on its side.

11. Make the final assembly. A number of procedures can be used. One method is to first glue the bottom onto the front and then glue one of the front corners as shown in Fig. 18-24. Be sure the bottom is centered. The side grooves are usually not glued. Turn the drawer on its side and glue the back into the assembled side. Finally glue on the remaining side as shown. Clamps or a few nails can be used to hold the joints together until the glue hardens.

12. Carefully check the drawer for squareness and then drive one or two nails through the bottom into the back. If the bottom was carefully squared, you should have little trouble with this operation. Wipe off the excess glue.

After the glue has cured, make a selective fit of each drawer to a particular opening. Trim and adjust drawer guides. Place an identifying number or letter on the underside of the bottom so it will be easy to return the drawer to its proper opening after sanding and finishing operations. The inside surfaces of quality drawers should always be sealed and waxed.

Shelves

Shelves are used extensively in cabinetwork, especially in wall units where drawers are not appropriate. In some designs, it may be necessary to fit and glue the shelves into dados cut in the sides of the structure to provide additional strength. When possible however, it is better to make the shelves adjustable, up and down, so they can be used for various purposes.

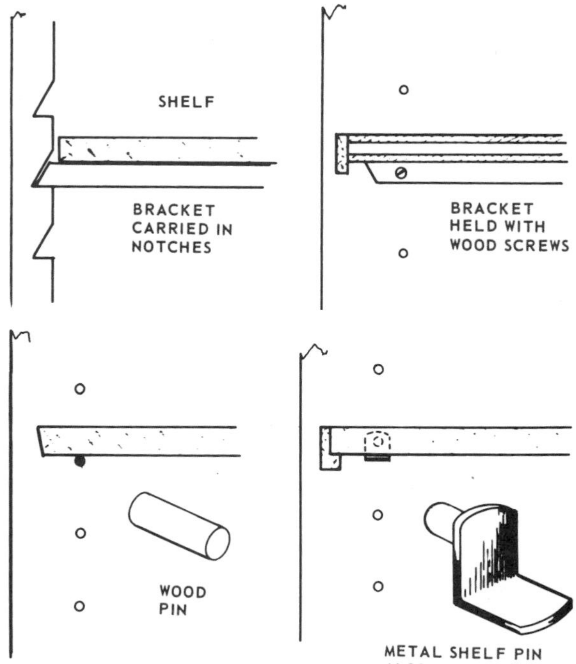

Fig. 18-25. Methods for supporting shelves. When plywood is used for the shelving material, it should be faced with solid stock.

Fig. 18-25 presents several methods of installing adjustable supports. Lay out the arrangement carefully so the shelves will be level. Usually it is best to do the cutting or drilling before the structure is assembled. This is essential when using one of the patented shelf standards where slotted strips are set in grooves cut in the sides or back. Fig. 18-26 shows an adjustable shelf support that mounts on the surface.

Fig. 18-26. Brackets are easily adjusted to provide various shelf spacing.

FIVE BASIC KITCHEN PLANS

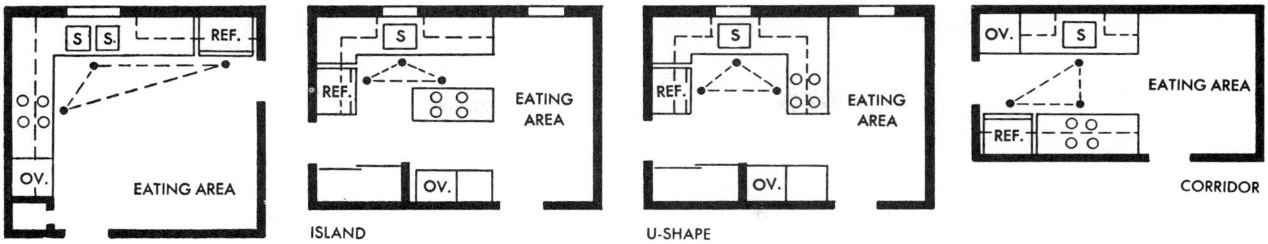

L-SHAPE   ISLAND   U-SHAPE   CORRIDOR

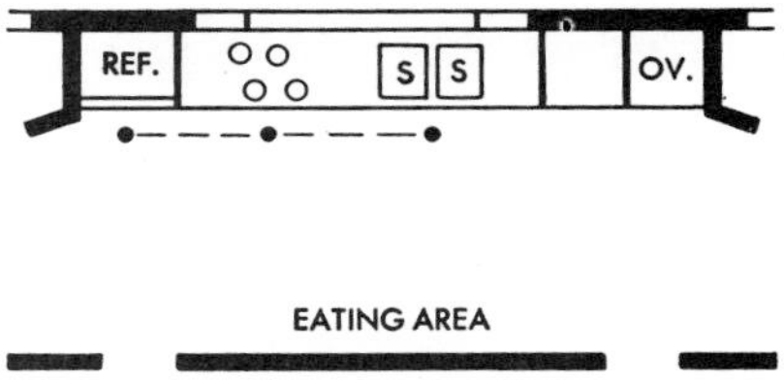

SINGLE WALL

The overall floor plan affects the shape, location and arrangement of the kitchen, so careful thought must go into its layout to insure sufficient wall space for the installation of cabinets and all necessary appliances. Five basic plans for kitchens of different shapes are shown above and at right.

Ideally, the kitchen should provide a continuous line of appliances linked by cabinets and counter tops. The U-shape does this most effectively. An L-shaped kitchen usually occupies two walls of the room and results in a triangular work area. If this triangle becomes too large, consider adding an island to bring the activity centers closer together. Single wall and corridor kitchens work well in long, narrow spaces, are best suited to small homes or apartments.

LOCATION OF DOORS AND WINDOWS AFFECTS PLANNING

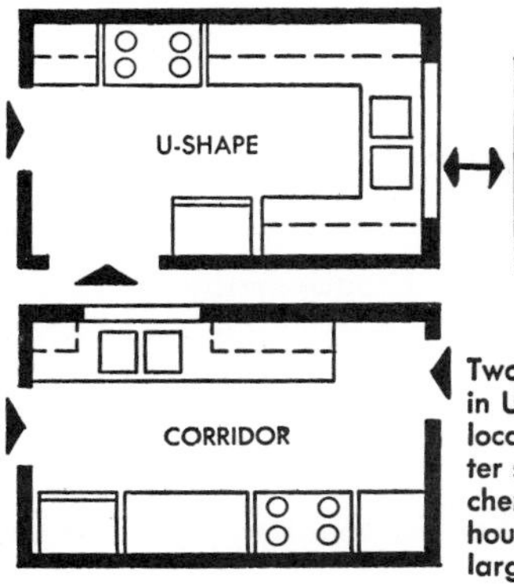

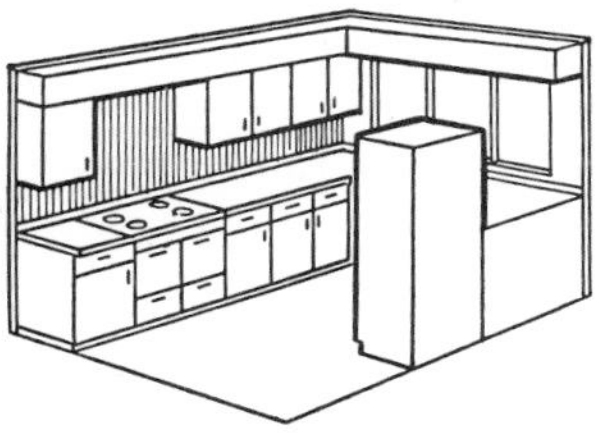

Two identical rooms at left must be laid out in U-shape or corridor plan because of the location of wall openings. At least one counter should be by a window. Combining kitchen with family room at front or rear of house (below) makes small kitchen appear larger.

KITCHEN FOR A SPLIT LEVEL

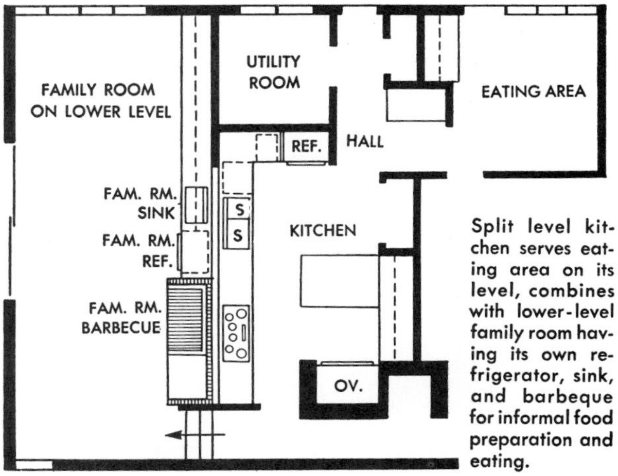

Split level kitchen serves eating area on its level, combines with lower-level family room having its own refrigerator, sink, and barbeque for informal food preparation and eating.

KITCHEN-FAMILY ROOM COMBINATIONS

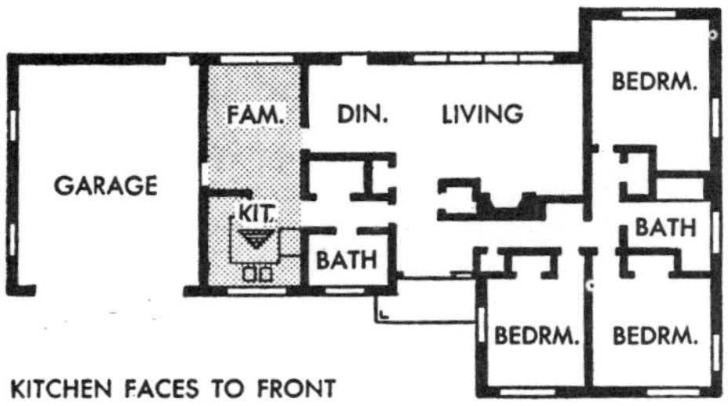

KITCHEN FACES TO FRONT

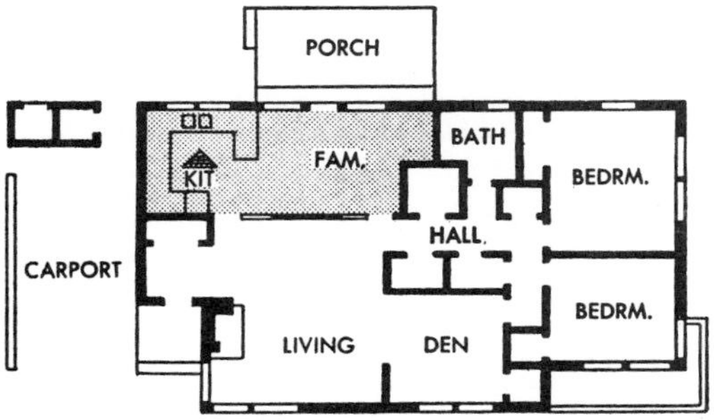

KITCHEN FACES TO REAR

PLAN ON DROP-INS FOR A BUILT-IN LOOK

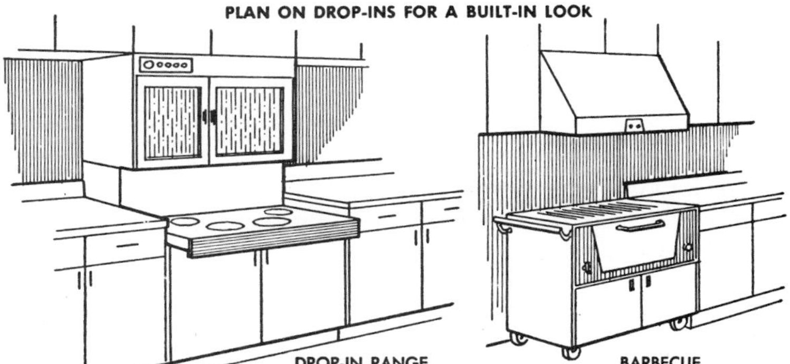

DROP-IN RANGE   BARBECUE

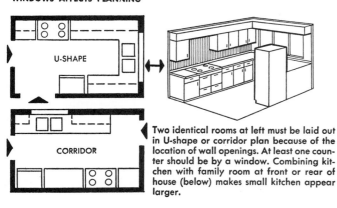

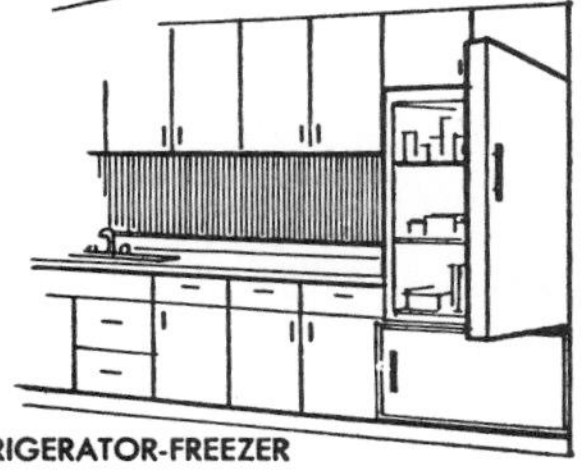

REFRIGERATOR-FREEZER

Latest trend: free-standing appliances that provide appearance of built-ins at lower installation cost.

Electric range has eye-level oven. Unit sets on a cabinet, requires no special installation. Refrigerator-freezer can be installed flush against wall. Doors swing within the dimensions of the unit, eliminating need for space between cabinet on hinged side. Hood makes portable barbeque look built in.

Kitchen Planning.

Fig. 18-27. Prefabricated cabinet doors. A–Flush door with overlay. B–Panel door with curved rails. C–Simulated leaded glass panel. D–Perforated metal panel. (Century Wood Products)

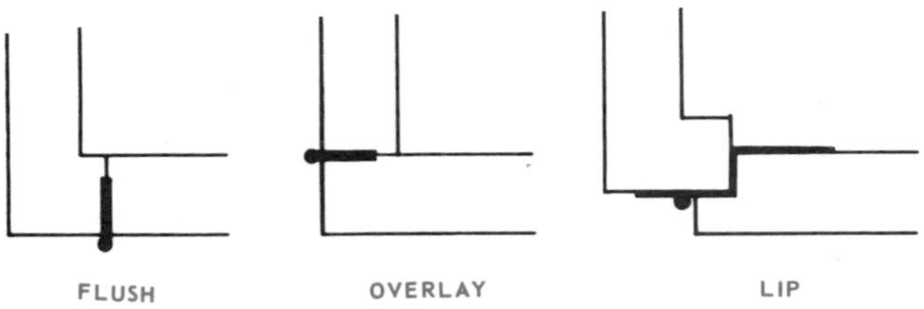

Fig. 18-28. Types of cabinet doors.

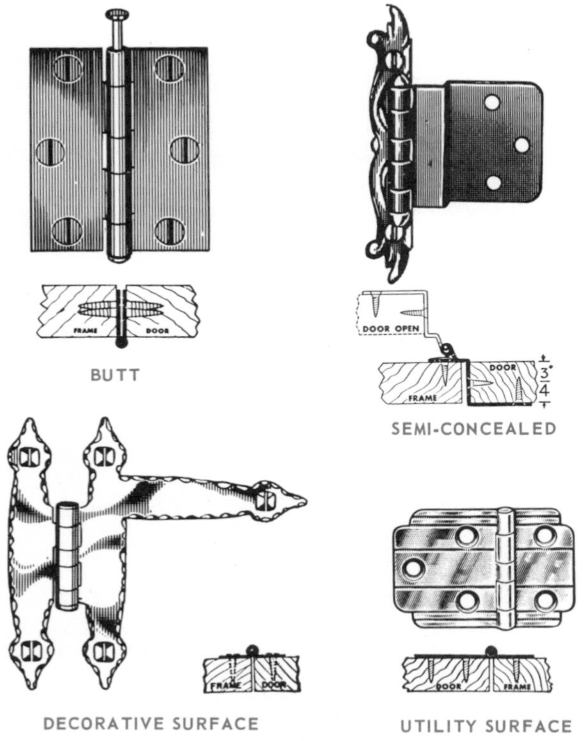

Fig. 18-29. Hinges for flush doors. (Amerock Corp.)

Standard shelving that is 3/4 in. thick should not be carried on supports that are spaced more than 42 in. apart. This applies especially to shelves that will carry books and other heavy loads. When plywood is used for the shelving material, the front edge should be overlaid with a strip of solid stock that matches the cabinetwork.

Doors

In built-in cabinetwork, the two general classifications of doors are swinging and sliding. For kitchen cabinets, the swinging door is most commonly used. Either type of door may be made of plywood (lumber core preferred) or consist of a frame with an inserted panel. Today, fine cabinet doors are often made of a frame covered with thin plywood on each surface -- similar to a regular hollow core door. Solid stock is seldom used, unless the door is quite small, because it will likely warp and excessive clearance must be provided to prevent it from sticking during humid weather.

Many carpenters use prefabricated cabinet doors which are available in various styles and in a wide range of stock sizes, Fig. 18-27. These are delivered to the job, cut to the exact dimensions with lipped edges (if specified) already machined.

HANDY SAYS:

"When specifying the size of the opening or the size of a cabinet door, always list width first and height second. This is standard practice among carpenters and is commonly followed by lumber supply stores and millwork plants."

Swinging doors may be hung flush with the opening, overlap the opening, or they can be inset part way like the lipped drawer front, Fig. 18-28. The flush door is usually installed with butt hinges. However, surface hinges, knife hinges or various invisible hinges can be used. Fig. 18-29 shows several types commonly selected for cabinetwork. Select a hinge that will be appropriate for the size of the door. The size of a hinge is determined by its length and width when open. On large wardrobe doors, three hinges should be used.

Flush doors must be carefully fitted to the opening with about 1/16 in. clearance on each edge. The total gain required for the hinge can be cut entirely in the door. However, for fine work, it is best to cut equally into each member. On large doors, it is practical to use the portable router for this operation.

Install the hinges on the door and then mount the door in the opening. Use only one screw in each hinge leaf. Make any necessary adjustments in the fit and then set the remaining screws. It may be necessary to plane a slight bevel on the edge of the door opposite the hinges.

Stops are set on the door frame so the door will be held flush with the surface of the opening when closed. The stops may be placed all around the opening or just on the lock or catch side.

Overlay doors provide an attractive appearance in some contemporary styles of cabinetwork. Butt hinges can be used; however, it is usually best to make the installation with a type that is designed for the purpose.

Lip doors are easier to cut and fit than flush doors because the clearance is covered. They must be installed with a special off-set hinge. See Fig. 18-30.

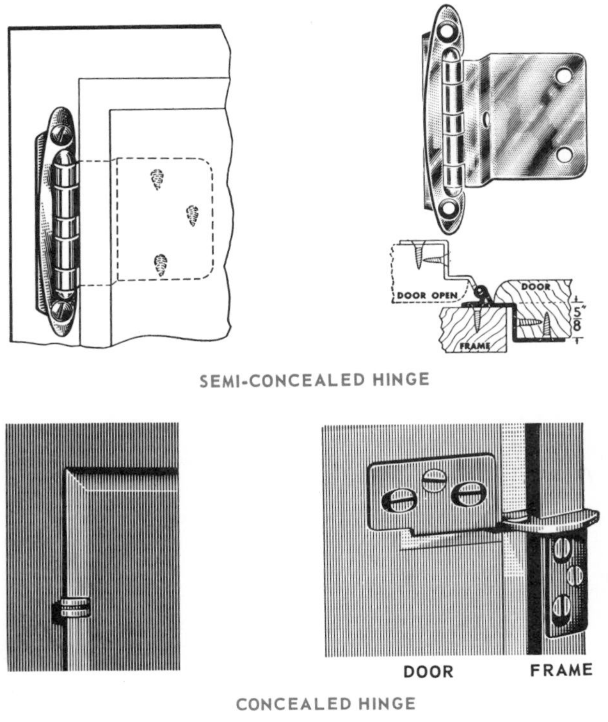

Fig. 18-30. Hinges for lip doors.

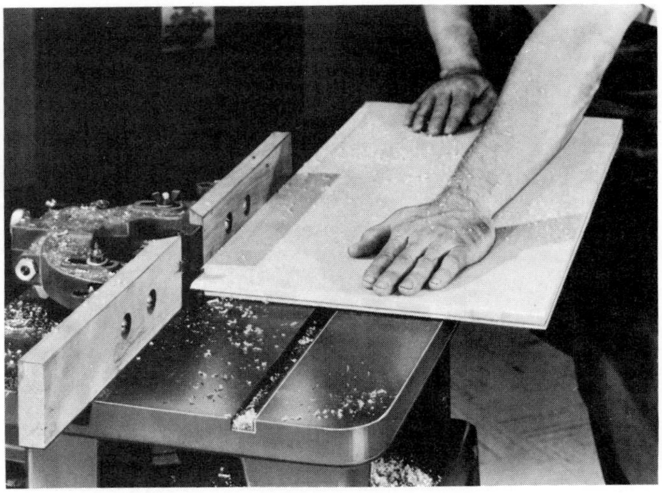

Fig. 18-31. Using a shaper to form the lip on a cabinet door. (Rockwell International)

After selecting the door blanks for a given series of openings (try to match grain and color when natural finish will be applied) cut the doors to the size of the openings plus the amount required for the lip. Clearance of about 1/16 in. is provided on each edge with an additional allowance for the hinges which are not gained into the door or frame. The lip is formed by cutting a rabbet along the edge. This can be accomplished on the table saw or with a shaper as shown in Fig. 18-31.

Check the fit of each door in its proper opening and mark its position in small letters on the back. Remove all machine marks with abrasive paper and soften edges and corners. Hardware may be prefitted and then removed during the finishing operations. Fig. 18-32 shows lip doors in place after finish has been applied.

Fig. 18-32. Lip doors installed on base cabinet.

Sliding Doors

Sliding doors are often used where the swing of regular doors would be awkward or cause interference. They are adaptable to various styles and structural designs, Fig. 18-33.

Fig. 18-33. Installing sliding doors in a storage cabinet.

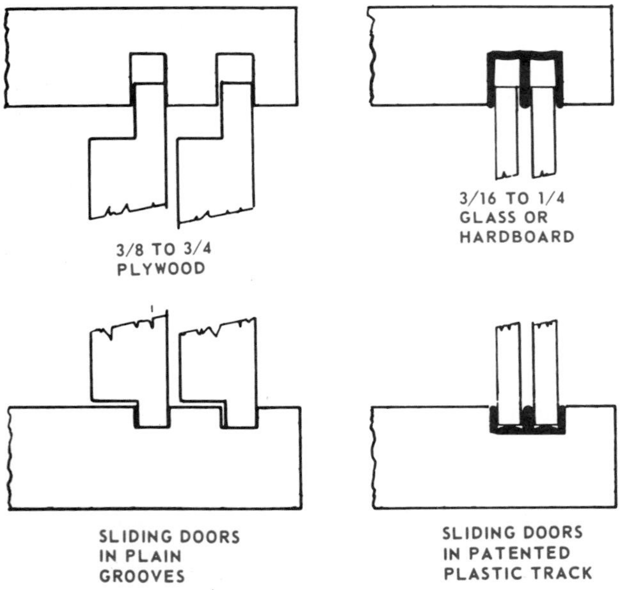

Fig. 18-34. Sliding door details.

Construction of a sliding door arrangement is shown in Fig. 18-34. Grooves are cut in the top and bottom of the cabinet (before assembly) and the doors are rabbeted so the edge formed will match the groove with about 1/16 in. clearance. Cut the top rabbet and groove deep enough that the door can be inserted or removed by simply raising it into the extra space. The doors will slide smoothly if the grooves are carefully cut, sanded, sealed and waxed. Excessive finish should

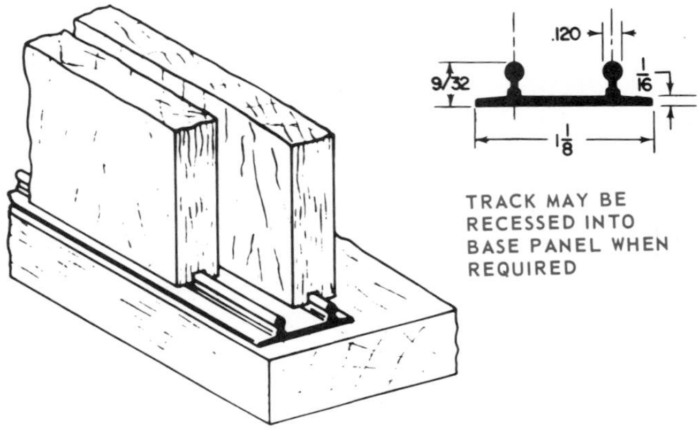

Fig. 18-35. Plastic door track installation. (Ronthor Reiss Corp.)

be avoided. Sliding glass doors are heavy and a special plastic or roller track should be used. Follow the manufacturer's recommendations for installation.

A wide range of sliding door track and rollers are available. Fig. 18-35 illustrates a self-lubricating plastic track that is easy to install and provides smooth operation for case and cabinet doors. Overhead track and rollers are used for large wardrobe doors and passage doors.

Sliding glass doors are adaptable to wall units and add variety to an installation. Fig. 18-36 shows bypass (1/4 in. plate glass) sliding doors in a cabinet in a kitchen-dinette area.

Fig. 18-36. Sliding plate glass doors featured in a wall cabinet section.

Counters and Tops

In modern cabinetwork, a high-pressure type of plastic laminate is commonly used as a surface for counters and tops. This material (usually 1/16 in. thick) offers high resistance to wear and is unharmed by boiling water, alcohol, oil, grease and ordinary household chemicals, Fig. 18-37.

Although the laminate is very hard, it does not possess great strength and is serviceable only when bonded to plywood, particle board or hardboard. This base or core material must be smooth and dimensionally stable. Hardwood plywood (usually 3/4 in. thick) makes a satisfactory base; however, some plywoods, especially fir, have a coarse grain texture which may telegraph (show through). Particle board which is less expensive than plywood provides a smooth surface and adequate strength, and is used extensively.

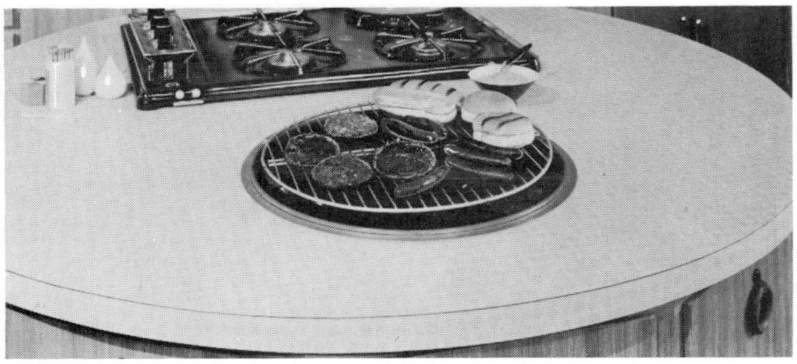

Fig. 18-37. Heat resistance factors make plastic laminates an ideal surface material for counters with built-in cooking units. (Formica Corp.)

When the core or base is free to move and is not supported by other parts of the structure, the laminated surface may warp. This can be counteracted by bonding a backing sheet of the laminate to the second face. It will minimize moisture penetration or loss and provide a balanced unit with identical materials on either side of the core. For a premium grade of cabinetwork, Architectural Woodwork Institute standards specify that a backing sheet be used on any unsupported area

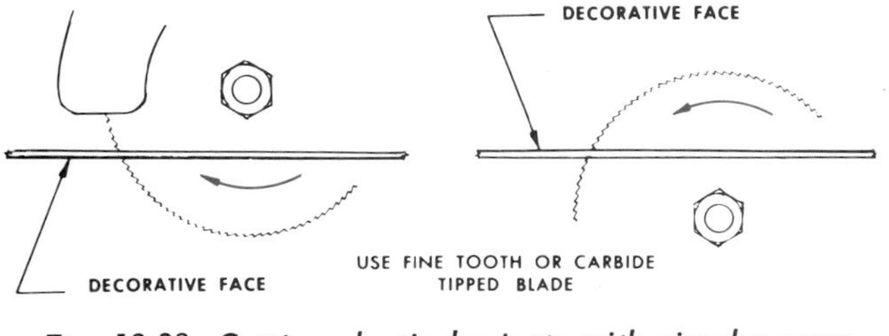

Fig. 18-38. Cutting plastic laminate with circular saws.

exceeding 4 square feet. Backing sheets are like the regular laminate without the decorative finish and are usually thinner. A standard thickness for use opposite a .060 in. (1/16) face laminate is .020 in.

Plastic laminates can be cut to rough size with a handsaw, tablesaw, portable saw or portable router, Fig. 18-38. Use fine-tooth blades and support the material close to the cut. Laminates 1/32 in. thick, which are used on vertical surfaces, can be cut with tin snips.

It is best to make the roughing cuts 1/8 in. to 1/4 in. oversize and then trim the edges after the laminate has been mounted. Handle large sheets carefully because they can be easily cracked or broken. Be careful not to scratch the decorative side.

Contact bond cement is used to apply the plastic laminate because no sustained pressure is required. It is applied with a spreader, roller, or brush to both surfaces to be joined. On large horizontal surfaces, it is desirable to use a spreader, Fig. 18-39. For soft plywoods, particle board or other porous surfaces, the spreader is held with the serrated edge perpendicular to the surface. On hard nonporous surfaces and plastic laminate, hold the edge at a 45 deg. angle as shown. A single coat should be sufficient.

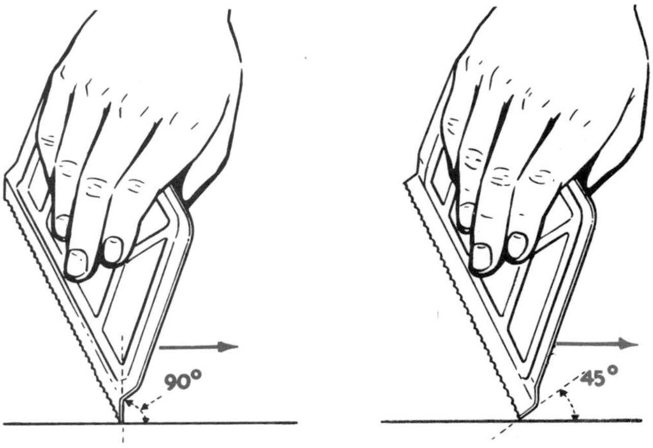

Fig. 18-39. Using a metal spreader to apply contact bond cement.

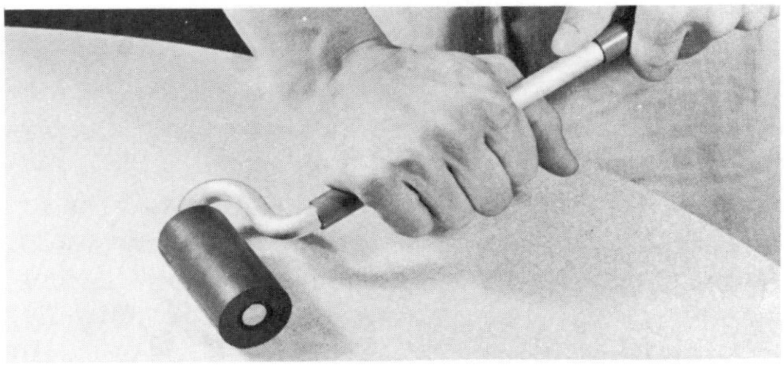

Fig. 18-40. Apply pressure to the laminate with a hand roller.

An animal hair or fiber brush may be used to apply the adhesive to small surfaces or those in a vertical position. Apply one coat, let it dry thoroughly and then apply a second. All of the surface should be completely covered with a glossy film. Dull spots, after drying, indicate that the application was too thin and that another coat should be applied.

Stir the adhesive thoroughly before using and follow the manufacturer's recommendations. Usually, brushes and applicators must be cleaned in a special solvent.

HANDY SAYS:

"Regular contact cement is extremely flammable. Keep it away from sparks, heat, and open flame. Be sure there is adequate ventilation and avoid breathing the vapor. Keep the container closed when not in use."

After the adhesive has been applied to both surfaces, let them dry (usually at least 15 minutes). You can test the dryness by pressing a piece of paper lightly against the coated surface. If no adhesive sticks to the paper, it is ready to be bonded. This bond can usually be made any time within an hour (time varies with different manufacturers). If the assembly cannot be made within this time, the adhesive can be reactivated by applying another thin coat.

Bring the two surfaces together in the EXACT POSITION required because they cannot be shifted once contact is made. When joining large surfaces, place a sheet of heavy wrapping paper (called a slip-sheet) over the base surface and then slide the laminate into position. Withdraw the paper slightly so one edge can be bonded and then remove the entire sheet and apply pressure.

Total bond is secured by the application of momentary pressure. Hand rolling provides satisfactory results if the roller is small (3 in. or less in length), Fig. 18-40. Long rollers apply less pressure per square inch. Work from the center to the outside edges and be certain to roll every square inch of surface. In corners and areas that are hard to roll, hold a block of soft wood on the surface and tap it with a rubber mallet.

Trimming and smoothing the edges is an important step in the application of a plastic laminate. A plane or file may be used but many carpenters prefer an electric router, equipped with an adjustable guide, Fig. 18-41. In making cutouts for sinks or other openings, an electric saber saw as illustrated in Fig. 18-42, is practical.

Fig. 18-41. Using an electric router to trim an overhanging edge. Note the guide that insures an accurate cut. (Black & Decker Mfg. Co.)

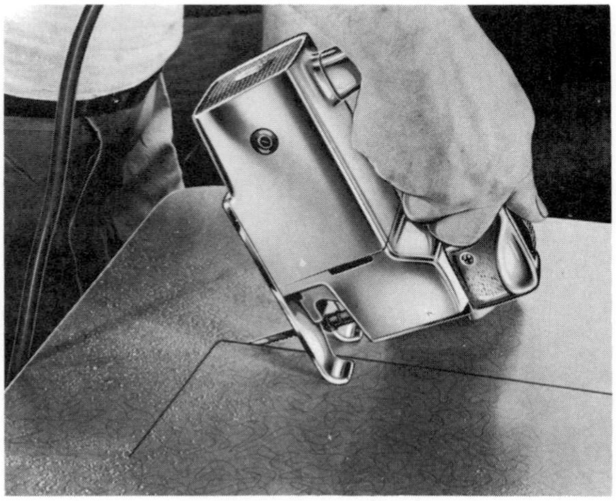

Fig. 18-42. Using an electric saber saw to cut an opening for a sink. View shows start of cut. (Stanley Power Tools)

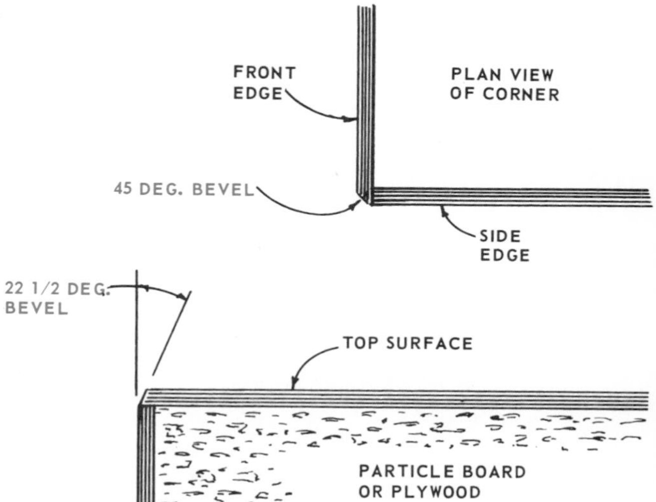

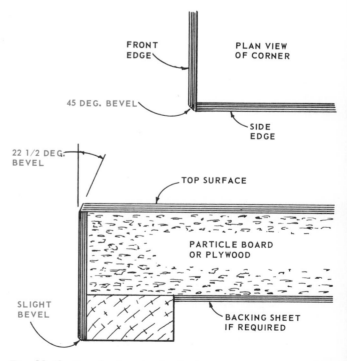

Fig. 18-43. Bevels for plastic laminate corners are very important in the production of quality work.

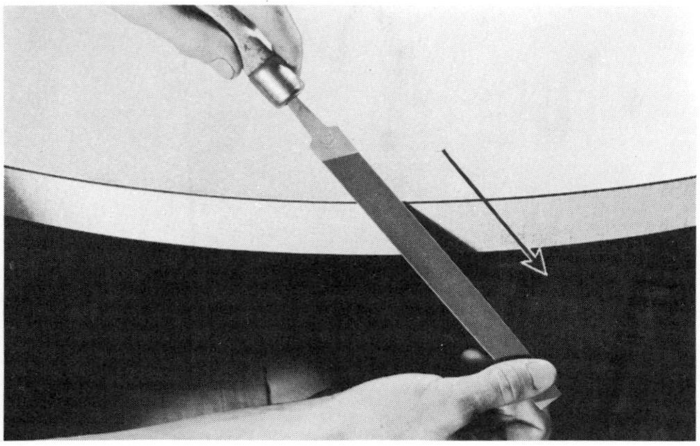

Fig. 18-44. Using a mill file to produce a beveled corner.

Fig. 18-45. Special router bit produces beveled corner. (Black & Decker Mfg. Co.)

Fig. 18-46. Completed plastic laminate counter surface for a bathroom built-in unit.

The corners of a plastic laminate application should be beveled, Fig. 18-43. This will make them smooth to touch and they will also wear better. The angle can be formed with a smooth mill file, Fig. 18-44. Stroke the file downward and be careful not to damage the surface of the edge trim strip. Some routers can be equipped with an adjustable guide and a special bit that will make this cut, Fig. 18-45. Final smoothing and a slight rounding of the bevel should be done with a 400 wet-or-dry abrasive paper.

When working with plastic laminates, be especially careful that files, edge tools or abrasive papers do not scratch or otherwise damage the finished surfaces. A completed countertop for the bathroom built-in unit pictured previously in constructional views is shown in Fig. 18-46. A special metal cove and cap strip has been used to trim the "back splash." FHA requires a minimum back splash height of 4 in. where kitchen counters join the wall surface.

Cabinet Hardware

After counters and tops have been covered and surface finishes applied, hardware is installed. This consists of knobs, pulls and various other metal fittings. Care should be used in the selection of an appropriate size, style, material and finish. Fig. 18-47 shows several designs of quality hardware for drawers and doors.

Some hardware, mainly hinges, should be prefitted before the final surface finish is applied. Pulls and catches can be installed afterward. Use care in the lay out and drill holes with sharp bits that will not splinter the surface. Drilling jigs, like the one shown in Fig. 18-48, save time and insure accuracy.

Drawer pulls usually look best when they are located slightly above the center line of the drawer

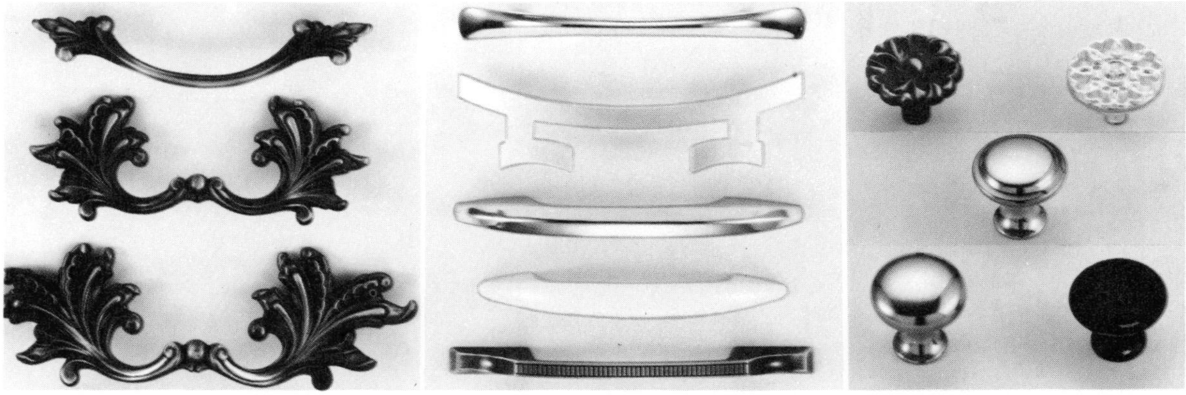

Fig. 18-47. Cabinet drawer and door hardware. (Amerock Corp.)

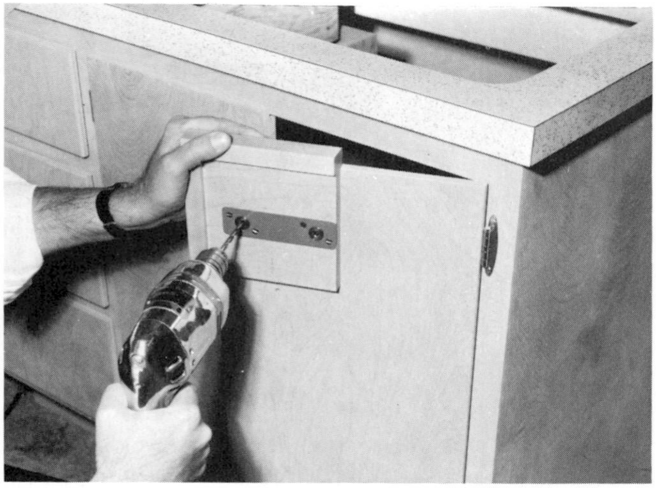

Fig. 18-48. *Using a drilling jig to locate holes for door pulls.*

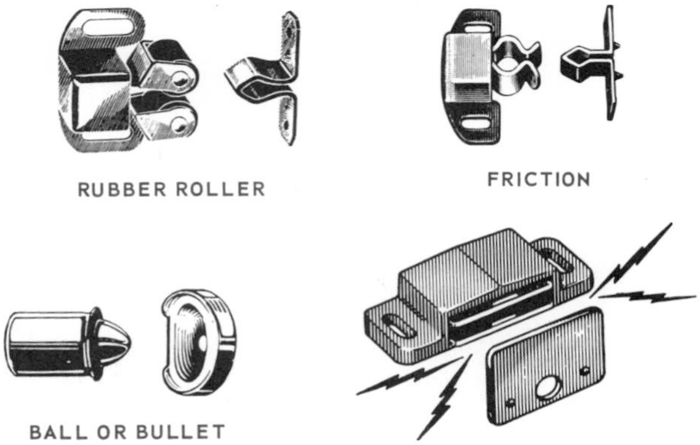

Fig. 18-49. *Types of cabinet door catches.*

front. Normally, they should be centered horizontally and in a level position. Door pulls are convenient when located on the bottom third of wall cabinets and the top third of base cabinets.

Swinging doors require some type of catch to keep them closed. Several common types are illustrated in Fig. 18-49. An important considera-

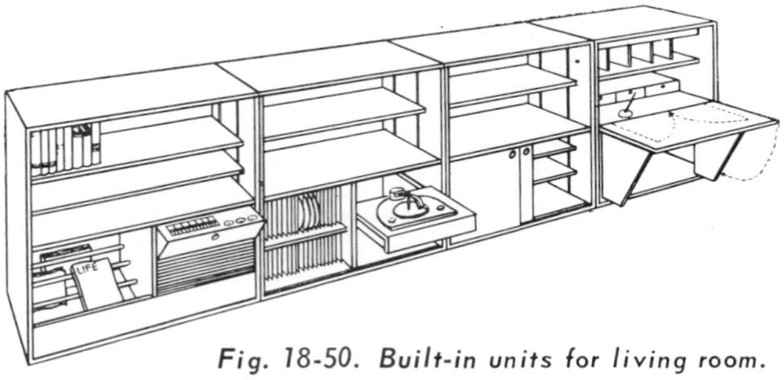

Fig. 18-50. *Built-in units for living room.*

Fig. 18-51. *Constructing a built-in unit that will serve as a wardrobe and room divider in a bedroom area. (American Plywood Assoc.)*

tion in their selection is the noise they produce. In general, the position of the catch should be as near as practical to the door pull. The package usually contains instructions for making the installation. Be sure to follow these carefully.

Fig. 18-52. Attaching facing strips to a bookcase that also serves as a room divider.

Other Built-In Units

Previous descriptions have been largely directed toward kitchen cabinets. The same general procedures can be applied to wardrobes, room dividers, and various built-in units for living rooms and family rooms. See Fig. 18-50.

The popularity of built-in features continues to grow in modern residential planning largely because of the demand for storage space and the general requirement that all space be organized and used as efficiently as possible. Another advantage is the attractive customized appearance that can result from well designed and carefully constructed units.

Fig. 18-53. Built-in bookcase, desk, and storage in a bedroom area. Note the door which features a woven wood pattern.

Fig. 18-54. Sequence of interior finish. Top. Basic framing for a built-in desk. Center. Facing and top complete. Bottom. Drawers fitted, plastic laminate applied, surface finish applied, and floor covering laid.

Study the details of the construction provided in the architectural plans or develop a carefully prepared working drawing using the actual dimensions secured from the wall, floor, and ceiling surfaces. When drawings are provided, check the space available to make sure that it corresponds with the requirements.

Wall and floor surfaces are seldom perfectly level or plumb and slight adjustments will be needed as the cabinetwork is constructed. Room corners or the corners of alcoves designed for built-in units will likely not be square and here again the cabinetwork will need to be trimmed or strips and wedges used to bring the work into proper alignment.

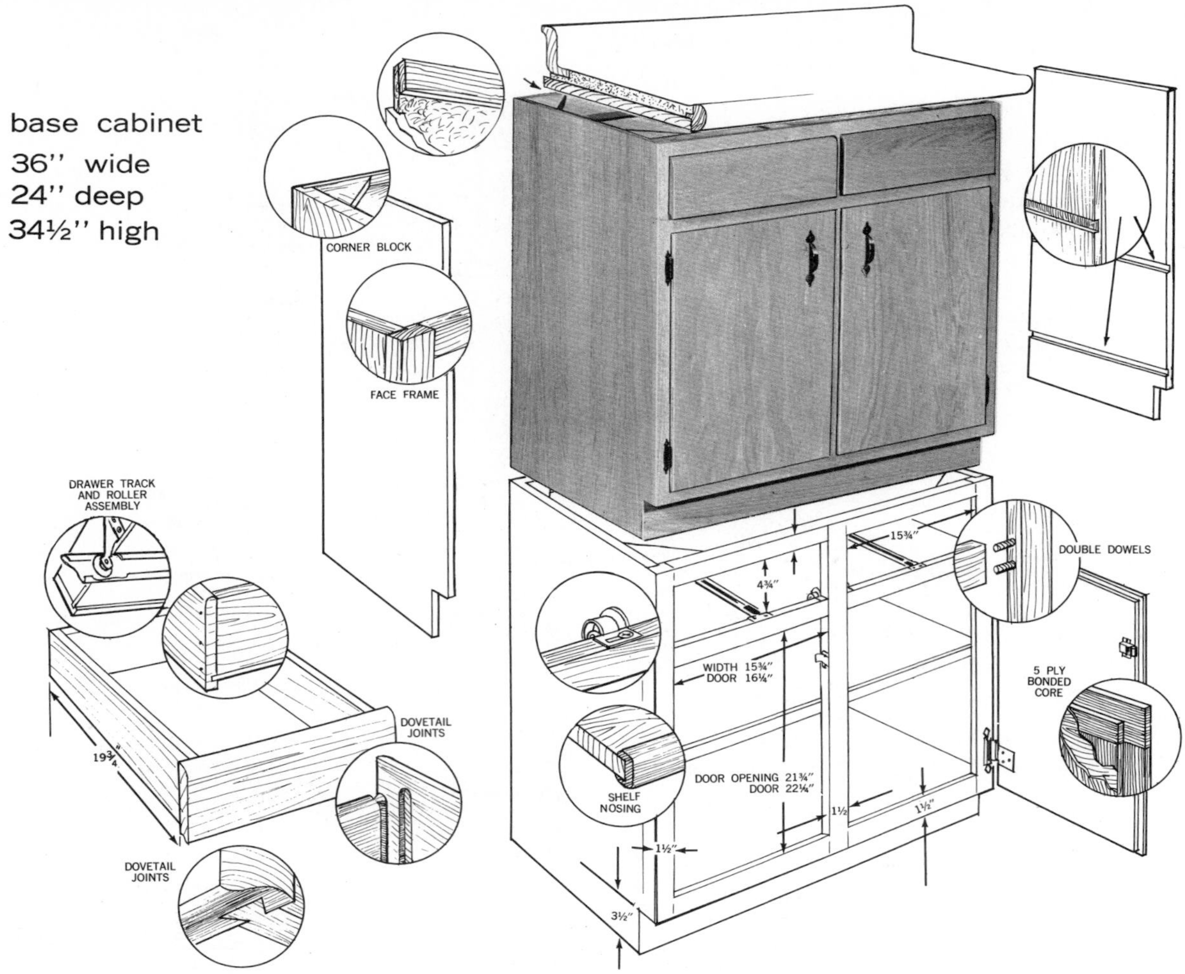

Fig. 18-55. Construction details of factory-built kitchen base cabinet. (Red Wing Wood Products)

Fig. 18-56. Construction details of factory-built wall cabinet.

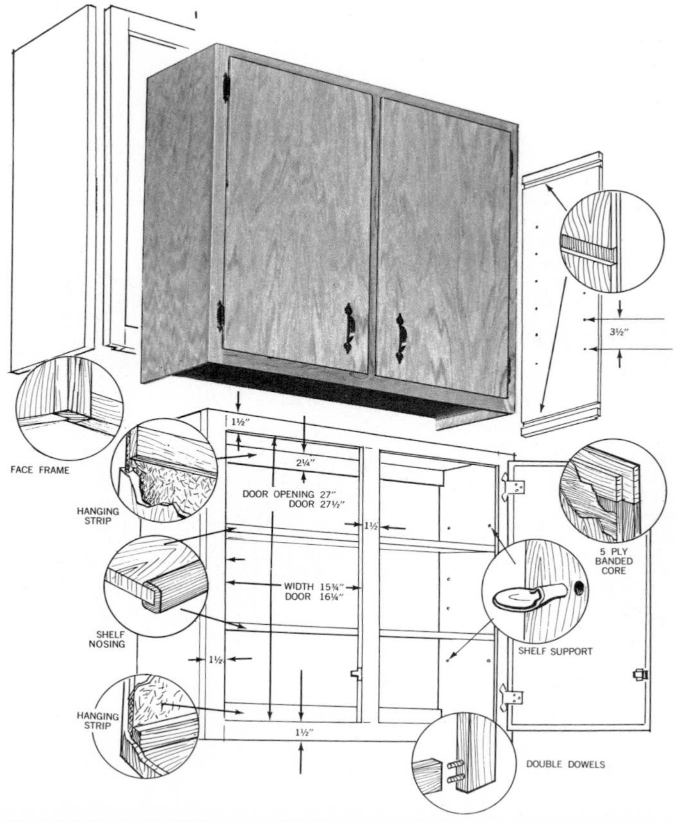

Fig. 18-57. Installation view of modern factory-built kitchen cabinets. (Kitchen Kompact Inc.)

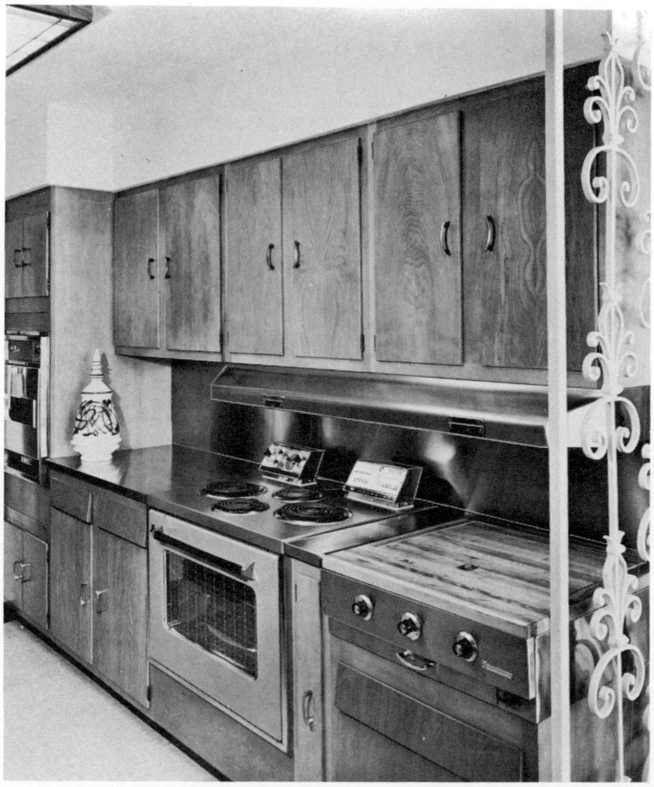

Keep a square and level constantly at hand, especially during the rough-in stages, Fig. 18-51. To carry inaccuracies from the wall or floor into the cabinet framework will make it difficult to apply the cabinet facing plumb and level. Errors, if permitted, will be a source of annoyance during the hanging of doors and fitting of drawers.

Built-in units vary in so many ways that no particular procedure can be recommended for their construction. For example: a bookcase that also serves as a room divider, Fig. 18-52, could be constructed during the regular interior finishing stages, as shown -- or it could be built as a detachable unit and installed just before interior painting and decorating. Fig. 18-53 shows a partial view of built-in units located in a bedroom area. Here the various components (headboard, bookcase, desk) were constructed and surface finishes applied in a basement area -- then moved into the room and attached to the walls after walls were painted and carpet laid.

Sequence of Interior Finish

Interior finish after wall, ceiling and floor surfaces are complete involves a wide variety of work as described in this Unit and Unit 17. The sequence in which the work is performed may vary considerably depending on the type of materials and construction used. The carpenter should give a great amount of attention to this -- organizing his work to prevent bottlenecks (delays). Appropriate times should be established for the delivery of materials. When delivered too far ahead, they will interfere with other work and may be subjected to damage. Work schedules need to be carefully planned and followed as closely as possible. The application of trim and the construction of cabinetwork generally proceed simultaneously.

Fig. 18-58. Mass production of cabinets in a factory. View shows door installation. (National Homes Corp.)

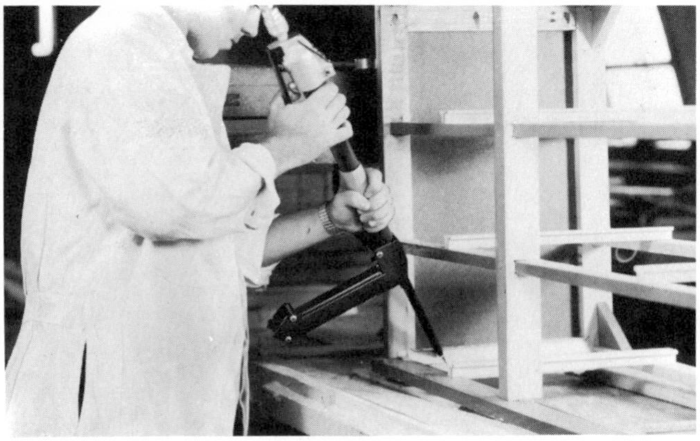

Fig. 18-59. Factory production of cabinetwork permits the use of specialized equipment. Here a long-nose stapler (air operated) is being used to attach drawer slides. (Duo-Fast Corp.)

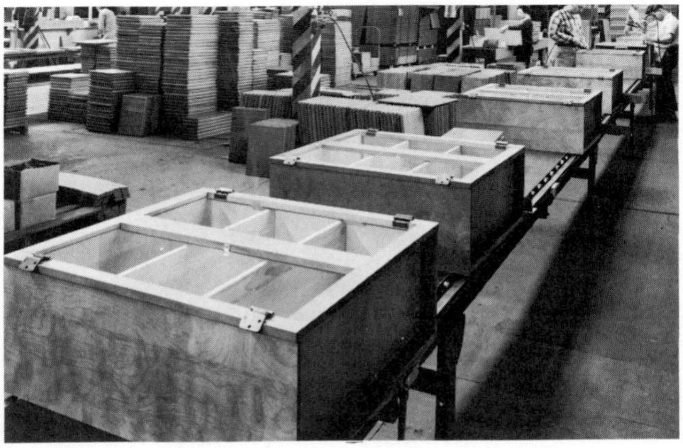

Fig. 18-60. Mechanized conveyor carries kitchen wall cabinets along assembly line in a modern woodworking plant. (Kitchen Kompact Inc.)

Fig. 18-61. Spraying a final coat of finish on base cabinets as they are carried along on an overhead conveyor.

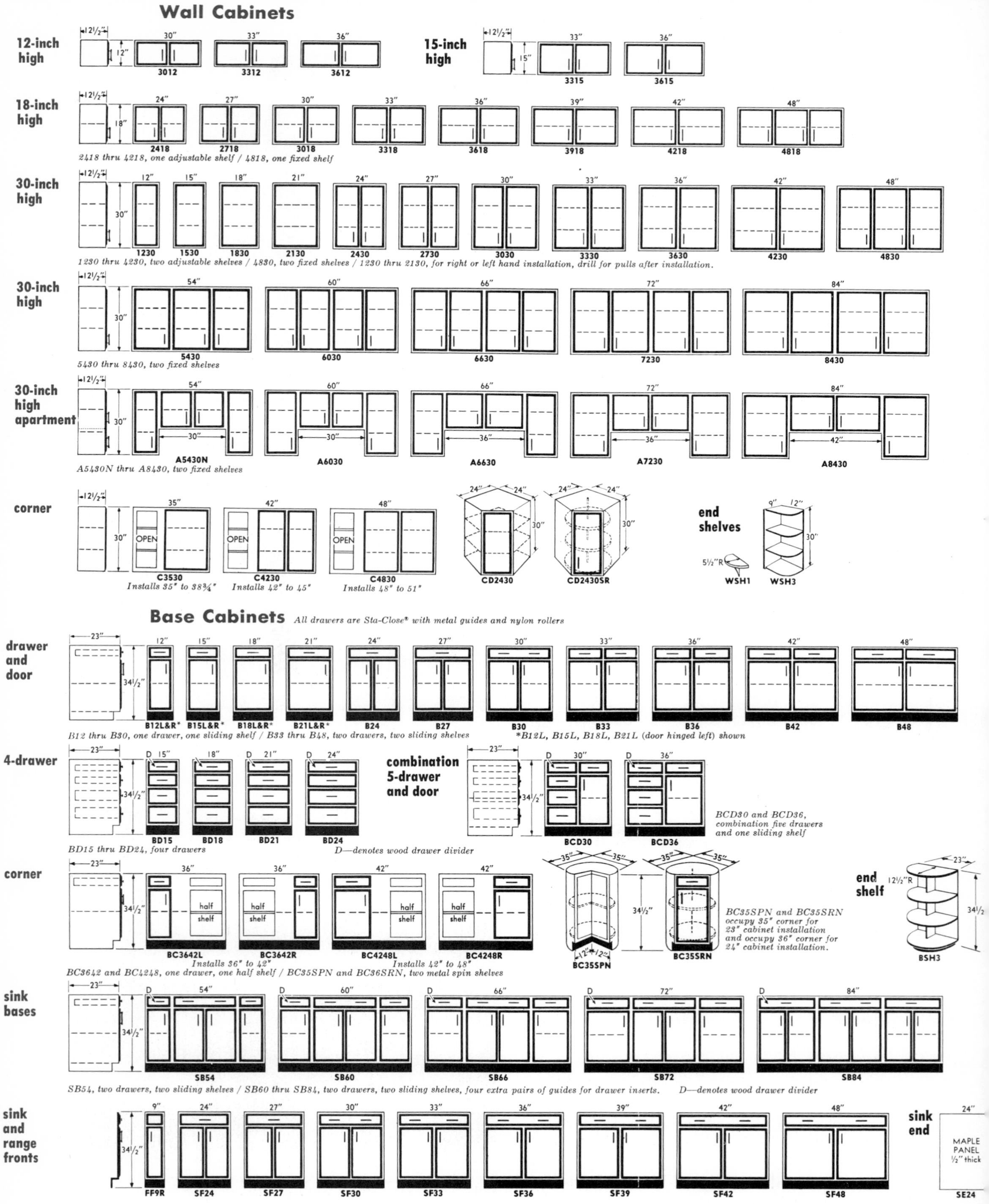

Fig. 18-62. Dimensions of base and wall units built by one manufacturer. (I-XL Furniture Co.)

Fig. 18-54 illustrates the sequence on a job where assembled-in-place cabinetwork was used. The top view shows the basic framework for a built-in desk in a kitchen area. It is installed on the underlayment and before the baseboard is set. In the center view, the facing and top have been added. Note that the baseboard is in position. Drawers are built and fitted and then surface finishes are applied to the desk, baseboard, and other trim. The plastic laminate top surface is installed, and when complete, the floor covering is laid. Finally, as shown in the bottom view, hardware is fitted and the base shoe is set in place.

Factory-Built Cabinets

Today, a major part of the cabinetwork for residential and commercial buildings is constructed in factories that specialize in this field. Modern production machines produce high quality work. Mass-produced parts are assembled with the aid of jigs and fixtures. See Figs. 18-55 to 18-60 incl.

Factory-built cabinets may be obtained in one of three forms: disassembled, assembled but not finished (in-the-white), or assembled and finished. Disassembled or "knocked-down" cabinets consist of parts cut to size and ready to be assembled on the job. The assembled but unfinished cabinet is ready to set in place. Hardware is included but not installed. All surfaces are sanded and ready for finish.

The latter type of cabinet (assembled and finished) is widely used today. Factory applied finishes, Fig. 18-61, which are durable, can be secured in various shades. The units are complete with hardware installed and are carefully packaged for shipment to the construction site.

Manufacturers of cabinetwork offer a variety of standard units, especially in the area of kitchen cabinets, Fig. 18-62. Most kitchen layouts can be made entirely from these units; however, when necessary, special custom built units can be ordered. Sometimes factory-built cabinets are combined with units constructed in custom cabinet shops, Fig. 18-63, or with various on-the-job built units. In addition to kitchen cabinets, various factory-built units are available for living rooms and other areas of the home, Figs. 18-64 and 18-65.

Manufacturers provide detailed instruction for the installation of their products. These directions should be studied and carefully followed. As previously stated, floors and walls are seldom exactly level and plumb, therefore shims and blocking must be used so the cabinets are not racked or

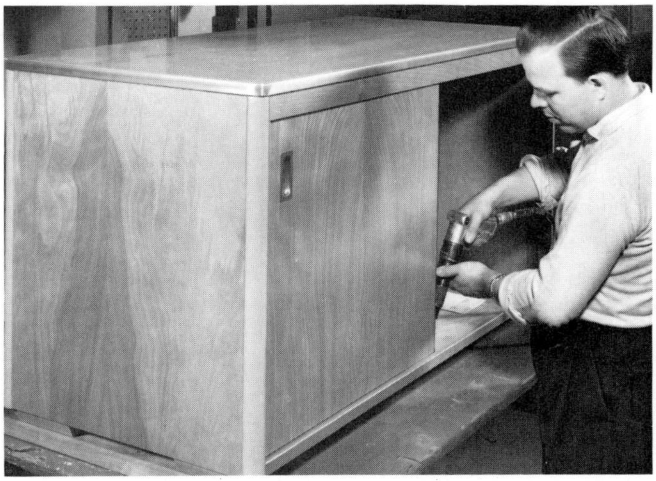

Fig. 18-63. Built-in storage cabinet with sliding doors constructed in a cabinet shop. (Weyerhaeuser Co.)

Fig. 18-64. Factory-built lavatory and vanity unit. (Kitchen Kompact Inc.)

Fig. 18-65. Modular cabinet units for a living room. (Herman Miller, Inc.)

twisted. Doors and drawers cannot be expected to operate properly if the basic cabinet is distorted by improper installation.

Screws should go through the hanging strips and into the stud framing. Toggle bolts are required when studs are inaccessible. Join units by first clamping them together and then, while they are perfectly aligned, install bolts and T-nuts. As each cabinet is installed, check from front to back and also along the front edge with a level. Be sure the front frame is plumb.

After the base cabinets are secured in place, the counter top is attached according to the manufacturer's specifications. It should be immediately covered with cardboard to protect it from damage.

Test Your Knowledge - Unit 18

1. The base unit of a standard kitchen cabinet is ____________ in. high.
2. When the carpenter draws a full-size sectional view of a cabinet, he generally refers to it as a ____________ ____________.
3. Assembled-in-place kitchen cabinets usually do not include ____________ panels.
4. When building assembled-in-place base cabinets, the partitions are installed ____________ (before, after) the bottom is secured to the base.
5. The vertical members used to face a cabinet are called ____________.
6. Three common types of drawer guides include: side guides, corner guides, and ______ ______.
7. Stock for kitchen cabinet drawer sides is usually ________ to ________ in. thick.
8. Standard shelving that is 3/4 in. thick should be carried on supports spaced no greater than ____________ in.
9. Three types of swinging doors that may be used in cabinetwork include: lip, flush and ____________.
10. When making a plastic laminate installation on a table or counter, it is recommended that a ____________ be used on any unsupported area greater than 4 square feet.
11. When exerting pressure to bond a plastic laminate by hand, it is best to use a ____________ (long, short) roller.
12. Drawer pulls are usually located slightly ______ (above, below) the center line.

Outside Assignments

1. Secure a set of architectural plans that include detail drawings of the kitchen cabinets. After a study of the details, prepare a master layout of a typical base cabinet. Since a complete view would be rather large, you may prefer to show only the front 4 - 6 in. Use a sharp pencil and show all the parts and clearances. Present your drawing to the class with explanations of the various structural parts.
2. If you have access to shop equipment, construct a typical cabinet drawer. Select appropriate joints that can be used for on-the-job built cabinets. Cut and fit the parts carefully. You may prefer to not glue the drawer together so that when you make a presentation to your class, it will be easier to show the joints. Another alternative is to glue the drawer, then cut it apart in several places so joint sections could be easily viewed.
3. Make a study of the various methods and devices used to install sinks in counter tops. From manufacturers' literature, prepare a sectional drawing about four times actual size that you can use in making a presentation to your class. Check into the possibility of borrowing sample rims and brackets from a building supply store that can be passed around to the group during your explanation.

Fig. 19-1. A fireplace creates a warm, cheerful atmosphere and provides an attractive interior design feature. (Majestic Company, Inc.)

Unit 19
CHIMNEYS AND FIREPLACES

A chimney is a vertical shaft through which smoke and gases from heating units, fireplaces, and incinerators are eliminated from the structure. When properly designed and constructed, it may also provide an architectural feature that will improve the exterior appearance.

The popularity of the fireplace in residences is high even though its efficiency as a source of heat is very low. The desire for a fireplace results from the cheerful, homelike atmosphere it creates and the decorative feature it contributes to the interior design. See Fig. 19-1.

Masonry Chimneys

Masonry chimneys are usually free-standing and constructed in such a way that they provide no support to, nor receive support from, the structural frame. Footings should extend below the frost line and project at least 6 in. beyond the sides. Walls of a chimney with clay-flue lining should be at least 4 in. thick. Foundations for a chimney or fireplace, especially when located on an outside wall, may be combined with those used for the building structure.

The size of a chimney will depend on the number, arrangement, and size of the flues (vertical openings in the chimney). The flue for a heating plant should be of sufficient cross-sectional area and height to create a good draft, thus permitting the equipment to develop its rated output. Always follow the heating equipment manufacturer's recommendations when deciding on flue sizes.

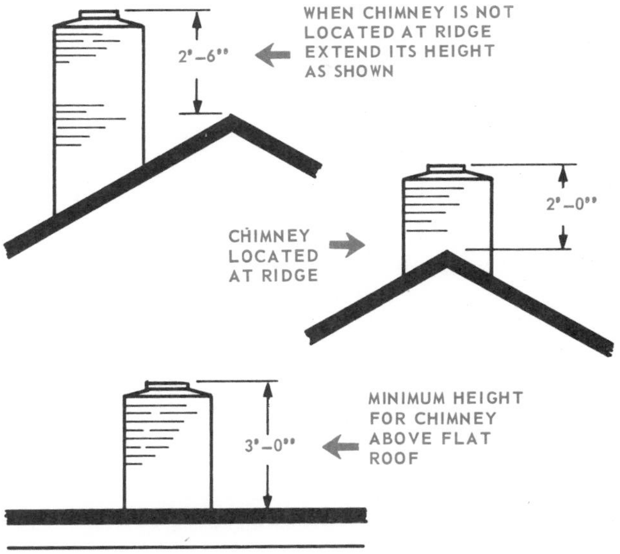

Fig. 19-2. Minimum chimney heights above roof. Check building codes in your area.

Building codes require that chimneys be constructed high enough to avoid downdrafts caused by the turbulence of wind as it sweeps past nearby obstructions or over sloping roofs. Minimum heights generally required are illustrated in Fig. 19-2. The chimney should always extend at least 2 ft. above any roof ridge that is within a 10 ft. horizontal distance.

Combustible materials such as wood framing members should be located at least 2 in. away from the chimney wall, Fig. 19-3. The open spaces between the framework and the chimney can be filled with mineral wool or other noncombustible material.

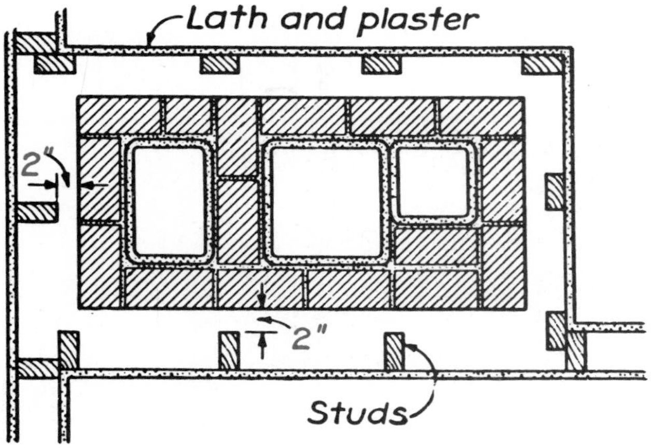

Fig. 19-3. Typical chimney cross section showing framing clearance. When two flues adjoin each other as shown, the joints of the flue lining should be staggered vertically a distance of at least 7 inches.

Flue Linings

The use of fireclay flue linings is recommended by the National Board of Fire Underwriters and is usually required by local building codes. Linings are available in square, rectangular, and round shapes and each shape is available in several sizes as listed in Fig. 19-4. Note that different methods of measurement are used for the three types: outside measurements for the old standard; nominal outside dimensions for the modular; and inside dimensions for the round linings. Small and medium flue linings are usually available in 2 ft. lengths. Flue rings are placed during the erection of the chimney and serve as a form around which the masonry is laid.

Although unlined fireplace chimneys with masonry walls (minimum thickness 8 in.) will operate satisfactorily, glazed flue lining is usually recommended because its smooth interior surface is less likely to attract pitch and tar which may eventually result in a restricted passage.

As a rule, a single flue should be used for only one heating unit. An exception often permitted is the vent from a gas-fired hot water heater which may be connected to a furnace flue -- usually with satisfactory results.

Construction

Permit the chimney footing to cure for several days so it will have adequate strength. Use care in starting the masonry work, making all lines level and plumb.

Each section of flue lining is set in place ahead of the chimney wall, serving somewhat as a guide for the brick work, Fig. 19-5. Joints in the flue lining are bedded in mortar or fireclay. Use considerable care in placing the lining units so they are square and plumb. Brick work is carried up along the lining and then another lining unit is placed.

STANDARD LINERS		MODULAR LINERS		ROUND LINERS	
Outside Dimensions	Area of Passage	Nominal Outside Dimensions	Area of Passage	Inside Diameter	Area of Passage
8½x 8½	52.56	8x12	57	8	50.26
		8x16	74		
8½x13	80.50			10	78.54
		12x12	87		
8½x18	109.69	12x16	120	12	113.00
13 x13	126.56	16x16	162	15	176.70
13 x18	182.84	16x20	208		
18 x18	248.06	20x20	262	18	254.40
		20x24	320	20	314.10
		24x24	385	22	380.13
				24	452.30

Fig. 19-4. Flue liner dimensions and clear cross-sectional areas in square inches. The National Building Code requires that the wall of the flue liner be a minimum of 5/8 in. thick.

Where offsets or bends are necessary in the chimney, they should be formed by mitering the ends of the abutting sections of lining. This prevents reduction of the flue area. The angle of offset should be limited to 60 deg. The center of gravity of the upper section must be located well within the limits described in Fig. 19-6.

Chimneys are often corbeled (extended outward) just before they project through the roof, Fig. 19-7. The larger exterior appearance is often more attractive and the greater amount of exposed masonry will usually withstand the weather for a longer period of time. Corbeling should not exceed a 1 in. projection in each course. The larger size should extend at least 6 in. below the roof framing.

Openings in the roof frame should be formed prior to the chimney construction. Refer to Unit

9 for information and procedure. Water leakage around the base can be prevented by proper flashing. Corrosion resistant metal such as sheet copper is commonly used for this purpose. The flashing is built into the roof surface and extends up along the masonry. Cap or counter flashing is bonded into the mortar joints and is then lapped down over the base flashing. Refer to Unit 10 for further details concerning the installation of flashing.

At the top of the chimney, the flue lining should project at least 4 in. above the top course or capping and should be surrounded with cement mortar at least 2 in. thick. See Fig. 19-8. Slope the cap so wind currents are directed upward and also to insure good water drainage from the top. When several flues are located in the same construction, it is best to extend them to different heights.

Prefabricated Chimneys

Various lightweight chimney units are available that require no masonry work. They demand little space and can be suspended from the structural members of the building. One type has a double metal wall with the space between filled with insulation. Other types consist of single walls made of cement-asbestos or vermiculite concrete, lined with some type of refractory material.

When this type of chimney is used for venting smoke and gases that may contain corrosive acids (incinerators or solid fuel boilers), the inside pipe should be made of stainless steel or have a porcelain coated surface.

HANDY SAYS:

"Prefabricated chimneys should show evidence that they have been tested and listed by the Underwriters' Laboratories, Inc. or other nationally recognized testing organizations. They must also conform to local building codes. Be sure to make the installation in strict accordance with the manufacturer's instructions."

Fireplace (Parts)

Fig. 19-9 shows a cutaway view that reveals the essential parts of a conventional masonry fireplace. The chimney and fireplace are combined in a single unit and often includes a flue for the regular heating equipment.

The hearth consists of two parts; one is located in front (front hearth) and the other is below the fire area (back hearth). The latter, along with

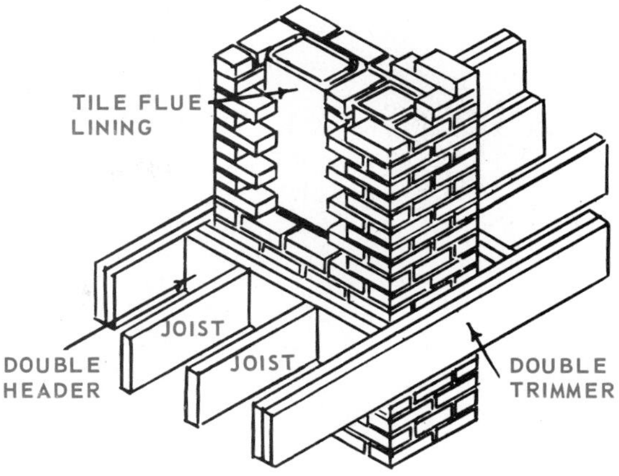

Fig. 19-5. Floor framing around chimney. Opening in brickwork shows flue lining.

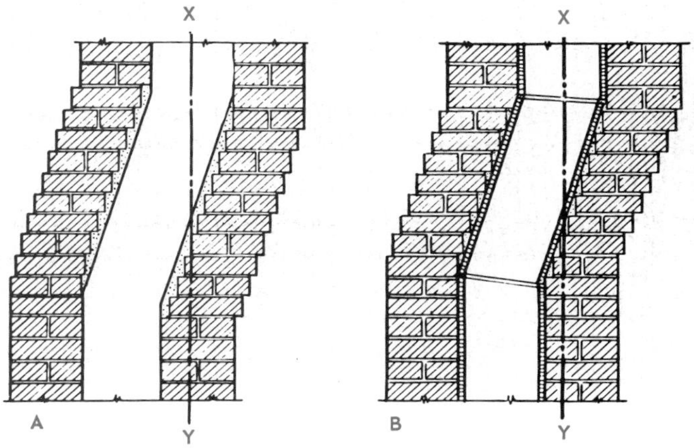

Fig. 19-6. For structural safety the amount of offset in a flue must be limited. The center line (XY) of the upper section must not fall beyond the center line of the wall in the lower section. In view A, the left wall of the unlined flue is started two courses higher than the right wall so the area of the sloping section will not be reduced after plastering. Note the method of cutting flue tile in view B.

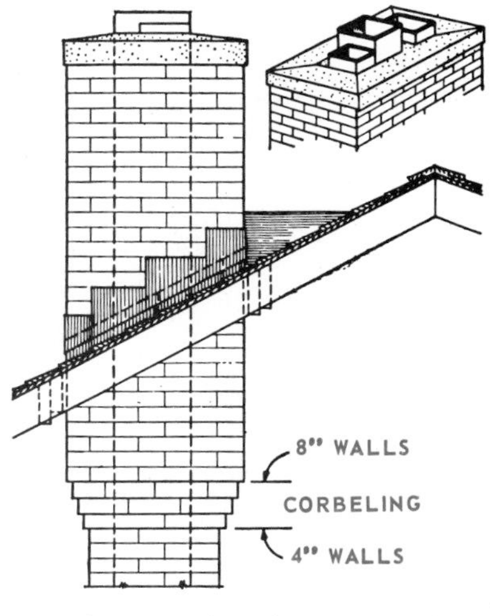

Fig. 19-7. Enlarging the top of a chimney results in greater resistance to weather and usually provides a more attractive appearance.

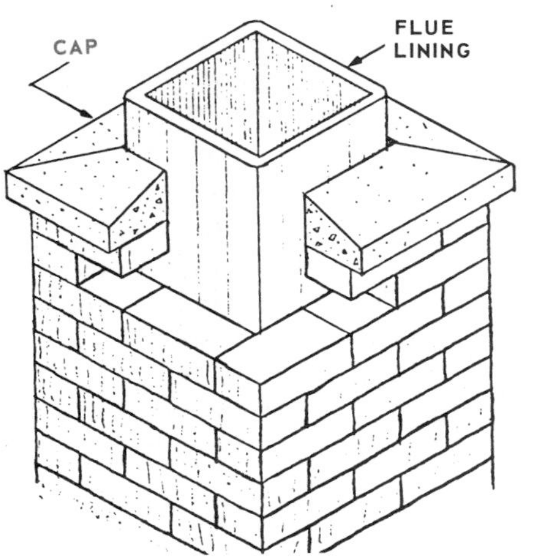

Fig. 19-8. Chimney cap made from cement mortar.

the sides and back wall are lined with firebrick which will withstand direct contact to flame. The side and back walls are usually sloped to reflect heat into the room.

A damper is located above the combustion area to control the fire and prevent loss of heat from the room when the fireplace is not being used.

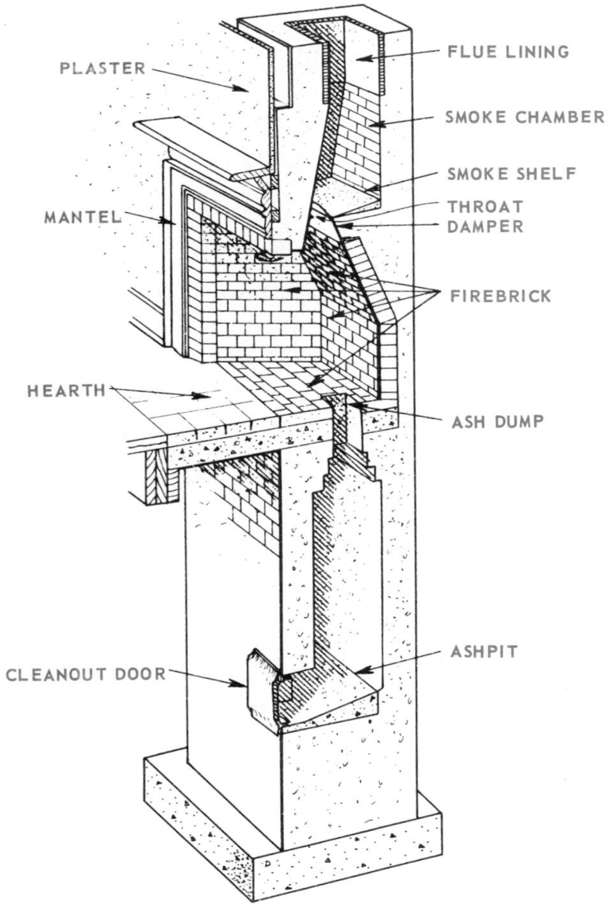

Fig. 19-9. Parts of a typical masonry fireplace.

The throat, smoke shelf, and smoke chamber are all important parts of the fireplace and must be carefully designed to insure proper operation.

Architectural plans often include details of the fireplace construction. See Fig. 19-10. The drawings include overall dimensions and can be scaled (measured) to secure others.

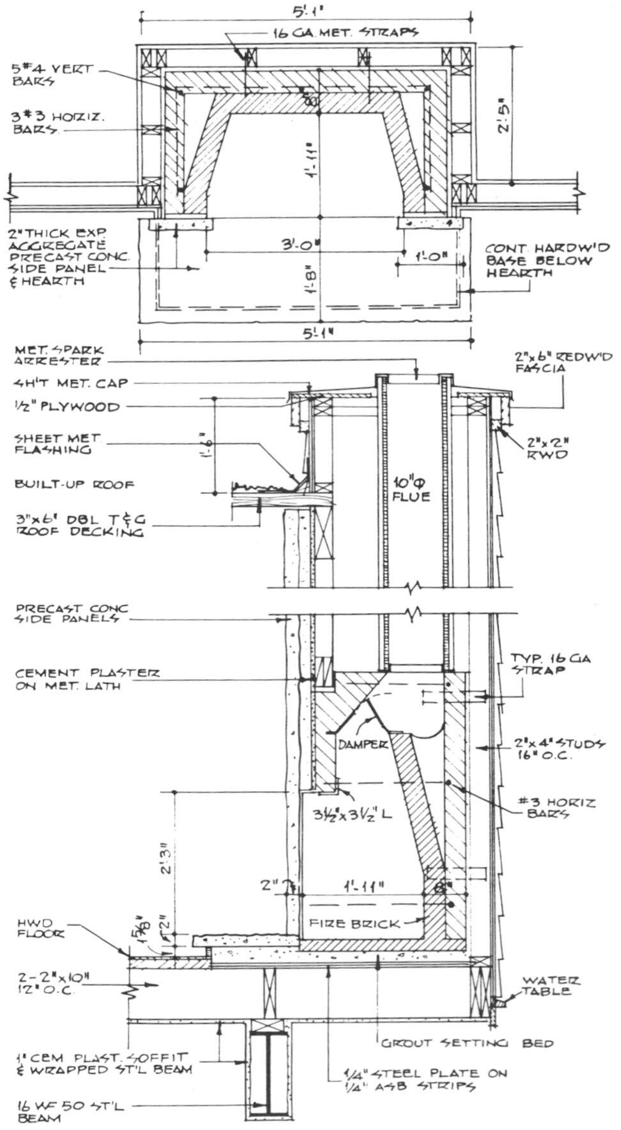

Fig. 19-10. Fireplace details are usually included in the architectural plans. In this design for a hillside house, the structure (and fireplace) is carried on 16 in. reinforced concrete piers and steel beams.

Construction Sequence

Masonry fireplaces are nearly always built in two separate stages. During the rough framing of the structure, masonry work is carried up from

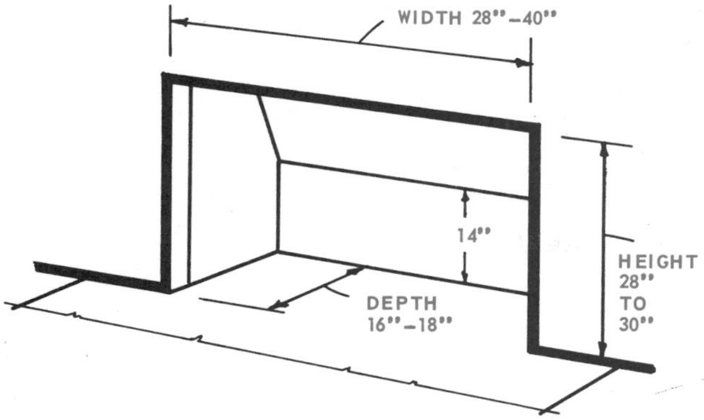

Fig. 19-11. Fireplace opening dimensions with range of sizes commonly used.

the foundation and the main walls of the fireplace are formed. After a steel lintel is set above the opening, the damper is installed and the smoke chamber built. Finally the chimney is carried on through the roof and the exterior masonry is completed -- usually before the roof surface is laid.

The second and finishing stage of the fireplace takes place after plastering or other type of wall surface is complete and during the application of interior trim. Decorative brick or stone may be set over the exposed front face. The finished surface of the front hearth is laid -- the reinforced

Fig. 19-12. A typical masonry fireplace illustrating details of construction. An alternate method of supporting the hearth is shown in the lower right-hand corner.

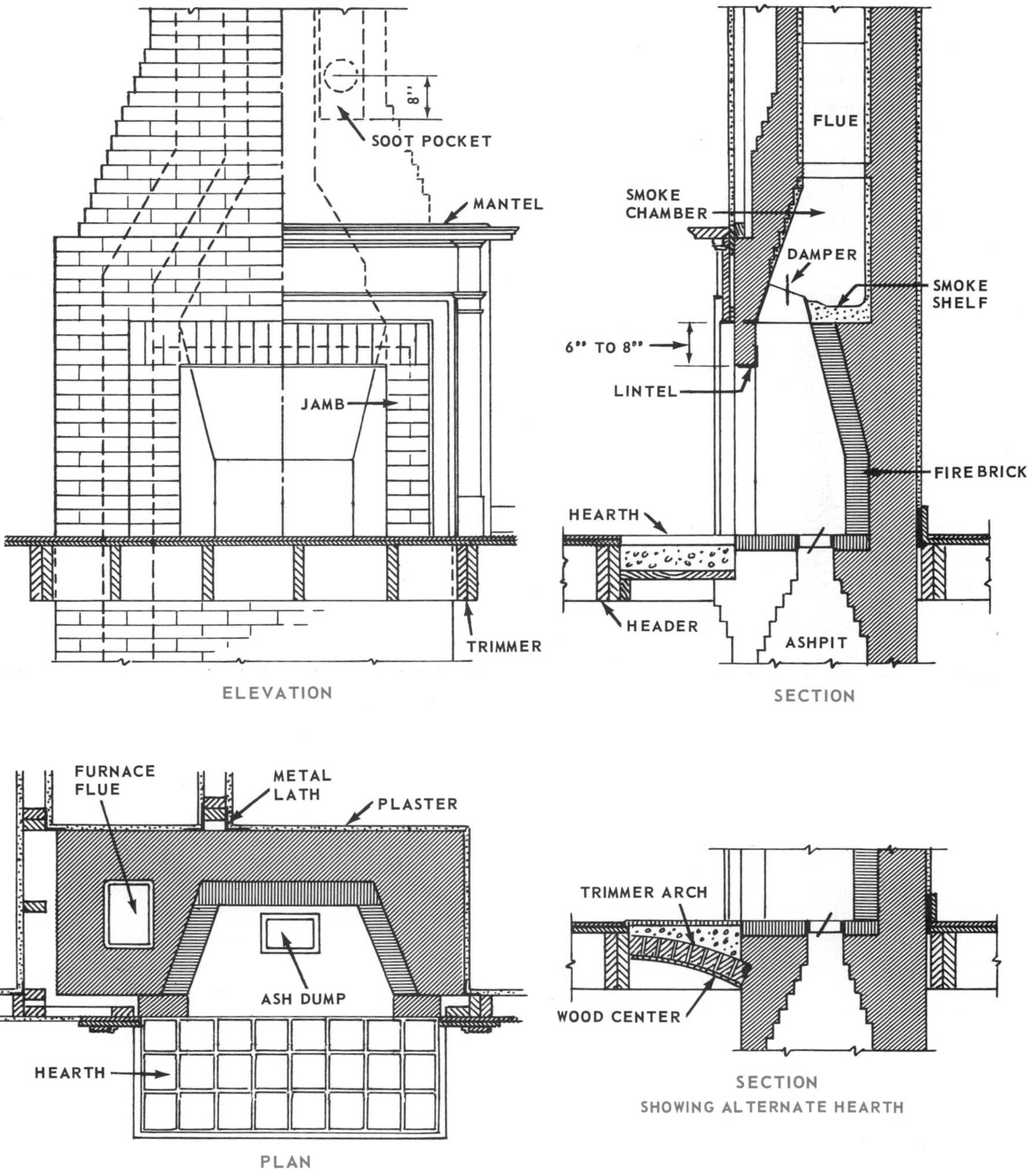

concrete base having been placed during the rough masonry construction. See opposite page for major steps in finishing a fireplace. Install wood trim (mantel) when masonry work is complete.

Design Details

The size of the fireplace opening should be based on the type of room and coordinated with the style of architecture. Some authorities suggest that it be of sufficient size to accomodate firewood 2 ft. long (standard cordwood cut in half). Fig. 19-11 illustrates standard procedure in listing fireplace dimensions.

More details of construction are shown in the detail drawing, Fig. 19-12. This particular design includes a flue for the furnace. The table in Fig. 19-13 lists some recommended dimensions for various openings and parts for a wide range of fireplace sizes.

Opening		Depth,	Minimum back (horizontal)	Vertical back wall,	Inclined back wall,	Outside dimensions of standard rectangular flue lining	Inside diameter of standard round flue lining
Width,	Height,						
Inches	*Inches*	*Inches*	*Inches*	*Inches*	*Inches*	*Inches*	*Inches*
24	24	16-18	14	14	16	8 1/2 by 8 1/2	10
28	24	16-18	14	14	16	8 1/2 by 8 1/2	10
24	28	16-18	14	14	20	8 1/2 by 8 1/2	10
30	28	16-18	16	14	20	8 1/2 by 13	10
36	28	16-18	22	14	20	8 1/2 by 13	12
42	28	16-18	28	14	20	8 1/2 by 18	12
36	32	18-20	20	14	24	8 1/2 by 18	12
42	32	18-20	26	14	24	13 by 13	12
48	32	18-20	32	14	24	13 by 13	15
42	36	18-20	26	14	28	13 by 13	15
48	36	18-20	32	14	28	13 by 18	15
54	36	18-20	38	14	28	13 by 18	15
60	36	18-20	44	14	28	13 by 18	15
42	40	20-22	24	17	29	13 by 13	15
48	40	20-22	30	17	29	13 by 18	15
54	40	20-22	36	17	29	13 by 18	15
60	40	20-22	42	17	29	18 by 18	18
66	40	20-22	48	17	29	18 by 18	18
72	40	22-28	51	17	29	18 by 18	18

Fig. 19-13. Recommended dimensions for an extensive range of fireplace sizes.

Hearth

The hearth, including the front section, must be completely supported by the chimney. This is accomplished by first building shoring (temporary support) and forms and then placing reinforced concrete (3 1/2 in. minimum thickness). In the best construction, a cantilevered design is secured by recessing the back edge of the total rough hearth into the rear wall of the chimney.

An ash dump may be provided in the rear hearth for clearing ashes, if there is space below for an ashpit. The dump consists of a metal frame with pivoted cover. The basement ashpit should be of tight masonry and include a clean-out door, Fig. 19-14.

When the floor structure consists of a slab-on-grade, an ashpit may be formed by a raised hearth

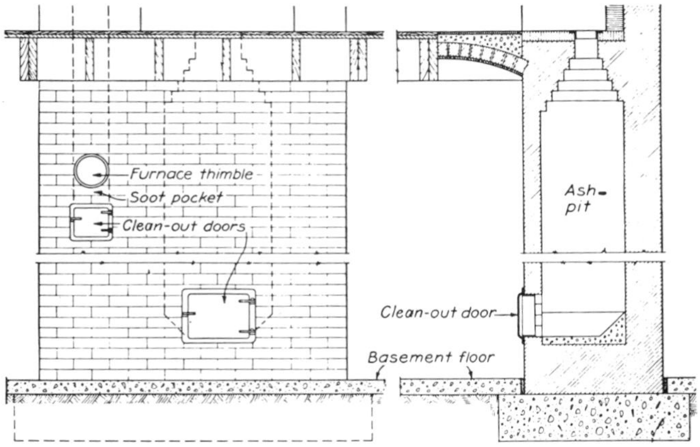

Fig. 19-14. Fireplace ashpit should include a tight fitting clean-out door. A clean-out for the furnace flue is also recommended but not required when gas fired equipment is used.

as illustrated in Fig. 19-15. This design is practical when the fireplace is located on an outside wall. In some designs, especially when no ashpit is included, the rear hearth is lowered several inches so ashes are easier to contain in this area.

Side and Back Walls

The side and back walls of the combustion chamber continue upward to the level of the damper. These must be lined with firebrick at least 2 in. thick; the firebrick being set in a special clay mortar that will withstand the heat. The total thickness of the walls, including the firebrick, should not be less than 8 in.

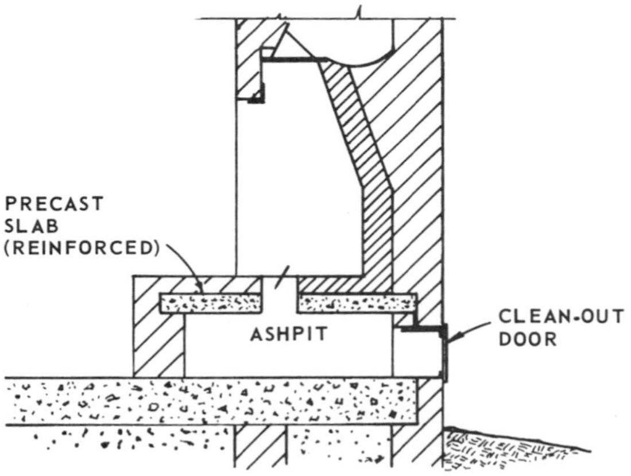

Fig. 19-15. Raised hearth provides space for ashpit in slab-on-grade construction.

Side walls are built at an angle for the purpose of reflecting heat into the room. This angle (also called "splay.") is usually laid out at 5 in. per foot. The back wall is extended vertically a distance

Rough-in for a corner fireplace. Lightweight concrete block have been laid up from concrete floor which was based on an adequate footing. Opening has been lined with firebrick. Steel lintel (arrow) supports masonry and damper above opening. Interior walls and ceilings have been finished, and ceiling has been spray painted. Note over-spray on top section of rough-in.

Skilled mason laying field stone gathered from building site and local area. Some of the stones must be held in place until the mortar sets. This is accomplished by using copper wires tied to galvanized nails driven into the blocks and mortar joints. After the joints have partially set, some of the mortar is raked out. Exposed wires are removed later. Note the temporary wood frame (arrow) used to form the edges of the opening. A steel lintel will be installed along the top of the frame (extending beyond the opening a minimum of 4 in. on each side) to support the stonework above the opening.

Stone facing complete and hearth installed. The mantel, a solid slab of wood, was laid in place with mortar and is an integral part of the masonry work. To anchor the mantel securely, galvanized nails were driven part way into the back edge of the wood slab. The stone facing has been cleaned with a mild solution of muriatic acid and then sealed with a thin coat of special finish to "bring out" the color of the stone.

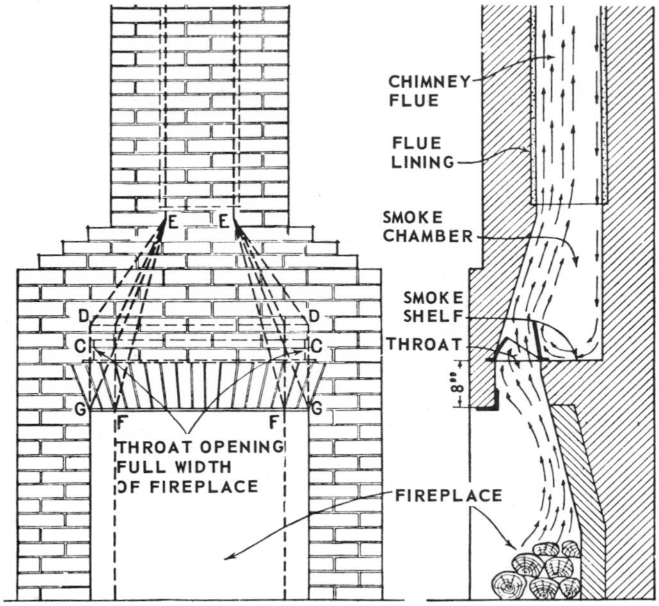

A B

Fig. 19-16. (View A) The top of throat damper is at DD, smoke shelf at CC. Side wall should not be drawn in until height of DD is passed. This assures full area. If drawing in is done as indicated by lines EF and EG, the width of the throat becomes less than the width of the opening and causes air currents to pile up in corners of the throat, resulting frequently in a smoky fireplace. (View B) Correct fireplace construction. Arrows indicate air and smoke currents.

slightly less than one-half of the opening height and then sloped forward. This slope directs the smoke into the throat of the fireplace and also forms an area for the smoke shelf which will be located above.

Throat

Fig. 19-16 shows the position of the throat, the narrow opening between the combustion area and the smoke chamber. Correct throat construction contributes much to efficiency of a fireplace. In the illustration, arrows indicate the upward flow of hot air and smoke which are directed forward through the throat and then pass into the front side of the smoke chamber. The rapid upward passage of hot gases creates a downward current on the opposite side. The downward flow is not strong but may be of such force that, if there is no throat to increase the velocity of the upward current, smoke may be forced out into the room. This explains why some fireplaces do not operate satisfactorily even though ample flue area has been provided.

Cross-sectional area of the throat should not be less than that of the flue. Its length should always be equal to the width of the fireplace opening with a vertical height of about 6 to 8 in. A length of heavy angle iron (called a lintel) is placed across the top of the opening to form the bottom edge of the throat and to carry the masonry above.

Damper

Part of the throat is formed by the damper, the size of which must be coordinated with the other aspects of the fireplace design. A vertical front flange permits it to rest snugly against the masonry of the front wall while the back flange is usually sloped to conform with the rear wall of the combustion chamber.

When installing the damper, be certain that masonry work above will not interfere with its full operation, also the ends should be provided with a slight clearance in the masonry to permit expansion. A standard damper design is shown in Fig. 19-17. Manufacturers will provide data and recommendations concerning model and size for specific installations.

Smoke Shelf and Chamber

To form the smoke shelf, the brickwork is set back at the top of the throat to the line of the flue wall. It should be made the full length of the throat. Depth of the shelf should not be less than 4 in., but it may vary from 4 in. up to 12 in. or more, depending on the depth of the fireplace.

Fig. 19-17. Top view of a standard fireplace damper. (Donley Brothers Company)

The purpose of the smoke shelf is to change the direction of the down draft so the hot gases at the throat will strike it approximately at a right angle instead of head-on. Therefore the shelf should usually be made as wide as possible. In the best construction, it is curved to reduce turbulence in the air flow.

The smoke chamber is the space extending from the top of the throat up to the bottom of the flue. The area at the bottom of the chamber is quite large, since its width includes that of the throat plus the depth of the smoke shelf. This space holds accumulated smoke temporarily if a gust of wind across the top of the chimney momentarily cuts off the draft. Without this chamber, smoke would likely be forced out into the room. A smoke chamber also lessens the force of the down draft by increasing the area through which it passes. Side walls are generally drawn inward one foot for each 18 in. of rise. All surfaces of the smoke chamber should be plastered smoothly with a 1/2 in. thickness of cement mortar.

Flue Size

Cross-sectional area of the flue is based on the area of the fireplace opening. A general rule states that the flue area should be at least 1/10th of the total opening when a lining is used. This applies to chimneys that are at least 20 ft. high. A somewhat larger flue may be required in the lower chimney heights normally used in modern single-story construction. The upward movement of smoke in a low chimney does not attain a high velocity -- thus a greater cross-sectional area is required.

One recommended method of calculating the area for a flue is to allow thirteen square inches of area for the chimney flue to every square foot of fireplace opening. For example: if the fireplace opening equaled 8.25 sq. ft., then the flue area should equal at least 107 sq. in. If the flue is to be built of brick and unlined, it would probably be made 8 in. x 16 in., or 128 sq. in., because brickwork can be laid to better advantage if the dimensions of the flue are multiples of 4 in. If the flue is lined, and lining is strongly recommended, the lining should have an inside area of at least 107 sq. in.

Special Designs

Contemporary styling in fireplaces quite frequently results in designs having openings on two or more sides. For these, observe the same principles in planning as previously described for conventional designs. When calculating the flue area, the SUM OF THE AREA OF ALL FACES must be used. Fig. 19-18 illustrates a corner fireplace design in which the flue area is based on the total of the front face opening, plus the end face opening. In this particular construction, the side walls are not splayed, however the rear wall is sloped in the usual manner.

Multi-face fireplaces must incorporate a throat and damper with requirements similar to standard designs. Special dampers with square ends and sides are available which are frequently used for two-way fireplaces that serve adjoining rooms.

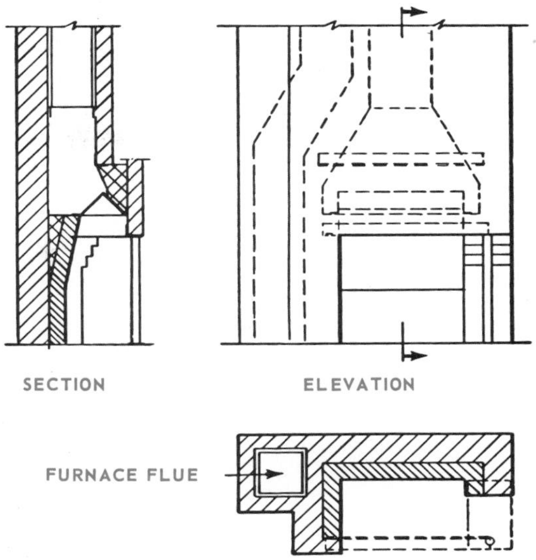

Fig. 19-18. General design for a projecting corner fireplace.

Built-In Circulators

The heating capacity of a fireplace can be increased by using a factory-built metal unit as shown in Fig. 19-19. The sides and back consist of a double wall within which air is heated. Cool air enters this chamber near the floor level and, when heated, rises and returns to the room through registers located at a higher level.

Modern built-in circulators (also called modified fireplaces) include not only the firebox and heating chamber, but also the throat, damper,

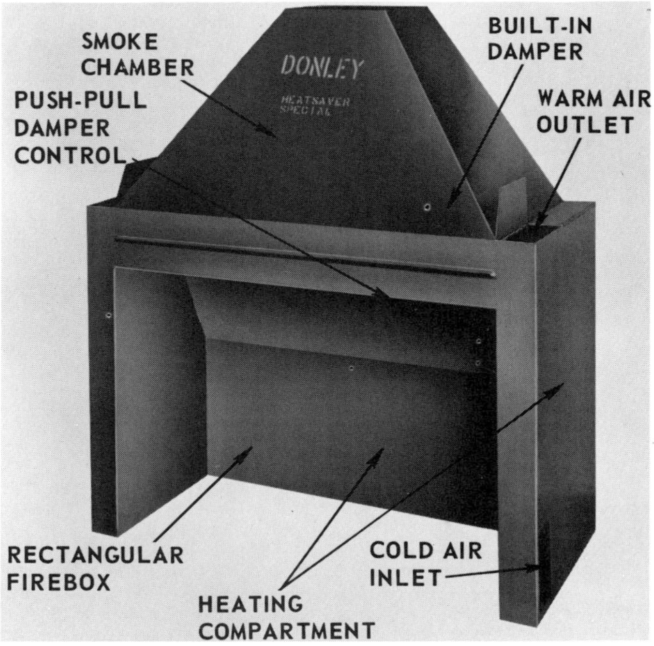

Fig. 19-19. Parts of a built-in fireplace circulator. Back and sides are constructed with a double steel wall.

smoke shelf, and smoke chamber. Since all of these parts are carefully engineered, satisfactory operation is assured when the installation is made according to the manufacturer's directions, and the flue size is adequate.

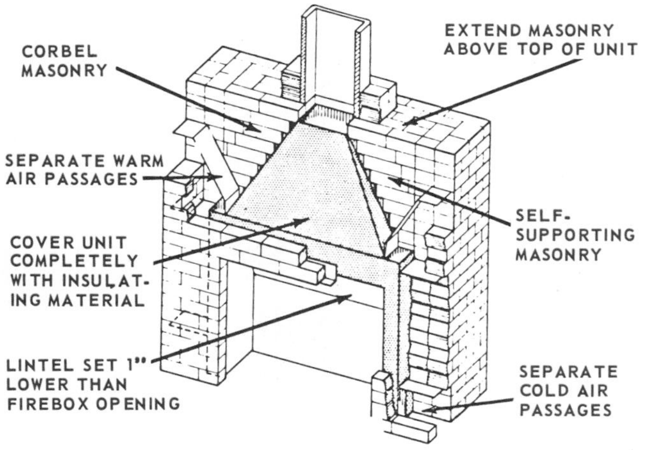

Fig. 19-20. General procedures for installing masonry around a circulator unit. (Donley Brothers Company)

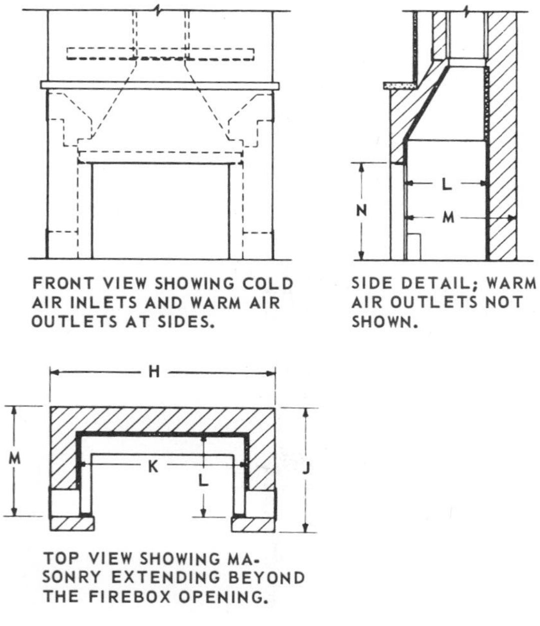

Model No.	H	J	K	L	M	N	Flue Sizes Modular	Flue Sizes Old Style
632	56	32	40	20	28	26½	12 x 12	8½ x 13
636	60	32	44	20	28	26½	12 x 12	8½ x 13
640	64	36	48	24	32	29¼	12 x 16	13 x 13
648	72	36	56	24	32	29¼	12 x 16	13 x 13

Fig. 19-21. Rough masonry dimensions for brick construction, including recommended flue sizes.

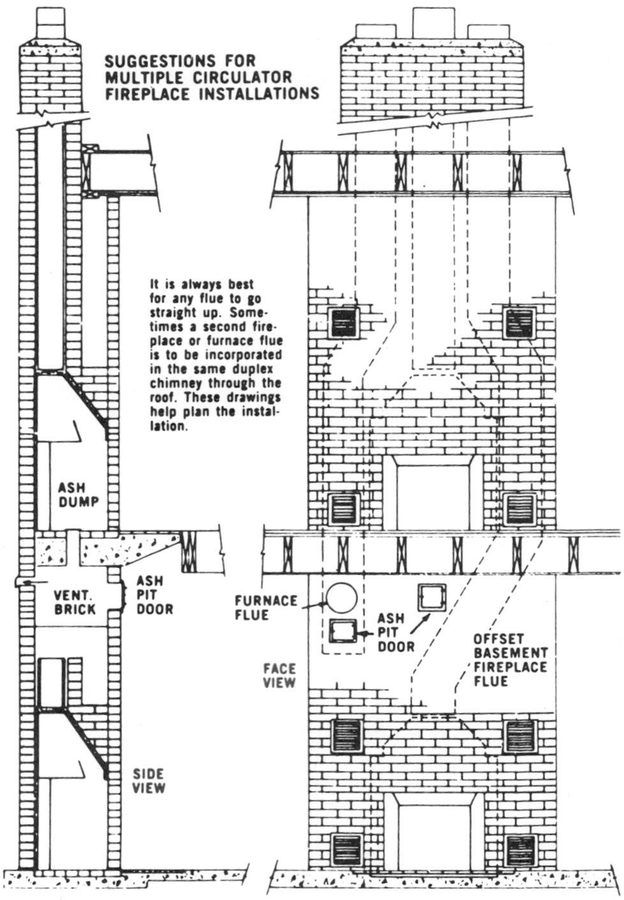

Fig. 19-22. Suggested chimney design for fireplaces located on two levels.

To install a circulator unit, first place it in position on the hearth and then build the brick and masonry work around the outside, Fig. 19-20. Steel angles (lintels) will be required across the top of the opening as shown and may be required in other locations to provide support. The unit itself should not be used for support of any masonry work.

When making an installation, follow specifications furnished by the manufacturer. Some type of fireproof insulating material is usually placed around the metal form not only to prevent the movement of heat but also to provide some expansion space between the metal and masonry.

Various sizes and types of built-in circulators are available. Fig. 19-21 lists sizes and model numbers manufactured by one firm. When units are located on different levels, the flues can be offset as illustrated in Fig. 19-22.

In any type of fireplace, proper operation will depend somewhat on an adequate flow of air into the building to replace that which is exhausted through the flue. Usually infiltration around doors and windows is sufficient; however, where weather

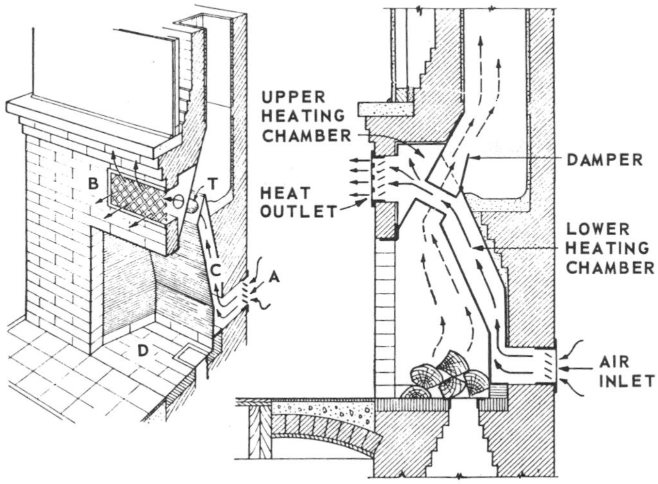

Fig. 19-23. In this modified fireplace, air enters the inlet (a) from the outside and is heated as it rises by natural circulation through the back chamber (c) and the tubes (t), being discharged into the room from the register (b). Air for supporting combustion is drawn into the fire at (d) and passes between the tubes and up the flue. A damper is provided to close the air inlet.

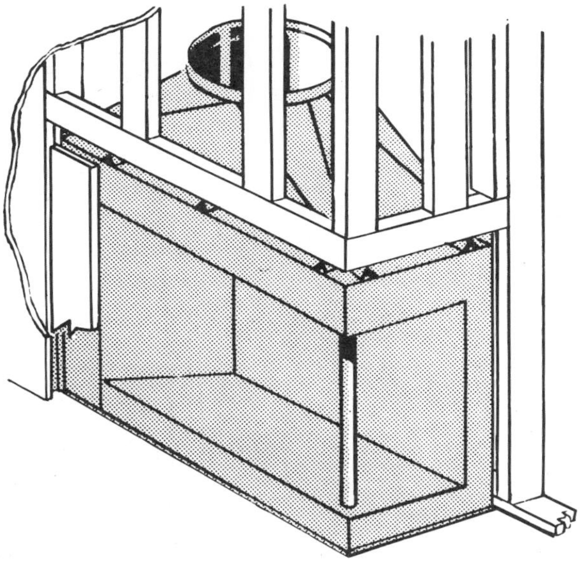

Fig. 19-24. Prefabricated fireplace unit especially designed for installation in regular wood framing. Specified clearances must be maintained and manufacturer's directions precisely followed.

stripping is tight, some inlet should be provided. Fig. 19-23 illustrates a special circulator design that uses fireplace heat to warm air brought in from the outside.

Prefabricated Fireplaces

Another type of factory-built fireplace is installed in a wood framework with little or no masonry work required, Fig. 19-24. The steel walls are doubled and usually include insulation that protects adjoining wood structures from excessive heat. This type of unit is sometimes installed with a stone or brick facing and hearth, Fig. 19-25, and provides an appearance similar to

Fig. 19-25. Modern prefabricated fireplace unit with masonry front facing and raised hearth.

Fig. 19-26. Packaged fireplace unit includes chimney and special roof projection.

Fig. 19-27. Contemporary styling applied to a free-standing factory-built fireplace. Includes screen and damper. Ashes accumulate in drawer underneath. (The Majestic Company, Inc.)

a conventional masonry fireplace. Fig. 19-26 illustrates a fireplace design available as a complete package which includes the chimney and roof housing of simulated masonry.

Free-standing, prefabricated fireplaces, Fig. 19-27, are appropriately used in various areas of the home. These are easily installed.

Modern prefabricated units provide freedom from fire hazards if the installation is made as specified by the manufacturer. All equipment should be approved by Underwriters' Laboratories, Inc. Also be sure to check requirements of the local building code.

Mantel

A mantel is a decorative member (usually wood) applied to the front of a fireplace. In addition to improving the appearance, it may provide a means to join the masonry of the fireplace with the interior wall surface. The design of the mantel should correspond with the style of architecture used in other parts of the interior, Fig. 19-28. According to FHA specifications, wooden parts should not be placed closer than 3 1/2 in. to the edge of the opening. Greater clearance is required when the parts project more than 1 1/2 in. For example: a wooden mantel shelf must be 12 in. above the opening.

Although simple mantel structures can be fabricated on the job, most of them are prefabricated in millwork plants; either in a completely assembled or knocked-down form, Fig. 19-29.

Fig. 19-28. Left. Fireplace mantel for traditional interiors. Right. Fireplace mantel surround that is adaptable to modern interiors. (Morgan Co.)

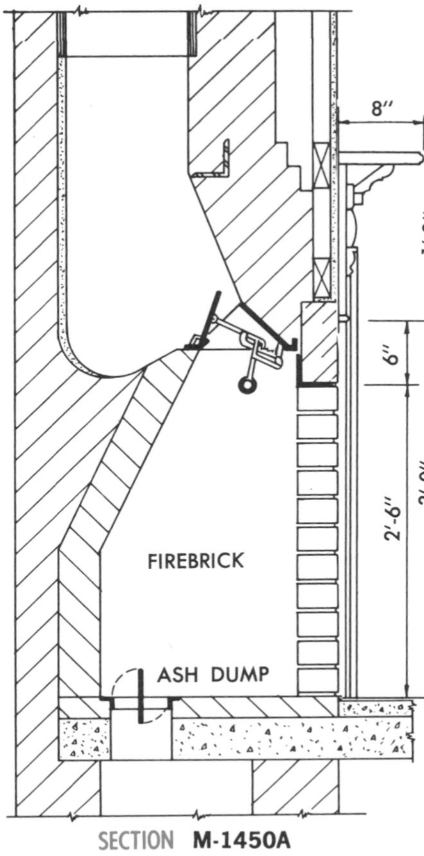

DIMENSIONS	M-1450A
Shelf Length	5'-11"
Shelf Width	8"
Breast Width	5'-2"
Height (overall)	4'-4 1/8"
Inside Measurement	3'-8" x 3'-0"

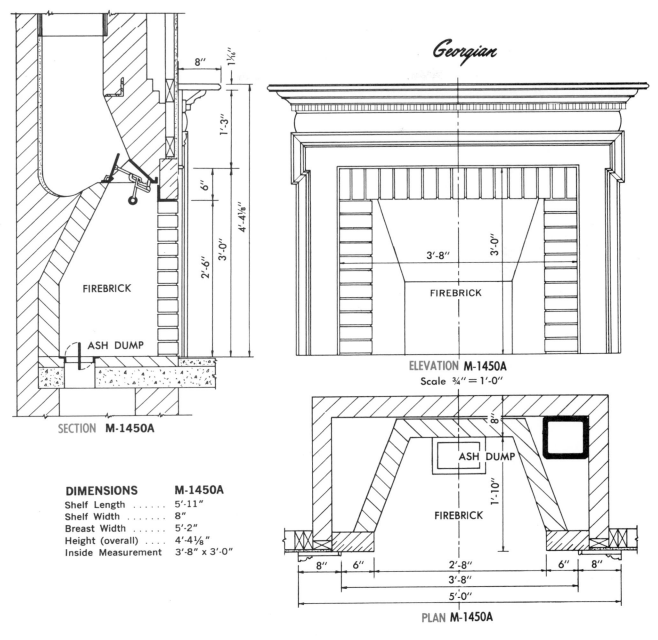

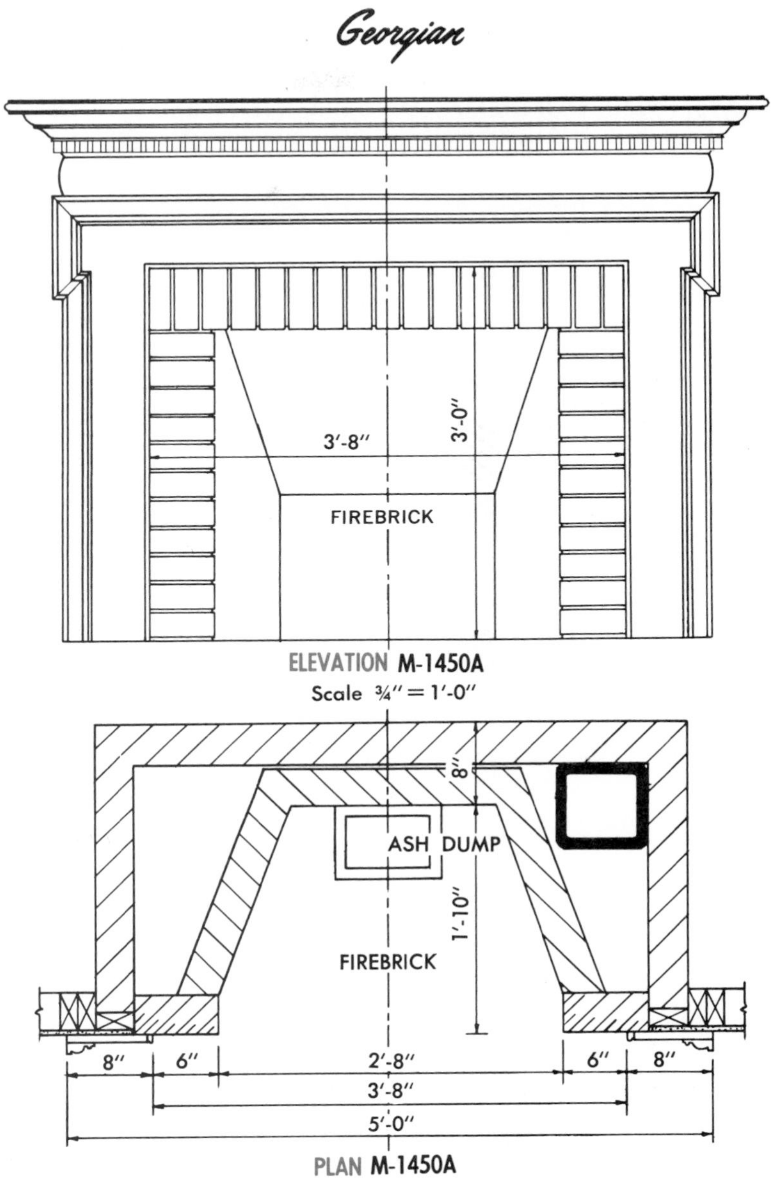

Fig. 19-29. Details of a factory-built mantel.

Test Your Knowledge - Unit 19

1. The overall size of a chimney will depend on the materials from which it is constructed and the number and size of the ___________.
2. Wood framing should be located at least _______ away from a masonry chimney.
3. The recommended minimum thickness of a chimney wall constructed without flue lining is ___________.
4. The offset angle in a chimney should be limited to _________ degrees.
5. When a prefabricated metal chimney is used for an incinerator, the inside pipe should be coated with porcelain or made of ___________.
6. To prevent heat loss from a room when the fireplace is not in operation, the fireplace must be equipped with a ___________.
7. The length of the throat should be equal to the ___________ of the fireplace opening.
8. All surfaces of the smoke chamber should be finished with a 1/2 in. thickness of __________.
9. As a general rule, the cross-sectional flue area should equal about _________ of the total area of the fireplace opening.

Outside Assignments

1. After a thorough study of reference books and such manufacturer's booklets as "Successful Fireplaces" distributed by the Donley Brothers Co., prepare a written report on the historical development of the fireplace. Start with the designs used by primitive man and highlight developments through the years. Place special emphasis on the colonial period of our country and explain how certain functional aspects of fireplaces of this era are incorporated into some modern designs.

2. Secure a copy of the building code that applies to your locality and study the sections that deal with chimneys, fireplaces, and venting systems. Make a list of the specific requirements concerning residential structures. Include information about masonry and prefabricated metal chimneys for heating plants, incinerators, and fireplaces. Also include requirements and restrictions for the use and installation of prefabricated fireplaces. Make a report to your class.

Fig. 20-1. Post-and-beam construction compared with conventional framing. Note that headers are not required for door and window openings that are located between the posts. (National Forest Products Assoc.)

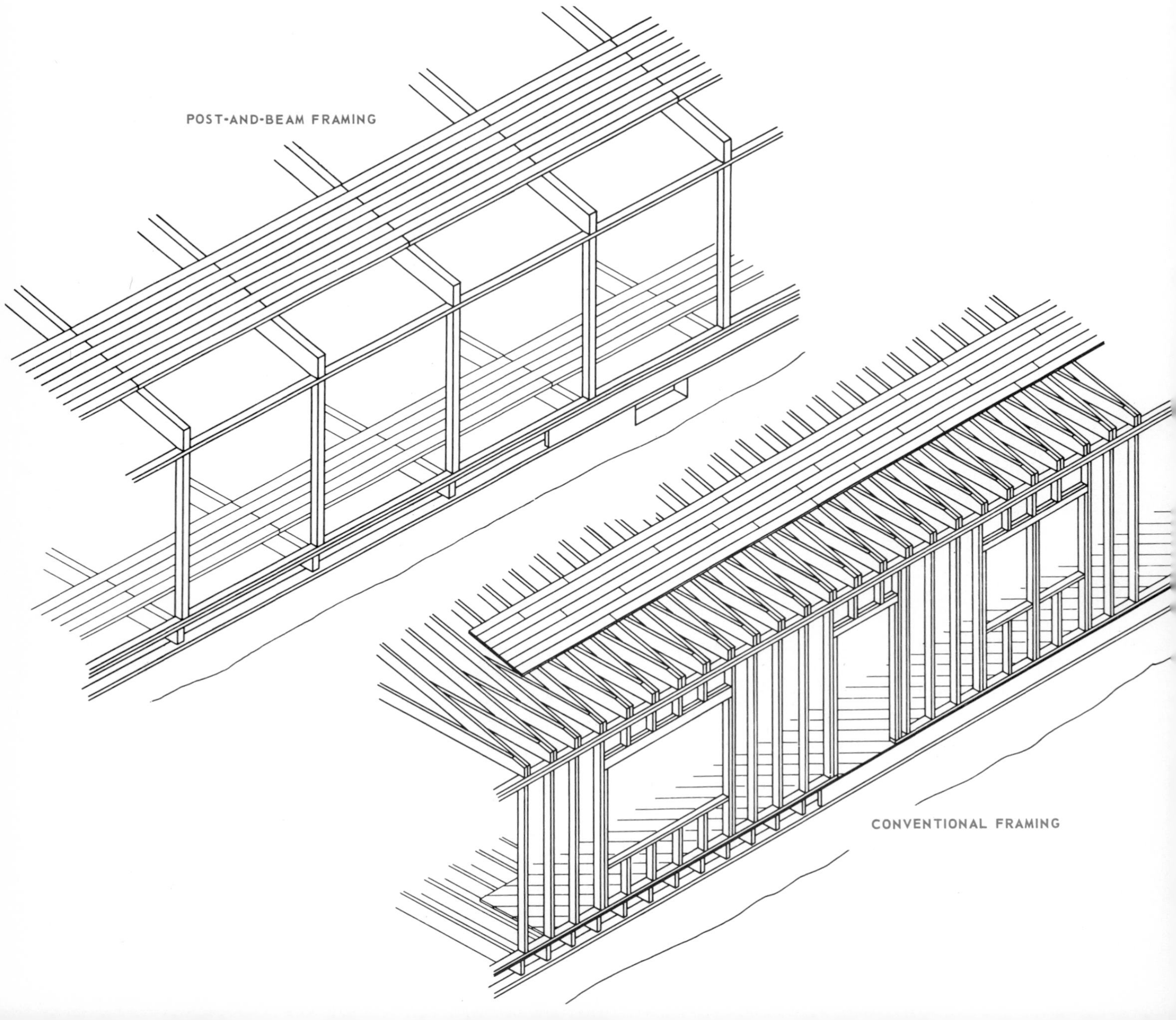

Unit 20

POST-AND-BEAM CONSTRUCTION

Post-and-beam construction consists of large framing members (posts, beams and planks) which are spaced farther apart than conventional framing members. This type of framing is similar to "mill construction" used for heavy timber buildings. It is often used today for residential work since it permits greater flexibility in contemporary and traditional designs than conventional framing methods. See Fig. 20-1.

Post-and-beam construction, also known as "plank-and-beam" construction, uses three components; posts, beams, and planks. Construction of this type is often combined with conventional framing. For example: the walls might be built with regular studs, plates, headers; and the roof framed with beams and planks. In such a case the term "plank-and-beam" could be applied to the roof structure. Similarly, it would be correct to refer to a floor structure as a "plank-and-beam" system.

Fig. 20-2. Exposed beams provide an architectural feature and added ceiling height.
(Western Wood Products Assoc.)

Fig. 20-3. Wide roof overhangs are easy to construct when posts, beams, and planks are used.

Some of the advantages of this construction are the distinctive architectural effect created by the exposed beams in the ceiling and the added height that normally results. See Fig. 20-2. The underside of the roof planks may serve as the ceiling surface thus providing a saving in material. Post-and-beam framing may also provide some saving in labor because the pieces are larger and fewer in number and can usually be assembled more rapidly than conventional framing.

One of the chief structural advantages is the simplicity of framing around door and window openings. Since the loads are carried by posts spaced at wide intervals in the walls, large openings can be framed without the need for headers. Window walls, characteristic of contemporary architectural styling, can be formed by merely inserting window frames between the posts. Still another advantage is the ease with which wide overhangs can be obtained by simply extending the heavy roof beams, Fig. 20-3.

In addition to the flexibility in design, post-and-beam construction also provides a high fire resistance factor. Wood beams do not transmit heat like unprotected metal beams which lose their strength and collapse under high tempera-

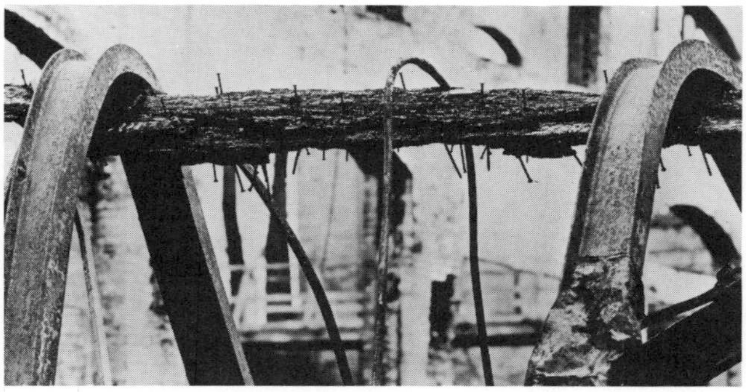

Fig. 20-4. After fire scene. Shows a wood beam supporting twisted steel I-beams. (Forest Products Laboratory)

tures, Fig. 20-4. Exposure of wood beams to flame results in a slow loss of strength. The wood is weakened in proportion to its slow reduction in cross section due to charring.

Most of the limitations of post-and-beam construction can be resolved through careful planning. The plank floors, for example, are designed for moderate uniform loads; therefore extra framing must be provided under bearing partitions, bath tubs, refrigerators, and other places where concentrated loads are likely to occur. Special provision must also be made to furnish lateral stability to the framework, especially the walls. While this might be provided with various types of bracing, it is common practice to enclose some of the wall area with large panels and use conventional stud constructions as shown in Fig. 20-5. Because of the absence of concealed spaces in outside walls and ceilings, electrical wiring, plumbing, and heating is somewhat more difficult to install.

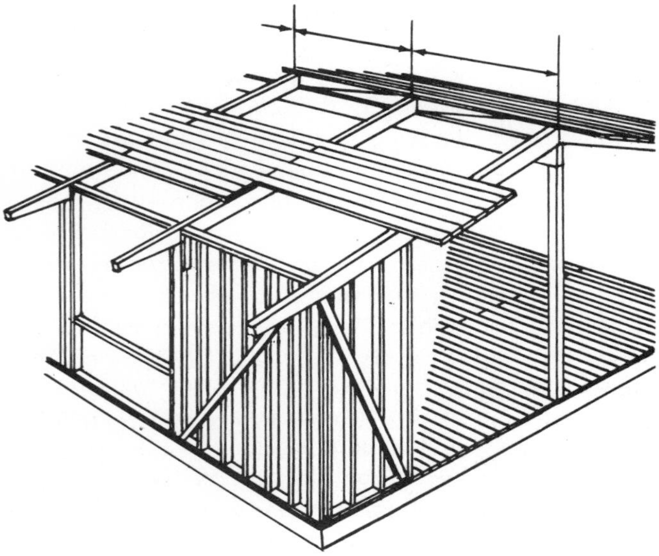

Fig. 20-5. Lateral bracing for a post-and-beam structure can be provided through the use of conventional braced-stud sections.

Posts

Foundations for post-and-beam framing may consist of continuous walls or simple piers located under each post. Either type of foundation must rest on footings that meet the requirements of local building codes. See Unit 6.

Posts must be strong enough to support the load and also large enough to provide full bearing surfaces for the ends of the beams. In general, posts should not be less than 4 x 4 in. nominal size. They may be made of solid stock or formed from 2 in. pieces that are securely fastened together. Where the ends of beams are joined over a post, the bearing surface should be increased with bearing blocks as shown in Fig. 20-6.

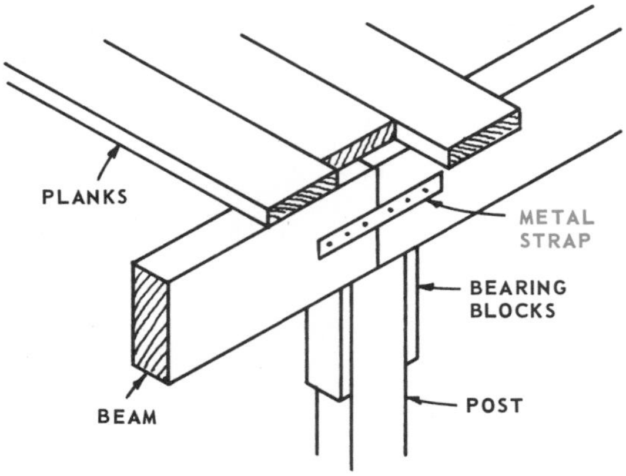

Fig. 20-6. Provide adequate bearing surface where beams are joined over a post. A heavy steel plate may replace the bearing blocks.

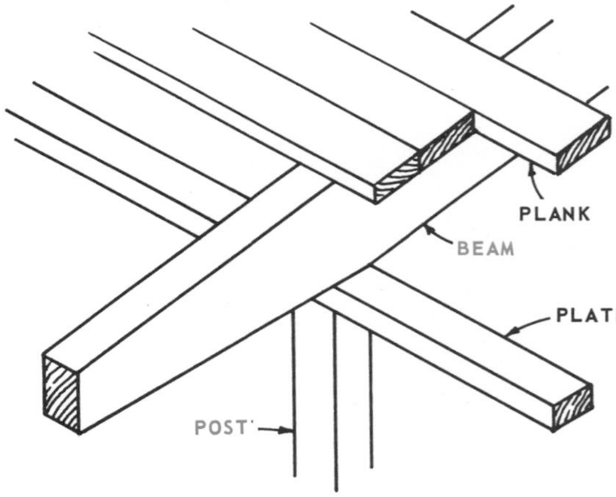

Fig. 20-7. Roof beams must be positioned directly over the supporting posts.

When posts extend upward any great distance without lateral bracing, a larger than usual cross-sectional area is required to prevent buckling. Requirements are usually listed in local building

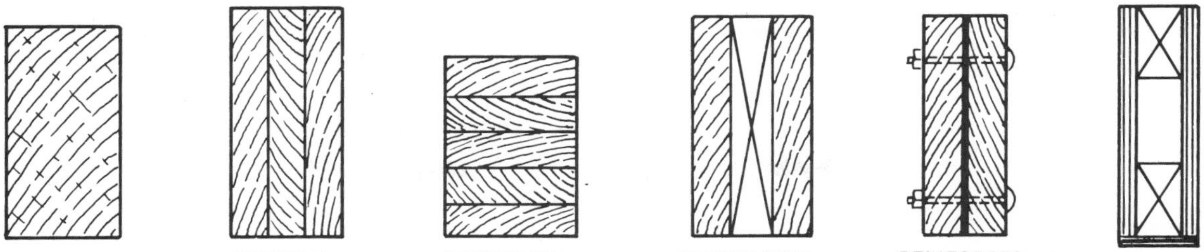

Fig. 20-8. Types of beams shown in cross section.

codes through a l/d ratio (also called slenderness ratio). The l represents the length in inches and the d stands for the smallest cross-sectional dimension (actual size). For example: this ratio for a 4 x 4 that is 8 ft. long would be about 27, which is within the limits usually prescribed.

The spacing of posts will be determined by the basic design of the structure which must be carefully engineered. Usually posts are spaced evenly along the length of the building and within the allowable free span of the floor or roof planks. With today's emphasis on modular coordination of building materials, the greatest economy will be gained when post-and-beam positions occur at standard increments of 16, 24, and 48 in.

In single story construction, a plate is attached to the top of the posts in about the same way as conventional framing. The roof beams are then positioned directly over the posts as shown in Fig. 20-7.

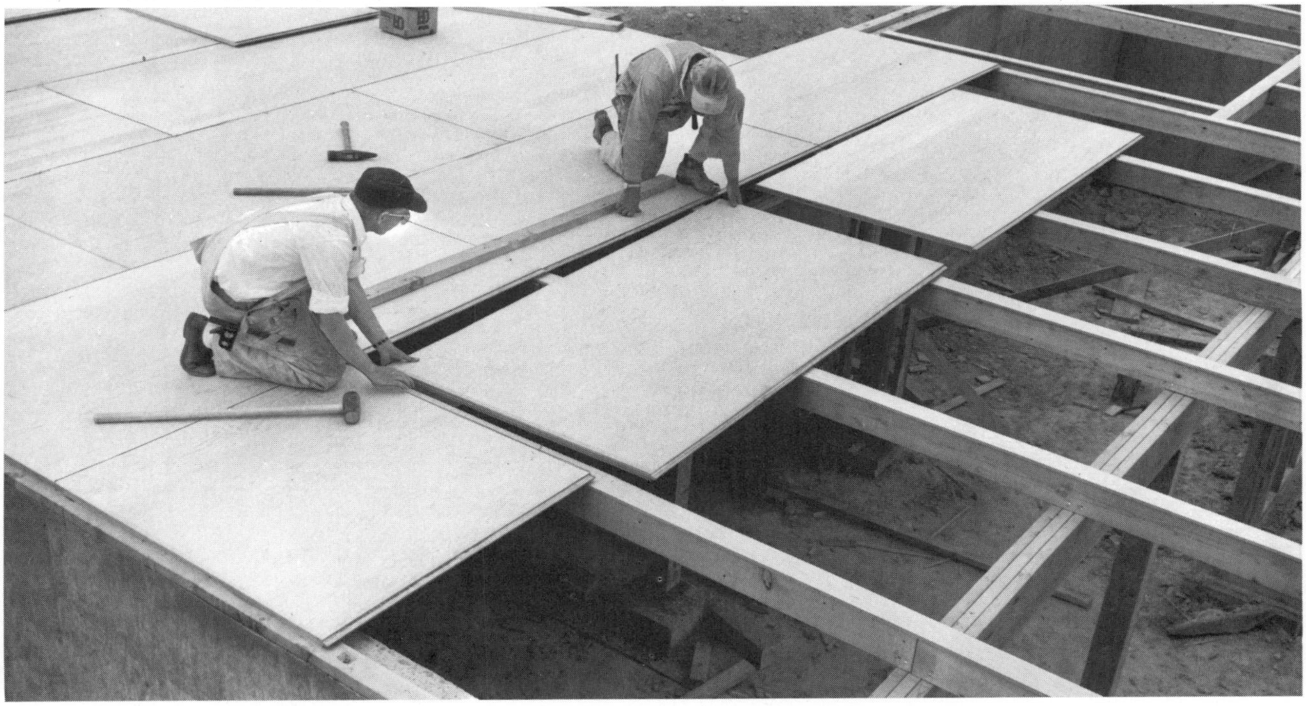

Fig. 20-9. Floor system with beams spaced 4 ft. O.C. Decking consists of 1 1/8 in. plywood with a tongue-and-groove edge. (American Plywood Assoc.)

Floor Beams

Beams for floor structures may be solid, glue-laminated, or built-up of several pieces securely nailed together, Fig. 20-8. Sometimes the built-up beams are formed with spacer blocks between the main members. Box-beams, described in another section of this Unit, may also be used.

For one story structures where under-the-floor appearance is not critical, standard dimension lumber can be securely nailed together to form the required size of beam. Fig. 20-9 illustrates how 2 x 8s have been nailed together and then installed 4 ft. on center. The decking being applied is 1 1/8 in. plywood with a special tongue-and-groove edge.

Sills for a plank-and-beam floor system can be designed similar to regular platform construction, Fig. 20-10. When it is desirable to keep the silhouette of the structure low (floor level near grade level) the beams can be carried in pockets in the foundation wall.

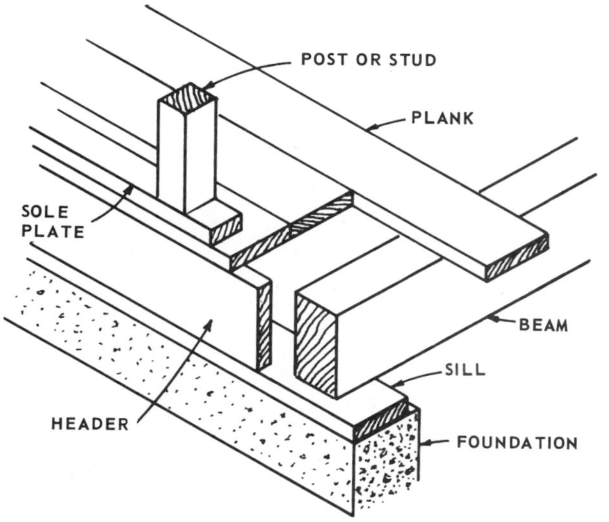

Fig. 20-10. Typical sill construction for post-and-beam frame. Blocking should be added under the post when it is not located directly over the beam.

Beam Descriptions

In general, it is best to use solid timbers when beam sizes are small and when a rustic architectural appearance is desired. Where high stress factors demand large sizes and a finished appearance is required, it is usually more economical to use laminated beams. They are manufactured in a wide range of sizes and finishes.

Solid timbers are available in a range of standard cross sections and in lengths of 6 ft. and longer, in multiples of 1 or 2 feet. These can be secured with a rough sawn surface or in a regular planed condition. When either solid or built-up beams are exposed, appearance becomes an important consideration. Fig. 20-11 illustrates on-the-job treatment that may be applied to exposed beams.

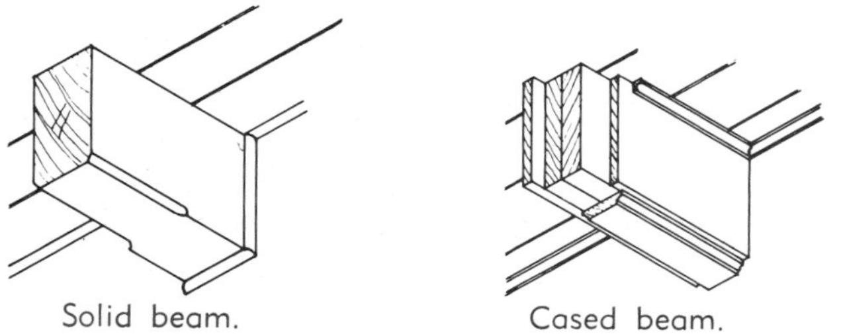

Solid beam. Cased beam.

Fig. 20-11. Common methods of finishing exposed beams.

SIZE (actual)	Wt. per lineal ft.	SIMPLE SPAN IN FEET													
		10	12	14	16	18	20	22	24	26	28	30	32	34	36
		LOAD BEARING CAPACITY—LBS. PER LINEAL FT.													
3" x 5¼"	3.7 lbs.	151	85	—	—	—	—	—	—	—	—	—	—	—	—
3" x 7¼"	4.9 lbs.	362	206	128	84	—	—	—	—	—	—	—	—	—	—
3" x 9¼"	6.7 lbs.	566	448	300	199	137	99	—	—	—	—	—	—	—	—
3" x 11¼"	8.0 lbs.	680	566	483	363	252	182	135	102	—	—	—	—	—	—
4½" x 9¼"	9.8 lbs.	850	673	451	299	207	148	109	—	—	—	—	—	—	—
4½" x 11¼"	12.0 lbs.	1,036	860	731	544	378	273	202	153	—	—	—	—	—	—
3¼" x 13½"*	10.4 lbs.	1,100	916	784	685	479	347	258	197	152	120	—	—	—	—
3¼" x 15"*	11.5 lbs.	1,145	1,015	870	759	650	473	352	267	206	163	128	104	—	—
5¼" x 13½"*	16.7 lbs.	1,778	1,478	1,266	1,105	773	559	415	316	245	193	154	124	101	—
5¼" x 15"*	18.6 lbs.	1,976	1,647	1,406	1,229	1,064	771	574	438	342	269	215	174	142	116
5¼" x 16½"*	20.5 lbs.	2,180	1,810	1,550	1,352	1,155	933	768	586	457	362	290	236	183	160
5¼" x 18"*	22.3 lbs.	2,378	1,978	1,688	1,478	1,308	1,113	918	766	598	474	382	311	254	204

* Horizontally Laminated Beams

TABLE I

example:
Clear Span = 18′ 0″
Beam Spacing = 8′ 0″
Dead Load = 8 lbs./sq. ft. (decking + roofing)
TABLE I — Live Load = 20 lbs./sq. ft. (snow)
ROOF BEAM — Total Load = (20 + 8) (8) = 224 lbs./lineal ft.
From Table I—Select 3″ x 11¼″ beam with capacity of 252 lbs./lin. ft.

SIZE (actual)	Wt. per lineal ft.	SIMPLE SPAN IN FEET													
		10	12	14	16	18	20	22	24	26	28	30	32	34	36
		LOAD BEARING CAPACITY—LBS. PER LINEAL FT.													
3" x 5¼"	3.7 lbs.	114	64	—	—	—	—	—	—	—	—	—	—	—	—
3" x 7¼"	4.9 lbs.	275	156	84	55	—	—	—	—	—	—	—	—	—	—
3" x 9¼"	6.7 lbs.	492	319	198	130	89	—	—	—	—	—	—	—	—	—
3" x 11¼"	8.0 lbs.	590	491	361	239	165	119	—	—	—	—	—	—	—	—
4½" x 9¼"	9.8 lbs.	738	479	298	196	134	96	—	—	—	—	—	—	—	—
4½" x 11¼"	12.0 lbs.	900	748	541	359	248	178	131	92	—	—	—	—	—	—
3¼" x 13½"*	10.4 lbs.	956	795	683	454	316	228	169	128	98	—	—	—	—	—
3¼" x 15"*	11.5 lbs.	997	884	756	626	436	315	234	178	137	108	—	—	—	—
5¼" x 13½"*	16.7 lbs.	1,541	1,283	1,095	732	509	367	271	205	158	123	96	—	—	—
5¼" x 15"*	18.6 lbs.	1,713	1,423	1,219	1,009	703	508	376	286	221	173	137	109	—	—
5¼" x 16½"*	20.5 lbs.	1,885	1,568	1,340	1,170	939	678	505	384	298	235	187	151	—	—
5¼" x 18"*	22.3 lbs.	2,058	1,710	1,464	1,278	1,133	886	660	503	391	309	247	200	—	—

* Horizontally Laminated Beams

TABLE II

example:
Clear Span = 18′ 0″
Beam Spacing = 5′ 0″
Dead Load = 7 lbs./sq. ft. (decking + covering)
TABLE II — Live Load = 40 lbs./sq. ft. (furniture, occupants, etc.)
FLOOR BEAM — Total Load = (40 + 7) (5) = 235 lbs./lineal ft.
From Table II—Select 4½″ x 11¼″ beam with capacity of 248 lbs./lin. ft.

Fig. 20-12. Tables for determining allowable spans for vertical glue-laminated beams. Table 1 is used for roof beams where a deflection of 1/240 of span is permitted. Table 2 is for floor beams and based on a deflection of 1/360 or less. (Weyerhaeuser Co.)

Beam sizes must be based on the span (spacing between supports), deflection permitted, and the load they must carry. Design tables are available from lumber manufacturers which may be used to determine sizes for simple buildings. See Fig. 20-12. Refer to Unit 7 for additional information on calculating beam loads.

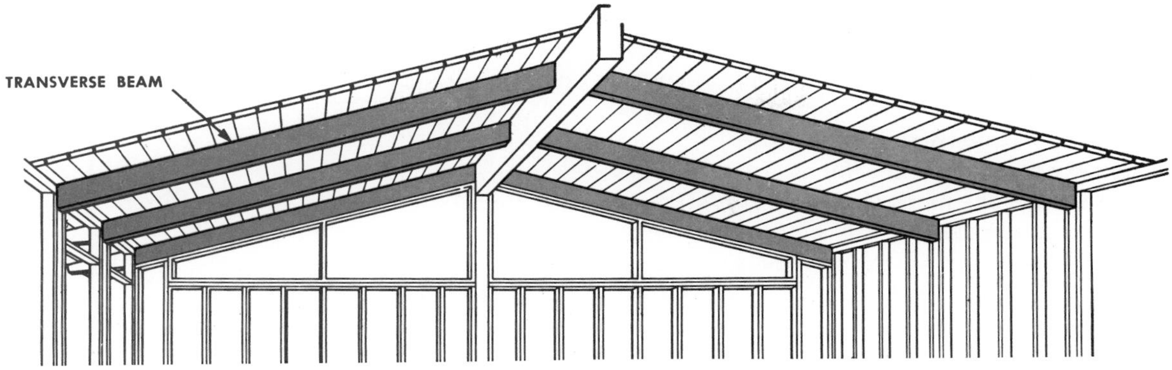

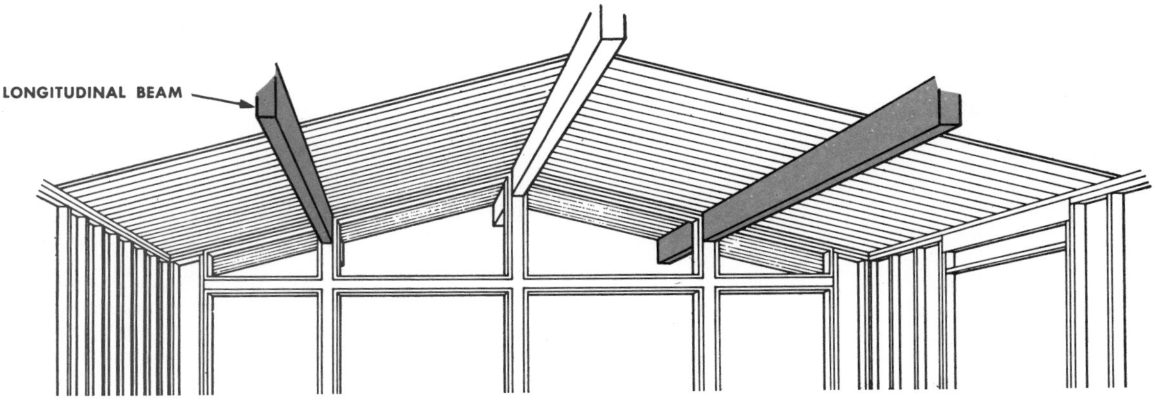

Fig. 20-13. Plank-and-beam roof constructions. Above. Transverse beams. Below. Longitudinal beams.

Roof Beams

Beam supported roof systems are of two basic types; transverse beams that are similar to exposed rafters on wide spacings; and longitudinal beams that run parallel to the supporting side walls and ridge beam. See Fig. 20-13.

Longitudinal beams (also called purlin beams) are usually larger in cross section than transverse beams because of the greater spans and heavier loads they must carry. The longitudinal beam permits many variations in endwall design with the extensive use of glass and extended roof overhangs as special features.

Either type of beam must be adequately supported either on posts or stud walls that incorpo-

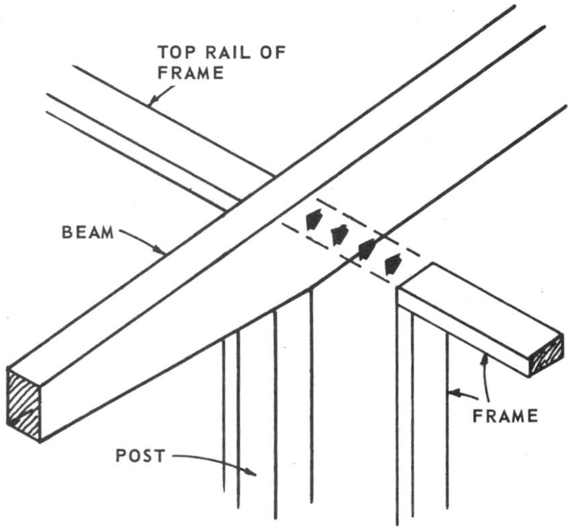

Fig. 20-14. Transverse beam bearing on post. Connection is reinforced by filler panel frame.

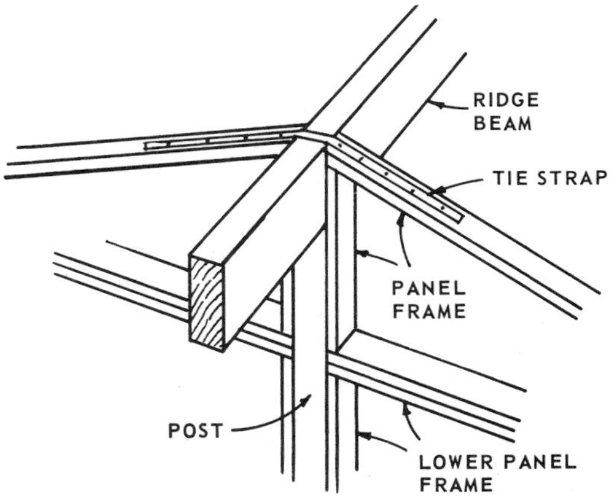

Fig. 20-15. Ridge beam held in position by panel frame and metal tie strap. Same detail of construction may be applied to transverse beams.

rate a heavy top plate. When supported on posts, the connection can be reinforced with a wide panel frame that extends to the top of the beam as illustrated in Fig. 20-14. A similar method of supporting a ridge beam is shown in Fig. 20-15.

Transverse beams can be joined to the sides of the ridge beam or supported on top as illustrated in Fig. 20-16. Metal tie plates, hangers, and straps are required to absorb the horizontal thrust.

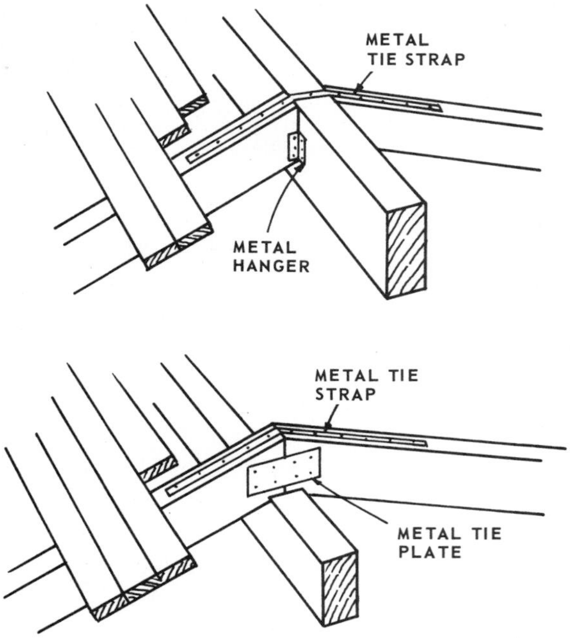

Fig. 20-16. Construction details of transverse beam and ridge beam. Above. Beam attached to side of ridge. Below. Beam supported on top of ridge.

Flat roof designs often consist of a plank-and-beam system. Details of construction are about the same as illustrated for low, sloping roofs.

Fasteners

A post-and-beam frame consists of a limited number of joints. Therefore the loads and forces exerted upon the structure are concentrated at these points. Regular nailing patterns used in conventional framing will usually not provide a satisfactory connection and the joints will need to be reinforced with special metal connectors. See Fig. 20-17. To increase the holding power of metal connectors, they should be attached with lag screws or bolts.

Since beam structures are usually exposed, some connectors will likely detract from the appearance and concealed devices will need to be

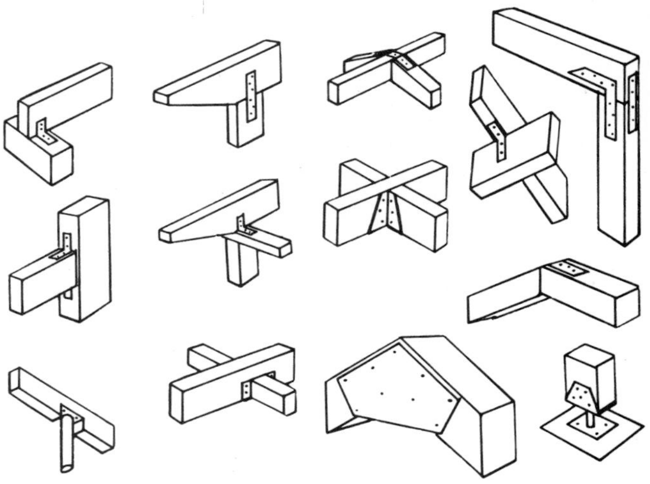

Fig. 20-17. Metal fasteners and connectors for post-and-beam construction. (Western Wood Products Assoc.)

substituted. Steel or wood dowel pins of appropriate size can be used. Notches and gains cut in the structural members may provide an interlocking effect or a recess for metal connectors.

HANDY SAYS:

"Use extra care when assembling exposed posts and beams. Tool and hammer marks will detract from the final appearance."

Partitions

Interior partitions are somewhat more difficult to construct under an exposed beam ceiling. Except for a load-bearing partition under a main ridge beam, it is usually best to make the installation after the beams and planks are in place. Partitions running perpendicular to a sloping ceiling should have regular top plates with filler sections installed between the beams.

Partitions parallel to transverse beams will have a sloping top plate. Sometimes it is best to construct these partitions in two sections; first building a conventional lower section the same height as the side walls, and then adding a triangular section above.

When nonbearing partitions run at a right-angle to a plank floor, no special framing is necessary. However, when nonbearing partitions run parallel, Fig. 20-18, additional support must be provided either with a small beam replacing the sole plate or located below the plank flooring as shown.

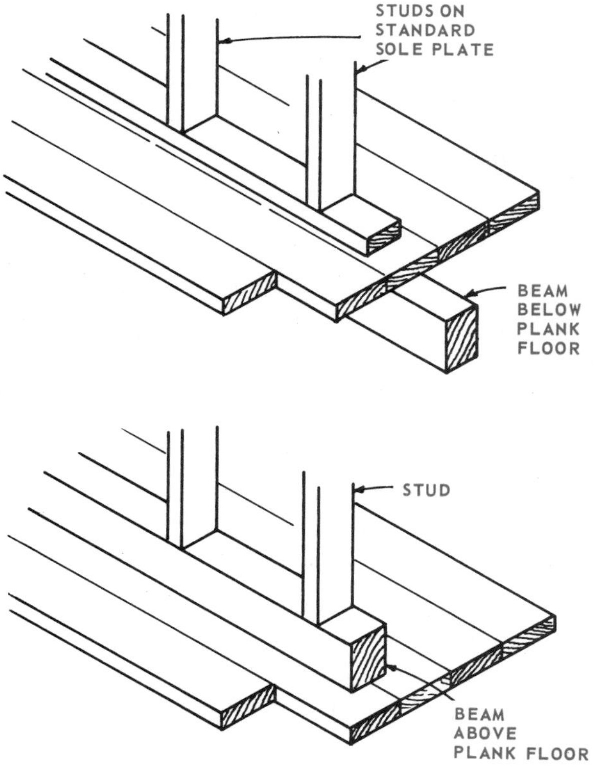

Fig. 20-18. Providing support for nonbearing partitions running parallel to flooring planks.

Planks

Planks for floor and roof decking vary from 2 to 4 in. in thickness depending on the span. The pieces usually have a tongue-and-groove edge or a groove for a spline joint that can be assembled into a tight, strong surface. Fig. 20-19 illustrates some standard designs with identification numbers as listed by the Western Wood Products Association. When planks are end-matched, Fig. 20-20, the joints need not occur over the beams.

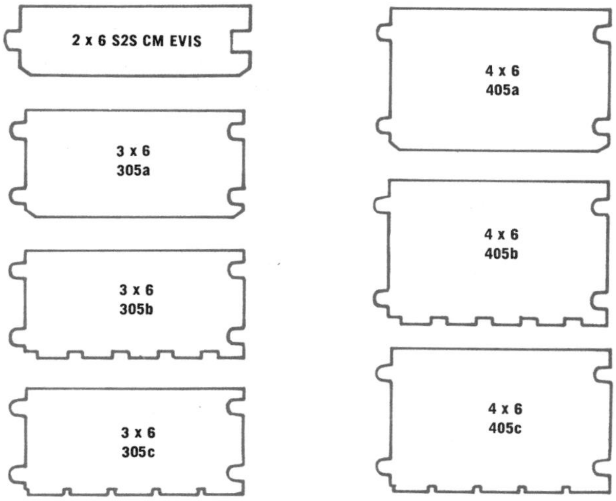

Fig. 20-19. Standard plank patterns with tongue-and-groove edge joints and machined faces.

In laying planks, added strength and stiffness will result if they are continuous over more than one span. This concept can also be applied to beams and other supports and is illustrated in Fig. 20-21.

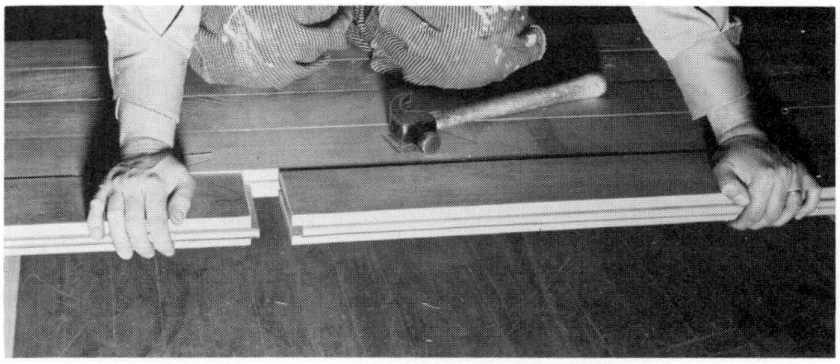

Fig. 20-20. End-matched planks being installed over floor beams. (Weyerhaeuser Co.)

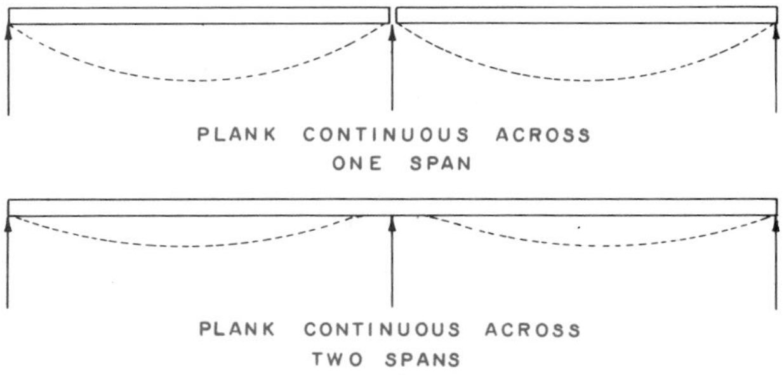

Fig. 20-21. How the stiffness of a plank over a single span compares with that of a plank continuous over two spans.

Roof planks should be selected carefully, especially when the faces will be exposed. Fig. 20-22 shows the application of planking with a V-joint along the edge which provides an attractive surface on the underside. Solid materials should have a moisture content that will correspond closely to the E.M.C. of the interior of the structure when it has been placed in service. Because of the large cross-sectional size of posts, beams, and planks, special precautions should be observed in selecting material with proper M.C. levels, otherwise difficulties due to excessive swelling or shrinkage may be encountered.

In cold climates, plank roof structures (directly over heated areas) require insulation and a vapor barrier. The thickness of the insulation will vary depending on the locality. The insulation should be a rigid type that will support the finished roof surface and the workmen. An approved vapor barrier should be installed between the planks and the insulation, Fig. 20-23.

Fig. 20-22. Laying 4 in. double tongue-and-groove planks on glue-laminated beams.

Today, several types of heavy structural composition board (2 to 4 in. thick) are available for roof decks. The panel sizes are large, and the material is relatively light in weight, thus permitting rapid application. See Fig. 20-24. Edges usually have some type of interlocking joint that provides a tight, smooth deck. When the underside (ceiling side) is prefinished, no further decoration is usually necessary. Always follow the manufacturer's recommendations when selecting and installing these materials.

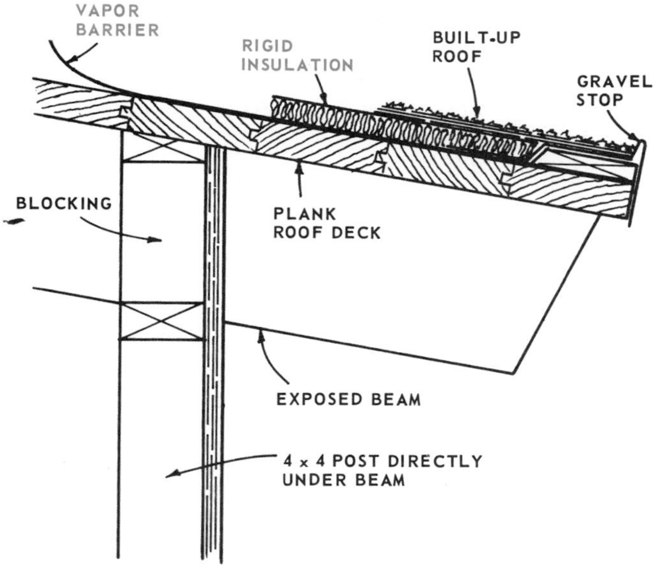

Fig. 20-23. Detail showing application of vapor barrier and insulation on a plank decking located directly over heated space.

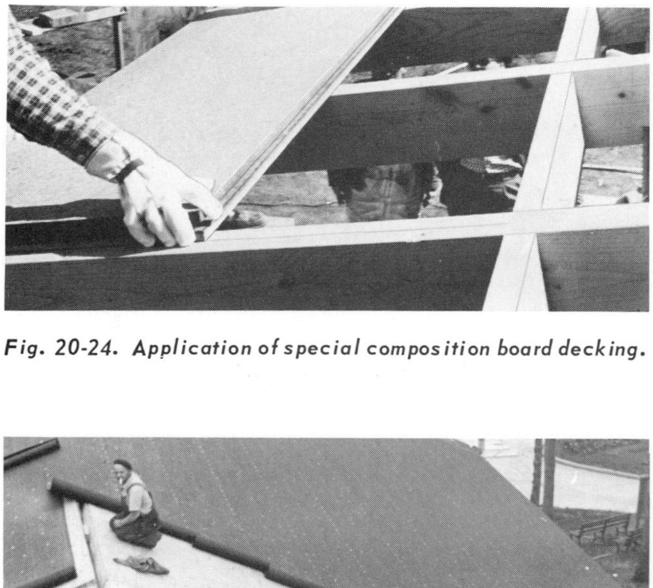

Fig. 20-24. Application of special composition board decking.

Fig. 20-25. Stressed skin panels form the roof decking. Note the size of the span. (Georgia-Pacific Corp.)

Stressed Skin Panels

Roof and floor decking, and also wall sections can be formed with plywood stressed skin panels. These can be designed to carry structural loads over wide spans as illustrated in Fig. 20-25.

Stressed skin panels are made by gluing sheets of plywood (skins) to longitudinal framing members or other core materials. They form a structural unit where the supporting action is similar to a series of built-up wooden I-beams as illustrated in Fig. 20-26.

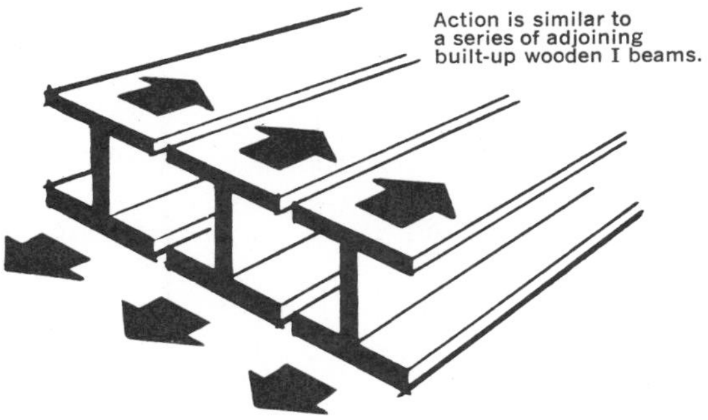

Fig. 20-26. How a stressed skin panel provides support.

These building components are usually produced in factories where rigid specifications in design and construction can be maintained. Fig. 20-27 provides general guidelines and constructional details. Note that insulation can be installed easily as a part of the manufacturing process.

Sandwich panels with plywood skins and cores

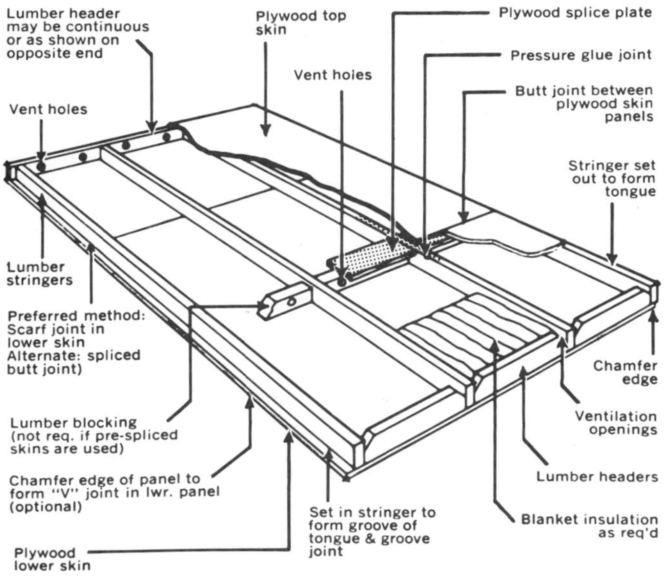

Fig. 20-27. General construction details for stressed skin panel. (Plywood Fabricator Service Inc.)

Fig. 20-28. Installing plywood box beams in a modern commercial building.

of such material as foamed polystyrene or paper honeycomb are similar to the stressed skin panel but do not provide as much rigidity and strength. These are used for curtain walls and various installations where the major support is carried by other components. Skins can be made from a wide variety of sheet materials including plywood, hardboard, plastic laminates, and aluminum.

Box Beams

Modern box beams made of plywood webs offer a structural unit that can span distances up to 120 ft. The high strength-to-weight ratio offers a tremendous advantage in commercial structures where wide, unobstructed areas are required, Fig. 20-28.

Basic design features of plywood box beams, illustrated in Fig. 20-29, consist of one or more vertical plywood webs which are laminated to seasoned lumber flanges. The flanges are separated at regular intervals by vertical spacers (stiffeners) which help distribute the load between the upper and lower flange and prevent buckling of the plywood webs. The strength of the unit depends to a large extent on the quality of the glue bond between the various members. Plywood box beams must be carefully designed and fabricated under controlled conditions.

Laminated Beams and Arches

Laminated wood beams and arches, available in many shapes and sizes, have opened new dimensions of design in modern construction. They offer strength, safety, economy, and permanence -- along with the natural beauty of wood, Fig. 20-30.

Most laminated structural members are made of softwoods. They are manufactured in industrial plants specializing in such production and then shipped, prefinished and ready for erection, to the

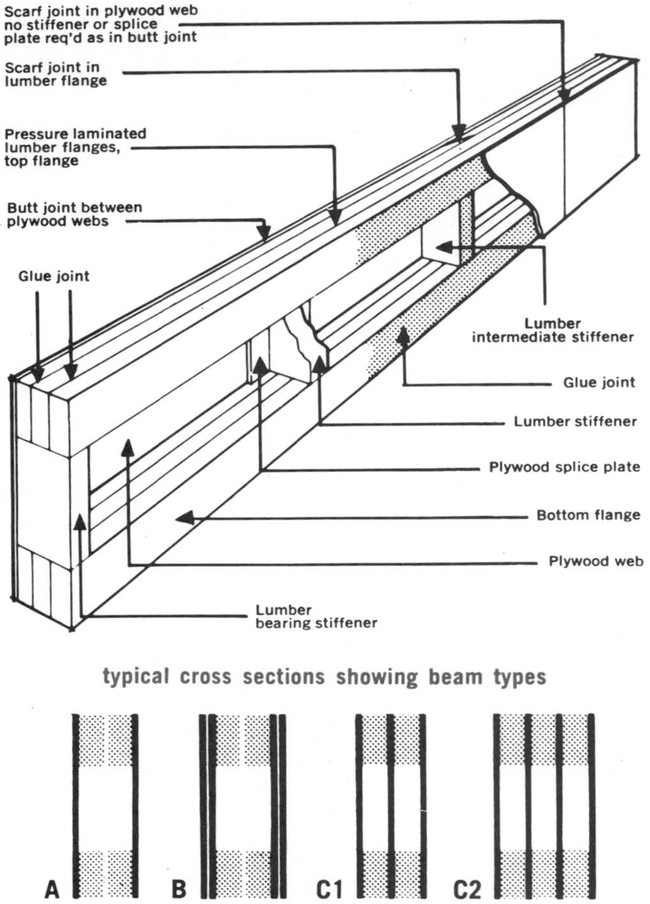

Fig. 20-29. Basic construction details used in fabricating plywood box beams. (Plywood Fabricator Service Inc.)

Fig. 20-30. Gracefully soaring parabolic arches of laminated wood used in church construction.

building site. In residential work, beams are usually straight or tapered, however in institutional and commercial buildings they are often formed into curves, arches and other complicated shapes. Some of the standard forms are illustrated in Fig. 20-31.

In the fabrication of beams and arches, lumber is carefully selected and machined to size. To secure the required length, pieces must often be end-joined. Since end-grain is hard to join, a

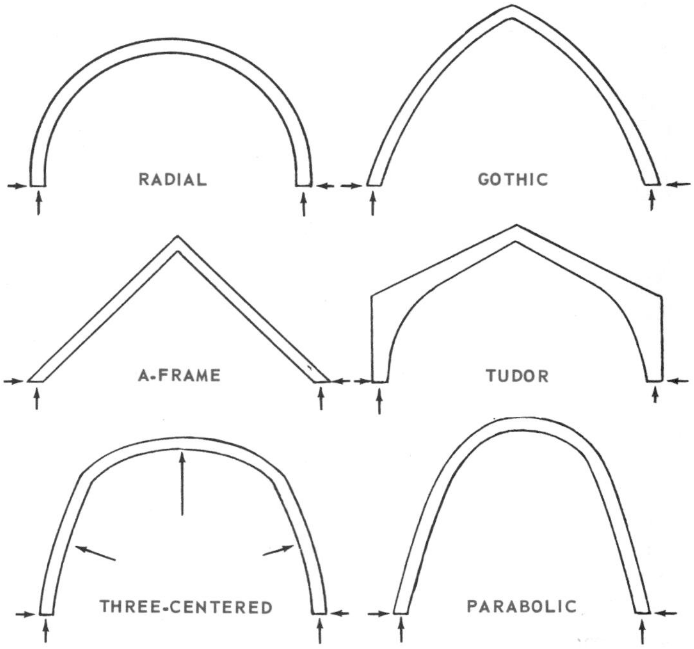

Fig. 20-31. Types of laminated wood arches. Arrows indicate the support and lateral thrust that must be provided.

special hooked-scarf joint may be used. A number of these joints may be required in each ply. The joints are staggered at least two feet from a similar joint in an adjacent layer. See Fig. 20-32.

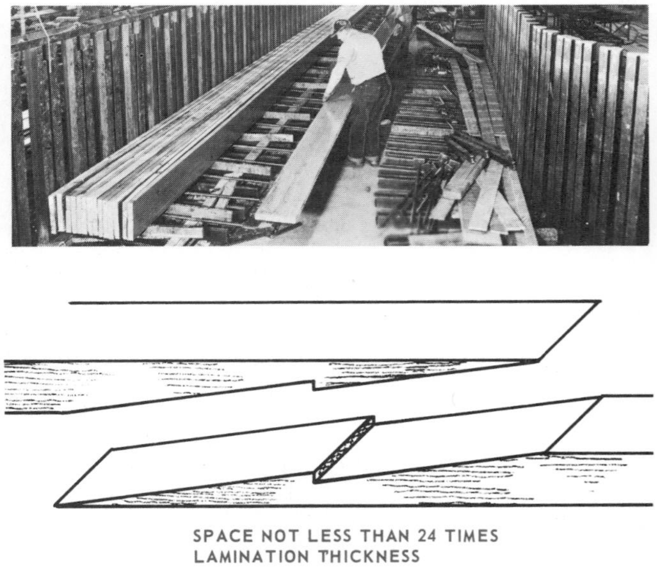

Fig. 20-32. Above. Assembling laminations for a straight beam 65 ft. long. Below. Hooked scarf joint used to join the ends of laminations.

Fig. 20-33. Gluing a laminated arch in a special clamping device. (Forest Products Laboratory)

Waterproof adhesives are applied and the layers are clamped in huge forms, Fig. 20-33. After the glue has cured, the edges are machined to size, and all surfaces are sanded. Today most of the units are prefinished to specifications that will match the interior of the completed building. Surfaces are carefully covered with special wrappings to provide protection during storage and shipment.

Test Your Knowledge - Unit 20

1. The slenderness ratio of a post compares the total height in inches with the ______________ (largest, smallest) dimension of its cross section.
2. The live load on a residential floor is usually figured at _____ lbs. per sq. ft. and represents the weight of furniture, occupants, and equipment.
3. Two common types of roof beams include longitudinal beams and ___________ beams.
4. Planks for floor and roof decking are available in thicknesses of 2 to ______ in.
5. A fabricated building component, similar to a stressed skin unit with a core of foamed plastic or paper honeycomb is called a _______ panel.
6. In plywood box beam construction, the flanges are separated at regular intervals with spacer blocks which distribute the load and prevent buckling of the ___________.

Outside Assignments

1. Visit a building supply center and secure descriptive literature about floor and roof decking that can be used in post-and-beam construction. Include both solid and laminated planks and composition panels. Be sure to obtain prices. Study these materials thoroughly -- qualities, characteristics and installation procedures. Prepare a written report or make an oral report to your class.
2. Build a mock-up of a stress skin panel in which you can experiment with the design and size of the various parts. Use a scale of about 3" = 1' - 0". Skins might be made of 1/8 in. plywood or 1/16 in. veneer. After you have tested the strength of your design, make up two units; one complete and one with a single skin so the inside structure can be observed. Present the mock-ups to your class along with explanations.

Installing prefabricated roof panels on a large commercial building. The panels (stacked in background) are supported by joists spaced 8 ft. apart. The joists are supported by large laminated beams (arrow). (American Plywood Assoc.)

Two story, four bedroom residence, designed and built by R. N. Blumstein and Thomas J. Markovic. Location, Flossmoor, Illinois.

PHOTOS SHOW CONSTRUCTION DETAILS OF HOUSE WITH EXPOSED BEAM CEILING

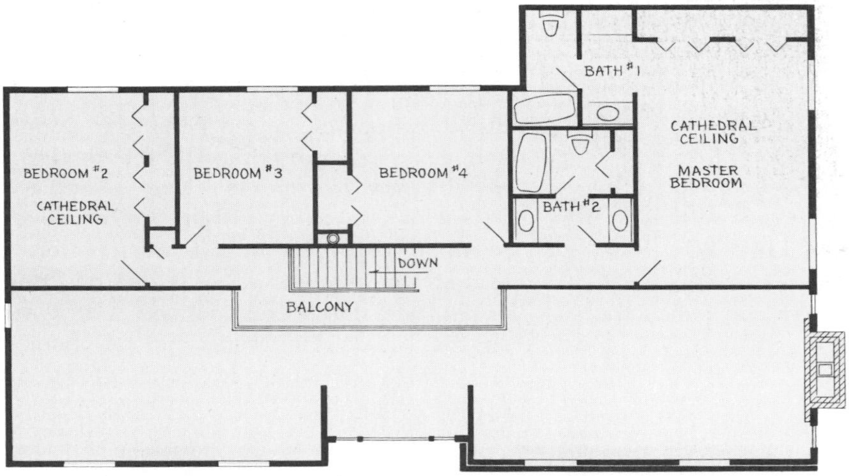

SECOND FLOOR PLAN

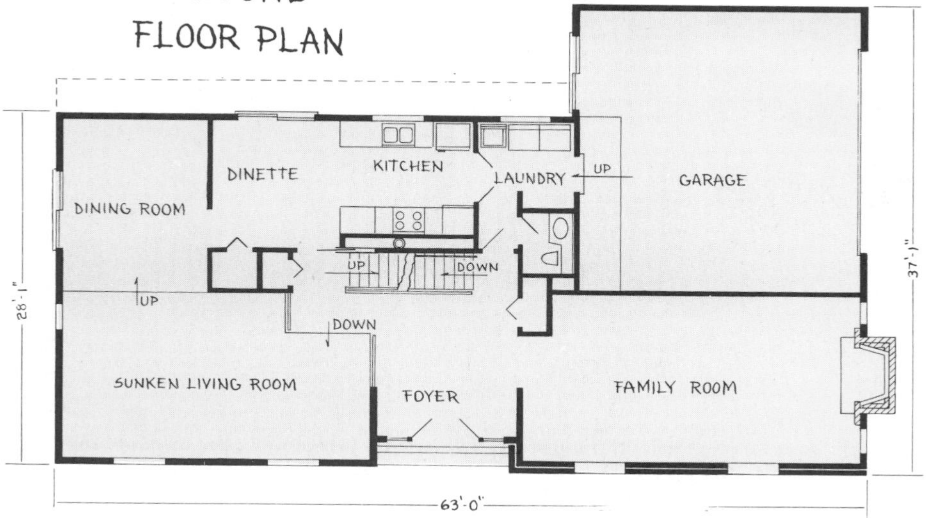

FIRST FLOOR PLAN

Above, left. 1/2 in. gypsum wall sheathing is placed between 1 x 3 furring strips spaced 16 in. O.C. Furring serves as nailing strips for vertical siding. Ceiling beams shown are 4 x 10, spaced 6 ft. O.C.

Above, right. Note exposed beam ceiling. The 2 x 6 tongue and grooved decking is covered with 15 lb. felt, 3/4 in. rigid insulation, and 380 lb. asphalt shingles. 235 lb. shingles are used on opposite side of roof.

First stages of wood construction. Floor joists are 2 x 10; wall studs 2 x 4. Joists and studs are 16 in. on center. Subfloor is 1 x 8.

On site assembly of panelized type of prefabricated (manufactured) home. The end wall panel being hoisted into position is 24 ft. long. Except for corner trim and bottom siding strip, the panel is completely finished inside and out. (Wausau Homes, Inc.)

Sectionalized (modular) home being placed on a treated wood foundation. Each section (commonly called a module) is completely finished inside and out, and includes all cabinetwork, built-in appliances, plumbing fixtures, and heating and electrical equipment. If you look closely, you can see a lighting fixture through the window, (arrow). (American Plywood Assoc.)

Unit 21

PREFABRICATION

The term prefabrication, as used in building construction, means to fabricate the parts, sections, or sub-assemblies in shops and factories, then transport them to the building site for final assembly. Prefabrication is also used in describing buildings that are partially or fully assembled before leaving the manufacturing plant.

Fig. 21-1. Automatic nailing machine with adjustable clamps, stops, and guides. Drives nails of 8–10 gauge and up to 5 in. long. (Frank Paxton Lumber Co.)

New developments in construction have resulted in improved fabrication techniques, many of which must be performed under controlled conditions with specialized equipment. For example: stressed-skin panels and plywood box beams require accurate glue applications, special presses, and handling equipment. Prefabrication in modern plants is accomplished with large production equipment, Fig. 21-1 and Fig. 21-2, which reduces the amount of human energy required and increases production.

Fig. 21-2. Roller coating machine applies finish to kitchen cabinet doors. (National Homes Corp.)

Components

In house construction today, whether the basic framework is erected by conventional methods or formed with various prefabricated assemblies, many factory-built components (parts) are used. Some of these are windows, door units, soffit systems, stairs, and built-in cabinetwork.

In the area of structural components, the roof truss is widely used. Manufacturing plants are usually able to furnish either simple Fink trusses (described in Unit 9) or coordinated units for a more complicated roof as shown in Fig. 21-3. Roof trusses must be carefully designed by structural engineers and fabricated according to exact specifications. High production double-end saws, Fig. 21-4, are employed to cut truss members to length and specially designed presses and jig tables are used in making the assembly with "gang-nail" connectors, Fig. 21-5. Completed roof trusses are shown being transported to the construction site in Fig. 21-6.

Floor trusses, Fig. 21-7, provide another example of prefabricated structural units that offer several advantages. Not only do they permit wide unsupported spans with a minimum of material, but they also allow open spaces for the installation

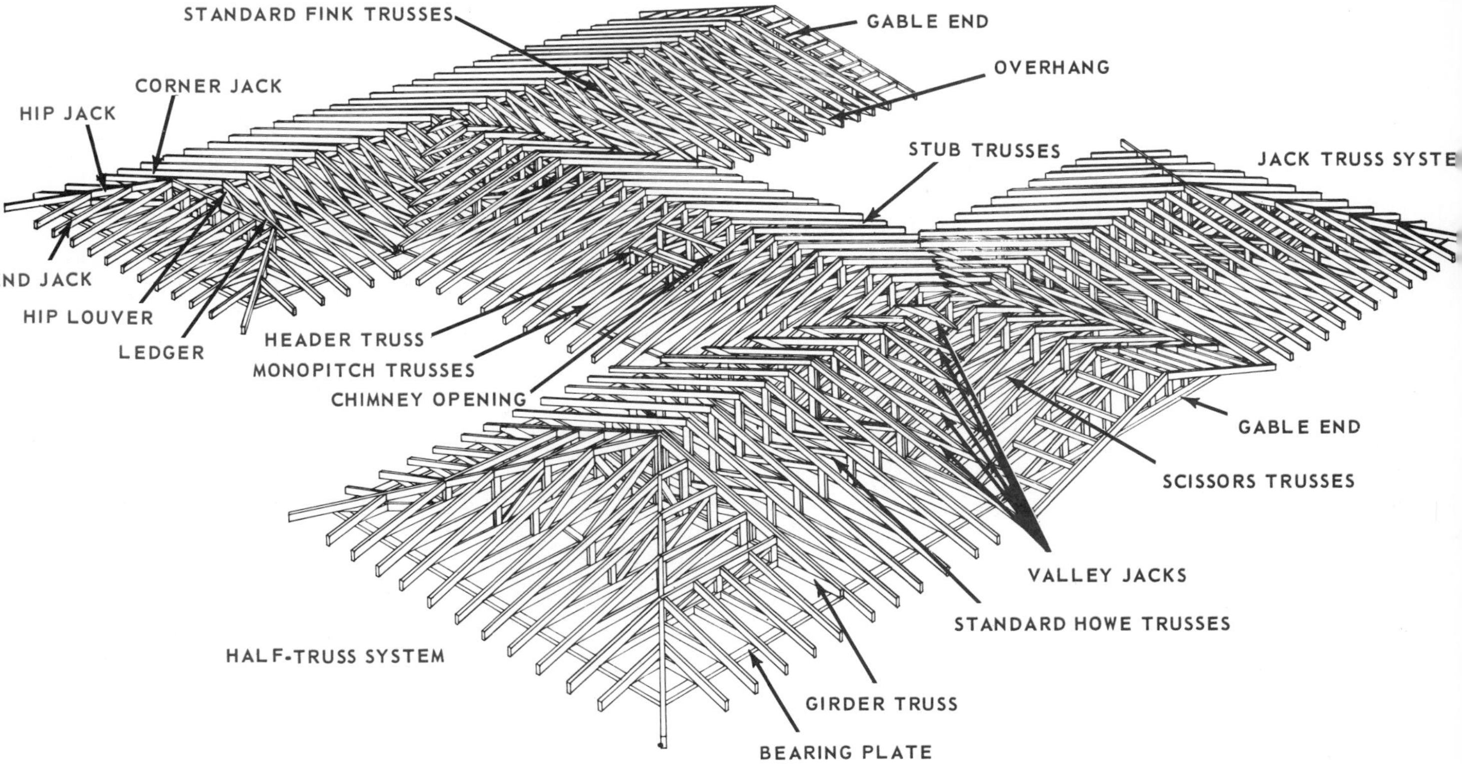

Fig. 21-3. Various types of roof trusses used to form hips, valleys, overhangs, and gable ends. (Automated Building Components, Inc.)

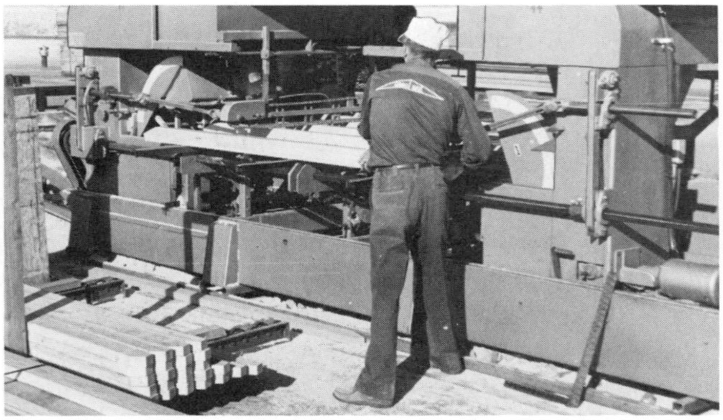

Fig. 21-4. Double-end saw with four cutting heads speeds production of truss members.

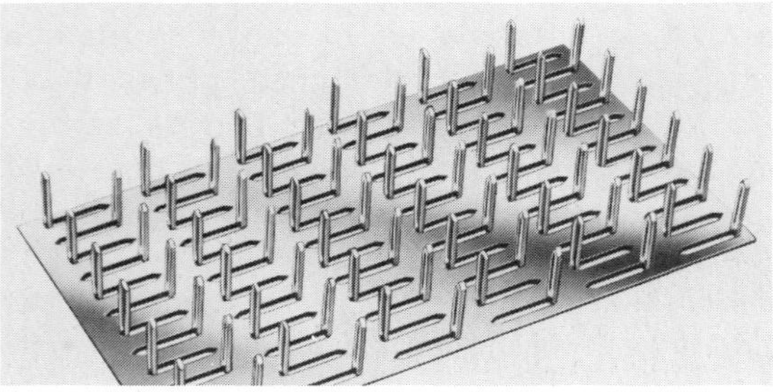

Fig. 21-5. "Gang-Nail" connector available in several sizes and gauges of steel. Must be applied with approved presses.

Fig. 21-6. Trussed roof system being transported to construction site. (Automated Building Components, Inc.)

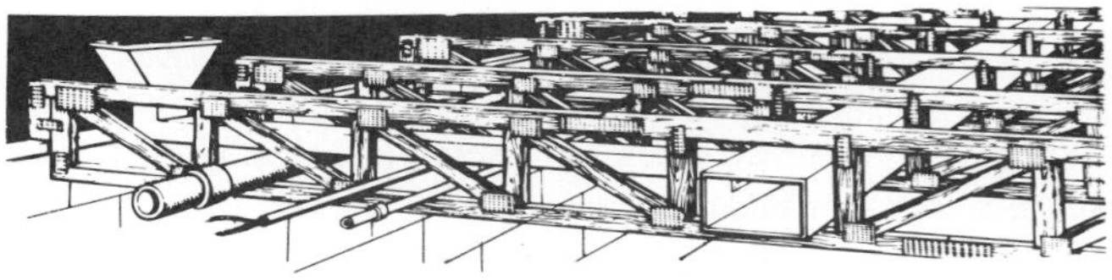

Fig. 21-7. Floor truss system permits wide spans and space for the installation of ducts.

of heating and air conditioning ducts, electrical conduit, and plumbing lines. Fig. 21-8 shows a truss of this type being fabricated.

Wall and roof panels are produced in various shapes and sizes and provide structural strength in addition to forming inside and outside surfaces. Fig. 21-9 shows the installation of curved stressed-skin panels constructed of lumber ribs and plywood surfaces. Flat panel construction is

Fig. 21-8. Hydraulic press (40 ton capacity) being used to install connectors in a flat girder truss 60 ft. long.

Fig. 21-9. Installing curved stressed-skin panels that will provide an unusual roof design and a vaulted ceiling effect. (Plywood Fabricators Service, Inc.)

Fig. 21-10. Starting component-built vacation home consisting of stressed-skin floor panels and sandwich wall panels.

described in Unit 20. Sandwich panels fabricated with a foamed plastic core provide sufficient strength for walls and partitions in some structures. See Fig. 21-10. These lightweight panels may be constructed with an aluminum outer skin and a hardboard or plywood interior surface. Special metal channels and angles are available which increase assembly speed.

Prefabricated Buildings

The total building in various stages of completion can be fabricated in factories and then shipped as a "package" to the building site. Residential structures constitute the major portion of this type of manufacturing; however small commercial buildings and farm structures are also being produced. See Fig. 21-11.

There are four basic types of prefabricated houses and other buildings; the precut, the panelized, the sectionalized, and the mobile-home. Combinations of these may be employed.

Fig. 21-11. Prefabricated home of Colonial design with three bedrooms, two baths, and a family room. Total living area 1600 sq. ft. (National Homes Corp.)

Precut

In the typical modern precut house, lumber is cut, shaped, and labeled to reduce labor and save time on the building site. Manufacturers of this type of house construction include materials necessary to complete the exterior and interior surfaces -- plus such millwork items as windows, doors, stairs and cabinets. Optional items may include electrical, plumbing, and heating equipment.

Fig. 21-12. Manufacturing plant where wall panels are being fabricated on production lines. (American Plywood Assoc.)

Panelized

In panelized prefabrication, flat sections of the structure are fabricated on assembly lines, Fig. 21-12, and hauled to the building site. Finish materials that are not applied to the panels in the plant are usually included in the "package" along with all the other items necessary to complete the structure. Today many home manufacturers utilize the panelized type of prefabrication.

Fig. 21-13. Control end of a double-end saw used to cut framing material in a plant that produces panelized structures.

Fig. 21-14. Using air-powered nailers to fasten wall frames. Nails are fed from overhead hoppers and carried by a stream of air through plastic hose to nailing mechanism. (Frank Paxton Lumber Co.)

Production woodworking machines are used to cut framing members to length and required angles, Fig. 21-13. These parts are stored and delivered to the assembly stations as needed. Wall panel frames are formed by placing the various studs, headers, and plates in positioning jigs on the production line and fastening them together with pneumatic powered nailers, Fig. 21-14.

As the completed frames move along conveyor lines, wall surface materials are placed in posi-

Fig. 21-15. A gang of pneumatic-powered nailers secure 1/2 in. plywood sheathing to a wall frame in a fraction of a minute. (National Forest Products Assoc.)

Fig. 21-16. Using production glue spreader to apply adhesive to wall panel. (National Homes Corp.)

tion and nailed with powered gang-nailers and/or staplers. See Fig. 21-15. Some materials are attached with an adhesive which is spread with large rollers, Fig. 21-16.

Farther along the line, special powered tables turn the panel over so the opposite side can be completed, Fig. 21-17. Insulation is installed

Fig. 21-17. Placing insulation between the studs. In background, workman controls a hydraulically operated mechanism that turns panels upside down.

Fig. 21-18. Applying prefinished aluminum siding to the exterior surface of a wall panel. (National Homes Corp.)

Fig. 21-19. Installing a window unit in a wall frame. Production lines require careful organization so correct components are at right place at right time.

along with electrical wiring or other facilities that may be required in a particular panel. Outside surface materials (siding) are attached, Fig. 21-18, either before or after window and door units are installed, Fig. 21-19. Some prefabricated

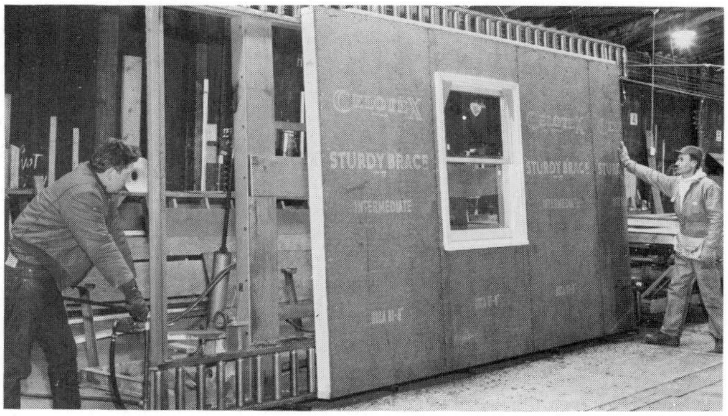

Fig. 21-20. Wall panel ready for shipment to construction site where siding will be applied after erection.

wall panels are assembled on the site before the siding is applied. Fig. 21-20 illustrates a completed wall panel of this type.

In other areas of the plant, roof framing components are prepared, Fig. 21-21. For post-and-

Fig. 21-21. Assembling roof trusses on special jig table equipped with a press for setting fastener plates. (National Forest Products Assoc.)

beam structures, unit panels that form both the roof and ceiling surfaces are commonly used. To save time after erection, the ceiling side is usually painted, Fig. 21-22.

Although a typical prefabricated house of the panelized type utilizes a first floor deck and foundation built by conventional methods, some manufacturers design and build floor panels. Full length joists are assembled with headers and then long plywood sheets, formed with scarf joints, are

Fig. 21-22. Painting the ceiling side of roof panels used in a post-and-beam type of prefabricated structure. (Wausau Homes Inc.)

Fig. 21-23. Prefabricated floor panels. Full length floor joists and scarf-glued plywood sheets are pressure glued to form units as large as 8 ft. by 24 ft.

glued in place. The resulting stressed-skin construction provides rigidity and strength and eliminates the possibility of nail pops and squeaks, Fig. 21-23.

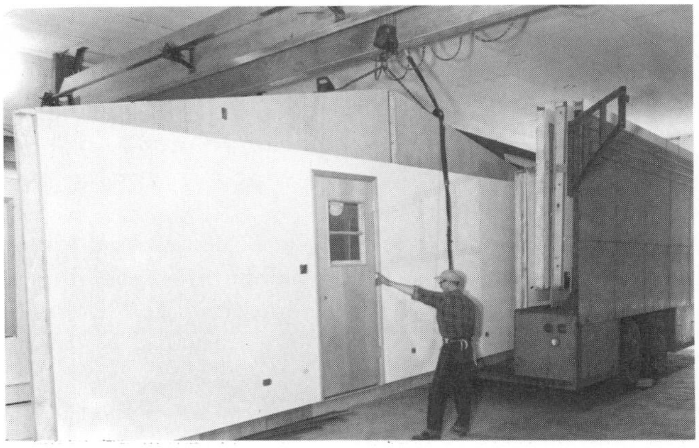

Fig. 21-24. Finished wall panel being loaded on trailer. Panel includes electrical wiring with outside door installed. Dry wall surface has been painted. (Wausau Homes Inc.)

As the various panels near the end of their respective production lines, they receive a final inspection. After inspection, each panel is numbered for assembly purposes and is then ready for storage or loading into a trailer, Fig. 21-24. When all materials and millwork to complete the "package" are loaded, the truck is dispatched to the building site. Kitchen cabinets and other built-in units are usually fabricated in plants specializing in these items and may be shipped either to the home fabricator or directly to the building site.

At the building site, the foundation should have been completed. The various prefabricated units were loaded on the trailer so they can be removed in the proper sequence required for assembly. After the floor deck is complete, side walls are placed in position as illustrated in Fig. 21-25.

Fig. 21-25. Erecting prefabricated wall panels at the construction site. (American Plywood Assoc.)

For large structures, cranes are often required to handle large panels or sections, Fig. 21-26. By the end of an eight hour day an experienced crew of men can usually erect and enclose an average size single-family house. Several weeks are usually required to complete the wiring, heating, plumbing and decorative work.

Fig. 21-26. Using a crane to install folded plate roof units made from lumber framing covered with plywood skins. (American Plywood Assoc.)

The panel type of prefabrication is not limited to any particular residential style or size. Multi-level, two story, and split-level types are available. Although houses usually constitute the major production, many manufacturing plants also supply panels and other components for school buildings, apartments, and small commercial structures.

Fig. 21-27. Applying siding to a modular home as it nears the end of plant production line.

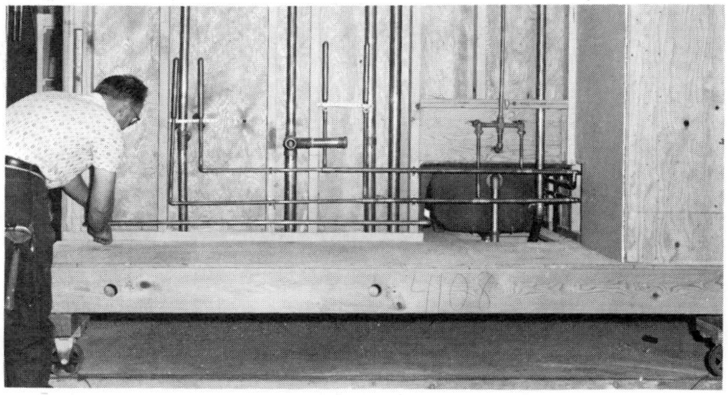

Fig. 21-28. Bathroom plumbing being installed as a sectionalized unit proceeds along a production line. Note the craftsman can perform all of the work in a standing position. (Wausau Homes, Inc.)

Sectionalized or Modular

In this type of prefabrication, entire sections (commonly called modules) of the structure are fabricated and finished in manufacturing plants, Fig. 21-27. The sections or modules are then transported to the site where they are assembled to form the complete structure. Because of limitations in transportation facilities, widths of a given module seldom exceed 14 feet.

An advantage of sectionalized construction results from the fact that nearly all of the detail finish work can be completed at the factory. Kitchen cabinets can be attached to the walls and other built-in features can be installed. Also wall, floor and ceiling surfaces can be applied and finished.

Fig. 21-29. Installing preassembled copper plumbing in a mechanical core. Factory conditions make it easy to check supply and waste lines with compressed air.

Electrical wiring, heating and air conditioning ducts, plumbing lines, and even plumbing fixtures can be installed under the controlled conditions that are possible in the manufacturing plants. Fig. 21-28 shows a section that includes a bathroom wall with tub in place and plumbing lines being completed as it moves along the production line.

A section which includes a concentration of plumbing and heating facilities is often included in a fabrication system that otherwise consists mainly of panels. Sections of this nature which include most of the utility hook-ups are called mechanical cores. Basic designs consist of grouping the kitchen, bath and utilities in one preassembled unit that requires only three connections

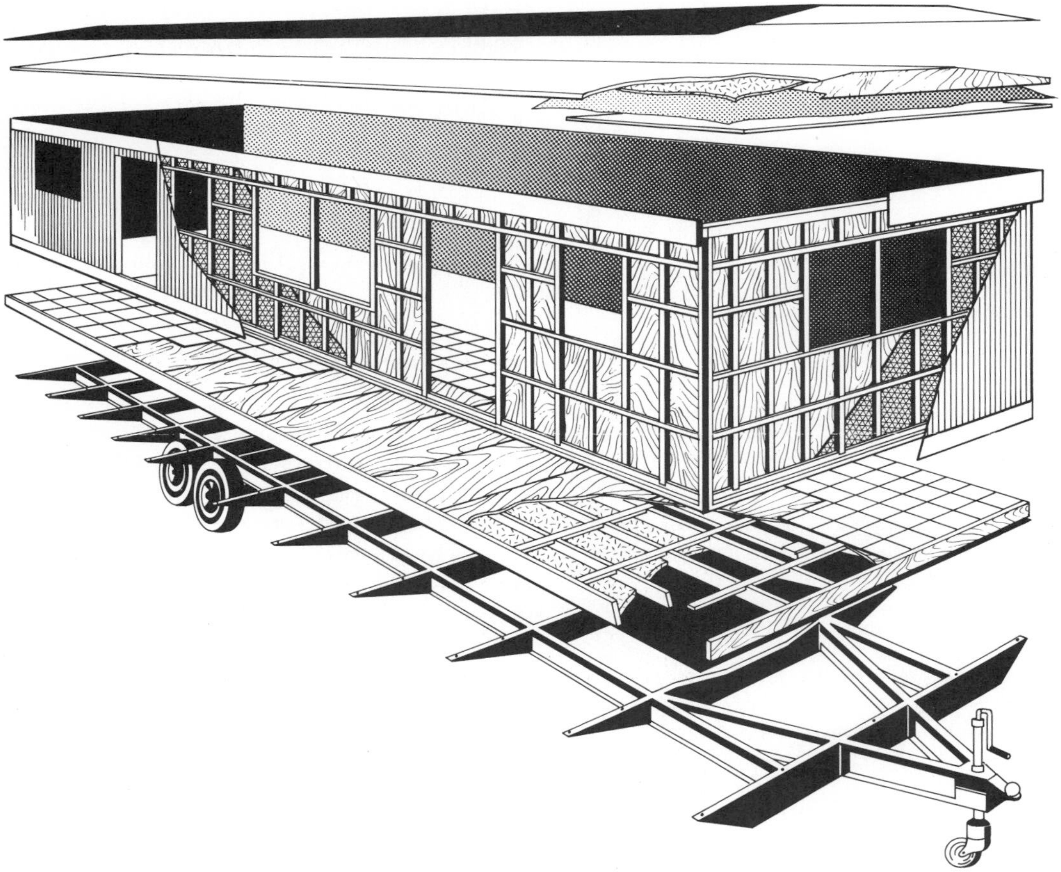

Fig. 21-30. Exploded view of a mobile home. All structural members above the steel chassis are wood. Plywood deck is securely screwed to floor frame which is assembled with glue. Outside walls consist of 2 x 3 studs with 1 x 2 horizontal rails glued in place. Inside surface is 3/16 prefinished plywood – outside prepainted aluminum panels. (Redman Industries, Inc.)

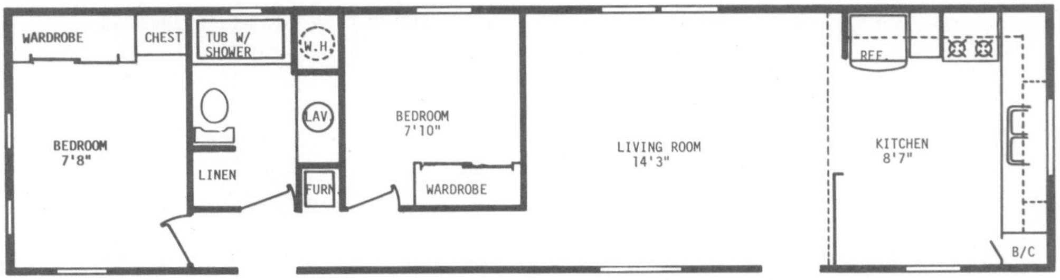

Fig. 21-31. Typical floor plan of a mobile home. Size shown is 50 x 12 feet and includes two bedrooms. (Skyline Corp.)

at the site. A mechanical core consisting of a bathroom and one kitchen wall is shown on an assembly line in Fig. 21-29. Some units are designed to include heating and air conditioning equipment as well as electrical and plumbing.

Disadvantages of the sectionalized type of prefabrication are the difficulties connected with storage and transportation, and the necessity of having power cranes available for handling and assembling the units at the construction site.

Mobile Homes

Mobile homes are constructed on a chassis, Fig. 21-30, and do not require a permanent foundation. They are generally defined by manufacturers as a trailer longer than 28 feet and heavier than 4500 lbs.

Widths of 12 and 14 ft. are considered standard. Lengths may be as great as 68 ft. A typical floor plan of a two bedroom model is shown in Fig. 21-31. Designs sometimes include an "expansible" feature that permits wider living rooms. The extended section is carried inside while the mobile home is being transported and is then attached to the side when the unit reaches its destination.

Production methods used in building mobile

Fig. 21-32. Wooden floor frame is attached to steel chassis. Floor must be fully insulated. Note plumbing and electrical lines. (Redman Industries, Inc.)

Fig. 21-33. Completed mobile homes ready to be towed to distributor or directly to parking area.

Fig. 21-34. Interior view of mobile home. (Marshfield Homes, Inc.)

homes are similar to those employed for standard prefabricated housing. Some of the structures however, are considerably different. For example, the floor must be one rigid unit and usually consists of a wood frame attached to a welded steel chassis, Fig. 21-32.

Mobile homes are completed and fully equipped at the factory, Fig. 21-33. They are then pulled by special trucks to the dealer or directly to trailer parks where they can be quickly prepared for occupancy. Interior furnishings include all major appliances, carpeting, drapes, furniture and lamps. Fig. 21-34 shows a typical interior view.

Test Your Knowledge - Unit 21

1. Prefabricated components are usually mass produced using conveyor lines and assembly ____________.
2. The type of prefabrication most commonly used by home manufacturers is the ____________.
3. A prefabricated section that includes a concentration of plumbing, heating, and other utility hookups is generally called a ____________.
4. The chief advantage of the __________ type of prefabrication is that most of the interior and exterior finish work can be completed at the factory.
5. Mobile homes are constructed on a metal chassis and do not require a permanent ______.

Outside Assignments

1. Visit a local builder or the manager of a building supply center that represents a "prefab" home manufacturer in your region. Secure descriptive literature and information concerning sizes, design, fabrication features, material quality, and erection procedures. Also secure prices for standard models and time required for delivery. Organize the information carefully and make an oral presentation to your class. Supplement your oral descriptions with appropriate visual aids.
2. Prepare a written report on the development of prefabrication as applied to building construction in this country. Highlight early experiments and devote most of your study to progress since World War II. Secure information from reference books, encyclopedias, and trade magazines.
3. Visit a mobile home sales center and secure information and descriptive literature concerning size, features, and price. Inspect models that are on display and give special attention to the quality of materials, workmanship, and finish. Also note the quality of equipment and furnishings. Compare prices with costs of prefabricated and conventionally built houses. Prepare carefully organized notes and make an oral report to your class.

Unit 22

SCAFFOLDS AND LADDERS

Carpenters use scaffolding (also called staging) to reach work areas which would normally be too high to work on while standing on the ground or floor deck. Scaffolding consists of rigid, elevated platforms used to support workers, tools, and materials with a high degree of safety. The height must enable the work to be performed with speed and accuracy, without undue strain on the worker.

The type of scaffold required depends on how many workers will be supported at one time, distance above the ground, and whether it will be required to support building materials.

Types of Scaffolding

Typical scaffolds are shown in Figs. 22-1, 22-2, and 22-3. In constructing wooden scaffolds, the uprights should be made of clear, straight-grain 2 x 4s. The lower ends should be placed on blocks or pads to prevent settling into the ground. Cross ledgers consist of 2 x 6's about 4 ft. long. Use at least three 16d nails at each end of the ledgers to fasten them to the uprights. In the case of the single-pole scaffold, one end of the ledger should be fastened to a 2 x 6 block securely nailed

Fig. 22-1. Installing an aluminum gutter. Workmen are supported by metal scaffolding. (ALCOA Building Products, Inc.)

to the wall. Braces may be made of 1 x 6 lumber, fastened to uprights with 10d nails, and with 8d nails where they cross. For the platform, 2 x 10 planks free from objectionable knots should be used. It is good practice to spike the planks to cross ledgers to prevent slipping.

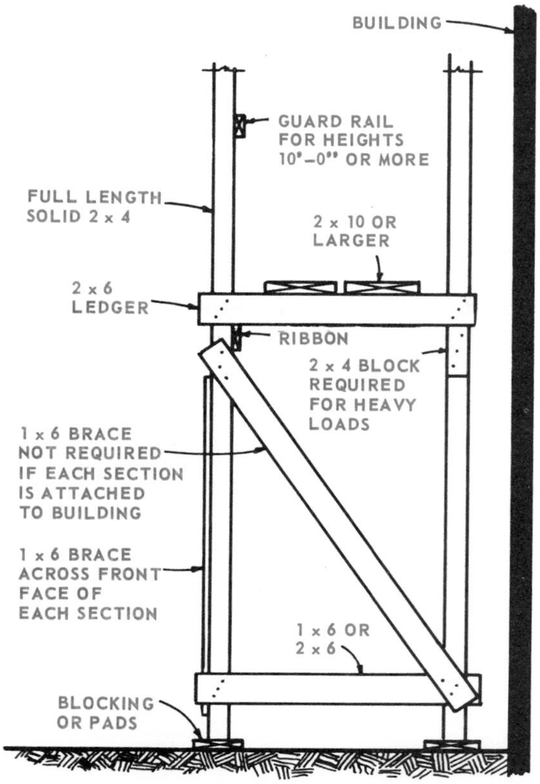

Fig. 22-2. Typical design for double-pole scaffold. Structure can be extended up to form several platforms. Maximum height should be limited to 18 feet.

Another scaffold type is the lean-to design which is so named because the upper end leans against the wall. Swinging scaffolds are suspended from the roof or other overhead structures. These are used mainly by painters and other craftsmen where only light equipment and materials are required.

Manufactured Scaffolding

In modern construction, many builders use sectional steel or aluminum scaffolding. The units are quickly and easily assembled from prefabricated frames. See Fig. 22-4. These units can be stacked and/or assembled horizontally to complete the required scaffold.

Sectional steel scaffolding may be obtained with adjustable legs and in sizes from 2 to 5 feet wide and heights from 3 to 10 feet. Various size

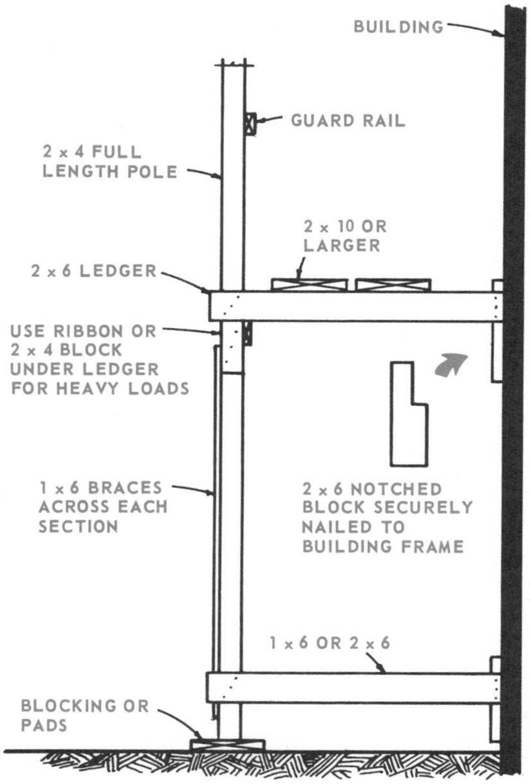

Fig. 22-3. Typical design for a single-pole scaffold. Horizontal distance between ledger sections should never exceed 10 feet.

bracing provides frame separations of 5 to 10 feet. The basic units are set up and joined vertically and horizontally. Fig. 22-5 illustrates the construction details of a sectional scaffold. The frames can be equipped with casters when a rolling scaffold is desired.

Brackets, Jacks, and Trestles

Metal wall brackets have gained wide acceptance in residential carpentry because they can be erected quickly, require little material, and are easily moved from one construction site to another, Fig. 22-6. Great care should be exercised when fastening brackets to a wall -- especially the type that is attached with nails. Use four 16d

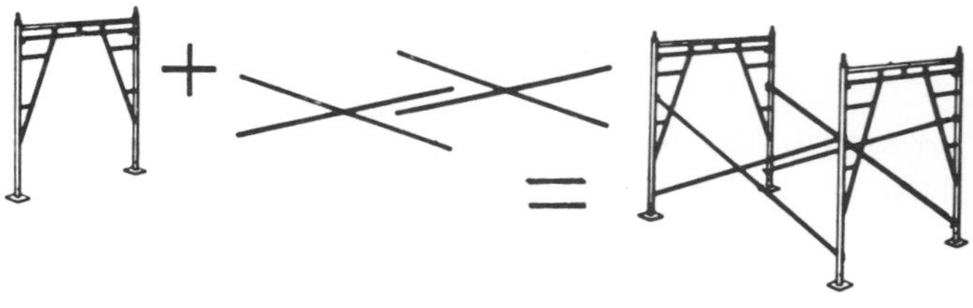

Fig. 22-4. Scaffold sections can be rapidly assembled from prefabricated trussed frames and diagonal braces.

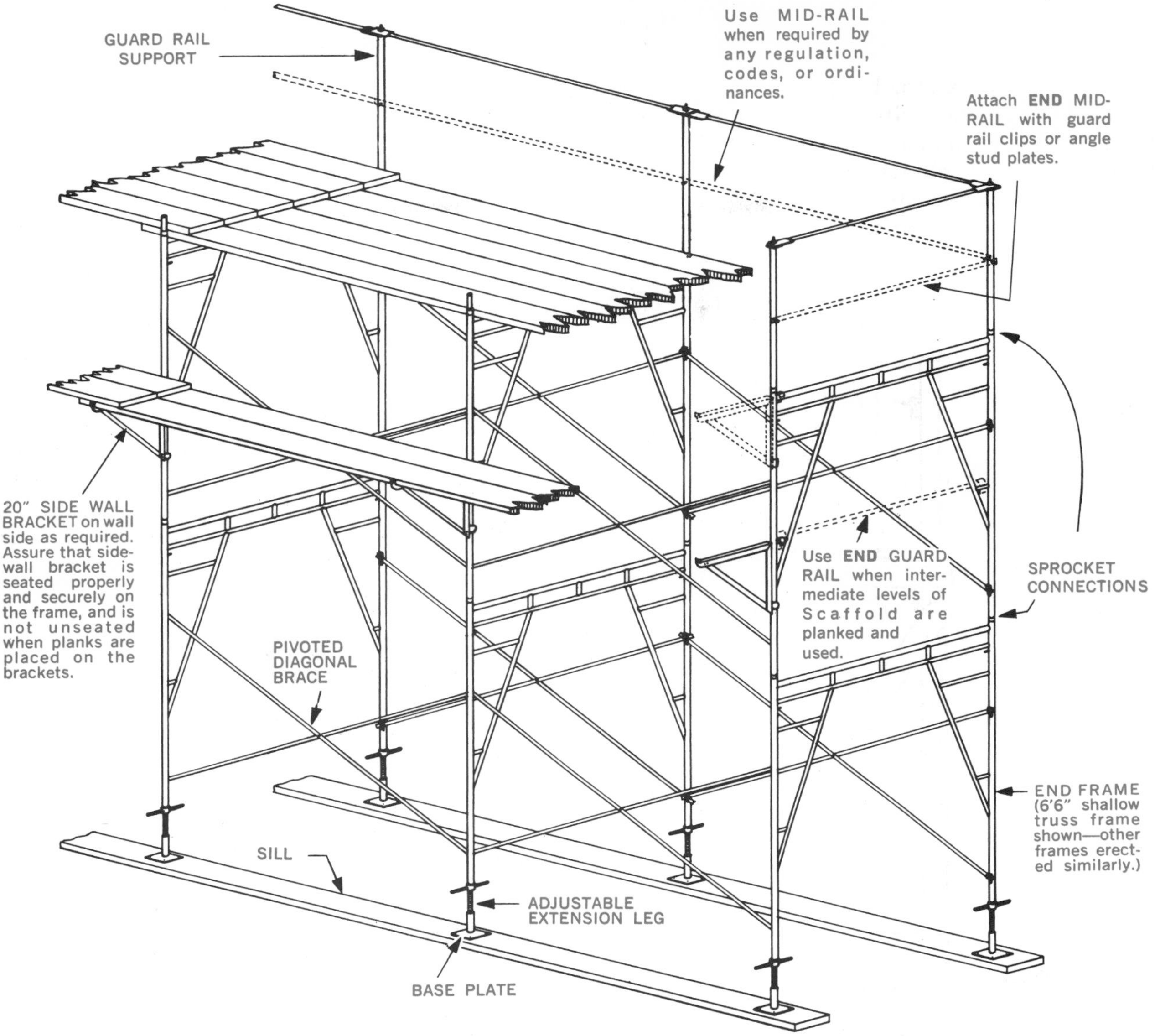
GUARD RAIL SUPPORT
Use MID-RAIL when required by any regulation, codes, or ordinances.
Attach END MID-RAIL with guard rail clips or angle stud plates.
20" SIDE WALL BRACKET on wall side as required. Assure that side-wall bracket is seated properly and securely on the frame, and is not unseated when planks are placed on the brackets.
PIVOTED DIAGONAL BRACE
Use END GUARD RAIL when intermediate levels of Scaffold are planked and used.
SPROCKET CONNECTIONS
END FRAME (6'6" shallow truss frame shown—other frames erected similarly.)
SILL
ADJUSTABLE EXTENSION LEG
BASE PLATE

Fig. 22-5. Scaffold assembly made from sections joined vertically and horizontally. (Patent Scaffolding Co.)

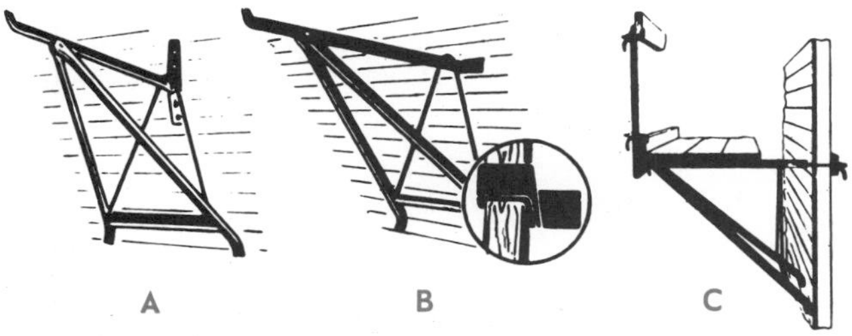
A
B
C

Fig. 22-6. Kinds of metal wall brackets. A–Attached with nails securely set in building frame. B–Hooked directly to studding. C–Bracket with detachable guard rail.

or 20d nails and be sure they penetrate sound framing lumber. Since the bracket is held largely by the nail heads, check to see that the heads have not been damaged by the hammer. Fig. 22-7 shows the use of metal brackets in forming scaffolds for a typical multi-level house.

Fig. 22-7. Scaffolding formed with metal wall brackets. When using this type to apply trim and siding, start at the top of the building and move downward.

Roofing brackets, Fig. 22-8, are convenient to use and provide a high degree of safety when working on steep slopes. The arm that supports the plank is adjustable so that a level position can be obtained on nearly any angle of roof surface.

Fig. 22-8. Roofing bracket. Slotted extension hooks on nails driven into the roof deck and frame. Arm that carries planking is adjustable. Overall length is 30 inches.

Fig. 22-9. Ladder jack. This type is held in position with hooks that fit around side rails. (Patent Scaffolding Co.)

Ladder jacks can be used to a good advantage in setting up a scaffold, particularly on repair jobs where a single carpenter is employed. Two sturdy ladders of the same size and a strong plank are required. Fig. 22-9 shows a jack that is suspended below the ladder and hooks to the side rails. Another type of jack, Fig. 22-10, is easily adjusted upward along the ladder.

Trestle jacks provide a practical method of supporting low platforms for interior work. They are assembled as illustrated in Fig. 22-11. In addition to selecting the proper size and weight of jack, the material used for the ledger (horizontal member to which the jacks are attached) must be sound and of adequate size to support the load.

Safety Rules for Scaffolding

1. All scaffolds should be constructed under the direction of an experienced craftsman.
2. Follow design specifications as listed in local and state codes. Inspect scaffolds daily before use.
3. Provide adequate pads or sills under scaffold posts.
4. Plumb and level scaffold members as the erection proceeds.
5. Equip planked areas with proper guard rails and toe boards when required.
6. Power lines near scaffolds are dangerous - consult power service company for advice and procedure.
7. Do not use ladders or makeshift devices on top of scaffold platforms to increase the height.
8. Be certain the planking is heavy enough to carry the load and span.
9. Planking should be lapped at least 12 in. and extend 6 in. beyond all supports.
10. Do not permit planking to extend an unsafe distance beyond supports.
11. Remove all materials and equipment from the platform of rolling scaffolds before moving.
12. The height of the platform on rolling scaffolds should not exceed by four times the smallest base dimension.

Ladders

Types of wood and aluminum ladders commonly used by the carpenter are illustrated in Fig. 22-12. Step ladders range in size from 4 to 20 ft. and regular single ladders are usually available in sizes of 8 to 26 ft. Extension ladders provide lengths up to 60 ft.

Quality wood ladders are fabricated from clear, straight-grained stock that is carefully seasoned.

Fig. 22-10. Ladder jacks that project on the front side. Left view shows method of raising position of jack while plank is supported on arm and shoulder. Use caution when performing this operation and do not raise the scaffold more than one rung at a time.

Ladders should be given a clear finish which permits easy visual inspection. When reconditioning a wood ladder, never use paint.

Basic care and handling of ladders is illustrated in Fig. 22-13. In addition, keep ladders clean and do not let grease, oil, or paint accumulate on the rails or rungs. On extension ladders, keep all fittings tight and lubricate locks and pulleys. Replace frayed or worn rope.

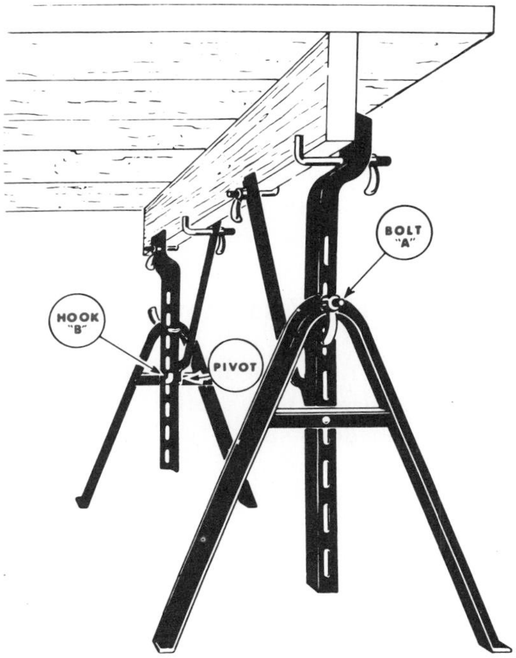

Fig. 22-11. Trestle jacks. To adjust height first loosen bolt A and then change hook B to another slot. (Patent Scaffolding Co.)

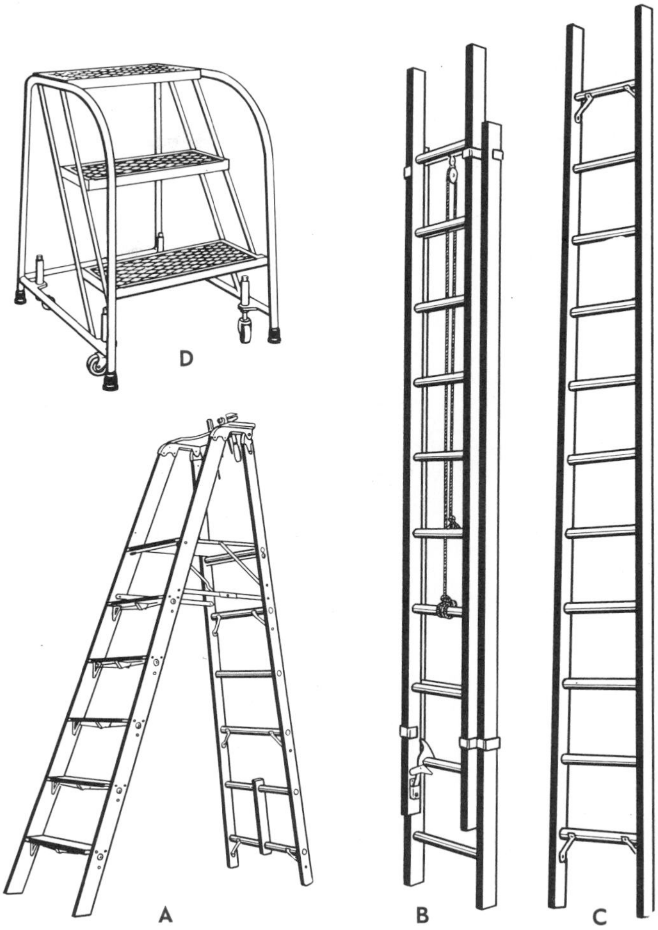

Fig. 22-12. Ladder types used by the carpenter. A–Stepladder with platform and toolholder. B–Extension ladder. C–Single ladder. D–Safety rolling ladder. Drops firmly to floor when mounted. Handy for inside trim work. (Patent Scaffolding Co.)

To erect a ladder, place the lower end against a solid base so it cannot slide. Raise the top end and walk toward the bottom end, grasping and raising the ladder rung by rung as you proceed. When vertical, lean against the structure at the proper angle (see Safety Rule 2 below) and make sure bottom ends of both rails rest on a firm base.

Safety Rules for Ladders

1. Always inspect a ladder before using it.
2. Place ladder so the horizontal distance from its lower end to the vertical wall is at least one-fourth the length of the ladder.
3. Before climbing the ladder, be sure both rails rest on solid footing.
4. Equip the rails with safety shoes, Fig. 22-14, when the ladder is used on surfaces that may permit the bottom to slip.

INSPECTION

LADDERS SHOULD BE INSPECTED FREQUENTLY AND THOSE WHICH HAVE DEVELOPED DEFECTS SHOULD BE EITHER REPAIRED OR DESTROYED.

CARRYING

ALWAYS CARRY A LADDER OVER YOUR SHOULDER WITH FRONT END ELEVATED. BE SURE NOT TO DROP OR LET FALL FOR SUCH IMPACT WEAKENS A LADDER.

STORAGE

STORE HORIZONTALLY ON SUPPORTS TO PREVENT SAGGING. DO NOT STORE NEAR HEAT OR EXPOSE TO ELEMENTS.

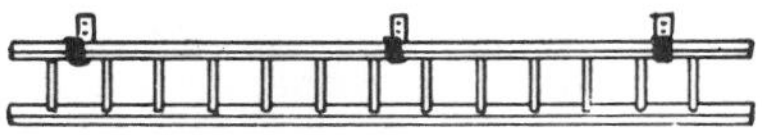

Fig. 22-13. Ladder care and handling.

5. Do not place ladder in front of doorway where the door can be opened toward the ladder.
6. Never place ladders on boxes or any unstable base to gain additional height.
7. Never splice together two short wood ladders to obtain a longer ladder.
8. Always face a ladder when climbing up or descending.

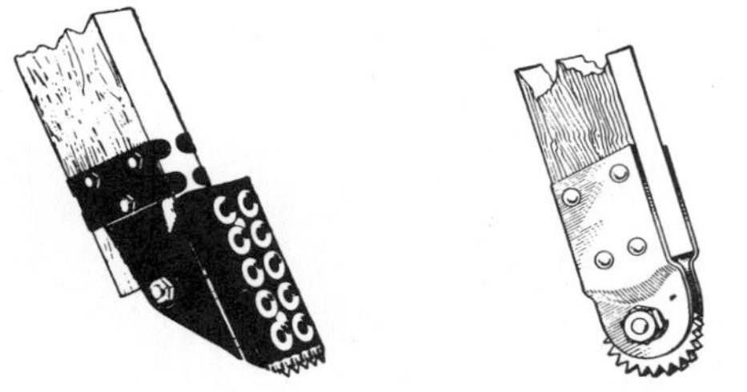

UNIVERSAL SPUR WHEEL

Fig. 22-14. Safety shoes for ladders. (Tilley Ladder Co.)

9. Place ladder in position so work can be done without leaning beyond either side rail.
10. Be certain extension ladders have sufficient lap between sections. A 36 ft. length should lap at least 3 ft.; a 48 ft., at least 4 ft.
11. When a ladder is used to gain access to a roof, it should extend above the roof at least 3 ft.
12. Both hands should be kept free when climbing a ladder. Use a hand line to raise or lower tools and materials.
13. Before mounting a step ladder, be sure it is fully open and locked, and all four legs are firmly supported.
14. Do not stand on either of the two top steps of a standard step ladder.
15. Do not leave tools on the top of a step ladder unless it is equipped with a special holder.
16. Never use metal ladders where there is a possibility of coming in contact with electric current.

Test Your Knowledge - Unit 22

1. Ledgers that support the planking of wood scaffolds should be made of material with a nominal cross section of _________ inches.
2. In wooden scaffolds, horizontal spacing of support sections should not exceed __________feet.
3. When the scaffolding consists of metal wall brackets, finish work must proceed __________ (upward, downward).
4. The minimum height of the platform of a rolling scaffold unit should not exceed ______times the smallest dimension of the base.
5. Regular single ladders are available in sizes ________ft. to __________ft.
6. The sides of a ladder are called _________ .
7. When a ladder is used to climb onto a roof, it should extend at least _______feet above the roofs edge.

Outside Assignments

1. Secure descriptive literature from a company that manufactures steel scaffolding. Check with your local building supply dealer or write directly to the company. From a study of this printed material, learn about the kinds and sizes of tubing that are used. Also learn how the various connecting devices operate. Try to secure approximate costs of this type of scaffolding. Prepare carefully organized notes and make an oral report to your class.
2. Make a study of the types of ladders used in construction and maintenance work and prepare a written report. Include information about the kind and quality of material used and how the ladders are assembled and finished. Also include commonly available sizes and list prices. Finally, develop a list of important considerations a homeowner should apply to the purchase of a ladder.

Fig. 23-1. An apprenticeship program provides on the job training. In the above view a carpenter explains the layout of a partition to an apprentice.

Unit 23
CARPENTRY—A CAREER

Carpentry offers a promising career for the individual who has an interest in and aptitude for working with tools and materials. It requires the development of manual skills (a combination of thinking and doing) and a thorough knowledge and understanding of basic principles and practices related to construction work.

Opportunities

The carpenter is an important craftsman in the broad field of building construction -- an industry which experts tell us will double in size during the next few years.

The carpenter needs a wide variety of skills. His versatility qualifies him for a broad range of job opportunities. Besides the construction of new buildings, he can handle work connected with remodeling, maintenance, and repair. Shops and factories that produce prefabricated buildings and components offer additional worthwhile opportunities.

The income earned by the carpenter is comparable to that received by other trades, and provides a good living. In addition, the carpenter experiences the pride and satisfaction that results from good craftsmanship... something that is not experienced in many other occupations.

After the carpenter gains experience, he may want to undertake a small construction contract. This may be a first step toward the general contracting business. The experienced carpenter is usually able to handle the work of the estimator who figures labor and material costs.

If the carpenter prefers the sales and service aspects of construction work, quite likely he can secure a position with a lumber yard, building supply center, or with a company that manufactures prefabricated structures.

Training

Basic to a career in carpentry and other areas of building construction, is a high school education. This is as true for the apprentice-to-be as for the technician or engineer. Take as many woodworking and building construction courses as possible. Other courses are also valuable -- especially drafting.

Students interested in the building trades sometimes minimize the importance of science and math. This is unfortunate because these studies are essential if you are to understand the technical aspects of modern methods and materials used in construction work. Social studies and English should also be included since everyone should be prepared to make a contribution to our society and be competent in reading, writing, and speaking.

After graduation from high school, you will probably be qualified to enter directly into an apprenticeship training program, Fig. 23-1. If circumstances permit, you may wish to enroll in a vocational-technical school in your area where you could take advanced courses in carpentry and related areas. Some high schools offer these vocational courses as a part of the regular programs before graduation. If possible, take classes in concrete work, bricklaying, plumbing, sheet-metal, and electric wiring. The carpenter usually works closely with tradesmen in these areas and a basic understanding of the methods and procedures will be very valuable. This is especially true if you desire to become a foreman or supervisor.

Apprenticeship

Our modern apprenticeship training program had its beginning in early times. Its origin is found in the father-son relationship whereby the knowledge and skill of a trade was passed on to succeeding generations. As society and the economic structure became more complex, the son often left his home and was placed under the guidance and direction of another master craftsman in the community. He was called an apprentice and learned the trade of his master.

During the period of apprenticeship (sometimes as long as seven years) the apprentice often lived in his master's household. He received no wages, but was given board, room, and clothing. When the training was complete, the apprentice was granted the status of journeyman and could then work for wages. The word journeyman was derived from the fact that the apprentice who had completed his training period was then free to "journey" to other places in search of employment. The term journeyman is still used to denote a tradesman who is fully qualified. After the journeyman had gained further experience by working with other masters, he often would establish his own business, and secure apprentices to work under his direction.

This form of apprenticeship declined rapidly with the advent of the industrial revolution. A new kind of system developed where the apprentice lived at home and received wages for his work. Under this system, the apprentice was often subjected to exploitation - becoming only a low paid worker who received little training. Such practices continued in varying degrees until federal legislation was enacted which established standards and specific requirements for apprenticeship training programs.

Today, apprenticeship programs are carefully organized and supervised. Local committees consisting of labor and management provide direct control, with assistance from schools, state and federal organizations and agencies, Fig. 23-2. The apprentice works under a signed agreement with an employer. The agreement includes the approval of local and state committees on apprenticeship training.

Applicants must be between the ages of 17 and 27 years and shall satisfy the local committee that they have the ability or aptitude to master the trade. The term of apprenticeship for the field of carpentry is normally four years. This may be adjusted for applicants with significant experience or those who may have completed certain advanced courses in vocational-technical schools.

In addition to the instruction and skills learned through regular work on the job, an apprentice attends school classes in subjects related to the trade. These classes are usually held in the evening and total at least 144 hours per year. They cover technical information about tools, machines, methods and processes; and provide practice in mathematical calculations, blueprint reading, sketching, layout work, and similar activities. A

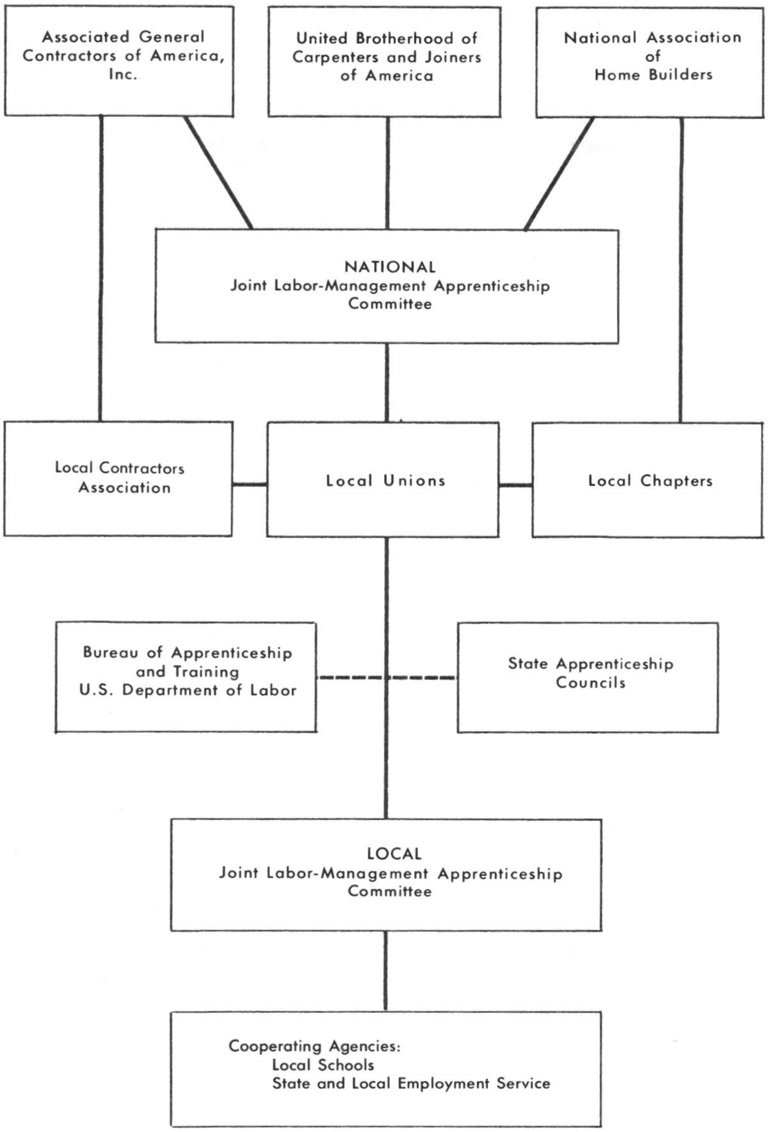

Fig. 23-2. Apprenticeship and training system for the carpentry trade. Additional information may be obtained from the bulletin National Carpentry Apprenticeship and Training Standards, prepared by the U. S. Department of Labor.

great deal of study is required to master the technical knowledge needed in modern carpentry work.

An apprentice is an employed worker and receives wages as he learns. The wage scale which is determined by the local committee, usually starts at about 50 percent of the amount received by a journeyman carpenter. This scale is advanced regularly and may approach 90 percent of the journeyman's pay during the last year.

When the training period is complete and the apprentice has satisfactorily passed a final examination, he becomes a journeyman carpenter. A certificate attesting to this status is issued and recognized throughout the country.

Personal Qualifications

To become a successful carpenter you must be physically able to perform the work and have a high ability or aptitude in working with the tools and materials of the trade. You need also sincere interest and enthusiasm that will intensify your efforts as you study and practice the skills and "know-how" required.

In addition, you must develop certain desirable character traits. Honesty in all your dealings is very important, especially in the quality and quantity of work performed. You must show courtesy, respect, and loyalty to those with whom you work.

Punctuality and reliability reflect your general attitude and are important not only during your training program but later when you enter regular employment.

The ability to cooperate and get along with others -- students, fellow craftsmen, supervisors, and employers -- is essential to success. Most individuals who fail in the carpentry trade do so because of a deficiency in personal characteristics - rather than inability to obtain a satisfactory level of performance.

The hazards associated with carpentry require that you develop a good attitude toward safety. This means that you must be willing to give time and attention to learning the safest way to perform your work -- that you will follow safety rules and regulations at all times.

Even after you complete your training program you must continue to be diligent in your efforts to perfect your skills and adjust to new methods and techniques. Each day brings new materials and improved procedures to the construction industry, and presents a special challenge to those in the carpentry trade. You should read and study new books, manufacturer's literature, and trade journals and magazines in the building construction field. Considerable information can be obtained at association meetings and conventions where new products are exhibited. To be a successful carpenter, you will also need to keep informed on code changes, new zoning ordinances, safety regulations and other aspects of construction work that apply directly to the local community in which you work.

During the next few years, the growth in building construction will result in a greatly increased demand for carpenters. Carpenters will be needed for the construction of homes, schools, churches, stores, public, institutional, and many other types of buildings.

For those who have ability, and are willing to work, the field of carpentry is unlimited.

Carpentry apprentices compete in a statewide contest sponsored by a carpenters union. Judging is based on the methods and procedures followed as well as the speed and accuracy of the work. (Master Builders of Iowa)

APPENDICES

ABBREVIATIONS COMMONLY USED IN BUILDING CONSTRUCTION

Term	Abbreviation
Acoustical Tile	AT
Aggregate	Aggr.
Air Conditioning	Air Cond.
Air dried	AD
Alternate	Alt.
Aluminum	AL
American Institute of Architects	A.I.A.
American Institute of Electrical Engineers	A.I.E.E.
American Society for Testing Materials	A.S.T.M.
American Standards Association, Inc.	A.S.A.
Apartment	Apt.
Approximate	Approx.
Architectural	Arch.
Asbestos	Asb.
Basement	Bsmt.
Bathroom	B
Beam	Bm.
Better	Btr.
Beveled	Bev.
Blocking	Blkg.
Board Foot	Bd. Ft.
Brick	Brk.
British thermal unit	Btu
Building	Bldg.
Bundle	Bdl.
Cabinet	Cab.
Casing	Csg.
Cement	Cem.
Cement Floor	Cem. Fl.
Cement Mortar	Cem. Mort.
Center Line	CL
Center Matched	CM
Closet	Cl or Clo.
Column	Col.
Common	Com.
Concrete	Conc.
Concrete Block	Conc. B
Conduit	Cnd.
Construction	Const.
Counter	Ctr.
Cubic Foot	Cu. Ft.
Cubic Yard	Cu. Yd.
Diagram	Diag.
Diameter	Dia. or Diam.
Dimension	Dim.
Ditto	Do.
Double Strength Glass	DSG
Drawing	Dwg.
Dressed and Matched	D & M
Edge	Edg.
Edge Grain	EG
Entrance	Ent.
Excavate	Exc.
Exterior	Ext.
Federal Housing Authority	FHA
Finish	Fin.
Fixture	Fix.
Flashing	Fl.
Flat Grain	FG
Flooring	Flg.
Fluorescent	Fluor.
Foot or Feet	Ft.
Footing	Ftg.
Foundation	Fdn.
Furring	Fur.
Gallon	Gal.
Galvanized Iron	GI
Glass	Gl.
Hardwood	Hdwd.
Horsepower	HP
Hose Bib	HB
Hot Water	HW
Hundred	C
Insulation	Ins.
Interior	Int.
Kiln-dried	KD
Kitchen	K
Lavatory	Lav.
Length	Lgth.
Light	Lt.
Linnen Closet	L Cl.
Linoleum	Lino.
Living Room	LR
Masonry Opening	MO
Material	Matl.
Maximum	Max.
Medicine Cabinet	MC
Minimum	Min.
Miscellaneous	Misc.
Modular	Mod.
Molding	Mldg.
Nosing	Nos.
On Center	OC
Opening	Opng.
Plaster	Plas.
Plate	Pl.
Plate Glass	Pl. Gl.
Plumbing	Plbg.
Precast	Prcst.
Prefabricated	Prefab.
Quart	Qt.
Random	Rdm.
Refrigerator	Ref.
Reinforcing	Reinf.
Revision	Rev.
Rough	Rgh.
Rough Opening	Rgh. Opng.
Schedule	Sch.
Screen	Scr.
Select	Sel.
Service	Serv.
Sheathing	Shthg.
Shelving	Shelv.
Shiplap	S/lap
Siding	Sdg.
Specifications	Spec.
Square	Sq.
Square Feet	Sq. Ft.
Stairway	Stwy.
Standard	Std.
Steel	St. or Stl.
Structural	Str.
Surfaced One Side	S1S
Surfaced Two Sides	S2S
Surfaced Four Sides	S4S
Surfaced One Side and Two Edges	S1S2E
Switch	Sw. or S
Temperature	Temp.
Thermostat	Thermo.
Thousand	M
Tongue and Groove	T & G
Typical	Typ.
Unexcavated	Unexc.
Ventilation	Vent.
Water Closet	WC
Water Heater	WH
Weight	Wt.
Wood	Wd.

COMPUTING BD.FT. OF FINISHED LUMBER (3 Percent Allowance For Waste)

STANDARD FLOORING: Multiply area (in square feet) by conversion factors given—

Nominal Thickness And Width	Net Thickness And Width	Conversion Factor
1 × 3	25/32 × 2/3/8"	1.301
1 × 4	25/32 × 3-1/4"	1.268
1 × 6	25/32 × 5-3/16"	1.192
1¼ × 3	1-1/16 × 2-3/8"	1.626
1¼ × 4	1-1/16 × 3-1/4"	1.585
1¼ × 6	1-1/16 × 5-3/16"	1.489

BEVEL SIDING: Compute area to be covered (in square feet) and deduct for openings in the wall. Multiply by conversion factors:

½ × 4" bevel siding, figuring lap cover of ¾" 1.498
½ × 6" bevel siding, figuring lap cover of 1" 1.373
½ × 8" bevel siding, figuring lap cover of 1½" 1.433

BUNGALOW SIDING:

¾ × 8" bungalow siding, figuring lap cover of 1½" 1.433
¾ × 10" bungalow siding, figuring lap cover of 1½" 1.329

1 x 6 DROP SIDING:

Pattern 105 lays with a net face width of 5-1/16" 1.220
Pattern 106 lays with a net face width of 5-3/16" 1.192

INCH BOARDS & SHIPLAP:

1 × 6" boards, S4S standard 25/32 × 5-5/8" 1.099
1 × 8" boards, S4S standard 25/32 × 7-1/2" 1.099
1 × 6" shiplap, finished 25/32 × 5-1/8" face width 1.206
1 × 8" shiplap, finished 25/32 × 7-1/8" face width 1.156

*For another allowance than 3%, divide factors given by 1.03 to obtain the net conversion factor for surfacing, and add any desired percentage for trim estimate.

STANDARD SIZES OF LUMBER (PRIOR TO 1970 RULES)

Type of Lumber	Nominal Size		Actual Size S4S At Comm. Dry Shp. Wt.	
	Thickness	Width	Thickness	Width
Dimension	2 in.	4 in.	1-5/8 in.	3-5/8 in.
	2 in.	6 in.	1-5/8 in.	5-5/8 in.
	2 in.	8 in.	1-5/8 in.	7-1/2 in.
	2 in.	10 in.	1-5/8 in.	9-1/2 in.
	2 in.	12 in.	1-5/8 in.	11-1/2 in.
Timbers	4 in.	6 in.	3-5/8 in.	5-1/2 in.
	4 in.	8 in.	3-5/8 in.	7-1/2 in.
	4 in.	10 in.	3-5/8 in.	9-1/2 in.
	6 in.	6 in.	5-1/2 in.	5-1/2 in.
	6 in.	8 in.	5-1/2 in.	7-1/2 in.
	6 in.	10 in.	5-1/2 in.	9-1/2 in.
	8 in.	8 in.	7-1/2 in.	7-1/2 in.
	8 in.	10 in.	7-1/2 in.	9-1/2 in.
Common Boards	1 in.	4 in.	25/32 in.	3-5/8 in.
	1 in.	6 in.	25/32 in.	5-5/8 in.
	1 in.	8 in.	25/32 in.	7-1/2 in.
	1 in.	10 in.	25/32 in.	9-1/2 in.
	1 in.	12 in.	25/32 in.	11-1/2 in.
Shiplap Boards	1 in.	4 in.	25/32 in.	3-1/8 in. face
	1 in.	6 in.	25/32 in.	5-1/8 in. face
	1 in.	8 in.	25/32 in.	7-1/8 in. face
	1 in.	10 in.	25/32 in.	9-1/8 in. face
	1 in.	12 in.	25/32 in.	11-1/8 in. face
Tongued and Grooved Boards	1 in.	4 in.	25/32 in.	3-1/4 in. face
	1 in.	6 in.	25/32 in.	5-1/4 in. face
	1 in.	8 in.	25/32 in.	7-1/4 in. face
	1 in.	10 in.	25/32 in.	9-1/4 in. face
	1 in.	12 in.	25/32 in.	11-1/4 in. face

BOARD FEET CONTENT

Size in Inches	Length in Feet								
	8	10	12	14	16	18	20	22	24
1 x 2	1-1/3	1-2/3	2	2-1/3	2-2/3	3	3-1/3	3-2/3	4
1 x 3	2	2-1/2	3	3-1/2	4	4-1/2	5	5-1/2	6
1 x 4	2-2/3	3-1/3	4	4-2/3	5-1/3	6	6-2/3	7-1/3	8
1 x 5	3-1/3	4-1/6	5	5-5/6	6-2/3	7-1/2	8-1/3	9-1/6	10
1 x 6	4	5	6	7	8	9	10	11	12
1 x 8	5-1/3	6-2/3	8	9-1/3	10-2/3	12	13-1/3	14-2/3	16
1 x 10	6-2/3	8-1/3	10	11-2/3	13-1/3	15	16-2/3	18-1/3	20
1 x 12	8	10	12	14	16	18	20	22	24
1 x 14	9-1/3	11-2/3	14	16-1/3	18-2/3	21	23-1/3	25-2/3	28
1 x 16	10-2/3	13-1/3	16	18-2/3	21-1/3	24	26-2/3	29-1/3	32
5/4 x 4	3-1/3	4-1/6	5	5-5/6	6-2/3	7-1/2	8-1/3	9-1/6	10
5/4 x 6	5	6-1/4	7-1/2	8-3/4	10	11-1/4	12-1/2	13-3/4	15
5/4 x 8	6-2/3	8-1/3	10	11-2/3	13-1/3	15	16-2/3	18-1/3	20
5/4 x 10	8-1/3	10-5/12	12-1/2	14-7/12	16-2/3	18-3/4	20-5/6	22-11/13	25
5/4 x 12	10	12-1/2	15	17-1/2	20	22-1/2	25	27-1/2	30
6/4 x 4	4	5	6	7	8	9	10	11	12
6/4 x 6	6	7-1/2	9	10-1/2	12	13-1/2	15	16-1/2	18
6/4 x 8	8	10	12	14	16	18	20	22	24
6/4 x 10	10	12-1/2	15	17-1/2	20	22-1/2	25	27-1/2	30
6/4 x 12	12	15	18	21	24	27	30	33	36
2 x 4	5-1/3	6-2/3	8	9-1/3	10-2/3	12	13-1/3	14-2/3	16
2 x 6	8	10	12	14	16	18	20	22	24
2 x 8	10-2/3	13-1/3	16	18-2/3	21-1/3	24	26-2/3	29-1/3	32
2 x 10	13-1/3	16-2/3	20	23-1/3	26-2/3	30	33-1/3	36-2/3	40
2 x 12	16	20	24	28	32	36	40	44	48
2 x 14	18-2/3	23-1/3	28	32-2/3	37-1/3	42	46-2/3	51-1/3	56
2 x 16	21-1/3	26-2/3	32	37-1/3	42-2/3	48	53-1/3	58-2/3	64
3 x 4	8	10	12	14	16	18	20	22	24
3 x 6	12	15	18	21	24	27	30	33	36
3 x 8	16	20	24	28	32	36	40	44	48
3 x 10	20	25	30	35	40	45	50	55	60
3 x 12	24	30	36	42	48	54	60	66	72
3 x 14	28	35	42	49	56	63	70	77	84
3 x 16	32	40	48	56	64	72	80	88	96
4 x 4	10-2/3	13-1/3	16	18-2/3	21-1/3	24	26-2/3	29-1/3	32
4 x 6	16	20	24	28	32	36	40	44	48
4 x 8	21-1/3	26-2/3	32	37-1/3	42-2/3	48	53-1/3	58-2/3	64
4 x 10	26-2/3	33-1/3	40	46-2/3	53-1/3	60	66-2/3	73-1/3	80
4 x 12	32	40	48	56	64	72	80	88	96
4 x 14	37-1/3	46-2/3	56	65-1/3	74-2/3	84	93-1/3	102-2/3	112
4 x 16	42-2/3	53-1/3	64	74-2/3	85-1/3	96	106-2/3	117-1/3	128
6 x 6	24	30	36	42	48	54	60	66	72
6 x 8	32	40	48	56	64	72	80	88	96
6 x 10	40	50	60	70	80	90	100	110	120
6 x 12	48	60	72	84	96	108	120	132	144
6 x 14	56	70	84	98	112	126	140	154	168
6 x 16	64	80	96	112	128	144	160	176	192
8 x 8	42-2/3	53-1/3	64	74-2/3	85-1/3	96	106-2/3	117-1/3	128
8 x 10	53-1/3	66-2/3	80	93-1/3	106-2/3	120	133-1/3	146-2/3	160
8 x 12	64	80	96	112	128	144	160	176	192

Lumber Calculation Tables. (Building Supply News)

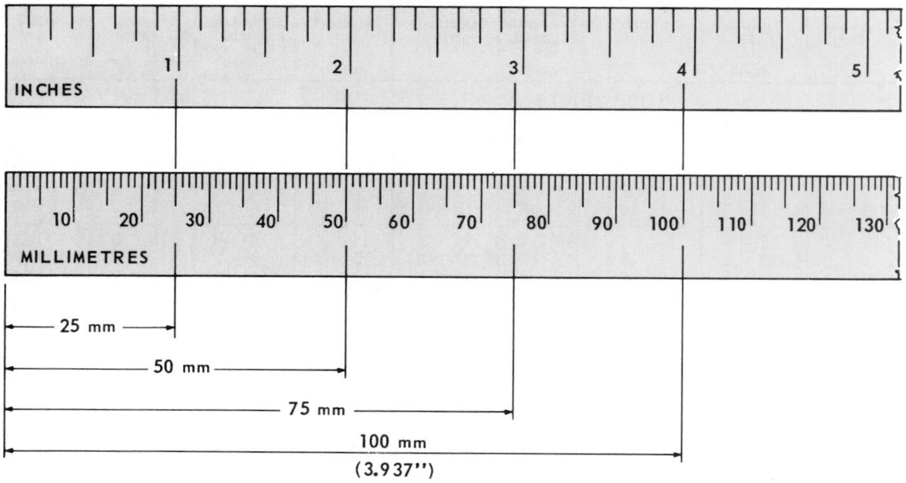

1/8" = 3.175 mm
1/4" = 6.350 mm
3/8" = 9.525 mm
1/2" = 12.700 mm
5/8" = 15.875 mm
3/4" = 19.050 mm
7/8" = 22.225 mm
1" = 25.400 mm

METRIC CONVERSION
(EXACT)

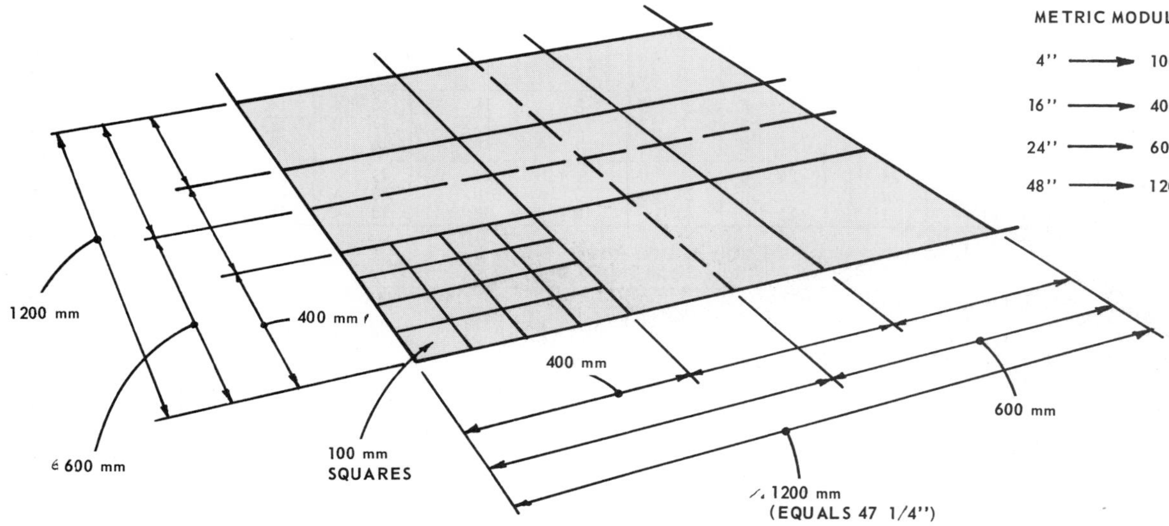

METRIC MODULES

4" → 100 mm
16" → 400 mm
24" → 600 mm
48" → 1200 mm

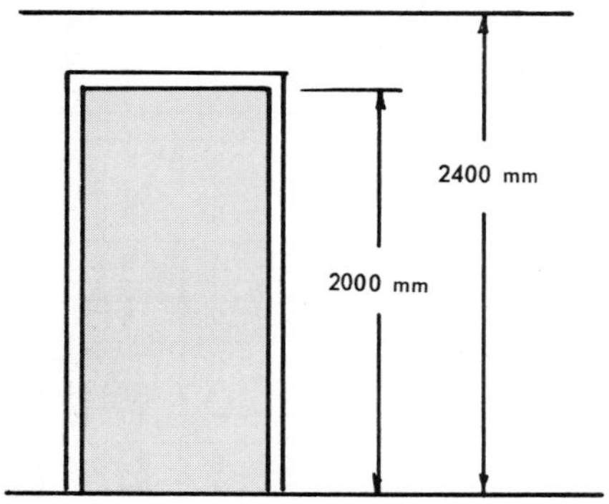

PASSAGE DOORS

WIDTHS

2'-2" → 650 mm
2'-6" → 750 mm
2'-8" → 800 mm
3'-0" → 900 mm

BUILDING PANELS

16" x 32" → 400 mm x 800 mm
48" x 96" → 1200 mm x 2400 mm
48" x 120" → 1200 mm x 3000 mm

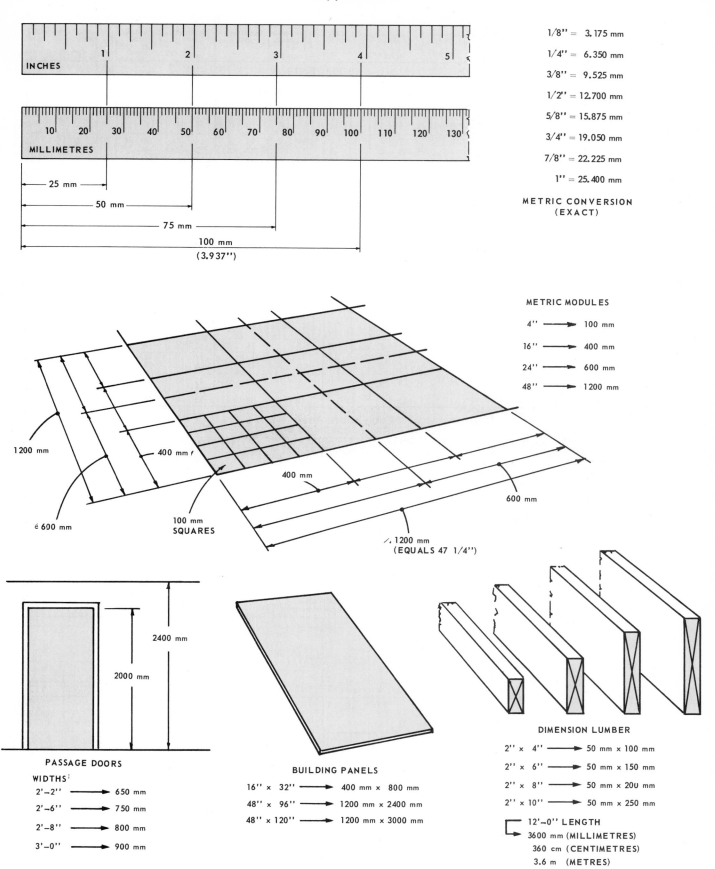

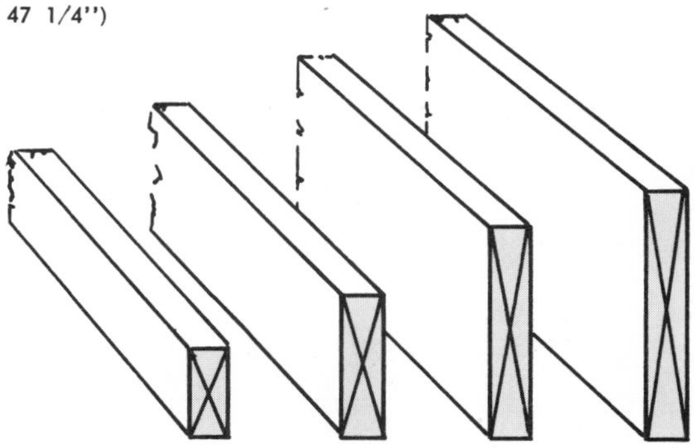

DIMENSION LUMBER

2" x 4" → 50 mm x 100 mm
2" x 6" → 50 mm x 150 mm
2" x 8" → 50 mm x 200 mm
2" x 10" → 50 mm x 250 mm

12'-0" LENGTH
→ 3600 mm (MILLIMETRES)
360 cm (CENTIMETRES)
3.6 m (METRES)

Metric modules for carpentry. Basic unit of four inches is converted to 100 millimetres (3.937 inches) as proposed by the National Forest Products Association. Note that arrows are used to indicate conversions made on this basis.

Nominal Size (inches)	Spacing (in.) o.c.	Construction Dense Constr. Sel. Struct. J&P MC-15	Construction J&P	Standard J&P MC-15	Standard J&P	Utility J&P MC-15	Utility J&P	Construction Dense Constr. Sel. Struct. J&P MC-15	Construction J&P	Standard J&P MC-15	Standard J&P	Utility J&P MC-15	Utility J&P
		FLOOR JOISTS											
		30-LB. LIVE LOAD						**40-LB. LIVE LOAD**					
		Ft. In.	Ft. In.	Ft. In.	Ft. In	Ft. In	Ft. In.	Ft. In.	Ft. In.	Ft. In.	Ft. In.	Ft. In.	Ft. In.
2x6	12	11 4	11 4	11 4	11 4	9 8	8 4	10 6	10 6	10 6	10 6	8 8	7 4
	16	10 4	10 4	10 4	10 4	8 6	7 2	9 8	9 8	9 8	9 8	7 6	6 4
	24	9 0	9 0	9 0	9 0	7 0	5 10	8 4	8 4	8 4	8 2	6 2	5 2
2x8	12	15 4	15 4	15 4	15 4	15 0	13 4	14 4	14 4	14 4	14 4	13 8	12 0
	16	14 0	14 0	14 0	14 0	13 0	11 6	13 0	13 0	13 0	13 0	11 8	10 4
	24	12 4	12 4	12 4	12 4	10 8	9 6	11 6	11 6	11 6	11 0	9 8	8 4
2x10	12	18 4	18 4	18 4	18 4	18 4	17 10	17 4	17 4	17 4	17 4	17 4	16 2
	16	17 0	17 0	17 0	17 0	17 0	15 6	16 2	16 2	16 2	16 2	16 0	14 0
	24	15 6	15 6	15 6	15 6	14 8	12 10	14 6	14 6	14 6	14 0	13 0	11 4
2x12	12	21 2	21 2	21 2	21 2	21 2	21 2	20 0	20 0	20 0	20 0	20 0	19 6
	16	19 8	19 8	19 8	19 8	19 8	18 8	18 8	18 8	18 8	18 8	18 8	16 8
	24	17 10	17 10	17 10	17 10	17 2	15 6	16 10	16 10	16 10	16 10	15 6	13 8
		CEILING JOISTS											
		NO ATTIC STORAGE						**LIMITED ATTIC SPACE**					
2x4	12	11 10	11 8	9 10	8 10	6 10	6 2	9 4	8 2	6 10	6 4	4 10	4 4
	16	10 10	10 0	8 6	7 8	6 0	5 4	8 0	7 2	6 0	5 6	4 2	3 10
	24	9 4	8 2	6 10	6 4	4 10	4 4	6 8	5 10	4 10	4 6	3 6	3 2
2x6	12	17 2	17 2	17 2	17 2	16 0	14 8	14 4	14 4	14 4	14 4	11 4	9 6
	16	16 0	16 0	16 0	16 0	13 10	11 8	13 0	13 0	13 0	12 10	9 8	8 4
	24	14 4	14 4	14 4	14 4	11 4	9 6	11 4	11 4	11 4	10 6	8 0	6 8
2x8	12	21 8	21 8	21 8	21 8	21 8	21 8	18 4	18 4	18 4	18 4	17 6	15 4
	16	20 2	20 2	20 2	20 2	20 2	18 10	17 0	17 0	17 0	17 0	15 0	13 4
	24	18 4	18 4	18 4	18 4	17 6	15 4	15 4	15 4	15 4	14 4	12 4	10 10
2x10	12	24 0	24 0	24 0	24 0	24 0	24 0	21 10	21 10	21 10	21 10	21 10	20 8
	16	24 0	24 0	24 0	24 0	24 0	24 0	20 4	20 4	20 4	20 4	20 4	18 0
	24	21 10	21 10	21 10	21 10	21 10	20 8	18 4	18 4	18 4	18 0	16 6	14 8
		LOW SLOPE ROOF JOISTS (Roof Slope 3 in 12 or Less)											
		NOT SUPPORTING FINISHED CEILING						**SUPPORTING FINISHED CEILING**					
2x6	12	14 4	14 4	14 4	14 4	11 4	9 6	13 8	13 8	13 8	13 8	10 6	8 10
	16	13 0	13 0	13 0	12 10	9 8	8 4	12 4	12 4	12 4	11 10	9 0	7 8
	24	11 4	11 4	11 4	10 6	8 0	6 8	10 10	10 8	10 8	9 8	7 4	6 2
2x8	12	18 4	18 4	18 4	18 4	17 6	15 4	17 8	17 8	17 8	17 8	16 2	14 2
	16	17 0	17 0	17 0	17 0	15 0	13 4	16 4	16 4	16 4	16 2	14 0	12 4
	24	15 4	15 4	15 4	14 4	12 4	10 10	14 6	14 6	14 6	13 2	11 4	10 0
2x10	12	21 10	21 10	21 10	21 10	21 10	20 8	21 0	21 0	21 0	21 0	21 0	19 0
	16	20 4	20 4	20 4	20 4	20 4	18 0	19 6	19 6	19 6	19 6	19 2	16 8
	24	18 4	18 4	18 4	18 0	16 6	14 8	17 8	17 8	17 8	16 8	15 8	13 6
2x12	12	24 0	24 0	24 0	24 0	24 0	24 0	24 0	24 0	24 0	24 0	24 0	23 0
	16	23 6	23 6	23 6	23 6	23 6	21 10	22 6	22 6	22 6	22 6	22 6	20 0
	24	21 2	21 2	21 2	21 2	20 0	19 6	20 4	20 4	20 2	20 2	18 6	16 4
		RAFTERS (Roof Slope Over 3 in 12)											
		LIGHT ROOFING						**HEAVY ROOFING**					
2x4	12	10 10	9 6	8 0	7 4	5 8	5 2	9 4	8 2	6 10	6 4	4 10	4 4
	16	10 0	8 4	7 0	6 4	5 0	4 6	8 0	7 2	6 0	5 6	4 2	3 10
	24	7 4	6 10	5 2	5 2	4 0	3 8	6 8	5 10	5 0	4 6	3 6	3 2
2x6	12	16 10	16 10	16 10	16 10	13 4	11 2	15 6	15 6	15 6	14 10	11 4	9 6
	16	15 8	15 8	15 8	15 0	11 6	9 8	14 4	14 0	14 2	12 10	9 8	8 4
	24	13 10	13 6	13 6	12 2	9 4	7 8	12 6	11 6	11 8	10 6	8 0	6 8
2x8	12	21 2	21 2	21 2	21 2	20 10	18 0	19 8	19 8	19 8	19 8	17 6	15 4
	16	19 10	19 10	19 6	19 0	17 8	15 8	18 4	18 4	18 4	17 6	15 0	13 4
	24	17 10	17 10	17 10	16 8	14 4	12 8	16 6	15 8	15 10	14 4	12 4	10 10
2x10	12	24 0	24 0	24 0	24 0	24 0	24 0	23 6	23 6	23 6	23 6	23 0	20 8
	16	23 8	23 8	23 8	23 8	23 8	21 0	21 10	21 10	21 10	21 10	20 4	18 0
	24	21 4	21 4	21 4	21 0	19 8	17 2	19 8	19 8	19 8	18 0	16 2	14 8

"MC 15" indicates the stock is surfaced at 15 per cent or less moisture content.

Maximum Spans for Visually Graded Douglas Fir. J and P Stand for Joist and Plank.

LENGTHS OF COMMON RAFTERS

FEET OF RUN	2 in 12	2½ in 12	3 in 12	3½ in 12	4 in 12	4½ in 12
	Inclination (set saw at) 9°-28'	Inclination (set saw at) 11°-46'	Inclination (set saw at) 14°-2'	Inclination (set saw at) 16°-16'	Inclination (set saw at) 18°-26'	Inclination (set saw at) 20°-33'
	12.17 in. per ft. of run	12.26 in. per ft. of run	12.37 in. per ft. of run	12.5 in. per ft. of run	12.65 in. per ft. of run	12.82 in. per ft. of run
4'	4' 0-11/16"	4' 1-1/32"	4' 1-15/32"	4' 2"	4'-2-19/32"	4' 3-9/32"
5	5' 0-27/32"	5' 1-5/16"	5' 1-27/32"	5' 2-1/2"	5' 3-1/4"	5' 4-3/32"
6	6' 1-1/32"	6' 1-9/16"	6' 2-7/32"	6' 3"	6' 3-29/32"	6' 4-15/16"
7	7' 1-3/16"	7' 1-13/16"	7' 2-19/32"	7' 3-1/2"	7' 4-9/16"	7' 5-3/4"
8	8' 1-3/8"	8' 2-3/32"	8' 2-31/32"	8' 4"	8' 5-3/32"	8' 6-9/16"
9	9' 1-17/32"	9' 2-7/16"	9' 3-11/32"	9' 4-1/2"	9' 5-27/32"	9' 7-3/8"
10	10' 1-23/32"	10' 2-19/32"	10' 3-23/32"	10' 5"	10' 6-1/2"	10' 8-7/32"
11	11' 1-7/8"	11' 2-7/8"	11' 4-1/16"	11' 5-1/2"	11' 7-5/32"	11' 9-1/32"
12	12' 2-1/32"	12' 3-1/8"	12' 4-7/16"	12' 6"	12' 7-13/16"	12' 9-27/32"
13	13' 2-7/32"	13' 3-3/8"	13' 4-13/16"	13' 6-1/2"	13' 8-15/32"	13' 10-21/32"
14	14' 2-3/8"	14' 3-21/32"	14' 5-3/16"	14' 7"	14' 9-3/32"	14' 11-1/2"
15	15' 2-9/16"	15' 3-29/32"	15' 5-9/16"	15' 7-1/2	15' 9-3/4"	16' 0-5/16"
16	16' 2-23/32"	16' 4-5/32"	16' 5-15/16"	16' 8"	16' 10-13/32"	17' 1-1/8"
INCHES OF RUN	These lengths are to be added to those shown above when run involves inches					
1/4"	1/4"	1/4"	1/4"	1/4"	1/4"	9/32"
1/2"	1/2"	1/2"	1/2"	17/32"	17/32"	17/32"
1"	1"	1"	1-1/32"	1-1/32"	1-1/16"	1-1/16"
2"	2-1/32"	2-1/32"	2-1/16"	2-3/32"	2-3/32"	2-1/8"
3"	3-1/16"	3-1/16"	3-3/32"	3-1/8"	3-5/32"	3-7/32"
4"	4-1/16"	4-3/32"	4-1/8"	4-5/32"	4-7/32"	4-9/32"
5"	5-1/16"	5-1/8"	5-5/32"	5-7/32"	5-9/32"	5-11/32"
6"	6-3/32"	6-1/8"	6-3/16"	6-1/4"	6-5/16"	6-13/32"
7"	7-3/32"	7-5/32"	7-7/32"	7-9/32"	7-3/8"	7-15/32"
8"	8-1/8"	8-3/16"	8-1/4"	8-11/32"	8-7/16"	8-17/32"
9"	9-1/8"	9-3/16"	9-1/4"	9-3/8"	9-15/32"	9-5/8"
10"	10-5/32"	10-7/32"	10-5/16"	10-7/16"	10-17/32"	10-11/16"
11"	11-5/32"	11-1/4"	11-11/32"	11-15/32"	11-19/32"	11-3/4"

FEET OF RUN	5 in 12	5½ in 12	6 in 12	6½ in 12	7 in 12	7½ in 12	8 in 12
	Inclination (set saw at) 22°-37'	Inclination (set saw at) 24°-37'	Inclination (set saw at) 26°-34'	Inclination (set saw at) 28°-27'	Inclination (set saw at) 30°-15'	Inclination (set saw at) 32°-0'	Inclination (set saw at) 33°-41'
	13.00 in. per ft. of run	13.20 in. per ft. of run	13.42 in. per ft. of run	13.65 in. per ft. of run	13.89 in. per ft. of run	14.15 in. per ft. of run	14.42 in. per ft. of run
4	4' 4"	4' 4-13/16"	4' 5-11/16"	4' 6-19/32"	4' 7-9/16"	4' 8-19/32"	4' 9-11/16"
5	5' 5"	5'-6"	5' 7-1/8"	5' 8-1/4"	5' 9-15/32"	5' 10-3/4"	6' 0-1/8"
6	6' 6"	6' 7-3/32"	6' 8-1/2"	6' 9-29/32"	6' 11-11/32"	7' 0-29/32"	7' 2-17/32"
7	7' 7"	7' 8-13/32"	7' 9-15/16"	7' 11-9/16"	8' 1-7/32"	8' 3-1/16"	8' 4-15/16"
8	8' 8"	8' 9-19/32"	8' 11-3/8"	9' 1-7/32"	9' 3-1/8"	9' 5-7/32"	9' 7-3/8"
9	9' 9"	9' 10-13/16"	10' 0-25/32"	10' 2-27/32"	10' 5"	10' 7-11/32"	10' 9-25/32"
10	10' 10"	11' 0"	11' 2-7/32"	11' 4-1/2"	11' 6-29/32"	11' 9-1/2"	12' 0-7/32"
11	11' 11"	12' 1-3/32"	12' 3-5/8"	12' 6-5/32"	12' 8-13/16"	12' 11-21/32"	13' 2-5/8"
12	13' 0"	13' 2-13/32"	13' 5-1/32"	13' 7-13/16"	13' 10-11/16"	14' 1-13/16"	14' 5-1/32"
13	14' 1"	14' 3-19/32"	14' 6-15/32"	14' 9-15/32"	15' 0-9/16"	15' 3-31/32"	15' 7-15/32"
14	15' 2"	15' 4-13/16"	15' 7-7/8"	15' 11-1/8"	16' 2-15/32"	16' 6-1/8"	16' 9-7/8"
15	16' 3"	16' 6"	16' 9-5/16"	17' 0-3/4"	17' 4-11/32"	17' 8-1/4"	18' 0-5/16"
16	17' 4"	17' 7-7/32"	17' 10-23/32"	18' 2-13/32"	18' 6-1/4"	18' 10-13/32"	19' 2-23/32"
INCHES OF RUN	These lengths are to be added to those shown above when run involves inches						
1/4"	9/32"	9/32"	9/32"	9/32"	9/32"	5/16"	5/16"
1/2"	17/32"	9/16"	9/16"	9/16"	9/16"	1-9/32"	5/8"
1"	1-3/32"	1-3/32"	1-1/8"	1-1/8"	1-5/32"	1-3/16"	1-7/32"
2"	2-5/32"	2-7/32"	2-1/4"	2-9/32"	2-5/16"	2-11/32"	2-13/32"
3"	3-1/4"	3-5/16"	3-3/8"	3-13/32"	3-15/32"	3-17/32"	3-19/32"
4"	4-11/32"	4-13/32"	4-15/32"	4-9/16"	4-5/8"	4-23/32"	4-13/16"
5"	5-13/32"	5-1/2"	5-19/32"	5-11/16"	5-25/32"	5-29/32"	6"
6"	6-1/2"	6-19/32"	6-23/32"	6-13/16"	6-15/16"	7-1/32"	7-7/32"
7"	7-9/16"	7-23/32"	7-13/16"	7-31/32"	8-1/8"	8-1/4"	8-13/32"
8"	8-21/32"	8-13/16"	8-15/16"	9-1/8"	9-1/4"	9-7/16"	9-5/8"
9"	9-3/4"	9-29/32"	10-5/32"	10-1/4"	10-13/32"	10-19/32"	10-13/16"
10"	10-27/32"	11"	11-3/16"	11-3/8"	11-9/16"	11-25/32"	1'0"
11"	11-29/32"	12-3/32"	1' 0-5/16"	1' 0-1/2"	1' 0-23/32"	1' 0-31/32"	1' 1-7/32"

Common Rafter Table. Subtract One-Half Ridge Board Thickness.
(Building Supply News)

PRODUCT	THICKNESS*	SIZES*	EDGES	MAJOR USES
Building Board	½"	4' x 8' 4' x 9' 4' x 10' 4' x 12'	Square	General purpose insulation board.
Insulating Roof Deck	Nominal / C Value 1½" 0.24 2" 0.18 3" 0.12	2' x 8'	Fabricated long edges, short edges interlocking or square.	Flat, pitched or shed type roofs.
Roof Insulation	Nominal / C Value ½" 0.72 1" 0.36 1½" 0.24 2" 0.19 2½" 0.15 3" 0.12	23" x 47" 24" x 48"	Varies	Built-Up roofs and under certain types of roofing on pitched roofs.
Wall Board	5/16" or ⅜"	4' x 8' 4' x 10'	Square	General purpose utility board.
Ceiling Tile: Plain or Perforated	½"	12" x 12" 12" x 24" 16" x 16" 16" x 32"	Fabricated or butt edges.	Decorative wall and ceiling finish.
Plank	½"	12" x 8' 12" x 10'	Fabricated long edges	Decorative wall and ceiling finish.
Sheathing, Regular Density	½", 25/32"	4' x 8' 4' x 9' 4' x 10' 4' x 12'	Square	Wall Sheathing for all types of wood framed construction.
	½", 25/32"	2' x 8'	Long edges fabricated short edges square.	
Sheathing, Intermediate	½"	4' x 8' 4' x 9'	Square	High Density product designed for use without supplementary corner bracing.
Sheathing, Nail-Base	½"	4' x 8' 4' x 9'	Square	High Density product designed for use in frame construction to permit the direct attachment of wood and asbestos-cement shingles
Shingle Backer	5/16" or ⅜"	11¾" x 48" 13½" x 48" 15" x 48"	Square	Undercoursing for wood or asbestos-cement shingles applied over insulation board sheathing.
Insulating Formboard	1", 1½"	24", 32" & 48" widths; 4' to 12' in length	Square	Used as a permanent form for reinforced gypsum or light-weight aggregate concrete poured-in-place roof construction.
Sound Deadening Board	½"	4' x 8' 4' x 9'	Square	In wall and floor assemblies to control Sound Transmission between units.

*For additional thicknesses and sizes, consult manufacturer.

Standard Insulation Board Products.

Cold Climate / Outside Design Temperature –30° F.

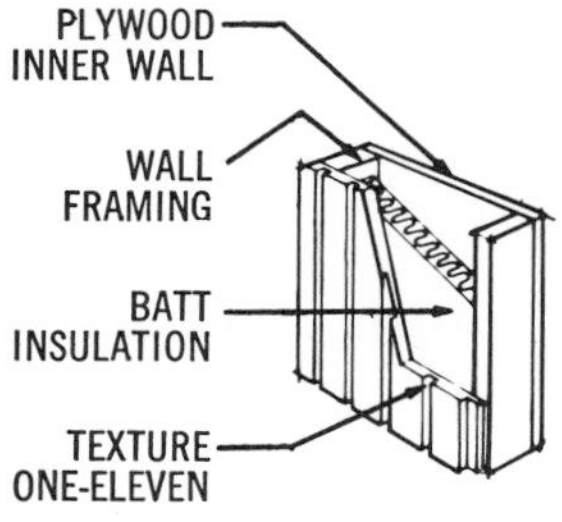

	Resistance
Outside air	0.17
Texture One-Eleven	0.58
Air space	0.97
2″ foil-faced glass-fiber batt	9.23
Air space	0.97
1/4″ plywood	0.25
Inside air	0.68
	R = 12.85

U = 0.078 Btu/hr./sq. ft./°F. temp. diff.

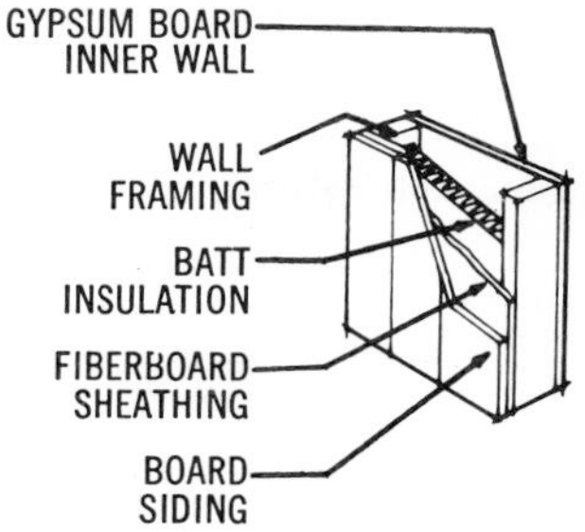

	Resistance
Outside air	0.17
1″-nominal board siding	0.74
25/32″ fiberboard sheathing	2.06
Air space	0.97
1-1/2″ glass-fiber batt	5.55
Air space	0.97
1/2″ gypsum board	0.45
Inside air	0.68
	R = 11.59
	U = 0.086

Warm Climate / Outside Design Temperature 15° F.

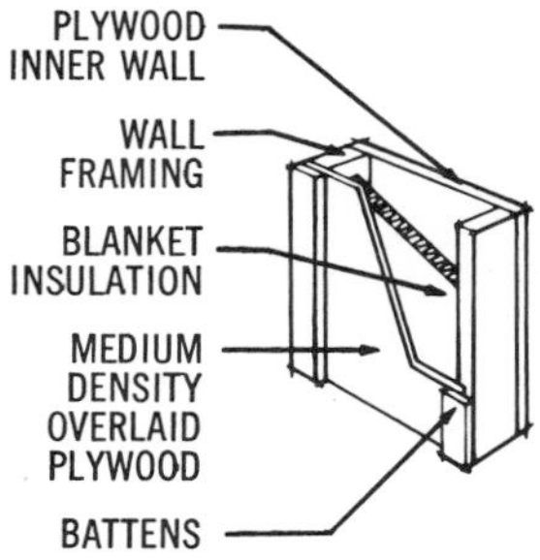

	Resistance
Outside air	0.17
3/8″ medium density plywood and battens	0.37
Air space	0.97
1″ wood fiber-blanket	4.00
Air space	0.97
1/4″ plywood	0.25
Inside air	0.68
	R = 7.41
	U = 0.14

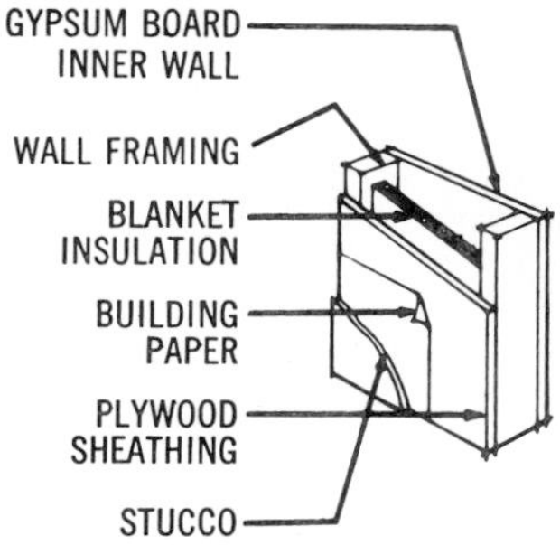

	Resistance
Outside air	0.17
Stucco (1″)	0.20
3/8″ plywood sheathing and paper	0.43
Air space	0.97
1″ wood-fiber blanket	4.00
Air space	0.97
1/2″ gypsum board	0.45
Inside air	0.68
	R = 7.87
	U = 0.13

Moderate Climate / Outside Design Temperature 0° F.

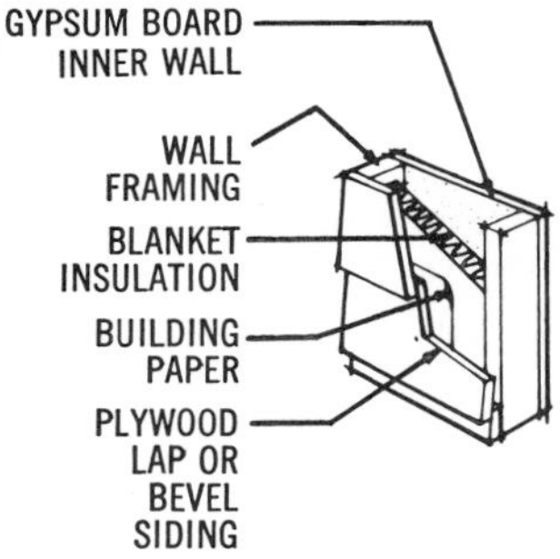

	Resistance
Outside air	**0.17**
3/8″ plywood lap or bevel siding and paper	**0.55**
Air space	**0.97**
1-1/2″ mineral wool blanket	**5.55**
Air space	**0.97**
1/2″ gypsum board	**0.45**
Inside air	**0.68**
	R = 9.34
	U = 0.11

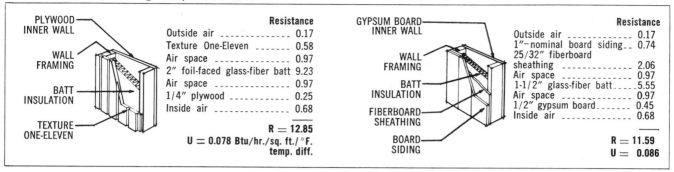

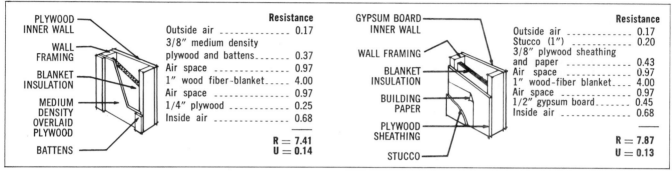

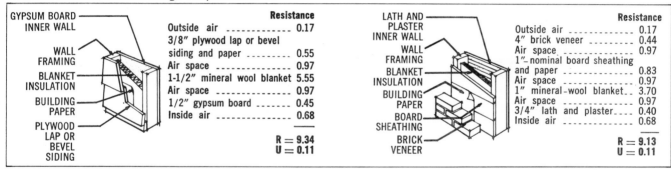

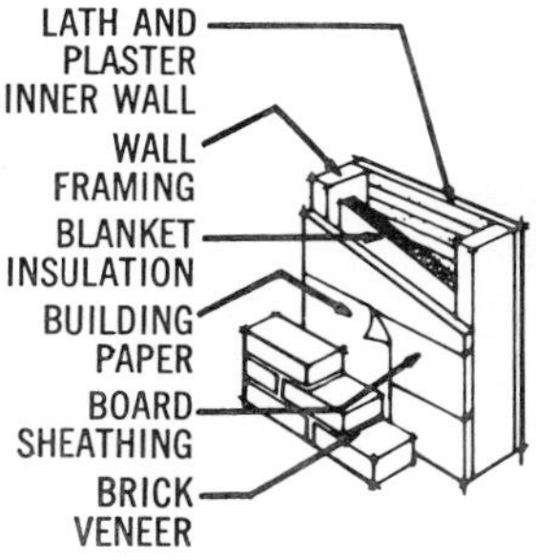

	Resistance
Outside air	0.17
4″ brick veneer	0.44
Air space	0.97
1″-nominal board sheathing and paper	0.83
Air space	0.97
1″ mineral-wool blanket	3.70
Air space	0.97
3/4″ lath and plaster	0.40
Inside air	0.68
	R = 9.13
	U = 0.11

Notes:

(1) Values in these examples are from ASHRAE Guide except that plywood and lumber values are based on information in Wood Handbook and in U. S. Forest Service Research Paper FPL-27.

(2) Average range of thermal resistance (R) and conductance (C) at 12% moisture content. Figures used for plywood and lumber in examples represent low portion of range, based on coefficient of thermal conductivity "k" for various species of wood. They therefore provide conservative values for the wall constructions.

(3) Thermal Conductance in Btu/hr./sq. ft./°F. temperature difference. Similar to "U" value, but numerically higher, because "U" value includes additional insulation value due to surface air effects.

SINGLE PLYWOOD PANELS

THICKNESS	R VALUE[2]	C VALUE[3]
1/4″	0.25 - 0.33	4.08 - 3.04
5/16″	0.31 - 0.41	3.26 - 2.44
3/8″	0.37 - 0.49	2.72 - 2.03
1/2″	0.49 - 0.66	2.04 - 1.52
5/8″	0.61 - 0.82	1.63 - 1.22
3/4″	0.74 - 0.99	1.36 - 1.01
1″	0.98 - 1.32	1.02 - 0.76
1-1/8″	1.10 - 1.49	0.91 - 0.68

Insulation Values for Various Wall Constructions. (American Plywood Assoc.)

SOUND ABSORPTION COEFFICIENTS OF GENERAL BUILDING MATERIALS AND FURNISHINGS

Complete tables of coefficients of the various materials that normally constitute the interior finish of rooms may be found in the various books on architectural acoustics. The following short list of materials give approximate values which will be useful in making simple calculations of the reverberation in rooms.

Materials	Coefficients					
	125 Hz	250 Hz	500 Hz	1000 Hz	2000 Hz	4000 Hz
Brick, unglazed	.03	.03	.03	.04	.05	.07
Brick, unglazed, painted	.01	.01	.02	.02	.02	.03
Carpet						
1/8" Pile Height	.05	.05	.10	.20	.30	.40
1/4" Pile Height	.05	.10	.15	.30	.50	.55
3/16" combined Pile & Foam	.05	.10	.10	.30	.40	.50
5/16" combined Pile & Foam	.05	.15	.30	.40	.50	.60
Concrete Block, painted	.10	.05	.06	.07	.09	.08
Fabrics						
Light velour, 10 oz. per sq. yd., hung straight, in contact with wall	.03	.04	.11	.17	.24	.35
Medium velour, 14 oz. per sq. yd., draped to half area	.07	.31	.49	.75	.70	.60
Heavy velour, 18 oz. per sq. yd., draped to half area	.14	.35	.55	.72	.70	.65
Floors						
Concrete or Terrazzo	.01	.01	.01	.02	.02	.02
Linoleum, asphalt, rubber or cork tile on concrete	.02	.03	.03	.03	.03	.02
Wood	.15	.11	.10	.07	.06	.07
Wood parquet in asphalt on concrete	.04	.04	.07	.06	.06	.07
Glass						
1/4", sealed, large panes	.05	.03	.02	.02	.03	.02
24 oz., operable windows (in closed condition)	.10	.05	.04	.03	.03	.03
Gypsum Board, 1/2" nailed to 2x4's 16" o.c., painted	.10	.08	.05	.03	.03	.03
Marble or Glazed Tile	.01	.01	.01	.01	.02	.02
Plaster, gypsum or lime,						
rough finish or lath	.02	.03	.04	.05	.04	.03
Same, with smooth finish	.02	.02	.03	.04	.04	.03
Hardwood Plywood paneling 1/4" thick, Wood Frame	.58	.22	.07	.04	.03	.07
Water Surface, as in a swimming pool	.01	.01	.01	.01	.02	.03
Wood Roof Decking, tongue-and-groove cedar	.24	.19	.14	.08	.13	.10
Air, Sabins per 1000 cubic feet @ 50% RH				.9	2.3	7.2

Sound Absorption Coefficients for General Materials Used in Light Construction. (Acoustical and Board Products Assoc.)

Interior Type

Use these terms when you specify plywood (2)	Description and Most Common Uses	Typical Grade-trademarks	Veneer Grade Face	Veneer Grade Back	Veneer Grade Inner Plies	Most Common Thicknesses (inch) (3)					
N-N, N-A, N-B INT-APA	Cabinet quality. For natural finish furniture, cabinet doors, built-ins, etc. Special order items.	N-N · G-1 · INT-APA · PS 1-74	N	N.A. or B	C						3/4
N-D INT-APA	For natural finish paneling. Special order item.	N-D · G-3 · INT-APA · PS 1-74	N	D	D	1/4					
A-A INT-APA	For applications with both sides on view. Built-ins, cabinets, furniture and partitions. Smooth face; suitable for painting.	A-A · G-4 · INT-APA · PS 1-74	A	A	D	1/4		3/8	1/2	5/8	3/4
A-B INT-APA	Use where appearance of one side is less important but two smooth solid surfaces are necessary.	A-B · G-4 · INT-APA · PS 1-74	A	B	D	1/4		3/8	1/2	5/8	3/4
A-D INT-APA	Use where appearance of only one side is important. Paneling, built-ins, shelving, partitions, and flow racks.	A-D GROUP 1 INTERIOR PS 1-74 000 APA	A	D	D	1/4		3/8	1/2	5/8	3/4
B-B INT-APA	Utility panel with two smooth sides. Permits circular plugs.	B-B · G-3 · INT-APA · PS 1-74	B	B	D	1/4		3/8	1/2	5/8	3/4
B-D INT-APA	Utility panel with one smooth side. Good for backing, sides of built-ins. Industry: shelving, slip sheets, separator boards and bins.	B-D GROUP 3 INTERIOR PS 1-74 000 APA	B	D	D	1/4		3/8	1/2	5/8	3/4
DECORATIVE PANELS—APA	Rough sawn, brushed, grooved, or striated faces. For paneling, interior accent walls, built-ins, counter facing, displays, and exhibits.	DECORATIVE GROUP 2 EXTERIOR PS 1-74 000 GROUP 1 FACE APA	C or btr.	D	D		5/16	3/8	1/2	5/8	
PLYRON INT-APA	Hardboard face on both sides. For counter tops, shelving, cabinet doors, flooring. Faces tempered, untempered, smooth, or screened.	PLYRON · INT-APA · PS 1-74			C & D				1/2	5/8	3/4

Exterior Type[7]

Use these terms when you specify plywood (2)	Description and Most Common Uses	Typical Grade-trademarks	Veneer Grade Face	Veneer Grade Back	Veneer Grade Inner Plies	Most Common Thicknesses (inch) (3)					
A-A EXT-APA (4)	Use where appearance of both sides is important. Fences, built-ins, signs, boats, cabinets, commercial refrigerators, shipping containers, tote boxes, tanks, and ducts.	A-A · G-3 · EXT-APA · PS 1-74	A	A	C	1/4		3/8	1/2	5/8	3/4
A-B EXT-APA (4)	Use where the appearance of one side is less important.	A-B · G-1 · EXT-APA · PS 1-74	A	B	C	1/4		3/8	1/2	5/8	3/4
A-C EXT-APA (4)	Use where the appearance of only one side is important. Sidings, soffits, fences, structural uses, boxcar and truck lining, farm buildings. Tanks, trays, commercial refrigerators.	A-C GROUP 4 EXTERIOR PS 1-74 000 APA	A	C	C	1/4		3/8	1/2	5/8	3/4
B-B EXT-APA (4)	Utility panel with solid faces.	B-B · G-1 · EXT-APA · PS 1-74	B	B	C	1/4		3/8	1/2	5/8	3/4
B-C EXT-APA (4)	Utility panel for farm service and work buildings, boxcar and truck lining, containers, tanks, agricultural equipment. Also as base for exterior coatings for walls, roofs.	B-C GROUP 2 EXTERIOR PS 1-74 000 APA	B	C	C	1/4		3/8	1/2	5/8	3/4
HDO EXT-APA (4)	High Density Overlay plywood. Has a hard, semi-opaque resin-fiber overlay both faces. Abrasion resistant. For concrete forms, cabinets, counter tops, signs and tanks.	HDO · A-A · G-1 · EXT-APA · PS 1-74	A or B	A or B	C or C plgd		5/16	3/8	1/2	5/8	3/4
MDO EXT-APA (4)	Medium Density Overlay with smooth, opaque, resin-fiber overlay one or both panel faces. Highly recommended for siding and other outdoor applications, built-ins, signs, and displays. Ideal base for paint.	MDO · B-B · G-4 · EXT-APA · PS 1-74	B	B or C	C		5/16	3/8	1/2	5/8	3/4
303 SIDING EXT-APA (6)	Proprietary plywood products for exterior siding, fencing, etc. Special surface treatment such as V groove, channel groove, striated, brushed, rough sawn.	303 SIDING 16 oc GROUP 1 EXTERIOR PS 1-74 000 APA	(5)	C	C			3/8	1/2	5/8	
T 1-11 EXT-APA (6)	Special 303 panel having grooves 1/4" deep, 3/8" wide, spaced 4" or 8" o.c. Other spacing optional. Edges shiplapped. Available unsanded, textured, and MDO.	303 SIDING 16 oc T 1-11 GROUP 1 EXTERIOR PS 1-74 000 APA	C or btr.	C	C					5/8	
PLYRON EXT-APA	Hardboard faces both sides, tempered, smooth or screened.	PLYRON · EXT-APA · PS 1-74			C				1/2	5/8	3/4
MARINE EXT-APA	Ideal for boat hulls. Made only with Douglas fir or western larch. Special solid jointed core construction. Subject to special limitations on core gaps and number of face repairs. Also available with HDO or MDO faces.	MARINE · A-A · EXT-APA · PS 1-74	A or B	A or B	B	1/4		3/8	1/2	5/8	3/4

(1) Sanded both sides except where decorative or other surfaces specified.
(2) Available in Group 1, 2, 3, 4 or 5 unless otherwise noted.
(3) Standard 4x8 panel sizes, other sizes available.
(4) Also available in Structural I (all plies limited to Group 1 species) and Structural II (all plies limited to Group 1, 2 or 3 species).
(5) C or better for 5 plies, C Plugged or better for 3 ply panels.
(6) Stud spacing is shown on grade stamp.
(7) For finishing recommendations, see Form V307.

Description and Use of Appearance Grades of Softwood Plywood. (American Plywood Assoc.)

"Nominal" is the size designation used by the *trade,* but it is not always the actual size. Sometimes the actual thickness of hardwood flooring is 1/32-inch less than the so-called nominal size. "Actual" is the *mill* size for thickness and face width, excluding tongue width. "Counted" size determines the board feet in a shipment. Pieces less than 1 inch in thickness are considered to be 1 in.

OAK

Nominal	Actual	Counted	Weights M Ft.
TONGUED AND GROOVED-END MATCHED			
25/32" x 3¼"	25/32" x 3¼"	1" x 4"	2300 lbs.
25/32" x 2¼"	25/32" x 2¼"	1" x 3"	2100 lbs.
25/32" x 2"	25/32" x 2"	1" x 2¾"	2000 lbs.
25/32" x 1½"	25/32" x 1½"	1" x 2¼"	1900 lbs.
3/8" x 2"	11/32" x 2"	1" x 2½"	1000 lbs.
3/8" x 1½"	11/32" x 1½"	1" x 2"	1000 lbs.
½" x 2"	15/32" x 2"	1" x 2½"	1350 lbs.
½" x 1½"	15/32" x 1½"	1" x 2"	1300 lbs.
SQUARE EDGE			
5/16" x 2"	5/16" x 2"	face count	1200 lbs.
5/16" x 1½"	5/16" x 1½"	face count	1200 lbs.

BEECH, BIRCH, HARD MAPLE AND PECAN

Nominal	Actual	Counted	Weights M Ft.
TONGUED AND GROOVED-END MATCHED			
25/32" x 3¼"	25/32" x 3¼"	1" x 4"	2300 lbs.
25/32" x 2¼"	25/32" x 2¼"	1" x 3"	2100 lbs.
25/32" x 2"	25/32" x 2"	1" x 2¾"	2000 lbs.
25/32 x 1½"	25/32" x 1½"	1" x 2¼"	1900 lbs.
3/8" x 2"	11/32" x 2"	1" x 2½"	1000 lbs.
3/8" x 1½"	11/32" x 1½"	1" x 2"	1000 lbs.
½" x 2"	15/32" x 2"	1" x 2½"	1350 lbs.
½" x 1½"	15/32" x 1½"	1" x 2"	1300 lbs.
SPECIAL THICKNESSES			
17/16" x 3¼"	33/32" x 3¼"	5/4" x 4"	2400 lbs.
17/16" x 2¼"	33/32" x 2¼"	5/4" x 3"	2250 lbs.
17/16" x 2"	33/32" x 2"	5/4" x 2¾"	2250 lbs.
Above Tongued and Grooved and End Matched			
JOINTED FLOORING — i. e., Square Edge			
25/32" x 2½"	25/32" x 2½"	1" x 3¼"	2250 lbs.
25/32" x 3¼"	25/32" x 3¼"	1" x 4"	2400 lbs.
25/32" x 3½"	25/32" x 3½"	1" x 4¼"	2500 lbs.
17/16" x 2½"	33/32" x 2½"	5/4" x 3¼"	2500 lbs.
17/16" x 3½"	33/32" x 3½"	5/4" x 4¼"	2600 lbs.

Prefinished hardwood flooring. When hardwood flooring is produced in unfinished form, sanding and other finishing operations are performed after the flooring is installed.

Some manufacturers produce flooring which is completely prefinished at the factory. It is ready for use immediately after being put down over subflooring.

All species and types of hardwood flooring might not be readily available in prefinished form, but a manufacturer equipped to make factory finished flooring usually can apply a finish to a species and type ordinarily produced unfinished.

Faster installation and elimination of finishing often can offset the higher initial cost of prefinished flooring.

ESTIMATING FLOORING

Figures below are board feet of various sizes of flooring required to cover square feet of floor space shown at left.

Square Ft. Floor Space	25/32 x 3¼	25/32 x 2¼	25/32 x 1½	½ x 2	½ x 1½	3/8 x 2	3/8 x 1½
10	13	14	16	13	14	13	14
20	26	28	31	26	28	26	28
30	39	42	47	39	42	39	42
40	52	56	62	52	56	52	56
50	65	70	78	65	70	65	70
60	78	83	93	78	83	78	83
70	91	97	109	91	97	91	97
80	104	111	124	104	111	104	111
90	117	125	140	117	125	117	125
100	129	139	155	130	139	130	139
200	258	277	310	260	277	260	277
300	387	415	465	390	415	390	415
400	516	553	620	520	553	520	553
500	645	692	775	650	692	650	692
600	774	830	930	780	830	780	830
700	903	969	1085	910	969	910	969
800	1032	1107	1240	1040	1107	1040	1107
900	1161	1245	1395	1170	1245	1170	1245
1000	1290	1383	1550	1300	1383	1300	1383
Counted as	1 x 4	1 x 3	1 x 2¼	1 x 2½	1 x 2	1 x 2½	1 x 2
Weight	2300	2100	2000	1350	1300	1000	1000
Size of nails used	8d	8d	8d	5d	5d	4d	4d
Nail spacing	10-12 in. on center			8 to 10 in. on center		6 to 8 in. on center	
No. pieces per bundle	8	12	12	18	18	24	24
To obtain B.M. from bundle multiply by	2⅔	3	2¼	3¾	3	5	4

COMPUTING FLOORING NEEDS

To determine the board feet of flooring needed to cover a given space, first find the area in square feet then add it to the percentage of that figure which applies to the size flooring to be used, as indicated below. The additions provide an allowance for side-matching plus an additional 5% for end-matching and normal waste.

55% for	25/32" x 1½"
42½% for	25/32" x 2"
38⅓% for	25/32" x 2¼"
29% for	25/32" x 3¼"
38⅓% for	3/8" x 1½"
30% for	3/8" x 2"
38⅓% for	½" x 1½"
30% for	½" x 2"

The above figures are based on laying flooring straight across the room. Where there are bay windows or other projections, allowance should be made for additional flooring.

NAIL SCHEDULE — OAK FLOORING

Tongued-and-grooved flooring. To be blind-nailed.

Flooring Dimen., In.	Type, Size, Spacing of Nails
3/8 x 2 3/8 x 1½	Bright casing nail — 4d wire or cut nail, or screw nail. 8 in. apart. Wood subfloor required.
25/32 x 3¼ 25/32 x 2¼ 25/32 x 1½	7d or 8d screw type or cut steel nail.* 10-12 in. apart.
½ x 2 ½ x 1½	5d screw type or cut steel or wire nail. 10 in. apart. Wood subfloor required.
Square-edge flooring. Face-nailed.	
5/16 x 2 5/16 x 1½	1-in. 15 gauge fully barbed flooring brad, preferably cement coated. 2 nails every 7 in. Wood subfloor required.

*If steel wire flooring nail is used, it should be 8d, preferably cement coated. Newly developed machine-driven barbed fasteners of the size recommended by the manufacturer are acceptable.

Hardwood Flooring.

— One-hour assembly — resilient channel ceiling system

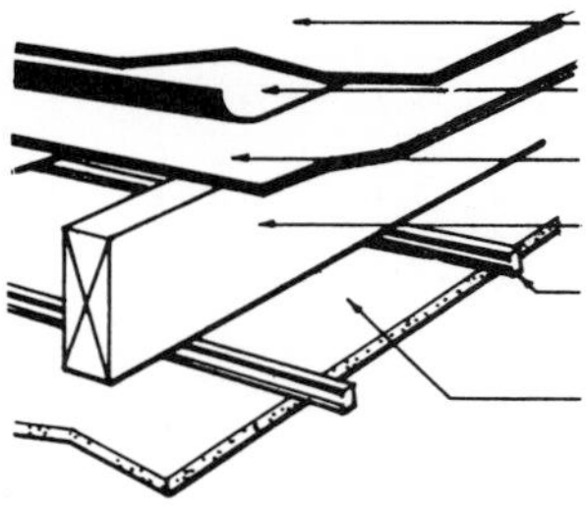

*Channels may be suspended below joists.

— One-hour assembly — T-bar grid ceiling system

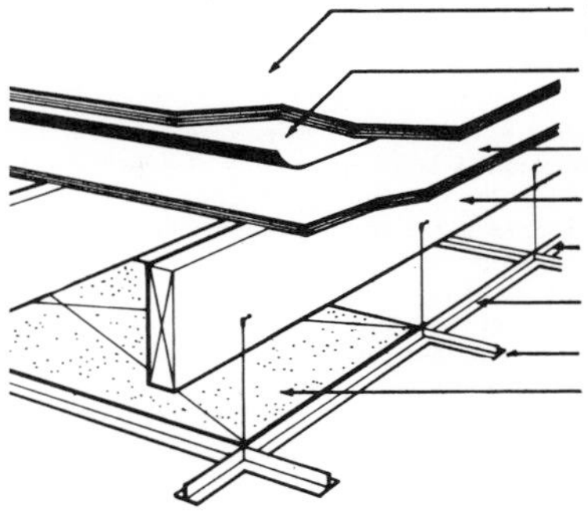

— One-hour interior shear wall construction

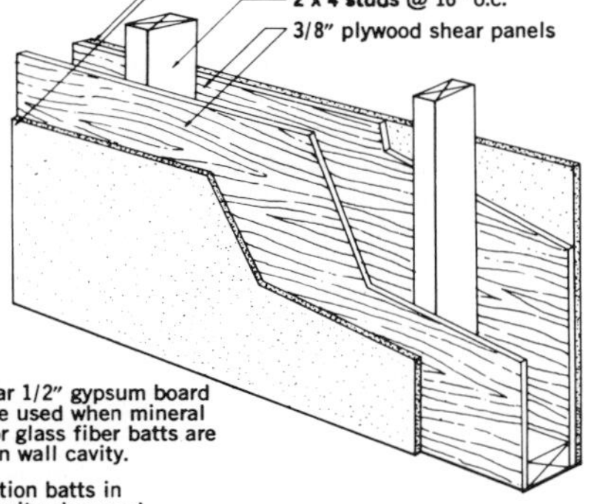

*Regular 1/2" gypsum board may be used when mineral wool or glass fiber batts are used in wall cavity.

Insulation batts in wall cavity also used for sound transmission control

— One-hour exterior wall construction

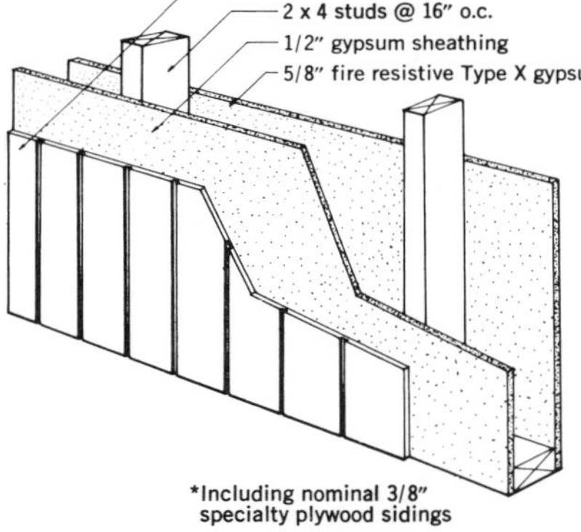

*Including nominal 3/8" specialty plywood sidings

— Treated stressed skin panel construction

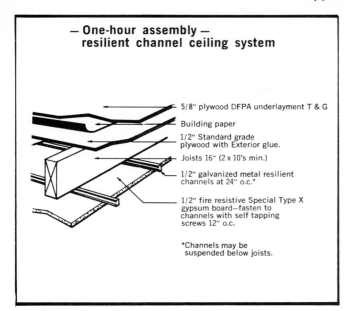

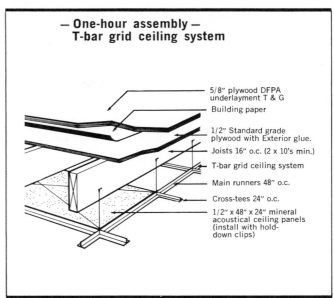

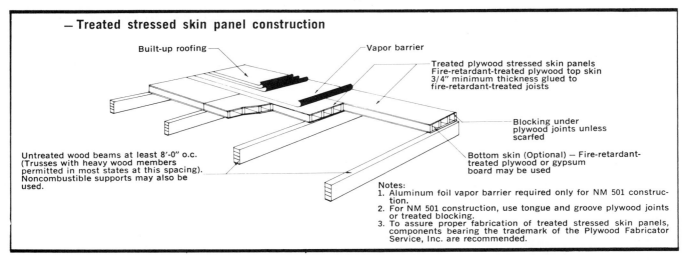

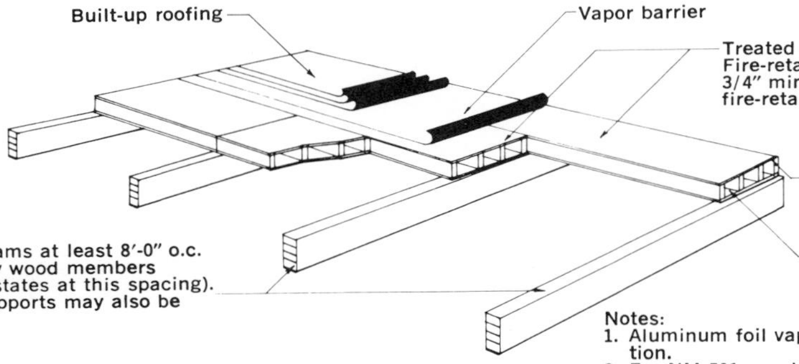

Notes:
1. Aluminum foil vapor barrier required only for NM 501 construction.
2. For NM 501 construction, use tongue and groove plywood joints or treated blocking.
3. To assure proper fabrication of treated stressed skin panels, components bearing the trademark of the Plywood Fabricator Service, Inc. are recommended.

Fire Resistant Construction -- All Assemblies Shown Provide a One-Hour Rating. (American Plywood Assoc.)

Better construction methods and materials used by builders today produce houses that are tighter, more draft-free than those built a half century ago.

In such snug shelters, moisture created inside the house often saturates insulation and causes paint to peel off exterior walls.

The use of a vapor barrier on the warm side of walls and ceilings will minimize this condensation, but adequate ventilation is needed to remove the moisture from the house.

FHA requires that attics have a total net free ventilating area not less than 1/150 of the square foot area, except that a ratio of 1/300 may be provided if:

(a) a vapor barrier having a transmission rate not exceeding one perm is installed on the warm side of the ceiling, or

(b) at least 50% of the required vent area is provided by ventilators located in the upper portion of the space to be ventilated, with the balance of the required ventilation provided by eave or cornice vents.

There are many types of screened ventilators available, including the handy miniature vents illustrated below, which can be used for problem areas or to supplement larger vents.

INSTALLING MINIATURE VENTS

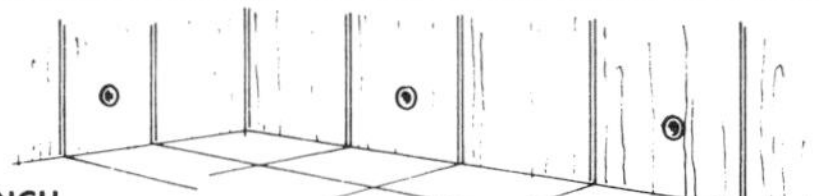

1-INCH
Install in paneling used in basement rooms. Can be painted over to match panel finish.

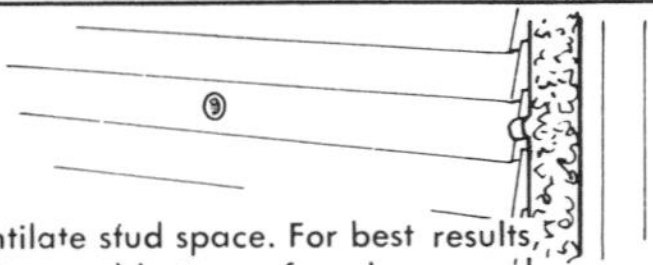

2-INCH
Use to ventilate stud space. For best results, install in top and bottom of each space.

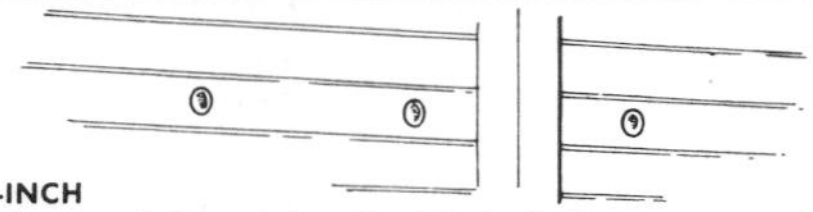

2½-INCH
Made especially to plug the 2½-in. hole cut to blow insulation between studs.

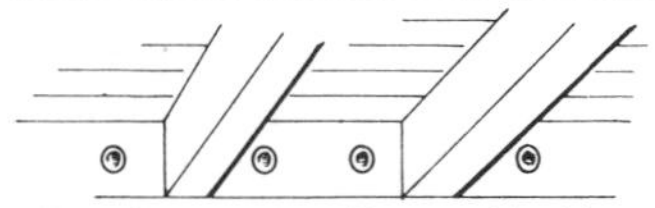

3-INCH
For ventilating rafter space in flat- roof buildings, or other jobs requiring fairly large free area.

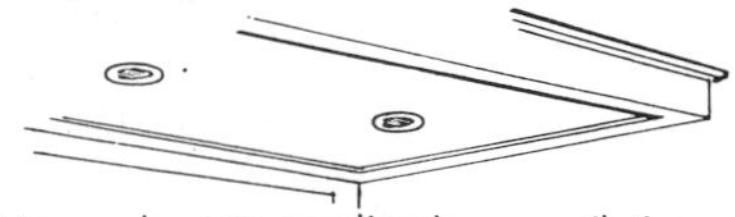

4-INCH
For hard-to-reach spots needing large ventilating area. Large enough for venting soffits.

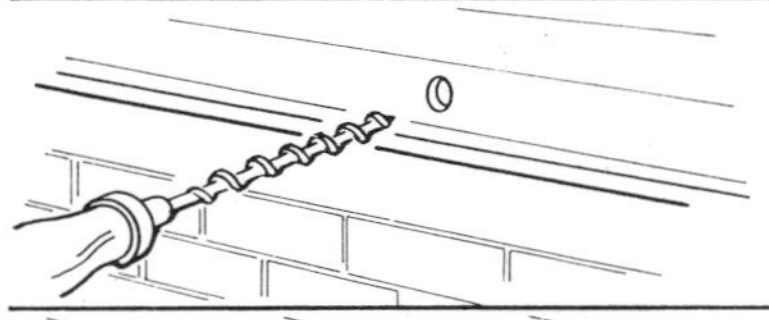

1. Drill or cut a hole the same size as the ventilator.

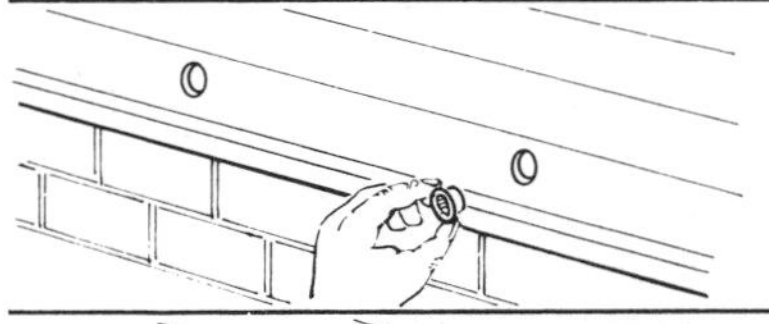

2. Insert the ventilator. Tension ridges hold ventilator in place . . . no nails or screws are needed.

3. Tap into place with a hammer, using a wood block to protect the margin. The louvers are recessed so there's no danger of damage during installation.

FREE AREA VENTILATION GUIDE
Square inches of ventilation required for attic areas

LENGTH (IN FEET) \ WIDTH (IN FEET)	20	22	24	26	28	30	32	34	36	38	40	42
20	192	211	230	250	269	288	307	326	346	365	384	403
22	211	232	253	275	296	317	338	359	380	401	422	444
24	230	253	276	300	323	346	369	392	415	438	461	484
26	250	275	300	324	349	374	399	424	449	474	499	524
28	269	296	323	349	376	403	430	457	484	511	538	564
30	288	317	346	374	403	432	461	490	518	547	576	605
32	307	338	369	399	430	461	492	522	553	584	614	645
34	326	359	392	424	457	490	522	555	588	620	653	685
36	346	380	415	449	484	518	553	588	622	657	691	726
38	365	401	438	474	511	547	584	620	657	693	730	766
40	384	422	461	499	538	576	614	653	691	730	768	806
42	403	444	484	524	564	605	645	685	726	766	806	847
44	422	465	507	549	591	634	676	718	760	803	845	887
46	442	486	530	574	618	662	707	751	795	839	883	927
48	461	507	553	599	645	691	737	783	829	876	922	968
50	480	528	576	624	672	720	768	816	864	912	960	1008

Using length and width dimensions of each rectangular or square attic space, find one dimension on vertical column, the other dimension on horizontal column. These will intersect at the number of square inches of ventilation required to provide 1/300th.

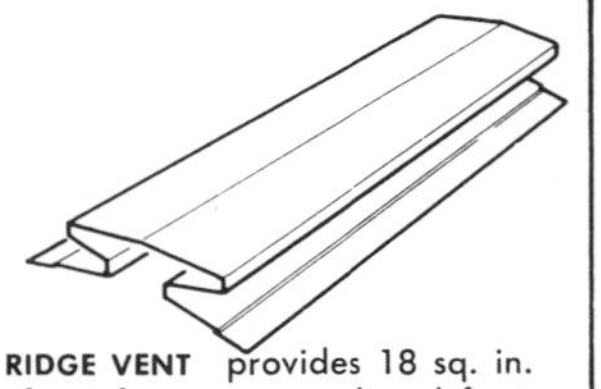

RIDGE VENT provides 18 sq. in. of net free area per lineal foot. Installed quickly over a 1½" gap in the sheathing at the ridge.

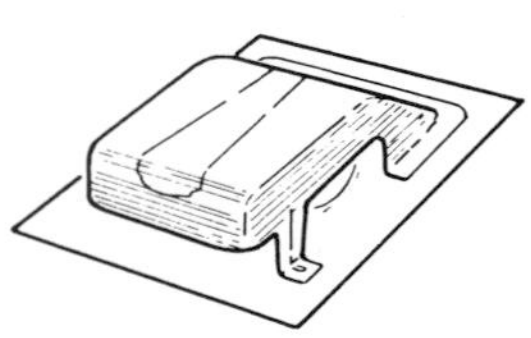

ROOF VENTS fit over openings cut between rafters to pull hot air out of attic.

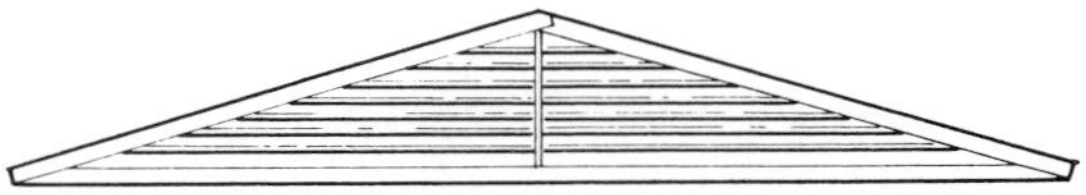

TRIANGLE VENT fits snugly under the roof gable to provide large vent areas at the highest point of the gable end.

ATTIC VENTS are installed in the gable end, usually above the level of the probable level of a future ceiling should the attic be finished later.

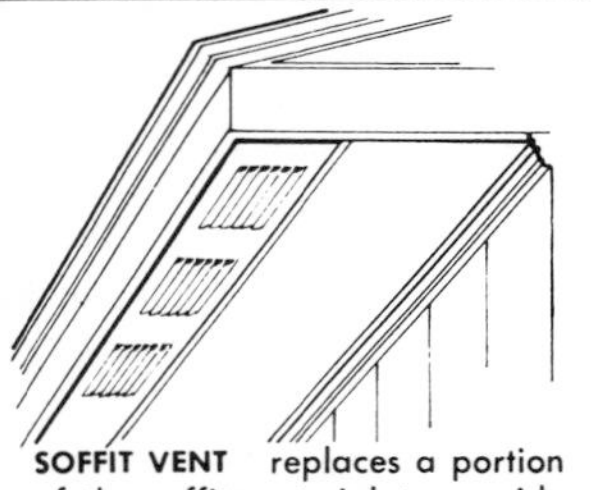

SOFFIT VENT replaces a portion of the soffit material to provide continuous ventilation along its entire length.

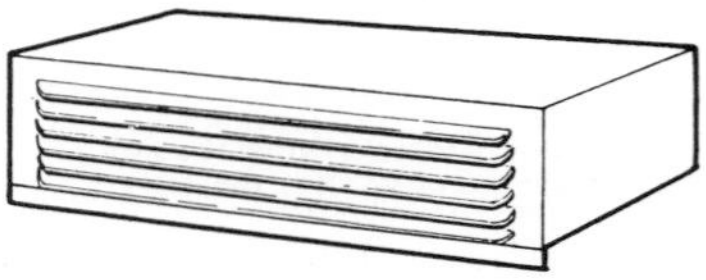

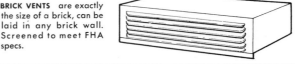

BRICK VENTS are exactly the size of a brick, can be laid in any brick wall. Screened to meet FHA specs.

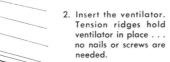

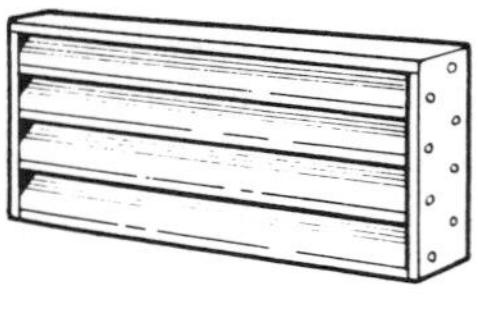

CEMENT BLOCK VENTS are designed to be mortared into the same space as an 8x16" cement block. At least four should be used to vent crawl space.

Modern Ventilators.

Acknowledgments

The author wishes to thank the individuals and organizations listed below for the valuable information, photographs, and other illustrations they so willingly provided. He is especially grateful for the assistance provided by Mr. Merlin Blais of the Western Wood Products Association and to Mr. R. Hugh Love of the American Plywood Association.

Acoustical and Board Products Association, Park Ridge, Ill.
Alcoa Building Products, Inc., Pittsburgh, Pa.
Aluminum Siding Association, Chicago, Ill.
Amerock Corporation, Rockford, Ill.
American Plywood Association, Tacoma, Wash.
Andersen Corporation, Bayport, Minn.
Architectural Woodwork Institute, Arlington, Va.
Armstrong Cork Company, Lancaster, Pa.
Asphalt Roofing Manufacturers Association, New York, N. Y.
Automated Building Components, Inc., Miami, Fla.
Barnes Builders Supply, Cedar Falls, Iowa.
Berger Scientific Supplies, Boston, Mass.
Bird and Son, East Walpole, Mass.
Black & Decker Mfg. Co., Towson, Md.
Bostitch, Inc., East Greenwich, R. I.
E. L. Bruce Co., Memphis, Tenn.
Building Dept., City of Cedar Falls, Cedar Falls, Iowa.
Building Supply News, Chicago, Ill.
Cedar Lumber Co., Cedar Falls, Iowa.
C-E Morgan Building Products, Oshkosh, Wisc.
Central Missouri State University, Warrensburg, Mo.
Century Wood Products, Kansas City, Mo.
Dennis C. Christensen, Contractor, Cedar Falls, Iowa.
Colonial Stair and Woodwork Company, Jeffersonville, Ohio.
Construction Craftsman Magazine, Washington, D.C.
Crown Aluminum Industries Corp., Pittsburgh, Pa.
Dexter Lock, Grand Rapids, Mich.
Donley Brothers Company, Cleveland, Ohio.
Duo Fast Corporation, Franklin Park, Ill.
Eaton Yale & Towne Inc., White Plains, N. Y.
Ekco Building Products Company, Canton, Ohio.
Ethyl Corporation, Baton Rouge, La.
Fine Hardwoods – American Walnut Association, Chicago, Ill.
Fir and Hemlock Door Association, Portland, Ore.
Fleet of America, Inc., Buffalo, N. Y.
Flintkote Company, New York, N. Y.
Follansbee Steel Corporation, Follansbee, W. Va.
Forest Products Laboratory, Madison, Wisc.
Formica Corporation, Cincinnati, Ohio.
Gamble Brothers, Inc., Louisville, Ky.
Garlinghouse Company, Inc., Topeka, Kan.
Georgia-Pacific Corporation, Portland, Ore.
Greenlee Brothers & Company, Rockford, Ill.
Gypsum Association, Evanston, Ill.
Herman Miller, Inc., Zeeland, Mich.
Home Planners, Inc., Detroit, Mich.
Ideal Company, Waco, Tex.
Independent Nail and Packing Co., Bridgewater, Mass.
International Conference of Building Officials, Whittier, Calif.
International Paper Company, Longview, Wash.
I-XL Furniture Company, Goshen, Ind.
Jordan Millwork Company, Sioux Falls, S. Dak.
Kitchen Kompact, Inc., Jeffersonville, Ind.
Libbey-Owens-Ford Glass Company, Toledo, Ohio.
Majestic Company, Inc., Huntington, Ind.
Maple Flooring Manufacturers Association, Oshkosh, Wisc.
Marshfield Homes, Inc., Marshfield, Wisc.
Masonite Corporation, Chicago, Ill.
Maywood, Inc., Amarillo, Tex.
Millers Falls Company, Greenfield, Mass.
Mineral Fiber Products Bureau, New York, N. Y.
Mohawk Flush Doors, Inc., South Bend, Ind.
National Forest Products Association, Washington, D.C.
National Gypsum Company, Buffalo, N. Y.
National Homes Corporation, Lafayette, Ind.
National Lock Company, Rockford, Ill.
National Oak Flooring Association, Memphis, Tenn.
National Woodwork Manufacturers Association, Chicago, Ill.
Nicholson File Company, Providence, R. I.
Nichols Wire and Aluminum Company, Davenport, Iowa.
Owens-Corning Fiberglas Corporation, Toledo, Ohio.
Patent Scaffolding Company, Long Island City, N. Y.
Frank Paxton Lumber Company, Des Moines, Iowa.
Pemko Manufacturing Company, Emeryville, Calif.
Perlite Institute, Inc., New York, N. Y.
Plywood Fabricator Service Inc., Tacoma, Wash.
H. K. Porter Company, Inc., Pittsburgh, Pa.
Portland Cement Association, Skokie, Ill.
PPG Industries, Inc., Pittsburgh, Pa.
Realist Inc., Menomonee Falls, Wisc.
Red Cedar Shingle and Handsplit Shake Bur., Seattle, Wash.
Redman Industries Inc., Dallas, Tex.
Red Wing Wood Products, Inc., Red Wing, Minn.
Richard Mehmen, Contractor, Cedar Falls, Iowa.
Richmond Screw Anchor Company, Brooklyn, N. Y.
Rock Island Millwork, Waterloo, Iowa.
Rockwell International, Power Tool Division, Pittsburgh, Pa.
Rolscreen Company, Pella, Iowa.
Ronthor Reiss Corporation, New York, N. Y.
Ruberoid Company, New York, N. Y.
Senco Products, Inc., Cincinnati, Ohio.
Shakertown Corporation, Cleveland, Ohio.
Simpson Timber Company, Seattle, Wash.
Skil Corporation, Chicago, Ill.
Skyline Corporation, Elkhart, Ind.
Southern Forest Products Association, New Orleans, La.
Spotnails, Inc., Rolling Meadows, Ill.
Stanley Works, New Britain, Conn.
Stenson and Warm, Inc., Waterloo, Iowa.
St. Regis Paper Company, Attleboro, Mass.
Timber Engineering Company, Washington, D. C.
Harold Truax, Mason, Waterloo, Iowa.
United States Gypsum Company, Chicago, Ill.
United States Steel Corporation, Pittsburgh, Pa.
Universal Form Clamp Company, Chicago, Ill.
Upson Company, Lockport, N. Y.
U. S. Plywood Corporation, New York, N. Y.
Vega Industries, Inc., Syracuse, N. Y.
Vermiculite Institute, Minneapolis, Minn.
Wagner Manufacturing Company, Cedar Falls, Iowa.
Wausau Homes Incorporated, Wausau, Wisc.
Western Wood Mldg. & Mlwk. Producers Assoc., Portland, Ore.
Western Wood Products Association, Portland, Ore.
Weyerhaeuser Company, Tacoma, Wash.
Whirlpool Corporation, Benton Harbor, Mich.
Wisconsin Knife Works, Beloit, Wisc.
Wood Conversion Company, St. Paul, Minn.

Badzinski, Stanley Jr., CARPENTRY IN RESIDENTIAL CONSTRUCTION, Prentice-Hall, Inc., Englewood Cliffs, New Jersey.

Durbahn, Walter E., FUNDAMENTALS OF CARPENTRY VOL. 1, American Technical Society, Chicago, Illinois.

Durbahn, Walter E. and Sundberg, Elmer W., FUNDAMENTALS OF CARPENTRY, VOL. 2, American Technical Society, Chicago, Illinois.

Feirer, John L., CABINETMAKING AND MILLWORK, Chas. A. Bennett Co., Peoria, Illinois.

Forest Products Laboratory, U. S. Department of Agriculture, WOOD HANDBOOK (1974 Edition) U. S. Government Printing Office, Washington, D.C.

Hornbostel, Caleb, MATERIALS FOR ARCHITECTURE, Reinhold Publishing Corp., New York City.

Irvin, Daniel W., POWER TOOL MAINTENANCE, McGraw-Hill Book Co., New York, New York.

Jones, Raymond P., FRAMING, SHEATHING AND INSULATION, Delmar Publishers, Inc., Albany, New York.

Kicklighter, Clois E., ARCHITECTURE - Residential Drawing and Design, Goodheart-Willcox Co., Inc., South Holland, Illinois.

Lloyd, William B., MILLWORK, PRINCIPLES AND PRACTICES, Cahners Publishing Co., Chicago, Illinois.

McDonnell, L. P., PORTABLE POWER TOOLS, Delmar Publishers, Albany, New York.

Muller, Edward J., ARCHITECTURAL DRAWING AND LIGHT CONSTRUCTION, Prentice-Hall, Inc., Englewood Cliffs, New Jersey.

Panshin, A. J. and others, FOREST PRODUCTS: THEIR SOURCES, PRODUCTION AND UTILIZATION, McGraw-Hill Book Co., New York City.

Reiner, Laurence E., METHODS AND MATERIALS OF CONSTRUCTION, Prentice-Hall, Inc., Englewood Cliffs, New Jersey.

Schmidt, John L. and others, CONSTRUCTION LENDING GUIDE, McGraw-Hill Book Co., New York City.

Smith, Ronald C., PRINCIPLES AND PRACTICES OF LIGHT CONSTRUCTION, Prentice-Hall, Englewood Cliffs, New Jersey.

Stegman, George K. and Stegman, Harry J., ARCHITECTURAL DRAFTING, American Technical Society, Chicago, Illinois.

Sundberg, Elmer W., BUILDING TRADES BLUEPRINT READING, American Technical Society, Chicago, Illinois.

U. S. Department of Agriculture, Forest Service, WOOD-FRAME HOUSE CONSTRUCTION, U.S. Government Printing Office, Washington, D.C.

Wagner, Willis H., MODERN WOODWORKING, Goodheart-Willcox Co., Inc., South Holland, Illinois.

Wass, Alonzo, METHODS AND MATERIALS OF RESIDENTIAL CONSTRUCTION, Reston Publishing Co., Inc., Reston, Virginia.

Watson, Don A., CONSTRUCTION MATERIALS AND PROCESSES, McGraw-Hill Book Co., New York, New York.

PERIODICALS

AUTOMATION IN HOUSING AND SYSTEMS BUILDING NEWS (bimonthly), 3740 Dempster Street, Skokie, Illinois 60076.

BUILDING SUPPLY NEWS, (monthly), 5 South Wabash Avenue, Chicago, Illinois 60603.

HOUSE AND HOME, (monthly), McGraw-Hill, Inc., 330 W 42nd Street, New York, New York 10036.

PROFESSIONAL BUILDER, (monthly), 5 South Wabash Avenue, Chicago, Illinois 60603.

WOOD AND WOOD PRODUCTS, (monthly), 300 W. Adams Street, Chicago, Illinois 60606.

GLOSSARY OF TERMS

ACOUSTICAL MATERIALS: Types of tile, plaster, and other materials which absorb sound waves. Generally applied to interior wall surfaces to reduce reverberation or reflection of the waves.

ADHESIVE: A substance capable of holding material together by surface attachment. A general term that includes glue, cement, mastic and paste.

AGGREGATE: Materials such as sand, rock, and gravel used to make concrete.

AIR CONDITIONING: Control of temperature, humidity, movement, and purity of air in buildings.

AIR DRIED: Wood seasoned by exposure to the atmosphere, in the open or under cover, without artificial heat.

ALTERATION: Any change in the facilities, structural parts, or mechanical equipment of a building which does not increase the cubic content.

ANCHOR BOLTS: Bolts embedded in concrete used to hold structural members in place.

ANNUAL RINGS: Rings or layers of wood which represent one growth period of a tree. In cross section the rings may indicate the age of the tree.

APRON: A piece of horizontal trim applied against the wall immediately below the stool. Conceals rough edge of plaster.

AREAWAY: An open space around a basement window or doorway. Provides light, ventilation, and access.

ASPHALT: A residue from evaporated petroleum. It is insoluble in water but is soluble in gasoline and melts when heated. Used for waterproofing roof coverings, exterior wall coverings, and flooring tile.

ASTRAGAL: An interior molding attached to one of a pair of doors or window sash in order to prevent swinging thru; also used with sliding doors to insure tighter fitting where doors meet.

BACKBAND: A narrow rabbeted molding applied to the outside corner and edge of interior window and door casing to create a "heavy trim" appearance.

BACKFILL: The replacement of earth after excavating.

BALUSTER: Squares of turned spindle-like vertical stair member which support the stair rail.

BALUSTRADE: A railing consisting of a series of balusters resting on a base, usually the treads, which supports a continuous stair or hand rail.

BASEMENT: The base story of a house, usually below grade.

BASE SHOE: Small narrow molding used around the perimeter of a room where the base meets the finish floor.

BATTEN: A strip of wood placed across a surface to cover joints.

BATTER: The slope, or inclination from the vertical, of a wall or other structure or portion of a structure.

BATTER BOARD: A temporary framework used to assist in locating corners when laying out a foundation.

BAY: One of the intervals or spaces into which a building plan is divided by columns, piers, or division walls.

BAY WINDOW: A rectangular, curved, or polygonal window, or group of windows usually supported on a foundation extending beyond the main wall of a building.

BEAM: A principal structural member used between posts, columns or walls.

BEARING PARTITION: A partition which supports a vertical load in addition to its own weight.

BEARING WALL: A wall which supports a vertical load in addition to its own weight.

BEDDING: A filling of mortar, putty, or other substance used to secure a firm bearing.

BED MOLDING: A molding applied where two surfaces come together at an angle. Commonly used in cornice trim especially between the plancier and frieze.

BENCH MARK: A mark on a permanent object fixed to the ground from which land measurements and elevations are taken.

BEVEL: To cut to an angle other than a right angle, such as the edge of a board or door.

BEVEL SIDING: Used as finish siding on the exterior of a structure. It is usually manufactured by "resawing" dry, square surfaced boards diagonally to produce two wedge-shaped pieces.

BID: An offer to supply, at a specified price: materials, supplies, and equipment; or the entire structure or sections of the structure.

BLEMISH: Any defect, scar, or mark that tends to detract from the appearance of wood.

BLIND STOP: A member applied to the exterior edge of the side and head jamb of a window to serve as a stop for the top sash and to form a rabbet for storm sash, screens, blinds and shutters.

BLUE STAIN: A stain caused by a fungus growth in unseasoned lumber -- especially pine. It does not affect the strength of the wood.

BOARD: Lumber less than 2 in. thick.

BOARD FOOT: The equivalent of a board 1 ft. square and 1 in. thick.

BRACKET: A projecting support for a shelf or other structure.

BRICK CONSTRUCTION: A type of construction in which the exterior walls are bearing walls made of brick.

BRICK MOLDING: A molding for window and exterior door frames. Serves as the boundary molding for brick or other siding material and forms a rabbet for the screens and/or storm sash or combination door.

BRICK VENEER CONSTRUCTION: A type of construction in which a wood-frame construction has an exterior surface of single brick.

BRIDGING: Pieces fitted in pairs from the bottom of one floor joist to the top of adjacent joists, and crossed to distribute the floor load. Sometimes pieces of solid stock of a width equal to the joist are used.

BUILT-UP ROOF: A roofing composed of several layers of rag felt or jute saturated with coal tar, pitch, or asphalt. The top is finished with crushed slag or gravel. Generally used on flat or low-pitched roofs.

BUTT: Type of door hinge. One leaf is fitted into space routed into the door frame jamb and the other into the edge of the door.

CABINET: Case or box-like assembly consisting of shelves, doors and drawers, used primarily for storage.

CABINET DRAWER GUIDE: A wood strip used to guide the drawer as it slides in and out of its opening.

CABINET DRAWER KICKER: Wood cabinet member placed immediately above and generally at the center of a drawer to prevent tilting down when pulled out.

CAMBER: A slight arch in a beam or other horizontal member which prevents it from bending into a downward or concave shape due to its weight or load it is to carry.

CANT STRIP: A triangular shaped strip of wood used under shingles at gable ends or under the edges of roofing on flat decks.

CASED OPENING: An interior opening without a door that is finished with jambs and trim.

CASEMENT: A window in which the sash swings on its vertical edge, so it may be swung in or out.

CASING: The trimming around a door or window, either outside or inside, or the finished lumber around a post or beam.

CAULK: To seal and make waterproof cracks and joints, especially around window and exterio- door frames. (also calk)

CHAIR RAIL: An interior molding applied along the wall of a room to prevent the chair from marring the wall.

CHAMFER: Corner of a board beveled at a 45 deg. angle. Two boards butt-jointed and with chamfered edges form a V joint.

CLEAT: A strip of wood fastened across a door to add strength. Also a strip fastened to a wall to support a shelf, fixture, or other objects.

CHECKRAILS: Meeting rails of a double-hung window which are made thicker to fill the opening between the top and bottom sash. They are usually beveled.

CLOSET POLE: A round molding installed in clothes closets to accommodate clothes hangers.

COLLAR BEAM: A tie beam connecting rafters considerably above the wall plate. It is also called a rafter tie.

COLUMN: Upright supporting member circular or rectangular in shape.

COMMERCIAL STANDARD: A voluntary standard that establishes quality, methods of testing, certification, rating, and labeling of manufactured items. It provides a uniform base for fair competition.

CONDUCTORS: Pipes for conducting water from a roof to the ground or to a receptacle or drain; downspout.

CONDUIT, ELECTRICAL: A pipe or tube, usually metal, in which wiring is installed.

CONVENIENCE OUTLET: Outlet into which may be plugged portable equipment such as lamps.

COPE: To cut or shape the end of a molded wood member so it will cover and fit the contour of an adjoining piece of molding.

CORBEL OUT: To extend outward from the surface of a masonry wall one or more courses to form a supporting ledge.

CORNER BEAD: Molding used to protect corners. Also a metal reinforcement placed on corners before plastering.

CORNER BRACES: Diagonal braces let into studs to reinforce corners of frame structures.

CORNICE: Exterior trim of a structure at the meeting of the roof and wall; usually consists of panels, boards, and moldings.

COUNTERFLASHING: Flashing used on chimneys at the roof-line to cover shingle flashing and prevent moisture entry.

COVE MOULDING: Moulding with a concave profile used primarily where two members meet at a right angle.

CUPOLA: Small vented four-sided structure installed on a roof. Adds decoration to the building

and provides ventilation for the attic.
CURTAIN WALL: A wall, usually nonbearing, between piers or columns.
DADO: A rectangular groove cut in wood across the grain.
DEAD LOAD: The weight of permanent, stationary construction included in a building.
DECAY: Disintegration of wood substance due to action of wood destroying fungi.
DIMENSION LUMBER: Lumber 2 to 5 in. thick, and up to 12 in. wide.
DIMENSIONAL STABILITY: The ability of a material to resist changes in its dimensions due to temperature, moisture and physical stress.
DOOR FRAME: An assembly of wood parts that form an enclosure and support for a door. Door frames are classified as exterior and interior.
DOOR STOP: A molding nailed to the faces of the door frame jambs to prevent the door from swinging through.
DORMER: A projecting structure built out from a sloping roof. Usually includes one or more windows.
DRIP CAP: A molding which directs water away from a structure to prevent seepage under the exterior facing material. Applied mainly over window and exterior door frames.
DRIP GROOVE: Semicircular groove on the underside of a drip cap or the lip of a window sill which prevents water from running back under the member.
DROP SIDING: Siding, usually 3/4 in. thick and machined into various patterns. Drop siding has tongue and groove or shiplap joints.
DRY ROT: A term loosely applied to many types of decay but especially to that which, when in an advanced stage, permits the wood to be easily crushed to a dry powder.
DRY WALL: Materials used for wall covering which do not need to be mixed with water before application.
EASED EDGE: Corner slightly rounded or shaped to a slight radius.
EAVES: The margin or lower part of a roof that projects over an exterior wall. Also called the overhang.
ELECTRIC MOISTURE METER: Meter used to determine the moisture content of wood. Action is based on electrical resistance or capacitance which varies with change in moisture content.
EQUILIBRIUM MOISTURE CONTENT: The moisture content at which wood neither gains nor loses moisture when surrounded by air at a given relative humidity and temperature.
ESCUTCHEON: In builders hardware, a protective plate or shield containing a key hole.
EXPANSION JOINT: A bituminous fiber strip used to separate blocks or units of concrete to prevent cracking due to dimensional change caused by shrinkage and variation in temperature.
FACADE: Main or front elevation of a building.
FACE NAIL: A nail driven perpendicular to the surface of a piece.
FACTORY AND SHOP LUMBER: Lumber intended to be cut up for use in further manufacture. It is graded on the basis of the percentage of the area which will produce a limited number of cuttings of a specified, or a given minimum, size and quality.
FASCIA: A wood member used for the outer face of a box cornice where it is nailed to the ends of the rafters and lookouts.
FENESTRATION: The placement or arrangement and sizes of the windows and exterior doors of a building.
FIBER BOARD: A broad term used to describe sheet material of widely varying densities; manufactured from wood, cane, or other vegetable fibers.
FIBER SATURATION POINT: The stage in the drying or wetting of wood at which the cell walls are saturated and the cell cavities are free from water. It is assumed to be 30 percent moisture content, based on oven dry weight and is the point below which shrinkage occurs.
FIRE STOP: A block or stop used in wall of building between studs to prevent the spread of fire and smoke through air space.
FIRE WALL: A wall which subdivides a building to restrict the spread of fire.
FLASHING: Sheet metal or other material used in roof and wall construction (especially around chimneys and vents) to prevent rain or other water from entering.
FLAT ROOF: A roof which is flat, or which is pitched only enough to provide for drainage.
FLUE: The space or passage in a chimney through which smoke, gas, or fumes rise. Each passage is called a flue, which with the surrounding masonry, makes up the chimney.
FLUSH: Adjacent surfaces even, or in same plane (with reference to two structural pieces).
FOOTING: The spreading course or courses at the base or bottom of a foundation wall, pier, or column.
FOUNDATION: The supporting portion of a structure below the first-floor construction, or grade, including the footings.
FRAMING: The timber structure of a building which gives it shape and strength; including interior and exterior walls, floor, roof and ceilings.
FURRING: Narrow strips of wood spaced to form a nailing base for another surface. Furring is used to level, to form an air space between the two surfaces and to give a thicker appearance to the base surface.
GABLE: That portion of a wall contained between the slopes of a double-sloped roof or that portion contained between the slope of a single-sloped roof and a line projected horizontally through the lowest elevation of the roof construction.
GAIN: Notch or mortise cut to receive the end of another structural member -- or a hinge and other hardware.
GIRDER: A large or principal beam used to support concentrated loads at particular points along its length.
GLAZING: The process of installing glass into sash and doors. Also refers to glass panes inserted in various types of frames.
GLAZING COMPOUND: A plastic substance of such consistency that it tends to remain soft and rubbery when used in glazing sash and doors.
GLUE BLOCK: A wood block, triangular or rectangular in shape, which is glued into place to reinforce a right-angle butt joint. Sometimes used at the intersection of the tread and riser in a stairs.
GROUND: A strip of wood assisting the plasterer in making a straight wall and in giving a place to which the finish of the room may be nailed.
GROUT: A thin mortar used in masonry work.
GUSSET: A panel or bracket of either wood or metal attached to the corners or intersections of a frame to add strength and stiffness.
GUTTER: Wood or metal trough attached to the edge of a roof to collect and conduct water from rain and melting snow.
FRIEZE: A boxed cornice wood trim member attached to the structure where the soffit (plancier) and wall meet.
HALF STORY: That part of a building situated wholly or partly within the roof frame, finished for occupancy.
HARDBOARD: A board material manufactured of wood fiber, formed into a panel having a density range of approximately 50 to 80 lbs. per cu. ft.
HEADER: Horizontal structural member that supports the load over an opening, such as a window or door. Also called a lintel.
HEADROOM: The clear space between floor line and ceiling, as in a stairway.
HEARTWOOD: The wood extending from the pith or center of the tree to the sapwood, the cells of which no longer participate in the life processes of the tree.
HEAT TRANSMISSION COEFFICIENT: Hourly rate of heat transfer for one square foot of surface when there is a temperature difference of one deg. F. of the air on the two sides of the surface.
HIP ROOF: A roof which rises from all four sides of a building.
HOLLOW-BACK: Removal of a portion of the wood on the unexposed face of a wood member to more properly fit any irregularity in bearing surface.
HOLLOW CORE DOOR: Flush door with a core assembly of strips or other units which support the outer faces.
HORN: The extension of a stile, jamb, or sill.
HOSE BIB: A water faucet that is threaded so a hose connection can be attached; a sill cock.
I BEAM: A steel beam with a cross section that resembles the letter I.
INCINERATOR: A device which consumes household waste by burning.
INTERIOR TRIM: General term for all the molding, casing, baseboard and other trim items applied within the building by finish carpenters.
INSULATION: (Thermal) Any material high in resistance to heat transmission that is placed in structures to reduce the rate of heat flow.
IN-THE-WHITE: Natural or unpainted; the natural unfinished surface of the wood.
JACK RAFTER: A short rafter framing between the wall plate and a hip rafter; or a hip or valley rafter and ridge board.
JALOUSIE: A series of small horizontal overlapping glass slats, held together by an end metal frame attached to the faces of window frame side jambs or door stiles and rails. The slats or louvers move simultaneously like a Venetian blind.
JAMB: The top and two sides of a door or window frame which contact the door or sash; top jamb and side jambs.
JOINERY: A term used by woodworkers when referring to the various types of joints used in a structure.
JOIST: One of a series of parallel framing members used to support floor and ceiling loads, and supported in turn by larger beams, girders, or bearing walls.
KERFING: Longitudinal saw cuts or grooves of varying depths (dependent on the thickness of the wood member) made on the unexposed faces of millwork members to relieve stress and prevent warping; members are also kerfed to facilitate bending.
KILN DRIED: Wood seasoned in a kiln by means of artificial heat, controlled humidity and air circulation.
KNOCKED DOWN: Unassembled; refers to structural units requiring assembly after being delivered to the job.
KNOT: Branch or limb embedded in the tree and cut through during lumber manufacture.
LALLY COLUMN: A cylindrically shaped steel member used to support beams and girders. Sometimes filled with concrete.
LATH: A building material of wood, metal, gypsum, or insulating board, fastened to frame of building to act as a plaster base.
LAZY SUSAN: A circular revolving cabinet shelf used in corner kitchen cabinet unit.
LEADER: A vertical pipe that carries rainwater from the gutter to the ground or a drain. Also downspout.
LEDGER: A strip attached to vertical framing or structural members to support joists or other horizontal framing. Similar to a ribbon strip.
LIGHT CONSTRUCTION: Construction generally restricted to conventional wood stud walls, floor and ceiling joists and rafters. Primarily residential in nature although it does include small commercial buildings.
LINEAL FOOT: Having length only, pertaining to a line one foot long -- as distinguished from a square foot or cubic foot.
LINTEL: A horizontal structural member which supports the load over an opening such as a door or window.
LIVE LOAD: The total of all moving and variable loads that may be placed upon a building.
LOCK BLOCK: A block of wood which is joined to the inside edge of the stile of a hollow core door and to which the lock is fitted. Flush doors have a lock block on each stile.
LOOKOUT: Structural member running between the lower end of a rafter and the outside wall. Used to carry the underside of the overhang; plancier or soffit.
LUMBER: The product of the saw and planing

mill not further manufactured than by sawing, resawing, passing lengthwise through a standard planing machine, and crosscutting to length. Some matching of ends and edges may be included.

MANSARD ROOF: A type of curb roof in which the pitch of the upper portion of a sloping side is slight and that of the lower portion steep. The lower portion is usually interrupted by dormer windows.

MASONRY: Stone, brick, hollow tile, concrete block or tile, and sometimes poured concrete and gypsum block, or other similar materials, or a combination of same, bonded together with mortar to form a wall, pier, buttress, etc.

MATCHED LUMBER: Lumber that is edge dressed and shaped to make a close tongue-and-groove joint at the edges or ends. Also generally includes lumber with rabbeted edges.

MECHANICAL EQUIPMENT: In architectural and engineering practice: All equipment included under the general heading of plumbing, heating, air conditioning, gasfitting, and electrical work.

MEDALLION: A raised decorative piece, sometimes used on flush doors.

MEETING RAIL: The bottom rail of the upper sash, and the top rail of the lower sash of a double-hung window. Also called a check rail.

MILLWORK: The term used to describe products which are primarily manufactured from lumber in a planing mill or woodworking plant; including moldings, door frames and entrances, blinds and shutters, sash and window units, doors, stairwork, kitchen cabinets, mantels, cabinets and porch work.

MODULAR COORDINATION: The dimensioning of a structure and use of building materials based on a common unit of measurement, namely a module.

MOISTURE CONTENT: The amount of water contained in wood. Expressed as a percentage of the weight of oven-dry wood.

MONOLITHIC: Term used for concrete construction poured and cast in one unit -- without joints.

MOULDER: A woodworking machine designed to run moldings and other wood members with regular or irregular profiles. Also called a sticker.

MOLDING: A relatively narrow strip of wood, usually shaped to a curved profile throughout its length. Used to accent and emphasize the ornamentation of a structure and to conceal surface or angle joints.

MULLION: A slender bar or pier forming a division between units of windows, screens, or similar frames -- generally nonstructural.

MUNTIN: Vertical member between two panels of the same piece of panel work. The vertical and horizontal sashbars separating the different panes of glass in a window.

NET FLOOR AREA: The gross floor area, less the area of the partitions, columns, and stairs and other floor openings.

NEWEL: The main post at the start of a stairs and the stiffening post at the landing; a stair newel.

NOMINAL SIZE: As applied to timber or lumber, the ordinary commercial size by which it is known and sold.

NONBEARING PARTITION: A partition extending from floor to ceiling which supports no load other than its own weight.

NOSING: The part of a stair tread which projects over the riser, or any similar projection; a term applied to the rounded edge of a board.

ON CENTER: A method of indicating the spacing of framing members by stating the measurement from the center of one member to the center of the succeeding one.

OPEN-GRAIN WOOD: Woods with large pores, such as oak, ash, chestnut, and walnut.

ORIEL WINDOW: A window that projects from the main line of an enclosing wall of a building and is carried on brackets, corbels, or a cantilever.

PARAPET: A low wall or railing along the edge of a roof, balcony or bridge. The part of a wall that extends above the roof line.

PARGETING: Thin coat of plaster applied to stone or brick to form a smooth or decorative surface.

PARTICLEBOARD: A formed panel consisting of particles of wood flakes, shavings, slivers, etc., bonded together with a synthetic resin or other added binder.

PARTITION: A wall that subdivides space within any story of a building.

PARTY WALL: A wall used jointly by two parties under easement agreement and erected at or upon a line separating two parcels of land that may be held under different ownership.

PENNY: Term used to indicated nail length; abbreviated by the letter d. Applies to common, box, casing, and finishing nails.

PIER: A column of masonry, usually rectangular in horizontal cross section. Used to support other structural members.

PILASTER: A part of a wall that projects not more than one-half of its own width beyond the outside or inside face of a wall. Chief purpose is to add strength -- may also be decorative.

PILE: A heavy timber, or pillar of metal or concrete, forced into the earth or cast in place to form a foundation member.

PITCH: Inclination or slope, as of roofs or stairs. Rise divided by the span.

PLAN: A drawing representing any one of the floors or horizontal cross sections of a building, or the horizontal plane of any other object or area.

PLANCIER: The underside of an eave or cornice, usually horizontal.

PLASTER: A mixture of lime, cement and sand, used to cover outside and inside wall surfaces.

PLAT: A map, plan, or chart of a city, town, section, or subdivision indicating the location and boundaries of individual properties.

PLATE: A horizontal structural member placed on a wall or supported on posts, studs, or corbels to carry the trusses of a roof or to carry the rafters directly. Also a sole or base member of a partition or other frame.

PLATFORM FRAMING: A system of framing a building where the floor joists of each story rest on the top plates of the story below (or on the foundation wall for the first story) and the bearing walls and partitions rest on the subfloor of each story.

PLUMB: Exactly perpendicular or vertical; at right angles to the horizon or floor.

PLUMBING: The work or business of installing in buildings the pipes, fixtures, and other apparatus for bringing in the water supply and removing liquid and water-borne wastes. This term is used also to denote the installed fixtures and piping of a building.

PLUMBING STACK: A general term for the vertical main of a system of soil, waste, or vent piping.

PORTICO: A porch or covered walk consisting of a roof supported by columns. A porch with a continuous row of columns.

PREFABRICATED CONSTRUCTION: Type of construction so designed as to involve a minimum of assembly at the site, usually comprising a series of large units manufactured in a plant.

PRESERVATIVE: Substance that will prevent the development and action of wood-destroying fungi, borers of various kinds, and other harmful insects that deteriorate wood.

PURLINS: Horizontal roof members used to support rafters between the plate and ridge board.

QUARTER ROUND: Small molding with a cross section of one-fourth of a cylinder.

QUARTER-SAWED: Lumber that has been cut at approximately a 90 deg. angle to the annular growth rings.

QUIT-CLAIM DEED: A deed whereby the grantor conveys without warranty, to the grantee, whatever interest he possesses in the property.

RABBET: A rectangular shape consisting of two surfaces cut along the edge or end of a board.

RAFTER: One of a series of structural members of a roof designed to support roof loads. The rafters of a flat roof are sometimes called roof joists.

RAIL: Cross or horizontal members of the framework of a sash, door, blind or other assembly.

RAKE: The trim members that run parallel to the roof slope and form the finish between the roof and wall at a gable end.

RAMP: Inclined plane connecting separate levels.

REINFORCED CONCRETE CONSTRUCTION: A type of construction in which the principal structural members, such as floors, columns, and beams, are made of concrete poured around steel bars or steel meshwork in such manner that the two materials act together to resist force.

RELATIVE HUMIDITY: Ratio of amount of water vapor in air in terms of percentage to total amount it could hold at the same temperature.

RESILIENT: The ability of a material to withstand temporary deformation, the original shape being assumed when the stresses are removed.

RETAINING WALL: Any wall subjected to lateral pressure other than wind pressures. Example -- a wall built to support a bank of earth.

RIBBON: A narrow board attached to studding or other vertical members of a frame that adds support to joists or other horizontal members.

RISER: The vertical stair member between two consecutive stair treads.

ROOFING: The materials applied to the structural parts of a roof to make it waterproof.

ROOF RIDGE: The horizontal line at the junction of the top edges of two roof surfaces where an external angle greater than 180 deg. is formed.

ROTARY CUT VENEER: Veneer cut on a lathe which rotates the log against a broad cutting knife. The veneer is cut in a continuous sheet much the same as paper is unwound from a roll.

ROUGHING-IN: The work of installing all pipes in the drainage system and all water pipes to the point where connections are made with the plumbing fixtures. Also applies to partially completed electrical wiring and other mechanical aspects of the structure.

ROUGH LUMBER: Lumber that has been cut to rough size with saws but which has not been dressed or surfaced.

ROUGH OPENING: The opening formed by the framing members.

SADDLE: A small gable type roof placed in back of a chimney on a sloping roof to shed water and debris.

SAPWOOD: The layers of wood next to the bark, usually lighter in color than the heartwood, that are actively involved in the life processes of the tree. More susceptible to decay than heartwood. Sapwood is not essentially weaker or stronger than heartwood of the same species.

SASH: The framework which holds the glass in a window.

SCAFFOLD: A temporary structure or platform used to support workmen and materials during building construction.

SCANTLING: Lumber with a cross section ranging from 2 in. x 4 in. to 4 in. x 4 in.

SCARFING: Joining the ends of stock together with a sloping lap-joint so they appear to be a single piece.

SCOTIA: A concave molding consisting of an irregular curve. Used under the nosing of stair treads and for cornice trim.

SCUTTLE: An opening in a ceiling which provides access to the attic.

SEASONING: Removing moisture from green wood to improve its serviceability.

SECOND GROWTH: Timber that has grown after the removal of a large portion of the previous stand.

SEPTIC TANK: A sewage-settling tank intended to retain the sludge in immediate contact with the sewage flowing through the tank, for a sufficient period to secure satisfactory decomposition of organic sludge solids by bacterial action.

SETTING BLOCK: A wood block placed in the glass groove or rabbet of the bottom rail of an insulating glass sash to form a base or bed for the glass.

SHAKES: Handsplit shingles.

SHEATHING: The structural covering. Consists of boards or prefabricated panels that are attached to the exterior studding or rafters of a structure.

SHEATHING PAPER: A building material used in

wall, floor, and roof construction to resist the passage of air.
SHIM: A thin strip of wood, sometimes wedge-shaped, for plumbing or leveling wood members. Especially helpful when setting door and window frames.
SHIPLAP: Lumber with edges that have been rabbeted to form a lap joint between adjacent pieces.
SHORING: Lumber and timbers used to prevent the sliding of earth adjoining an excavation. Also the timbers used as temporary bracing against a wall or under a platform.
SHUTTER: A wood assembly of stiles and rails to form a frame which encloses panels used in conjunction with door and window frames. Also may consist of vertical boards cleated together.
SIDE OF TRIM: Trim required to finish one side of a door or window opening.
SIDING: The finish covering of the outside wall of a frame building. Many different types are available.
SILL: The lowest member of the frame of a structure, usually horizontal, resting on the foundation and supporting the uprights of the frame. Also the lowest member of a window or outside door frame.
SLEEPER: A timber laid on or near the ground to support floor joists and other structures above. Also wood strips laid over or embedded in a concrete floor to which finish flooring is attached.
SOFFIT: The underside of the members of a building, such as staircases, cornices, beams and arches. Relatively minor in size as compared with ceilings. Also called drop ceiling and furred-down ceiling.
SOFTWOODS: The botanical group of trees that have needle or scalelike leaves and are evergreen for the most part, cypress, larch, and tamarack being exceptions. The term has no reference to the actual hardness of the wood. Softwoods are often referred to as conifers, and botanically they are called gymnosperms.
SOIL STACK: A general term for the vertical main of a system of soil, waste, or vent piping.
SOLAR ORIENTATION: Placement of a structure on a building site to obtain the maximum benefits of sunlight.
SPAN: The distance between structural supports such as walls, columns, piers, beams, girders, and trusses.
SPECIFICATION: A written document stipulating the kind, quality, and sometimes the quantity of materials and workmanship required for a construction job.
SPECIFIC GRAVITY: The ratio of the weight of a body to the weight of an equal volume of water at some standard temperature.
SPLASH BLOCK: A small masonry block laid with the top close to the ground surface to receive roof drainage and carry it away from the building.
SPLINE: A small strip of wood that fits into a groove or slot of both members to form a joint.
SQUARE: Unit of measure -- 100 square feet -- usually applied to roofing material and to some types of siding.
STAIRWAY, STAIR, OR STAIRS: A series of steps, with or without landings, or platforms, usually between two or more floors of a building.
STAIRWELL: The framed opening which receives the stairs.
STEEL-FRAME CONSTRUCTION: A type of construction in which the structural parts are of steel or are dependent on a steel frame for support.
STICKERS: Strips of wood used to separate the layers in a pile of lumber so air can circulate.
STILE: The upright or vertical outside pieces of a sash, door, blind or screen.
STOOL: A molded interior trim member serving as a sash or window frame sill cap. Stools may be beveled-rabbeted or rabbeted to receive the window frame sill.
STOOP: A small porch, veranda, or platform, or a stairway, outside an entrance to a building.
STORY: That part of a building comprised between any floor and the floor or roof next above.
STORY POLE: A strip of wood used to lay out and transfer measurements for door and window openings, siding and shingle courses, and stairways. Also called a rod.
STRAIGHTEDGE: A straight strip of wood or metal used to lay out or check the accuracy of work.
STRESSED SKIN: Two facings, one glued to one side and the other to the opposite side of an inner structural framework to form a panel. Facings may be of plywood or other suitable material.
STRIKE PLATE: A metal piece mortised into or fastened to the face of a door frame side jamb to receive the latch or bolt when the door is closed.
STRUCTURAL WINDOW WALL PANEL: A window unit framed into a wall panel at the factory. Also called a factory-assembled structural wall panel.
STUD: One of a series of vertical wood or metal structural members in walls and partitions. Plural -- studs or studding.
SUBFLOOR: Boards or panels laid directly on floor joists over which a finished floor will be laid.
SURFACED LUMBER: Lumber that is dressed or finished by running it through a planer.
TAIL BEAM: A relatively short beam or joist supported by a wall on one end and by a header on the other.
TERMITE SHIELD: A shield, usually made of sheet metal, placed in or on a foundation wall or around pipes to prevent the passage of termites into the structure.
TERRAZZO FLOORING: A floor produced by embedding small chips of marble or colored stone in concrete and then grinding and polishing the surface.
THERMOSTAT: An instrument that controls automatically the operation of heating or cooling devices by responding to changes in temperature.
THREE-WAY SWITCH: A switch designed to operate in conjunction with a similar switch to control one outlet from two points.
THRESHOLD: A wood member, beveled or tapered on each side, used to close the space between the bottom of a door and the sill or floor underneath. Sometimes called a saddle.
TIE BEAM (collar beam): A beam so situated that it ties the principal rafters of a roof together and prevents them from thrusting the plate out of line.
TIMBERS: Lumber 5 inches or larger in least dimension.
TOENAILING: To drive a nail at a slant with the initial surface in order to permit it to penetrate into a second member.
TOE SPACE: A recessed space at the floor line of a base kitchen cabinet or other built-in units. Permits one to stand close without striking the vertical surface with his toes.
TRANSFORMER: A device for transforming the voltage characteristics of an electric current.
TRANSOM: A small opening above a door separated by a horizontal member (transom bar). Usually contains a sash or a louver panel hinged to the transom bar.
TRAP: A plumbing fitting or device designed to provide a liquid trap seal which will prevent the sewer gases from passing through and entering a building.
TREAD: The horizontal part of a step on which the foot is placed.
TRIM: The finish materials in a building, such as moldings applied around openings (window trim, door trim) or at the floor and ceiling of rooms (baseboard, cornice, picture molding).
TRIMMER: The beam or floor joist into which a header is framed. Adds strength to the side of the opening.
TRUSS: A structural unit consisting of such members as beams, bars, and ties; usually arranged to form triangles. Provides rigid support over wide spans with a minimum amount of material.
UNPROTECTED-METAL CONSTRUCTION: A type of construction in which the structural parts are of metal unprotected by fireproofing.
VALLEY: The internal angle formed by the two slopes of a roof.
VALLEY RAFTER: A rafter which forms the intersection of an internal roof angle.
VAPOR BARRIER: A watertight material used to prevent the passage of moisture or water vapor into or through structural elements (floors, walls, ceilings).
VENEERED WALL: A frame building wall with a masonry facing (example -- single brick). A veneered wall is nonload bearing.
VENT: A pipe installed to provide a flow of air to or from a drainage system or to provide a circulation of air within such system to protect trap seals from siphonage and back pressure.
VENTILATION: The process of supplying and removing air by natural or mechanical means. Such air may or may not have been conditioned.
VERMICULITE: Mineral closely related to mica, with the faculty of expanding on heating to form lightweight material with insulating qualities. Used as bulk insulation, also as aggregate in insulating and acoustical plaster, and in insulating concrete floors.
WAINSCOT: A lower interior wall surface (usually 3 to 4 ft. above the floor) that contrasts with the wall surface above. May consist of solid wood or plywood.
WALLBOARD: Wood pulp, gypsum, or other materials made into large rigid sheets that may be fastened to the frame of a building to provide a surface finish.
WALL TIE: Metal strip or wire used to bind tiers of masonry in cavity wall construction, or to bind brick veneer to a wood frame wall.
WARP: Any variation from a true or plane surface. Warp includes bow, crook, cup, and twist, or any combination thereof.
WATER REPELLENT: A solution, primarily paraffin wax and resin in mineral spirits, which upon penetrating wood retards changes in its moisture content.
WATER TABLE: A ledge or slight projection at the bottom of a structure which carries the water away from the foundation.
WEATHERING: The mechanical or chemical disintegration and discoloration of the surface of wood. It can be caused by exposure to light, the action of dust and sand carried by winds, and the alternate shrinking and swelling of the surface fibers that come with the continual variation in moisture content brought by changes in the weather. Weathering does not include decay.
WEATHERSTRIP: Narrow strips of metal, vinyl plastic or other material, so designed that when installed at doors or windows, they will retard the passage of air, water, moisture, or dust around the door or window sash.
WEEPHOLE: A small hole, as in a retaining wall, to drain water to the outside. Commonly used at the lower edges of masonry cavity walls.
WET WALL: An interior wall finish surface usually consisting of 3/8 in. gypsum plaster lath and 1/2 in. gypsum plaster applied to the lath surface.
WHALER (also waler): A horizontal member used in concrete form construction.
WIND ("i" pronounced as in kind): A term used to describe the warp in a board when twisted (winding). It will rest upon two diagonally opposite corners, if laid upon a perfectly flat surface.
WINDOW UNIT: Consists of a combination of the frame, window, weatherstripping and sash activation device. May also include screens and/or storm sash. All parts are assembled as a complete operating unit.
WING: A lateral extension of a building from the main portion thereof or one of two or more coordinate portions of a building which extends from a common junction.
WIRE GLASS: Glass having a layer of meshed wire incorporated approximately in the center of the sheet.
WP SERIES MOLDING PATTERNS: A recent molding series of more than 500 profiles which was a joint venture of the former West Coast Lumbermen's Association and the Western Pine Association now merged as Western Wood Products Association.

INDEX

D

E

H

I

J

K

L

R

S

T

U